AF332942

OPTOELECTRONICS – DEVICES AND APPLICATIONS

Edited by **Padmanabhan Predeep**

Optoelectronics – Devices and Applications
Edited by Padmanabhan Predeep

CBS, Edition **2016**

Published by InTech
Janeza Trdine 9, 51000 Rijeka, Croatia

Publishing Process Manager Mirna Cvijic

Technical Editor Teodora Smiljanic

Cover Designer InTech Design Team

Image Copyright john austin, 2010. Used under license from Shutterstock.com

Additional hard copies can be obtained from orders@intechopen.com

Optoelectronics – Devices and Applications, Edited by Padmanabhan Predeep
p. cm.
ISBN 978-953-307-576-1

Contents

To my father; but for his unrelenting efforts I would not have made it to this day.

Preface

Optoelectronics - Devices and Applications is the second part of an edited anthology on the multifaceted areas of optoelectronics by a selected group of authors including promising novices to experts in the field, where are discussed design and fabrication of device structures and the underlying phenomena. Many of the optoelectronic and photonic effects are integrated into a vast array of devices and applications in numerous combinations, and more are in fast development. New branches of optoelectronics continues to sprout up such as military optoelectronics, medical optoelectronics etc. The field of optoelectronics and photonics was originally aimed at applying light to tasks that could previously only be solved through electronics, such as in data transfer technology. Optoelectronics, being graduated to photonics seeks to continue this endeavor and to expand upon it by searching for applications for light. At any rate the optics related electronic and photonic phenomena, where the closely connected players like electrons and photons, often refuse to be demarcated into water tight compartments. With applications touching everyday life and consumer electronic gadgets, optoelectronics is emerging as a popular technology and draws from and contributes to several other fields, such as quantum electronics and modern optics.

There are many aspects of light and its behavior that are important to those studying electronics for scientific or industrial purposes. Light sensing is particularly important in photonics, as the light involved in experiments and tests often needs to be quantified and may not even be visible and electrons invariably helps in this. The role of lasers in increasing the quality of life in modern times is unique. It is a lifesaving source of light that enormously helped in medicine as in military technology and even in entertainment, data storage, and holography.

The wide range of such applications in the field of optoelectronics and photonics ensures that it is generally a well-funded and thriving area of scientific research and upcoming researchers are sure to find it extremely encouraging. In the global energy front also optics and photonics hold the hope of harnessing light to provide safe energy and power especially in the light of the hidden dangers of nuclear power as an alternative. I am sure that this collection of articles by experts from the field would help them enormously to understand the underlying principles, design and fabrication philosophy behind this wonderful technology. The first part of this set presents recent trends in the development of materials and techniques in optoelectronics and the readers are suggested to have a look into that as well in the InTech websites.

July 2011

P. Predeep
Professor
Laboratory for Unconventional Electronics & Photonics
Department of Physics
National Institute of Technology Calicut
India

Part 1

Optoelectronic Devices

Organic Light Emitting Diodes: Device Physics and Effect of Ambience on Performance Parameters

T.A. Shahul Hameed[1], P. Predeep[1] and M.R. Baiju[2]
[1]Laboratory for Unconventional Electronics and Photonics, National Institute of
Technology, Calicut, Kerala,
[2]Department of Electronics and Communication, College of Engineering,
Trivandrum, Kerala,
India

1. Introduction

Research in Organic Light Emitting Diode (OLED) displays has been attaining greater momentum for the last two decades obviously due to their capacity to form flexible (J. H. Burroughes et al, 1990) multi color displays. Their potential advantages include easy processing, robustness and inexpensive foundry (G.Yu & A.J.Heeger, 1997) compared to inorganic counterparts. In fact, this new comer in display is rapidly moving from fundamental research into industrial product, throwing many new challenges (J. Dane and J.Gao, 2004; G. Dennler et al, 2006) like degradation and lifetime. In order to design suitable structures for application specific displays, the studies pertaining to the device physics and models are essentially important. Such studies will lead to the development of accurate and reliable models of performance, design optimization, integration with existing platforms, design of silicon driver circuitry and prevention of device degradation. More over, a clear understanding on the device physics (W.Brutting et al, 2001) is necessary for optimizing the electrical properties including balanced carrier injection (J.C.Scott et al, 1997: A.Benor et al., 2010) and the location of the emission in the device. The degradation (J.C.Scott et al, 1996; J. Dane and J.Gao, 2004) of the device is primarily caused by the moisture , which poses questions to the reliability and life of this promising display. How the device responds to different temperature ambience (T.W.Lee and O.Park, 2000) also attracts attention of researchers since its applications at cryogenic temperature are yet to be explored. The basic device physics and modeling philosophies based on the mathematical formulations of its physical behavior are revisited in this article. Also it reviews the prominent ambient studies and the efforts to enhance the reliability of the device by new fabrication methods with inexpensive ways of encapsulation, making it suitable for long life display applications.

2. Principle and physics of organic LEDs

2.1 Device structure, principle

The simplest structure of OLED is shown in fig 1. The Tris(8-hydroxyquinolinato) aluminium (Alq3) is an evaporated emissive layer on the top of spun cast hole transport layer Poly-(3,4-ethyhylene dioxythiophene):poly-(styrenesulphonate) (PEDOT:PSS). Indium Tin Oxide (ITO) and aluminium are the anode and cathode respectively. Charge injection, transport and recombination (I.H.Campbell et al,1996) occur in the light emitting conductive layer of organic light emitting diodes and its features influence efficiency and color of emission from the device. Besides the characteristics of light emitting organic layer, interface interactions (P.S.Davids et al, 1996) of this layer with other layers in OLED play important role in defining the characteristics of the display. There have been innumerable studies on different aspects of PEDOT: PSS (L.S.Roman et al,1999;S.Alem et al,2004) enhancing the performance of photo cells and light emitting diodes. In practical implementations, more layers for carrier injection and transport are normally incorporated.

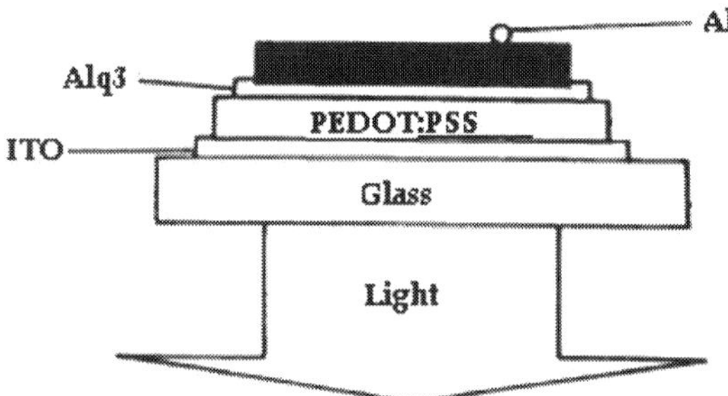

Fig. 1. Structure of Organic Light Emitting Diode.

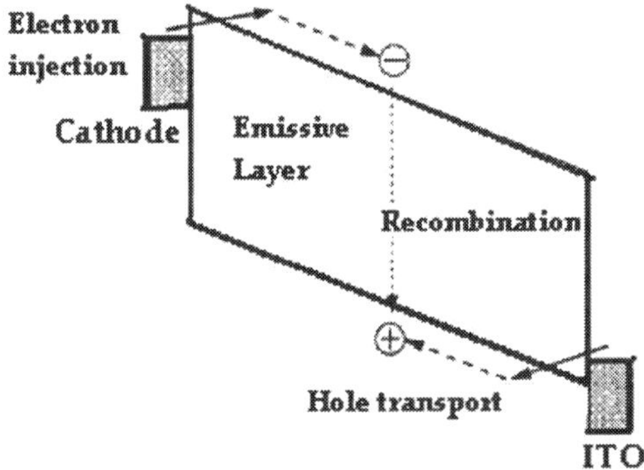

Fig. 2. Injection, Transport and Recombination in PLED[15].

In Polymer Light Emitting Diodes(PLED), conducting polymers like Poly (2-methoxy, 5-(2-ethylhexoxy)-1, 4-phenylene-vinylene (MEH- PPV) are used as the emissive layer in which dual carrier injection takes place (Fig. 2). Electrons are injected from cathode to the LUMO of the polymer and holes are injected from anode to HOMO of the conducting polymer and they recombine radiatively within the polymer to give off light (Y.Cao et al,1997). The fabrication of the device is easy through spin casting of the carrier transport layer and Electro Luminescent layer (MEH-PPV) for thickness in A^{o} range.

2.2 Device physics

For OLEDs, it is more often a practice to follow many concepts derived from inorganic semiconductor physics. In fact, most of the organic materials used in LEDs form disordered amorphous films without forming crystal lattice and hence the mechanisms used for molecular crystals cannot be extended. Detailed study on device physics of organic diodes based on aromatic amines (TPD) and aluminium chelate complex (Alq) was carried out by many research groups (W.Brutting et al,2001).Basic steps in electroluminescence are shown in fig. 3 where charge carrier injection, transport, exciton formation and recombination are accounted in presence of built-in potential. Built-in potential(Vbi) across the organic layers is due to the different work functions between anode and cathode (I.H.Campbell et al,1996).

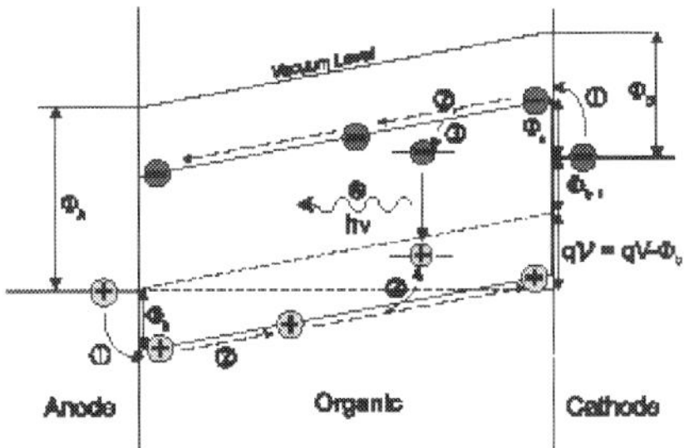

Fig. 3. Basic Steps of Electroluminescence with Energy Band[4].

Built-in potential (Vbi) found out by photovoltaic nulling method, where OLED is illuminated and an external voltage is applied till photocurrent is equal to dark current (J.C.Scott et al,2000). Its physical significance is that it reduces the applied external voltage V such that a net drift current in forward bias direction can only be achieved if V exceeds built in voltage.Carrier injection is described by Fowler-Nordheim tunneling or Richardson-Schottky thermionic emission, described by the equations

$$j_{FN} = \frac{A^* q^2 F^2}{\phi_B \alpha^2 K_B^2} \exp(-\frac{2\alpha\phi_B^{3/2}}{3qF}) \qquad (1)$$

$$j_{RS} = A^* T^2 \exp(-\frac{\phi_B - \beta_{RS}\sqrt{F}}{K_B T}) \qquad (2)$$

The current is either space charge limited (SCLC) or trap charge limited (TCLC).The recombination process in OLED has been described by Langevin theory because it is based on a diffusive motion of positive and negative carriers in the attractive mutual Coulomb field. To be more clear, the recombination constant (R) is proportional to the carrier mobility (W.Brutting et al,2000).

$$R = [q / \varepsilon\varepsilon_0][\mu_h + \mu_e] \qquad (3)$$

Apart from the discussion on the dependence of current on voltage and temperature, the current has a direct dependence on the thickness of the organic layer and it was observed that thinner the device better will be the current output. Similar observations were also made by the group on J-V and luminance characteristics of ITO/TPD/AlQ/Ca hetero junction devices for different organic layer thickness. The thickness dependence of current at room temperature leads to the inference that the electron current in Alq device is predominantly space charge limited with a field dependent charge carrier mobility and that trapping in energetically distributed states is additionally involved at low voltage and especially for thick layers. The temperature dependence of current in Al/Alq/Ca device (from 120 K to 340K) indicates that device is having a less turn-on current at higher temperature and recombination in OLED to be bimolecular process following the Langevin theory. The mathematical analysis of the device, considering traps and temperature has been a new approach in device physics.

Towards the search of highly efficient device, the combining of Alq and NPB, with a thickness of 60nm for the Alq layer has been determined to yield higher quantum efficiency whereas thickness variation of NPB layer doesn't show any measurable effect.

The field and temperature dependence of the electron mobility in Alq leads to the delay equation (W.Brutting et al,2000) as

$$t_d = \frac{d}{\mu F} \tag{4}$$

where

$$F = \frac{V - V_{bi}}{d} \;.$$

The behavior of hopping transport in disordered organic solids has been better explained by Gaussian Disorder Model (H.Bassler,1993). The quantitative model for device capacitance with an equivalent circuit of hetero layer device gives more insight into interfacial charges and electric field distribution in hetero layer devices.

The transport behavior in polymer semiconductor has been a matter of active debate since many theories were put forwarded by different groups. Charge transport is not a coherent motion of carriers in well defined bands - it is a stochastic process of hopping between delocalized states, which leads to low carrier mobilities $(\mu << 1cm^2 / Vs)$ (W.Brutting et al,1999). Trap free limit for dual carrier device was studied by Bozano et al,1999. Space charge limited current was observed above moderate voltages (>4V), while zero field electron mobility is an order of magnitude lower than hole mobility. Balanced carrier injection is one of the pre requisites for the optimal operation of single layer PLEDs. Balanced carrier transport implies that injected electrons and holes have same drift mobilities. In fact, it is difficult to achieve in single layer devices due to the predominance of one of the carriers and hence bi-layer devices are used to circumvent the problem. ITO/PPV/TPD: PC/ Al devices fabricated where ITO/PPV is an ideal hole injecting contact for the trap-free MDP TPD: PC. Here ITO/PPV contact acts as an infinite, non depletable charge reservoir, which is able to satisfy the demand of the TPD: PC layer under trap-free space-charge-limited (TFSCL) conditions (H.Antoniadis et al,1994). Trap free space charge limited current (TFSL) [L.Bozano et al,1999) can be expressed as

$$J_{TFSL} = \frac{9}{8}\varepsilon\varepsilon_0\mu E^2 / d \tag{5}$$

where ε_0 is the permittivity of vacuum, ε is the permittivity of the polymer, μ is the mobility of holes in trap free polymer, d is inter electrode distance(M. A. Lampert and P. Mark ,1970). Trapping is relatively severe at low electric fields and in thick PPV layers. At high electric fields, trapping is minimized even for thick PPV layers.

The carrier drift distance x at a given electric field E before trapping occurs is given by $x = \mu\tau E$ where τ is the trapping time. The electron deep trapping product $\mu\tau$ determines the average carrier range per applied electric field before they get immobilized in deep traps. It is imperative that the difference in $\mu\tau$ values of electrons and holes in PPV (10^{-12} and $10^{-9} cm^2 / v$ respectively) reflects their discrepancy in transport. In fact, not the structure of PPV contributes to this difference, but oxygen related impurities in PPV (P.K. Konstadinidis et al,1994) with strong electron accepting character and reduction potential lower than PPV may act as the predominant electron traps and limit the range of electrons.

The study of temperature dependence of current density versus electric field for single carrier (both electron dominated and hole dominated) and dual carrier devices at temperatures 200K and 300K exhibits interesting results (L.Bozano et al,1999). In both temperatures, the reduction in space charge due to neutralization contributes to significant enhancement in current density in dual carrier devices . Also it was deduced that the electric field dependence of the mobility is significantly stronger for electrons than for holes. The electric field coefficient γ is related to temperature as per the empirical relation $\gamma = (1 / kT - 1 / kT_0)B$ where B and T_0 are constants (W.D.Gill,1972). In MEH-PPV devices, charge balance will be improved by cooling which in turn leads to enhanced quantum efficiency. By adjusting barrier heights, at the level of 0.1eV, quantum efficiency close to theoretical maximum can be achieved. In order to limit the space charge effects and hence to enhance the performance in terms of current density, the intrinsic carrier mobility to be taken care by modifying dielectric constant or electrically pulsing the device at an interval greater than recombination time. The other means of improvisation is aligning of polymer backbone, but such efforts may lead to quenching (L.Bozano et al,1998)

2.3 Device models

Device modeling is useful in many ways like optimization of design, integration with existing tools, prediction of problems in process control and better understanding of degradation mechanism. By modeling PLEDs current-voltage -luminance behavior, with which quantum and power efficiencies can be analytically seen, this in turn normally has to be subjected to experimental validation.

Both band based models and exciton based models were proposed to explain the electronic structure and operation of polymer devices. Out of the two, there are more supportive arguments for band based model. I.D.Parker examined (I.D.Parker,1994) the factors that control carrier injection with a particular reference to tunneling, by experimenting on ITO/MEH-PPV/Ca device. The thickness dependability of current density with respect to bias and field strength are shown in fig.4 and 5 respectively. It is obvious from these figures that the device operating voltage shall be reduced by reducing the polymer thickness. The field dependence of I-V behavior points to the tunneling model of carrier injection, in which carriers are field emitted through a barrier at electrode/polymer interface (fig.4).

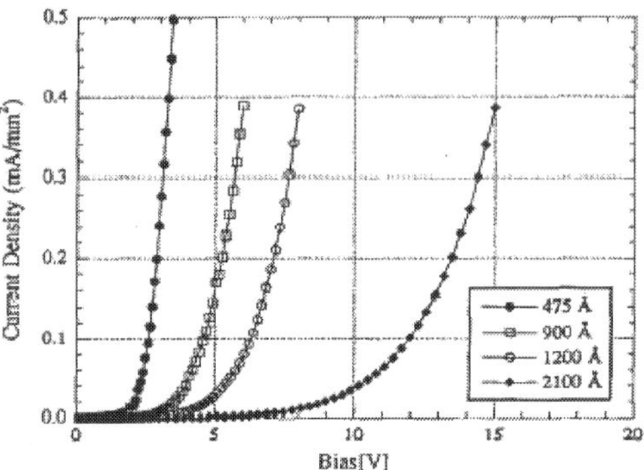

Fig. 4. Thickness Dependence of the I-V Characteristics in ITO/MEH-PPV/Ca Device
(I.D.Parker,1994).

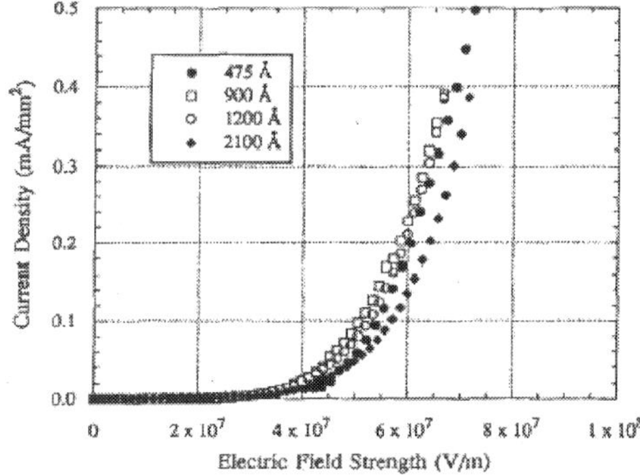

Fig. 5. Field v Current Dependence for ITO/MEH-PPV/Ca Device ((I.D.Parker,1994).

For a clear understanding of the device physics and models, it is customary to fabricate
single carrier and dual carrier devices. On replacing Ca, having low work function (2.9eV)
with higher work function metals like In (4.2eV), Au (5.2eV), hole only devices can be made.
This increases the offset between Fermi energy of cathode and LUMO of polymer which
causes a substantial reduction in injected electrons and holes become dominant carriers. It is
apparent that the external quantum efficiency reduces in single carrier devices. The current
characteristics show only a slight dependence with temperature which is predicted by
Fowler-Nordheim tunneling.

$$I \propto F^2 \exp(\frac{-k}{F}) \tag{6}$$

where F is the field strength The constant k is defined by

$$k = \frac{8\pi\sqrt{2m^*}\phi^{3/2}}{3qh} \tag{7}$$

where ϕ is the barrier height and m^* is the effective mass of the holes(S.M.Sze,1981).
A rigid band model better explains experimental results where holes and electrons tunnel into the polymer when applied electric field tilts the polymer bands to present sufficiently thin barriers. Fig.6 clearly indicates how this model envisages tunneling of holes.

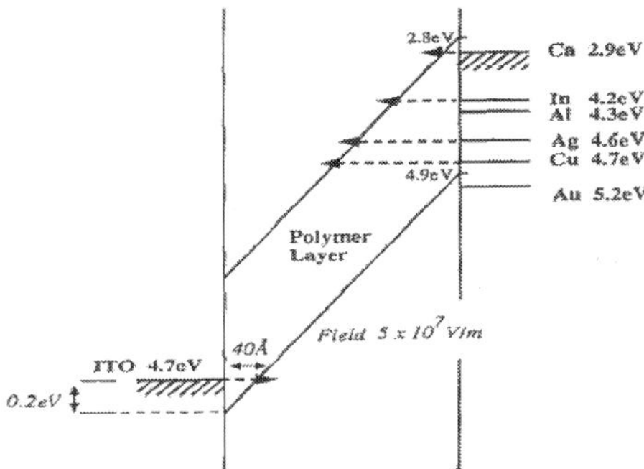

Fig. 6. Band Diagram (in Forward Bias) for Model, indicating positions of Fermi Level for different electrode materials (I.D.Parker,1994).

From the band based model and characterization, the improvements in device performance was suggested by I.D. Parker. Of the devices he made, ITO/MEH-PPV/Ca devices exhibit better results due to the reasons explained elsewhere. The device turn – on happens at a flat band condition and it is in fact the voltage required to reach the flat-band condition and it depends on the band gap of the polymer and work-function of electrodes. The operating voltage of the device is sensitive to barrier height whereas the turn-on voltage is not.
From the equations mentioned before, an approximation for the current can be made as

$$I \propto \exp(-\frac{\gamma\phi^2}{V}) \tag{8}$$

where V is the applied voltage and ϕ is the barrier height. This prediction of barrier height dependence of operating voltage has been supported by experimental credentials.
Efficiency of the device is a function of current density due to minority carriers, increasing barrier height leads to an exponential decrease in current and efficiency, which is shown in fig.7.Parker had suggested the suitable combination of electrode materials and polymers so that low turn-on voltage and operating voltage can be achieved.
J.C.Scott et al(J.C.Scott et al,2000) contributed to unveil the phenomena like built in potential, charge transport, recombination and charge injection with a numerical model to calculate the recombination profile in single and multilayer structures. 'Essentially trap free' transport, Langevin mechanism for recombination and model of thermionic injection with Schottkey barrier at metal organic interface are the important features used by them. It is to be highlighted that charge trapping is neglected in the analysis and transport is described in terms of trap free space charge limited currents. Fowler-Nordheim mechanism was used to explain the injection, but by analytical methods and simulations, thermionic injection (G.G. Malliaras ,1998) is said to best suit for explaining the injection in organic diodes.

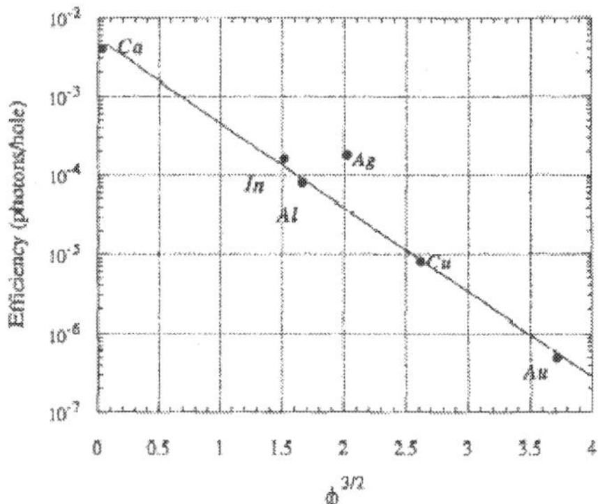

Fig. 7. Device Efficiency v (Barrier Height)$^{3/2}$ [I.D.Parker,1994].

There are remarkable efforts (P.W.M.Blom & Marc J.M,1998) in characterization and modeling of polymer light emitting diodes. Their experiments on PPV devices, both single carrier and dual carrier devices, paved the way to the better understanding of mobility of electrons and holes. Electron only devices are fabricated by a PPV layer sandwiched between two Ca electrodes whereas hole only devices with an evaporated Au on top. For hole only devices, current density depends quadratically on voltage.

$$J = \frac{9}{8}\varepsilon_0\varepsilon_r\mu_p\frac{V^2}{L^3} \tag{9}$$

where μ_p is hole mobility and L is the thickness of the device. Transport properties of the single carrier devices are described in detail with analytical expressions. Hole only device is having effect of space charge holes and electron only devices show trapping of electrons. For double carrier device, two additional phenomenon becomes important-recombination and charge neutralization. Recombination is bimolecular since its rate is directly proportional to electron and hole concentration. Without traps and field dependent mobility, the current in double carrier device is

$$J = \left(\frac{9\pi}{8}\right)^{1/2}\varepsilon_0\varepsilon_r\left(\frac{2q\mu_p\mu_n(\mu_p+\mu_n)}{\varepsilon_0\varepsilon_r B}\right)^{1/2}\frac{V^2}{L^3} \tag{10}$$

where B is bimolecular recombination constant. (P.W.M.Blom & Marc J.M,1998).
In PLEDs, conversion efficiency is dependent on applied voltage whereas in conventional LEDs, it is not. Temperature dependence of charge transport in PLEDs is investigated by performing J-V measurements on hole only and double carrier devices. Carrier transport strongly dependent on temperature (P.W.M.Blom et al, 1997) and the fig.8 explains the variation of current density with respect to applied voltage for different temperature.
Also, the plot of bimolecular recombination constant B for different temperatures (fig.9) sheds light into the fact that recombination is Langevin type [31] and mathematically it is expressed in terms of mobility

$$B = \frac{e}{\varepsilon_0\varepsilon_r}(\mu_n+\mu_p) \tag{11}$$

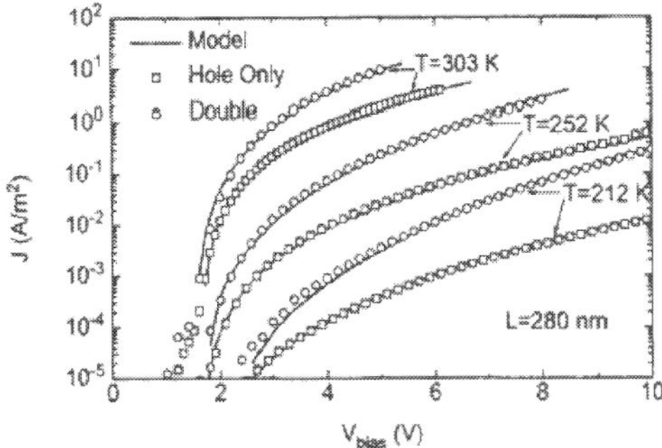

Fig. 8. Experimental and Calculated (Solid lines) J-V characteristics in hole only (squares) and double carrier (circle) for different thickness (P.W.M.Blom & Marc J.M,1998).

The enhancement of maximum conversion efficiency is by decreasing non radiative recombination and by use of electron transport layer which shifts recombination zone away from metallic cathode.

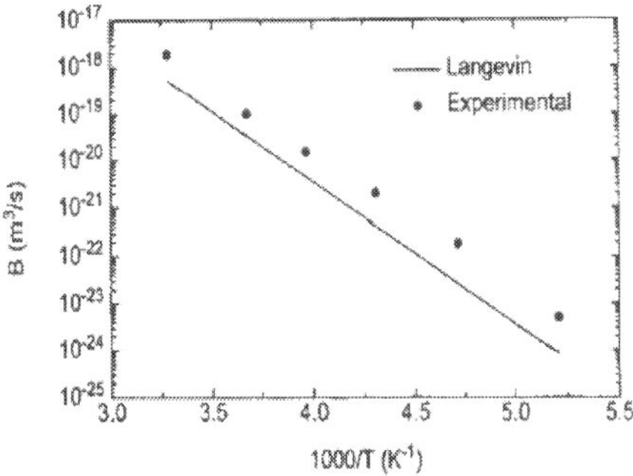

Fig. 9. Temperature Dependence of Bimolecular Recombination Constant (P.W.M.Blom & Marc J.M,1998).

Device model based on Poisson's equation and conservation of charges was more a traditional presenattion (Y.Kawabe et al,1998) in organic electronic devices. By assuming that recombination rate is proportional to collision cross section A, electric field, sum of mobility values of electrons and holes and the product of carrier densities, charge conservation equation has been rewritten as

$$\frac{dJ_{h,e(x)}}{dx} = \pm A(\mu_h + \mu_e)E(x)n_h(x)n_e(x) \tag{12}$$

where + and – signs indicate electron and hole currents.
By conservation law of the total current

$$J_{h(x)} + J_{e(x)} = eE(x)n_h(x)\mu_h + eE(x)n_e(x)\mu_e = J_0 , \tag{13}$$

with the boundary conditions given by current injection at both electrodes (Y.Kawabe et al,1998) .

Besides, current density, relative quantum efficiency was calculated by the model equation

$$\eta = \frac{J_{h(0)} - J_{h(d)}}{J_0} = \frac{J_{e(d)} - J_{e(0)}}{J_0} \tag{14}$$

Here numerical values of the parameters are used to simulate J-V and quantum efficiency characteristics .Two devices-one with semiconducting polymer (BEH-PPV) and the other with dye doped polymer ($PVK : AlQ_3$) were fabricated by spin casting techniques and characterized. The results validate the model for the single layer devices and its suitability for complex devices is yet to be tested.

The model is having the advantages of incorporating   charged traps as shown in equation below

$$\frac{dE(x)}{dx} = \frac{e}{q}[n_h(x) - n_e(x) \pm n_t(x)] \tag{15}$$

where $\pm$ indicates positive ad negative charges respectively. This sends limelight to the causes of degradation process in real devices due to the accumulation of electrons in the vicinity of the cathode. The inferences include low barrier height for low voltage operation, high mobility for high brightness devices and low electron mobility confines the emission region near the cathode and should be avoided to prevent electrode quenching.

3. Ambient studies of organic light emitting diodes

The temperature dependence of current density versus bias voltage exhibits interesting results in organic light emitting diodes. The studies made on four sets of devices namely Device A: ITO/PEDOT-PSS/MEH-PPV/Al, Device B: ITO/PEDOT-PSS/MEH-PPV/LiF/Al, Device C: ITO/PEDOT-PSS/Alq3/Al and Device D: ITO/PEDOT-PSS/Alq3/LiF/Al show the effects of temperature variation in their performance. The OLEDs were fabricated on ITO coated glass of surface resistivity in the range of tens of ohms. The standard cleaning procedure (J W. H. Kim et al,2003) in deionized water, acetone and isopropyl alcohol were carried out. PEDOT:PSS and MEH:PPV were spun cast on ITO coated glass for polymer devices. For fabricating small molecule based OLEDs, Tris(8-hydroxyquinolinato) aluminium (Alq3) was vacuum evaporated at 10^{-6} torr by physical vapor deposition. The buffer layer of LiF was also vacuum evaporated in the devices where such caps were used to enhance the injection of carriers. The metallic cathode was also vacuum evaporated in all the four sets of devices.The J-V characteristics were plotted by using a Keithley 2400 Source meter interfaced to a computer. Impedance versus frequency behavior was studied using Electrochemical workstation IM6 ex from Zahner, Germany. It also gives the plots of real versus imaginary impedances. The measurements from cryogenic temperature to room temperature were taken with the help of cryostat. The thickness of the evaporated as well as spun cast layers and refractive index of PEDOT:PSS film on ITO were measured by Sopra make Spectroscopic Ellipsometer. The luminance behavior was observed with the help of a fibre optic spectrometer Avantes.

3.1 Current density versus bias voltage

The variation of current density with respect to the applied voltage explains the turn on phenomena of the device. Figures 10 and 11 show the J-V characteristics of devices A, B, C and D respectively at a temperature varying from very low value of 100K to room temperature. The devices A and B are having MEH:PPV as the emissive layer and their J-V characteristics are shown in figure 10a and 10b respectively. The devices C and D in which the emissive material is small molecule Alq3 exhibits a current variation as shown in figure 11a and 11b respectively.

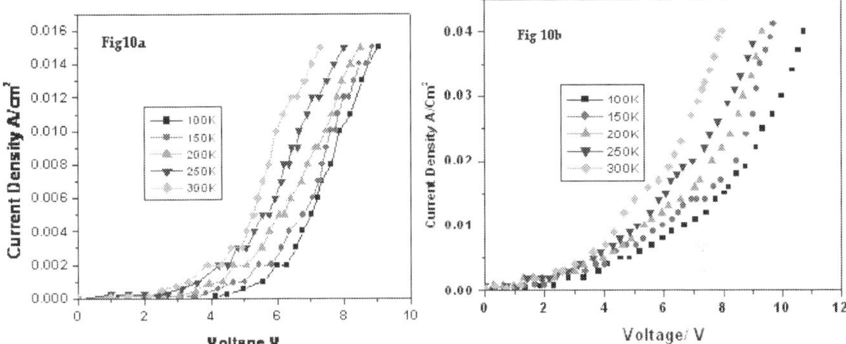

Fig. 10. JV characteristics of Device A and B at different temperatures.

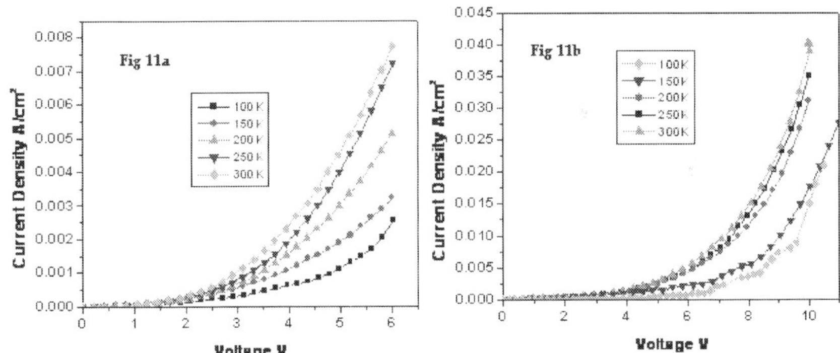

Fig. 11. J-V Characteristics of Device C and D at different temperatures.

The lowest voltage required [26] for the start of tunneling and hence the light emission is the 'turn on' voltage. At very small forward voltage, tunneling doest not occur and it begins at the flat band condition. In fact, 'flat band voltage' is the energy gap minus the two energy offsets. The turn on voltage is a function of the energy levels of the polymer and considered to be independent of the polymer thickness. The emission from the device starts to occur at a point where the current starts to increase rapidly when plotted in linear axis. This is the 'operating voltage' at which light emission becomes visible to the naked eye and it is a function of the thickness of the emissive layer.

In the device A, 'turn on' happens at 4volts at 100K and it gradually comes down at every fall of 50k and finally it reaches 2.2 volts at 300K. For the device where LiF buffer layer (device B) is used to catalyze the carrier injection, 'turn on' occurs bit earlier than device A-2.8 volts at 300K and falls to 2.1 volts at 100K. It is obvious that the rate of this fall in device A is more than that of device B. The operating voltage also experiences a similar shift due to the variation in temperature-5.8 volts at 100K to 3.9 volts at 300K in device A and in the case of device B, it is 4 volts at 100K to 2.9 volts at 300K. A similar variation can be seen in organic light emitting devices also where Alq3 is the emissive material.

In all the four sets of devices, it was observed that the 'turn on' occurs at smaller values of applied bias voltage in room temperature. As the temperature goes on decreasing, the turn on becomes slower and it becomes worst at the lowest temperature of 100K. At lower forward bias Fowler Nordheim tunneling contributes to the device current whereas at higher bias voltages, space charge limited current (SCLC) governs the current. The current density in dual carrier device is a direct function of the product and the sum of the mobilities of electrons and holes (P.W.M.Blom et al,1998) , which is clear from the eqn.10.

$$J = \left(\frac{9\pi}{8}\right)^{1/2} \varepsilon_0\varepsilon_r \left(\frac{2q\mu_p\mu_n(\mu_p + \mu_n)}{\varepsilon_0\varepsilon_r B}\right)^{1/2} \frac{V^2}{L^3} \qquad [10]$$

On increasing the recombination constant B, the neutralization decreases which brings down the current density.The mobilities at lower temperatures substantially come down which contribute to the slower 'turn on' process at lower temperatures.

3.1.2 Impedance characteristics

Impedance spectroscopy is a powerful tool (Shun-Chi Chang et al,2001) to investigate the behavior of OLEDs when applied with an alternating input having a frequency ranging from tens of hertz to several hundreds of kilohertz with a small ac input signal like 100mV peak to peak and this can be performed in the presence or absence of a superimposing DC voltage. The use of lower excitation voltage could assure the quasi-equilibrium condition needed to carry out such experiments and probe charged states in the bulk. Further, small ac voltage without a superimposing DC voltage would ensure clear separation of bulk effects from interfacial effects. By using spectroscopic investigations, real versus imaginary impedance can be derived which helps to evolve the electrical models of the device.

The equivalent circuit of OLED is normally represented by a series resistance with a parallel combination of resistance and capacitance in the case of single layer devices. More RC layers to be included when more layers are added in the device. This is normally deduced from the real and imaginary impedance obtained through impedance spectroscopy. The resistance and capacitance can be computed by fixing the points of series resistance (Rs) and the parallel resistance as shown in the figure 12. From the measurements of imaginary impedance (Z″) the frequency corresponding to its maximum value can be equated as ω = 1/(Rp.Cp). From this equation value of Cp can be computed and the equivalent circuit is drawn as shown in figure 13. It is to be highlighted that when PEDOT:PSS is used as hole transport layer in organic or polymer devices, the impedance spectra resembles to that of a single layer device giving only one semicircle in the Cole-Cole plot or only one peak in the imaginary impedance measurements. In the real versus imaginary impedance plot (Cole-Cole plot), frequency is always an implicit variable.

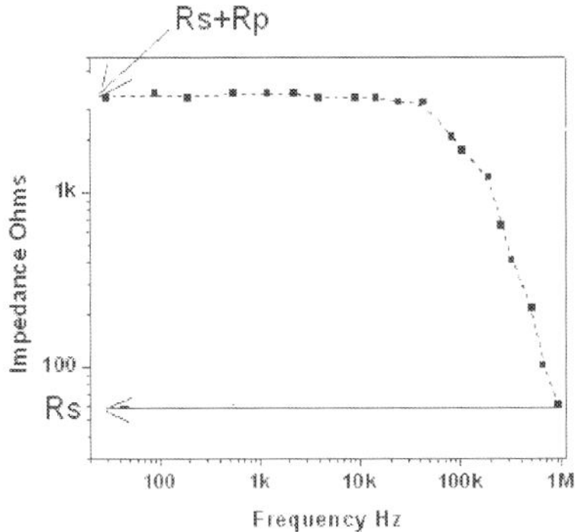

Fig. 12. Deducing equivalent circuit from impedance plots.

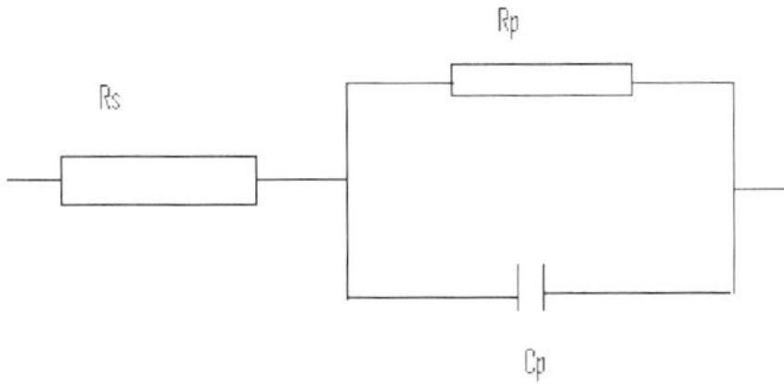

Fig. 13. Equivalent Circuit.

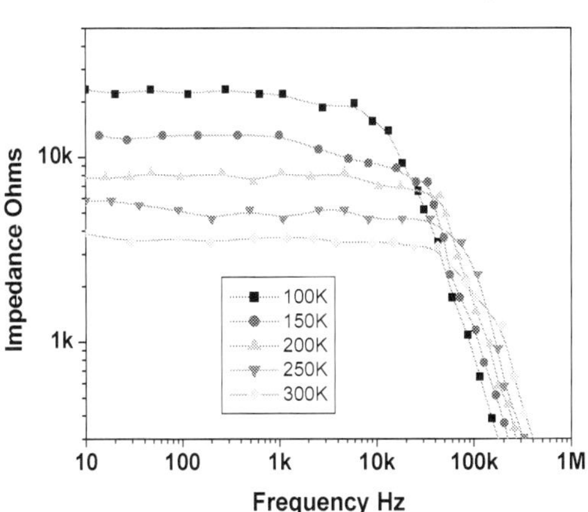

Fig. 14a. Impedance spectra of ITO/PEDOT:PSS/MEH:PPV/Al.

The qualitative difference between the device behaviors when subjected to a small excitation of 100mV peak to peak with no superimposing DC voltages at different temperature is an interesting case to be analyzed. Figures 14 and 15 show the impedance spectra of the devices which use the polymer and small molecule electroluminescent layers in dual carrier injection devices. It is clear from the figure 14a and 14b that the device in which a buffer layer of LiF is used (device B) offers more impedance at the same frequency than the one without it (device A).

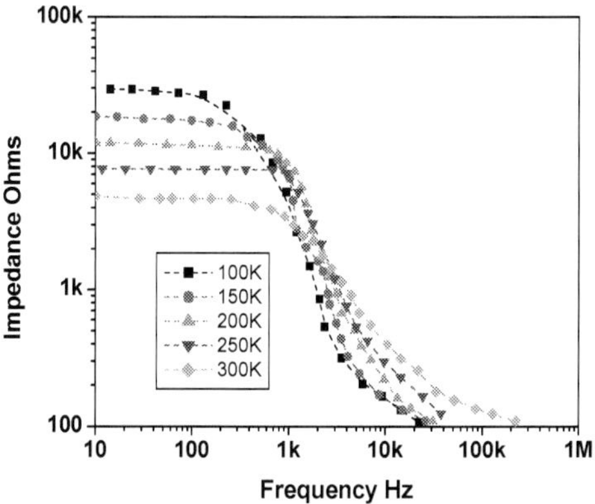

Fig. 14b. Impedance spectra of ITO/PEDOT:PSS/MEH:PPV/LiF/Al.

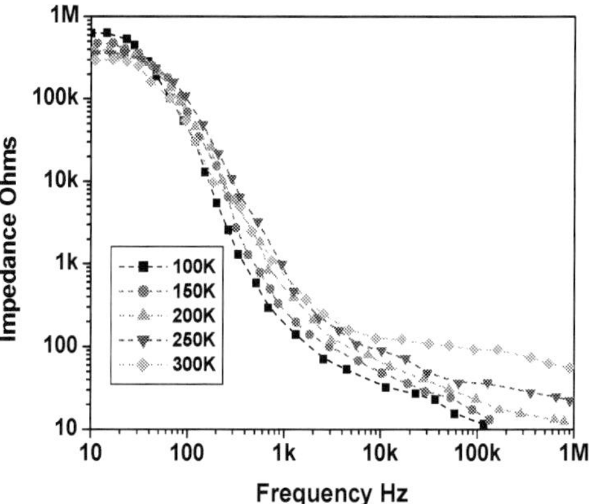

Fig. 15a. Impedance spectra of ITO/PEDOT:PSS/Alq3/Al.

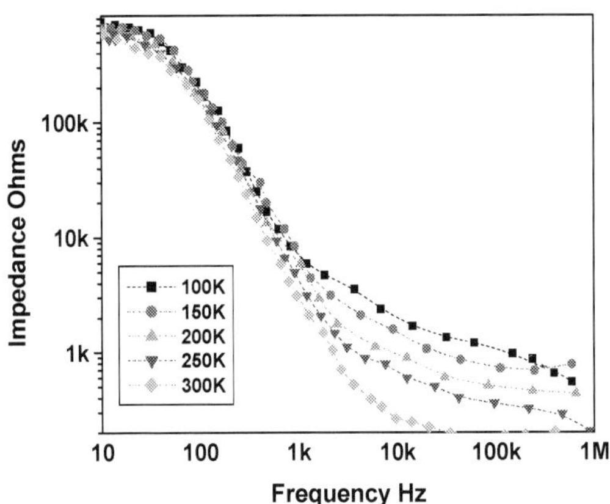

Fig. 15b. Impedance spectra of ITO/PEDOT:PSS/Alq3/LiF/Al.

In both cases, impedance remains high for higher value of frequency and it comes down as temperature is lowered. The impedance falls at lower frequencies in device B than A.
In the case of the devices which use small molecule Alq3 as emissive layer exhibits a higher impedance than that of MEH:PPV device in identical thickness of the layers. Here in low frequency regime of the spectra, the impedance remains constant for a smaller span of frequencies than that of the polymer devices. It is worth mentioning that the organic light emitting device which does not use a buffer layer of LiF offers less impedance than its counterpart which uses a buffer layer. At room temperature, the fall of impedance is faster for the device D.

Temp.	Value of Rs	Value of Rp	Value of Cp
100K	20Ω	23.35 KΩ	725 pF
150K	40Ω	13.1 KΩ	16.6 nF
200K	72Ω	7.78 KΩ	5.76nF
250K	50Ω	5.8KΩ	9.68nF
300K	60Ω	3.89 KΩ	144.6nF

Table 1. Values of the Parameters in the Equivalent Circuit.

A sample computation of the parameters in the equivalent circuit of the device shown in fig. 14a is given in Table1.

4. Encapsulation and reliability enhancement

Ever since efficient organic light emitting diodes were reported (C.W.Tang et al,1989), there has been unending efforts for devising full color displays with the least degradation. Evolution of dark spots and consequent decay of device luminance were the reported (P.E.Burrows et al,1994) phenomena in degradation studies of organic luminescent devices.

No doubt, the degradation due to moisture poses threat in lifetime and performance and this problem is worse in devices having flexible substrates since they are more permeable to moisture and oxygen to which organic materials are sensitive too.

The first systematic study in this respect was from Burrows et al (P.E.Burrows et al,1994) and they had proposed encapsulation as a means of circumventing the decay of life time. Large area devices when operated for extended life, there has been occurrence of short circuits. Once the device is applied with a voltage, current in the range of tens of milli amperes is allowed to send through it and short circuit begins to develop. If a high current is applied for a short period, again it causes short circuit between electrodes. The formation of microscopic conduction paths through organic layers leads to burn out when high current is applied. These paths exist initially due to the non planarity at the interfaces, eventually leading to the formation of short circuits. Considering the sustainable features of the OLED devices, the encapsulation material (G.Dennler et al,2006) should be having low moisture absorption, low curing temperature, short curing time and transparent to visible light. A structure with an encapsulation proposed by Burrows et al is shown in figure 16. The device fabricated by conventional cleaning and coating procedures to be transferred from vacuum to a glove box in nitrogen ambience. A thin bead of epoxy adhesive to be applied through syringe around the edges with care that adhesive doesn't get in contact with the active layers. A clean glass of suitable dimension to be used for covering the top and UV curing can be used to ensure proper adhesion and connections from electrodes are to be taken out with thin Au bonding connected with colloidal silver solution.

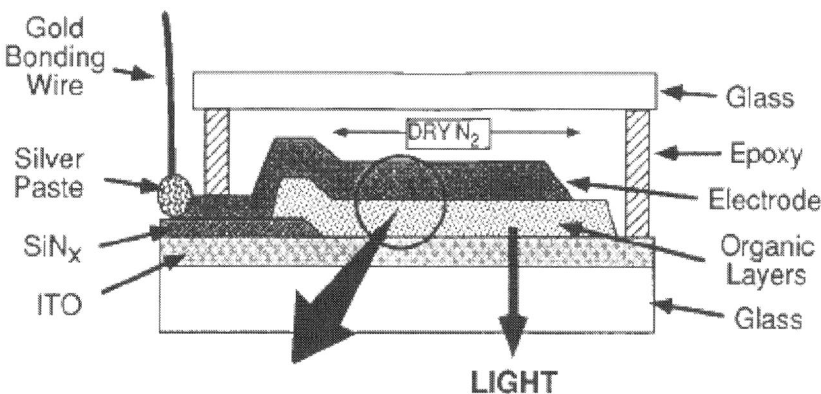

Fig. 16. Schematic on Encapsulation of OLEDs (P.E.Burrows et al,1994).

Another method of encapsulation is based on physical lamination (Tae-Woo Lee et al,2004) of thin metal electrodes supported by elastomeric layer against an electroluminescent organic is shown in figure 17. This method relies only on van der Waals interactions to establish spatially homogeneous, intimate contacts between the electrodes and the organic layers.

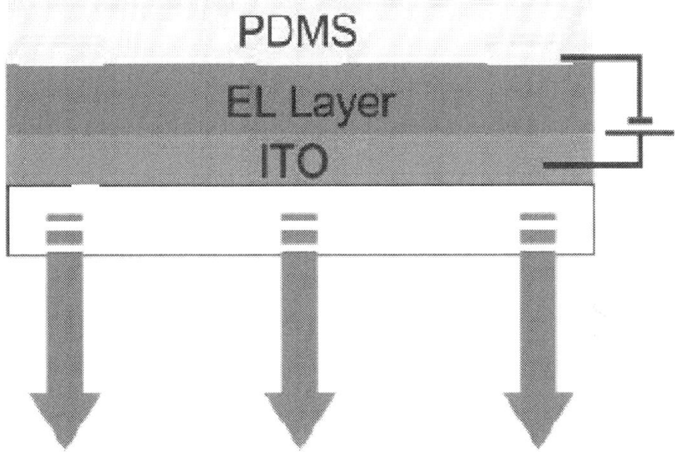

Fig. 17. Soft Contact Lamination of OLEDS [Tae-Woo Lee et al,2004).

The disruption at the electrode-organic interface can be substantially minimized with a high degree of protection against pinhole defects. This is better termed as soft contact lamination, which is intrinsically compatible with soft contact lithography which could very well be used for devices in nanometric regime. The encapsulation methods, however precise they are, induce morphological, physical or chemical changes in organic layers, which could be minimized by this soft contact lamination technique.

5. Perimeter leakage

Computation of leakage current is necessary for taking measures to gain control over it in order to enhance the performance of organic light emitting diodes. Not many number of studies have been reported so far in this regard and Garcia-Belmonte et al (Germa` Garcia-Belmonte et al,2009) has made some pioneering works considering the structural aspects of organic light emitting diodes. No doubt, leakage current has a momentous role in the stand-by life of the battery operated device and hence tracing its physical origin is equally important. The ohmic behavior in reverse biased and forward biased region (till built in voltage) is assumed to be linked to leakage current component too and hence total current density can be equated as $J_{tot}= J_{oper}+J_{leakage}$ .The current till built in voltage has a predominant leakage component and after this point it is due to the applied potential. The surface roughness of Indium Tin Oxide Layer (K.B. Kim et al,2003) and the local damage of the organic layer induced during radio frequency sputtering of cathodes (J H. Suzuki & M. Hikita ,2003;L.S.Liao et al,1999) are assumed to have links with the leakage paths.

6. References

Amare Benor, Shin-ya Takizawa, C. Pérez-Bolivar, and Pavel Anzenbacher(2010), *Energy barrier, charge carrier balance, and performance improvement in organic light-emitting diodes,*Applied Physics Letters 96, 243310

C. W. Tang, S. A. VanSlyke, and C. H. Chen (1989) *Electroluminescence of doped organic thin films,* Journal of Applied Physics 65, 3610

G. Dennler , C. Lungenschmied , H. Neugebauer , N.S. Sariciftci , M. Latre`che ,

G. Czeremuszkin , M.R. Wertheimer(2006). *A new encapsulation solution for flexible organic solar cells ,*Thin Solid Films 511 – 512 ,349 – 353

G.G. Malliaras, J.R. Salem, P.J. Brock, J.C. Scott (1998),*Current limiting mechanisms in polymer diodes,* Physical Review.B 58 R13411.

G.Yu and Alan J.Heeger(1997). *High efficiency photonic devices made with semiconducting polymers,*Synthetic Metals 85, 1183

Germa` Garcia-Belmonte , Jose´ M. Montero , Yassid Ayyad-Limonge , Eva M. Barea ,Juan Bisquert and Henk J. Bolink(2009), *Perimeter leakage current in organic light emitting diodes ,* Current Applied Physics 9 , 414–416

H. Suzuki, M. Hikita, (1996*) Organic Light emitting diodes with radio frequency sputter deposited electron injecting eelctrodes,* Applied Physics Letters 68 , 2276

H.Antoniadis. M.A.Abkowitz and B.R.Hsieh (1994), *Carrier deep – trapping mobility life time products in poly (p-phenylene vinylene),* Applied Physics Letters 65 (16) 2030

H.Bassler(1993*). Charge Transport in Disordered Organic Photoconductors,* Physica status solidi (b) 175,. 15

I.D. Parker(1994*), Carrier tunneling and device characteristics in polymer light emitting diodes,* Journal of Applied Physics. 75, 1656.

I.H. Campbell, T.W. Hagler, D.L. Smith, J.P. Ferraris(1996) *Direct Measurement of Conjugated Polymer Electronic Excitation Energies Using Metal/Polymer/Metal Structures* Physical Review Letters. 76 11,1900-1903.

J. C. Scott, J. H. Kaufman, P. J. Brock, R. DiPietro, J. Salem, and J. A.Goitia(1996), *Degradation and failure of MEH-PPV light-emitting diodes,* Journal of Applied Physics 79, 2745

J. H. Burroughes, D. D. C. Bradley, A. R. Brown, R. N. Marks, K. Mackay,R. H.Friend, P. L. Burns & A.B.Holmes. (1990) *Light-emitting diodes based on conjugated polymers,* Nature, 347, 539 .

J.C.Scott, Philip J.Brock, Jesse R.Salem, Sergio Ramos, George G.Malliaras, Sue A .Carter and Luisa Bozano (2000). *Charge transport processes in organic light emitting devices,* Synthetic Metals 111-112, 289-293

J.C.Scott, S.Karg and S.A.Carter(1997). *Bipolar charge and current distributions in organic light-emitting diodes* Journal of Applied Physics 82(3) , 1454-60

Justin Dane and Jun Gao(2004).*Imaging the degradation of polymer light emitting diodes* Applied Physics Letters, 85, 3905

K.-B. Kim, Y.H. Tak, Y.-S. Han, K.-H. Baik, M.-H. Yoon, M.-H.Lee (2003). *Relationship between Surface Roughness of Indium Tin Oxide and Leakage Current of Organic Light-Emitting Diode* Japanese Journal of Applied Physics, Vol.42,part 2, No.4B-letters 438-440

L. S. Roman, M. Berggren, and O. Ingana(1999).*Polymer diodes with high rectiification*: Applied Physics Letters. 75, 3557–3559

L.Bozano, S.A.Carter, and P.J.Brock(1998), *Temperature-dependent recombination in polymer composite light-emitting diodes* Applied Physics Letters 73, 3911

L.Bozano,S.A.Carter, J.C.Scott, G.G.Malliaras and P.J.Brock(1999) , *Temperature-and Field-dependent electron and hole mobilities in polymerlight-emitting diodes*(1999),Applied Physics Letters Volume 74, Number 8

L.S. Liao, L.S. Hung, W.C. Chan, X.M. Ding, T.K. Sham, I. Bello,C.S. Lee, S.T. Lee(1999). *Ion-beam-induced surface damages on tris-(8-hydroxyquinoline) aluminum*, Applied Physics Letters 75, 1619.

M. A. Lampert and P. Mark(1970) *Current Injection in Solids* ~Academic, NewYork, 1970P. E. Burrows, V. Bulovic, S. R. Forrest, L. S. Sapochak, D. M. McCarty, and M. E. Thompson(1994), *Reliability and degradation of organic light emitting diodes*, Applied Physics Letters 65 (23)

P. S. Davids, Sh. M. Kogan, I. D. Parker, and D. L. Smith (1996). *Charge injection in organic light emitting diodes: Tunneling into low mobility materials*, Appl. Phys. Lett., vol. 69, pp. 2270- 2272

P.W.M.Blom and Marc J.M.de jong (1998), *Electrical characterization of polymer lightemitting diodes,* IEEE journal of selected topics in quantum electronics, vol. 4, no. 1

P. W. M. Blom, M. J. M. De Jong, and S. Breedijk (1997). *TemperatureDependentElectron-HoleRecombinationinPolymerLight-EmittingDiodes,*Applied Physics Letters., vol. 71, pp. 930-932

Papadimitrakopoulos, K. Konstadinidis, T. Miller, R. Opila, E. A. Chandross and M. E. Galvin(1994).*Quantum Efficiencies of Poly(Paraphenylene vinylenes)*, Chemistry of materials 6, 1563

S. Alem, R. de Bettignies, J. M. Nunzi, and M. Cariou(2004) *Efficient polymer based interpenetrated network photovoltaic cells ,*Applied Physics Letters. 84, 2178–2180

S. M. Sze(1981), *Physics of Semiconductor Devices* (Wiley, New York, Shun-Chi Chang and Yang Yang, Fred Wudl, Gufeng He and Yongfang Li(2001), *AC impedance characteristics and modeling of polymer solution light emitting devices,* Journal of physical chemistryB , *105,* 11419-11423

Tae-Woo Lee and O Ok Park(2000) *The Effect of Different Heat Treatments on the Luminescence Efficiency of Polymer Light-Emitting Diodes,* Advancecd Materials. 12, No. 11

Tae-Woo Lee, Jana Zaumseil, Zhenan Bao, Julia W. P. Hsu, and John A. Rogers(2004) *Organic light-emitting diodes formed by soft contact lamination,* PNAS , 101 (2),429–433

W. H. Kim,G. P. Kushoto, H.Kim, Z.H.Kafafi(2003) *Effect of annealing on the electrical properties and morphology of a conducting polymer used as anode in organic light emitting devices,* Journal of Polymer Science: Part B: Polymer Physics, Vol. 41,21, 2522–2528

W.D.Gill(1972), *Drift mobilities in amorphous charge transfer complexes of trinitrofluorenone and poly-n-vinylcarbazole,* Journal of Applied Physics 43, 5033

Wolfgang Brutting, Stefan Berleb and Anton G.Muckl (2001). *Device physics of organic light-emitting diodes based on molecular materials,* Organic Electronics 2, 1-36

Y. Cao, G. Yu, C. Zhang, *R. Menon,* and A.J. Heeger(1997), *Polymer light emitting diodes with polyethylene dioxythiophene polystyrene sulfonate as the transparent anode,*Synthetic Metals,,87, 171

Y.Kawabe, M.M.Morrell, G.E.Jabbour, S.E.Shaheen, B.Kippelen and .Peyghambarian(1998) *A numerical study of operational characteristics of organic light-emitting diodes.*Journal of Applied Physics 84(9)

2

Integrating Micro-Photonic Systems and MOEMS into Standard Silicon CMOS Integrated Circuitry

Lukas W. Snyman
*Silicon Photonics Group, Laboratory of Innovative Electronic Systems,
Department of Electrical Engineering, Tshwane University of Technology (TUT),
Pretoria,
South Africa*

1. Introduction

Various researchers have highlighted the integration of small-dimension, optical communication- and micro-systems into mainstream silicon fabrication technology (Bourouina et al 1996; Clayes, 2009; Fitzgerald & Kimerling, 1998; Gianchanadni, 2010; Robbins , 2000; Soref , 1998). The realization of sufficiently efficient light-emitters have, been a major technological challenge.

A research area, known as "Silicon Photonics", has emerged in recent years (Kubby and Reed , 2005-2010; Savage 2002; Wada, 2004). This technology offers the advanced processing of data at ultra high speeds and provides advanced optical signal processing. It can analyse diverse optical data directly on chip, and may even contribute towards solving the interconnect density problem, associated with current microprocessor systems. Until now, this technology has been primarily established at 1550 nm. The reason is to conform with the main long haul and low loss telecommunication bands. The realization of waveguides, modulators, resonators, filters etc. on silicon platforms has been achieved until now with relative ease by using mainly Silicon-on-Insulator (SOI) technology.

Two main application fields have been developed, namely (1) high speed optical communication with modulation speeds and bandwidths reaching up to THz , utilizing Si-Ge technology , and (2) the so-called "Lab on chip" approach, where the main goal is the realization of an optical micro-system, which can perform certain analysis of the environment or attached media.

In the absence of an efficient light source at 1550 nm on a chip, these systems operate currently with external light sources. They also incorporate Si-Ge detectors, which are not compatible with mainstream silicon technology (Beals at al, 2008; Lui et al, 2010; Wada, 2004). The integration of germanium into silicon structures requires the addition of complex and very expensive processing procedures. Recently, a Ge-on- Si laser source was announced by Lui et al. in 2010. This technology provides coherent optical emission on a chip, but utilizes quite complex strained Si-Ge layer technology.

Making use of adequately emitting Complementary Metal Oxide Semiconductor (CMOS) optical sources, together with good silicon detectors, shows good potential to manufacture diverse new optical communication and integrated systems directly onto CMOS silicon chips. The optical communication bandwidth of these systems may not necessarily compete with that of Si-Ge technology, but it still could take a substantial market share when the benefits of an all silicon and CMOS compatible systems are considered. These benefits are, mainly, (1) lower complexity of the technology; (2) lower cost of fabrication; (3) ease of integration into the mainstream CMOS technology; and (4) higher system integration capabilities.

The realization of micro-photonic systems on CMOS chips can lead to many new products and markets in the future. Achieving these goals can lead to low cost "all-silicon" opto-electronic based technologies and so-called "smarter" and more " intelligent" CMOS chips. Envisaged systems could range from CMOS based micro-systems, analyzing environmental or bio-logical substances to sensors on chips, which can detect vibration,, inertia and acceleration. Whole new products, aimed at the medical and biological market could be developed and sensor systems, which could measure colour, optical intensities, absorption, and distances (including metrology) . Such a new field could be appropriately nomenclated "Silicon CMOS Photonics".

Propagation wavelength (nm)	Optical source	Waveguide technology	Detectors	Complexity (10)	Estimated cost to implement (10)
450	Available	Not available	Available	10	8
750 – 850	Available	Challenges	Available	3	2
1100	Available	Available	Not available	8	10
1500	Available	Available	Not available	8	10

Table 1. Comparison of optical source, waveguide and optical detector technologies for generating new micro-photonic systems in CMOS integrated circuit technology.

Table 1 summarizes the current options for integrating photonic systems into CMOS technology with regard to optical source, waveguide, detector technology and complexity. The composition of the table is based on the presentation of results in the field at recent international conferences (SPIE Photonic West 2009,2010).

The analysis shows, that if efficiently enough waveguides could be developed in the wavelength regime of 750-850 nm, both optical sources and detectors could be completely compatible with CMOS technology. The waveguide technology at these wavelengths faces some major challenges, as very little research and development work has been done in this field. The operating wavelength would be about one half of that of 1550 nm, which is currently the wavelength for long haul communication systems. This wavelength could still very effectively link with the current wide bandwidth 850 nm local area network technology.

In this chapter, research results are presented with regard to the following : (1) The optical compatibility of silicon CMOS structures. (2) The current state of the art technology of optical sources at submicron wavelengths, that are compatible with mainstream CMOS technology. (3) Development capabilities of waveguides in the 750 – 850 nm wavelength regime utilising CMOS technology. (4) "Proof of concept" of optical communication systems that utilize "all silicon CMOS components ". (5) Finally, the development of CMOS based micro-photonic systems using CMOS technology in the 650 – 850 nm wavelength regime.

2. Optical compatibility of CMOS technology

First, the capability of CMOS technology is evaluated to accommodate optical propagation in micro-photonic systems. An investigation of the CMOS structure, as in Fig. 1, (Fullin et al., 1993), shows that the field oxide, the inter-metallic oxide, and the silicon nitride (Si_3N_4) passivation layer are all optical transparent and can serve as "optical propagation and/or optical coupling structures " in CMOS integrated circuitry.

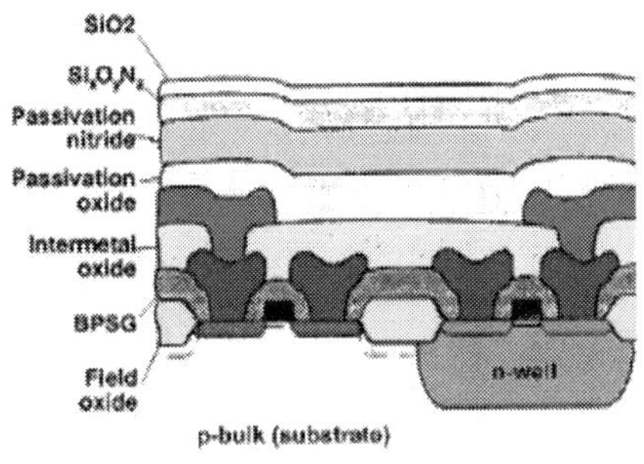

Fig. 1. Schematic diagram displaying a typical structure used in field oxide based CMOS integrated circuit technology. Layers that are optically transparent below 1 µm are shown in yellow, green and grey . Bright yellow: Native silicon dioxide; Yellow: intermetallic oxide; White: Passivation oxide ; Green: Silicon nitride.

Field oxide, used for electrical isolation between MOSFET transistors by older CMOS processes, is formed by oxidation of silicon. This results in a high quality "glassy" layer of superb optical transmission with a refractive index of 1.46. A drawback is that this layer is bonded at the bottom to a highly absorptive silicon substrate with a refractive index of 3.5 and a very high absorption coefficient for all optical radiation below 950 nm. It is anticipated to use this layer as medium to transport optical radiation vertically outward from Si Avalanche based Light Emitting Diodes (Si AvLEDs), which are situated at the silicon-overlayer interface (Snyman et al. , 2009). The specific structure associated with the field oxide, favours simple convex lensing for outward directed vertical optical radiation. Fig. 2 demonstrates the concept obtained by structural analysis and ray tracing.

The inter-metallic oxide, positioned between metallic layers, are CVD plasma deposited and mainly used as electrical isolation between the metal layers. Literature surveys (Beals et al , 2008, Gorin et al, 2008 show that, even so, these layers are porous, they offer suitable propagation for the longer mid infra-red wavelengths, where structural defects, such as porosity, grain boundaries and side wall scattering due to roughness play a lesser role . The metallic layers bonding to the oxide inter-metallic oxide layers can be used as effective reflectors or optical confinement layers.

The oxide which is deposited on top of the metal layers serves as a pre-passivation step prior to the final passivation by silicon nitride. This layer could be used for the propagation of mid infrared wavelengths.

The silicon nitride layer possesses interesting optical properties. One main advantage is that the refractive index is higher than that of the surrounding plasma oxide layers. Depending on the composition and deposition technology, its refractive index can be varied between 1.9

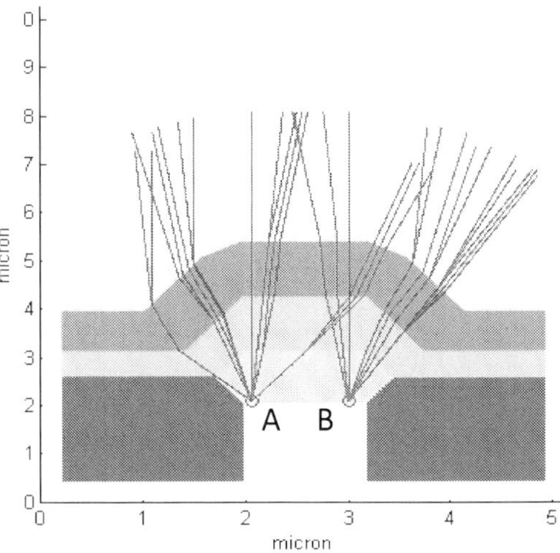

Fig. 2. Optical propagation phenomenon at 750nm in CMOS over layers using simple ray tracing techniques. The layer color indexing is the same as in Fig.1.

and 2.4. This layer, when surrounded by silicon oxide, is ideal for waveguiding optical radiation laterally in the CMOS structure. Since this layer, too, is created by the CVD process, is porous and has a rough surface. Therefore, it is anticipated to use this layer for the propagation of longer wavelengths. Fig. 3 demonstrates this concept.

Optically transparent layers made of polymer or silicon oxi-nitride can be deposited on top of the CMOS layers with relative ease by means of suitable post processing procedures. Since these layers are deposited at low temperatures, they can be subjected to further procedures to generate sloped or lens like structures in the final outer layer CMOS structure. Recently, at the SPIE Photonic West Trade Show in San Francisco, it was reported that RF etching and other technologies exist to pattern such layers with up to 150 steps using appropriate software and process technology (Tessera, 2011).

CMOS processes below 350 nm utilize a planarization process after the MOSFET transistor fabrication. They deposit up to six metal layers on top of these layers, where sloping of these layers is caused by the thicker outer metal layers (Foty, 2009, Sedra, 2004). However, this technology uses trench isolation for electrically isolating n- and p MOSFETS laterally in the CMOS structure. The trench- isolation technology opens up interesting optical properties. Trenches are spatially defined. This implies that light emitters can be fabricated in the CMOS structure at certain areas which are laterally bounded by isolation trenches or deep crevasses in the silicon. It hence follows that, if these trenches could be filled with an optical material of higher refractive index, optical radiation emitted from the silicon-overlayer interface, could then be coupled with high efficiency directly into adjacent optical channels. The current CMOS technology can create a thin oxidation layer that is used as isolation layer in the trench technology. If this layer can be enhanced and is followed by a layer of high refractive index material such as silicon nitride, interesting lateral optical conductors or waveguides can be constructed at the silicon–overlayer interface. Some of these concepts are illustrated in Fig. 4 and Fig 5 (Snyman 2010d) (Snyman and Foty, 2011a).

Since certain optical sources can only be fabricated at the silicon–overlayer interface in the CMOS structure (such as Si-avalanche LED technology), coupling of optical radiation from the silicon-overlayer interface to the outer CMOS surface layers needs to be investigated. Analysis conducted has shown, that by applying special CMOS layer definition techniques and positioning these layers under 45 degree, structures can be generated which couple the optical radiation from the silicon substrate to the over layers. Fig. 3 illustrates this concept. Additional structures can be designed to ensure nearly 100 % coupling into the silicon nitride layer.

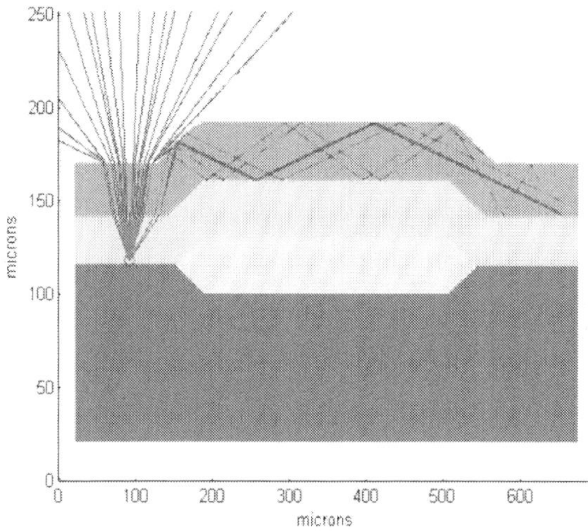

Fig. 3. Optical propagation phenomenon at 750nm in CMOS over layers using simple ray tracing techniques. The layer color indexing is the same as in Fig.1. Waveguiding of radiation along the silicon nitride overlayer is demonstrated.

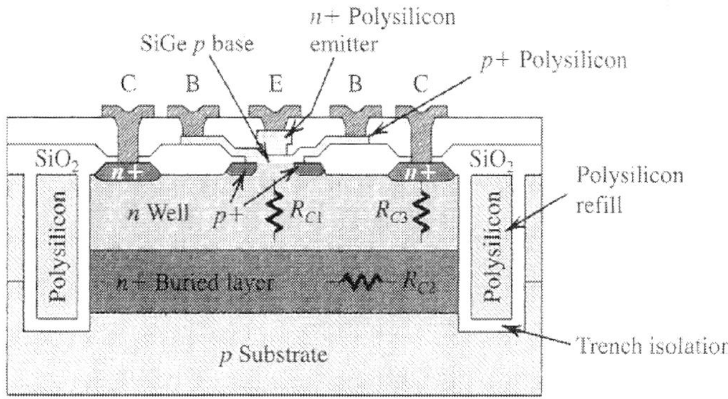

Fig. 4. Components and structural layout of the latest CMOS processes utilising isolation trench based technology (Sedra, 2004).

This implies that photonic system structures can be generated in CMOS technology which incorporate so called " multi-planing", where optical radiation can be coupled from one plane to the next. Obviously, the concepts described here are still in its infancy, and further research is necessary. Both standard CMOS as well as Silicon-on-Insulator (SOI) technology are suitable to realise some of the concepts.

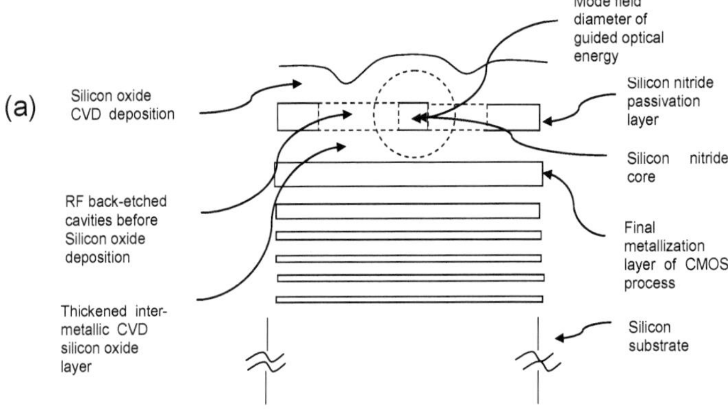

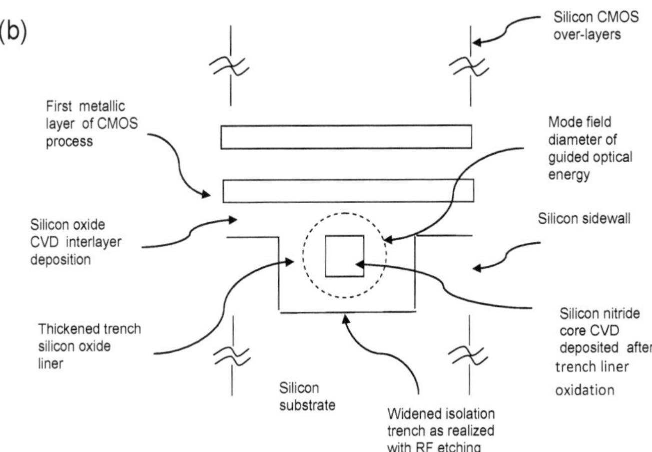

Fig. 5. Cross-sectional profiles of possible wave guides that can be constructed by modification of existing CMOS processes. (a) Waveguide structure fabricated by post processing procedures in the overlayers. (b) Trench based waveguide structure with silcon nitride embedded in silcon oxide.

3. Viable optical sources for all- silicon CMOS technology

The availability of optical sources suitable for integration into CMOS technology is evaluated. A survey reveals that a number of light emitters have been developed since the nineties that can be integrated into mainstream silicon technology. They range from forward biased Si p-n LEDs which operate at 1100 nm (Green et al, 2001; Kramer et al 1993; Hirschman et al 1996); avalanche based Si LEDs which operate in the visible from 450 – 650 nm (Brummer et al, 1993; Kramer et al 1993; Snyman et 1996- 2006); organic light emitting diodes (OLED) incorporated into CMOS structures which also emit in the visible (Vogel et al., 2007); to, strained layer Ge-on-ilicon structures radiating at 1560 nm (Lui, 2010). Fig. 6 illustrates the spectral radiance versus wavelength for a number of these light sources as found in various citations.

Forward biased p-n junction LEDs and Ge-Si hetero-structure devices emit between 1100 and 1600 nm. This wavelength range lies beyond the band edge absorption of silicon, and all silicon detectors respond only weakly or not at all to this radiation. Hence, these technologies are not viable for the development of only silicon CMOS photonic systems. The Ge-Si hetero-structure can be realized in Si–Ge CMOS processes, but increases complexity and costs.

Organic based Light Emitting Diodes (OLED) utilize the sandwiching of organic layers between doped silicon semiconductor layers with high yields between 450 and 650 nm (Vogel et al , 2007). In spite, the incorporation of foreign organic materials through post-processes this technology is a viable option. The photonic emission levels are quite high, up to 100 cd m^{-2} at 3.2 V and 100 mA cm^{-2}. The organic layers must be deposited and processed at low temperature. This technology is, therefore, particularly suited for post processing, and as optical sources in the outer layers of the CMOS structures. A major uncertainty with regard to this technology is the high speed modulation capability of these devices.

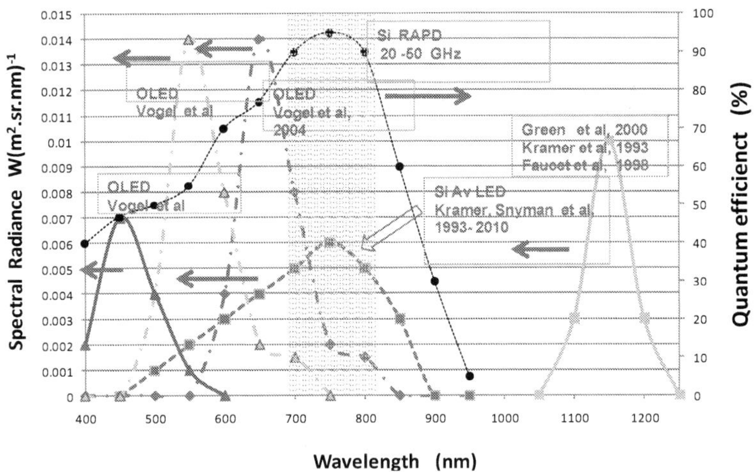

Fig. 6. Spectral radiance characteristics of Organic Light Emitting Devices (OLEDs and Si avalanche-based light emitting device (Si Av LED), and comparison with the spectral detection range of reach through avalanche detector (RAPD) devices.

Si avalanche light emitting devices in the 450 – 650 nm regime have been known for a long time (Newman 1955; Ghynoweth et al, 1956)]. The fabrication of these devices is high temperature compatible and can be used in standard silicon designs. Viable CMOS compatible avalanche Si LEDs (Si CMOS Av LEDs) have emerged since the early 1990's. Kramer & Zeits (1993) were the first to propose the utilization of Si Av LEDs inside CMOS technology. They illustrated the potential of this technology. Snyman et al (1998-2005) have realized a series of very practical light emitting devices in standard CMOS technology, such as micro displays and electro-optical interfaces, which displayed higher emission efficiencies as well as higher emission radiances (intensities). Particularly promising results have been obtained regarding efficiency and intensity, when a combination of current density confinement, surface layer engineering and injection of additional carriers of opposite charge density into the avalanching junction, were implemented (Snyman et al., 2006 - 2007). These devices showed three orders of increase in optical output as compared with previous similar work. However, increases in efficiency seemed to be compromised by higher total device currents; because of loss of injected carriers, which do not interact with avalanching carriers. Du Plessis and Aharoni have made valuable contributions by reducing the operating voltages associated with these devices (2000, 2002).

Fig. 7 presents an example of an electro-optical interface that was developed by Snyman et al. in association with the Kramer- Seitz group in 1996 in Switzerland and which offered very high radiance intensity (approximately 1 nW) in spot areas as small as 1 μm^2.

The latest analysis of the work of Kramer et al and Snyman et al (Snyman et al, 2010), shows that, particularly, the longer wavelength emissions up to 750 nm can be achieved by focusing on the electron relaxation techniques in the purer n-side of the silicon p-n avalanching junctions. This development has a very important implication. The spectral radiance of this device compares extremely well with the spectral detectivity of the silicon reach through avalanche photo detector (RAPD) technology. A particular good match is obtained between the emission radiance spectrum of this device and the detectible spectrum of a RAPD (see Fig. 6).

 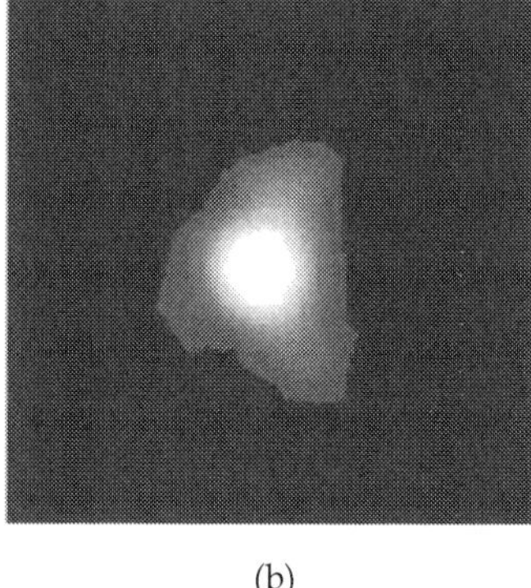

(a)   (b)

Fig. 7. Si avalanche-based light emitting device (Si Av LED) and electro-optical interfaces realized in 1.2 μm Si CMOS technology with standard CMOS design and processing procedures (Snyman, 1996). (a) Top view with bright field optical microscopy. (b) Optical emission characteristics in dark field conditions

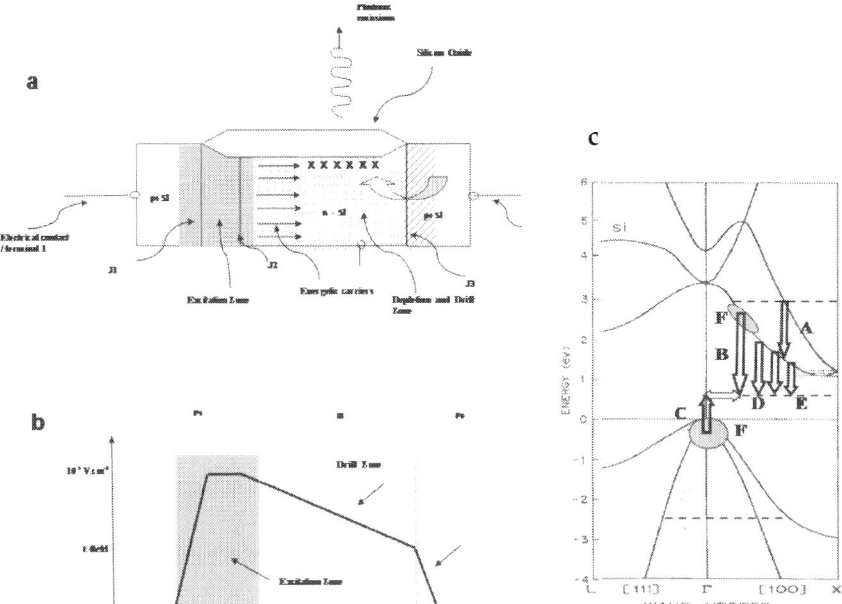

Fig. 8. Schematic diagram showing the operation principles of a Si avalanche-based light emitting device (Si Av LED) and electro-optical interface. (a) Structure of the device. (b) Electric field profile through the device and , (c), nature of photonic transitions in the energy band diagram for silicon .

Fig. 8 represents some of the latest in house designs with regard to a so called "modified E-field and defect density controlled Si Av LED". Only a synopsis is presented here and more details can be found in recent publications (Snyman and Bellotti, 2010a). The device consists of a p+-i-n-p+ structure with a very thin lowly doped layer between the p+ and the n layer. The purpose of this layer is to create a thin but elongated electric field region in the silicon that will ensure a number of diffusion multiplication lengths in the avalanche process. The excited electrons loose their energies mainly in the n–type material through various intra-band and inter-band relaxation processes. If the p⁺n junction at the end of the structure is slightly forward biased and a large number of positive low energy holes is injected into the n-region, these holes can then interact with these high energy electrons . This enhances the recombination probability between high energetic electrons and low energy holes.

The recombination process can be further enhanced by inserting a large number of surface states at the Si-SiO$_2$ interface in the n-region. This can cause a "momentum spread" in the n-region for both, the energetic electrons as well as the injected holes. Fig. 8 (c) presents the photonic transitions that are stimulated by this design. Excited energetic electrons from high up in the conduction band may relax from the second conduction band to the first conduction band. Energetic electrons excited by the ionization processes may interact and relax to defect states which are situated in the mid-bandgap level between the conduction band and the valence band. The maximum density distribution (electrons per energy levels) is around 1 to 1.8 eV (Snyman 2010a) , and relaxation to mid-bandgap defect states will cause a spread of light emission energies from 0.1 eV to 2.3 eV , with maximum transition possibilities between 1.5 eV and 2.3 eV. By controlling the defect density in this device, one can favour either the 650 nm or 750 nm emissions. Total emission intensities of up to 1 μW per 5 μm^2 area at the Si-SiO$_2$ interface have recently been observed (Snyman and Bellotti, 2010a). Further improvement is currently underway in order to increase particularly the longer wavelength emissions associated with these structures.

In summary, particularly promising about the application of Si Av LEDs into CMOS integrated systems, is the following :

- Si Av LEDs can emit an estimated 1 μW inside silicon and at compatible CMOS operating voltages and currents (3-8 V, 0.1- 1 mA) they can emit up to 10 nW / μm^2 at 450 -750 nm (Snyman and Bellotti, 2010a; Snyman 2010b; Snyman 2010c).
- They can be realized with great ease by using standard CMOS design and processing procedures , vastly reducing the cost of such systems.
- The emission levels of the Si CMOS Av LEDs are 10^{+3} to 10^{+4} times higher than the detectivity of silicon p-i-n detectors, and hence offer a good dynamic range in detection and analysis.
- These types of devices can reach very high modulation speeds, greater than 10 GHz, because of the low capacitance reverse biased structures utilised (Chatterjee, 2004).
- They can be incorporated in the silicon-CMOS overlayer interface, because they are high temperature processing compatible.
- They can emit a substantial broadband in the mid infrared region (0.65 to 0.85 μm) . Particularly, p$^+$n designs emit strongly around 0.75 μm (Kramer 1993, Snyman 2010a).

4. Development of CMOS optical waveguides at 750nm

The development of efficient waveguides at submicron wavelengths in CMOS technology faces major challenges, particularly due to alleged higher absorption and scattering effects at submicron wavelengths.

A recent analysis shows that both, silicon nitride and Si oxi-nitride, transmitting radiation at low loss between 650 and 850 nm (Daldossa et al., 2004; Gorin et al., 2008). Both, Si O$_x$ N$_y$ and Si$_x$ N$_y$ possess high refractive indices of 1.6 - 1.95 and 2.2 - 2.4 respectively, against a background of available SiO$_2$ as cladding or background layers in CMOS silicon .

Subsequently, a survey was conducted of the optical characteristics of current CVD plasma deposited silicon nitrides that can be easily integrated in CMOS circuitry. In Fig. 9, the absorption coefficients versus wavelength are given for three types of deposited silicon nitrides. The first curve corresponds to the normal high frequency deposition of silicon nitride used in CMOS fabrication. The results were published by Daldossa et al. , 2004. The second curve corresponds to a low frequency deposition process as recently developed by Gorin et al (2008). The third curve corresponds to a special low frequency process followed by a low temperature "defect curing" technique as developed by Gorin et al. This process offers superb low loss characteristics. These results are extremely promising , and calculations show that, with this technology , very low propagation losses of 0.5 dB cm^{-1} at around 750 nm can be achieved when combined with standard CMOS technology. This wavelength falls into the maximum detectivity range of state-of-the-art reach- through avalanche silicon photo detectors (Si-RAPDs).

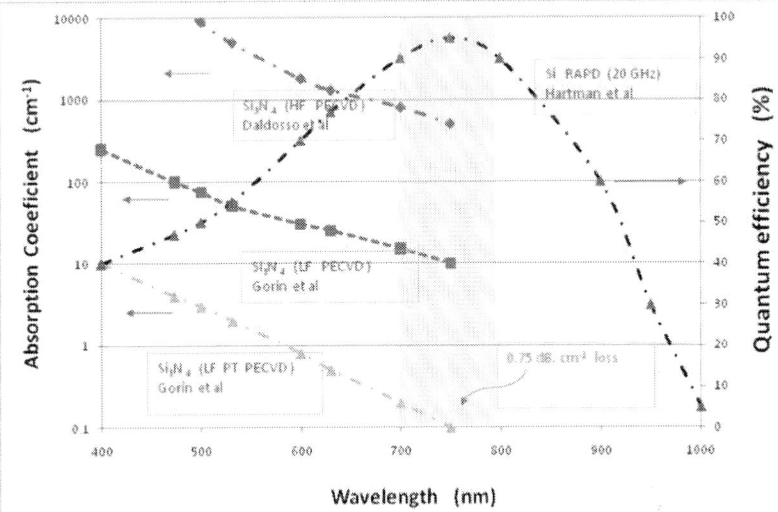

Fig. 9. Analyses of the loss characteristics of plasma deposited silicon nitride versus propagation wavelength and comparison with the detectivity of CMOS compatible reach through avalanche detectors.

Optical simulations were performed with RSOFT (BeamPROP and FULL WAVE) to design and simulate specific CMOS based waveguide structures operating at 750 nm, using CMOS materials and processing parameters. First, simple lateral uniform structures were investigated with no vertical and lateral bends and with a core of refractive index ranging from n = 1.96 (oxi-nitride) to n = 2.4 (nitride). The core was surrounded by silicon oxide (n = 1.46).

The analysis showed that both, multimode as well as single mode waveguiding can be achieved in CMOS structures. Fig. 10 and Fig.11 illustrate some of the obtained results.

Fig. 10 shows a three dimensional view of the electrical field along the 0.6 µm diameter silicon nitride waveguide. Multi-mode propagation with almost zero loss is demonstrated as a function of distance over a length of 20 µm. Multi-mode propagation in CMOS micro-systems has the following advantages: (1) a large acceptance angle for coupling optical radiation into the waveguide; (2) exit of light at large solid angles at the end of the waveguide; (3) allowing narrow curvatures in the waveguides; and (4) more play in dimensioning of the waveguides. (1) and (2) are particularly favourable for coupling LED light into waveguides.

Fig. 11 shows the simulation of a 1 µm diameter trench-based waveguide with an embedded core layer of 0.2 µm radius silicon nitride in a SiO_2 surrounding matrix. The two dimensional plot of the electrical field propagation along the waveguide as shown in Fig. 11 (a) reveals single mode propagation. The calculated loss curve in the adjacent figure (b), shows almost zero loss over a distance of 20 µm in Fig 11(b). Fig. 12(a) displays the transverse field in the waveguide perpendicular to the axis of propagation. Using the value of the real part of the propagation constant, as derived in the simulation, an accurate energy loss could be calculated using conventional optical propagation. With the imaginary part of the refractive index, as predicted by RSOFT, a low loss propagation of 0.65 dB cm⁻¹ is found, taking the material properties into account, as used by the RSOFT simulation program.

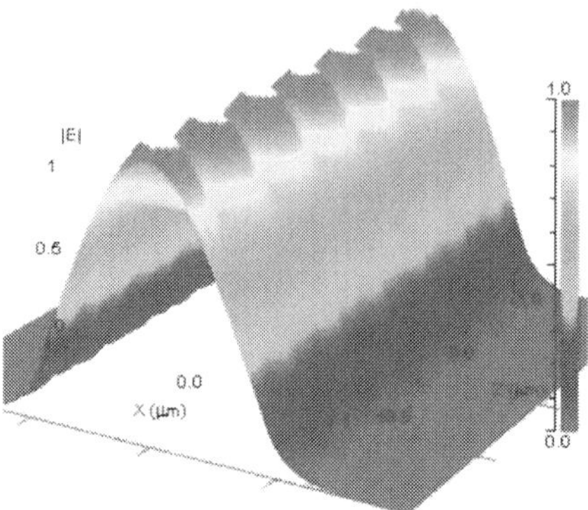

Fig. 10. Advanced optical simulation of the electrical field propagation in a 0.6 μm wide silicon nitride layer embedded in SiO_2 in CMOS integrated circuitry. Multimode optical propagation at 750 nm is demonstrated over 20 μm with a loss of less than 1 dB cm⁻¹.

Single mode propagation, where the light is more difficult to couple into the waveguide, results in low modal dispersion loss along the waveguide, as well as in extreme high modulation bandwidths.

It is important to note that waveguide mode converters can be designed to convert multimode into single mode.

In Fig 12 (b) , the same simulation was performed as in Fig. 11, but with a silicon oxi-nitride core of 0.2 μm embedded in a silicon oxide cladding. The mode field plot shows a slight increase in the fundamental mode field diameter, and less loss of about 0.35 dB cm⁻¹. This suggests that a larger proportion of the optical radiation is propagating in the silicon oxide cladding.

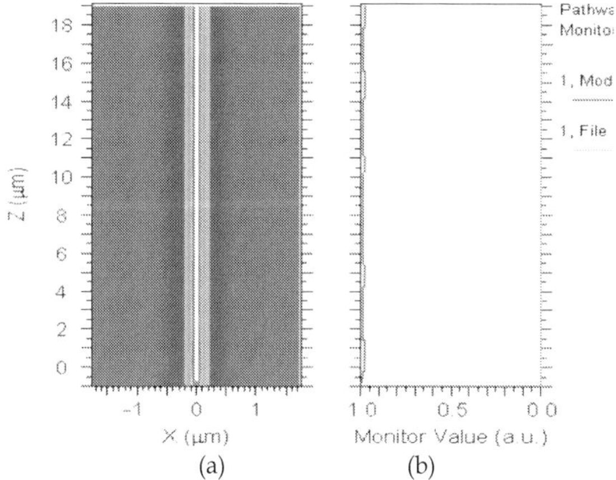

Fig. 11. (a) and (b): Advanced simulation of the electrical field propagation in a silicon nitride layer within CMOS integrated circuitry. Single mode propagation is demonstrated at 750 nm over a distance of 20 μm for a 0.2 μm wide silicon nitride waveguide , embedded in SiO_2.

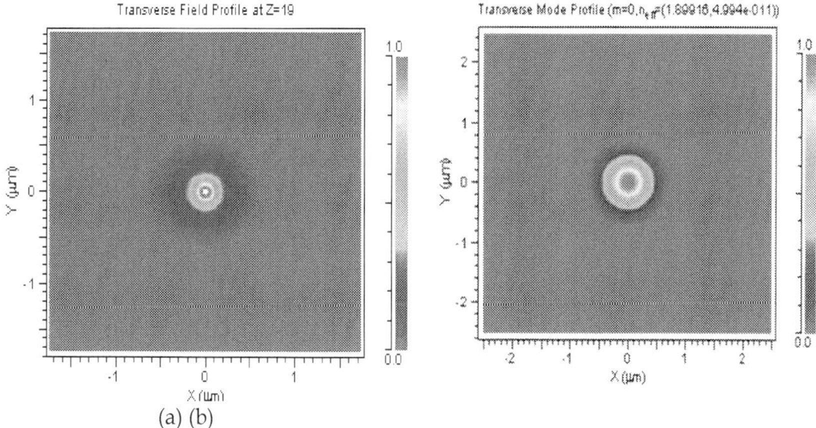

Fig. 12. (a) Transverse field profile prediction for a silicon nitride based CMOS waveguide. The core of the silicon nitride is 0.2 μm in diameter and is embedded in a 1 μm diameter SiO_2 cladding. (b) Transverse mode field profile for a 0.3 μm oxi-nitride layer embedded in SiO_2.

Subsequently, a modal dispersion analysis was conducted on these structures. The calculations reveal a maximum dispersion of 0.5 ps cm^{-1} and a bandwidth-length product of greater than 100 GHz-cm for a 0.2 μm silicon nitride based core. A maximum modal dispersion of 0.2 ps cm^{-1} and a bandwidth-length product of greater than 200 GHz-cm was found for a 0.2 μm silicon-oxi-nitride core which was embedded in a 1 μm diameter silicon-oxide cladding. Due to the lower refractive index difference between the core and the cladding, a larger transverse electric field of about 0.5 μm radius, as well as lower modal dispersion, is achieved with a silicon oxi-nitride core. The material dispersion characteristic was estimated at approximately 10^{-3} ps nm^{-1} cm^{-1}, which is much lower than the maximum predicted modal dispersion for the designed waveguides.

5. CMOS optical link - proof of concept

The photo-micrographs in Fig. 13 illustrate results which have been achieved with a CMOS opto-coupler arrangement, containing a CMOS Av-based light-emitting source, an 5 x 1 x 150 μm silicon over-layer waveguide and a lateral incident optimized CMOS based photo-detector (Snyman &Canning 2002, Snyman et al, 2004). The waveguide was fabricated in CMOS similar to that as shown in Fig. 5 (b).

Fig. 13 (a) shows an optical microscope picture of the structure under normal illumination conditions with the Si LED source, the waveguide and the elongated diode detector. Fig.13 (b) shows the structure as it appeared under subdued lighting conditions. At the end of the silicon oxide structure, some leakage of the transmitted light was observed (feature B). This observation is quite similar to light emission observed at the end of a standard optical fibre, and it confirms that good light transmission occurs along the waveguide.

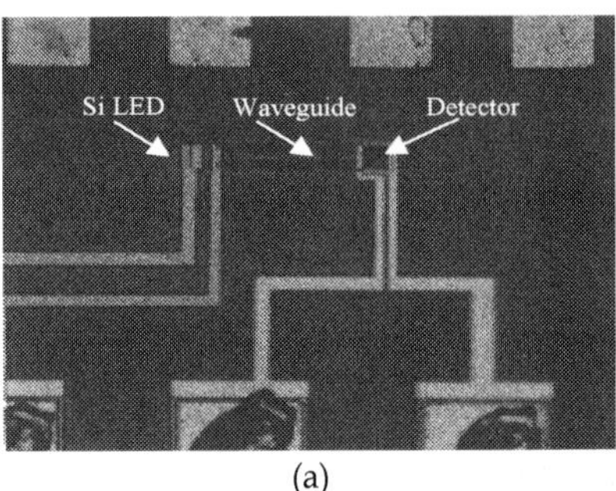

(a)

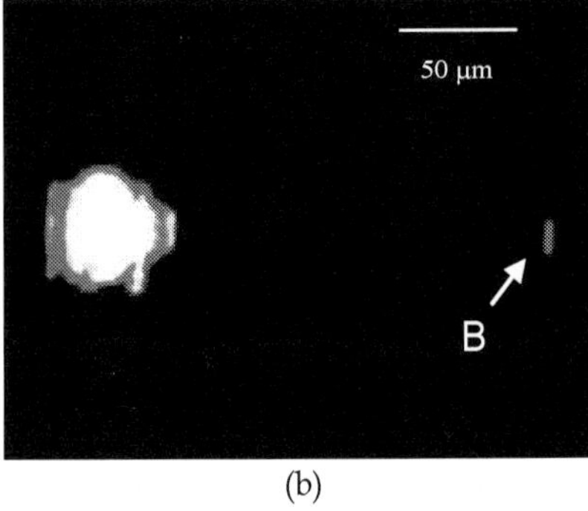

(b)

Fig. 13. Photomicrographs of a CMOS opto-coupler arrangement consisting of a CMOS Av-based light-emitting source, an optically waveguide and a CMOS lateral incident photo-detector. (a) shows a bright field photo-micrograph of the arrangement, and (b) shows the optical performance as observed under dark field conditions (Snyman et al, 2000, 2004).

Signals of 60 – 100 nA could be observed for 0 to +20 V source pulses and +10 V bias at the elongated diode detector. When the detector was replaced with a n^+pn photo-transistor detector (providing some internal gain at the detector at appropriate voltage biasing), signals of up to 1 μA could be detected.

The arrangement showed good electrical isolation of larger than 100 MΩ between the Si LED and the detector for voltage variations between the source and the detector of 0 to +10V on either side when no optical coupling structures were present . This was mainly due to the p^+n and n^+p reversed biased opposing structures utilised in the silicon design. Once an avalanching light emitting mode was achieved at the source side, a clear corresponding current response was observed at the detector. Detailed test structures are currently investigated.

6. Proposed CMOS and SOI waveguide-based optical link technology

Building on the optical source and waveguide concepts, as outlined in the preceding sections, optical source based systems may be designed which optimally couple light into the core of an adjacently positioned optical waveguide. Similarly, the core of the waveguide can laterally couple light into an adjacent RAPD based photo diode. It follows that interesting high speed source- detector optical communication channels and systems can be implanted in CMOS technology as illustrated in Fig. 14 (Snyman , 2010d, 2011a). The proposed isolation trench waveguide technology as outlined in Section. 2 is particularly well suited in order to create such configurations in CMOS technology. However, OLED surface layer structures together with CMOS technology and Si Av LED and SOI technologies may also generate such structures.

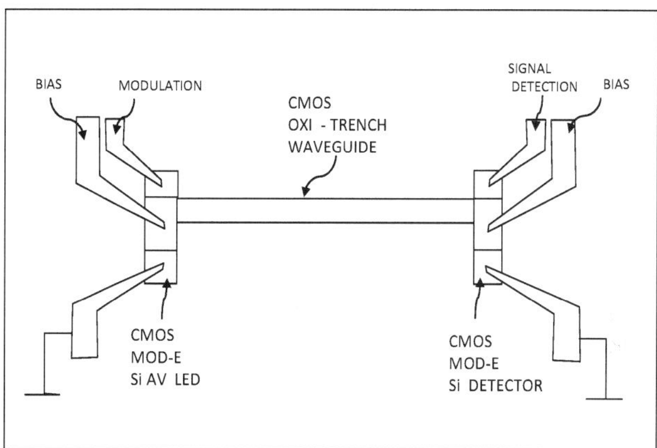

Fig. 14. Conceptual optical link design using a optical source arrngement as in Fig . 8 , a CMOS trench based waveguide and a RAPD photo detector arrangement.Bi-directional optical communication may be realised with the structure.

Using a Si Av LED optical source, an optical p+npn source, as outlined in Fig. 8 can be designed, with its optical emission point aligned with a lateral propagating CMOS based waveguide. Similarly, lateral incident detectors can be designed that take advantage of the carrier multiplication and high drift concept of reach through avalanche based diodes (RAPD). This can be combined with the proposed CMOS trench- waveguide systems. This implies that a similar lateral n+pp-p+ structure could be designed, such that with suitable voltage biasing, a high carrier generation adjacent to a high carrier drift region is formed. By placing an appropriate contact probe in the high drift region, varying voltage signals could be detected as a function of drift current. Silicon detector technology has been quite well established during the last few decades. These devices enerate up to 0.6 A W^{-1} and reach up to 20 GHz (Senior, 2008).

The generic nature of these designs open up numerous and diverse types of optical communication and optical signal processing devices realized in CMOS technology. Transmitter-receiver arrangements can be designed that will enable full bi-directional optical communication. The concepts, outlined here are not final , and there is scope for further improvement.

A drawback of these designs is the fact that the optical source needs to be driven by direct modulation methods. OLEDs have the advantage of low modulation current or voltage. However, they may be limited by forward biased diffusion capacitance effects. Si Av LEDs require low modulation voltage, but high driving currents. Since the driving current needs to be supplied by CMOS driver circuitry, this implies large area CMOS driving PMOS and NMOS transistors with high capacitance. Through the incorporation of localized hybrid technologies, appropriate waveguide based modulators can be designed , that are either based on the electro-optic (Kerr) effect or the charge injection effect.. It is envisaged to reach modulation speeds, orders of magnitude higher (reaching far into the GHz range), with much less driving currents (Snyman, 2010d).

7. Optical coupling efficiencies and optical link power budgets

Obtaining good coupling efficiencies with Si Av LEDs and OLEDs when incorporated into CMOS structures presents a major challenge. It is estimated that the optical power emitted from the Si Av LEDs is in the order of 100 – 1000 nW (for typical driving powers of 8 V and 10 µA). Since most of the emission occurs inside the silicon with a refractive index of 3.5, it implies that only about 1 % of this optical power can leave the silicon because of the small critical angle of only 17 degrees inside the silicon. After leaving the silicon the light spreads over an angle of 180 degrees (Fig.15 (a)). When a standard multimode optical fibre with a numerical aperture of 0.3 is placed close to such an emission point, only 0.3 % of the forward emitted optical power enters the fibre.

Our research has shown that remarkable increases in optical coupling efficiencies can be achieved by means of two techniques : (1) concentrating the current that generates the light as close as possible to the surface of the silicon (for Si Av LEDs) ; and, (2), maximizing the solid angle of emission in the secondary waveguide.

By displacing the metal contacts that provide current to the structure as shown in Fig 15 (b) , the current is enforced on the one side surface facing the core of the waveguide. Since mainly surface emission is generated, about 50 % of the generated optical power enters the waveguide (Snyman 2010d, Snyman 2011 a). A silicon nitride core with a silicon oxide cladding could then ensure an acceptance angle of up to 52.2 degrees within the waveguide. The total coupling efficiency that can be achieved with such an arrangement is of the order of 30%. This is an 100 fold increase in coupling efficiency from the point of generation to within the waveguide as achieved in Fig 15 (a) (Snyman, 2011c).

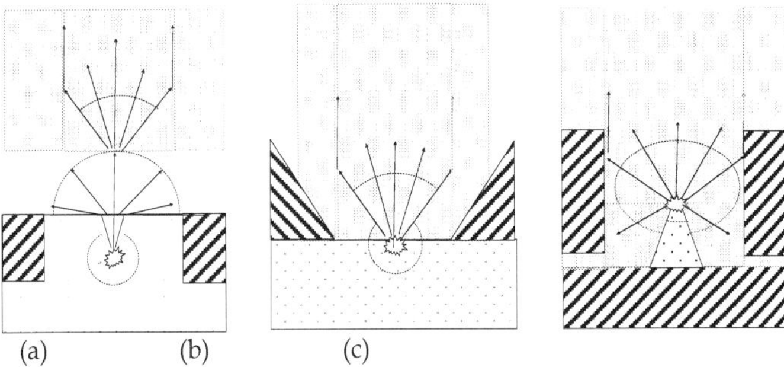

Fig. 15. Demonstration of optical coupling between a Si Av LED optical source and the silcon nitride CMOS based optical waveguide.

Fig. 15 (c) shows a further optimized design. Here a thin protrusion of doped silicon material is placed inside the core of a silicon nitrate core CMOS based waveguide (Snyman 2011a). Such a design is quite feasible with standard layout techniques of
CMOS silicon provided that the side trenches surrounding the silicon protrusion are effectively filled with silicon nitride through the plasma deposition process. The core is surrounded by trenches of silicon oxide. The optical power that is generated at the tip of the protrusion and radiates in a solid angle of close to a full sphere inside the waveguide. Simulation studies show that up to 80 % of the emitted light is now coupled into the silicon nitride core. Reflective metal surfaces at the sides and the back of this waveguide may further improve the forward propagation.

The optical radiation produced inside these waveguides will be highly multimode. The diameters of these waveguides may be bigger than the ones suggested for single mode propagation in Section 4. However, in such cases, standard type waveguide mode converters can reduce the number of modes or even generate single mode propagation.

With an optical power source of 1 µW at the silicon surface, one can achieve a coupling efficiency between source and waveguide of 30 to 50 %, assuming a coupling loss of only 3 dB between source and waveguide. With a 0.6 dB cm^{-1} wave guide loss, the loss in the 100 µm waveguide itself is estimated to be 0. 01 dB. Since the whole radiation propagating in the waveguide can be delivered with almost 100 % coupling efficiency, one can expect about 500 nW of optical power reaching the detector. With an 0.3 A per Watt conversion efficiency of the detector, current levels of about 100 nA (0.1 µA) can be sensed with a 10 x 10 µm detector. Values for OLEDs together with surface CMOS waveguides could be much higher. The low frequency detection limit of silicon detectors of such dimensions is of the order of pico-Watt. For low frequencies and low optical level detection , a dynamic range of about 10^3 to 10^4 is achievable. At high modulation speeds, the achievable bit error rates will obviously increase.

The optical powers quoted above are much lower when compared with current LASER, LED and optical fiber link "macro" technology. However, we are addressing a new field of "micro-photonics" with micrometer and nano-meter dimensions, and power levels as well as other parameters should be scaled down accordingly. Furthermore, our research showed that the optical intensities determine the achievable bit error rates rather than absolute intensities. As stated earlier, the calculated intensity levels with some of our Si Av LEDs are as high as 1 nW µm^{-2} .

8. Connecting with the environment

We present only two viable ways of communicating with the outside chip environment:, i.e optical communication vertically outward from the chip, and optical communication via lateral waveguide connections.

In the first case, silicon oxide and silicon nitride are used as well as trench technology as outlined in Section 2 in order to increase the vertical outward radial emission (Snyman, 2011c). Fig. 16 illustrates the concept. By placing a thin layer of silicon adjacent to two semi-circular trenches, the solid angle of the optical outward emission is increased within the silicon from about 17 to almost 60 degrees. Filling up the trenches with silicon oxide and placing of thin layer of silicon oxide increases the critical emission angle from the silicon from 17 to 37 degrees. The thin layer of silicon nitride can be appropriately shaped with post processing RF etching techniques such that all emitted light can be directed vertically upward. It is estimated that a total optical coupling efficiency from silicon to fibre of up to 40 % can be achieved in this way.

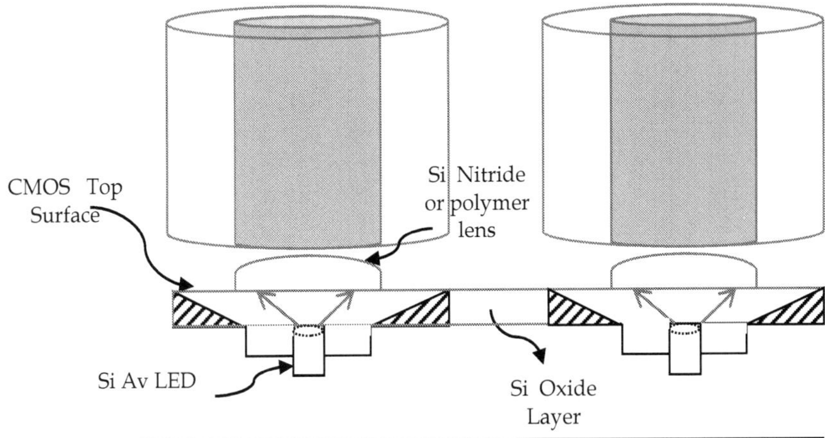

Fig. 16. Vertical outward coupling of optical radiation into optical fiber waveguides using trench based and overlayer post processing technology.

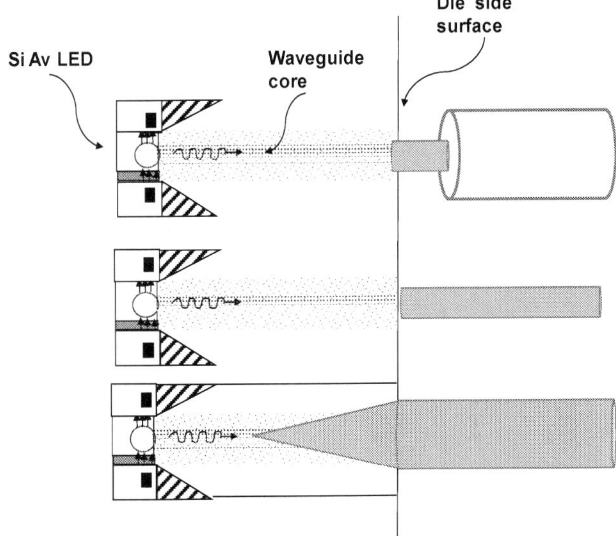

Fig. 17. Lateral out coupling using CMOS waveguide based optical coupling with optical fibers alligned at the side surface of the chip.

In the second case, optical coupling is achieved via lateral wave guiding (Snyman, 2010 d, 2011 c, 2011d). Fig. 17 illustrates this. (1) The lateral coupling from the optical source can be as high as 80 % as demonstrated in the previous section in Fig. 15 (c). (2) The optical radiation can be converted from multi-mode propagation to single mode propagation by waveguide mode converters; (3) Single mode radiation at the side surface ensures high collimation. This assures a coupling efficiency of almost 100 % at the side surface. In total, an optical coupling efficiency of up to 80 % can be achieved. This is much higher than achievable with vertical coupling. (4) Our analysis shows that the far field pattern of the optical radiation emitted from the waveguides can be manipulated by either adiabatic expansion or by tapering the core near the end of the waveguide. In this way the mode field diameter extends into the silicon oxide cladding, and the radiation couples more efficiently into the core from an externally positioned optical fibre.

In conclusion, the analysis shows that combining CMOS compatible sources effectively with on- chip lateral extending waveguide technology, offers major advantages , like increased coupling efficiencies, increased optical power link budgets, lower achievable bit error rates in data communication , and better coupling with the external environment.

9. Proposed first iteration CMOS micro-photonic systems

The on-chip optical and signal processing applications have been already highlighted in Section 6. A particular interesting design , made possible with the CMOS waveguide technology, is a so called H-configuration waveguide that can be used for optical clocks in very large CMOS micro-processor systems (Wada, 2004).

The realization of diverse other CMOS and waveguide based micro-photonic systems as well as the incorporation of a whole range of micro-sensors into CMOS technology is possible. The advantages are, (1), high levels of miniaturization; (2), higher reliability levels; (3), a vast reduction in technology complexity and, (4), a drastic reduction in production costs. The proposed waveguide technologies, particularly in this chapter, offer high optical coupling between Si Av LEDs or OLEDS and CMOS based waveguides, with diverse applications in optical interconnect and future on chip micro-photonic systems.

Fig. 18 to 20 illustrate some applications, as proposed here, for CMOS based micro-photonic systems (Snyman 2008a, 2009a, 2010c, 2011b, 2011c).

In Fig. 18, a hybrid approach is demonstrated. A mechanical module is added to an existing CMOS package creating a CMOS-based micro-mechanical optical sensor (CMOS MOEMS), capable of detecting diverse physical parameters such as vibration , pressure, mechanical osscillation etc. Optical radiation is coupled from the CMOS platform to the mechanical platform. The mechanical platform returns optical signals which contain information about the deflection (Snyman, 2011 c).

Fig. 19 shows a monolithic approach of creating CMOS MOEMS involving only post-processing procedures. A cantilever is fabricated in part of the CMOS IC die, by post processing procedures. Si Av LED or OLEDs couple optical radiation into a slanted waveguide track, transmit the optical radiation laterally across the die, collimate the radiation through the crevasse onto the one side of the cantilever. Optical radiation is reflected from the cantilever and detected by a series of p-i-n photo- detectors arranged laterally along the crevasse side surface. The accumulated signals are processed by adjacent

CMOS analogue and digital processing circuits. Such a structure can detect vibrations, rotations and accelerations (Snyman, 2011 c).

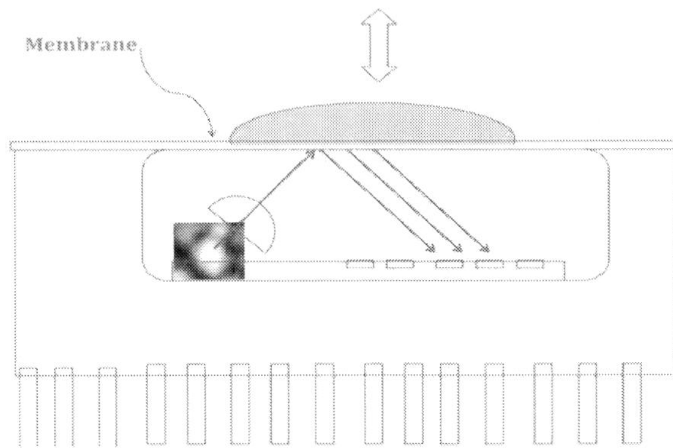

Fig. 18. Schematic diagram of a hybrid CMOS-based micro-photonic system that can be realized by placing a mechanical- module on top of a optically radiative and detector active CMOS platform using standard packaging technology.

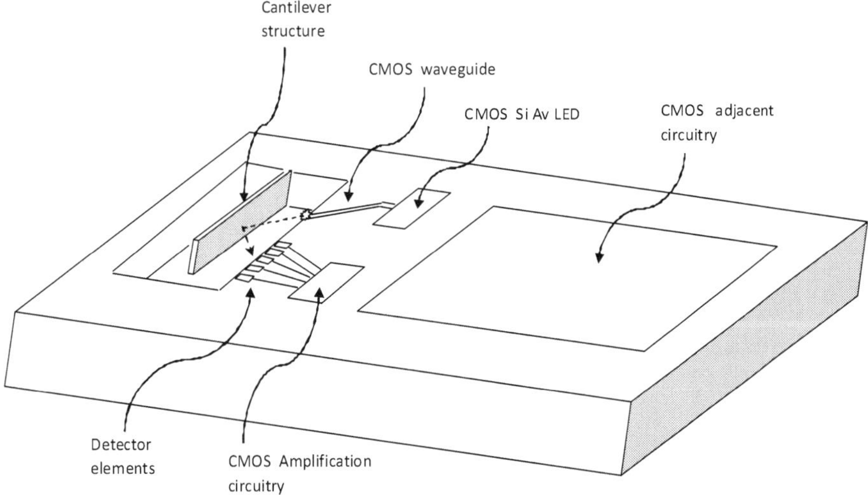

Fig. 19. Schematic diagram of an example CMOS-based Micro-Mechanical-Optical Sensor (MOEMS) device that can be realized with conventional CMOS integrated design with additional post processing procedures. Key constituents of such a device is an effective CMOS on-chip optical source , coupling of the source to a waveguide, CMOS competible optical waveguiding and optical collimation and detection circuitry.

The immunity to electromagnetic induced noise of these systems is a major advantage. Key components of such the systems are an effective CMOS compatible optical source, CMOS compatible optical wave guiding, effective optical coupling into the waveguide, and optical collimation circuitry. The sensitivity and functionality of these systems are a function of the waveguide design.

Fig. 20 explores a more complete and more advanced waveguide based micro-photonic system design including ring resonators, filters and an unbalanced Mach-Zehnder interferometer. By selectively opening up a portion of the waveguide in the one arm of the interferometer to the environment, molecules or gases can be absorbed and both, phase and intensity changes can be detected by the interferometer. Sensors can be designed which detect the absorption spectra of liquids (Snyman 2008a, 2009a, 2010c, 2011b, 2011c,).

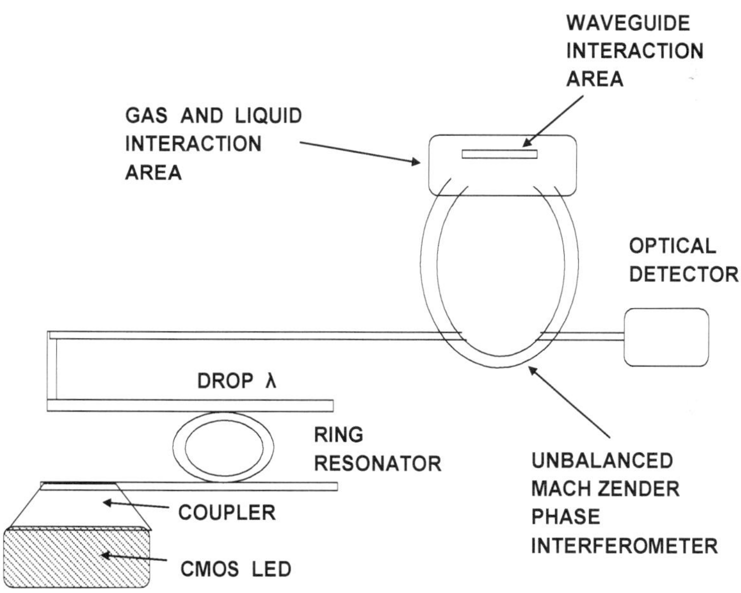

Fig. 20. Schematic diagram of a CMOS-based micro-photonic system that can be realized using an on chip Si Av LED, a series of waveguides, ring resonators and an unbalanced Mach-Zehnder interferometer. A section of the waveguide is exposed to the environment and can detect phase and intensity contrast due to absorption of molecules and gases in the evanescent field of the waveguide.

Obviously, a great variety of diverse other types of CMOS based micro-photonic systems are possible, each incorporating specific optical micro-sensors and waveguides. It is anticipated to implement future CMOS based micro-photonic systems in micro-spectro-photometry, micro-metrology, and micro- chemical absorption analysis.

10. Conclusions

It is evident that the analyses as presented in this study with regard to Si Light Emitting Devices operating at 650 – 850 nm and lateral optical waveguides can lead to the generation of diverse photonic micro-systems systems in standard CMOS integrated circuitry. The generation of lateral waveguides in CMOS technology operating in this wavelength regime poses particular challenges. However, enough evidence has been obtained from our analyses and first iteration experimental realisations that this technology is indeed feasible. The proposed sub-technologies has major advantageous for the generation of complete new families of photonic micro-systems on CMOS chip avoiding the more complex Si Ge or III-V hybrid technology. The following serves as brief summaries of results and statements made:

1. The potential of CMOS technology was analysed and evaluated for sustaining the generation of optical micro-photonic systems in CMOS integrated circuitry. Particularly, the silicon dioxide "field " oxide , inter-metallic oxides and passivation nitride and added polymer over-layer structures show good potential to be utilised as "building blocks" in new generation CMOS based micro-photonic systems.

2. It was shown that a variety of optical source technologies currently already exists that can be utilised for the generation of 650- 850nm optical sources on chip. OLEDs offers high irradiance in this wavelength regime. There are however challenges with regard to incorporation of the hybrid organic based technologies into CMOS technology and with regard to achieving high modulation speeds. Silicon avalanche-based Si LEDs can be integrated into CMOS integrated circuitry with relative ease, they offer high modulation bandwidth , and can be integrated particularly at the silicon-overlayer interface, and offer both vertical and lateral optical coupling possibilities. Their power conversion efficiency is lower, but analysis show that the power levels is enough to offer adequate power link budgets , with high modulation bandwidth. Particularly, they can generated micron size optical emission points, with high irradiance levels, offering unique application possibilities with regard to generating of micro-structured photonic devices.

3. Analyses and simulation results as presented in this study, show that it is possible to design waveguides with CMOS technology at 650-850 nm. Particularly, the generation of waveguides with small dimension silicon nitride cores embedded in larger silicon dioxide surrounds seems particularly attractive. The utilisation of lateral CMOS waveguides increase coupling efficiencies, improve optical link power budgets, and supports numerous designs with regard to the generation of micro-photonic structures in CMOS integrated circuitry. These aspects are all beneficial for generating lateral layouts of micro-photonic systems on chip and offers viable options for interfacing optically with the environment..

4. The technology as proposed, may not necessarily compete with the ultra high modulation speeds offered by Si-Ge based and SOI based technologies currently

operating at above 1100nm. However, Si Av LEDs, waveguides detectors, demonstrated in this study, support the generation of micro-photonic systems in standard CMOS technology, offering modulation speeds of up to 10 GHz at the added advantage of ease of integration into standard integrated circuit technology. Direct driving of the sources in CMOS may reduce modulation speeds. Particularly, the use of waveguide-based modulators may produce higher modulation speeds (Snyman 2010e). Several advances in this area could still be made.

5.   Lastly, a few designs are proposed for the realisation of first iteration micro-photonic sensor systems on chip. Both mechanically as well as adhesion and waveguide based sensors systems are proposed and the application possibilities of each were presented. Particularly, the proposed technologies offers the realisation of a complete new family of source–sensor based micro-photonic systems where bandwidth is not the essential parameter, but rather the capability to add to the integration level, intelligence level, and the interfacing level of the processing circuitry with the environment.

11. Acknowledgements

The hypotheses, analyses, first iteration results and research interpretations as presented in this study were generated by means of South African National Research Foundation grants FA200604110043 (2007-2009) and NRF KISC grant 69798 (2009-2011) and SANRF travel block grants (2007,2008, 2009). The utilization of facilities at the Carl and Emily Fuchs Institute for Microelectronics for confirmation of some of the experimental results is gratefully acknowledged. The provision and use of advanced software facilities at the TUT is acknowledged. The final proof reading of the script by Dr. D. Schmieder is especially acknowledged.
Selected topics of this article forms the subject of recent PCT Patent Application PCT/ZA2010/00032 of 2010, (Priority patents: ZA2010/00200, ZA2009/09015, ZA2009/08833, ZA2009/04508) ; PCT Patent Application PCT/ZA2010/00033 of 2010 (Priority patents: ZA 2008/1089, ZA2009/04509, ZA2009/04665, ZA2009/04666, ZA2009/05249, ZA2009/08834, ZA2009/0915), ZA2010/08579, ZA2011/03826; and PCT Patent Application PCT/ZA2010/00031 of 2010" (Priority patents: ZA 2010/02021, ZA 2010/00201, ZA2010/00200, ZA 2009/07233, ZA2009/07418, ZA 2009/04164, ZA200904163, ZA2009/04161). These all deal with our latest technology definitions with regard to OLED and Si Av LED CMOS based optical communication systems, Si Av LED design, CMOS waveguide design, CMOS modulator and switch design, CMOS based data transfer systems, CMOS micro-photonic system and Micro-Optical Mechanical Sensors (MOEMS) design. The purpose of these patents is to secure intellectual property protection on investments made already, and to secure licensing of certain key components of the technology as already developed. However, the opportunities in this field is so extensive, that numerous further investment opportunities with interested further investors exist.

12. References

Aharoni, H. & Du Plessis, M., "The Spatial Distribution of Light from Silicon LED's", *Sensors and Actuators*, Vol. A 57 No. 3, (1996), pp. 233-237.

Beals, M., Micheal, J., Liu, F.J, Ahn, D.H, Sparacin, D. , Sun, R., Hong, C.Y. & Kimerling, L.C., (2008) , "Process flow innovations for photonic device integration in CMOS", *Proceedings of SPIE 6898*, pp. 6898 O4 – O14.

Bourouina, T. , Bosseboeuf, A. & Grandchamp, J-P, (1996) . "Design and Simulation of an Electrostatic Micro-pump for Drug-Delivery Applications". *MicroMechanics Europe Workshop (MME'96)*, pp 152-155, Barcelone .

Brummer, J. C., Aharoni, H. ,& du Plessis, M. , "Visible Light from Guardring Avalanche Silicon Photodiodes at Different Current Levels", *South African Journal of Physics* Vol. 16 (1/2),(1993), pp 149-152.

Bude, J., Sano, & N., Yoshii, A. , "Hot carrier Luminescence in Silicon", *Physical Review* Vol. B 45, No. 11, (1992), pp. 5848 – 5856.

Carbone, L. , Brunetti, R., Jacoboni, C. , Lacaita, A. & Fischetti, M. , "Polarization analysis of hot-carrier emission in silicon, " *Semicond.Sci. Technol.* Vol. 9, (1994), pp. 647-676.

Chatterjee, B., Bhuva & Schrimpf, R., "High-speed light modulation in avalanche breakdown mode for Si diodes", *IEEE Electron Device Letters* , Vol. 25, No. 9,(2004), pp. 628-630.

Claeys, C., (Fellow, IEEE), (2009). "Trends and Challenges in More Moore and More than Moore Research", *Proceedings of the South African Conference on Semiconductors and Superconductors (SACSST)*, pp 1-6 , ISBN 978-0-620-43865-0, Cape Town, South Africa.

Daldosso,N., Melchiorri, M., Riboli, F., Sbrana, F., Pavesi, L., Pucker,G., Kompocholis, C., Crivellari,M., Belluti, P., & Lui, A., "Fabrication and optical characterization of thin two-dimensional Si_3N_4 waveguides" *Mater. Sci. Semicond. Process.* Vol. 7, (2004), pp. 453-458.

Du Plessis, M ., Aharoni, H. & Snyman, L.W., "A silicon trans-conductance light emitting device (TRANSLED)." , *Sensors and Actuators* , Vol. A 80, No. 3, (2003), pp 242-248.

Du Plessis, M., Aharoni, H. , & Snyman, L.W., " Spatial and intensity modulation of light emission from a silicon LED matrix", *IEEE Photonics Technology Letters* , Vol. 14, No. 6, (2002), pp. 768-770.

Du Plessis, M., Snyman, L.W. , & Aharoni, H. (2003). " Low-Voltage Light Emitting Devices in Silicon IC Technology", *Proc. of the IEEE International Symposium on Industrial Electronics (ISIE 2005)*, Vol. 3 , ISBN 0-7803-8739-2 , pp. 1145- 1149, Dubrovnik, Croatia.

Fauchet, P. M., "Progress Toward Nano-scale Silicon Light Emitters", *IEEE Jour. Selected Topics in Quantum Electron* Vol.4, (1998) , pp. 1020-1028.

Fitzgerald, E.A. ,& Kimerling, L.C. , "Silicon-Based Technology for Integrated Opto-electronics", *MRS Bulletin*, (1998), pp. 39-47.

Foty, D., "Digital CMOS Electronics", Gilgamesh Associates, LLC, Fletcher, Vermont, USA. (2009). Private communication .

Fullin, E., Voirin,E., Lagos, A., Moret, J.M., "MOS-Based Technology for Integrated Opto-electronic Circuits", *Scientific and technical report, CSEM, (1993)*, pp. 26.

Ghynoweth, W.G. & McKay, K.G. , "Photon emission from avalanche breakdown in silicon ", *Physical Rev.* Vol. 102 , (1956), pp. 369-376.

Gianchandani , Y. B. , (Fellow, IEEE), (2010) " Emerging Research in Micro and Nano Systems: Opportunities and Challenges for Societal Impact" , In: *Microfluidics, BioMEMS, and Medical Microsystems VIII,* edited by Holger Becker and Wanjun Wang, Proc. of SPIE, available at doi: 10.1117/12.848216 (Plenery lecture).

Gorin, A., Jaouad, A., Grondin, E., Aimez, V. & Charette, P. , " Fabrication of silicon nitride waveguides for visible-light using PECVD: a study of the effect of plasma frequency on optical properties". *Optics Express,*Vol. l16, No. 18, (2008), pp 13509-13516.

Green, M.A., Zhao, J., Wang, A. , Reece, P. & Gal , M., "Efficient silicon light-emitting diodes" *Nature* , Vol. 412, (2001), pp 805-808.

Gunn ,G., Narasimha, A., Analui, B. , Liang, Y. & Sleboda, T.J. (2007) "A 40Gbps CMOS Photonics Transceiver",. In: *Silicon Photonics II* , *Proc SPIE* , Vol. 6477, (ISBN 9780819465900), pp. 64770N1- 8.

Hirschman, K. D., Tsybeskov, L., Duttagupta, S. P. & Fauchet, P. M. , "Silicon-based visible light-emitting devices integrated into microelectronic circuits," *Nature,* Vol. 384, (1996), pp. 338–341.

Kramer, J. , Seitz, P. , Steigmeier, E. F. , Auderset, H., and Delley, B., "Industrial CMOS technology for the integration of optical metrology systems," *Sensors and Actuators A* Vol. A37–38,(1993), pp. 527–533.

Kubby, J. A . & Reed, G. (2007). Introductory remarks, In: *Silicon Photonics II* , *Proc. SPIE,* Vol. 6477. (ISBN 9780819465900), pp. xi.

Kubby, J. A.& Reed, G. (2010). Introductory remarks. In: *Silicon Photonics V* , *Proc SPIE* Vol. 6477, ISBN 9780819480026, pp xi.

Lacaita, K., Zappa, F, Bigliasrdi, S. & Manfredi, M. , "On the Brehmstrahlung origin of hot-carrier-induced photons in silicon devices", *IEEE Trans. Electron Devices*, Vol. ED40, (1993), pp. 577-582.

Liu, J., Sun, X. , Rodolfo Camacho-Aguilera, R., Kimerling, L.C. and Michel, J. , " Ge-on-Si Laser operating at room temperature" *Optics Letters* Vol. 35, No. 5, (2010), pp. 679-681.

Newman, R. , "Visible light emission from a silicon p-n junction", *Phys. Rev.* Vol. 100, (1955), pp. 700 – 703.

Ogudo, K. A. , Snyman , L.W. , Du Plessis, M. & Udahemuka, G. (2009). "Simulation of Si LED (450nm-750nm) light propagation phenomena in CMOS integrated circuitry for MOEMS applications" , *Proceedings of the South Africa Conference on Semiconductor and Superconducting Technologies (SACSST 2009)*, ISBN 978-0-620-43865-0, pp. 168-182.

Robbins, D.J. , Editorial comments . (2000). "Silicon Opto-electronics" , *Proc. of SPIE,* Vol. 3953, pp. vi, San Jose, California, USA.

Savage, N., "Linking with Light", *IEEE Spectrum,* Vol. 39, (2002), pp. 32–36.

Schneider, K., and Zimmerman, H. , In: *Highly sensitive optical receivers* , (2006) , Springer, ISBN 13978-3-546-29613-3.

Sedra , A.S. & Smith, K.C. (2004). "VLSI Fabrication Technology" In:*Microelectronics Circuits*, Oxford University Press, New York, ISBN 13:978-0-19-514252-5, pp. A 12.

Senior, J. M. , "Optical Detectors", In: *Optical Fiber Communications, Principles and Practice,* (3rd Edition), (2008). Prentice Hall. pp 444..

Snyman , L. W. , Ogudo , K.A. , Du Plessis M. & Udahemuka , G. , (2009). "Application of Si LED's (450nm-750nm) in CMOS integrated circuitry based MOEMS - Simulation and Analyses", *Proc. SPIE,* Vol. 7208, pp. 72080C (ISSN 0277-786X) Available at http://dx.doi.org/10.1117/12.808551.

Snyman L. W. "Hybrid and monolithic Microsystems and MOEMS", (2011). RSA Patent application 2011/03826 of 25th May 2011.

Snyman L. W., "CMOS LED MOEMS Devices", (2008), RSA Patent Appl. No. 2008/1089, of 1 February 2008 .

Snyman, L . W. , Du Plessis, M.,& Aharoni, H . , (2005). "Three terminal optical sources (450nm - 750nm) for next-generation silicon CMOS OEIC's" , *12th International Conference of Mixed Design of Integrated Circuits and Systems (MIXDES'2005),* pp. 737 – 747. Krakov, Poland (Invited plenary session paper).

Snyman, L. W. , & Du Plessis, M. (2008). "Increasing The emission intensity of p+np+ CMOS LED's (450 – 750 nm) by means of depletion layer profiling and reach-through techniques" ; In: *Silicon Photonics III* , Edited by Joel A Kubby and Graham T. Reed, *Proc. SPIE* , Vol. 6898 , pp. OE 1- 12, ISSN: 0277-786X, ISBN 9780819470737SPIE, Bellingham, W.A.

Snyman, L. W. , Aharoni, H. and Du Plessis, M, "Two order increase in the quantum efficiency of silicon CMOS n+pn avalanche-based light emitting devices as a function of current density" *IEEE Photonic Technology Letters*, Vol. 17, (2005), pp. 2041-2043.

Snyman, L. W. , du Plessis, M. & Aharoni, H. (2006). "Two order increase in the optical emission intensity of CMOS integrated circuit LED's (450 – 750 nm). Comparison of n+pn and p+np designs ", *Proc of the SPIE* , Vol. 5730, pp. 59-72.

Snyman, L. W. , Du Plessis, M. & Aharoni, H. , "Injection-avalanche based n+pn Si CMOS LED's (450nm . 750nm) with two order increase in light emission intensity - Applications for next generation silicon-based optoelectronics", *Jpn. J. Appl. Physics* Vol. 46 No. 4B, (2007), pp. 2474-2480.

Snyman, L. W. "CMOS waveguide based transceiver ", (2010). RSA Patent Application 2010 /06804 of 23 September 2010.

Snyman, L. W. "OLED and CMOS based micro-photonic systems", (2010). RSA Patent Application 2010 /08579 of 30 November 2010.

Snyman, L. W., du Plessis, M., & Bellotti, E. , "Increasing the emission intensity of p+np+ CMOS LED's (450 – 750 nm) by means of depletion layer profiling and defect engineering techniques", *IEEE Journal of Quantum Electronics* , Vol. 46 , No. 6, (2010), pp. 906-919.

Snyman, L. W., Ogudo, K. D. and Foty, D. (2011). "Development of a 0.75 micron wavelength CMOS optical communication system ", Proc SPIE Vol 7943, edited by

Joel A. Kubby, Graham T. Reed, CCC code: 0277-786X/11/$18 doi: 10.1117/ 12.873202, pp 79430K-1 to 12.

Snyman, L. W., T. Okhai, T. Bourouina and W. Noell, (2011) "Development of on-CMOS chip Micro-Photonic and MOEMS Systems " Proc SPIE Vol 7930. MOEMS and Miniaturized Systems X, edited by Harald Schenk, Wibool Piyawattanametha, CCC code: 0277-786X/11/$18 · doi: 10.1117/12.873353, pp 79300Z-1 to 11.

Snyman, L.W. (2010d)., "Optical communication system". PCT Patent Application PCT/ZA2010/00032.

Snyman, L.W. , Aharoni, H ., du Plessis, M. , & Gouws, R.B.J. , "Increased efficiency of silicon light emitting diodes in a standard 1.2 micron complementary metal oxide semiconductor technology", *Optical Engineering* , Vol. 37, (1998), pp. 2133 – 2141.

Snyman, L.W. , Aharoni, H., du Plessis, M. , Marais, J.F.K , van Niekerk, D. & Biber, A., "Planar light emitting electro-optical interfaces in standard silicon complementary metal oxide semiconductor integrated circuitry", *Optical Engineering* , Vol. 41, (2002), pp 3230 – 3240.

Snyman, L.W. , Auderset, H., Derendinger, M. , Patterson, B.D. , & Von Lanthen, A. , "Efficient electroluminescence from 2- and 3-junction silicon structures", In: *Annual Report 1996/Annex.111B* of the Paul Scherrer Institute, Switzerland, Vol. 28 , (1996).

Snyman, L.W. , Bogalecki, A. , Canning, L. , Du Plessis, M., & Aharoni, H. (2002). "High frequency optical integrated circuit design and first iteration realization in standard silicon CMOS integrated circuitry" *Proc. of the 10th IEEE International Symposium on Electron Devices and Optoelectronic Applications (EDMO 2002)*, IEEE Cat No. 02TH8629, pp. 77-82, Manchester, United Kingdom.

Snyman, L.W. , Du Plessis, M. & Aharoni, H. (2005), "Three Terminal n+ppn Silicon CMOS Light Emitting Devices (450nm – 750nm) with Three Order Increase in Quantum Efficiency", *Proc. of the IEEE International Symposium on Industrial Electronics (ISIE 2005)* , Vol. 3, pp. 1159 – 1169. (Special session paper).

Snyman, L.W. , du Plessis, M., Seevinck, E. , & Aharoni, H. , "An efficient, low voltage, high frequency silicon CMOS light emitting device and electro-optical interface", *IEEE Electron Device Letters* , Vol. 20,(1999), pp. 614-617.

Snyman, L.W. "MOEMS sensor device", (2010). *PCT Patent Application* PCT/ZA2010/000 33.

Snyman, L.W. "Wavelength Specific Si CMOS LEDs", (2010). *PCT Patent Application* PCT/ZA2010/00031 (Priority patents: ZA 2009/07233, ZA2009/07478, ZA2009/07233, ZA2010/00201, ZA2010/02021) .

Snyman, L.W., & Biber, A. (1999) , "Enhanced light emission from a Si n+pn CMOS structure" , *Proceedings of the 1999 IEEE SoutheastCon*, ISBN 0-7803-5237-8, . pp. 242 – 245, Conference held at Lexington, Kentucky, U.S.A.,

Snyman, L.W., "Increased coupling efficiencies with CMOS-based waveguides ", (2011). *RSA Patent* (submitted August 2011).

Snyman, L.W., and Bellotti, E. , (2010). "New Interpretation of Photonic Yield Processes (450-750 nm) in Multi-junction Si CMOS LEDs : Simulation and Analyses " , in

Silicon Photonics IV , edited by Joel A Kubby, Graham T. Reed, Proceedings of SPIE Vol. 7606, ISBN 9780819480026 , ISSN 0277-786X, pp. 760613-15 , as available at doi: 10.1117/12.843134).

Snyman, L.W., Canning, L.M., Bogalecki, A. , Aharoni, H., & du Plessis, M. (2004). "200-Mbps optical integrated circuit design and first iteration realizations in 0.8 micron bi-CMOS silicon integrated circuitry", *Proc SPIE.* Vol. 5357, p. 6.

Soref, R. , "Applications of Silicon-Based Optoelectronics", In : *MRS Bulletin,* (1998), pp. 20-47.

Tessera, Charlotte, USA, (2011), as available at www.tessera.com.

Vogel, U., Kreye, D., Reckziegel, S, Torker, M. , Grillberger, C ,& Amelung, J. (2007) "OLED-on-CMOS Integration for Optoelectronic Sensor Applications, In: *Silicon Photonics II,* edited by Joel Kubby, G Reed, *Proc. of SPIE* , Vol. 6477, pp. 647703-1 to 8 (2007).

Wada, K. (2004). "Electronics and Photonics convergence on Silicon CMOS Platforms" *Proc. of SPIE* Vol. 5357, pp. 16 .

Zimmermann, H. (1997). "Improved CMOS-integrated Photodiodes and their Application in OEICs", *Proc. IEEE 1997 Workshop on High Performance Electron Devices for Microwave and Opto-electronic Applications,* pp.346-351. King's College, London.

3

SPSLs and Dilute-Nitride Optoelectronic Devices

Y Seyed Jalili
Science Research Campus, Islamic Azad University
Iran

1. Introduction

Currently the main concern in GaAs-based dilute nitride research is the understanding of their material properties. There are many contradictory conclusions specially when it comes to the origin of the luminescence efficiency in these systems. different ideas have been put forward some more plausible than others. However there is a lack of new ideas to overcome the differences. This chapter will address such issues and then finally we will study SPSL structures as an alternative to the the random alloy quaternary GaInNAs for more efficient growth, design and manufacture of optoelectronic devices based on these alloys.

One of the major issues in current studies of GaInNAs is the metastability of the material. To overcome the rather low solubility of N in GaAs or GaInAs, non-equilibrium growth conditions are required, which can be realized only by molecular-beam epitaxy (MBE) Kitatani et al. (1999); Kondow et al. (1996) or metal-organic vapour phase epitaxy (MOVPE) Ougazazaden et al. (1997); Saito et al. (1998). Growing off thermal equilibrium implies a certain degree of metastability. The aim of growing GaInNAs, emitting at the telecommunication wavelengths of 1.3 μm and, also 1.55 μm, is only possible by incorporating nearly 40% In and several per cent of N. These concentrations are at the limits of feasibility in MBE and MOVPE growth on GaAs substrates. The emission wavelength of such GaInNAs layers was strongly blue-shifted when, after the growth of the actual GaInNAs layer, the growth temperature was raised for growing AlGaAs-based top layers (such as distributed Bragg reflectors in vertical-cavity surface-emitting laser (VCSEL) structures or for confinement and guiding in edge emitting laser structures). This led to a number of annealing studies which yield somewhat contradictory results Bhat et al. (1998); Francoeur et al. (1998); Gilet et al. (1999); Kitatani et al. (2000); Klar et al. (2001); Li et al. (2000); Pan et al. (2000); Polimeni et al. (2001); Rao et al. (1998); Spruytte et al. (2001a); v H G Baldassarri et al. (2001); Xin et al. (1999). This, ofcourse, is partly due to the different annealing conditions and growth conditions used, but is also a strong manifestation of the metastability of this alloy system. The full implications of the metastability are just evolving and different mechanisms causing a blue shift of the band gap have been suggested Grenouillet et al. (2002); Mussler et al. (2003); Spruytte et al. (2001b); Tournie et al. (2002); Xin et al. (1999). Nevertheless, all discussions and investigations, so far, have suggested that GaInNAs material system is a very promising candidate for telecoms and in particular datacom applications. However, for both GaNAs and GaInNAs material systems, the higher the nitrogen incorporation, the weaker the alloy luminescence efficiency. A key to the utilization of nitride-arsenide for long wavelength optoelectronic devices is obtaining low defect materials with long non-radiative lifetimes. Therefore currently, these materials must be annealed to obtain device quality material. Photoluminescence and capacitance-voltage measurements indicate the presence of a trap associated with excess nitrogen hsiu Ho & Stringfellow (1997); Spruytte et al. (2001a). Therefore the likely defect responsible for the low luminescence efficiency is associated with excess nitrogen. It is believed that the effect of thermal annealing on the PL properties of these structures is generally attributed to the elimination of non-radiative centers and improved uniformity. Non-radiative centers are considered to originate from phase separation and/or plasma damage from the N radicals Kitatani et al. (2000).

Interest in the tertiary material system GaNAs had been waned in favour of the quaternary GaInNAs due to its inability to reach the long wavelengths required for commercial applications. However, its new-found use in diffusion-limiting layers and in short-period superlattice structures, and ofcourse being the simpler, ternary, dilute nitride equivalent of GaInNAs and therefore, probably, easier to investigate and understand means that its material properties and behaviour upon annealing are not only important but useful considerations Gupta et al. (2003); Sik et al. (2001). The post-growth rapid thermal annealing (RTA) is usually performed on these ternary Francoeur et al. (1998) and quaternary alloys Spruytte et al. (2001b). Rapid thermal anneal strongly improves the photoluminescence (PL) efficiency. This increase in PL intensity is usually accompanied with a blue shift of the PL peak. In the following section, we focus on the effect of emission energy changes in the photoluminescence (PL) spectrum with annealing of the GaNAs material system and try to elucidate the controversy over its origin.

2. Annealing effects

2.1 Annealing of the ternary GaAs-based dilute nitride: GaNAs

In order to investigate the effect of annealing on this ternary dilute nitride, the sample structure shown in figure 1 was devised. It consists of a 5×8 nm MQW structure, which would provide a good PL signal, and that 8 nm wells (a few nm smaller than the critical thickness for GaNAs layers) would prevent strain relaxation-related defects. Another reason for using an 8 nm well was that a model of emission from a GaNAs MQW structure used to compute emission energies for different well thicknesses and different nitrogen concentrations indicates that. As the well width increases, the nitrogen concentration has increasingly less influence on the bandgap, and so slight growth-rate-related variations in well thickness have will have less of an effect on emission.

Samples with nitrogen concentrations of 1.0% and 2.5% were grown for our annealing studies. The lower limit of 1.0% was chosen because it had been suggested theoretically (and has since been demonstrated experimentally) that up to about 1.0%, the coexistence of strongly perturbed host states (PHS) and localized cluster states (CS) of an isoelectronic nitrogen impurity is observed, reflecting the non-amalgamation character of the band formation process Kent & Zunger (2001a;b); Klar et al. (2003). In other words, GaNAs begins to act as a 'dilute nitride' at around $y = 1.0\%$. Samples with 2.5% nitrogen were also grown, as this is approximately the upper limit at which XRD data reflects the total nitrogen content of the sample. It was also thought that if nitrogen out-diffusion was to be responsible for the changes seen as a result of annealing, the sample with higher-nitrogen concentration might, or should, illustrate this more clearly than the sample with lower-nitrogen content.

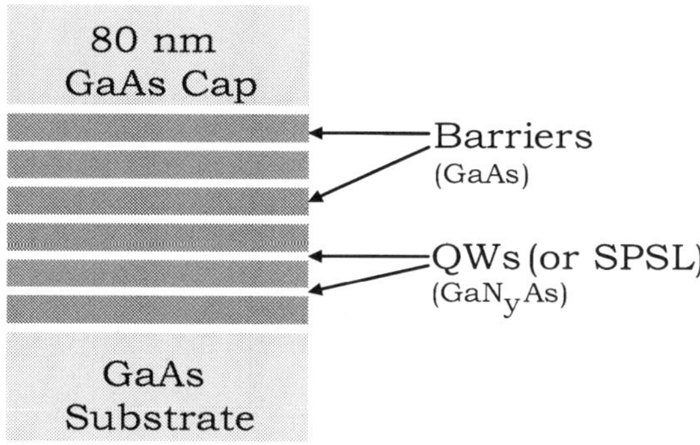

Fig. 1. Schematic nominal GaNAs/GaAs MQW structure used for annealing studies.

Sample Name	Nominal N-Concentration	RTA Round 1	RTA Round 2	Total
GaNAs21	1%	15 sec	30 sec	45 sec
GaNAs22	2.5%	15 sec	30 sec	45 sec
GaNAs23	1%	30 sec	30 sec	60 sec
GaNAs24	2.5%	30 sec	30 sec	60 sec

Table 1. Table showing the RTA times for different samples at 800°C.

PL measurements were made on the as-grown samples at 15 K and also after two ex-situ, RTA treatments, see figures 2 and 3, which were performed at 800°C in ambient Ar using a GaAs (001) insulating substrate proximity cap. Table 1 shows how the first and second rounds of annealing were carried out so that the maximum amount of information could be extracted from only three treatments. In this way, PL could be measured for two different nitrogen concentrations and for five different annealing times, 0 s (as-grown), 15 s, 30 s, 45 s and 60 s. Upon annealing, the peak wavelength of the 1.0% nitrogen samples blue shifted from 1.340 to 1.356 eV (at approx. 0.3 meV s^{-1}), and the full-width half-maximum (FWHM) decreased from 65 to 23 meV (see figures 2). For the 2.5% nitrogen samples, the peak wavelength blue shifted from 1.176 to 1.207 eV (at approx. 0.5 meV s-1), and the FWHM decreased from 38 to 22 meV, see figure 3. Blue shifting, increased peak intensity and decreased FWHM are all effects typical of a post-growth annealing treatment,the changes observed here are in agreement with those reported by Buyanova *et al* Buyanova, Pozina, Hai, Thinh, Bergman, Chen, Xin & Tu (2000) for similar MQW samples and annealing conditions. The fact that the rate of blue shifting for the 2.5% sample is greater than (almost double) that of the 1.0% sample suggests that the underlying mechanism may be N-dependent, but further work would be needed to verify this.

The main changes that occur due to thermal annealing, i.e. a blue shift in peak wavelength and an improvement in integrated intensity and FWHM, have proved rather difficult to explain

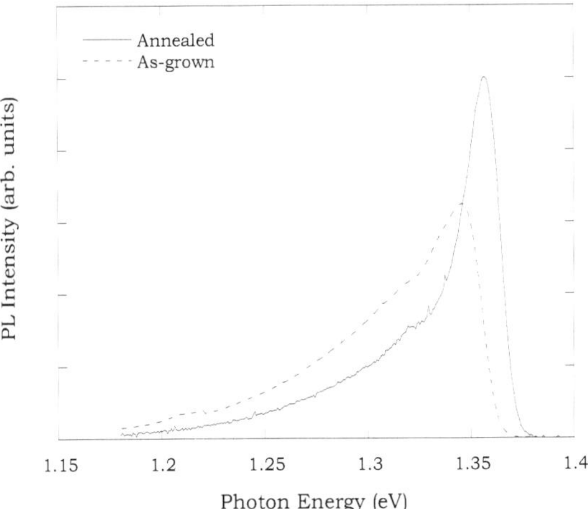

Fig. 2. 15K PL spectra for a five-quantum-well $GaN_{0.025}As/GaAs$ structure grown by SS-MBE and annealed at $800^{o}C$ for different lengths of time.

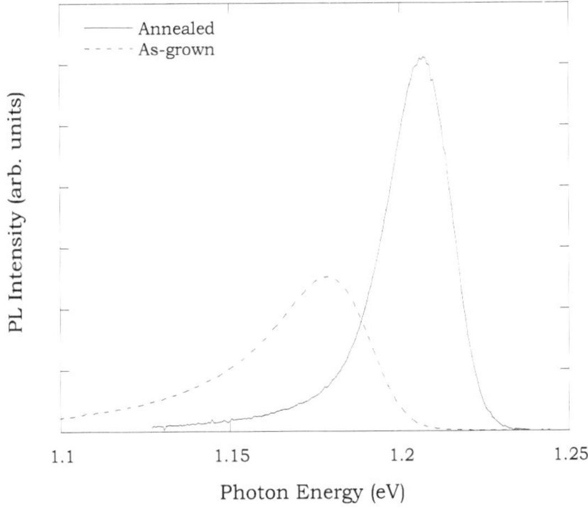

Fig. 3. 15K PL spectra for a five-quantum-well $GaN_{0.025}As/GaAs$ structure grown by SS-MBE and annealed at $800^{o}C$ for different lengths of time.

in terms of the physical properties of the alloy, and definitive explanations remain elusive, due, as already mentioned, in part to the sometimes contradictory nature of published results Grenouillet et al. (2002); Li, Pessa, Ahlgren & Decker (2001).

Two possible explanations have, so far, been proposed to account for the observed blue shift of GaNAs PL spectra with annealing. Li *et al* Li et al. (2000) observed a RTA-induced blue shift in the low temperature photoluminescence (LTPL) spectrum of a single GaNAs quantum well and explained it quantitatively by nitrogen diffusion out of the quantum well. On the other hand, Buyanova *et al* Buyanova, Hai, Chen, Xin & Tu (2000) performed low temperature optical studies of both GaNAs multi-quantum wells and thick epilayers and showed that annealing could induce a blue shift of the PL spectra without necessarily changing the photoluminescence excitation (PLE) spectra energy, that is, the peak PL emission wavelength. They therefore suggested that the change in the PL maximum was related to improvement of the alloy uniformity and that RTA decreased the value of the localization potential. This implied that nitrogen preferentially reorganized in the GaNAs layers rather than diffused into the GaAs barriers. Further investigations by Grenouillet *et al* Grenouillet et al. (2002), confirms the explanation by Buyanova *et al*, that nitrogen reorganizes into the narrow band gap GaNAs material rather than escapes out of it. However it is important, ofcourse, to be aware that this peculiar behaviour reflects the interplay of growth conditions, which is a whole research area of its own, as well as metastability in this system.

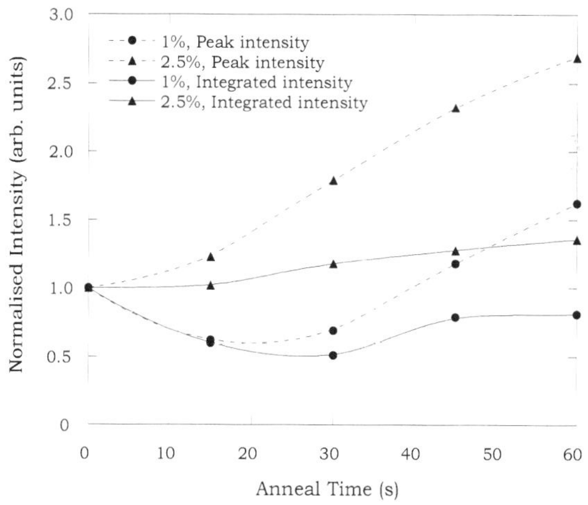

Fig. 4. 15K PL peak intensity and integrated intensity data for the 1.0% and 2.5% GaNAs MQW samples for different anneal times.

Something which seems to have received less attention in the published literature, but which is also an important consideration, is the limit to which annealing can improve PL efficiency. If the degradation of PL intensity is related to defect density, then performance enhancement through the annealing-out of defects is certainly limited. Additionally, the model suggested by Grenouillet et al. Grenouillet et al. (2002) to explain optical performance based on composition fluctuations has been demonstrated both theoretically and experimentally Grenouillet et al. (2002); Pan et al. (2000) to result in a limit beyond which improvement is negligible, although some papers have shown that 'extreme' annealing can also cause samples to degrade after passing through an 'optimal' state see Hierro Hierro et al. (2003), Gupta *et al* Gupta et al. (2003) and Xin *et al* Xin et al. (2000). The presence of hydrogen (which pacifies optically-active centers) can also confuse matters, since its incorporation during 'gas-source' and 'metal-organic' growth and subsequent out-diffusion during annealing can augment perceived improvements in optical efficiency Klar et al. (2003); v H G Baldassarri et al. (2001).

The work and the data presented here is insufficient to comment on annealing to 'extremes', although the behaviour of both peak and integrated intensity, illustrated in figure 4, seem to indicate a fall-off in the rate of increase with anneal time. The initial drop in both peak and integrated intensity for the 1.0% samples might also suggest that localised excitonic emission is the main source of emission in the as-grown 1.0% samples, but is quickly annealed out in favour of band-edge emission from the uniform alloy. However, the underlying problems with the morphology and defect density of both sets of samples are still sufficient to prevent RT emission, see figure 5.

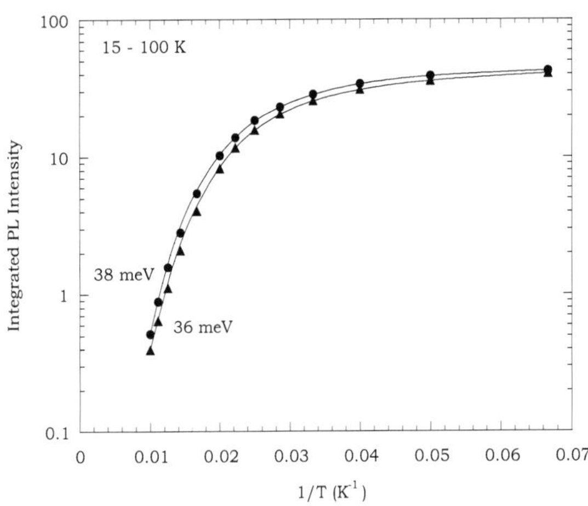

Fig. 5. Arrhenius plot for a GaN0.025As MQW sample after annealing for 30s (•) and for 60s (△) at 800°C.

The Arrhenius plot of the 2.5% nitrogen samples shows that emission begins to fall off at around 35 K, dropping by around two orders of magnitude by 100 K. The activation energy of the thermal loss mechanism is 38 meV after 30s and 36 meV after 60s, and a decrease from 41 to 38 meV is also observed for the 1.0% samples after a similar anneal. These values are comparable with those given for both GaNAs and GaInNAs MQW samples Pomarico et al. (2002); Toivonen et al. (2003a), although relatively few papers analyse GaNAs samples before and after annealing in such a manner.

The observed consistent drop in the activation energy upon annealing might indicate a loss of nitrogen from the wells. If we were to assume that the blue shift is entirely due to a change in overall nitrogen composition of the wells then, according to figure 6, such a blue shift would be consistent with a decrease in nitrogen composition of about 0.1%. This result is more than two orders of magnitude smaller than the same result given by Wang et al Wang et al. (2002). But this slight reduction in N-concentration (around 0.1%) is unlikely to be the cause of the large blue shifting and improvements in emission. In any case, the evidence published on the diffusion of nitrogen (or lack thereof) from dilute nitride layers is not conclusive, and seems to be strongly-dependent on the composition/miscibility of the alloy and defect density Albrecht et al. (2002); Loke et al. (2002); Peng et al. (2003).

Illustrated in figure 7 is the XRD rocking curves from samples GaNAs21-24, which shows no evidence of significant changes in the two structures after the rapid thermal annealing

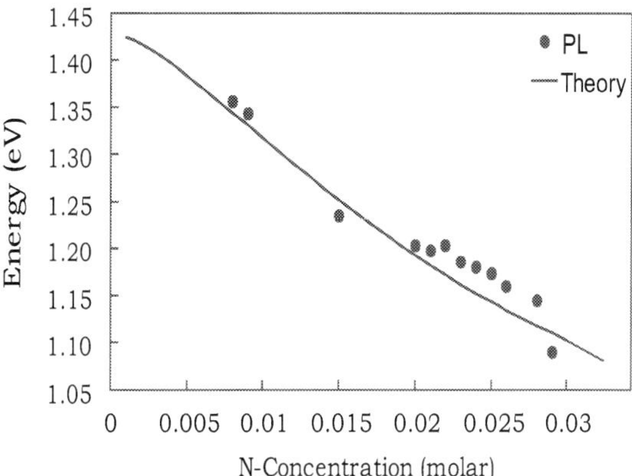

Fig. 6. Theoretical (Solid line), for a well width of 70Å and experimental (Circles) optical transitions in GaN_yAs_{1-y} MQW annealed samples, with varying nitrogen concentrations determined from low temperature PL measurements.

(within the errors of the model). The main finding was that the structures grown were smaller than had been intended due to a lower than expected growth rate. This suggests that neither nitrogen out-diffusion from the wells, nor changes in well thickness, are likely to be responsible for the blue shifting and intensity enhancement demonstrated by this set of samples as a result of annealing. However, there is insufficient information here to comment on whether the changes are related to a general reduction in defect density and improvement in alloy uniformity, or to an improvement in morphology and/or compositional uniformity at the interfaces Li, Pessa, Ahlgren & Decker (2001); Toivonen et al. (2003a;b).

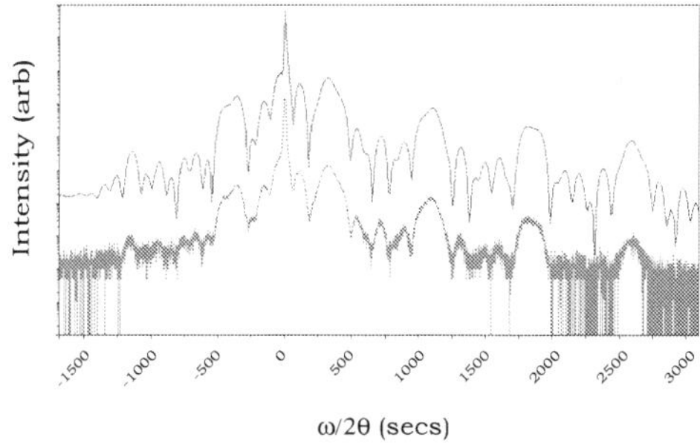

Fig. 7. Measured (lower) and simulated (upper) XRD rocking curves for as-grown 'GaNAs24'. The simulation is based on a five-period $GaN_0.025As/GaAs$ MQW structure with 6.5 nm-thick QWs, 18.8 nm-thick cladding layers and a 70 nm GaAs cap.

Another interesting feature of the PL spectra for the 1.0% samples (at low T) is that of the prominent low-energy tail, observed by both Buyanova et al. Buyanova, Pozina, Hai, Thinh, Bergman, Chen, Xin & Tu (2000) for MQW samples and by Wang et al. Wang et al. (2003) for 100 m-thick $GaN_{0.0145}As$ epilayers. Attempts have been made to explain the origins of this feature with particular reference to localised excitons (LEs), initially due to the exponential shape of the tail, even though the origin of this localisation is not fully understood. In some cases, PL spectra measured as T is increased from ~10 to 300 K display two peaks, one of which diminishes with increasing T (characteristic of LE emission) and the other of which increases with T (characteristic of free exciton (FE) emission) Buyanova et al. (2002); Mair et al. (2000); Shirakata et al. (2002). In addition, the S-shaped temperature dependence of the peak emission wavelength for GaNAs samples also suggest that localised excitons dominate recombination at low T in dilute nitrides Hierro et al. (2003); Mazzucato et al. (2003); Pomarico et al. (2002).

To date, the reasons offered for localisation at low T relate to compositional fluctuations within the lattice Buyanova et al. (2003); Grenouillet et al. (2002); Hong & Tu (2002); Kent & Zunger (2001a), or to the presence of point defects. These point defects can take the form of low-level contaminants and vacancy defects Li, Pessa, Ahlgren & Decker (2001); Toivonen et al. (2003b), and of N-related defects such as interstitials and complexes Ahlgren et al. (2002); Li, Pessa & Likonen (2001); Masia et al. (2003) (which are shown to increase with N-concentration) and ion-induced damage at the interfaces Ng et al. (2002); Pan et al. (2000). What is clear from published material is that the annealing of GaInNAs samples removes interstitial nitrogen and other non-radiative centers, thereby improving alloy homogeneity, enhancing PL efficiency and reducing the prevalence localised excitons.

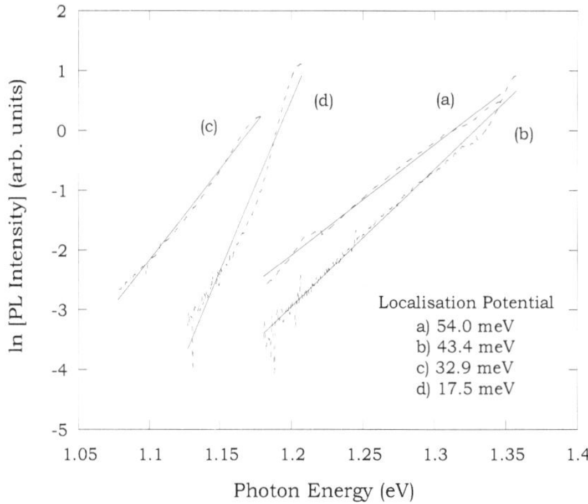

Fig. 8. Natural-log plots of low-E halves of PL spectra (dotted lines) and linear fits (solid lines) for (a) GaNAs23 (as grown), (b) GaNAs23 (60 s anneal), (c) GaNAs24 (as grown), and (d) GaNAs24 (60 s anneal). An estimate of the localisation potential is given by the reciprocal of the gradient.

In order to estimate the localisation potential responsible for the exponential tails seen in figures 2 and 3, the low-E side of each spectrum was plotted on a natural log scale, see figure 8. These estimates are very close to one made by Buyanova et al. for a virtually-identical structure Buyanova et al. (1999), and would seem to indicate that the localisation potential decreases with a 60 s RTA at 800°C for both samples, although the data for GaNAs24 does not fit as well as for that of GaNAs23. This might be expected, looking at figures 2 & 3, since the exponential tail is much more prominent for GaNAs23 than for GaNAs24.

2.2 Annealing of the quaternary GaAs-based dilute nitride: GaInNAs

Even though by nature a more complex system than GaNAs, being a quaternary RTA in GaInNAs and its mechanisms are better understood and the general interpretation of experimental observations are less contradictory. The presence of In seems to be the key in this alloy. During the growth process, chemical bonding aspects dominate at the surface which favour GaŰN bonds instead of InŰN bonds Kurtz et al. (2001). This surface state is frozen in during the non-equilibrium growth process. In contrast to the surface, In-rich nn-configurations of N are favoured in bulk at equilibrium due to the dominance of local strain effects. Therefore, the frozen non-equilibrium bulk state can be transformed into the equilibrium bulk state by annealing under appropriate conditions. Annealing GaInNAs leads to a rearrangement of the N-sites favouring In-rich nn-environments. Depending on the growth conditions, as well as annealing procedure, this presents one of the main contributions to the large blue shift after annealing, which is observed in the PL of $Ga_{1-y}In_yN_xAs_{1-x}$ structures grown either by MOVPE or by MBE Masia et al. (2003); Moison et al. (1989). Further experimental and theoretical evidence for this process was given in Wagner et al. (2003) and Seong et al. (2001). Combining Monte Carlo and pseudo-potential supercell studies, Kim and Zunger found that annealing of GaInNAs causes changes in the nn-configurations of N towards In-rich environments and results in a blue shift of the band gap. Kurtz et al showed by FT-IR vibrational spectroscopy that annealing of $Ga_{0.94}In_{0.06}As_{0.98}N_{0.02}$ converts nn-environments of N from 4Ga to 3Ga and 1In. X-ray photoelectron spectroscopy have to date revealed that nitrogen exists in two bonding configurations in not-annealed material, a GaŰN bond and another nitrogen complex in which N is less strongly bonded to gallium atoms. Annealing removes this second nitrogen complex. A combined nuclear reaction analysis and channeling technique showed that not annealed GaNAs contains a significant concentration of interstitial nitrogen that disappears upon anneal. It is believed that this interstitial nitrogen is responsible for the deviation from VegardŠs law and the low luminescence efficiency of not annealed GaNAs and GaInNAs quantum wells.

The low luminescence efficiency in not-annealed GaInNAs, again, indicates the existence of nonradiative recombination centers or traps. Therefore annealing increases the luminescence efficiency by decreasing the concentration of these centers. Saito *et al* Saito et al. (1998) Xin *et al* Xin et al. (1999) and Geisz *et al* Geisz et al. (1998) postulated that this trap was due to hydrogen impurities. They observed that during the growth of nitride-arsenides by metal-organic chemical vapor deposition (MOCVD) or gas source molecular-beam epitaxy hydrogen supplied in the group V gas sources is incorporated. They also observed changes in the hydrogen concentration profile when annealing. However, Spruytte *et al* Spruytte et al. (2001b) group and other people growing nitride-arsenides by solid source MBE (such as the group I am involved with) using an rf plasma Miyamoto et al. (2000) observed the same increase of luminescence efficiency with annealing with almost no hydrogen present. Hence, despite the recent laser successes, there is still significant work remaining to tame GaAs-based dilute nitride materials system in order to realize the full wavelength range of high-performance optoelectronic emitters. Most critically the strong impact of annealing on the material properties clearly indicate that, still, the epitaxial growth of these alloys is not yet fully mastered. Some aspects such as, to cite but a few, the effect of plasma species on the crystal quality during MBE growth, N segregation, phase separation, ordering, remain to be explored. For all growth techniques there appear to be a need to further work on N-sources. The fine structure of the band gap of GaInNAs, and the metastability caused by different N-environments, requires further studies. The implications on the band alignment of heterostructures containing GaInNAs, as well as on the properties of lasers containing this quaternary alloy, need to be discussed. Also transport properties of these alloys should be investigated in more details in connection with epitaxial growth conditions. Most work to date has focused on low N-content alloys. the growth of high-quality high-N-content alloys for fundamental as well as applied purposes remains a real challenge.

2.3 Alternative dilute nitride emitting structures: SPSL structures

Further to our studies above, in order to gain a better understanding of the issues raised above we have also started to investigate short period super lattice (SPSL) structures of GaInNAs system. The SPSL growth method provides a simple way to tune the N concentration in ternary, GaNAs QWs as well as the In concentration in the quaternary, GaInNAs QWs. The low temperature PL spectra of a $7(GaN_{0.025}As)_5$ $6(GaAs)_{6.4}$ SPSL and the equivalent 70 $GaN_{0.01}As/GaAs$ QW, where the parameters stated are determined through XRD measurements, low temperature PL is illustrated in figure 9.

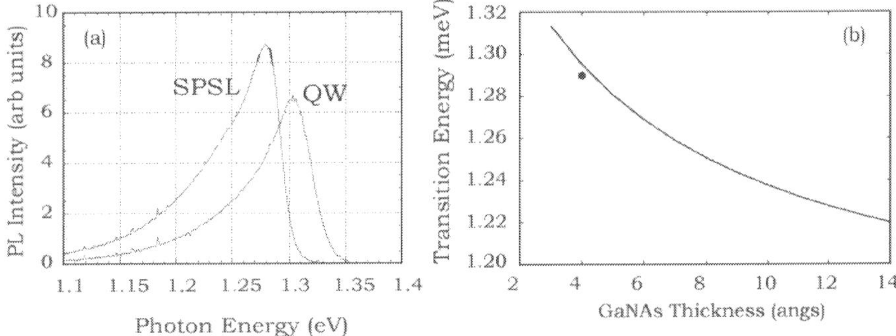

Fig. 9. (a) 15K PL of $7(GaN_{0.025}As)_5$ $6(GaAs)_{6.4}$ SPSL and 80 $GaN_{0.01}As/GaAs$ QW samples. (b) The calculated transition energy of $7(GaN0.025As)_m$ $6(GaAs)_n$ SPSL structure vs. $GaN_{0.025}As$ thickness, m. The $7(GaN_{0.025}As)_5$ $6(GaAs)_{6.4}$ SPSL measured PL-peak is shown by the dark circle.

The PL intensity of the SPSL sample is stronger than that of the bulk layer, as reported also by Hong *et al* Hong et al. (2002; 2001). The PL-peak position, however, is slightly smaller than that of the bulk layer. This is in accordance with our transition energy calculations of figures 6 and 9(b), by comparing their relative positions on figures 6 and 9(b) respectively. The SPSL calculation of figure 9(b) is based on the propagation matrix algorithm Jalili et al. (2004), see chapter 5.

Therefore a better way to improve the luminescence efficiency (material quality) of $III-N_y-V_{1-y}$ alloys, would be that instead of growing the alloy GaInNAs with a random spatial distribution of atoms of the group III elements, a super-lattice (SL) based on the binary compound InAs and the ternary GaNAs should be grown. This would result in a precise arrangement of group III elements and separation of In and N into distinct, separate, layers

Hong et al. (2001). There would, of course, be significant strain to be accommodated due to the lattice mismatch of $\sim 6\%$ between InAs and GaAs and this would restrict the thickness of the layers forming the super-lattice Gerard et al. (1989); Hasenberg et al. (1991); Jang et al. (1992); Moreira et al. (1993); Toyoshima et al. (1990; 1991). An obvious way to reduce the strain and hence remove the bound on the layer thickness would be to use two ternary alloys GaInAs and GaNAs. Certainly, this could be done. The advantage of the binary/ternary combination is that we remove the need to control both In and nitrogen, needing only to control the nitrogen.

2.3.1 (Short-period) Super-lattice structures

The concept of the semiconductor "super-lattice" (SL) was introduced by Leo Esaki,Esaki & Tsu (1970) to describe a crystalline structure with a periodic one-dimensional structural modification. This was achieved by growing, epitaxially and sequentially, multiple thin layers of two semiconductor materials of similar crystal structure but distinctly different energy band-gaps. In this paper we wish to develop and apply techniques to predict the optical properties of such super-lattices (SL) particularly short period super-lattices (SPSL). We aim to use this information to design and fabricate SPSL structures which are optically equivalent to structures formed using the quaternary III-N-V random alloy systems, Hong *et al* Hong et al. (2002; 2001), which will be our main topic for the rest of the current chapter.

3. The idealised super-lattice and its limitations

The electronic and optical properties of SPSL structures have usually been studied by considering an infinite number of identical layers stacked within the same semiconductor structure, and positioned periodically to form a super-lattice Sai-Halasz et al. (1977). The basic SL structure is illustrated in Fig. 10(a), where each period, or unit cell, consists of one well region with width, d_A, and a barrier region of width, d_B. The period of the super-lattice is $d = d_A + d_B$. The host materials, labeled A and B, are the binary (InAs) and ternary (GaNAs) semiconductors respectively. Other configurations using the ternary, GaInAs, and the ternary GaNAs, structures are also possible Hong et al. (2002; 2001).

The Kronig-Penny model of the super-lattice, which is based on the Bloch theorem, is an idealized model. Any practical structure will have a finite number of period and, as shown in Fig. 10(b), is often bound by a wider band-gap semiconductor (labeled 'C' in Fig. 10(b)). Indeed it is quite possible to consider the structure shown in Fig. 10(b) to be one element of a structure in which the structure is repeated in an analogous way to the formation of multiple quantum wells. We will consider the modeling of this (short-period) super-lattice later. At this point, however, we will discuss the model of

the idealized structure, considering it as a limiting case which our finite model must tend to as the number of periods becomes large. To determine the band-structure of the SL structure (a), we consider the potential within a period $0 < z < d$ to be given by

$$V(z) = \begin{cases} V_A & 0 < z < d_A \\ V_B & d_B < z < d \end{cases} \tag{1}$$

and

$$V(z + nd) = V(z), \quad \text{for any integer } n. \tag{2}$$

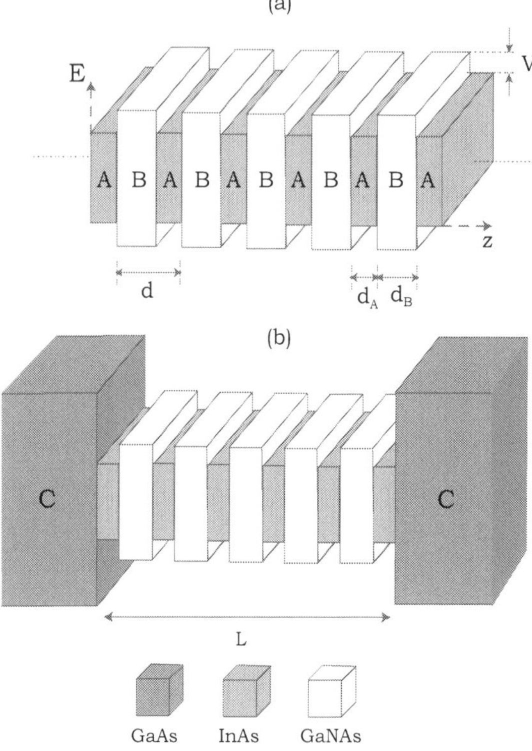

Fig. 10. Schematic representation of the conduction and valence band edge profile of (a) a SL structure of semiconductors A and B (or in-short $(A)_{d_A}(B)_{d_B}$), showing wells of width d_A alternating periodically with barriers of width d_B and differential height V_o, to form a SL of period $d = d_A + d_B$. (b) SL structure of finite length L, confined by semiconductor C, GaAs, i.e. SPSL.

Within such a structure the electrons and holes experience a periodic potential, which is unbounded in the same way as we assume for a bulk crystal. This implies that the electron or hole wave-functions are no longer localised but extend through-out the lattice. Electrons are therefore equally likely to be found in any of the wells in the super-lattice. Electrons in such a structure are said to occupy "Bloch states".

$$\Psi_l(z + d) = e^{iqd}\Psi_l(z) \tag{3}$$

The energy dispersion curve for the SL, $E(q)$, will be restricted to the first Brillouin zone of the SL, i.e. $-\pi/d \leq q \leq \pi/d$. The SL (or Bloch) wavevector, q, is orientated along the crystal growth direction, which we have taken to be the z-axis. The SL dispersion curves representing the energy E of a particle as a function of its wavevector, q, are obtained from the Kronig-Penny expression.

$$\cos(qd) = \cos(k_A d_A) \cos(k_B d_B) - \frac{k_A^2 + k_B^2}{2 k_A k_B} \sin(k_A d_A) \sin(k_B d_B) \qquad (4)$$

We use the nomenclature $(InAs)_m$ $(GaN_yAs)_n$ to specify a SL in which m mono-layers of the binary alloy and n mono-layers of the ternary constitute the basic lattice which is repeated to form the super-lattice. Calculations of the band-structure are shown below in Fig. 11(a) and (b) for m=4, n=4; m=4, and n=9; and for two different nitrogen compositions.

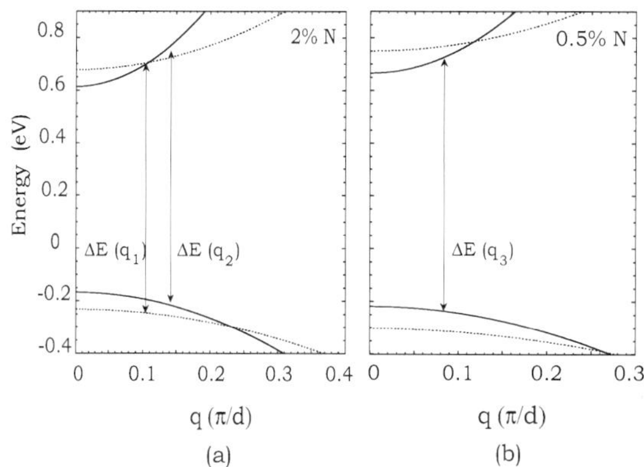

Fig. 11. Band structure calculations of $(InAs)_4$ $(GaNyAs)_4$, solid line, and $(InAs)_4$ $(GaNyAs)_9$, dashed line, SLs. (a) GaNAs with 2% N (b) GaNAs with 0.5% N.

By analogy with bulk and QW energy dispersion relations, we should be able to predict the thickness of a SPSL structure by analysing the energy wave-vector dispersion plot of an equivalent SL-structure, i.e. the same unit cell structure, in the reciprocal space. This can be done from the definition of k-vector, in the k-space the propagation vector, which is basically, assuming $\Psi(z) \to 0$ at end points, the simple quantum mechanical expression

$$
\begin{aligned}
E &= \frac{1}{2m^*(E)} \hbar^2 |q|^2 \\
&= \frac{1}{m^*(E)} \hbar^2 \frac{2\pi^2}{L^2} (n_x^2); \quad n_x = 0, \pm 1, \pm 2 \ldots
\end{aligned}
\qquad (5)
$$

Relating the wave-vector, $k(\equiv q)$, and the transition energy, $\Delta E(k)$. Pictorially, by using the energy wave-vector dispersion diagram of a SL structure, illustrated schematically in Fig. 10(a), we would want to estimate the thickness, L, of a SPSL structure, illustrated schematically in Fig. 10(b), which have the same unit cell and transition energy. The energy wave-vector dispersion of $(InAs)_4(GaN_{0.02}As)_9$ SL structures for two different N compositions of 0.5% and 2% are illustrated in Fig. 11(a) and (b) respectively. The two transition energies, $\Delta E(q_1)$ and $\Delta E(q_2)$ both correspond to 1.3 μm emission, shown in Fig. 11(a). However as the wave vector in the latter transition is larger, then the corresponding SPSL structure will have a smaller overall length in comparison. This is deduced from Eq 5.

4. The short period super-lattice

Before we proceed to the details of the propagation matrix calculation of the optical transitions in a SPSL, we need to consider the properties of the binary (InAs) and ternary (GaNAs) alloys that form the SPSL. There are two distinct issues we need to address. The first is simply the modeling of the bulk properties of these materials, particularly when there is considerable strain at the interface with say, GaAs. The second issue is the need to determine the band off-set arrangement both between InAs and GaNAs and GaNAs and GaAs.

We start with the binary alloy InAs, this is a narrow gap III-V semiconductor, so when considering optical transitions, band non-parabolicity is important and must be accounted for. We used the 3-band Kane Hamiltonian to investigate the band-edge electronic description of bulk InAs-GaAs, where the lattice mismatch is accommodated by biaxial compressive strain within the InAs layer. To model the GaNAs, we use the band-anti-crossing (BAC) model to explain the unusually strong band-gap reduction of GaAs through the replacement of only a few percent of the arsenic atoms by nitrogen Shan et al. (2001); Weyers et al. (1992). The good agreement between a 5-band (10 including spin) k.p and the BAC model, confirms the validity of this two band model at the band-edge O'Reilly et al. (2002). In this model, the nitrogen atoms form a flat and almost dispersionless band, resonant with the GaAs conduction band, but, which also interacts with the GaAs conduction band minimum. This interaction leads to the formation of two bands of mixed GaAs-nitrogen character. The lower band is shifted downwards with respect to the GaAs conduction band-edge by > 100 meV for each percent increase of nitrogen concentration. The new conduction band minimum is strongly non-parabolic. To take account of this, and the non-parabolicity of the valence band, we have used the modified Kane Hamiltonian, with the inclusion of the nitrogen.

The Hamiltonian, which includes the off-diagonal matrix elements linking the ψ_N, ψ_C and ψ_{lh}, ψ_{so} basis states, is written as

$$
\begin{bmatrix}
E_N & V_{NC} & P_N k_z & 0 \\
V_{NC} & E_C & -i\sqrt{\tfrac{2}{3}}P_K k_z & i\sqrt{\tfrac{1}{3}}P_K k_z \\
P_N k_z & i\sqrt{\tfrac{2}{3}}P_K k_z & \delta E_s & -\sqrt{\tfrac{1}{2}}\delta E_s \\
0 & -i\sqrt{\tfrac{1}{3}}P_K k_z & -\sqrt{\tfrac{1}{2}}\delta E_s & E_{so}
\end{bmatrix}
\tag{6}
$$

Where the nitrogen dependant terms are defined as

$$
E_C = E_g^{GaAs} - (1.55 - 3.88)y
$$
$$
E_N = 1.65 - (3.89 - 3.88)y
\tag{7}
$$
$$
V_{NC} = -2.4\sqrt{y}
$$

The nitrogen concentrations, y, in molar units, are extremely small, only a few percent. δE_s, and E_{so} are the light-hole and the spin-orbit splitting band-edges respectively. The heavy-hole band is decoupled from the rest of the bands. The heavy-hole band-edge is set at zero. The coupling term, the matrix element P_N, in the Hamiltonian above is $\sim 10\%$ of P_K, the Kane matrix element, as reported by O'Reilly et al O'Reilly et al. (2002). However, others suggest that P_N is about an order of magnitude smaller than this Linsay (2002). Therefore in view of there being no consensus over the value of P_N and also the general difficulty in determining its value, it is therefore best to set the k-dependant N-related terms to zero. The determinant equation $|\ H_{Kane,N} - EI\ |$ then gives the following dispersion relation within the vicinity of the band edges

$$k_z^2 = \frac{3}{P_K^2} \frac{(E - E_N)(E - E_C)[(E - \delta E_s)(E - E_{so}) - \frac{\delta E_s^2}{2}]}{(E - E_N)[3E + \delta E_s - 2E_{so}]}$$

$$\hspace{6cm}(8)$$

$$- \frac{V_{NC}^2[(E - \delta E_s)(E - E_{so}) - \frac{\delta E_s^2}{2}]}{(E - E_N)[3E + \delta E_s - 2E_{so}]}$$

The lattice mismatch between bulk GaNAs and GaAs is accommodated by biaxial tensile strain within the GaNAs layer. Plots of band structures of InAs-GaAs and $GaN_{0.02}As$ are illustrated in Fig. 12(a) and (b) respectively.

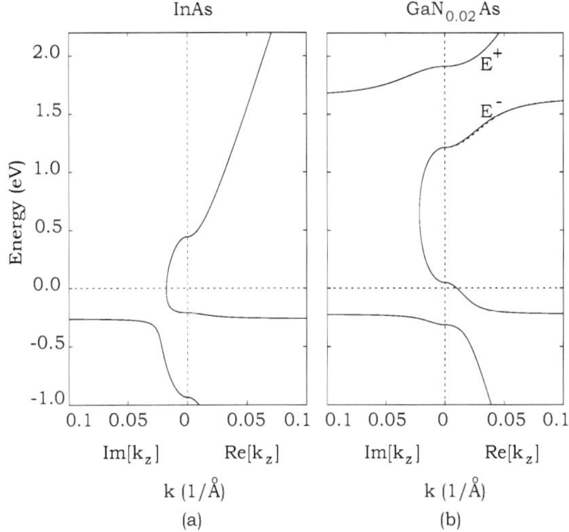

Fig. 12. Illustration of calculated, Kane description, dispersion relations of real and imaginary wave-vectors in InAs and $GaN_{0.02}As$ semiconductors, corresponding to 6.67% compressive and 0.53% tensile strains respectively. The dashed line is based on the BAC Shan et al. (1999) model.

Except for the band-gap energy, the electron effective mass, Skierbiszewski et al. (2000) the lattice constant and the valence band deformation potential, Raja et al. (2002) most parameters for GaN_yAs_{1-y} are obtained by a linear interpolation between the parameters of the relevant binary semiconductors given in table 2.

Moving to the discussion of the band off-set issue, we note that the presence of nitrogen atoms is assumed to influence predominantly the host conduction band, with the valence bands remaining largely undisturbed. Following this model one might therefore expect the band-gap mismatch between the well and barrier material in GaN_yAs_{1-y} /GaAs heterostructure to show up predominantly in the conduction band off-set, with the valence-band offset being zero or negative. However, recent experiments suggest that

	GaAs	InAs	GaN	InP
$E_{g,\,RT}$ (eV)	1.424	0.36	3.2	1.344
$E_{g,\,LT}$ (eV)	1.519	0.41	3.39	1.424
E_p (eV)	25.7	22.2	25.4	16.7
Δ (eV)	0.34	0.32	0.015	0.11
a_0 (Å)	5.6533	6.0584	4.503	5.8687
a_{gap} (eV)	-8.3768	-6.08	-7.8	-6.31
b (eV)	-1.7	-1.8	-1.9	-1.7
c_{11} (GPa)	181.1	83.29	293	10.11
c_{12} (GPa)	53.2	101.1	159	5.61
m_e^*/m_0	0.067	0.023	0.22	0.08
m_{hh}^*/m_0	0.5	0.4	0.8	0.6
m_{lh}^*/m_0	0.087	0.026	0.19	0.089
m_{so}^*/m_0	0.15	0.16		

Table 2. Parameters of GaAs (de Walle (1989)), InAs (de Walle (1989)), InP and GaN (de Walle (1989); Fan et al. (1996); Kim et al. (1996); Pearton (2000); Persson et al. (2001)) used in the calculations.

this might not be true Buyanova, Pozina, Hai & Chen (2000); Egorov et al. (2002). Latest results and measurements have determined that the band lineup in GaNAs-GaAs (for low N compositions) is of type I Buyanova, Pozina, Hai & Chen (2000); Egorov et al. (2002); Klar et al. (2002). InAs and GaAs are also expected to show a type I band line up de Walle (1989). However, to authors' knowledge, there is currently no information on band lineup of InAs-GaNAs heterostructure. Therefore, with GaAs band edges as the reference, we have lined up InAs and GaNyAs by lining up InAs/GaAs (60:40 in favour of CB) and GaNyAs/GaAs ($y = 0.02$, $\sim$ 90:10 in favour of CB). The band diagram is illustrated in Fig. 13.

5. SPSL design & calculations: The propagation matrix approach

In this section we show how the Schrödinger equation for a one dimensional potential profile with an arbitrary shape, a SPSL in this case, (Fig. 10(b)), can be solved using a propagation matrix approach, similar to that used in electromagnetic wave reflection or guidance in a multilayered medium Kong (1990); Yeh & Yariv (1984). An arbitrary profile, V(z) can always be approximated by a piece-wise step profile. In contrast to the conventional method, described above, where the Bloch condition must be employed to obtain the required translational symmetry of the problem Bastard & Brum (1986), a propagation matrix approach produces a more accurate energy band solution of a SPSL structure Ram-Mohan *et al* Ram-Mohan et al. (1988), since it provides an extra degree of freedom in space, z, to vary the, layer dependant, potential $V_l(z)$ accordingly. This is in contrast to the conventional method in which the layer dependant potential remains fixed and repeats itself infinitely over the SL unit cell Bastard (1988). Of course, a SPSL structure should tend towards a SL structure with increasing period. One way of discovering exactly how many quantum wells or unit cells are required before a finite structure resembles an infinite one (i.e. a SL) would be to look at the ground state energy as a function of the number of periods within a SPSL structure.

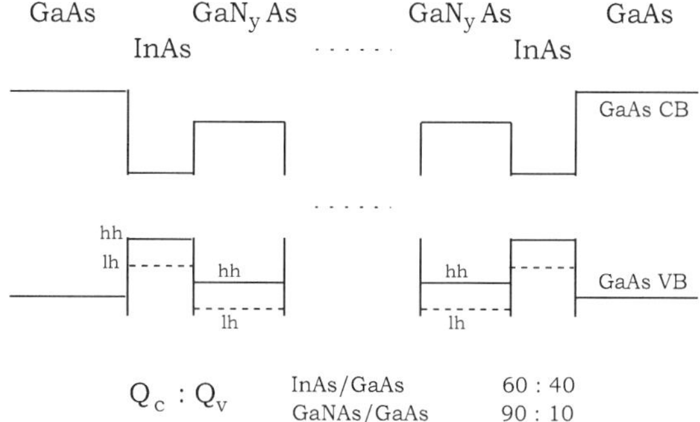

Fig. 13. Energy band line-up profile of GaAs/M(InAs)$_{d_A}$N(GaN$_{0.02}$As)$_{d_B}$/GaAs SPSL structure. lh and hh denote light and heavy holes, the conduction and valence band offsets, denoted by Q_C and Q_V respectively is determined to be 50:50 for InAs/GaN$_{0.02}$As.

A SPSL structure may consist of a number of heterostructures. As we know QWs are fabricated by forming heterojunctions between different semiconductors. From an electronic viewpoint, semiconductors are different because they have different band structures and, hence, band gaps. Apart from the bandgap, there are other properties which are also different in semiconductors, such as the dielectric constant, the lattice constant and, what is considered as the next most important quantity, the effective mass. In general the calculation of static energy levels within QWs should account for the variation in the effective mass across the heterojunction under appropriate boundary conditions. The boundary conditions on the envelope functions can be obtained by integration of the coupled differential equations across an interface. This problem has been previously addressed in the literature, Johnson et al, also Bastard *et al* and Taylor *et al*, these can be stated as requiring the continuity of both

$$\psi_{l,n}(E,z) \quad \text{and} \quad \frac{1}{m_{l,n}^*(E,z)}\frac{\partial}{\partial z}\psi_{l,n}(E,z) \tag{9}$$

across a heterostructure interface. Where $m_{l,n}^*(E,z)$ and $\psi_{l,n}(E,z)$ are the energy dependant effective mass and envelope coefficients in layer 'l' and band 'n', which we will looked at in detail in chapter 2.

Within a multilayer structure, for a given energy, the envelope function in each layer can be written as a sum of forward and backward traveling waves

$$\psi_l(z) = A_l e^{ik_l(z-z_l)} + B_l e^{-ik_l(z-z_l)}, z_{l-1} \leq z \leq z_l \tag{10}$$

Where the wavevector k_l is given by the auxiliary equation for each layer and is expressed as

$$k_l^2 = \frac{2m_l^*(E,z)(E-V_l(z))}{\hbar} \tag{11}$$

Applying the boundary conditions of Eq. (9) a general form of a propagating matrix at an interface between layers l and l' is obtained as

$$\Pi_{l \to l'}(E) = P_l D_l^{-1} D_{l'} P_{l'} \tag{12}$$

where D_l, the transition matrix, which is obtained from the boundary conditions, and P_l, the propagating matrix, are defined as

$$D_l = \begin{pmatrix} 1 & 1 \\ \dfrac{k_l}{m_l^*(E,z)} & -\dfrac{k_l}{m_l^*(E,z)} \end{pmatrix}, \quad P_l = \begin{pmatrix} e^{ik_l z_l} & 0 \\ 0 & e^{-ik_l z_l} \end{pmatrix} \tag{13}$$

Therefore the matrix $P_l D_l$ takes us from layer $l + 1$ to l. For a large number of layers, the propagation matrix for each layer can be linked forming the propagation matrix of the whole structure

Imposing the relevant boundary conditions, (i.e. $\psi(z) \to 0$ as $z \to \pm\infty$) the wavefunction must tend toward zero into the outer barriers (GaAs in this case), and the coefficients of the growing exponentials must be zero. Therefore

$$\begin{pmatrix} A_N \\ 0 \end{pmatrix} = \begin{pmatrix} \Pi_{11}(E) & \Pi_{12}(E) \\ \Pi_{21}(E) & \Pi_{22}(E) \end{pmatrix} = \begin{pmatrix} 0 \\ B_0 \end{pmatrix} \tag{14}$$

Eq. (14) implies that for nontrivial solutions, we must have the eigenequation above satisfy,

$$\Pi_{22}(E) = 0 \tag{15}$$

SPSLs are conventionally defined by the number M of interfaces of InAs grown upon GaNAs, and the number N of interfaces of GaNAs grown upon InAs, with the numbers m and n, measured in Angstrom (Å),which define thicknesses of InAs and GaNAs layers respectively.Jang et al. (1992); Moreira et al. (1993); Toyoshima et al. (1990) The short form of the expression is $M(InAs)_m \, N(GaNAs)_n \equiv N[(InAs)_m(GaNAs)_n] + (InAs)_m$. Therefore a SPSL of N periods consists of 'N' GaN_yAs layers and 'N+1' (=M) InAs layers. In this case the SPSL structure equivalent to a single quantum well of the random alloy 8.5 nanometers thick and composition $GaIn_{0.35}N_{0.015}As/GaAs$, is obtained by imposing the molar composition ratios of $III-N_y-V_{1-y}$ constituents onto SPSL layer thicknesses which would give us the numbers M, N, m and n, written in short form, as $7(InAs)_46(GaN_{0.015}As_9)$ or $14(InAs)_213(GaN_{0.015}As_{4.5})$. Hence the overall length of the SL is limited to what would be the typical dimensions of an 8.5 nm $GaIn_{0.35}N_{0.015}As$-GaAs QW, which places the band-edge (0.95eV) at around 1.3 μm.

Fig. 14, illustrates, in plots (a) and (b), transition energies of $7(InAs)_46(GaN_{0.02}As)_n$ and $14(InAs)_213(GaN_{0.02}As)_n$ SPSLs, surrounded on either side by GaAs, as a function of barrier layer thickness, n, respectively. As expected, the transition energy is larger for the structure with smaller QW thickness, and for both structures, the transition energy levels off with increasing barrier thickness. Note that the points that correspond to the $7(InAs)_46(GaN_{0.02}As)_9$ and $14(InAs)_213(GaN_{0.02}As_{4.5})$ SPSL structures in Fig. 14 plots (a) and (b) respectively, are indicated with a dashed line. As was predicted, the band edge transition is at 1.3 μm. Plot (a) of Fig. 15 is transition energy of an $M(InAs)_3N(GaN_{0.02}As)_2$ SPSL as function of period, N, and plot (b) is that of the $M(InAs)_4N(GaN_{0.03}As)_2$ SPSL, where in the latter case we have effectively lowered the barrier height by increasing the nitrogen concentration to 3% and the well thickness to 4 Å respectively.

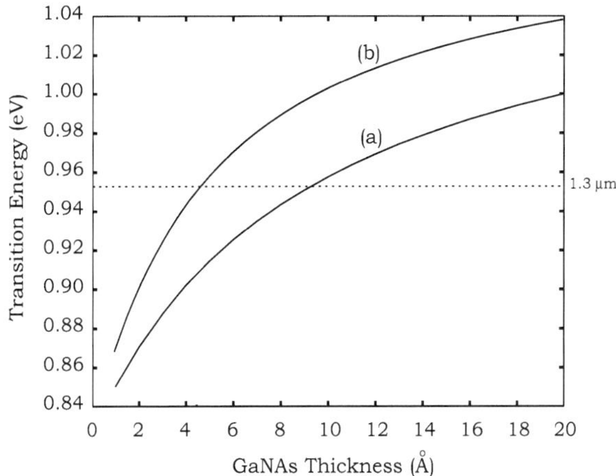

Fig. 14. Energy gap of InAs/GaN$_{0.02}$As SPSL structure as function of varying GaNAs (barrier) layer thickness (a) $7(InAs)_4 6(GaNAs)_n$ configuration (b) $14(InAs)_2 13(GaNAs)_n$ configuration.

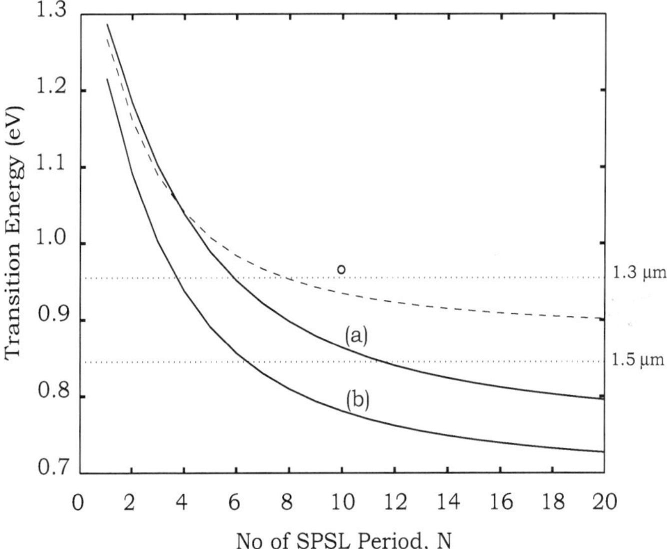

Fig. 15. The calculated transition energy plots of SPSL structures as function of SPSL-period, N. (a) M(InAs)$_3$N(GaN$_{0.02}$As)$_2$ and (b) M(InAs)$_4$N(GaN$_{0.03}$As)$_2$. The dotted line is the numerical result for the M(InAs)$_3$N(GaN$_{0.023}$As)$_{6.2}$ SPSL structure. The circle (o) is from Hong et al(needs reference in here)and is the experimental result for 10(InAs)$_3$9(GaN$_{0.023}$As)$_{6.2}$ SPSL annealed structure.

Therefore varying by the number of periods and/or barrier height within a SPSL structure, the position of the band edge can be modified significantly. For the plots it is clear that a structure which would absorb or emit at the important telecommunication wavelength of 1.5 μm can be achieved. We could equally reduce the potential barrier height of the cladding layer (GaAs in this case) by incorporation of In, in order to reduce the band edge to 1.5 μm, since, due to limitations of strain, the InAs layer thickness, with a critical thickness, $h_c \leq 5$ Angstroms cannot be varied arbitrarily. As expected a larger number of SPSL periods, N, reduces the transition energy. The same pattern holds with a reduction in potential barrier height.

The following plots illustrate contour plots for various SPSL structures which emit or absorb light at 1.3 μm. The contours in Fig. 16(i) indicate that by reducing dB, tunneling across the barriers increases and leads to a reduction of the carrier energy within the wells. Therefore to make up for this reduction we need to increase the barrier height, Vo, or we must reduce the N concentration since the number of unit cells and the well width, d_A, are fixed. The two contour lines in the figure imply that if SPSL-period, N, is reduced in going from solid line contour to the dashed line contour, then the carrier energy is increased. Therefore thinner barriers or more nitrogen, are required to lower the barrier height, since d_A is fixed. Further more, for nitrogen concentrations of 0.5-1.5% the contour curvature is negligible with respect to N concentrations. This is particularly so for smaller numbers of periods, N. This is very significant considering that band gap variation in III-(N)-V systems is nonlinear with respect to the nitrogen concentration and is therefore very difficult to control even by sophisticated epitaxial growth techniques. Fig. 16(ii) illustrates 1.3 μm contour plots for fixed nitrogen concentration and well thickness. In this case an increase in barrier thickness, dB, reduces the carrier energy within the wells, and therefore, to make up for this we would have to increase the number of periods. Going from the contour represented by a dashed line to the one represented in dotted line, the nitrogen concentration increases from 0.5% to 2% respectively. For higher nitrogen concentrations the barrier height V_o, is lowered implying that the carrier energy decreases. Therefore we would have to reduce the number of periods to make up for the carrier energy reduction. In Fig. 16(iii) the contours indicate that, since increase in number of periods lowers the carrier energy, the barrier height needs to be raised as d_A and d_B are both kept fixed. This is achieved by reducing the nitrogen concentration. The same pattern holds when barrier width, d_B, is reduced, as shown by the solid line of Fig. 16(iii). Again, as with contours of Fig. 16(i), the transition energy is not very sensitive to variations in nitrogen concentration for the smaller barrier width particularly for 2-3% nitrogen concentrations. This is in contrast to structures with comparatively larger barrier width (dashed line of Fig. 16(iii)) which leads to better control over nitrogen concentration in growth. These results, which are based on numerical models are in agreement with the predictions based on the SL model.

The results are very encouraging for design and fabrication of short period superlattices suitable for devices which emit or absorb light at 1.3μm and also 1.5 μm of GaAs-based dilute nitrides. Specifically, more degrees of freedom are available for the design of nanostructure optoelectronic devices based on a given choice of materials. Structures can be engineered to vary the SPSL energy gap, by suitable choice of layer thicknesses, which can be atomically controlled using thin film crystal growth techniques such as MBE, as well as varying the number of SL period and layer composition. The proposals to use dilute nitride SPSL structures results in the separation of In and N and would over-come some of the key material issues limiting growth of III-N_y-V_{1-y} alloys. The growth of the binary and ternary configuration of GaInNAs SPSL should also provide better compositional control since the

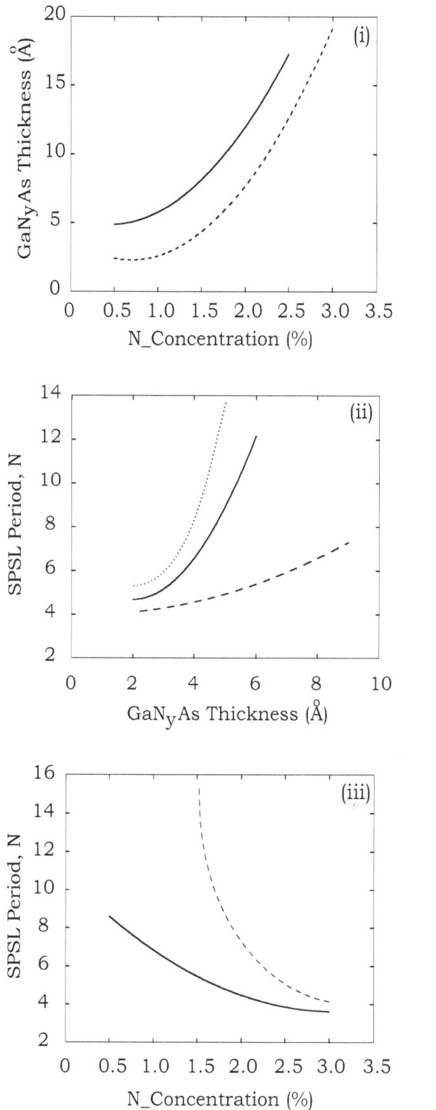

Fig. 16. 1.3 μm contour plots of (i) $4(InAs)_4 13(GaN_yAs)_n$, solid line, and $7(InAs)_4 6(GaN_yAs)_n$, dashed line, SPSLs vs. barrier width, n, and N-concentration,y. (ii)$M(InAs)_4 N(GaN_{0.005}As)_n$, dotted line, $M(InAs)_4 N(GaN_{0.01}As)_n$ solid line, and $M(InAs)_4 N(GaN_{0.02}As)_n$, dashed line, SPSLs vs. number of periods and barrier width. (iii) $M(InAs)_4 N(GaN_yAs)_9$, dashed line, and $M(InAs)_4 N(GaN_yAs)_4$, solid line, SPSL structures as function of number of periods, N, and N-concentration, y.

incorporation of nitrogen will involve only one group III-element in each period of the structure. Also, since in SPSL structures the well/barrier width and therefore the period are in effect reduced to less than the electron mean free path, the entire electron system will enter a quantum regime of reduced dimensionality in the presence of nearly ideal interfaces, resulting in improved mobility within these structures. Therefore, design and growth of more efficient optoelectronic devices based on III-N_y-V_{1-y} systems should be possible. The current work on SPSL dilute nitride structures is very scarce. To authors knowledge apart from our group only one other has produced such work without any proper theoretical back up tough. Therefore the potential is tremendous in this field with many possible directions in obtaining a better understanding of the important GaAs-based dilute nitride systems.

If dilute nitride materials are to prove their worth, then it must be demonstrated that they can be used to produce durable optoelectronic devices for use at 1.3-1.55 m applications. Unfortunately, a full understanding of the fundamental nature and behaviour of nitride alloys, especially during the annealing treatments that are required for optimum performance, continues to elude researchers. Certain trends have been identified qualitatively, such as that optimum anneal conditions depend on composition, and more specifically on (2D/3D) growth mode Hierro et al. (2003), on nitrogen content Francoeur et al. (1998); Loke et al. (2002), and on indium content for GaInNAs Kageyama et al. (1999), but 'optimum' annealing treatments continue to vary widely, according to growth method, growth conditions, structure and composition. We believe that SPSL structures have an important role to play in such studies. Therefore the priority should be to repeat the previous annealing study and try to obtain more information about the improvements seen during annealing. This could be done by measuring more-comprehensively the relationship seen in Arrhenius plots of integrated PL intensity vs. 1/T. Additionally, a series of experiments designed to find the optimum combination, duration and temperatures for in-situ and/or ex-situ annealing should be carried out, and repeated for SPSL active layers to determine whether such dilute nitride structures are capable of outperforming more-primitive MQW structures. These experiments should also provide another opportunity to investigate the optical performance of nitrides.

We made use of the transfer matrix algorithm based on the envelope function approximation (EFA). The results obtained demonstrated excellent agreement with those obtained experimentally so far, to authors knowledge, Hong *et al* Hong et al. (2001). Since the transfer matrix method is based on the EFA, it has the corresponding advantage that the input parameters are those directly determined by experimentally measured optical and magneto-optical spectra of bulk materials. The effect of additional perturbations, such as externally applied fields, built in strain in superlattices are easily incorporated into the **k.p** Hamiltonian with no additional analysis in the transfer matrix method. Furthermore the transfer matrix method provides a simple procedure to obtain the wavefunctions, which are particularly useful in evaluating transition probabilities.

6. References

Ahlgren, T., Vainonen-Ahlgren, E., Likonen, J., Li, W. & Pessa, M. (2002). Concentration of interstitial and substitutional nitrogen in GaN$_x$As$_{1x}$, *Applied Physics Letters* 80: 2314–2316.

Albrecht, M., Grillo, V., Remmele, T., Strunk, H. P., Egorov, A. Y., Dumitras, G., Riechert, H., Kaschner, A., Heitz, R. & Hoffmann, A. (2002). Effect of annealing on the In and N distribution in InGaAsN quantum wells, *Applied Physics Letters* 81: 2719–2721.

Bastard, G. A. (1988). *Wave mechanics applied to semicindcutor heterostructures*, Les Editiond de Physique, Paris.

Bastard, G. & Brum, J. (1986). Electronic states in semiconductor heterostructures, *Quantum Electronics, IEEE Journal of* 22: 1625–1644.

Bhat, R., Caneau, C., Salamanca-Riba, L., Bi, W. & Tu, C. (1998). Growth of GaAsN/GaAs, GaInAsN/GaAs and GaInAsN/GaAs quantum wells by low-pressure organometallic chemical vapor deposition, *Journal of Crystal Growth* 195: 427–437.

Buyanova, I. A., Chen, W. M., Pozina, G., Bergman, J. P., Monemar, B., Xin, H. P. & Tu, C. W. (1999). Mechanism for low-temperature photoluminescence in GaNAs/GaAs structures grown by molecular-beam epitaxy, *Applied Physics Letters* 75: 501–504.

Buyanova, I. A., Chen, W. M. & Tu, C. W. (2002). Magneto-optical and light-emission properties of III-As-N semiconductors, *Semiconductor Science and Technology* 17: 815–822.

Buyanova, I. A., Chen, W. M. & Tu, C. W. (2003). Recombination processes in N-containing IIIŰV ternary alloys, *Solid State Electronics* 47: 467–475.

Buyanova, I. A., Hai, P. N., Chen, W. M., Xin, H. P. & Tu, C. W. (2000). Direct determination of electron effective mass in GaNAs/GaAs quantum wells, *Applied Physics Letters* 77: 1843–1845.

Buyanova, I. A., Pozina, G., Hai, P. N. & Chen, W. M. (2000). Type I band alignment in the GaN_xAs_{1-x}/GaAs quantum wells, *Physical Review B* 63: 033303–033307.

Buyanova, I. A., Pozina, G., Hai, P. N., Thinh, N. Q., Bergman, J. P., Chen, W. M., Xin, H. P. & Tu, C. W. (2000). Mechanism for rapid thermal annealing improvements in undoped GaN(x)As(1-x)/GaAs structures grown by molecular beam epitaxy, *Applied Physics Letters* 77: 2325–2327.

de Walle, C. G. V. (1989). Band lineups and deformation potentials in the model-solid theory, *Physical Review B* 39: 1871–1883.

Egorov, A. Y., Odnobludov, V. A., Zhukov, A. E., Tsatsul'nikov, A. F., Krizhanovskaya, N. V., Ustinov, V. M., Hong, Y. G. & Tu, C. W. (2002). Valence band structure of GaAsN compounds and band-edge line-up in GaAs/GaAsN/InGaAs hetorosructures, *Molecular Bean Epitaxy, 2002 International Conference on*, IEEE, pp. 269–270.

Esaki, L. & Tsu, R. (1970). Superlattice and negative differential conductivity in semiconductors, *IBM J. Res. Development* 14: 61–65.

Fan, W. J., Li, M. F., Chong, T. C. & Xia, J. B. (1996). Electronic properties of zinc-blende GaN, AlN, and their alloys $Ga_{1-x}Al_xN$, *Journal of Applied Physics* 79: 188–194.

Francoeur, S., Sivaraman, G., Qiu, Y., Nikishin, S. & Temkin, H. (1998). Luminescence of as-grown and thermally annealed GaAsN/GaAs, *Applied Physics Letters* 72: 1857–1859.

Geisz, J. F., Friedman, D. J., Olson, J. M., Kurtz, S. R. & Keyes, B. M. (1998). Photocurrent of 1 eV GaInNAs lattice-matched to GaAs, *Journal of Crystal Growth* 195: 401–408.

Gerard, J. M., Marzin, J. Y., Jusserand, B., Glas, F. & Primot, J. (1989). Structural and optical properties of high quality InAs/GaAs short-period superlattices grown by migration-enhanced epitaxy, *Applied Physics Letters* 54: 30–32.

Gilet, P., Chevenas-Paule, A., Duvaut, P., Grenouillet, L., Hollinger, P., Million, A., Rolland, G. & Vannuffel, C. (1999). Growth and characterization of thick GaAsN epilayers and GaInNAs/GaAs multiquantum wells, *Physica-Status-Solidi a* 176: 279–283.

Grenouillet, L., Bru-Chevallier, C., Guillot, G., Gilet, P., Ballet, P., Duvaut, P., Rolland, G. & Million, A. (2002). Rapid thermal annealing in GaN_xAs_{1x}/GaAs structures: Effect of nitrogen reorganization on optical properties, *Journal of Applied Physics* 91: 5902–5908.

Gupta, J. A., Barrios, P. J., Aers, G. C., Williams, R. L., Ramsey, J. & Wasilewski, Z. R. (2003). Properties of 1.3 μm InGaNAs laser material grown by MBE using a N_2/Ar RF plasma, *Solid State Electronics* 47: 399–405.

Hasenberg, T. C., McCallum, D. S., Huang, X. R., Dawson, M. D., Boggess, T. F. & Smirl, A. L. (1991). Linear optical properties of quantum wells composed of all-binary InAs/GaAs short-period strained-layer superlattices, *Applied Physics Letters* 58: 937–939.

Hierro, A., Ulloa, J.-M., Chauveau, J.-M., Trampert, A., Pinault, M.-A., Tournié, E., Guzmán, A., Sánchez-Rojas, J. L. & Calleja, E. (2003). Annealing effects on the crystal structure of GaInNAs quantum wells with large In and N content grown by molecular beam epitaxy, *Journal of Applied Physics* 94: 2319–2324.

Hong, Y. G., Egorov, A. Y. & Tu, C. W. (2002). Growth of GaInNAs quaternaries using a digital alloy technique, *Journal of Vaccum Science Technology B* 20: 1163–1166.

Hong, Y. G. & Tu, C. W. (2002). Optical properties of $GaAs/GaN_xAs_{1-x}$ quantum well structures grown by migration-enhanced epitaxy, *Journal of Crystal Growth* 242: 29–34.

Hong, Y. G., Tu, C. W. & Ahrenkiel, R. K. (2001). Improving properties of GaInNAs with a short-period GaInAs/GaNAs superlattice, *Journal of Crystal Growth* 227: 536–540.

hsiu Ho, I. & Stringfellow, G. B. (1997). Solubility of nitrogen in binary III-V systems, *Journal of Crystal Growth* 178: 1–7.

Jalili, Y. S., Stavrinou, P. N., Jones, T. & Parry, G. (2004). GaAs-based $III-N_y-V_{1-y}$ active regions based on short-period super-lattice structures, to be submitted to, *Physical Review B* .

Jang, J. G., Miller, D. L., Fu, J. & Zhang, K. (1992). Study of $(InAs)_m(GaAs)_n$ short-period superlattice layers grown on gaas substrates by molecular-beam epitaxy, *Journal of Vaccum Science Technology B* 10: 772–774.

Kageyama, T., Miyamoto, T., Makino, S., Koyama, F. & Iga, K. (1999). Thermal annealing of GaInNAs/GaAs quantum wells grown by chemical beam epitaxy and its effect on photoluminescence, *Japanese Journal of Applied Physics* 38: L298–L300.

Kent, P. R. C. & Zunger, A. (2001a). Evolution of III-V Nitride alloy electronic structure: The localized to delocalized transition, *Physical Review Letters* 86: 2613–2616.

Kent, P. R. C. & Zunger, A. (2001b). Theory of electronic structure evolution in GaAsN and GaPN alloys, *Physical Review B* 64: 115208–115231.

Kim, K., Lambrecht, W. R. L. & Segall, B. (1996). Elastic constants and related properties of tetrahedrally bonded BN, AlN, GaN, and InN, *Physical Review B* 53: 16310–16326.

Kitatani, T., Kondow, M., Nakahara, K., Larson, M. C., Yazawa, Y., Okai, M. & Uomi, K. (1999). Nitrogen incorporation rate, optimal growth temperature, and AsH_3-flow rate in GaInNAs growth by gas-source MBE using N-radicals as an N-source, *Journal of Crystal Growth* 201-202: 351–354.

Kitatani, T., Nakahara, K., Kondow, M., Uomi, K. & Tanaka, T. (2000). Mechanism analysis of improved GaInNAs optical properties through thermal annealing, *Journal of Crystal Growth* 209: 345–349.

Klar, P. J., Gruning, H., Chen, L., Hartmann, T., Golde, D., Gungerich, M., Heimbrodt, W., Koch, J., Volz, K., Kunert, B., Torunski, T., Stolz, W., Polimeni, A., Capizzi, M., Dumitras, G., Geelhaar, L. & Riechert, H. (2003). Unusual properties of metastable (Ga, In)(N,As) containing semiconductor structures, *in* IEEE (ed.), *Optoelectronics, IEE Proceedings-*, Vol. 150, IEEE, pp. 28 –35.

Klar, P. J., Gruning, H., Heimbrodt, W., Weiser, G., Koch, J., Volz, K., Stolz, W., Koch, S. W., Tomic, S., Choulis, S. A., Hosea, T. J. C., O'Reilly, E. P., Hofmann, M., Hader, J. & Moloney, J. V. (2002). Interband transitions of quantum wells and device structures containing Ga(N, As) and (Ga, In)(N, As), *Semiconductor Science and Technology* 17: 830–842.

Klar, P. J., Gruning, H., Koch, J., Schafer, S., Volz, K., Stolz, W., Heimbrodt, W., Saadi, A. M. K., Lindsay, A. & OŠReilly, E. P. (2001). (Ga, In)(N, As)-fine structure of the band gap due to nearest-neighbor configurations of the isovalent nitrogen, *Physical Review B* 64: 121203(R)–121207(R).

Kondow, M., Uomi, K., Kitatani, T., Watahiki, S. & Yazawa, Y. (1996). Extremely large N-content (up to 10%) in ganas grown by gas-source MBE, *Journal of Crystal Growth* 164: 175–179.

Kong, J. A. (1990). *Electromagnetic wave theory*, 2nd edn, Wiley, New York.

Kurtz, S., Webb, J., Gedvilas, L., Friedman, D., Geisz, J., Olson, J., King, R., Joslin, D. & Karam, N. (2001). Structural changes during annealing of GaInAsN, *Applied Physics Letters* 78: 748–750.

Li, L. H., Pan, Z., Zhang, W., Lin, Y. W., Zhou, Z. Q. & Wu, R. H. (2000). Effects of rapid thermal annealing on the optical properties of GaN_xAs_{1-x}/GaAs single quantum well structure grown by molecular beam epitaxy, *Journal of Applied Physics* 87: 245–248.

Li, W., Pessa, M., Ahlgren, T. & Decker, J. (2001). Origin of improved luminescence efficiency after annealing of Ga(In)NAs materials grown by molecular-beam epitaxy, *Applied Physics Letters* 79: 1094–1096.

Li, W., Pessa, M. & Likonen, J. (2001). Lattice parameter in GaNAs epilayers on GaAs: Deviation from Vegard's law, *Applied Physics Letters* 78: 2864–2866.

Linsay, A. (2002). PhD Thesis, University of Surrey.

Loke, W. K., Yoon, S. F., Wang, S. Z., Ng, T. K. & Fan, W. J. (2002). Rapid thermal annealing of GaN_xAs_{1x} grown by radio-frequency plasma assisted molecular beam epitaxy and its effect on photoluminescence, *Journal of Applied Physics* 91: 4900–4903.

Mair, R. A., Lin, J. Y., Jianga, H. X., Jones, E. D., Allerman, A. A. & Kurtz, S. R. (2000). Time-resolved photoluminescence studies of In(x)Ga(1-x)As(1-y)N(y), *Applied Physics Letters* 76: 188–190.

Masia, F., Polimeni, A., Baldassarri, G., von Högersthal, H., Bissiri, M., Capizzi, M., Klar, P. J. & Stolz, W. (2003). Early manifestation of localization effects in diluted Ga(AsN), *Applied Physics Letters* 82: 4474–4476.

Mazzucato, S., Potter, R. J., Erol, A., Balkan, N., Chalker, P. R., Joyce, T. B., Bullough, T. J., Marie, X., Carrère, H., Bedel, E., Lacoste, G., Arnoult, A. & Fontaine, C. (2003). S-shaped behaviour of the temperature-dependent band gap in dilute nitrides, *Physica E* 17: 242–244.

Miyamoto, T., Kageyama, T., Makino, S., Schlenker, D., Koyama, F. & Iga, K. (2000). CBE and MOCVD growth of GaInNAs, *Journal of Crystal Growth* 209: 339–344.

Moison, J. M., Guille, C., Houzay, F., Barthe, F. & Rompay, M. V. (1989). Surface segregation of third-column atoms in group iii-v arsenide compounds: ternary alloys and heterostructures, *Physical Review B* 40: 6149–6162.

Moreira, M. V. B., Py, M. A. & Ilegems, M. (1993). Molecular-beam epitaxial growth and characterization of modulation-doped field-effect transistor heterostructures using InAs/GaAs superlattice channels, *Journal of Vaccum Science Technology B* 11: 601–609.

Mussler, G., Däweritz, L., K H Ploog, J. W. T. & Talalaev, V. (2003). Optimized annealing conditions identified by analysis of radiative recombination in dilute Ga(As,N), *Applied Physics Letters* 83: 1343–1345.

Ng, T. K., Yoon, S. F., Wang, S. Z., Loke, W. K. & Fan, W. J. (2002). Photoluminescence characteristics of GaInNAs quantum wells annealed at high temperature, *Journal of Vaccum Science Technology B* 20: 964–968.

O'Reilly, E. P., Lindsay, A., Tomic, S. & Kamal-Saadi, M. (2002). Tight-binding and **k.p** models for the electronic structure of Ga(In)NAs and related alloys, *Semiconductor Science and Technology* 17: 870–879.

Ougazazaden, A., Bellego, Y. L., Rao, E. V. K., Juhel, M. & Leprince, L. L. (1997). Metal organic vapor phase epitaxy growth of GaAsN on GaAs using dimethylhydrazine and tertiarybutylarsine, *Applied Physics Letters* 70: 2861–2863.

Pan, Z., Li, L. H., Zhang, W., Lin, Y. W. & Wu, R. H. (2000). Effect of rapid thermal annealing on GaInNAs/GaAs quantum wells grown by plasma-assisted molecular-beam epitaxy, *Applied Physics Letters* 77: 1280–1282.

Pearton, S. J. (2000). *Optoelectronic properties of semiconductors and superlattices: GaN and related materials II*, Gordon and Breach Science Publishers.

Peng, C. S., Pavelescu, E. M., Jouhti, T., Konttinen, J. & Pessa, M. (2003). Diffusion at the interfaces of InGaNAs/GaAs quantum wells, *Solid State Electronics* 47: 431–435.

Persson, C., da Silva, A. F., Ahuja, R. & Johansson, B. (2001). Effective electronic masses in wurtzite and zinc-blende GaN and AlN, *Journal of Crystal Growth* 231: 397–406.

Polimeni, A., v H G Baldassarri, H., Bissiri, M., Capizzi, M., Fischer, M., Reinhardt, M. & Forchel, A. (2001). Effect of hydrogen on the electronic properties of $In_xGa_{1x}As_{1y}N_y$/GaAs quantum wells, *Physical Review B* 63: 201304(R)–201308(R).

Pomarico, A., Lomascolo, M., Cingolani, R., Egorov, A. Y. & Riechert, H. (2002). Effects of thermal annealing on the optical properties of InGaNAs/GaAs multiple quantum wells, *Semiconductor Science and Technology* 17: 145–149.

Raja, M., Lloyd, G. C. R., Sedghi, N., Eccleston, W., Lucrezia, R. D. & Higgins, S. J. (2002). Conduction processes in conjugated, highly regio-regular, high molecular mass, poly(3-hexylthiophene) thin-film transistors, *Journal of Applied Physics* 92: 1446–1449.

Ram-Mohan, L. R., Yoo, K. H. & Aggarwal, R. L. (1988). Transfer matrix algorithm for the calculation of band structure of semiconductor super lattices, *Physical Review B* 38: 6151–6159.

Rao, E. V. K., Ougazzaden, A., Begello, Y. L. & Juhel, M. (1998). Optical properties of low band gap $GaAs_{1-x}N_x$ layers: Influence of post-growth treatments, *Applied Physics Letters* 72: 1409–1411.

Sai-Halasz, G. A., Tsu, R. & Esaki, L. (1977). A new semiconductor superlattice, *Applied Physics Letters* 30: 651–653.

Saito, H., Makimoto, T. & Kobayashi, N. (1998). MOVPE growth of strained InGaAsN/GaAs quantum wells, *Journal of Crystal Growth* 195: 416–420.

Seong, M. J., Hanna, M. C. & Mascarenhas, A. (2001). Composition dependence of Raman intensity of the nitrogen localized vibrational mode in GaAs(1-x)N(x), *Applied Physics Letters* 79: 3974–3976.

Shan, W., Walukiewicz, W., III, J. W. A., Haller, E. E., Geisz, J. F., Friedman, D. J., Olson, J. M. & Kurtz, S. R. (1999). Band anticrossing in GaInNAs alloys, *Physical Review Letters* 82: 1221–1224.

Shan, W., Walukiewicz, W., Yu, K. M., III, J. W. A., Haller, E. E., Geisz, J. F., Friedman, D. J., Olson, J. M., Kurtz, S. R., Xin, H. P. & Tu, C. (2001). Band anticrossing in III-N-V alloys, *Physica-Status-Solidi B* 223: 75–85.

Shirakata, S., Kondow, M. & Kitatani, T. (2002). Temperature-dependent photoluminescence of high-quality GaInNAs single quantum wells, *Applied Physics Letters* 80: 2087–2089.

Sik, J., Schubert, M., Leibiger, G., Gottschalch, V. & Wagner, G. (2001). Band-gap energies, free carrier effects, and phonon modes in strained GaNAs/GaAs and GaNAs/InAs/GaAs superlattice heterostructures measured by spectroscopic ellipsometry, *Journal of Applied Physics* 89: 294–305.

Skierbiszewski, C., Perlin, P., Wisniewski, P., Knap, W., Suski, T., Walukiewicz, W., Shan, W., Yu, K. M., Ager, J. W., Haller, E. E., Geisz, J. F. & Olson, J. M. (2000). Large, nitrogen-induced increase of the electron effective mass in $In_yGa_{1-y}N_xAs_{1-x}$, *Applied Physics Letters* 76: 2409–2411.

Spruytte, S. G., Coldren, C. W., Harris, J. S., Wampler, W., Krispin, P., Ploog, K. & Larson, M. C. (2001a). Incorporation of nitrogen in nitride-arsenides: Origin of improved luminescence efficiency after anneal, *Journal of Applied Physics* 89: 4401–4406.

Spruytte, S. G., Coldren, C. W., Harris, S. J., Wampler, W., Krispin, P., Ploog, K. & Larson, M. C. (2001b). Incorporation of nitrogen in nitride-arsenides: Origin of improved luminescence efficiency after anneal, *Journal of Applied Physics* 89: 4401–4406.

Toivonen, J., Hakkarainen, T., Sopanen, M. & Lipsanen, H. (2003a). Effect of post-growth laser treatment on optical properties of Ga(In)NAs quantum wells, *in* IEEE (ed.), *Optoelectronics, IEE Proceedings-*, Vol. 150, IEE, pp. 68 –71.

Toivonen, J., Hakkarainen, T., Sopanen, M. & Lipsanen, H. (2003b). Observation of defect complexes containing Ga vacancies in GaAsN, *Applied Physics Letters* 82: 40–42.

Tournie, E., Pinault, M.-A. & Guzman, A. (2002). Mechanisms affecting the photoluminescence spectra of GaInNAs after post-growth annealing, *Applied Physics Letters* 80: 4148–4150.

Toyoshima, H., Anan, T., Nishi, K., Ichihashi, T. & Okamoto, A. (1990). Growth by molecular-beam epitaxy and characterization of $(InAs)_m(GaAs)_n$ short period superlattices on InP substrates, *Journal of Applied Physics* 68: 1282–1286.

Toyoshima, H., Onda, K., Mizuki, E., Samoto, N., Kuzuhara, M., Itoh, T., Okamoto, A., Anan, T. & Ichihashi, T. (1991). Molecular-beam epitaxial growth of InAs/GaAs superlattices on GaAs substrates and its application to a superlattice channel modulation-doped field-effect transistor, *Journal of Applied Physics* 69: 3941–3949.

v H G Baldassarri, H., Bissiri, M., Polimeni, A., Capizzi, M., Fischer, M., Reinhardt, M. & Forchel, A. (2001). Hydrogen-induced band gap tuning of (InGa)(AsN)/GaAs single quantum wells, *Applied Physics Letters* 78: 3472–3474.

Wagner, J., Geppert, T., Köhler, K., Ganser, P. & Maier, M. (2003). Bonding of nitrogen in dilute GaInAsN and AlGaAsN studied by raman spectroscopy, *Solid State Electronics* 47: 461–465.

Wang, S. Z., Yoon, S. F., Loke, W. K., Liu, C. Y. & Yuan, S. (2003). Origin of photoluminescence of GaAsN/GaN(001) layers grown by plasma-assisted solid source molecular beam epitaxy, *Journal of Crystal Growth* 255: 258–265.

Wang, S. Z., Yoon, S. F., Ng, T. K., Loke, W. K. & Fan, W. J. (2002). Molecular beam epitaxial growth of $GaAs_{1-X}N_X$ with dispersive nitrogen source, *Journal of Crystal Growth* 242: 87–94.

Weyers, M., Sato, M. & Ando, H. (1992). Red shift of photoluminescence and absorption in dilute GaAsN alloy lasers, *Japanese Journal of Applied Physics* 31: L853–L857.

Xin, H. P., Kavanagh, K. L. & Tu, C. W. (2000). Gas-source molecular beam epitaxial growth and thermal annealing of GaInNAs/GaAs quantum wells, *Journal of Crystal Growth* 208: 145–152.

Xin, H. P., Tu, C. W. & Geva, M. (1999). Annealing behaviour of p-type GaInAsN grown by GSMBE, *Applied Physics Letters* 75: 1416–1418.

Yeh, P. & Yariv, A. (1984). *Optical waves in crystals*, John Wiley and Sons, New York.

4

Optoelectronic Plethysmography for Measuring Rib Cage Distortion

Giulia Innocenti Bruni[1], Francesco Gigliotti[1] and Giorgio Scano[1,2]
[1]Fondazione Don Carlo Gnocchi, Pozzolatico Firenze
[2]Department of Internal Medicine,
Section of Clinical Immunology and Respiratory Medicine
Italy

1. Introduction

The pressure acting on the part of the Rib Cage that is apposed to the costal surface of the lung is quite different from that acting on the part apposed to the diaphragm. The non uniformity of pressure distribution led Agostoni and D'Angelo (1985) to suggest that the rib cage could be usefully regarded as consisting of two compartments mechanically coupled to each other (Agostoni & D'Angelo, 1985; Jiang et al., 1988, Ward, 1992): the pulmonary rib cage (RCp), and the abdominal rib cage (RCa). The magnitude of the coupling determines the resistance to distortion and is an important parameter in the mechanics of breathing. Unitary behaviour of the rib cage was thought to be dictated by rigidity and the restrictive nature of rib articulations and interconnection. Nonetheless, important distortion of the rib cage from its relaxation configuration has been described in asthma (Ringel et al., 1983) quadriplegia (Urmey et al., 1981) and also in health individual during a variety of breathing pattern (quiet breathing, hyperventilation, single inspiration, involuntary breathing acts, such as phrenic nerve stimulation); (Crawford et al., 1983; McCool et al., 1985; Ward et al., 1992; D'Angelo, 1981; Roussos et al., 1977). In summarizing these results Crawford et al., (1983) and more recently McCool et al., (1985) concluded that the maintenance of rib cage shape needs not be attributed to inherent stiffness but may be the consequence of apparently coordinated activity of the different respiratory muscles. Under circumstance such as lung hyperinflation or when mechanical coupling between the upper rib cage (RCp) and the lower rib cage (RCa) is very loose rib cage muscle recruitment is essential to prevent paradoxical (inward) rib cage displacement. (Ward et al., 1992). Moreover the deformation of the chest wall (CW) occurring during hyperventilation and while breathing through a resistance implies that the work of breathing in these conditions is slight larger than that calculated only the basis of the volume-pressure diagram. And indeed part of the force exerted by the respiratory muscles is expended to change the shape of the chest wall relative to that occurring at the same lung volume during relaxation (Agostoni & Mognoni 1966).

Most of what is known about the kinematics of the chest wall i,e., the thoraco-abdomen compartment comes from studies (Sackner, 1980; Gilbert et al., 1972) using RIP (Respitrace®). However, the RIP method is subject to error, the volume being inferred from cross-sectional area changes. Also, evaluation of the breathing pattern with RIP is reliable only when the rib cage and abdomen behave with a single degree of freedom such as during quiet breathing. The validity of the calibration coefficient obtained experimentally to convert one or two dimensions to volume is limited to the estimation of tidal volume under conditions matched with those during which the calibration was performed (Henke et al., 1988). Conversely, OEP has been proven able to evaluate, without any assumptions regarding degree of freedom, changes in compartmental volume of the chest wall. (Pedotti et al., 1995; Cala et al., 1996; Kenyon et al., 1997; Sanna et al., 1999; Aliverti et al., 1997; Duranti et al., 2004; Romagnoli et al., 2004a; Romagnoli et al., 2004b; Romagnoli et al., 2006; Binazzi et al., 2006; Filippelli et al., 2001; Lanini et al., 2007; Gorini et al., 1999; Filippelli et al., 2003). The precise assessment of changes in thoraco-abdominal volumes, combined with pressure measurements, allows a detailed description of the action and control of the different respiratory muscle groups. That is the reason why the accurate computation of thoraco-abdominal volume changes is needed. It is well known that methods actually in use for the computation of thoraco-abdominal volume displacement are affected by several limitations. The most used devices able to compute dynamic changes of the thoraco-abdominal wall are magnetometers and inductance plethysmography (Respitrace®). Both these systems are based on the assumption that the thoraco-abdominal wall has only two degrees of freedom but it is well known that changes in both antero-posterior diameter and changes in cross-sectional area of thoracic and abdominal compartments are not linearly related to their respective volumes. Furthermore both devices are strongly influenced by artefacts due to the subject's posture that limit their utilization in dynamic conditions (e.g. exercise).

An ideal system able to measure movements and volumes of the respiratory system should have the following characteristics as much as possible:

1. Accurate computation of volume changes without using a mouthpiece that may alter the normal breathing pattern (Gilbert et al., 1972).
2. Necessitating of a simple, stable and repeatable calibration.
3. Possibility of use in non collaborating subjects (during sleep, or in unconscious patients).
4. Permitting the analysis in different postures.
5. Permitting the analysis in dynamic conditions such walking, or cycling.
6. Allowing high frequency response in order to accurately describe rapid phenomena (i.e. electric or magnetic stimulation of phrenic nerves).
7. Allowing the analysis of movements and volume changing of the different compartments of the chest wall: the upper thorax, lower thorax, and abdomen).
8. Allowing the analysis of movements and volume changing of the two halves (left and right) of the chest wall.
9. Being non-invasive, non-joining and safe for the patient.

An OEP device able to track the three-dimensional co-ordinates of a number of reflecting markers placed non-invasively on the skin of the subject satisfies many of these characteristics. The simultaneous acquisition of kinematic signals with pleural and gastric pressures during a relaxation manoeuvre allows the representation of pressure-volume plots describing the mechanical characteristics of each compartment. The OEP system was developed in 80's by the Bioengineering Department of the University of Milano in order to overcome as many of the previous limitations as possible (Pedotti et al., 1995; Cala et al., 1996; Kenyon et al., 1997)

Here we quantify distortion in healthy and diseased rib cage using a method that requires an accurate measurement of absolute volumes of upper and lower rib cage .

2. Methods

2.1 Subjects and experimental protocol

We studied non-smoking healthy subjects experienced in physiological studies and in performing respiratory manoeuvres, and patients with a number of respiratory disorders.

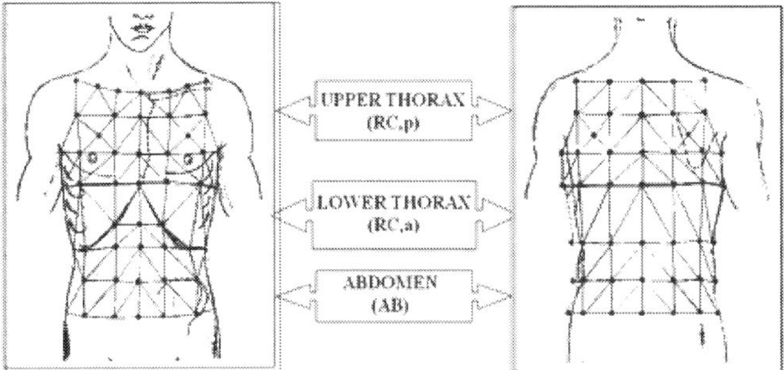

Fig. 1. Eighty-nine markers model.

Eighty-nine reflecting markers are placed in front and back over the trunk from the clavicles to the anterior superior iliac spines along predefined vertical and horizontal lines. To measure the Vcw compartments from the surface markers, we define the following: (i) the diaphragm border confirmed by percussion at end-expiration in sitting position, (ii) the boundaries of the upper rib cage (RCp) as extending from the clavicles to a line extending transversely around the thorax corresponding to the top of the area of the apposition of the diaphragm to the rib cage; (iii) the boundaries of lower rib cage (RCa) as extending from this line to the lower costal margin anteriorly, and to the level of the lowest point of the lower costal margin posteriorly, and (iv) the boundaries of the abdomen as extending caudally from the lower rib cage to a horizontal line at the level of the anterior superior iliac spine.

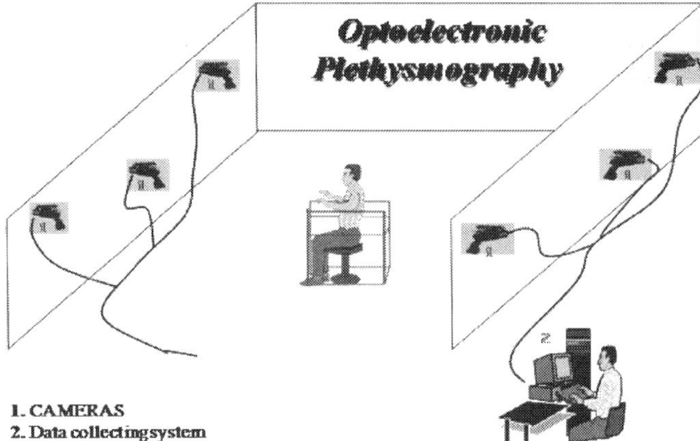

Fig. 2. The coordinates of the landmarks were measured with a system configuration of six infrared television cameras, three placed 4 m behind and three placed 4 m in front of the subject at a sampling rate of ≥60 Hz.

2.2 Compartmental volume measurements

Volumes of the different chest wall compartments were assessed by using the ELITE system, which allows computation of the 3-dimensional coordinates of 89 surface markers applied on the chest wall surface with high accuracy (Cala et al., 1996). The markers, small hemispheres (5 mm in diameter) coated with reflective paper, were placed circumferentially in seven horizontal rows between the clavicles and the anterior superior iliac spine. Along the horizontal rows, the markers were arranged anteriorly and posteriorly in five vertical columns, and there was an additional bilateral column in the mid-axillary line. In agreement with Cala et al., (1996), the anatomic landmarks for the horizontal rows were 1) the clavicular line, 2) the manubrio-sternal joint, 3) the nipples, 4) the xiphoid process, 5) the lower costal margin, 6) umbilicus, and 7) anterior superior iliac spine. The landmarks for the vertical rows were 1) the midlines, 2) both anterior and posterior axillary lines, 3) the midpoint of the interval between the midline and the anterior axillary lines, and 4) the midaxillary lines. To measure volume of chest wall (Vcw) compartments from the surface markers, we defined the following: (i) confirmed by percussion at end-expiration in sitting position, the diaphragm border in the mid clavicular line was always below the anterior end of the seventh rib, (ii) the boundaries of the upper rib cage (RC, p) as extending from the clavicles to a line extending transversely around the thorax corresponding to the top of the area of the apposition of the diaphragm to the rib cage, (iii) the boundaries of lower rib cage (RC, a) as extending from this line to the lower costal margin anteriorly, and to the level of the lowest point of the lower costal margin posteriorly, and (iv) the boundaries of the abdomen as extending caudally from the lower rib cage to a horizontal line at the level of the anterior superior iliac spine.

The coordinates of the landmarks were measured with a system configuration of six infrared television cameras, three placed 4 m behind and three placed 4 m in front of the subject at a sampling rate of 25-100 Hz. Starting from the marker coordinates, the thoraco-abdominal volumes were computed by triangulating the surface. For closure of surface triangulation, additional phantom markers were constructed as the average position of surrounding points at the center of the caudal and cephalad extremes of the trunk. Volumes were calculated from the surface triangulation between the marker points.

2.3 Pressure measurements

Pes and Pga were measured by using conventional balloon-catheter systems connected to two 100-cmH$_2$O differential pressure transducers (Validyne, Northridge, CA). Pes was used as an index of pleural pressure and Pga as that of abdominal pressure (Pab). From the pressure signals, we measured the following: Pes and Pga at end inspiration (PesEI and PgaEI, respectively) and end expiration (PesEE and PgaEE, respectively) at zero-flow points. The transdiaphragmatic pressure (Pdi) was obtained by subtracting Pes from Pga. Pdi at end expiration during quiet breathing was assumed to be zero. The difference between PgaEI and PesEI was defined as active Pdi and that between PgaEE and PesEE as passive Pdi. ΔPdi was defined as the difference between passive Pdi and active Pdi. Pressure and flow signals were synchronized to the kinematics signals of the OEP system and sent to an IBM-compatible personal computer through an RTI 800 analogue-to-digital card for subsequent analysis.

2.4 Rib cage and abdomen relaxation measurements

Relaxation characteristics of the chest wall were studied at rest. The subjects, in a sitting position, inhaled to total lung capacity and then relaxed and exhaled through a high resistance. Relaxation manoeuvres were repeated until curves were reproducible, pressure at the mouth returned to zero at functional residual capacity (FRC), and Pdi was zero throughout the entire manoeuvre. The best relaxation curve was retained. To assess rib cage relaxation characteristics, volume of pulmonary rib cage (Vrc,p) was plotted against Pes. The best fitting linear ($y = ab + x$) regression for the Vrc,p-Pes curve was constructed to obtain a relaxation curve of RC, p. The relaxation curve of the abdomen was obtained by plotting Pga vs. Vab from end-expiratory volume of abdomen (Vab) to end-inspiratory Vab during quiet breathing; we found a curvilinear relationship to which we fitted a second-order polynomial regression (Sanna et al., 1999; Aliverti et al., 1997). (This was extrapolated linearly from higher and lower values of Vab). This method was preferred to the actual data obtained during relaxation because the latter were reliably obtained only at values of Vab greater than at FRC.

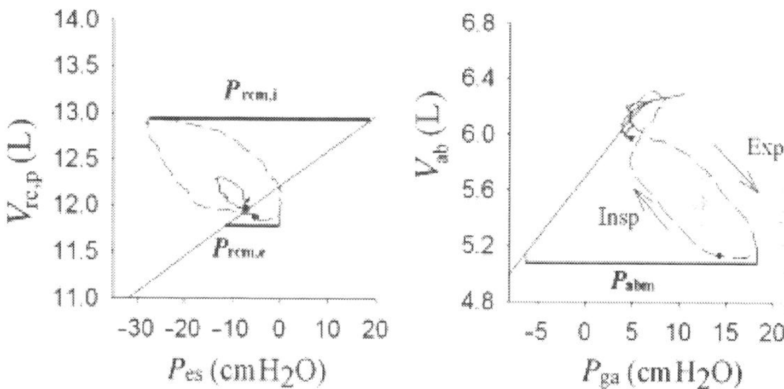

Fig. 3. Schematic representation of relationship between oesophageal pressure (Pes) and volume of pulmonary rib cage(Vrc,p) during quiet breathing (continuous loop) and at 50 L min-1 of VE (dashed loop) assuming minimal rib cage distortion during leg exercise. The thin line is the relaxation line. The closed circle is end expiratory volume. Measurement of pressure generated by rib cage muscles at Vrc,p (thin line) is obtained from horizontal distance between dynamic loop and the relaxation line at that volume during inspiration (Prcm,i) and expiration (Prcm,e). Right: schematic representation of the relationship between gastric pressure (Pga) and volume of the abdomen (Vab) at rest and at 50 L min-1 of ventilation (VE). The thin line is the relaxation line. Measurement of pressure generated by abdominal muscles (Pabm) at that Vab is obtained from the horizontal distance between gastric pressure and the relaxation line at that volume.

2.5 Cardiopulmonary exercise testing

Flow was measured with a mass flow sensor (Vmax 229; SensorMedics; 70 ml dead space) near the mouthpiece and lung volume changes were obtained by integrating the flow signal. A gas mixture (inspiratory oxygen fraction of 0.50 balanced with nitrogen) was inspired by the patients from a Douglas bag through a two-way non-rebreathing valve (mod 27900; Hans-Rudolph, Kansas City, MO, USA, 115 ml dead space). The flow into the Douglas bag was constant and patients breathed the gas mixture at the rate they demanded. We carefully reduced the impedance of the tubing by increasing its width and minimizing its length. To ascertain the linearity of the analyzer we used a 0.50 oxygen calibration cylinder. During the test flow rate at the mouth and gas exchange were recorded breath-by-breath (Vmax 229,

SensorsMedics). Expired gas was analyzed for oxygen uptake (V'_{O2}), and carbon dioxide production (V'_{CO2}). Cardiac frequency was continuously measured using a 12-lead electrocardiogram and oxygen saturation was measured using a pulse oxymeter (NPB 290; Nellcore Puritan Bennett, Pleasanton, CA, USA). The equipment was calibrated immediately before each test. V'_{CO2} and V'_{O2} were expressed as standard temperature, pressure and dry. The flow signal was synchronized to that of the motion analysis used for OEP and sent to a personal computer for subsequent analysis.

3. Analysis of the data

3.1 Operational chest wall volume measurements

To measure the Vcw compartments from the surface markers, we defined the following: (i) confirmed by percussion at end expiration in sitting position, the diaphragm border in the mid clavicular line was always below the anterior end of the seventh rib, (ii) the boundaries of the upper rib cage (RCp) as extending from the clavicles to a line extending transversely around the thorax corresponding to the top of the area of the apposition of the diaphragm to the rib cage; (iii) the boundaries of lower rib cage (RCa) as extending from this line to the lower costal margin anteriorly, and to the level of the lowest point of the lower costal margin posteriorly, and (iv) the boundaries of the abdomen as extending caudally from the lower rib cage to a horizontal line at the level of the anterior superior iliac spine. The arrangement of the chosen markers and the geometric model allow the computation of the contribution of rib cage and abdomen to tidal volume (VT). The difference between the end-inspiratory and end-expiratory volumes of each compartment was calculated as the VT. The OEP calculates absolute volumes. The absolute volume of each compartment at functional residual capacity (FRC) in control conditions was considered as the reference volume. Volumes are reported either as absolute values or as changes from the volume at FRC in control conditions. The total chest wall volume (Vcw) was modeled as the sum of volume of the upper rib cage, i.e., the rib cage apposed to the lung (Vrc,p), volume of the lower rib cage, i.e., the rib cage apposed to the abdomen (Vrc,a) and volume of the abdomen (Vab). Thus, the Vcw was calculated as Vcw = Vrc+Vab and changes (Δ) in Vcw were calculated as ΔVcw = ΔVrc+ΔVab. The time course of the volume of each region (Vrc,p, Vrc,a and Vab) along their sum (Vcw) was processed to obtain a breath-by-breath assessment of both ventilatory pattern and operational chest wall volume . From VT and respiratory frequency, VE was calculated. VT was simultaneously measured by using a mass flow sensor (sVT). The volume accuracy of the OEP system was tested by comparing VToep to sVT. All respiratory cycles at rest and during walking were pooled for each subject.

The time course of the volume of each region (Vrc,p, Vrc,a and Vab) and their sum (Vcw) was processed to obtain a breath-by-breath assessment of both ventilatory pattern and operational chest wall volume (Johnson et al., 1999; Gorini et al., 1999).

3.2 Rib cage distortion measurements

a) The undistorted rib cage configuration was defined by plotting Vrc,p against Vrc,a during relaxation. Rib cage distortion was evaluated by comparing Vrc,p-Vrc,a at rest and during exercise to the undistorted rib cage configuration, according to the method of Chihara et al (1996). Thus we measured the perpendicular distance of the distorted configuration away from the relaxation line and divided it by the value of Vrc,p at the insertion point. This results in a dimensionless number, which, when multiplied by 100, gives percent distortion.

b) Because most patients were unable to relax their respiratory muscles enough to yield accurate and meaningful relaxation volume-pressure curves of the thorax, the presence of rib cage distortion was established by: 1. comparing the time courses of Vrc,p *vs* Vrc,a and 2. the phase shift between Vrc,a and Vrc,p when these two volumes were plotted against each other. This was measured as the ratio of distance delimited by the intercepts of Vrc,p versus Vrc,a dynamic loop on line parallel to the X-axis at 50% of RCp tidal volume (m), divided by RCa tidal volume (s), as $\theta = \sin^{-1}$ (m s⁻¹), as previously adopted approach (Agostoni & Mognoni, 1996; Aliverti et al., 2009) (Fig. 5). In this system a phase angle of zero represents a completely synchronous movement of the compartments and 180° total asynchrony. Rib cage to abdomen displacement was assessed by the ratio of changes in Vrc to change in Vab.

The rest signals were recorded over a 3-min period after a 10-min period of adaptation to equipment. In each patients, the volume tracings were normalized with respect to time to allow ensemble averaging over three reproducible consecutive breaths chosen within the period of interest (rest, warm-up, each minute of exercise) and to derive an average respiratory cycle over each of the data acquisition periods. Inspiratory and expiratory phases of the breathing cycles were derived from the Vcw signal.

3.3 Respiratory muscle pressure measurements

The pressure developed by inspiratory and expiratory rib cage muscles (Prcm,i and Prcm,e, respectively) and that developed by the abdominal muscles (Pabm) were measured as the difference between the Pes-Vrc,p loop and the relaxation pressure-volume curve of RCp and between the Pga-Vab loops and the relaxation pressure-volume curve of the abdomen, respectively, according to the method of Aliverti et al. (1997).

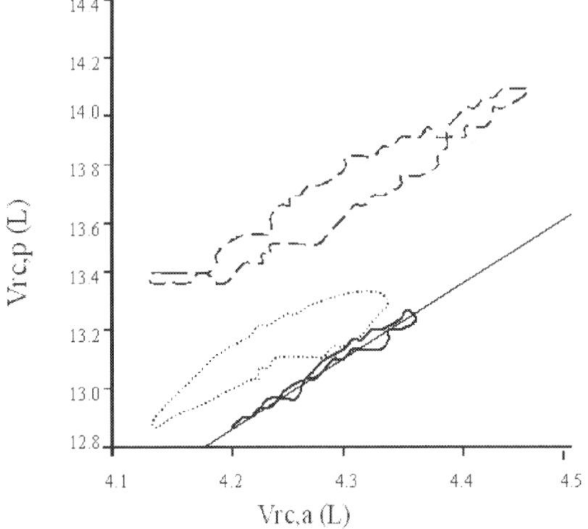

Fig. 4. The undistorted rib cage configuration is defined by plotting Vrc,p against Vrc,a during relaxation. Rib cage distortion is evaluated by comparing Vrc,p-Vrc,a at rest and during exercise to the undistorted rib cage configuration. Individual Vrc,p–Vrc,a plot at quiet breathing (QB) and at 50 L of VE In a representative subject. Continuous lines: relaxation lines. Continuous loops: respiratory cycle at QB. Dotted loops: respiratory cycle during leg exercise; dashed loops: respiratory cycle during arm exercise.

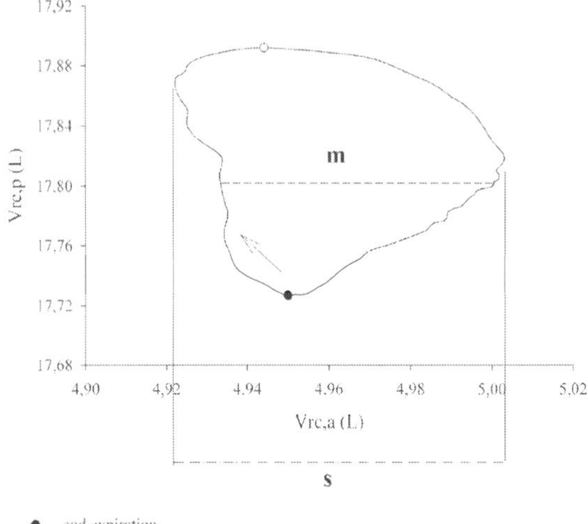

Fig. 5. In patients unable to relax their respiratory muscles enough to yield accurate and meaningful relaxation volume-pressure curves of the thorax, the presence of rib cage distortion is established by: 1. comparing the time courses of Vrcp, Vrca and 2. the phase shift between Vrc,a and Vrc,p when these two volumes are plotted against each other. This is measured as the ratio of distance delimited by the intercepts of Vrcp, Vrca dynamic loop on line parallel to the X-axis at 50% of RCp tidal volume (m), divided by RCa tidal volume (s), as $\theta = \sin^{-1}$ (m s^{-1}). In this system a phase angle of zero represents a completely synchronous movement of the compartments and 180° total asynchrony.

4. Results and discussion

4.1 OEP vs pneumotachograph volume

We compared change in Vcw during inspiration obtained by OEP (ΔVcw) with inspired volumes obtained by integration of flow (ΔVm). Also, the linear regression analysis between ΔVcw and ΔVm calculated simultaneously over a period of 20s yielded the following equation: r:0.94, Y= -0.103+1.093X. The small discrepancy we found between VToep and VTm may be explained as follows. While pneumotachograph measures the volume of the lung OEP measures the volume of the trunk. This includes volume changes in the mouth, gas compression and decompression in the lung, and blood shift between trunk and extremity.

4.2 Physiology
4.2.1 Effect of exercise

Studies concerning chest wall mechanics during exercise or walking in normal humans (Kenyon et al., 1997; Aliverti et al., 1997; Sanna et al., 1999; Duranti et al., 2004) have used OEP to investigate a new aspect of respiratory mechanics: the rib cage distortion, that is due to the different pressure acting on the volumes of the lower (abdominal) and upper rib cage i,e., the non diaphragmatic inspiratory/expiratory muscles acting on volume of the upper rib cage, and diaphragm and abdominal muscles acting on volume of the lower rib cage. The volume distortion surprisingly is <1% (Kenyon et al., 1997; Aliverti et al., 1997; Sanna et al., 1999). Thus, during exercise, the diaphragm, rib cage and abdominal muscles are coordinated so that rib cage distortion, although measurable, is minimised. In particular, the progressive relaxation of abdominal muscles observed during inspiration could prevent volume of the lower rib cage from an unbalanced expansion with respect to volume of the upper rib cage. (Aliverti et al., 1997; Sanna et al., 1999; Duranti et al., 2004)

4.2.2 Effect of coughing

The three-compartment model of the chest wall dictates that contraction of the abdominal muscles has both a deflationary action on the lower rib cage via their insertional components (the rectus and obliquus muscles), and an inflationary action via their non-insertional components (the trasversus muscle), the net effect being that upper rib cage deflation is commensurate with lower rib cage deflation (Kenyon et al., 1999). However, if forces applied to the upper rib cage are out of proportion with those applied to the lower rib cage, distortion might ensue during fits of coughing. In this way the abdominal rib cage is exposed to greater positive abdominal pressure at the end of expiration during cough (Man et al., 2003). Lanini et al., (2007) therefore hypothesized that uneven distribution of operating forces may results in rib cage distortion during coughing. The results obtained in 12 healthy subjects during voluntary single and prolonged coughing efforts at functional residual capacity and after maximal inspiration (max) showed that the three chest wall compartments contributed to reducing end expiratory volumes of the chest wall during coughing at functional residual capacity and prolonged maximum coughing, with the latter resulting in the greatest chest wall deflation. Mean rib cage distortion, did not differ between men and women, but tended to significantly increase from single to prolonged coughing maximum. Lanini et al. (2007) therefore concluded that rib cage distortion may ensue during coughing, probably as a result of uneven distribution of forces applied to the rib cage.

4.3 Pathophysiology
4.3.1 Neuromuscular diseases (NMD)

NMD are characterized by progressive loss of muscle strength resulting in cough ineffectiveness with deleterious effects on the respiratory system (Bach, 1993; Bach, 1997). Assessment of cough effectiveness is therefore a prominent component of the clinical evaluation and respiratory care in these patients. Owing to uneven distribution of muscle weakness in neuromuscular patients (De Troyer & Estenne 1995). Lanini et al., (2008) hypothesized that forces acting on the chest wall may have impact on the compartmental distribution of gas volume resulting in a decrease in cough effectiveness. The current authors have shown that unlike controls patients were unable to reduce end expiratory chest wall volume and exhibited greater rib cage distortion during cough. Peak cough flow was negatively correlated with rib cage distortion, the greater the former the smaller the latter, but not with respiratory muscle strength. Therefore, insufficient deflation of chest wall compartments and marked rib cage distortion resulted in cough ineffectiveness in these neuromuscular patients.

4.3.2 Pathology of the rib cage

Few detailed physiological studies have been carried out in young pectus excavatum PE subjects either preoperatively or postoperatively (Mead et al., 1985); it has been suggested however that the depression of the sternum limits the movement of the ribs especially in the lower ones, thus preventing the expansion of the lower thoracic cross-sectional area (Koumbourlis, 2009). On theoretical grounds uncoordinated displacement of chest wall compartments is not unexpected in these patients, considering that a non-uniform distribution of pressure over the different parts may distort the rib cage (Crawford et al., 1983; McCool et al., 1985; Chihara et al., 1996; Ward et al., 1992; Kenyon et al., 1997). By contrast, recent studies (Kenyon et al., 1997, Aliverti et al., 1997; Sanna et al., 1999; Romagnoli et al., 2006) have shown that the expiratory action of the abdominal muscles plays a key role in minimizing rib cage distortion during sustained ventilatory effort in healthy subjects. Moreover, a normal swing in abdominal pressure with a normal abdominal pressure-volume loop is associated with normal rib cage mobility during increased ventilation in PE patients (Mead et al., 1985). In keeping with these data, the preliminary results of our laboratory (Binazzi et al., 2009) indicate a normal reduction in end-expiratory abdominal volume (suggestive of phasic expiratory activity) during hyperventilation in PE patients. Collectively these data allow us to hypothesize that a coordinated motion of upper to lower rib cage prevents distortion during ventilatory tasks in PE patients. It has been suggested that the rib cage fails to move up and out during inspiration (Whol et al., 1995). Available data, however, argue against this possibility (Koumbourlis, 2009; Mead et al., 1985). Plotting of upper rib cage volume (Vrc,p) vs lower rib cage volume (Vrc,a) we were able to find a normal phase angle degree at QB and through maximal voluntary ventilation in control subjects and in a few PE patients.

4.3.3 Asthma

The mechanics of the chest wall was studied in asthmatic patients before and during histamine-induced bronchoconstriction. The volume of the chest wall (Vcw), pleural and gastric pressures were simultaneously recorded. Vcw was modeled as the sum of the volumes of the pulmonary-apposed rib cage (Vrc,p), diaphragm-apposed rib cage (Vrc,a), and abdomen (Vab). During bronchoconstriction, hyperinflation was due to the increase in end-expiratory volume of the rib cage, whereas change in Vab was inconsistent because of phasic recruitment of abdominal muscles during expiration. Changes in end-expiratory Vrc,p and Vrc,a were along the rib cage relaxation configuration, indicating that both compartments shared proportionally the hyperinflation. Vrc,p-Ppl plot during bronchoconstriction was displaced leftward of the relaxation curve, suggesting persistent activity of rib cage inspiratory muscles throughout expiration. Changes in end-expiratory Vcw during bronchoconstriction did not relate to changes in airway obstruction or time and volume components of the breathing cycle. We concluded that during bronchoconstriction in asthmatic patients: (1) rib cage accounts largely for the volume of hyperinflation, whereas abdominal muscle recruitment during expiration limits the increase in Vab; (2) hyperinflation is influenced by sustained postinspiratory activity of the inspiratory muscles; (3) this pattern of respiratory muscle recruitment seems to minimize volume distortion of the rib cage at end-expiration and to preserve diaphragm length despite hyperinflation (Gorini et al., 1999).

4.3.4 Chronic obstructive pulmonary disease (COPD)

Lung hyperinflation in patients with chronic obstructive pulmonary disease (COPD) places the respiratory muscles at a mechanical disadvantage and impairs their force generation capacity (De Troyer, 1997). Clinical evidence of the diaphragm's vulnerability in the effect of hyperinflation is abundant (Sharp, 1985). One indicator of diaphragm dysfunction is Hoover's sign (Hoover, 1920) consisting of inward movement of the lower lateral rib cage during inspiration. However, the basis of abnormal rib cage motion and the effect of hyperinflation on rib cage distortion have not been systematically examined in patients with COPD. Some factors argue against a close relationship between Hoover's sign and hyperinflation: (i) the primary factor contributing to rib cage distortion in COPD patients is an abnormal alteration of the forces applied to the pulmonary and abdominal rib cage, with hyperinflation making only a minor contribution (Jubran & Tobin, 1992); (ii) hyperinflation produces a decrease in airway resistance (Briscoe & Dubois, 1958) which may account for the greater degree of abnormal CW motion observed with resistive loading (Tobin et al., 1987) than with hyperinflation (Jubran & Tobin, 1992); (iii) hyperinflation is closely linked to expiratory flow limitation which at least theoretically, can be due entirely to loss of lung elastic recoil and tracheo or bronchomalacia. We therefore asked whether hyperinflation would produce rib cage distortion *per se*. We hypothesized that lung hyperinflation and rib cage distortion (Binazzi et al., 2008) could independently define the functional conditions of COPD patients. We based the hypothesis on the following observations: (i) a remarkable directed correlation has been found between abdominal rib cage compliance and distortability (Chihara et al., 1996), and (ii) passive tension in the abdominal muscles exerts an important deflationary action on abdominal rib during tidal inspiration (Kenyon et al., 1997). Rib cage distortion associated with Hoover's sign was indicated by the negative values of Vrc,p/Vrc,a which contrasted with the positive values in patients without Hoover's sign. Most importantly, the fact that we found a lack of any significant relationship between quantitative indices of Hoover's sign and functional residual capacity validates the starting hypothesis that rib cage volume distortion cannot be fully explained by static hyperinflation in patients with COPD. Chihara et al. (1996) have also speculated that when rib cage distortion is present the greater the distortability the greater the degree of recruitment of inspiratory rib cage muscles and the greater the predisposition to dyspnea for a given load and strength (Ward & Macklem 1992). On the other hand, the role of hyperinflation on abnormal chest movement is questionable (Binazzi et al., 2008; Hoover, 1920; Aliverti et al., 2009; Joubran & Tobin, 1992; O'Donnell et al., 1997; O'Donnell et al., 2001). By contrast, Aliverti et al., (2009) have shown that lower rib cage paradox yields to an early onset of dynamic hyperinflation as a likely explanation for the increased dyspnea during incremental exercise in these patients. Contradicting this interpretation we have shown that, neither rib cage distortion nor, despite being reduced, dynamic lung hyperinflation do not contribute to oxygen-induced decrease in dyspnea in exercising normoxic COPD patients.

The coordinated respiratory muscle action translates into proportional changes in the volume of the CW compartments when human beings cycle, run or walk (Sanna et al., 1999; Aliverti et al., 1997; Duranti et al., 2004). This complex interaction between the diaphragm, inspiratory rib cage muscles, and abdominal muscles is poorly understood during unsupported arm exercise [UAE]. Comparing UAE with leg exercise [LE] in normal subjects Celli et al. (1988) found that UAE resulted in less ventilatory contribution of inspiratory muscles of the rib cage and more contribution by the diaphragm and abdominal muscles. In a two compartment rib cage model this shift in dynamic work results in rib cage distortion (Kenyon et al., 1997). Romagnoli et al., (2006) therefore hypothesized that a decrease in pressure contribution of the rib cage inspiratory muscles, and increase in pressure production of the diaphragm and abdominal muscles with UAE might be associated with rib cage distortion as opposed to undistorted configuration during LE at comparable ventilation. The results showed that unlike LE, with UAE inspiratory pressure production of the rib cage muscles did not significantly increase from quiet breathing. However, no clear-cut differences in rib cage distortion were found between UAE and LE. What is the clinical relevance of this study? Based on the present results and those in patients with ankylosing spondylitis and rib cage rigidity (Romagnoli et al., 2004) we speculate that diverting rib cage muscles from ventilatory function to postural function limits inspiratory rib cage expansion more than some degree of rib cage rigidity does. This may have negative ventilatory consequences in severely hyperinflated patients with chronic obstructive pulmonary disease (COPD) who unlike the subjects of the present study are not able to deflate the rib cage and abdominal compartments to maintain an adequate tidal volume when using rib cage muscles for daily living activities.

4.3.5 Rib cage distortion and dyspnea

Both the joint and muscles receptors should sense the deformation of the chest wall occurring during hyperventilation and breathing through resistance. The information sent by these receptors could alter the pattern of activity of some respiratory muscles. Because of the phase shift between the change of lung volume and those of some parts of the rib cage there might be a phase shift between the different impulses from the lung receptors and the those of the rib cage. This paradoxical information contributes to a sensation of dyspnea (Agostoni & Mognoni, 1996). More recently Chihara et al., (1996) have speculated that when rib cage distortion is present the greater the distortability the greater the degree of recruitment of inspiratory rib cage muscles and the greater the predisposition to dyspnea for a given load and strength. However our recent data shown that BORG score on air did not differ between patients with and without rib cage distortion, and that changes in BORG with oxygen associated with no change in phase shift do not provide evidence that rib cage distortion plays a major role in the perceived sense of breathlessness. But that does not mean that it could not contribute as we do not have any evidence that phase shift accurately reflects the different pressures acting on lower and upper rib cage (Chihara et al., 1996; Kenyon et al., 1997), or energy wasted during inspiration on rib cage distortion. Further studies in these patients are needed to assess the relationship between changes in the applied muscle pressures, displacement of chest wall compartment, rib cage phase shift, and dyspnea during exercise, on air and oxygen.

Either dyspnea or leg effort, or both may be the principal complaints for stopping exercise in patients with COPD (O'Donnell et al., 1997; O'Donnell et al., 2001) Regardless of whether patients dynamically hyperinflated or deflated the chest wall, or distorted the rib cage, was dyspnea the primary symptom limiting exercise. These data are in keeping with those of Iandelli et al., (2002) who have found that externally imposed expiratory flow limitation does not necessarily lead to dynamic hyperinflation, nor to impaired exercise performance in subjects who do not hyperinflate the chest wall, and does not contribute to dyspnea in subjects who hyperinflate until the highest expiratory flow limitation exercise level is reached. Collectively these data are not in line with a previous report (Aliverti et al., 2009) showing that an early onset of dynamic hyperinflation of the chest wall is the most likely explanation of predominance of dyspnea in patients with rib cage distortion, and that when paradox is absent the sense of leg effort becomes a more important symptom limiting exercise. The effort-dependent nature of different exercise tests, underlying multifactorial mechanisms, and subjective nature of dyspnea scaling might account for these different results. In conclusion, the rib cage paradox, changes in chest wall dimension and dyspnea do not appear to be closely interrelated phenomena during exercise in COPD patients.

4.3.6 Study limitations

The limitations of OEP in assessing the relative changes in Vrc,p and Vrc,a might be the changes in the cephalic margin of the zone of apposition, i.e., in the area over which the rib cage is effectively exposed to abdominal pressure (Chihara et al., 1996). Inasmuch as the area of apposition is diminished in patients with COPD, the abdominal rib cage is considerably smaller than normal. However, to the extent that abdominal displacement is the principal determinant of the upper boundary of the area of apposition (Kenyon et al., 1997), the similarity of this displacement at end inspiration during rest and exercise suggests that its caudal excursion during inspiration is not greatly affected by exercise. We therefore believe that an error, if any exists, introduced by defining the boundary of the upper limit of the area of apposition on the dynamics of abdominal rib cage and pulmonary rib cage would not qualitatively affect our results.

5. References

Agostoni, E. & D'Angelo E. (1985). Statics of the chest wall. In: *The Thorax*, Editor, Roussos, Macklem, pp. 259-295, Publisher, Marcel Dekker, New York

Jiang, TX. Demedts, M. & Decramer, M. (1988). Mechanical coupling of upper and lower canine rib cage and its functional significance. *Journal of Applied Physiology*, Vol. 64, pp. 620-626

Ward, MEJW. & Macklem, PT. (1992). Analysis of human chest wall motion using a two compartment rib cage model. *Journal of Applied Physiology*, Vol.72, pp. 1338-1347

Ringel, ER. Loring, SH. Mcfadden, ER. &. Ingram, RH. (1983). Chest wall configurational changes before and during acute obstructive episodes in asthma. *American Review of Respiratory Dis*ease, Vol.128, pp. 607-610

Urmey, WF. Loring, SH. Mead, J. Brown, R. Slutsky, AS. Sarkarati, MS. & Rossier, A. (1981). Rib Cage (Rc) Mechanics In Quadriplegic Patients. *Physiologist* , Vol. 24, No. 4, pp. 97

Crawford, ABH. Dodd, D. & Engel, LA. (1983). Changes in the rib cage shape during quiet breathing, hyperventilation and single inspiration. *Respiratory Physiology*, Vol 54, pp. 197-209

McCool, FD. Loring, SH. & Mead J. (1985). Rib cage distortion during voluntary and involuntary breathing acts. *Journal of Applied Physiology*. Vol. 58, No5, pp. 1703-1712

D'Angelo, E. (1981). Cranio-caudal rib cage distortion with increasing inspiratory airflow in man. *Respiratory Physiology*. Vol. 44 pp. 215-237

Roussos, CS. Fixley, M. Geneset, J. Cosio, M. Kelly, S. Martin, RR. & Engel, LA. (1977). Voluntary factors influencing the distribution of inspired gas. *American Review of Respiratory Disease*. Vol. 116, pp. 457-467, ISSN

Agostoni, E. & Mognoni, PJ. (1966). Deformation of the chest wall during breathing efforts. *Journal of Applied Physiology*. Vol. 21, No. 6, pp. 1827-1832.

Sackner, MA. (1980). Monitoring of ventilation without a physical connection to the airway. In: *Diagnostic Techniques in Pulmonary Disease, Part I Volume 1*. pp. 508, Publisher: Marcel Dekker, New York.

Gilbert, R. Auchincloss, JH. Brodsky, J. & Boden, W. (1972). Changes in tidal volume, frequency and ventilation induced by their measurement. *Journal of Applied Physiology*, Vol. 33, pp. 252-254

Henke, KG. Sharratt, M. Pegelow, D. & Dempsey, JA. (1988). Regulation of end expiratory lung volume during exercise. *Journal of Applied Physiology*, Vol. 64, pp. 135-146

Pedotti, A. & Ferrigno, G. (1995). Opto-electronics based systems. In *Three-Dimensional Analysis of Human Movement, Human Kinetics, 1st Ed*, Editors, Allard, P. Stokes, IA. Bianchi, J. pp. 57-78, Publishers, Human Kinetics. Champaign, USA.

Cala, SJ. Kenyon, CM. Ferrigno, G. Carnevali, P. Aliverti, A. Pedotti, A. Macklem, P.T. & Rochester, DF. (1996). Chest wall and lung volume estimation by optical reflectance motion analysis. *Journal of Applied Physiology*, Vol. 81, pp.2680-2689

Kenyon, CM. Cala, SJ. Yan, S. Aliverti, A. Scano, G. Duranti, R. Pedotti, A. & Macklem, P.T. (1997). Rib cage mechanics during quiet breathing and exercise in humans. *Journal of Applied Physiology*, Vol. 83, pp. 1242–1255

Sanna, A. Bertoli, F. Misuri, G. Gigliotti, F. Iandelli, I. Mancini, M. Duranti, R. Ambrosino, N. & Scano, G. (1999). Chest wall kinematics and respiratory muscle action in walking healthy man. *Journal of Applied Physiology*, Vol. 87, pp. 938-946

Aliverti, A. Cala, SJ. Duranti, R. Ferrigno, G. Kenyon, CM. Pedotti, A. Scano, G. Sliwinsky, P. Macklem, PT. & Yan, S. (1997). Human respiratory muscle actions and control during exercise. *Journal of Applied Physiology*, Vol. 83, pp.1256-1269

Duranti, R. Sanna, A. Romagnoli, I. Nerini, M. Gigliotti, F. Ambrosino, N. & Scano, G. (2004). Walking modality affects respiratory muscle action and contribution to respiratory effort. *Pflugers Archives*, Vol.448, pp.222-230

Romagnoli, I. Gigliotti, F. Lanini, B. Bianchi, R. Soldani, N. Nerini, M. Duranti, R. & Scano, G. (2004). Chest wall kinematics and respiratory muscle coordinated action during hypercapnia in healthy males. *European Journal of Applied Physiolology*, Vol. 91, pp.525-533

Romagnoli, I. Gigliotti, F. Galarducci, A. Lanini, B. Bianchi, R. Cammelli, D. & Scano, G. (2004) Chest wall kinematics and respiratory muscle action in ankylosing spondylitis patients. *European Respiratory Journal*, Vol. 24, pp. 453-460

Romagnoli, I. Gorini, M. Gigliotti, F. Bianchi, R. Lanini, B. Grazzini, M. Stendardi, L. & Scano, G. (2006). Chest wall kinematics, respiratory muscle action and dyspnoea during arm vs. leg exercise in humans. *Acta Physiologica* (Oxf), Vol. 188, pp. 63-73

Binazzi, B. Lanini, B. Bianchi, R. Romagnoli, I. Nerini, M. Gigliotti, F. Duranti, R. Milic-Emili, J. & Scano, G. (2006). Breathing pattern and kinematics in normal subjects during speech, singing and loud whispering. *Acta Physiologica* (Oxf), Vol. 186, pp. 233-246.

Filippelli, M. Pellegrino, R. Iandelli, I. Misuri, G. Rodarte, J.R. Duranti, R. Brusasco, V. & Scano, G. (2001). Respiratory dynamics during laughter. *Journal of Applied Physiology*, Vol. 90, pp. 1441-1446

Lanini, B. Bianchi, R. Binazzi, B. Romagnoli, I. Pala, F. Gigliotti, F. & Scano, G. (2007). Chest wall kinematics during cough in healthy subjects. *Acta Physiologica* (Oxf), Vol. 190, pp. 351-358

Gorini, M. Iandelli, I. Misuri, G. Bertoli, F. Filippelli, M. Mancini, M. Duranti, R. Gigliotti, F. & Scano, G. (1999). Chest wall hyperinflation during acute bronchoconstriction in asthma. *American Journal of Respiratory and Critical Care Medicine*, Vol. 160, pp. 808-816

Filippelli, M. Duranti, R. Gigliotti, F. Bianchi, R. Grazzini, M. Stendardi, L. & Scano, G. (2003). Overall contribution of chest wall hyperinflation to breathlessness in asthma. *Chest*, Vol. 124, pp. 2164-2170

Johnson, BD. Weisman, IM. Zeballos, RJ. & Beck, KC. (1999). Emerging concepts in evaluation of ventilatory limitation during exercise: the excessive tidal flow-volume loop. *Chest,* Vol. 116, pp. 488-503

Chihara, K. Kenyon, CM. Macklem, PT. (1996). Human rib cage distortability. *Journal of Applied Physiology.* Vol. 81, No.1, pp. 437-447.

Aliverti, A. Quaranta, M. Chakrabarti, B. Albuquerque, ALP. & Calverley, PM. (2009). Paradoxical movement of the lower ribcage at rest and during exercise in COPD patients. *European Respiratory Journal,* Vol. 33, pp. 49-60

Man, WD. Kyroussis, D. Fleming, TA. Chetta, A. Harraf, F. Mustfa, N. Rafferty, GF. Polkey, MI. & Moxham, J. (2003). Cough gastric pressure and maximum expiratory mouth pressure in humans. *American Journal of Respiratory and Critical Care Medicine*, Vol. 68, pp. 714-717

Bach, JR. (1993). Mechanical insufflation-exsufflation: comparison of peak expiratory flows with manually assisted and unassisted coughing techniques. *Chest,* Vol. 104, pp. 1553-1562

Bach, JR. Ishikawa, I. & Kim, H. (1997). Prevention of pulmonary morbidity for patients with Duchenne muscular dystrophy. *Chest,* Vol.112, pp. 1024–1028

De Troyer, A. & Estenne, M. (1995). Respiratory system in neuromuscular disorders. In: *The Thorax Part C,* Editor, Roussos, C. pp. 2177–2212. Publisher, Marcel Dekker, New York, NY

Lanini, B. Masolini, M. Bianchi, R. Binazzi, B. Romagnoli, I. Gigliotti, F. & Scano G. (2008). *Respiratory Physiology and Neurobiology,* Vol. 161, No. 1, pp. 62-68

Mead, J. Sly, P. Le Souefm, P. Hibbert, M. & Phelan, P. (1985). Rib cage mobility in Pectus Excavatum. *American Review of Respiratory Disease.* Vol. 132 pp.1223-1228

Koumbourlis, C. (2009). Pectus excavatum: Pathophysiology and clinical characteristics. *Paediatric Respiratory Reviews,* Vol.10 pp. 3-6

Binazzi, B. Coli, C. Innocenti Bruni, G. Gigliotti, F. Messineo, R. Lo Piccolo, R. Romagnoli, I. Lanini, B. & Scano, G. (2009). Breathing pattern and chest wall kinematics in patients affected by pectus excavatum. *European Respiratory Journal,* Vol. 34 No. 53

Whol, ME. Stark, A. & Stokers, DC. (1995). Thoracic disorders in Childhood. *In The Thorax Part C, Disease,* Editor, Roussos, C. pp. 2035-2069. Publisher, Marcel Dekker New York, NY

De Troyer, A. (1997). Effect of hyperinflation on the diaphragm. *European Respiratory Journal,* Vol.10, pp. 708–713

Sharp, JT. (1985). The chest wall and respiratory muscles in airflow limitation. In: *The Thorax: Vital pump. Lung Biology in Healthy and Disease Series,* Editors, Macklem, PT. Roussos, C. & Lenfant, C. pp. 1155–1202, Publisher, Marcel Dekker, NewYork, NY

Hoover, CF. (1920). The diagnostic significance of inspiratory movements of the costal margins. *American Journal. Medical Sciences.* Vol. 159, pp. 633–646

Jubran, A. & Tobin, MJ. (1992). The effect of hyperinflation on rib cage-abdominal motion. *American Review of Respiratory Disease,* Vol. 146, pp. 1378–1382

Briscoe,WA. & Dubois, AB. (1958). The relationship between airway resistance, airway conductance and lung volume in subjects of different age and body size. *Journal of Clinical Investigation, Vol.* 37, pp. 1279–1285

Tobin, MJ. Perez, W. Guenther, SM. Lodato, RF. & Dantzker, DR. (1987). Does rib-cage abdominal paradox signify respiratory muscle fatigue? *Journal of Applied Physiology,* Vol. 63, pp. 851–860

Binazzi, B. Bianchi, R. Romagnoli, I. Lanini, B. Stendardi, L. Gigliotti, F. & Scano G. (2008). Chest wall kinematics and Hoover's sign. *Respiratory Physiology and Neurobiology,* Vol. 160 No.3, pp. 325-333

O'Donnell, DE. Bain, DJ. & Webb, K. (1997). Factors contributing to relief of exertional breathlessness during hyperoxia in chronic air flow limitation. *American Journal of Respiratory and Critical Care Medicine,* Vol.155 pp. 530-535

O'Donnell, DE. D'arsigny, C. & Webb, K. (2001). Effects of hyperoxia on ventilatory limitation during exercise in advanced chronic obstructive pulmonary disease. *American Journal of Respiratory and Critical Care Medicine,* Vol. 163 pp. 892-898

Celli, BR. Criner, G. & Rassulo, J. (1988).Ventilatory muscle recruitment during unsupported arm exercise in normal subjects. *Journal of Applied Physiology,* Vol. 54, pp. 1936-1941

Iandelli, I. Aliverti, A. Kayser, B. Dellacà, R. Cala, SJ. Duranti, R. Kelly, S. Scano, G. Sliwinski, P. Yan, S. Macklem, PT. & Pedotti, A. (2002). Determinants of exercise performance in normal men with externally imposed expiratory flow limitation. *Journal of Applied Physiology,* Vol. 92, pp. 1943-1952

5

Development of Cost-Effective Native Substrates for Gallium Nitride-Based Optoelectronic Devices via Ammonothermal Growth

Tadao Hashimoto and Edward Letts
SixPoint Materials, Inc.
U.S.A.

1. Introduction

Since the realization of blue light emitting diodes (LEDs) in 1991 (Nakamura et al., 1991), GaN-based optoelectronic devices have been widely developed and commercialized. Adding blue and green colors to the existing red LEDs enabled full-color displays and white light sources. Blue LEDs were also combined with yellow phosphor to achieve solid-state white light sources. Another major application of GaN-based materials is blue laser diodes (LDs). After the first lasing in 1996 (Nakamura et al., 1996), blue LDs were intensively developed to commercialize high-density digital video disks (HD-DVDs). Starting from its applications to indicators, traffic lights, cell phone back-lighting, and HD-DVDs, GaN-based optoelectronic devices are now used in general lighting and replacing conventional electric light sources using vacuum tubes.

Due to the lack of native substrates (e.g., GaN or AlN) in the 1980s and early 1990s, GaN-based devices were developed on heterogeneous crystalline substrates such as c-plane sapphire and c-plane silicon carbide. After the commercialization of free-standing GaN substrates grown by hydride vapor phase epitaxy (HVPE) on gallium arsenide (Motoki et al., 2007), blue LDs have been developed on GaN substrates because highly dislocated devices on heteroepitaxial substrates were susceptible to degradation caused by high current density (1~10 kA/cm^2) in laser devices. Nevertheless, people have not switched to GaN substrates for LED applications because of the following reasons: 1) extremely high cost of GaN wafers ($5,000~$10,000 per 2″ in 2008) does not fit the cost structure of LEDs; and 2) dislocations do not have deteriorating effects on LEDs owing to the low current density. However, to realize solid-state lighting which replaces existing light bulbs, HB-LEDs are required to handle higher current than conventional LEDs. State-of-the-art HB-LED chips carry 1 A in 1 mm^2 chip (Cree, 2009), which corresponds to 0.1 kA/cm^2. Although power saturation at high current injection called "droop" is not an easy problem to solve (Stevenson, 2009), each HB-LED chip will be required to handle even higher current densities to achieve significant reduction of lumen cost in the near future. When the current density becomes comparable to that of LDs, GaN substrate will be indispensable to ensure the reliability of the device. With an increasing demand for low-cost GaN substrates, several methods such as HVPE (Hanser et al., 2008; Fujito et al., 2009), high-pressure solution growth (Porowski, 1999; Inoue et al., 2001), flux method (Yamane et al., 1998; Kawamura et al., 2003) and ammonothermal method (Dwiliński et al., 1998; Ketchum et al., 2001; Purdy et al., 2002) have been researched to grow bulk GaN. Among these methods, the ammonothermal growth of bulk GaN has a potential to provide cost competitive GaN wafers for lighting LEDs due to its excellent scalability.

This chapter first explains the ammonothermal growth including its development history and current major research activities. Second, a physicochemical aspect of the method is described. The ammonothermal growth of GaN is categorized into the basic and acidic ammonothermal growth. Owing to their physicochemical properties, these two conditions exhibit a significant difference in terms of the phase selection, solubility, and chemical compatibility with reactor materials. The third section covers recent results of GaN growth by the basic ammonothermal method. Crystal quality characterized with x-ray diffraction (XRD), transmission electron microscopy (TEM), Nomarski differential interface contrast microscopy (Nomarski microscopy), secondary ion mass spectroscopy (SIMS), photo absorption spectroscopy, and four-probe measurement is discussed. The last section concludes with a summary and future prospect. The progress in the ammonothermal growth promises cost-effective GaN wafers for solid-state lighting LEDs.

2. Ammonothermal growth

2.1 Fundamentals of ammonothermal growth

A method of growing crystals in supercritical solvent is generally called solvothermal growth. The solvothermal growth utilizes an autoclave which can contain supercritical solvent at high-pressure and high-temperature. The most successfully commercialized solvothermal growth is hydrothermal growth of α-quartz, in which silica is dissolved in supercritical water and recrystallized on seed crystals. The solvothermal growth has an advantage of scalability over other crystal growth methods. Nowadays, a hydrothermal quartz autoclave can grow over 1,000 crystals in one batch. Ammonothermal growth is one type of solvothermal growth using ammonia as a solvent, which is mainly used to grow nitride materials. In general, growth mechanism of GaN in ammonothermal growth is explained as follows: 1) nutrient such as metallic Ga or polycrystalline GaN placed in a nutrient region dissolves in supercritical ammonia under high-pressure (100~400 MPa) and high-temperature (400~600 °C); 2) the dissolved solute is transported by the convective flow of solvent (i.e. supercritical ammonia) to a seed region; 3) the solvent becomes supersaturated in the seed region under different temperature and/or pressure condition from the nutrient region; and 4) GaN crystallizes on seed crystals. Schematic drawings of typical ammonothermal growth setups are shown in Fig. 1. An autoclave heated by external heaters houses internal components and source materials. The heater is divided into two or more regions; the lower region is usually maintained at higher temperature than the upper region. The configuration differs depending on the acidity because temperature dependence of solubility varies with the acidity of the supercritical ammonia. In the case of basic ammonothermal growth, the nutrient is located in the upper region of the autoclave and the seed crystals are located in the lower region of the autoclave as shown in Fig.1 (a). By contrast, the seed crystals are located in the upper region and the nutrient is located in the lower region for the acidic ammonothermal growth as shown in Fig.1 (b).

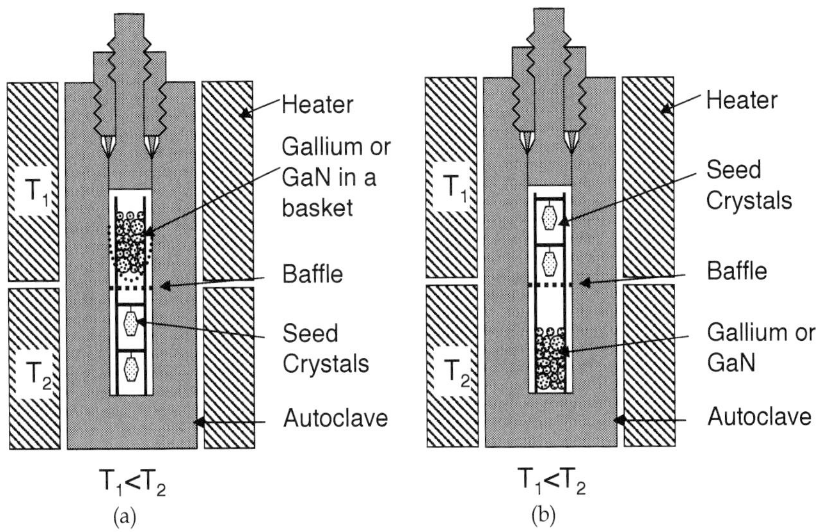

Fig. 1. Schematic drawings of ammonothermal growth setups. (a) Configuration for basic ammonothermal growth, (b) configuration for acidic ammonothermal growth. Note the difference of nutrient and seed positioning.

2.2 Development history and current activities of the ammonothermal GaN growth

The first ammonothermal growth of group III nitride crystals was synthesis of AlN reported in 1990 (Peters, 1990). Surprisingly, the result in this report has already implied the retrograde solubility of group III nitride materials in ammono-basic solutions. In 1995, synthesis of GaN in the ammono-basic environment was reported (Dwiliński et al., 1995), followed by reports in the ammono-acidic environment (Purdy, 1999) and the ammono-basic environment (Ketchum et al., 2001). After these pioneering works of microcrystalline GaN synthesis, the research started to focus on seeded growth of GaN via fluid transport (Callahan et al., 2004). Retrograde solubility of GaN in ammono-basic solution was reported in a technical journal in 2005 (Hashimoto et al., 2005) as well as in a patent application (Dwiliński et al., 2001), while research on ammono-acidic conditions revealed normal dependence of solubility on temperature (Kagamitani et al., 2006). In 2007, the main focus of the ammonothermal GaN growth has shifted from thick-film growth to bulk growth (Hashimoto et al., 2007) and an exceptionally high-quality GaN bulk crystal was demonstrated in 2008 (Dwiliński et al., 2008). Since these reports on bulk-shaped GaN were published, the ammonothermal growth has drawn more attention in the research body of GaN crystal growth.

Currently, there are several companies and institutes researching and developing the ammonothermal growth of GaN such as Ammono z o. o. in Poland, SixPoint Materials, Inc. in the U.S.A., Mitsubishi Chemical Corp. in Japan, Soraa, Inc. in the U.S.A., the University of California, Santa Barbara (UCSB) in the U.S.A., the Air Force Research Laboratory (AFRL) in the U.S.A., Tohoku University in Japan, and Asahi Chemical Corp. in Japan. There are mainly three different approaches to grow GaN in supercritical ammonia: 1) basic ammonothermal with external heaters; 2) acidic ammonothermal with external heaters; and 3) acidic high-temperature ammonothermal with internal heaters. Table 1 summarizes typical growth configurations, conditions and research groups.

	Basic	Acidic	Acidic high-temperature
Growth configuration	Seeds: lower region Nutrient: upper region	Seeds: upper region Nutrient: lower region	Seeds: upper region Nutrient: lower region
Temperature range (°C)	400~600	400~600	600~1500
Pressure range (MPa)	100~400	100~400	100~400
Mineralizers	Alkali metals and their compounds	Halide compounds	Halide compounds
Research institutes	Ammono z o. o. AFRL Mitsubishi Chemical Corp. SixPoint Materials, Inc. UCSB	Tohoku University	Asahi Chemical Corp. Soraa, Inc.

Table 1. Summary of current major activities of the ammonothermal growth.

3. Physicochemical properties of the ammonothermal growth

The ammonothermal method is divided into two categories in terms of acidity of supercritical ammonia: the basic ammonothermal method and the acidic ammonothermal method. These two methods exhibit quite different physicochemical properties. A chemical additive called a mineralizer determines the acidity of the fluid. Basic mineralizers are alkali metals and their compounds such as potassium (K), sodium (Na), lithium (Li), potassium amide (KNH_2), sodium amide ($NaNH_2$), lithium amide ($LiNH_2$), and sodium azide (NaN_3). Acidic mineralizers are halide compounds such as ammonium fluoride (NH_4F), ammonium chloride (NH_4Cl), ammonium bromide (NH_4Br), ammonium iodide (NH_4I), and gallium iodide (GaI_3). In addition to controlling acidity of fluid, mineralizers play an important role of increasing the solubility of GaN in the supercritical ammonia. Mineralizers also have an influence on phase selection of GaN; that is, formation of hexagonal wurtzite GaN or metastable cubic zinc-blende GaN. In this section, physicochemical properties of ammonothermal growth are discussed in terms of phase selection, solubility, and impact on reactor material.

3.1 Phase selection of GaN in ammonothermal synthesis

In addition to the stable wurtzite structure, GaN represents two other phases: zinc-blende structure and rocksalt structure. Rocksalt GaN is only observed under an extremely high pressure (Xia et al., 1993), whereas zinc-blende structure is a low-temperature metastable phase and more frequently observed for GaN grown at relatively low temperature (Wu et al., 1996) or grown on zinc-blende templates (Mizuta et al., 1986). Since the growth temperature of GaN in the ammonothermal method falls in the temperature range at which metastable zinc-blende GaN is frequently formed, it is of great interest to investigate the phase of GaN synthesized in the ammonothermal growth. Fig. 2 illustrates X-ray 2θ-ω scan for microcrystalline GaN powers synthesized by the ammonothermal method with various mineralizers. As shown in the figure, microcrystalline GaN powder synthesized with acidic or neutral mineralizers contained both wurtzite and zinc-blende phases. On the contrary, GaN powder synthesized with basic mineralizers contains negligible amount of zinc-blende GaN. Similar result was reported in literature (Dwiliński et al., 1995; Purdy et al., 2002). Considering the temperature range of GaN syntheses in this experiment, it is quite natural that the microcrystalline GaN contains zinc-blende phase. Therefore, we argue that some kinetics have a dominant influence on the selection of wurtzite phase in the basic environment. From an industrial point of view, basic mineralizers have a great advantage because a potential inclusion problem of metastable phase can be avoided. Ammonothermal growth with acidic mineralizers requires a delicate optimization of growth condition to eliminate phase inclusions (Ehrentraut et al., 2008a).

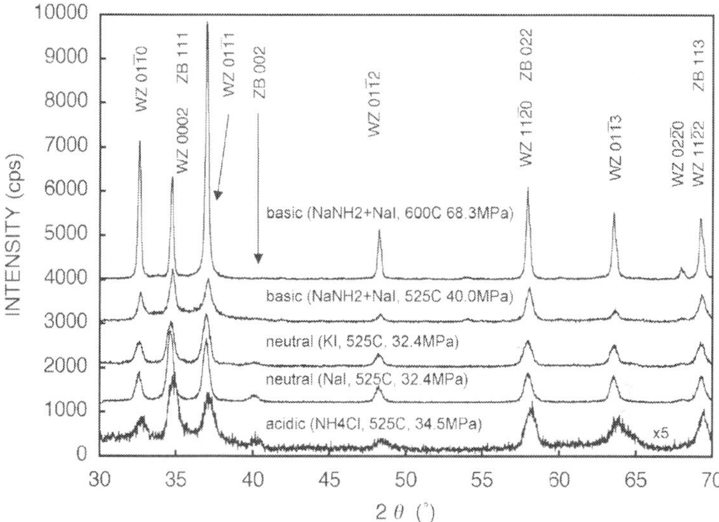

Fig. 2. X-ray 2θ-ω scan for microcrystalline GaN powers synthesized with various mineralizers. Both wurtzite (WZ) and zinc-blende (ZB) GaN were synthesized with acidic or neutral mineralizers, while wurtzite GaN was synthesized with basic mineralizers.

3.2 Solubility of GaN in supercritical ammonia

It is indispensable to understand the solubility of GaN in supercritical ammonia for successful development of ammonothermal GaN growth. Several groups have reported solubility data; however, the reported data are scattered possibly due to different conditions, immaturity of measurement technique, and measurement errors. Fig. 3 shows one example of solubility curve as a function of temperature. GaN was dissolved in supercritical ammonia at 76±12 MPa with 1.5 mol% of NaNH$_2$ as a mineralizer for 120 hours, and the weight loss of GaN was measured after a sudden release of ammonia. The solubility increases as a function of temperature until around 500 °C, and then, decreases at higher temperature. The increase in solubility can be explained with increased kinetic energy, while the retrograde solubility can be explained with one or combination of the following three reasons. The first reason is the dissociation of NH_3 at high temperature. The solubility decreased in the temperature range at which NH_3 starts to dissociate into N_2 and H_2. If NH_3 dissociates into N_2 and H_2, the total amount of NH_3 to dissolve GaN decreases, causing underestimation of solubility in the experiment. The second reason is preferred formation of gallium amide or gallium imide compounds from GaN at lower temperature. The third possible reason is the effect of entropy in dissolution process. The standard free energy of solution, ΔG is expressed as follows:

$$\Delta G = \Delta H - T\,\Delta S,$$

where ΔH and ΔS are changes in enthalpy and entropy associated with dissolution, and T is temperature. In dissolution of ionic salts, an entropy contribution to the free energy cannot be neglected since the enthalpy changes associated with dissolution process are usually comparatively small. It is reported that the entropy of solution is exceedingly negative when the anion and cation that would be formed on dissociation are both triply charged (Johnson, 1968). This might be the case in GaN dissolution; i.e., the entropy of GaN dissolution into NH_3 is negative, resulting in higher free energy at higher temperature. In the experiment shown in Fig. 3, up to 1.2 wt% of GaN dissolved in supercritical ammonia with 1.5 mol% of $NaNH_2$ at 76 MPa and 550-600 °C. Other solubility value for basic supercritical ammonia available in literature is 3 mol% ($\approx$ 15 wt%) with KNH_2:NH_3 = 0.07 at 340 MPa and 400 °C (Dwiliński et al., 2001), 13 wt% at 350 °C and 1 wt% at 560 °C for 2-4 molar KNH_2 solutions (Callahan et al., 2006). Unlike ammono-basic conditions, GaN does not show retrograde solubility for ammono-acidic conditions. It was reported that up to 0.25 mol of GaN dissolved in supercritical ammonia with 0.127 mol NH_4Cl at 67.7~100.9 MPa and 550 °C (Ehrentraut et al., 2008b). Considering the temperature limitation of the autoclave materials, the retrograde solubility for basic conditions is beneficial; seed temperature can be maintained as high as possible to grow better quality GaN crystals.

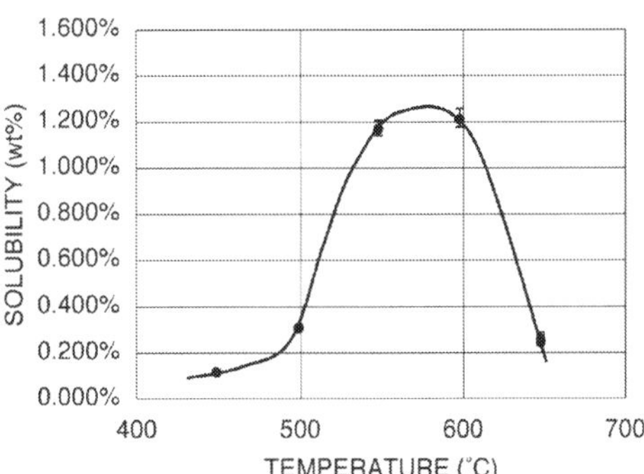

Fig. 3. Solubility of GaN in supercritical ammono-basic solutions as a function of temperature. The pressure was 76±12 MPa (11000±1800 psi) and the mineralizer was 1.5 mol% of $NaNH_2$.

3.3 The impact of acidity on the autoclave material

Ammonothermal autoclaves are usually constructed with Ni-Cr based superalloy because growth temperature higher than 500 °C is preferred to grow GaN crystals. Although Ni-Cr superalloy presents excellent corrosion resistance against water and salt water, it is seriously corroded when exposed to acidic supercritical ammonia. To avoid corrosion of the autoclave, the acidic ammonothermal method employs Pt-based liner (Kagamitani et al., 2006). In addition to extreme cost of the liner, it is quite challenging to cover all wet surfaces of high-pressure components such as valves, safety rupture discs, and pressure transducers. In the case of acidic high-temperature ammonothermal method, all growth components are housed in a capsule made of a noble metal to protect internal heater and high-pressure enclosure (D'Evelyn et al., 2004). Conversely, Ni-Cr based superalloy withstands corrosion against supercritical ammono-basic solutions. From an industrial point of view, basic ammonothermal method is advantageous over acidic conditions for simpler and inexpensive autoclave design.

4. Characteristics of GaN crystals grown by the basic ammonothermal method

Recently, the crystal quality of GaN grown by the ammonothermal method has become sufficient for successive device fabrications. In our group, GaN crystals are usually grown on c-plane GaN seed crystals prepared by a vapor phase method. Given the polar nature of wurtzite GaN, crystals grown on the Ga-polar side exhibit different characteristics from that on the N-polar side. Here, structural characteristics and surface morphology are compared between crystals on the Ga-side and the N-side, followed by impurity analysis, optical characterization, and electrical characterization for crystals grown on the N-side.

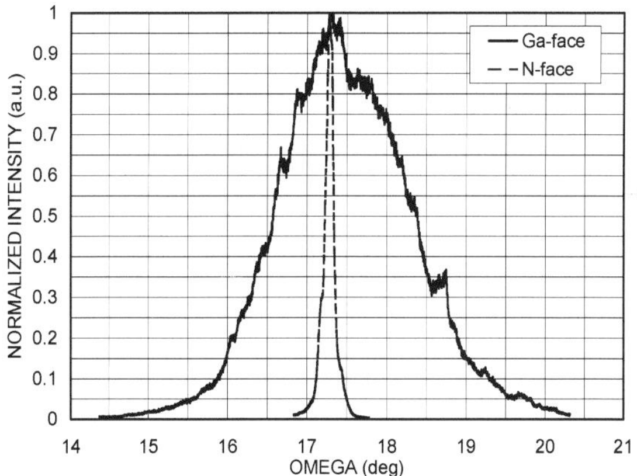

Fig. 4. XRD rocking curves of 002 reflection from crystals grown on Ga-side and N-side with an unoptimized condition.

4.1 Structural characteristics and surface morphology

In the basic ammonothermal method, crystals grown on the N-side tend to present better structural quality than that on Ga-side. Fig. 4 shows an example of XRD rocking curve measured for 002 reflection from Ga-polar and N-polar crystals. The crystal grown on the Ga-side in this sample contained numerous poorly aligned grains, although structural quality was acceptable on the N-side. It was demonstrated, however, that high-quality crystals can be grown on both sides with optimized conditions as presented in Fig. 5. The full width half maximum (FWHM) of the 002 XRD rocking curve became as narrow as ~60 arcsec. Surface morphology on the Ga-side has a strong correlation with FWHM of the 002 XRD rocking curve. Fig. 6 (a) and (b) show Nomarski microscopy images of the surface on the Ga-side for the unoptimized and optimized growth conditions, respectively. The Ga-polar surface exhibits smooth humps when 002 XRD FWHM became 200 arcsec or less. By contrast, the N-polar surface is covered with large hillocks [Fig. 7 (a)] with microscopic dots [Fig. 7 (b)] regardless of 002 XRD FWHM. Threading dislocation density of crystals grown on N-polar surface is typically under the detection limit of TEM, which corresponds to the order of 10^6 cm^{-2}. Generation of new dislocations at the interface between the seed crystal and the grown crystal was not observed.

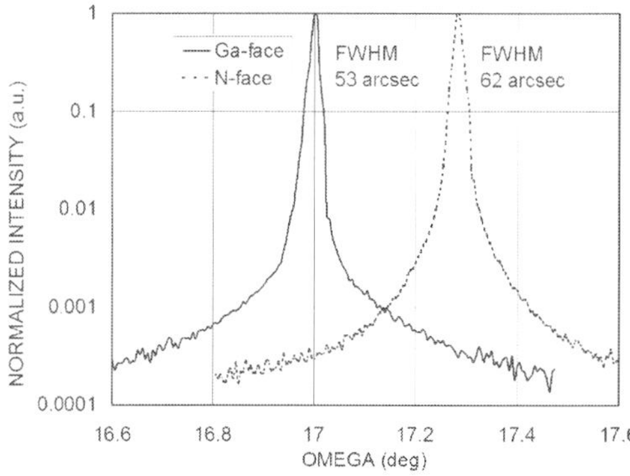

Fig. 5. XRD rocking curve of 002 reflection from crystals grown on Ga-side and N-side with an optimized condition.

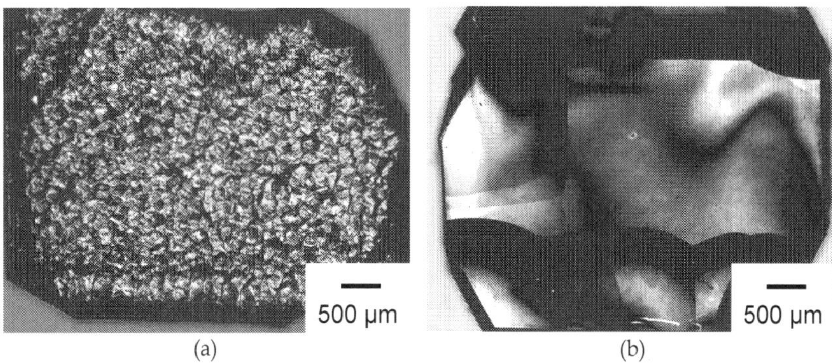

Fig. 6. Morphology of Ga-polar surface observed with Nomarski microscopy for crystals grown with (a) an unoptimized condition and (b) an optimized condition.

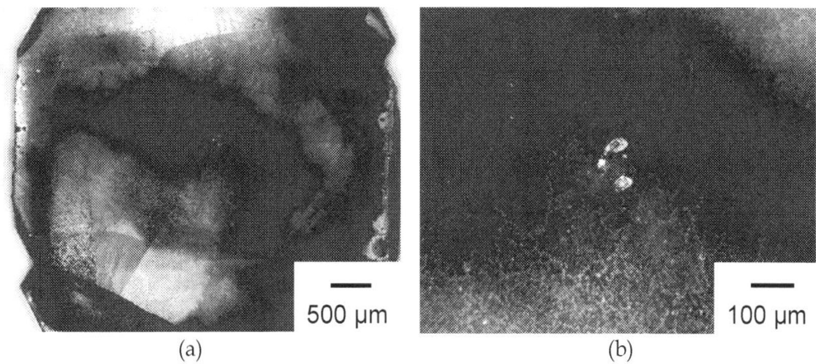

Fig. 7. Morphology of N-polar surface observed with Nomarski microscopy for crystals grown with an optimized condition, (a) lower and (b) higher magnification.

4.2 Impurity analysis, characterization of optical and electrical property

GaN grown by the basic ammonothermal method contains relatively higher impurities than crystals grown by other methods. The main reasons are as follows: 1) lower growth temperature than other methods; 2) oxygen contamination caused by oxidation of mineralizers; and 3) incorporation of heavy metal elements from the reactor wall. Table 2 presents an example of impurity concentration in a GaN crystal on the Ga-face and N-face. The crystal grown on the Ga-side contains more Na and heavy metals than that on the N-side, whereas the Ga-side incorporates less O than the N-side. It is surmised that internal polarization field attracts cation on the Ga-side and anion on the N-side. We also demonstrated that annealing of the crystals at 1200 °C induced accumulation of alkali metal impurities on the Ga-polar surface. This phenomenon is tentatively explained by thermally enhanced drifting of alkali metals under internal electric field. Some amount of alkali metal impurities can be removed through annealing followed by lapping of the top Ga-polar surface. Note that annealing at 1100 °C or lower did not cause the migration of alkali metals, which implies that the alkali metals do not migrate into device layers during the epitaxial growth process on the ammonothermal GaN wafers.

	Na	O	Heavy metals (Ni, Cr, etc.)
Ga-face	2×10^{18} cm^{-3}	7×10^{18} cm^{-3}	$10^{15} \sim 10^{17}$ cm^{-3}
N-face	2×10^{16} cm^{-3}	2×10^{19} cm^{-3}	$10^{15} \sim 10^{16}$ cm^{-3}

Table 2. Example of impurity concentrations in a GaN crystal grown by the basic ammonothermal method.

Although the ideal GaN crystal is transparent, GaN grown by the ammonothermal method looks yellowish, brownish, or blackish, depending on growth conditions. Fig. 7 (a) shows one example of bulk GaN crystal. The origin of color is attributed to impurities and/or vacancies. Brownish crystals usually yield yellowish wafers after slicing [Fig. 7 (b)]. In order to use GaN wafer for white LEDs, optical absorption at around 450 nm must be minimized to maximize optical output. To date, the smallest optical absorption coefficient at 450 nm obtained from sliced and polished GaN wafer is 8 cm^{-1}.

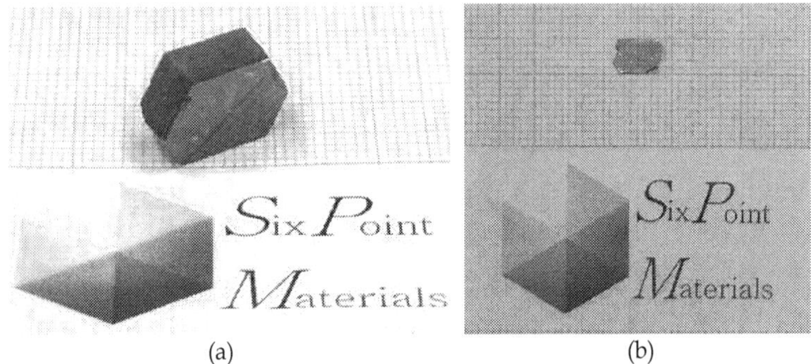

(a)   (b)

Fig. 7. Photographs of (a) GaN bulk crystal, and (b) GaN wafer.

The crystal is electrically conductive with n-type conduction. Resistivity measured by the four-probe method was 10^{-2} Ω cm or less, which is low enough for LED and LD applications. Fig. 8 demonstrates LED devices fabricated on a GaN wafer sliced from an ammonothermally grown bulk GaN crystal. GaN wafers produced by the basic ammonothermal method have sufficient quality for successive device fabrication.

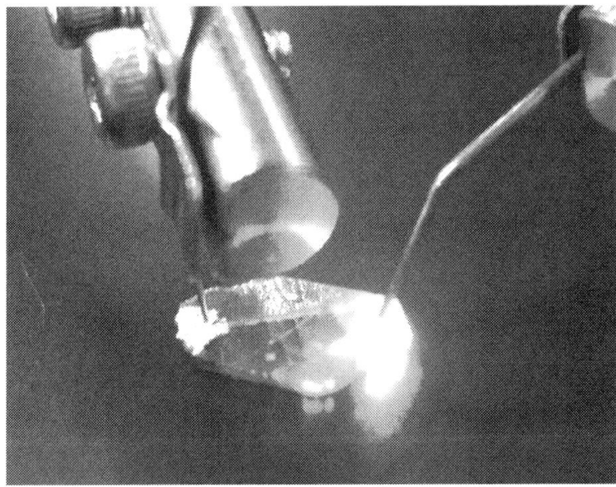

Fig. 8. Photograph of LEDs on a GaN wafer produced by the ammonothermal method.

5. Conclusion

Motivated by the successful mass production of quartz crystals by the hydrothermal growth, we have been developing the ammonothermal growth of GaN. It has been demonstrated that the ammonothermal method can produce device-quality GaN wafers. Fundamental research for over a decade has revealed general physicochemical properties of the ammonothermal GaN growth. Structural quality of GaN crystals evaluated with XRD FWHM and TEM has been improved to the acceptable level for successive device fabrications. However, there remain several challenges to bring this technology into industrial mass production of GaN wafers. First, more detailed knowledge on chemistry and crystal growth mechanism is indispensable. In particular, solubility reported in literature seems highly scattered, although it is one of the most important parameters to optimize crystal growth conditions. A more comprehensive study on solubility for various conditions is strongly demanded. Since the reaction occurs in a "black box," it is quite challenging to obtain an idea of what is happening in the reactor. Accurate control of the crystal growth process will require fundamental knowledge on the relationship among pressure, volume, and temperature for supercritical ammonia with mineralizers. Growth mechanisms and kinetics are also poorly understood. It is preferable to develop an *in-situ* technique to analyze or observe the growth process. Second, the grown crystals still contain a large amount of impurities and point defects, which is the cause of colored crystals. Optical absorption coefficient as low as 1~2 cm^{-1} is requested for LED applications. Purification of the process as well as searching for an effective dopant are necessary to achieve colorless GaN wafers. Third, seed crystals with large surface area are needed to achieve large diameter wafers because lateral expansion is quite slow in the ammonothermal growth. It would be necessary to combine other growth methods such as HVPE and flux methods to obtain high-quality, large-area seed crystals. Lastly, more time and efforts are required to develop large-capacity autoclaves. Development of large diameter superalloy billets is one of the keys for the success of the ammonothermal mass production. Also, complete safety assessment against potential blast hazards and toxic gas hazards becomes a critical issue for large-capacity autoclaves. Despite the above-mentioned challenges, the outstanding scalability of the ammonothermal method makes it one of the most promising pathways to achieve low-cost, high-quality GaN wafers. The entire GaN industry strongly expects successful development of the ammonothermal method for mass production of GaN wafers.

6. References

Callahan, M.; Wang, B.; Bouthillette, L.O.; Wang, S.Q.; Kolis, J.W.& Bliss, D.S. (2004). *Material Research Society Symposium Proceeding*, Vol. 798, (2004), pp. Y2.10.1.

Callahan, M.; Wang, B.G.; Rakes, K.; Bliss, D.; Bouthillette, L.; Suscavage, M. & Wang, S.Q. (2006). *Journal of Materials Science*, Vol.41, (2006), pp.1399.

Cree product brochure (2009). www.cree.com/products/pdf/CPR3EC.pdf.

D'Evelyn, M.P.; Narang, K.J.; Park, D.S.; Hong, H.C.; Barber, M.; Tysoe, S.A.; Leman, J.; Balch, J.; Lou, V.L.; LeBoeuf, S.F.; Gao, Y.; Teetsov, J.A.; Codella, P.J.; Tavernier, P.R.; Clarke, D.R. & Molnar, R.J. (2004), *Materials Research Society Symposium Proceedings*, Vol.798, (2004), pp.Y2.3.1.

Dwiliński, R.; Wysmolek, A.; Baranowski, J. & Kamińska, M. (1995). *Acta Physics Polonica A*, Vol.88, (1995), pp.833.

Dwiliński, R.; Doradziński, R.; Garczyński, J.; Sierzputowski, L.; Palczewska, M.; Wysmolek, A. & Kamińska, M. (1998). *MRS Internet Journal of Nitride Semiconductor*, Res. 3, (1998), pp.25.

Dwiliński, R.; Doradziński, R.; Garczyński, J.; Sierzputowski, L.P. & Kanbara, Y. (2001). Poland Patent Application, P347918.

Dwiliński, R.; Doradziński, R.; Garczyński, J.; Sierzputowski, L.P.; Puchalski, A.; Kambara, Y.; Yagi, K.; Minakuchi, H. & Hayashi, H. (2008). *Journal of Crystal Growth*, Vol.310, (2008), pp.3911.

Ehrentraut, D.; Kagamitani, Y.; Fukuda, T.; Orito, F.; Kawabata, S.; Katano, K. & Terada, S. (2008a). *Journal of Crystal Growth*, Vol.310, (2008), pp.3902.

Ehrentraut, D.; Kagamitani, Y.; Yokoyama, C. & Fukuda, T. (2008b). *Journal of Crystal Growth*, Vol.310, (2008b), pp.891.

Fujito, K.; Kubo, S.; Nagaoka, H.; Mochizuki, T.; Namita, H. & Nagao, S. (2009). *Journal of Crystal Growth*, Vol.311, (2009), pp.3011.

Hanser, D.; Liu, L.; Preble, E.A.; Udwary, K.; Paskova, T. & Evans, K.R. (2008). *Journal of Crystal Growth*, Vol.310, (2008), pp.3953.

Hashimoto, T.; Fujito, K.; Haskell, B.; Fini, P.T.; Speck, J.S. & Nakamura, S. (2005). *Journal of Crystal Growth*, Vol.275, (2005), pp.e525.

Hashimoto, T.; Wu, F.; Speck, J.S. & Nakamura, S. (2007). *Japanese Journal of Applied Physics*, Vol.46, (2007), pp.L889.

Inoue, T.; Seki, Y.; Oda, O.; Kurai, S.; Yamada, Y. & Taguchi, T. (2001). *Physica Status Solidi (b)* Vol.223, (2001), pp.15.

Johnson, D.A. (1968). *Some thermodynamic aspects of inorganic chemistry –2nd ed.*, Chapter 5, (1968, 1982), Cambridge University Press.

Kagamitani, Y.; Ehrentraut, D.; Yoshikawa, A.; Hoshino, N.; Fukuda, T.; Kawabata, S. & Inaba, K. (2006). *Japanese Journal of Applied Physics*, Vol.45, (2006), pp.4018.

Kawamura, F.; Morishita, M.; Omae, K.; Yoshimura, M.; Mori, Y & Sasaki, T. (2003). *Japanese Journal of Applied Physics*, Vol.42, (2003), pp.L879.

Ketchum, D.R. & Kolis, J.W. (2001). *Journal of Crystal Growth*, Vol.222, (2001), pp.431.

Mizuta, M.; Fujieda, S.; Matsumoto, Y. & Kawamura, T. (1986). *Japanese Journal of Applied Physics*, Vol.25, (1986), pp.L945.

Motoki, K.; Okahisa, T.; Hirota, R.; Nakahata, S.; Uematsu, K. & Matsumoto, N. (2007). *Journal of Crystal Growth*, Vol.305, (2007), pp.377.

Nakamura, S.; Mukai, T. & Senoh, M. (1991). *Japanese Jounal of Applied Physics Part II*, Vol.30, (1991), pp.L1998.

Nakamura, S.; Senoh, M.; Nagahama, S.; Iwasa, N.; Yamada T.; Matsushita, T.; Kiyoku, H. & Sugimoto, Y. (1996). *Japanese Journal of Applied Physics Part II*, Vol.35, (1996), pp.L74.

Peters, D. (1990). *Journal of Crystal Growth*, Vol.104, (1990), pp.411.

Porowski, S. (1999). *MRS Internet Journal of Nitride Semiconductor*, Res. 4S1, (1999), pp.G1.3.

Purdy, A.P. (1999). *Chemistry of Material*, Vol.11, (1999), pp.1648.

Purdy, A.P.; Jouet, R.J. & George, C.F. (2002). *Crystal Growth Design*, Vol.2, (2002), pp.141.

Stevenson, R. (2009). The LED's Dark Secret, *IEEE Spectrum*, August (2009), www.specturm.ieee.org/semiconductors/optoelectronics/the-leds-dark secret.

Wu, X.H.; Kapolnek, D.; Tarsa, E.J.; Heying, B.; Keller, S.; Keller, B.P.; Mishra, U.K.; DenBaars, S.P. & Speck, J.S. (1996). *Applied Physics Letter*, Vol.68, (1996), pp.1371.

Xia, H.; Xia, Q.; & Ruoff, A.L. (1993). Physical Review B, Vol.47, (1993), pp.12925.

Yamane, H.; Shimada, M.; Sekiguchi, T. & DiSalvo, F.J. (1998). *Journal of Crystal Growth*, Vol.186, (1998), pp.8.

6

Computational Design of A New Class of Si-Based Optoelectronic Material

Meichun Huang

Dept . of Physics ,Xiamen University ,Xiamen
CCAST (World Laboratory), PO Box 8730, Beijing,
China

1. Introduction

Computational materials science, as an intersection of theoretical physics, condensed matter physics and material science, its main tasks, in general, are firstly, research and calculate the physical and chemical properties for known materials. This would help us to better understand the microscopic mechanism of some material properties and open up new fields of applications. Secondly, computational design of new materials, that would help us to explore and design new materials to meet some proposed requirements. As we known, new material design and development is the basis of high-tech material and its device applications.

The research history of computational materials science can actually be traced back to the mid 20s of last century. Immediately after the birth of quantum mechanics theory, many excellent theoretical works were applied to the study of physical properties of solid materials. In the early years of the computational materials science, it is mainly to adopt semi-empirical method. This is because the object being studied is a many-particle system which contains very large amounts of atoms and electrons. One cannot find the rigorous analytical solutions, but the empirical values of the measurable parameters are available, Therefore, in semi-empirical method, the number of experience parameters is at least as much as that of material types, hence, most studies are pointed only to the computing research category of physical properties. Until the mid 60s of last century, Hohenberg-Kohn (Hohenberg & Kohn. 1964) and Kohn-Sham (Kohn & Sham 1965) proposed the Density Functional Theory (DFT) and Local Density Approximation (LDA), to get rid of the semi-empirical method restriction, and hence causing the birth of so-called *ab initio* or the first principles method that do not depend on experimental parameters. This method has now become a core technology of the computational material science, computational chemistry and computational condensed matter physics and is widely used.

What is a First Principles for the physical world, no one seems to be specified clarify its meaning. Many physicists believe that the most precise description of atoms and smaller than the atomic system is the theory of quantum electrodynamics (QED). Therefore, QED is a theory called first principles of the micro world. In the field of condensed matter physics and materials science, within the framework of the density functional theory, if one do not consider the high-energy radiation correction, one needs only the electromagnetic theory, Schrödinger equation and/or Dirac equation and a few of parameters unrelated to specific materials, namely, electron mass m and charge e; atomic mass M and charge Z; Bohr radius a_0; Planck constant h and fine structure constant a, so as to describe the various properties of many-particle system. This is referred to the First Principles in this article. In other words, the first principles in the condensed matter physics, in contrast with the basic theory of theoretical physics, it does not involve the physical basis of a more profound question, such as where the electronic mass comes from, why the fine structure constant is approximately equal to 1/137 and whether it changes over time, etc., and consequently, the Principles only consider the 7 parameters to be fixed natural constants.

This chapter is not to simply introduce an *ab initio* calculation for a particular Si-based optoelectronic materials, but also the computational design of the Si-based new materials is presented here. It mainly consists of two steps: at first, we should point out the functional requirements on these new materials such as optoelectronic materials, and put forward the design of the new materials and its basic properties for the requirements: analyze the factors needed to achieve these material requirements, a new material composition, atomic choice and lattice type are designed. Then, we should calculate its electronic structure by using the first principles method and examine whether the results meet the pre-made requirements. Obviously, the semi-empirical methods are often useless because there are no available experimental parameters. If the result is not satisfactory, the designed model, calculation methods and accuracy should be adjusted and improved in next steps.

In this chapter, the design idea, atomic model of materials, and their electronic structure for a new class of silicon-based optoelectronic materials are presented. A summary of the research background, recent progress and problems concerning the Si-based materials is briefly given in Sec.2. Based on this analysis, we propose design of a new silicon-based material with direct band gap is one of the better solutions. However, how to design direct band-gap semiconductor new materials, there is no ready-made principle to be followed. Therefore, a design idea that easier to operate is proposed based on the widely analysis and synthesis of band-gap types for existing semiconductor materials. In fact, the design idea contains only simple physics principle. Though it can be excavated out from piles of existing experimental and theoretical calculated data, up to now, no one has been able to extract it out and give a clear elaboration. The design idea is presented in Sec.3. Our design of new Si-based light emitting materials model and preliminary results are described in Sec.4. The result shows that the design has more clear physical basis and is more compatible with silicon technology. We note that, *ab initio* calculation is good for material properties, but material design needs more advanced physical model to solve the involved physical problem, thus many unnecessary calculation workload may be avoided. In Sec.5, a simple summary is presented.

Our computational design and preliminary results still need experimental validation. There is no doubt that the designed material is an ideal structure coming out from the theoretical calculations, its crystal growing is not an easy task. It will definitely encounter many practical problems to be overcome. Experimental topics will not be discussed in this chapter.

2. Research progress and problems

In this section, the main progress and the existing problems in the field of silicon-based optoelectronic material will be briefly reviewed. It is well known, the development of silicon microelectronics technology in the 20th century is one of the most compelling high-tech world wide achievements. It has aroused changes almost to all kinds of technology and even most people's daily life. Now, when the Si microelectronics technology becomes more and more close to its quantum limit, there are great challenges on the transmission rate of information and communication technology, also developing ultra-high speed, large capacity optoelectronic integration chip. Thus, the development and research of Si-based optoelectronic materials has become the must topic of major concern in the scientific world.

Since crystal silicon is an indirect band gap semiconductor, the conduction band bottom is located at near X point in the Brillouin zone that has an O_h point group symmetry. The indirect optical transition must have other quasi-particle participation, such as the phonons, so as to satisfy the quasi momentum conservation. We know that ordinary crystal silicon could not be an efficient light emitter, since the indirect transition matrix element is much less than that of the direct transition. For more than 20 years, people have been seeking methods to overcome the shortcomings of silicon yet unsuccessful. However, in recent years, researches show it is possible to change the intrinsic shortcomings of Si-based material. The main strategies include: (a) use of Brillouin zone folding principle (Hybertsen & Schlüter 1987), selecting appropriate number of layers m and n, the super lattices $(Si)_m/(Ge)_n$ can become a quasi-direct band gap materials; (b) synthesis of silicon-based alloys. such as $FeSi_2$, etc. (Rosen ,et al. 1993), the electronic structure also has a quasi-direct band gap; (c) in silicon, with doped rare earth ions to act the role of luminescent centers (Ennen. et al 1983); (d) use of a strong ability of porous silicon (Canham 1990; Cullis & Canham. 1991; Hirschman et al.1996); (e) use of the optical properties of low-dimensional silicon quantum structures, such as silicon quantum wells, quantum wires and dots, may avoid indirect bandgap problem in Si (Buda. et al, 1992); (f) use of silicon nano-crystals (Pavesi, et al. 2000; Walson ,et al 1993) ; (g) silicon/insulator superlattice (Lu et al. 1995) and (h) use of silicon nano-pillars (Nassiopoulos,et al 1996). All these methods are possible ways to achieve improved properties of silicon-based optoelectronic materials.

Recently, an encouraging progress on the experimental studies of the silicon-based optoelectronic materials and devices has been achieved. The optical gain phenomenon in nanocrystalline silicon is discovered by Pavesi's group. (Pavesi, et al. 2000). They give a three-level diagram of nano-silicon crystal to describe the population inversion. The three levels are the valence band top, the conduction band bottom and an interface state level in the band gap, respectively. Absorbed pump light (wavelength 390 nm) enables electronic transitions from the valence band top to the conduction band bottom, and then fast (in nanosecond scale) relaxation to interface states under the conduction band bottom. The electrons in interface states have a long lifetime, therefore can realize the population inversion. As a result the transition from the interface states to the valence band top may lead stimulated emission. In short, the optical gain of silicon nanocrystals in the short-wave laser pump light has been confirmed by Pavesi's experiment.

However, that is neither the procession of minority carriers injected electroluminescence, nor the coherent light output. In fact, nano-crystalline silicon covered with SiO_2 still retains certain features of the electronic structure of bulk Si material with indirect band gap. It is not like a direct band gap material, such as GaAs, that achieves injection laser output. In addition, light-emitting from the interface states of silicon nanocrystals is a slow (order of 10 microseconds) luminous process, much slower than that of GaAs (magnitude of nanoseconds). It indicates that the competition between heat and photon emission occurs during the luminous process. Therefore, the switching time for such kind of silicon light-emitting diode (LED) is only about the orders of magnitude in MHz, whereas the high-speed optical interconnection requires the switching time in more than GHz. It is still at least 3 to 4 magnitudes slower.

Another development of the Si-based LED is the use of a c-Si/O superlattice structure by Zhang Qi etc (Zhang Q ,*et al.* 2000). They found that it has a super-stable EL visible light (peak of ~2 eV) output. The published data indicates that the device luminous intensity had remained stable, almost no decline for 7 months.This feature is obviously much better than that of porous silicon, and reveals an important practical significance for the developing of silicon-based optoelectronic-microelectronic integrated chips. They believe that if an oxygen monolayer is inserted between the nanoscale silicon layers, it may cause electrons in Si to undergo the quantum constraint. But a theoretical estimation indicates that the quantum confinement effect is very small, and even can be ignored in this case, because the thickness of the oxygen monolayer is too small (less than 0.5 nm). Therefore, the green electroluminescent mechanism in this LED still needs further study.

In addition, an important work from Homewood's group, they investigated a project called dislocation engineering which achieved effective silicon light-emitting LED at room temperature (Ng ,et al. 2001). They used a standard silicon processing technology with boron ion implantation into silicon. The boron ions in Si-LED not only can act the role of *pn* junction dopant, and also can introduce dislocation loops. In this way the formation of the dislocation array is in parallel with the *pn* junction plane. The temperature depending peak emission wavelength of the device (between 1.130-1.15μm) , has an emitting response time of ~18μs, and the device external quantum efficiency at room temperature $~2\times10^{-4}$. As it's at the initial stage of development, it is a very prospective project worth to be investigated.

After a careful analysis of the luminous process of the above silicon-based materials and devices, it is not hard to find that many of them are concerned with the surface or interface state, from there the process is too slow to emit light. It causes the light response speed to become too slow to satisfy the requirements of ultra-high speed information processing and transmission technology. To fully realize monolithic optoelectronic integrated (OEIC), it needs more further explorations, and more fundamental improvement of the performance of silicon-based optoelectronic materials.

To solve these problems, from the physical principles point of view, there are two major kinds of measures: namely, To try to make silicon indirect bandgap be changed to direct bandgap, and to make full use of quantum confinement effect to avoid the problem of indirect bandgap of silicon. Recently, a large number of studies on quantum wires and dots, quantum cascade lasers and optical properties are presented.

This article is based on the exploration of the band modification. The main goal is to design the direct band gap silicon- based materials, hoping to avoid the surface states and interface states participation in luminous process and to have compatibility with silicon microelectronic process technology.

One of the research targets is looking for the factors that bring out direct bandgap and using them to construct new semiconductor optoelectronic materials. Unfortunately, Although the "band gap" concept comes from the band theory, the modern band theory does not clearly give the answers to the question whether the type of bandgap for an unknown solid material is direct or indirect. To clarify the type of bandgap of the material we should precede a band computation. In fact, the research around semiconductor bandgap problems has been long experienced in half of a century. A summary from the chemical bond views for analysis and prediction the semiconductor band gap has been given in early 1960s by Mooser and Pearson (Mooser & Pearson .1960). In the 1970s, the relations between the semiconductor bond ionicity and its bandgap are systematically analyzed by Phillips in his monographs (Phillips. 1973). Over the past 20 years, in order to overcome the semiconductor bandgap underestimate problems in the local density approximation (LDA), various efforts have been taken. The most representative methods are the development of quasi-particle GW approximation method (Hybertsen & Louie. 1986 ; Aryasetiawan & Gunnarsson. 1998; Aulbur et al.2000) and sX-LDA method (Seidl , *et al.* 1996), their bandgap results are broadly consistent with the experimental results. Recently, about the time-dependent density functional theory (TDDFT) (Runge & Gross 1984; Petersilka et al. 1996) and its applications have been rapidly developed and become a powerful tool for researching the excited state properties of the condensed system. All of the above important progress have provided us with semiconductor bandgap sources, the main physical mechanism and estimation of bandgap size. They have a clearer physical picture and are considered to be main theoretical basis in the current bandgap engineering.

However, these efforts are mainly focused in the prediction and correction of the band gap size, they almost do not involve the question whether the bandgap is direct or indirect. From the perspective of material computational design, a very heavy and complicated calculation in a "the stir-fries type" job and choosing the results to meet the requirements are unsatisfactory. In order to minimize the tentative calculation efforts, physical ideas must be taken as a principle guidance before the band structure calculations are proceeded. In next Section, a design concept and the design for new material model will briefly be presented

3. Computational design: principles

The complexity in the many-body computation of the actual semiconductor materials rises not only from without analytical solution of the electronic structure, but also lack of a strictly theory to determine their bandgap types. Nevertheless, we believe that the important factors determining a direct band gap must be hidden in a large number of experimental data and theoretical band structure calculations. We comprehensively analyze the band structure parameters for about 60 most commonly used semiconductor, including element semiconductor, compound semiconductor and a number of new semiconductor materials. It was found that there are three major factors deciding bandgap types, namely, the core state effect, atomic electronegativity difference effect and crystal symmetry effect (Huang M.C 2001a; Huang & Zhu Z.Z. 2001b,c, Huang et al. 2002; Huang 2005). Actually, these three effects belong to the important component in effective potential that act on valence electrons. The first two effects have also been pointed out in literature on some previous band calculation, but the calculations did not concern on material design as it's goal. A more detailed description will be given in the following

3.1 Core states effect

First of all, let us consider the element semiconductors Si, Ge and α-Sn. Their three energy at the conduction band bottom relative to the valence band top (set it as a zero energy) with the increase in core state shell in atom, the variation rules are as follows:

1. When going from Si to Sn, the conduction band bottom energy X_1 at X-point , does not have obvious changes.
2. The conduction band bottom energy L_1 at L-point constantly decreases, when going from Si to Sn, the reduction rate is about 1. 5 eV.
3. It is noteworthy that the Γ-point conduction band bottom's energy $\Gamma_{2'}$ shows the trend of rapid decline with the increase of core state shell, the decline rate is about 4 eV.

The changing tendency of the three conduction band bottom energy not only indicates the Si, Ge and Sn conduction band bottom are located at (near) X, L and Γ point (α-Sn is already a zero band gap materials) and more, it indicates the importance of core states effects for the design of direct band gap materials. With the core states increases, the indirect band gap materials will be transformed to a direct band gap material. In the design of a direct band gap group IV alloys, selection of the heavier Sn atoms as the composition of materials will be inevitable. Recently, the electronic structures of SiC, GeC and SnC with a hypothetical zincblende-like structure have been calculated by Benzair and Aourag (Benzair & Aourag (2002)), the results also show that the conduction band bottom energy Γ_1 will reduced rapidly with the Si, Ge, Sn increasing core state, and eventually led to that SnC is a direct band gap semiconductor. From another perspective, the effect of the lattice constant on the band structure is with considerable sensitivity, which is a well-known result. Even if the identical material, as the lattice constant increases, the most sensitive effect is also contributed to rapid reduction of the conduction band bottom energy Γ (Corkill & Cohen (1993)). Therefore, for a composite material under normal temperature and pressure, a natural way to achieve larger lattice parameter is to choose the substituted atom with larger core states. From this point of view, the core states effect and the influence of lattice constant on the band structure have a similar physical mechanism. Figure 1(a) shows the core states effect, the size of the core states is indicated by a core-electron number $Z_c = Z - Z_v$, where Z is atomic number and Z_v the valence electron number.

3.2 Electronegativity difference effect

In the compound semiconductor, there are two kind of atoms which were bonded by so-called polar bond or partial polar bond, and this is directly related to their interatomic electronegativity difference. In pseudopotential theory, that is included in the antisymmetric part of the crystal effective potential. The variation trend of three conduction band bottom energies at Γ-, L- and X- point for two typical zinc blende semiconductors, Ga-V and III-Sb, with their interatomic electronegativity difference is shown in Figure 1 (b) and (c). Note that here the Pauling electronegativity scale (see Table 15 in Phillips. 1973) was selected, because it is particularly suitable for sp^3 compound semiconductors. It can be seen from the Figure 1(b-c), the Γ conduction band bottom energy will be rapidly reduced as the electronegativity difference decrease and then get to close to the Γ valence band top, so that GaAs, GaSb, and InSb in these two series compounds are of direct band gap semiconductors, whereas GaP and AlSb are the indirect band gap material due to a larger electronegativity difference. However, there is no theory available at present to quantitatively explain this change rule, moreover we note, using of other electronegativity scale (for example, Phillips's scale) , the variation rule is not so obvious. For all of these, the change tendency of semiconductor conduction band bottom energy under the Pauling electronegativity scale can still be taken as a reference to design the direct band gap material model.

The above two effects, core states and electronagativity difference effect, indicate that the direct and indirect bandgap properties in semiconductor within the same crystal symmetry have the characteristic change trend as follows:

- An atom with bigger core state is more advantageous to the composition of semiconducting material having a direct band gap.
- The compounds by atoms with a smaller electronegativity difference, are conducive to compound semiconductor transformation from indirect band gap to direct band gap.

These results may give us a sense that choosing the atomic species makes a design reference, but they cannot explain the existing data completely. For example the above two typical III-V series, have important exception:

1. For the series of AlN (d) AlP (ind) AlAs (ind) AlSb (ind) , only AlN is a direct gap semiconductor, but it has a largest electronegativity difference and a smallest core states, which are mutually contradictory with the first two effects. .

2. For the series of GaN (d) GaP (ind) GaAs (d) GaSb (d), the GaN is a direct band gap material, although the electronegativity difference is larger than that of GaP and the core states is smaller.

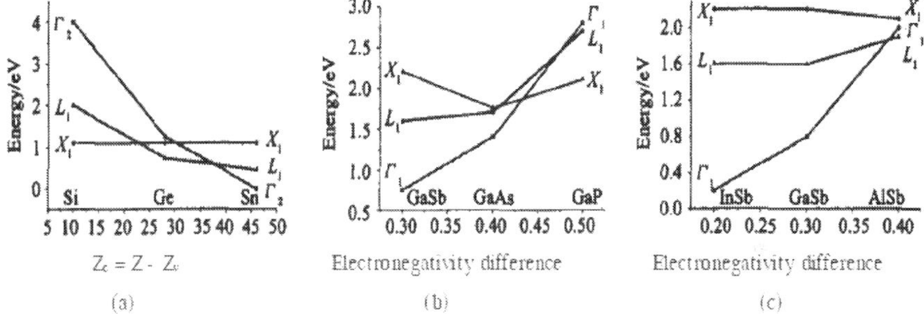

Fig. 1. The energies (Γ, X, L) at conduction band bottom vs (a) the electron number in core states for element semiconductors, and vs (b and c) the electronegativity difference between the component atoms in compound semiconductors.

This fact shows that the direct-indirect variation tendency of the band structure for these two series semiconducting material has another mechanism which needs be further ascertained.

3.3 Symmetry effect

In fact, the band gap type of AlN and GaN is different from their corresponding materials in that series, one of the important reasons is that they have different crystal symmetry. What kind of crystal symmetry can help the formation of a direct band gap of electronic structure in solids? This is the issue to be discussed in this section. In general, the electronic structure in solids depends on the electron wave function and crystal effective potential, in which the symmetry of the crystal unit cell is concealed. In order to reveal the connection between band gap type and crystal symmetry, we consider that now we can only use statistical methods to reveal the relationship, because there is no theoretical description for this issue at present. In Table 1, we list out both the point group symmetry and bandgap type for about 50 most common semiconductors. A careful observation will find out that some of variation tendency which so far has not been clearly revealed in this very ordinary table:

1. The unit cells of the main semiconductor materials have O_h, T_d, and C_{6v} point group symmetry, also they do not exclude other symmetry, such as D_{6h}, D_2 and so on. Let us make a simple statistical distribution for the crystal symmetry vs band-gap type. It can be seen that the materials have an O_h cubic symmetry and are all of indirect band gap, including II-VI group's CdS and CdS having a stable cubic structure O_h under high pressure (Benzair & Aourag 2002), although they have a C_{6v} symmetry and a direct

bandgap in normal pressure. In addition, I-VII group Ag halide, AgCl and AgBr have O_h symmetry though they are indirect band gap material. The only exception is α-Sn, but it is the zero direct band gap, which does not belong to semiconducting material in strict sense.

2. The materials which have hexagonal symmetry C_{6v} and D_2 symmetry, including the new super-hard materials BC_2N (Mattesini & Matar 2001), all have a direct band gap.

Group IV			II-V			II-VI			I-VII			Others		
SC	PG	d/i	SC	PG	d/i	SC	PG	d/i	SC	PG	d/i	SC	PG	d/i
C	O_h	i	BN	D_{6h}	i	ZnO	C_{6v}	d	CuCl	T_d	d	MnO	O_h	i
Si	O_h	i	BP	T_d	i	ZnS	C_{6v}	d	CuBr	T_d	d	NiO	O_h	i
Ge	O_h	i	BAs	T_d	i	ZnS	T_d	d	CuI	T_d	d	GaSe	D_{6h}	d
Sn	O_h	d/0	AlN	C_{6v}	d	ZnSe	T_d	d	AgCl	O_h	i	InSe	D_{6h}	d
SiC	T_d	i	AlP	T_d	i	ZnTe	T_d	d	AgBr	O_h	i	BC_2N	D_2	d
GeC	T_d	i	AlAs	T_d	i	CdS	C_{6v}	d	AgI	T_d	d			
SnC	T_d	d	AlSb	T_d	i	CdS	O_h	i	AgI	C_{6v}	d			
CSi_2Sn_2	D_2	d	GaN	C_{6v}	d	CdSe	C_{6v}	d						
CGe_3Sn	D_2	d	GaP	T_d	i	CdSe	O_h	i						
			GaAs	T_d	d	CdTe	T_d	d						
			GaSb	T_d	d	HgS	T_d	d						
			InN	T_d/C_{6v}	d	HgSe	T_d	d						
			InP	T_d	d	HgTe	T_d	d						
			InAs	T_d	d									
			InSb	T_d	d									

Table 1. Point-group symmetry and band-gap type of crystals. Where SC=semiconductor, PG=point group and d/i=direct or indirect gap.

3. The materials which have zinc-blende structure symmetry, T_d and D_{6h} symmetry, are kind of between two band gap types, direct- and indirect gap, in which HgSe and HgTe reveal only a small direct band gap. If the relativistic corrections are included, they will be the semi-metal (Deboeuij et al. 2002). Now we temporarily ignore these facts. In the materials which have T_d and D_{6h} symmetry, there are an estimated ~75% belonging to direct bandgap semiconductors.

For convenience, we use the group order g of the point group of the crystal unit cell to describe the crystal symmetry, in which the point group T_d and D_{6h} have a same group order g (=24), and call it 'same symmetry class'. Let F_d be the percentage of direct band gap materials accounted for the material number of the same symmetry class. Statistical dependence of the F_d vs the group order g is an interesting diagram scheme, as shown in Figure 2. In this case, $F_d=1$ for the direct bandgap and $F_d=0$ for the indirect bandgap. This diagram indicats very explicitly that reducing the crystal symmetry or, the points group's operand is advantageous to the design and synthesis of the direct band gap semiconducting material. In fact, the Brillouin zone folding effect can also be seen as an important effect of lowering the symmetry of the crystal. For example, lower the symmetry from T_d to C_{6v}, the face-centered cubic Brillouin zone length Γ- L is equal to twice the Γ-A line of hexagonal Brillouin zone. In this case, the conduction band bottom L of T_d will be folded to the conduction band bottom Γ of C_{6v}, leading to a direct band gap. We note that the band gap type will also be determined by the other factors, for example, the symmetry of electronic wave function at the conduction band bottom and the valence band top. Nevertheless, the main features of both the electronic structure and the band gap type are dominantly determined by crystal structure and their crystal potentials and charge density distribution that should be understandable.

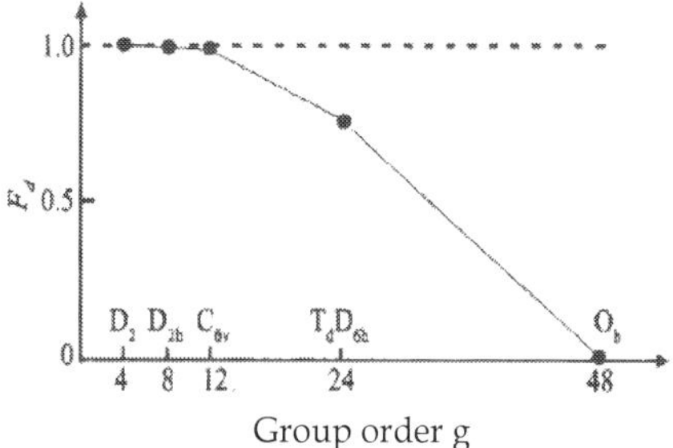

Group order g

Fig. 2. A relationship between crystal symmetry and band gap type.

Note that the main statistical object in Fig.2 is sp^3 and sp^3-like hybridization semiconductor; it also includes some of ionic crystals and individual magnetic ion oxide compounds. It does not exclude increasing other more complex semiconducting material in the Table 1. However, we believe that the general changing trend of F_d has no qualitative differences. In other words, reducing the crystal symmetry is conducive to gain direct bandgap semiconductors. In addition, the semi-magnetic semiconductors, most of the magnetic materials and the transition metal oxides have a more complex mechanism. To determine their band gap type also needs to consider the spin degree of freedom, the strongly correlation effect, more complex effects and other factors. The topic needs to be investigated in the future.

4. Computational design: model

The design requirements are: the new material must be compatible with Si microelectronics technology; it contains Si to achieve lattice matching, and the material is of direct band gap so as to avoid the light-emitting process involving surface and/or interface state, so that the devices to provide the required functions for ultra-high-speed applications.

As stated above, in order to meet these requirements, the reduced symmetry principle can provide the direction of the crystal geometry design. We carry out energy band structure computation beforehand, so that the ascertainment on the crystal structure model has a reliable basis. There are two available essential methods to reduce the crystal symmetry:

Method 1: in the Si lattice, insert some non-silicon atoms to substitute part of silicon atoms, or produce silicon compounds (alloy), so as to reduce the crystal from O_h point group symmetry to T_d point group symmetry, or to D_{4h}, D_{2h} and other crystal structures with a lower symmetry.

Method 2: in the Si lattice, by using periodic insertion of non-silicon atom layer or Si alloy layer to obtain the lower symmetry materials.

The above two methods may realize the modification for the Si bandgap type. Among them, the method 2 is more suitable for the growth process requirements on Si(001) surface. for example, in order to obtain a Si-based superlattice with symmetry lower than silicon crystal, the non-silicon atom monolayer can be grown on the silicon (001) surface, and then silicon atoms are grown, Repeatedly proceed this process by using Molecular Beam Epitaxy (MBE), Metal-Organic Chemical Vapour Deposition (MOCVD) or Ultra-high vacuum CVD (UHV-CVD), a new Si-based superlattice can be synthesized. In this way, we can not only reduce the symmetry of the silicon-like crystal, but also modify the bandgap type. This is a primarily method for the computational design.

On intercalated atoms choice, from the theoretical point of view, an inserted non-silicon atoms layer can lower the symmetry. The kinetics of crystal growth requires careful selection of insertion atoms, we consider here, the bonding nature of the Si atom with the inserting non-Si atoms. A natural selection on the insertion atoms is the IV-group atoms (C, Ge, Sn), the same group element with silicon, and the VI-group atoms (O, S, Se), due to the fact that they and Si atoms can form a stable thin film similar to SiO_2 film

We have performed a detailed study on electronic structure of two series of silicon based superlattice materials, which include $(IV_xSi_{1-x})_m/Si_n$ (001) superlattices (Zhang J L . et al. 2003; Chen et al.2007; Lv & Huang. 2010) and $VI(A)/Si_m/VI(B)/Si_m$ (001) superlattice series (Huang 2001a; Huang & Zhu . 2001b,c, Huang et al. 2002; Huang 2005).

4.1 $(Sn_xSi_{1-x})_m/Si_n$ (001) superlattices

The $(Sn_xSi_{1-x})_m/Si_n$ (001) superlattices we designed is composed of Sn_xSi_{1-x} alloy layer and Si layer, alternatively grown on Si (001) substrates. The unit cells of the $(Sn_xSi_{1-x})_m/Si_n$ (001) superlattices are shown in Figure 3 (a,b,c) for atomic layer mumber m=n=5 and x=0.125, 0.25, 0.5, respectively. Where Si_5 is a cubic unit cell which includes 5 Si atomic layers on Si(001) substrate. Similarly, the $(Sn_xSi_{1-x})_5$ is also a cubic Sn_xSi_{1-x} alloy on Si(001) surface. Although the Si and IVSi alloy are cubic crystals, the $(IV_xSi_{1-x})_5/Si_5$ (001) superlattices is a tetragonal crystal, the unit cell has a D_{2h} symmetry that is lower than cubic point group O_h. Note that the unit cell of this superlattice contains nine atomic layer along the [001] direction (c-axis) , because two cubes (IVSi)$_5$ and (Si$_5$) have common crystal faces. For simplicity, we present it in the following:

This structure will be named as $IV_xSi_{1-x}/Si(001)$. The equilibrium lattice constants after lattice relaxation of the superlattices and pure silicon have been obtained by means of total energy calculation within the DFT-LDA framework.

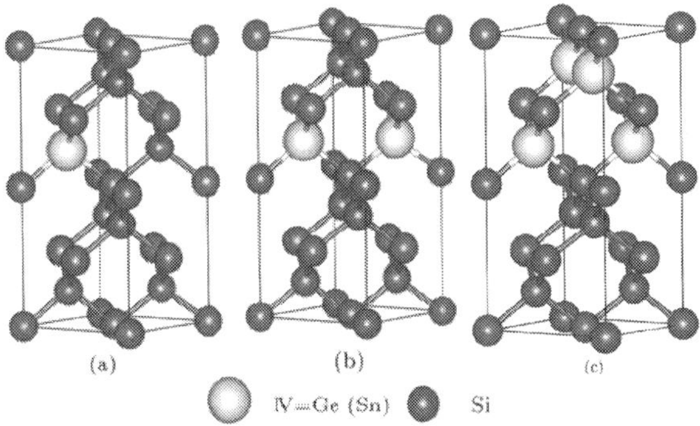

Fig. 3. The unit cell of ($IV_xSi_{1-x})_5/Si_5$ (001) superlattices. (a) x=0,125, (b) x=0.25, (c) x=0.5.

The results are shown in Table 2. From Table 2 we can find obviously that these superlattices have the reasonable lattice matching with the silicon. The lattice mismatch is less than 3% for a smaller IV component, e.g. for $x < 0.25$. The result indicates that epitaxy alloy (IVSi) on silicon (001) surface, (a IV-atom doped homogeneous epitaxy alloy), will be much easier to form than the heterogeneous epitaxy III-V compounds on silicon surface. The detailed calculation study shown that, although (IVSi) alloy is probably an indirect bandgap material, yet the IV_xSi_{1-x}/Si (001) superlattice composed of the Si and (IV_xSi_{1-x}) alloys might be a direct bandgap semiconductor with smallest bandgap located at Γ-point in Brillioun zone. Their electronic properties will be discussed in section 5.

Materials	$a=b$	c
Si	10.26	20.52
$Sn_{0.125}Si_{0.875}/Si$ (001)	10.49	20.92
$Sn_{0.25}Si_{0.75}/Si$ (001)	10,58	21.30
$Sn_{0.5}Si_{0.5}/Si$ (001)	10.79	21.90
$Ge_{0.125}Si_{0.875}/Si$ (001)	10.36	20.71
$Ge_{0.25}Si_{0.75}/Si$ (001)	10,39	20.79
$Ge_{0.5}Si_{0.5}/Si$ (001)	10.47	20.92

Table 2. The theoretical equilibrium lattice constants (in a.u.) of superlattices (IV_xSi_{1-x})$_5$/Si$_5$ (001) and a pure silicon.

4.2 VI(A)/Si $_m$/ VI(B)/Si $_m$ (001) superlattices

Another new Si-based semiconductor we designed is VI(A)/Si_m/VI(B)/Si_m (001) superlattice, here VI(A) and VI(B) are VI-group element monolayer grown on silicon (001) surface, VI(A or B) =O , S or Se. In token of Si_m, index m is the silicon atomic layer number. The superlattice structure can be grown epitaxially on silicon (001) surface, layer by layer, and then a VI-group atomic monolayer is epitaxially grown as an inserted layer. In the epitaxial growth process, the location of VI-group atoms is dependent on the silicon (001) reconstructed surface (i.e., dimerization) mode, while the surface atoms of the dimerization are also dependent on the number of silicon layers. For example, in the case of m=6 or even number, it has a simple (2x1) dimerization (Dimer) structure, whereas in m=5 or odd number, a (2x2) dimerization (Dimer) structure will be obtained. Therefore, we have two unit cells with different symmetry; they are orthogonal and tetragonal superlattice, respectively. The unit cell models for m=5 and m=10 are shown in Figure 4. It can be shown that the two structures models have been avoided dangling bonds in bulk. From the perspective of chemical bonds, each silicon atom has four nearest neighbor bonds, whereas each VI atom has two nearest neighbour Si-VI bonds. They form a stable structure, and prevent the participation of interface states. The designed models of superlattice unit cells, VI(A)/Si_5/VI(B)/Si_5 and VI(A)/Si_{10}/VI(B)/Si_{10} are shown in Figure 4, in which the inserted VI atoms layer is a periodic monolayer and the dimer reconstruction on surface has been considered. Note that the primitive lattice vectors of the superlattices are different from the (Sn_xSi_{1-x})$_5$/Si$_5$ (001) due to the Si(001) surfaces having been restructured. During the first-principles calculations, the distance between the VI-atoms and Si-atoms, the positioning of the VI-atoms parallel to the interface with respect to the Si (001) surface and the lattice parameters of the superlattice cell can be varied. After the relaxations are finished, the total energy of the relaxed interface system is at the lowest, then a stable unit cell will be obtained. The theoretical equilibrium lattice constants (in a.u.) of the superlattices are given in Table 3. It can be seen that the $a \approx b$ for tetragonal structure superlattice VI(A)/Si$_5$/VI(B)/Si$_5$(001) with (2x2) dimer, whereas the VI(A)/Si$_6$/VI(B)/Si$_6$(001) is an orthogonal structure superlattice with (2x1) dimer. In all cases, these superlattices formed by alternating a VI-atom monolayer and diamond structure Si along to [001] direction, their lattice parameters are increased with the core states of inserted VI-atoms increased.

Materials	a	b	c
Se/Si$_5$/O/Si$_5$(001)	14,62	14.53	33.07
Se/Si$_5$/S/Si$_5$ (001)	14.64	14.59	34.28
Se/Si$_5$/Se/Si$_5$ (001)	14.66	14.66	34.79
Se/Si$_6$/O/Si$_6$(001)	14,42	7.31	38.57
Se/Si$_6$/S/Si$_6$ (001)	14.47	7.33	39.80
Se/Si$_6$/Se/Si$_6$ (001)	14.53	7.33	40.27

Table 3. The theoretical equilibrium lattice constants (in a.u.) of the superlattices VI(A)/Si$_m$/VI(B)/Si$_m$ (001).

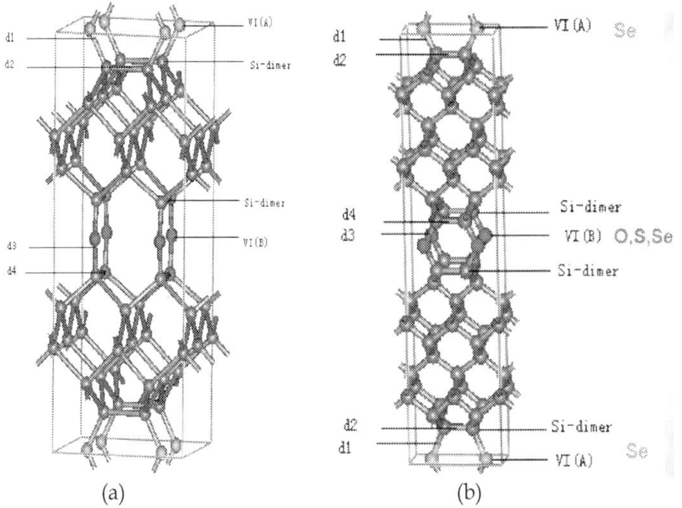

Fig. 4. The model of designed superlattice unit cell. The inserted VI atoms layer is a monolayer, the dimer reconstruction on surface has been considered. (a) VI(A)/Si$_5$/VI(B)/ Si$_5$(001). (b) VI(A)/Si$_{10}$/VI(B)/Si$_{10}$(001).

5. Results and discussion

According to our computational design principle, the theoretical superlattices IV$_x$Si$_{1-x}$ / Si (001), (IV=Ge,Si; x=0.125,0.25,0.5) and VI(A)/Si $_m$/ VI(B)/Si $_m$ (001) (VI=O,S.Se; m=5.6.10) have been investigated. In our calculations, the band structures based on the density functional theory (DFT) and local density approximation (LDA) are performed first. The purpose is to find and demonstrate the direct bandgap materials. On this basis, in order to correct the Kohn-Sham band gap which is always underestimate due to the LDA limitation, a representative quasiparticle band structure calculation in Hedin's GW approximation was carried out. The calculation in details and main results are described below.

5.1 Electronic structure of IV_xSi_{1-x} / Si (001) superlattices

The DFT-LDA calculation for these new superlattices is based on a total energy pseudopotential plane-wave method. The wavefunctions are expressed by plane waves with the cutoff energy of $|k+G|^2 \leq 450$ eV. The Brillouin zone integrations are performed by using 6x6x3 k-mesh points within the Monkhorst-Pack scheme. The convergence with respect to both the energy cutoff and the number of k-point has been tested. With a larger energy cutoff or more k points, the change of the total energy of the system is less than 1 meV. Calculated equilibrium lattice constants after lattice relaxation are given in Table 2, and it is very closely Vegard's law for different IV component.

The Band structures of Ge_xSi_{1-x}/ Si (001) and Sn_xSi_{1-x}/ Si (001) superlattices are shown in Fig.5(a,b) for x=0.125, 0.25 and 0.5., respectively. It can be seen that the Ge_xSi_{1-x}/ Si (001) (x=0.125 and 0.25) and Sn_xSi_{1-x}/ Si (001) (x=0.125) are the superlattices with a direct gap at Γ-point. Although the dispersion relation of the valence band is quite similar in all cases, the lowest conduction band revealed great differences in the dispersion. The reason is that both the Ge and Sn have a larger core states and hence larger lattice parameters than that of Si, Their perturbation potential will change the Kohn-Sham effective potential V_{eff} and eigenvalues $E^{KS}(k)$. As Corkill-Cohen has pointed out (Corkill & Cohen M.L (1993)), the result is that the lowest conduction band (Γ-band edge of Si) will continue to lower with the increase of lattice constant. This feature can lead to an above three Γ- point direct band gap superlattice, of course, also there is a greater possibility in transforming them to direct band gap material due to the lower symmetry of the unit cells. In the same way, with the Sn superlattice band gap becoming small compared with the Ge is understandable.

We note that the selection of superlattice primitive cell is not unique. If the location of alternative atoms Ge, Sn are chosen symmetrically for the unit cell center, a D_{4h} symmetry superlattice can be obtained. In order to examine the energy band structure in this case, the band structures of Sn_xSi_{1-x}/ Si (001) superlattices are calculated again. In the same time, as a comparison, the band structure of pure Si (in D_{4h}) is also given in Figure 6(a). The results show that silicon is still an indirect band gap semiconductor, the conduction band bottom is in Γ-X and Γ-Z line, and only Sn_xSi_{1-x}/ Si (001) (x = 0.125) is a direct band gap material. The results excellently agree with Figure 5 (b). The shift of conduction band edge for these systems is also clearly visible when we inspect going from Si to $Sn_{0.5}Si_{0.5}$/Si(001) superlattice. First of all, the energy of Γ-band edge is reduced and hence the direct gap superlattice $Sn_{0.125}Si_{0.875}$/Si(001) is formed. Then, the reduction of Z-band edge exceeds that of the Γ-band edge (if Sn component increased), the indirect gap superlattices are obtained, with smaller relevant band gap.

The Kohn-Sham band gap E_g^{KS} of the superlattices are summarized in Table 4, the data is corresponding to different model and exchange-correlation approximation quasi-particle energy E^{QP} and quasi-particle wavefunction ψ^{QP}, the key-point is calculated. In order to correct the Kohn-Sham band gap E_g^{KS} of the superlattices, the quasiparticle band structure within Hedin's GW method (GWA) is performed by using PARATEC and ABINIT packages, for a representative superlattice $Sn_{0.125}Si_{0.875}$/Si(001), where G is a one-particle Green function, W is a dynamic screening Coulomb interaction. The quasi-particle energy

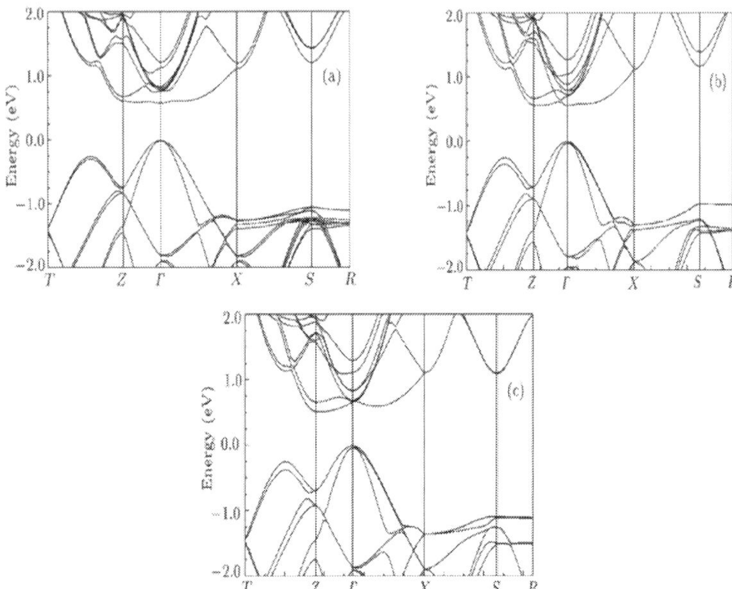

Fig. 5(a). Band structure of Ge$_x$Si$_{1-x}$/ Si (001) superlattices. (a)x=0.125, (b) x=0.25, (c)x=0.5

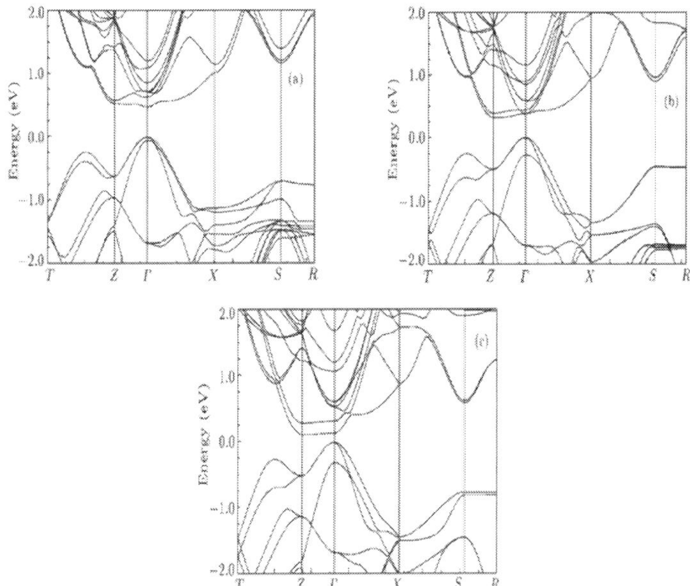

Fig. 5(b). Band structure of Sn$_x$Si$_{1-x}$/ Si (001) superlattices. (a)x=0.125, (b) x=0.25, (c)x=0.5

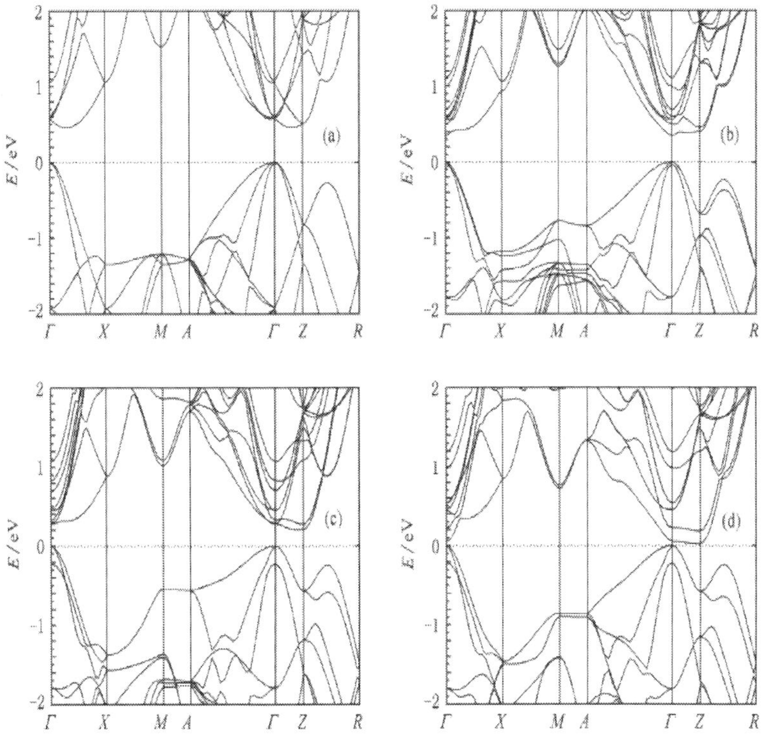

Fig. 6. DFT-LDA band structures of Si and Sn_xSi_{1-x}/ Si (001) superlattice in D_{4h} symmetry.(a) Si, (b,c,d) superlattices for x=0.125, 0.24, 0.5, respectively.

E^{QP} and quasi-particle wavefunction ψ^{QP} are solutions of quasi-particle equation which contains an electron self-energy operator Σ. One of key-points is to calculate the Σ. In Hedin's GWA, Σ = iGW, it does not consider vertex corrected. The extensive research points out (Hybertsen M.S.& Louie S.G. 1985, 1986), the quasi-particle wave function ψ^{QP} is almost completely overlapped with the Kohn-Sham wave function ψ^{KS}, the overlap range exceeds 99.9%. Therefore, in our GWA calculation, we will assume that can use Kohn-Sham wave function as quasi-particle wave function of zero-level approximation. Therefore, we can construct the Green function G that employ the Kohn-Sham wave function ψ^{KS}, based on the Kohn-Sham equation solutions. The dynamic screening Coulomb interaction W depends on the bare Coulomb interaction v and dielectric function matrix χ. The dielectric matrix calculation is also a difficult task, we adopt the simpler RPA approximation. In this way, based on the KS equation solutions, we could solve the quasi-particles equation and obtain the quasi- particle band structure of the superlattice. As a representative result of IV_xSi_{1-x}/Si(001) superlattices, a quasi-particle band structure is given in Figure 7, which is quite similar to its LDA band structure in Figure 6(b). The main difference is that the direct band gap increases from E_g^{LDA} = 0.35 eV to E_g^{QP} = 0.96 eV. In other words, the quasi-particle bandgap correction of this system is 0.61 eV. Although G and W has not carried out self-consistent calculation in present work, one can see that the result is quite accurate and reliable,

Materials	$E_g^{KS}(D_{2h}, GGA)$	$E_g^{KS}(D_{4h}, LDA)$	$E_g^{QP}(D_{4h}, G_0W_0)$
Si	0.58	0.46	
$Sn_{0.125}Si_{0.875}/Si(001)$	0.47 (Γ-Γ)	0.35 (Γ-Γ)	0.96 (Γ-Γ)
$Sn_{0.25}Si_{0.75}/Si(001)$	0.31 (Γ- Z)	0.21 (Γ- Z)	
$Sn_{0.5}Si_{0.5}/Si(001)$	0.10 (Γ- Z)	0.03 (Γ- Z)	
$Ge_{0.125}Si_{0.875}/Si(001)$	0.57 (Γ-Γ)		
$Ge_{0.25}Si_{0.75}/Si(001)$	0.55 (Γ-Γ)		
$Ge_{0.5}Si_{0.5}/Si(001)$	0.51 (Γ- Z)		

Table 4. Band gap E_g (in eV) of the $IV_xSi_{1-x}/Si(001)$ superlattices. (Γ-Γ) stands for direct gap at Γ, GGA the generalized gradient approximation, LDA the local density approximation.

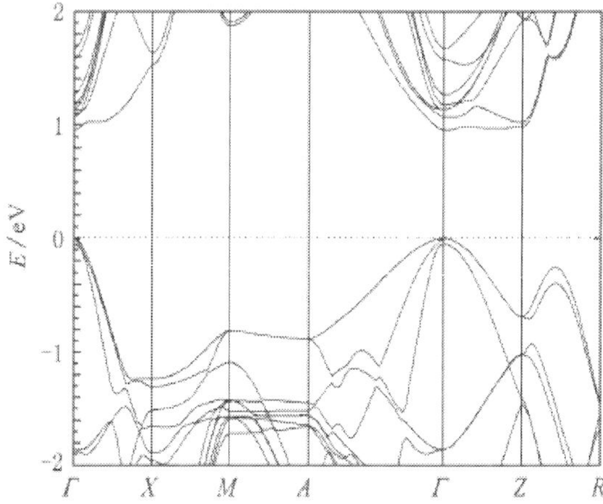

Fig. 7. Quasiparticle energy band of $Sn_{0.125}Si_{0.875}/Si(001)$ superlattice.

In fact, this approach is often called G_0W_0 method in the literature. But by this method, results obtained are better for the *sp* semiconductors even than partial self-consistent method G_0W and GW_0 as well as the complete self-consistent method GW. There are already some works studying the reasons for these facts (e.g. see Ishii et al. 2010).

5.2 Electronic structure of VI(A)/Si $_m$/ VI(B)/Si $_m$ (001) superlattices

Another series of computational designed silicon-based superlattice is VI(A)/Si $_m$/ VI(B)/Si $_m$ (001), which includes $O/Si_m/O/Si_m(001)$, S/Si $_m$/S/Si $_m$ (001), Se/Si $_m$/Se/Si $_m$ (001), Se/Si $_m$/O/Si $_m$ (001), and Se/Si $_m$/ S/Si $_m$ (001) etc for m=5,6,and 10. The results show that, for the cases of selected VI (A) = Se, VI (B) = O, S, Se, the direct band gap superlattices can be formed. Two unit cell structure models, tetragonal and orthogonal structure for m=5 (or odd number) and m=10 (or even number) are shown in Figure 4. These stable lattice structure models and their equilibrium lattice constants, the VI-Si bond length and

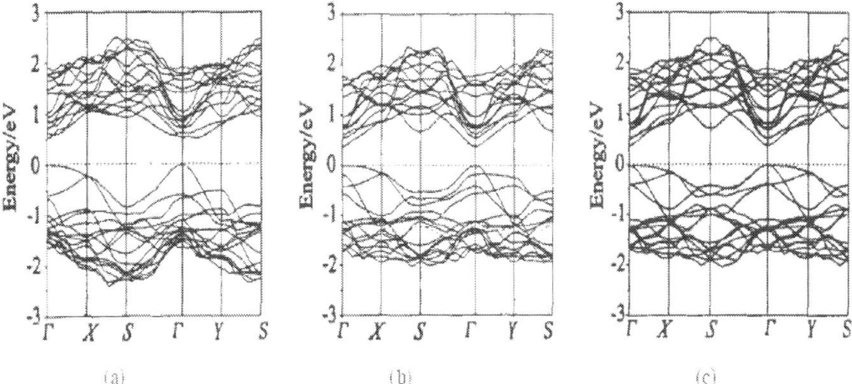

Fig. 8. Band structures of Si-based superlattices with odd number layers Si and tetragonal structure. (a) Se/Si$_5$/O/Si$_5$(001) , (b) Se/Si$_5$/S/Si$_5$(001) (c) Se/Si$_5$/Se/Si$_5$(001).

bond angle are calculated by using the first principles total energy method. The DFT-LDA band structure calculation of the Si-based superlattices use mixed-basis pseudopotential method with norm- conservation pseudopotential (Hamann et al 1979) and VASP package with ultra-soft pseudopotential (Kresse & Furthmüller 1996) and Ceperly- Alder Exchange-correlation potential (Ceperley & Alder 1980), respectively. The wavefunctions are expanded by plane waves with the cutoff energy of 12 Ry which has been optimized via total energy tolerance $\Delta E=1$ meV.

The band structures of Se/Si$_5$/VI(B)/Si$_5$(001) (VI(B)=O,S,Se) superlattices with a tetragonal structure are shown in Figure 8. It is shown that the materials are the potential Si-based optoelectronic semiconductors with Γ-point direct gap.

The band structures of O/Si$_5$/VI(B)/ Si$_5$(001) (VI(B)=O,S) which only involve the VI-atoms of smaller core-states, are also studied and found that are the quasi direct gap materials with the X-point valence band top (Huang 2001a; Huang & Zhu 2001b,c, Huang et al. 2002; Huang 2005)., although their smallest direct band gap is still at Γ- point.

To investigate the influence of Se/Si$_m$/VI(B)/Si$_m$(001) (VI(B)=O,S,Se) with even number layers silicon that have the orthogonal structures on the electronic properties and band gap type, the Se/Si$_m$/VI(B)/Si$_m$(001) (VI(B)=O,S,Se; m=6,10) are calculated with the same method. The results indicate that they are also direct band gap superlattices as shown in Figure 9. In other words, band-gap type and number of layers of silicon in Se/Si$_m$/VI(B)/Si$_m$(001) (VI(B)=O,S,Se) are not sensitively dependent. However, choosing the appropriate size of the VI atoms, such as Se, is important. Using Se and O or S periodic cross intercalation in Si(001), the desired results can be achieved more satisfactorily (Zhang J.L. Huang M.C. et al, (2003)) due to the core states effect and the smaller electronegativity difference. The LDA band gap of these Si-based materials is listed in Table 5. For the tetragonal structure material (m=5), its band gap is a little bit bigger than that of the orthogonal structure situation (m=6,10). As well known, the LDA band gap is not a real material band gap, since the exchange correlation potential in DFT-LDA equation can not correctly describe the excited states properties. In order to revise LDA band gap, we can use GWA methods or screen-exchange-LDA (sX-LDA) method to solve the quasiparticle equation. The existed research shows that this energy gap revision is quite large, for example, for

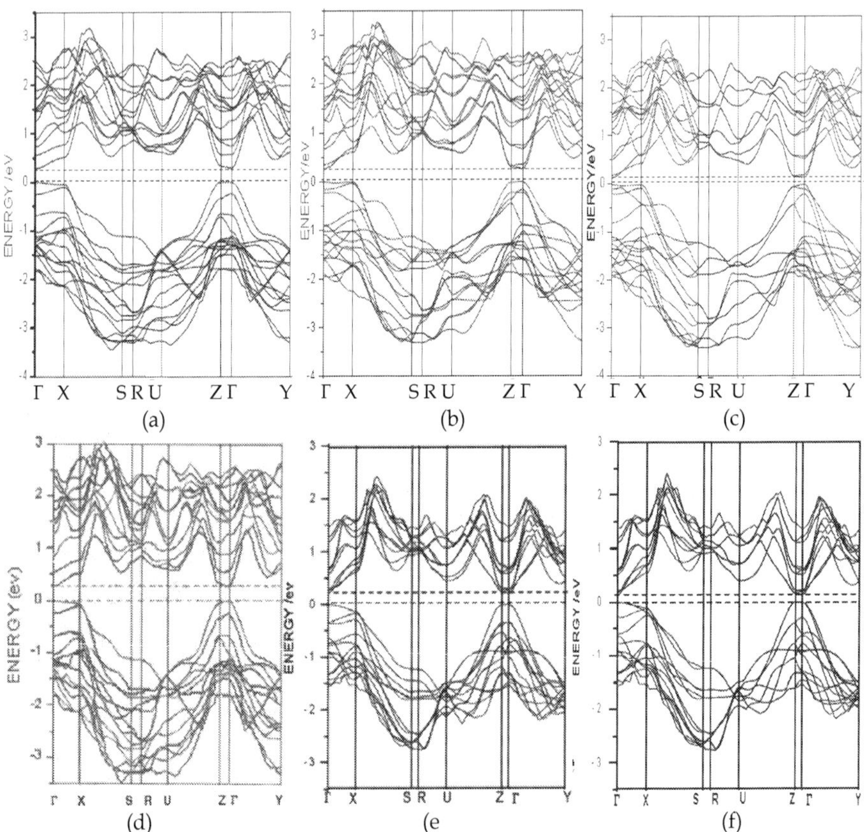

Fig. 9. Band structures of Si-based superlattices with even number layers Si in orthogonal structure. (a) $Se/Si_6/O/Si_6(001)$, (b) $Se/Si_6/S/Si_6(001)$, (c) $Se/Si_6/Se/Si_6(001)$, (d) $Se/Si_{10}/O/Si_{10}(001)$, (e) $Se/Si_{10}/S/Si_{10}(001)$, (f) $Se/Si_{10}/Se/Si_{10}(001)$.

Materials	$E_g{}^{KS}$(LDA, Tet)	$E_g{}^{KS}$(LDA,Orth.)
$Se/Si_5/O/Si_5(001)$	0.50	
$Se/Si_5/S/Si_5(001)$	0.40	
$Se/Si_5/Se/Si_5(001)$	0.35	
$Se/Si_6/O/Si_6(001)$		0.30
$Se/Si_6/S/Si_6(001)$		0.25
$Se/Si_6/Se/Si_6(001)$		0,20
$Se/Si_{10}/O/Si_{10}(001)$		0.30
$Se/Si_{10}/S/Si_{10}(001)$		0.25
$Se/Si_{10}/Se/Si_{10}(001)$		0,20

Table 5. Band gap E_g (in eV) of the $Se/Si_m/VI(B)/Si_m(001)$ (VI(B)=O,S,Se) superlattices.

silicon and germanium, It is about 0.7 and 0.75 eV, respectively (Hybertsen M.S. and Louie S.G. (1986)). Our GWA calculation for IVSi/Si superlattice has the band gap revision of 0.61 eV, which is near to Silicon. Taking into account the quasi-particle band gap correction, for example, 0.61 eV, the band gap of these si-based materials is in the region of 1,11-0.81 eV, which is corresponding to the infrared wavelength of 1.12-1.53μm, just matching to the windows of lower absorption in the optical fiber. Therefore they are potentially good Si-based optoelectronic materials.

Similar to our computation cited above, MIT's research group (Wang et al 2000) had provided a class of semiconductors, in which a particular suitable configuration, $(ZnSi)_{1/2}P_{1/4}As_{3/4}$, is identified that lattice constant matched to Si and has a direct band gap of 0.8 eV. Although this material has good performance, but its complex structure, involving the four elements in the heteroepitaxy on silicon substrates, the crystal growth may have much more difficulties.

Another well-known computational design is proposed by Peihong Zhang etc (Zhang P.H, et al. 2001). They suggest two IV-group semiconductor alloys CSi_2Sn_2 and CGe_3Sn that have body-centered tetragonal (bct) structure, the lattice matched with Silicon. Among them, CSi_2Sn_2 has a direct band gap located at X point in the BZ, and CGe_3Sn has a Γ- point direct band gap, because its lattice is slightly distorted from b.c.t,, the crystal symmetry of CGe_3Sn is lower than that of CSi_2Sn_2. Their GW band gap is in 0.71-0.9eV range. Anyway, they are also potential contenders of Si-based optoelectronic materials. The heterogeneous epitaxy of these IV group alloys on silicon substrate is not an easy task, because the positions of the component atoms have to meet some particular requirements in these alloys. In contrast, we use of periodic atomic intercalation method to have more practical application prospect. The symmetry reduction principle, core states effect and electronegativity difference effect can be used not only for silicon-based materials but also can be extended to other indirect band gap semiconductor systems, such as AlAs, diamond and other materials, to realize the energy band modification. They also have a significant research and development prospect. We designed Si-based optoelectronic materials can natural be realized lattice matched with silicon substrate. The growth process on the MBE, MOCVD or UHV-CVD might easier to control. Once the experimental research of these materials is brokenthrough, OEIC technology will have a significant development.

6. Conclusion

This chapter has given an overview of our works on the computational design of a new class of Si-based optoelectronic materials. A simple effective design idea is presented and discussed. According to the design ideas, two series models of superlattice are constructed and calculated by the first principles method. It is found that the superlattices $Ge_xSi_{1-x}/Si(001)$ (x=0.125,0.25), $Sn_xSi_{1-x}/Si(001)$ (x=0.125), $Se/Si_m/VI/Si_m/Se(001)$ (VI=O,S,Se; m=5,6.10) are the Γ-point direct energy gap Semiconductors, moreover, they can be realized lattice matched with silicon substrate on (001) surface. These new materials have the band gap region of 0.63-1.18 eV under the GW correction that is corresponding to infrared wavelength of 1.96-1.05 μm and are suited for the applications in the optoelectronic field. An open question for all kind of Si-based new materials is what and how to do to achieve them under the experimental research.

7. Acknowledgments

This work was supported by the Chinese National Natural Science Foundation in the Project Code: 69896260, 60077029, 10274064, 60336010. Author wishes to thank Dr. T.Y. Lv, Dr. J. Chen and Dr. D.Y.Chen for their calculation efforts successively in these Projects. We also are grateful to Prof. Q.M.Wang and Prof. Z.Z. Zhu for many fruitful discussions. Finally, author want to express his thanks to Prof. Boxi Wu for reading the Chapter manuscript and gave valuable comments.

8. References

Aryasetiawan F , Gunnarsson O. (1998), The GW method .*Rep. Prog. Phys.* ,61 :p.237.

Aulbur W G, Jonsson L ,Wilkins J W. (2000), Quasiparticle calculations in solids. *Solid State Physics* ,54 :p.1-231.

Benzair A ,Aourag H. (2002), Electronic structure of the hypothetical cubic zincblende-Like semiconductors SiC, GeC and SnC . *Phys. Stat . Sol.* (b) , 231 (2) : p.411-422.

Buda F. et al, (1992), *Phys. Rev. Lett.*,69,p.1272

Canham L T. (1990), Silicon quantum wire array fabrication by electrochemical dissolution of wafer . *Appl. Phys. Lett* . 57 :p.1 045 - 1 048.

Ceperley D M ,Alder B I. (1980), Ground state of the elect ron Gas by a stochastic method. *Phys. Rev. Lett* .45 :p.566.

Chen J. Lv T,Y. Huang M.C.(2007), Electronic structure of $Si_{1-x}IV_x/Si$ superlattices on Si(001), *Chin.Phys. Lett.* 74,p.811.

Corkill J.L and Cohen M.L (1993), Band gaps in some group-IV materials: A theoretical analysis, *Phys. Rev*.B47,p.10304

Cullis A G, Canham L T. (1991), Visible light emission due to quantum size effects in highly porous crystalline silicon . *Nature* , 353 :p.335 - 338.

Deboeuij Y P L , Kootstra F J, Snijders G. (2002), Relativistic effects in the optical response of HgSe by time-dependent density functionals theory. *Int, J. Quantum Chem.* , 85 :p.450 - 454.

Ennen H, Schneider J, Pomerenke G,Axmann A. (1983) 1.54μm luminescence of erbium-implanted III-V semiconductors and silicon, *Appl. Phys. Lett.* 43, p.943.

Hamann D R , Schluter M , Chiang C. (1979), Norm-conserving pseudopotentials . *Phys. Rev. Lett* .,43, p.1 494.

Hirschman K D, Tsybeskov L, Duttagupta S P, et al. (1996), Silicon-based light emitting devices integrated into microelectronic circuits . *Nature* , 384 :p.338 - 340.

Hohenberg P.and Kohn W. (1964). Inhomogeneous electron Gas .*Phys. Rev.* 136 p.B864,:.

Huang M.C. (2001a), The new progress in semiconductor quantum structures and Si-based optoelectronic materials, . *J. Xiamen Univ.(Natural Sci.)* , 40 , p. 242 -250.

Huang M.C. and Zhu Z. Z. (2001b), An exploration for Si-based superlattices structure with direct-gap . *The 4th Asian Workshop on First principles Electronic Structure Calculations..* p.12.

Huang M.C. and Zhu Z. Z.(2001c), A new Si-based superlattices structure with direct band-gap. *Proc. of 5th Chinese Symposium on Optoelectronics.* p.44 - 47.

Huang M.C, Zhang J L ,Li H P ,et al. (2002), A computational design of Si-based direct bandgap materials. *International J . of Modern Physics* B 16 :p.4 279 - 4 284.

Huang M.C,(2005), An ab initio Computational Design of Si-based Optoelectronic Materials, *J. Xiamen Univ.(Natural Sci.)* , 44 , p. 874

Hybertsen M.S. and Louie S.G. (1985). First-Principles Theory of Quasiparticles: Calculation of Band Gaps in Semiconductors and Insulators, *Phys. Rev. Lett.* 55, p.1418

Hybertsen M S ,Louie S G. (1986), Electron correlation in semiconductors and insulators . *Phys. Rev.* B34 :p.5 390 - 5 413.

Hybertsen M.S. and Schluter M. (1987), Theory of optical transitions in Si/Ge(001) strained-layer superlattices, *Phys. Rev.* B36, p.9683

Ishii S., Maebashi H. and Takada Y. (2010), Improvement on the GWΓ Scheme for the Electron Self-Energy and Relevance of the G_0W_0 Approximation from this Perspective. *arXiv [cond-mat,mtrl-sci]* 1003.3342v2

Kohn W AND Sham L J. (1965), Self-consistent equations including exchange and correlation effects. Phys. Rev. 140 : p.1 133.

Kresse G.and Furthmüller J. (1996), Efficient iterative schemes for *ab initio* total-nenrgy calculations using a plane-wave basis set. *Phys.Rev.* B54, p.11169

Lu Z H ,Lockwood D J ,Baribeau J M. (1995), Quantum confinement and light emission in SiO_2/Si superlattice . *Natur,* 378 :p.258 - 260.

Lv T.Y. Chen J. Huang M.C. (2010), Band structure of Si-based superlattices $Si_{1-x}Sn_x$/Si. *Acta Phys. Sinica,* 59,p.4847

Mattesini M ,Matar S F. (2001), Search for ultra-hard materials :theoretical characterisation of novel orthorhombic BC_2N crystals . *Int . J . Inorganic Materials* , 3 : p.943 -957.

Mooser E ,Pearson W B. (1960),The chemical bond in semiconductors. *Progress in Semiconductors* ,5 : p.141 -188.

Nassiopoulos A G, Grigoropoulos S , Papadimit riou S. (1996),Electroluminescent device based on silicon nanopillars . *Appl. Phys. Lett* .69 :p.2 267 - 2 269.

Ng W.L ,Lourenc M A ,Gwilliam R M et al. (2001), An efficient room-temperature silicon-based light emitting diode .*Nature* , ,410 :p.192 - 194.

Pavesi L ,Dal Negro L ,Mazzoleni C ,et al. (2000), Optical gain in silicon nanocrystals. *Nature* , ,409 :440 - 444.

Petersilka M , Gossmann U I , Gross E K U. (1996), Excitation energies from time-dependent desity functional theory . *Phys. Rev. Lett* .76 :p.1 212.

Phillips J C. (1973) *Bonds and Bands in Semiconductors.* USA : Academic Press ,.

Rosen Ch H ,Schafer B ,Moritz H ,et al. (1993) Gas source molecular beam epitaxy of $FeSi_2$ / Si (111) heterostructures . *Appl. Phys. Lett* .,62 :p.271.

Runge E , Gross E K U. (1984), Density functional theory for time-dependent systems . *Phys. Rev. Lett* .,52 :p.997.

Seidl A, Gorling A, Vogl P, et al. (1996), Generalized Kohn-Sham schemes and the band-gap problem . *Phys. Rev.*,B53 :p.3 764.

Walson W L, Szajowski P F, Brus L E. (1993), Quantum confinement in size-selected surface-oxidised silicon nanocrystals . *Science* , 262 :p.1 242 - 1 244.

Wang T, Moll N, Kyeongjae Cho, Joannopoulos J. D. (2000) Computational design of compounds for monolithic integration in optoelectronics, *Phys. Rev.* B63, p.035306

Zhang J.L.,Huang M.C. et al, (2003), Design of direct gap Si-based superlattice VIA/Si_m/VIB / Si_m/ViA. *J. Xiamen Univ.(Natural Sc.,)*,42 :p.265 - 269.

Zhang P.H, Crespi V H, Chang E, et al. (2001), Computational design of direct-bandgap semiconductors that lattices-match silicon. *Nature* , 409 :p.69 - 71.

Zhang Q , Filios A ,Lofgren C ,et al. (2000), Ultrastable visible electroluminescence from crystalline c-Si/O superlattice. *Physica* , E8 :p.365 - 368.

Part 2

Optoelectronic Sensors

7

Coupling MEA Recordings and Optical Stimulation: New Optoelectronic Biosensors

Diego Ghezzi
Istituto Italiano di Tecnologia
Italy

1. Introduction

In the last twenty years the efforts in interfacing neurons to artificial devices played an important role in understanding the functioning of neuronal circuity. As result, this new *brain technology* opened new perspectives in several fields as neuronal basic research and neuro-engineering. Nowadays it is well established that the functional, bidirectional and real-time interface between artificial and neuronal living systems counts several applications as the brain-machine interface, the drug screening in neuronal diseases, the understanding of the neuronal coding and decoding and the basic research in neurobiology and neurophysiology. Moreover, the interdisciplinary nature of this new branch of science has increased even more in recent years including surface functionalisation, surface micro and nanostructuring, soft material technology, high level signal processing and several other complementary sciences.

In this framework, Micro-Electrode Array (MEA) technology has been exploited as a powerful tool for providing distributed information about learning, memory and information processing in cultured neuronal tissue, enabling an experimental perspective from the single cell level up to the scale of complex biological networks. An integral part in the use of MEAs involves the need to apply a local stimulus in order to stimulate or modulate the activity of certain regions of the tissue. Currently, this presents various limitations. Electrical stimulation induces large artifacts at the most recording electrodes and the stimulus typically spreads over a large area around the stimulating site.

Compound optical uncaging is a promising strategy to achieve high spatial control of neuronal stimulation in a very physiological manner. Optical uncaging method was developed to investigate the local dynamic responses of cultured neurons. In particular, flash photolysis of caged compounds offers the advantage of allowing the rapid change of concentration of either extracellular or intracellular molecules, such as neurotransmitters or second messengers, for the stimulation or modulation of neuronal activity. This approach could be combined with distributed MEA recordings in order to locally stimulate single or few neurons of a large network. This confers an unprecedented degree of spatial control when chemically or pharmacologically stimulating complex neuronal networks.

Starting from this point, the main objective of this chapter is the discussion of an integrated solution to couple the method based on optical stimulation by caged compounds with the technique of extracellular recording by using MEAs.

2. Scientific background

In the second half of the last century the functional properties of neurons, e.g. receptor sensitivity and ion channel gating, have been investigated providing a detailed picture of the neuronal physiology. In fact, some peculiar behaviors, e.g. plasticity, have been deepened down to the different molecular mechanisms underlying this function. Nowadays the high level of knowledge about single neuron functioning does not reflect an high level of understanding of the complex way of intercommunication between neurons in neuronal networks. The need of learning the neuronal language and the desire to bidirectionally communicate with neurons encouraged the development of new technologies, as MEA devices, focused to this purpose.

MEAs have been proposed more than thirty years ago (Gross, 1979; Pine, 1980; Thomas et al., 1972) for the study of electrogenic tissues, i.e. neurons, heart cells and muscle cells. Nowadays, they represent an emerging technology in such studies. In the last thirty years, MEAs have been exploited with various preparations such as dissociated cell cultures (Marom & Shahaf, 2002; Morin et al., 2005), organotypic cultures (Egert et al., 1998; Hofmann et al., 2004; Legrand et al., 2004) and acute tissue slices (Egert et al., 2002; Kopanitsa et al., 2006) for a large variety of applications, such as the study of functional activity of larger biological networks (Tscherter et al., 2001; Wirth & Lüscher, 2004), as well as applications in the fields of pharmacology and toxicology (Gross et al., 1997; 1995; Natarajan et al., 2006; Reppel et al., 2007; Steidl et al., 2006). Recently, MEA biochips have also been used as *in vitro* biosensors to monitor both acute and chronic effects of drugs and toxins on heart/neuronal preparations under physiological conditions or pathological conditions modelling human diseases (Stett et al., 2003; Xiang et al., 2007).

Referring to neuronal preparations, a major distinguishing feature of the nervous system is its ability to inter-connect regions that are relatively distant from each other, via synaptic connectivity and complex circuits/networks. Consequently, when studying the nervous system and its complex circuitry *in vitro*, it is necessary and desirable to be able to provide a given stimulus (typically electrical or chemical/pharmacological in nature) at a well-defined point of the circuit and subsequently monitor how it propagates through the circuit. The MEA technology provides key advantages for carrying out such studies. It allows the possibility to record electrical activity at multiple sites simultaneously, thereby providing information about the spatio-temporal dynamics of the circuit. Moreover the usefulness of MEAs comes also from the possibility to electrically stimulate cells cultured on top of them.

However MEA applicability in cell culture/tissue electrical stimulation could not be simple as it sound. Usually the amplitude of stimulation is at least an order of magnitude bigger than the cell spiking activity, thus making impossible the detection of activity during the stimulation. Moreover, the stimulation produces large electrical artefact lasting on most channels for milliseconds after the real stimulus, making uncertain the interpretation of data in the first period after stimulation. Some attempts to remove the stimulus artifacts from the recordings have been recently proposed using off-line or on-line blanking methods (Jimbo et al., 2003; Wagenaar & Potter, 2002) partially solving this problem.

Another important disadvantage is related to the poorly controlled spatial distribution of the electrical stimuli. In fact, it has been demonstrated that electrical stimuli spread to the whole biological preparation with amplitude decreasing with the square of the distance from the stimulation site (Heuschkel et al., 2002); in fact, electrical stimuli can directly activate a large number of cells distributed in a quite large area (hundreds of microns) around the stimulation electrode also in the presence of synaptic blockers (Darbon et al., 2002). The reason of that is unknown, but probably due to several axons passing through the region of the stimulating electrode. Varying the stimulation protocol (i.e. amplitude, polarity, waveform or duration of the pulse) the number of cells directly responding to the electrical stimulus could be adjusted, however the classification of responses detected at different electrodes surrounding the stimulating electrode in directly elicited or due to synaptic transmission remains uncertain. Finally, electrical stimulation needs care to use voltages or current densities that do not harm the electrode.

Some attempts have been done in order to keep down the extension of electrical stimulation. Clustering structures have been proposed (Berdondini et al., 2006) showing a clear difference in the Post-Stimulus Time Histogram (PSTH) between traditional and clustered MEAs. Whereas the traditional MEA shows a the dominance of the early responses (mean latency of 10 ms), the different clusters show a great variability in mean latency (from 10 ms to 100 ms). Unfortunately, the use of clustering structures as well as network patterning structured PDMS layers or neurocages (Erickson et al., 2008) can relatively limiting the random nature of the network and its functional plasticity.

Another method commonly used to stimulate or modulate in vitro neuronal preparations is the application of chemical or pharmacological compounds, e.g. neurotransmitters, ion-channel blockers etc. The problem here is that the chemical/pharmacological compound traditionally is applied over the whole culture preparation through bath addiction, and thus affects almost the entire culture/circuit. Local drug delivery has been proposed in several fashions, from the use of glass pipettes placed near the target cell to dedicated Lab On Chips (LOCs). Glass pipettes are widely used in neuroscience for the local delivery of chemical compounds, but this method is limited by the time needed for the pipette placing and the impossibility to perform parallel multipoint delivery. On the contrary, several publications report on microfluidic devices making possible to transport molecules to cells in a spatially resolved way, i.e. multiple laminar flows (Takayama et al., 2003). Unfortunately, a few systems have been reported where MEAs were combined with microfluidic devices for the testing of toxins (DeBusschere & Kovacs, 2001; Gilchrist et al., 2001; Pancrazio et al., 2003) but without efforts towards the localization of the delivery or complete characterization. A dispensing system for localised stimulation was recently designed to be combined with a MEA chip (Kraus et al., 2006) but not yet completely implemented.

A useful method to combine local neuronal stimulation and local drug delivery involve the use of optical techniques. In principle, different works report on methods for optical stimulation of neurons (Callaway & Yuste, 2002), including direct (Fork, 1971) or dye-mediated laser stimulation (Farber & Grinvald, 1983), direct two-photon excitation (Hirase et al., 2002), endogenous expression of molecules sensitive to light (Zemelman et al., 2002) and caged neurotransmitter activation (Callaway & Katz, 1993). Among the above, the use of caged compounds seems to be the most physiologically suitable approach for the coupling of light with either neuronal excitation, e.g. with caged glutamate (Wieboldt et al., 1994), or modulation, e.g. with caged intracellular second messengers (Nerbonne, 1996).

Caged compounds are characterized by the presence of a blocking chemical group that can be removed by ultra-violet (UV) light pulses (Ellis-Davies, 2007). In this manner, a rapid increase in the concentration of the desired molecule can be obtained by switching the caged analogue into its active form through the cleavage of its blocking group. However, while the process of compound uncaging can be well controlled temporally, the spatial control of this process is limited by the width of the light beam and by light diffraction effects between the light source and the biological preparation, as well as by compound diffusion in the medium around the site of stimulus application.

In these years, various methodological solutions have been adopted to optically stimulate neurons by caged compounds going from the use of UV sources (e.g. xenon flash lamps) coupled to the port of an epifluorescent microscope (Callaway & Katz, 1993), to the use of laser scanning approaches (Shoham et al., 2005) or digital holographic microscopes (Lutz et al., 2008). Moreover, also external devices such as optical fibres (Bernardinelli et al., 2005) or semiconductor UV light-emitting diodes (Venkataramani et al., 2007) have been used.

The idea of coupling MEAs and optical uncaging has been explored (Ghezzi et al., 2008) using a micro-actuated optical fibre that is able to activate a single site of a cultured neuronal network. In that work, the evaluation of the compound diffusion and of the uncaging efficiency confirmed the applicability of this approach to the local excitation of a selected region of a neuronal network cultured on a MEA device.

3. Evaluation of photostimulation with PhotoMEA

The novel PhotoMEA platform (Ghezzi et al., 2007) combines the standard MEA features, i.e. electrical monitoring, with local chemical stimulation through compound uncaging. In comparison to electrical stimulation, where the electrical stimulus spreads over the whole biological preparation with an amplitude decreasing with the square of the distance from the stimulation site (Heuschkel et al., 2002), the optical stimulation scheme of caged compounds is limited to areas that are exposed to light pulses with sufficient energy to uncage the compounds and diffusion of the compounds in the medium, i.e. uncaging takes place only in a well defined volume (Ghezzi et al., 2008). In order to allow local chemical stimulation, the spatial control of light, i.e. the propagation of light through MEA biochips, has to be defined carefully to reduce the stimulation area to an electrode location and its close surroundings.

Thus, in order to achieve local chemical stimulation, the PhotoMEA platform introduces two novel features to conventional MEA-based data acquisition systems, i.e. the use of specific MEA biochips that integrate a metal shadow mask and the addition of an optical fibre bundle specially designed to fit to the PhotoMEA electrode layout. This allows unprecedented highly localised chemical stimulation at a single electrical recording site, while monitoring the overall culture preparation. Moreover, the use of a multiple-fibre bundle system gives some advantages respect to the use of a single fibre. In fact, it avoids the movement of an optical fibre above the culture plane, thus protecting the culture from possible damages and reducing the experimental time needs for the alignment. In addition, the automatic alignment of multiple fibres to the electrode layout allows patterned stimulation at each electrode of the PhotoMEA biochip, improving the ability to release compounds in multiple sites in parallel.

3.1 The PhotoMEA biochip

Basically, the PhotoMEA biochip is based on a glass substrate, transparent Indium-Tin Oxide (ITO) recording electrodes, a titanium shadow mask that blocks light and thus prevents chemicals from uncaging in undesired regions, and an SU-8 epoxy insulation layer.

Fabrication of the PhotoMEA biochip is made using micro-fabrication technologies, i.e. positive and negative photolithography and wet chemical etching. It is built from a float glass wafer (diameter: 4 inch, thickness: 700 μm) covered with 100 nm ITO and 100 nm titanium (Fig. 1A1,B1). First, the titanium layer is patterned using Microposit S1805 positive tone photoresist (Shipley, Marlborough, USA) and wet chemical etching in 1 % HF solution for 5 s (Fig. 1A2,B2). This step defines the shadow mask of the PhotoMEA (opening diameter of 80 μm). The Microposit S1805 photoresist is then stripped away. The ITO layer is then patterned

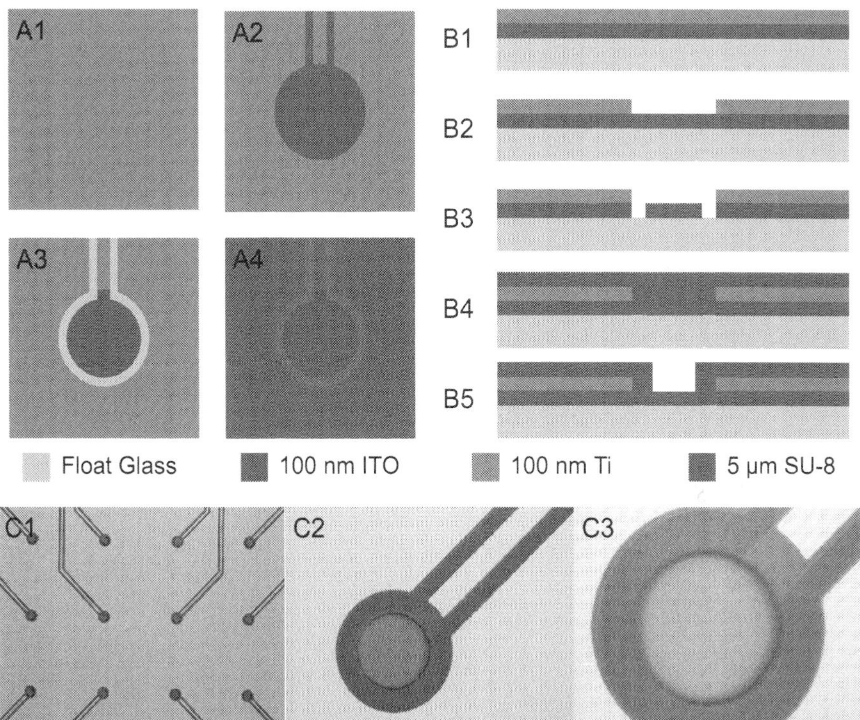

Fig. 1. (**A**) Top view of the fabrication process regarding electrodes and conductive leads. The basic design consists of ITO leads and electrodes (Blue) on a glass substrate (Cyan) covered by a titanium shadow mask (Orange). The titanium mask covers the entire region between electrodes (**A1**). To avoid shorts between the leads and the metal mask, titanium (**A2**) and ITO (**A3**) are patterned creating two small separations along each electrode lead. The insulating layer is composed of SU-8 epoxy covering all the area except for electrodes and contact pads (**A4**). (**B**) Cross-section view of the electrode fabrication process. (**B1**) Cleaning of the glass wafer covered with 100 nm ITO and 100 nm titanium. (**B2**) Patterning of the titanium layer by lift-off. (**B3**) Patterning of the ITO layer by lift-off. (**B4**) Deposition of the SU-8 epoxy insulation layer by spin-coating. (**B5**) Photolithography and opening of the insulator layer by lift-off. (**C**) Partial view of the PhotoMEA biochip workspace. Transparent electrode sites allow both local chemical stimulation and electrical readout. The space around the electrodes is covered by an integrated thin film titanium shadow mask in order to avoid unwanted uncaging of compounds (**C1**). High magnifications of one electrode site show the opening into the shadow mask (**C2**) and the real ITO electrode (**C3**). The scale bar is 200 μm in **C1** and 40 μm in **C2,3**.

using also Microposit S1805 photoresist and wet chemical etching in 37 % HCl for 150 s. This step defines the locations and wires of the ITO electrodes (diameter of 55 μm). The Microposit S1805 photoresist is then stripped away (Fig. 1A3,B3). The next step is the fabrication of SU-8 epoxy insulation layer. SU-8 GM1060 negative tone resist (Gersteltec, Pully, Switzerland) is coated (5000 rpm for 40 s) and baked for solvent evaporation (15 min at 95 °C) in order to obtain a 5 μm thick layer (Fig. 1B4). It is then exposed to UV light at 365 nm (120 mJ/cm^2) and cross-linking of the illuminated SU-8 parts is achieved by a polymerisation bake (15 min at 95 °C). Unexposed parts, i.e. effective electrode areas (diameter of 50 μm) and connection pads, are released in SU-8 developer (poly-glycol-methyl-ether-acetate) for 1 min (Fig. 1A4,B5). An oxygen-plasma (500 W, 1 min) and a hard bake (2 hrs at 140 °C) insure well definition and good adhesion of the SU-8 insulation layer. Finally, the PhotoMEA chips were released by wafer dicing.

The obtained PhotoMEA chips are then assembled onto a printed circuit board using silver-epoxy glue E212 (Epotecny, Levallois Perret, France) and are sealed using EPO-TEK 302-3M epoxy (Epoxy Technology Inc., Billerica, USA). A glass ring (internal diameter of 19 mm, external diameter of 24 mm and height of 6 mm) defining the culture chamber is finally mounted on top of the PhotoMEA assembly using Sylgard 184 silicone elastomer (Dow Corning, Seneffe, Belgium).

The electrode layout is based on an 8x8 matrix without corner electrodes, whit an electrode spacing of 500 μm (Fig. 1C1). The space between the recording-sites is covered with titanium in order to avoid unwanted chemical stimulation. The electrode leads are also made of ITO covered with titanium in order to limit the area where light can pass through the PhotoMEA biochip. To avoid shorts between the electrode leads and the titanium mask, ITO and titanium are patterned creating to small separations along the electrode lead (Fig. 1C2). The electrode shape is circular with a diameter of 50 μm and an opening in the metal shadow mask with a diameter of 80 μm defines the actual chemical stimulation area (Fig. 1C3).

3.2 The PhotoMEA platform set-up

The basic idea of the PhotoMEA platform is the combination of a standard MEA data acquisition system (Multi Channel Systems MCS GmbH, Reutlingen, Germany) with an optical fibre bundle (Ceramoptec, Bonn, Germany) coming from its bottom side (Fig. 2A). The fibre bundle is composed of 64 optical fibres (UV 50/120/150, NA0.12) arranged in an 8x8 square matrix (Fig. 2B). This arrangement was designed to match the exact geometry of the PhotoMEA biochip electrode layout. A spacing of 500 μm was achieved by positioning the fibres in a special mount made in Arcap AP1D alloy (Fig. 2C) where 160 μm holes were done by precise mechanical drilling (Fig. 2D). Each fibre was glued in the corresponding hole using the semi-rigid optical grade epoxy resin Epo-Tek 305 (Epoxy Technology Inc.).

A TTL controllable 375-nm UV laser source (Coherent Italia, Milano, Italy) was used to generate UV pulses coupled to the selected optical fibre via a 20x objective lens (Thorlabs Inc., Newton, USA). The laser can be alternative coupled to every optical fibre by moving the input side of the fibre bundle through a M105.3 DC motorised 3-axes micropositioning stage (Physik Instrumente SrL, Milano, Italy), controlled by a LabView (Teoresi SrL, Torino, Italy) custom application. On the other side, the exact alignment between the fibres of the bundle and the electrodes of the PhotoMEA biochip was obtained using another M105.3 DC motorised 3-axes micropositioning stage (Physik Instrumente SrL) controlled by the same LabView custom application. The alignment was optimized exactly matching at least 4 optical fibres with the corresponding electrode site (Fig. 2E).

3.3 Neuronal cultures

Low-density primary cultures of hippocampal neurons were prepared from embryonic day 18 rat embryos (Charles River Laboratories Italia SrL, Calco, Italy), essentially as previously

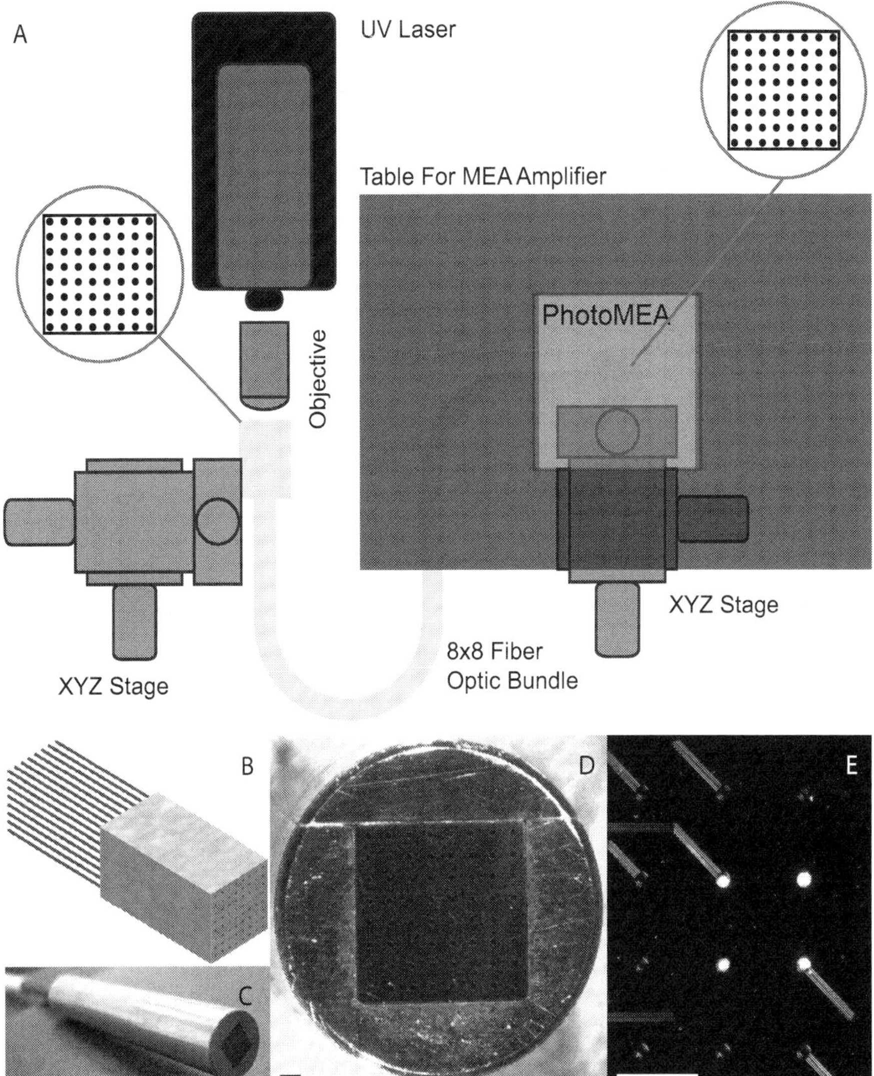

Fig. 2. (**A**) Scheme of the PhotoMEA platform set-up. (**B**) Drawn of the fiber arrangement in the bundle bundle. (**C**) Picture of the bundle head. (**D**) Magnified picture of the bundle head. (**E**) Fiber bundle aligned with the PhotoMEA biochip.

described (Kaech & Banker, 2006). Some modifications were introduced to adapt the method to the PhotoMEA biochip (Ghezzi et al., 2008).
Rat hippocampal neurons can be cultured over the MEA and PhotoMEA biochip for up to several weeks, making large neuronal network characterized by dense synaptic

interconnections and huge spontaneous electrical activity detected at MEA electrode sites. Using conventional transparent MEAs, images of neurons can be easily acquired either

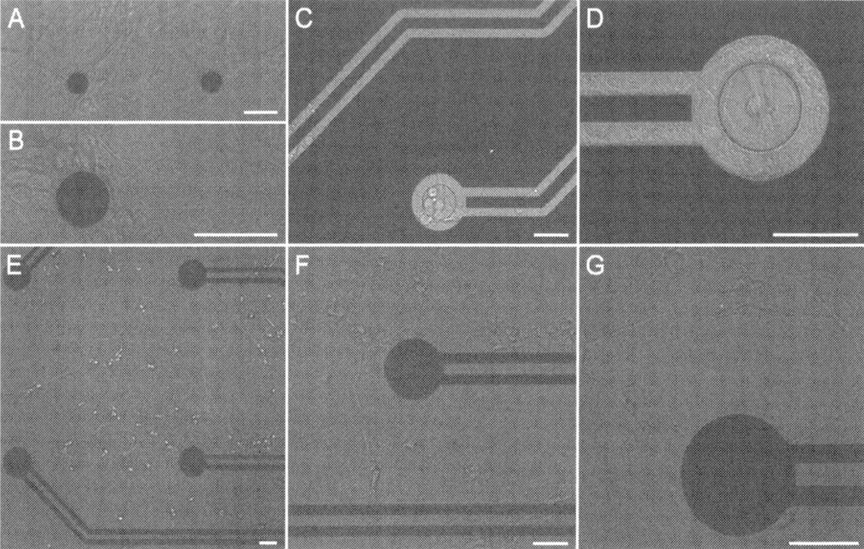

Fig. 3. **(A)** Transmitted light microscopy picture of neurons cultured on a commercially available ThinMEA. **(B)** Enlargement of an electrode site of the ThinMEA. **(C)** Transmitted light microscopy picture of neurons cultured on the PhotoMEA biochip. **(D)** Enlargement of an electrode site of the PhotoMEA. **(E-G)** Reflected light microscopy image of a portion of the PhotoMEA biochip covered with hippocampal neurons at different magnifications. Scale bar is 50 μm.

using inverted or upright microscope in transmitted light microscopy (Fig. 3A,B). Images of neuronal cultures were taken by an Axiovert 200 inverted epifluorescence microscope (Carl Zeiss SpA, Arese, Italy) positioned over an anti-vibration table and equipped with a 20x/0.8NA Plan-Apochromat short distance objective lens, a 40x/1.3NA EC Plan-Neofluar oil immersion objective lens and a an ORCAII CCD camera (Hamamatsu Photonics Italia SrL, Arese, Italy).

On the contrary, the titanium mask of the PhotoMEA biochip hampers the observation of the entire network in transmitted light microscopy. In fact, only neurons at the electrode sites (not covered by the titanium mask) are clearly visible (Fig. 3C,D).

Working with non-transparent substrates, neurons cultured on top of them can be observed using an upright microscope in reflected light mode. Images of neuronal cultures were taken by a FN1 upright microscope (Nikon Instruments SpA, Calenzano, Italy) positioned over an anti-vibration table and equipped with a 4x/0.1NA and a 10x/0.25NA long distance objective lenses, a Brightfield filter (Chroma Technology Corporation, Rockingham, USA) and a an ImagEM CCD camera (Hamamatsu Photonics Italia SrL). This method allows us to observe the entire network cultured covering the PhotoMEA chip with the exception of the transparent electrode sites (Fig. 3E,G). In conclusion, the titanium mask does not block the possibility to observe the development and the vitality of the neurons in culture.

3.4 Electrical properties

To completely characterize our fabricated PhotoMEA chips, we measured characteristic 1 kHz impedances of all electrodes. Mean measured electrode impedance is 1015 kΩ ± 112 kΩ, with a minimum value of 780 kΩ and a maximum of 1420 kΩ. Moreover, electrical recordings were performed in the culturing medium at 37 °C using the MEA1060 system (Multi Channel Systems MCS GmbH). Data recorded at 25 kHz/ch from the 60 channels were then filtered from 10 Hz to 3 kHz and spikes were sorted using a threshold algorithm included in the MC Rack software (Multi Channel Systems MCS GmbH). The threshold was defined as a multiple of the standard deviation of the biological noise computed during the first 500 ms of the recording (-5 * SD$_{noise}$). PhotoMEA electrodes showed a noise level appropriate to spike detection (Fig. 4) during recordings.

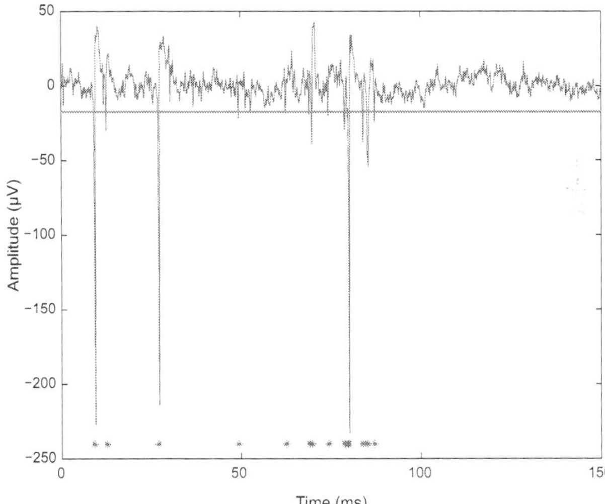

Fig. 4. Neuronal spikes can be easily detected with PhotoMEA electrodes, where root-mean square basal noise was measured as ±5.4 μV in a trace of 1 s without spiking activity. Threshold for detection was fixed to -17.5 μV by the software. Markers highlight the detected spikes after band-pass filtering.

3.5 Electrical stimulation

Hippocampal neurons (18 DIV) cultured on PhotoMEA biochips were electrically stimulated in order to illustrate the electrical stimulation disadvantages using conventional MEA technology. Biphasic, positive then negative, voltage pulses (amplitude of ±100 mV and pulse-width of 100 μs/phase) were applied to the neuronal network through one electrode of an PhotoMEA biochip (Fig. 5B). The electrical recording performed by the MEA system on all biochip electrodes shows that the stimulus spreads to the entire area of the culture, in spite of the large electrode spacing (Fig. 1). It results that the whole neuronal network could be electrically stimulated with an amplitude decreasing with the square of the distance from the stimulation site, affecting data quality as it is not known if the evoked responses detected at other electrode sites (Fig. 5C) correspond to direct cell stimulation due to the electrical stimulus (the responses follow the electrical artefact) or to network activity (i.e. signals that were propagated within the cell culture by synaptic transmission). Moreover, often the stimulated electrode remains not available for a long time after the stimulus application.

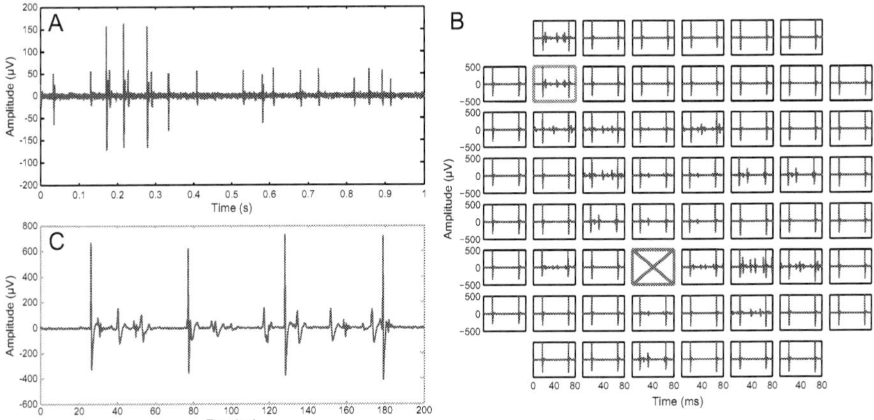

Fig. 5. (**A**) Recorded trace from one electrode of the PhotoMEA biochip showing a period of spontaneous spiking activity. (**B**) Applying a biphasic, positive then negative, voltage pulse (amplitude ±100 mV and pulse duration 100 μs/phase) to electrode 46 (red cross) the stimulus artefact is recorded by all electrodes of the biochip. Evoked responses can be found at several electrodes of the recording space. (**C**) High resolution trace recorded at an electrode (green box in **B**) far from the recording site (red cross), when a train of four pulses is applied. The trace shows electrically evoked spikes directly coupled to the electrical artifacts.

3.6 Optical stimulation

The optical stimulation approach was first evaluated in its ability to stimulate a small region surrounding a recording electrode and then in its efficiency in stimulating neurons.

In order to demonstrate the compound uncaging principle in a small volume at an electrode location, optical pulses with different pulse duration were delivered to a fluorescent caged compound (CNB-caged fluorescein, Invitrogen SrL, Milano, Italy). Fluorescence images were taken with an MZ16F stereomicroscope (Leica Microsystems, Wetzlar, Germany) equipped with a Moticam1000 CMOS camera (Motic, Xiamen, China), a x-cite120 metal halide fluorescence illuminator (EXFO, Quebec, Canada) and a Leica blue filter set (ex: BP470/40, em: LP515). The stereomicroscope was positioned over the PhotoMEA experimental setup.

The resulting fluorescence intensity due to compound uncaging increased with the light pulse duration, indicating that the amount of uncaged compound increased with the energy delivered to the sample (Fig. 6A,B). On the other hand, the stimulated area measured as Full Width at Half Maximum (FWHM) also increased along with the pulse duration, but it did not spread widely over the size of the hole in the metal mask, even for long stimuli (Fig. 6C). The final experiment was aimed at demonstrating that neuronal activity can be locally evoked using the PhotoMEA platform. A UV light pulse was applied to cultured hippocampal neurons at 14 DIV (Fig. 7A) in the presence of MNI-caged-L-glutamate (Tocris Bioscience, Bristol, UK) at a concentration of 100 μM. When neurons were stimulated with

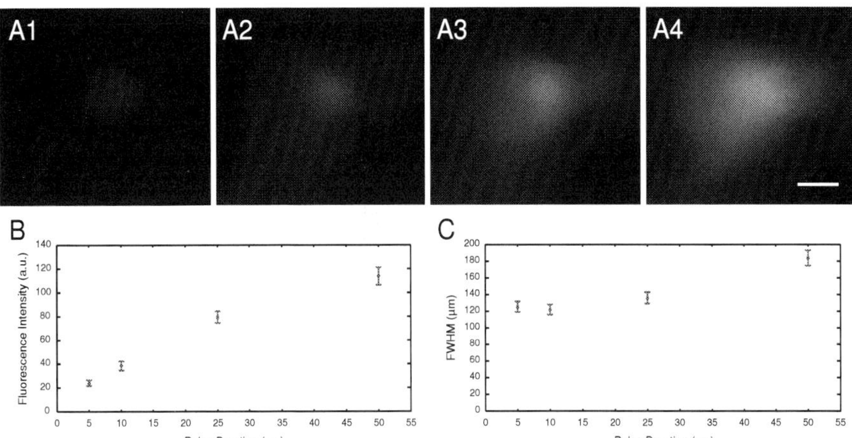

Fig. 6. (A) Pictures of the CMNB-caged fluorescein dissolved in glycerol at a final concentration of 100 μM activated by four optical pulses differing in their pulse duration: (A1) 5 ms, (A2) 10 ms, (A3) 25 ms and (A4) 50 ms. Measured maximum fluorescence intensity (B) and FWHM (C) of uncaged CMNB-caged fluorescein obtained by UV-light pulses with the same pulse durations as in A. The mean values ś standard deviations of the computed values are reported for every pulse duration (n = 5). Scale bars are 100 μm.

a UV pulse of 15 ms, evoked spikes were electrically detected at the stimulation electrode (Fig. 7B). Unfortunately, during the pulse, an interaction between the UV light and the stimulated electrode site was found. Similarly to what happens with electrical stimulation, the stimulated electrode site presented a stimulus artefact due to the optical pulse, which was not found on all other electrodes of the PhotoMEA biochip (Fig. 7C). These artifact seem to be related to the energy transferred by the UV light pulse as their duration is approximately twice the duration of the light pulse and increases with increasing pulse duration (mean $\pm$ standard deviation for n = 20 subsequent stimulation repeated for all pulse durations; 8.1 ms $\pm$ 0.59 ms at 5 ms, 18.8 ms $\pm$ 0.73 ms at 10 ms, 28.4 ms $\pm$ 0.62 ms at 15 ms, 35.2 ms $\pm$ 0.36 ms at 20 ms, 41.8 ms $\pm$ 0.61 ms at 25 ms, 1414.5 ms $\pm$ 215.16 ms at 50 ms and 2819.89 ms $\pm$ 274.53 ms at 100 ms). The physical nature of these artifacts and their possible influence on the neuronal activity are currently under investigation. However, the optical stimulation was found to work at the stimulated electrode site as evoked biological responses followed the chemical stimulation (Fig. 7B). At some other electrode sites, spontaneous and/or evoked activity appearing with a large delay and probably in response to a plastic effect of the network linked to the chemical stimulation were detected (Fig. 7C). We exclude that the activity at the other electrode sites can be evoked because of either direct local uncaging or diffusion of the glutamate. In fact, it has been demonstrated that the uncaging is localized to the close surrounding of the stimulated electrode (Fig. 6). Moreover, based on previous evaluation of the diffusion rate of the glutamate (Ghezzi et al., 2008), we can exclude that free glutamate affects electrodes far from the stimulation site.

As shown by a temporal representation of the network activity after stimulus (Fig. 7D), the activity is initially evoked at the stimulated site and after few tens of milliseconds it spreads

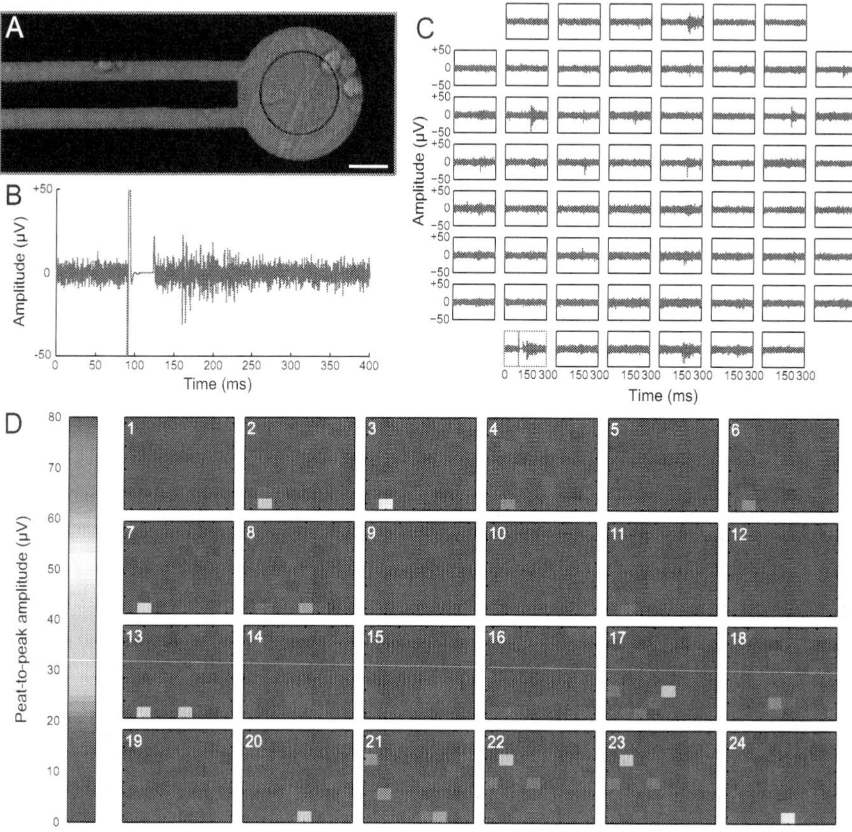

Fig. 7. (**A**) Picture of the optically stimulated neurons close to the ITO electrode of the PhotoMEA biochip. (**B**) Activity recorded at the stimulated site after an optical pulse of 15 ms. The artefact induced by the optical pulse and the following biological activity evoked by the glutamate uncaging is shown. (**C**) Activity recorded from the entire network after the UV pulse. The red box highlights the stimulated site. (**D**) Graphical representation of electrical activity spreading in the network after the optical pulse. Every frame, acquired with a sample rate of 1 kHz, represents a measure of the activity at every site. The color map represents a color representation of the peak-to-peak amplitude at every recorded site. Scale bar is 25 μm.

to other regions of the culture, thus revealing the interconnection between the different parts of the network. This also supports our conclusion concerning the localization of the stimulus.

4. Conclusion

Electrical stimulation on MEA presents certain experimental limitations, as it is difficult to prevent the electrical stimulus from spreading over the whole culture. Thus, the induction of evoked responses within the whole cell culture masks functional and network characteristics, and makes difficult the proper evaluation of signal propagation. The same problem arises for chemical stimulation, as the chemical compound spreads throughout the culture medium, thereby also limiting the results to proper network behavioural observations. The goal of the development of the PhotoMEA platform was to generate a tool and method that would allow local chemical stimulation, in order to stimulate only a small portion of the biological preparation. It was expected that such device would facilitate the acquisition of more precise information about functional processes within complex biological networks.

In the current work, through the use of the novel PhotoMEA biochips combined with UV-light pulse stimulation, local chemical compound uncaging and hence local chemical stimulation was successfully achieved. The optical stimulation performed through the PhotoMEA platform limits the activation of the stimulus only to the area surrounding the electrodes, thus allowing the possibility to have a better defined study of information processing in neuronal networks with several independent inputs and outputs. Beside caged neurotransmitters, virtually every kind of signalling molecule or second messenger has already been caged, from protons to proteins, including also inositols, nucleotides, peptides, enzymes, mRNA and DNA (Ellis-Davies, 2007). This considerably widens the scope and potential impact of the PhotoMEA tool in cell signalling, systems biology and complex biological cultures, and makes it also amenable to use with non-neuronal cultures. In the field of pharmacology, the features of the PhotoMEA platform improve the possibility to create in a more controlled manner a spatial map of the drug effect's on the tissue preparation, in order to improve the evaluation of the drug's specificity, a result that cannot be easily achieved using conventional methods for the drug application, e.g. pipetting.

An important feature is that the PhotoMEA technology can be readily scaled up for higher throughput applications, and thus may provide opportunities in drug screening applications, especially for central nervous system (CNS) disorders. The CNS drug discovery industry currently has several high throughput tools (e.g. planar patch-clamp) for monitoring and testing drugs on single isolated cells. However, there are no suitable tools and methods, especially high-throughput, to evaluate drug activity on synaptic biology, i.e. at the network level (Dunlop et al., 2008) and on real neurons, thereby presenting an excellent opportunity for PhotoMEA tools in CNS drug screening community. In addition to caged compounds, the PhotoMEA system is also expected to rise considerable interest for applications using photosensitive tissue preparations, such as retinal tissue. There are already several studies that have used MEAs with retinal explants for electrical recording and stimulation (Puchalla et al., 2005; Segev et al., 2004). The combination of the standard MEA electrical recording feature with the PhotoMEAA capability to optically uncage a signalling molecule and/or optically stimulate light-sensitive retinal neurons, promise to provide an unparalleled information-rich paradigm for investigating the complex information processes that take place in the mammalian retina.

5. References

Berdondini, L., Chiappalone, M., van der Wal, P., Imfeld, K., de Rooij, N., Koudelka-Hep, M., Tedesco, M., Martinoia, S., van Pelt, J., Masson, G. L. & Garenne, A. (2006). A microelectrode array (mea) integrated with clustering structures for investigating in vitro neurodynamics in confined interconnected sub-populations of neurons., *Sensors and Actuators B* 114: 530–541.

Bernardinelli, Y., Haeberli, C. & Chatton, J.-Y. (2005). Flash photolysis using a light emitting diode: an efficient, compact, and affordable solution., *Cell Calcium* 37(6): 565–572.

Callaway, E. M. & Katz, L. C. (1993). Photostimulation using caged glutamate reveals functional circuitry in living brain slices., *Proc Natl Acad Sci U S A* 90(16): 7661–7665.

Callaway, E. M. & Yuste, R. (2002). Stimulating neurons with light., *Curr Opin Neurobiol* 12(5): 587–592.

Darbon, P., Scicluna, L., Tscherter, A. & Streit, J. (2002). Mechanisms controlling bursting activity induced by disinhibition in spinal cord networks., *Eur J Neurosci* 15(4): 671–683.

DeBusschere, B. D. & Kovacs, G. T. (2001). Portable cell-based biosensor system using integrated cmos cell-cartridges., *Biosens Bioelectron* 16(7-8): 543–556.

Dunlop, J., Bowlby, M., Peri, R., Vasilyev, D. & Arias, R. (2008). High-throughput electrophysiology: an emerging paradigm for ion-channel screening and physiology., *Nat Rev Drug Discov* 7(4): 358–368.

Egert, U., Heck, D. & Aertsen, A. (2002). Two-dimensional monitoring of spiking networks in acute brain slices., *Exp Brain Res* 142(2): 268–274.

Egert, U., Schlosshauer, B., Fennrich, S., Nisch, W., Fejtl, M., Knott, T., Müller, T. & Hämmerle, H. (1998). A novel organotypic long-term culture of the rat hippocampus on substrate-integrated multielectrode arrays., *Brain Res Brain Res Protoc* 2(4): 229–242.

Ellis-Davies, G. C. R. (2007). Caged compounds: photorelease technology for control of cellular chemistry and physiology., *Nat Methods* 4(8): 619–628.

Erickson, J., Tooker, A., Tai, Y.-C. & Pine, J. (2008). Caged neuron mea: a system for long-term investigation of cultured neural network connectivity., *J Neurosci Methods* 175(1): 1–16.

Farber, I. C. & Grinvald, A. (1983). Identification of presynaptic neurons by laser photostimulation., *Science* 222(4627): 1025–1027.

Fork, R. L. (1971). Laser stimulation of nerve cells in aplysia., *Science* 171(974): 907–908.

Ghezzi, D., Menegon, A., Pedrocchi, A., Valtorta, F. & Ferrigno, G. (2008). A micro-electrode array device coupled to a laser-based system for the local stimulation of neurons by optical release of glutamate., *J Neurosci Methods* 175(1): 70–78.

Ghezzi, D., Pedrocchi, A., Menegon, A., Mantero, S., Valtorta, F. & Ferrigno, G. (2007). Photomea: an opto-electronic biosensor for monitoring in vitro neuronal network activity., *Biosystems* 87(2-3): 150–155.

Gilchrist, K. H., Barker, V. N., Fletcher, L. E., DeBusschere, B. D., Ghanouni, P., Giovangrandi, L. & Kovacs, G. T. (2001). General purpose, field-portable cell-based biosensor platform., *Biosens Bioelectron* 16(7-8): 557–564.

Gross, G. W. (1979). Simultaneous single unit recording in vitro with a photoetched laser deinsulated gold multimicroelectrode surface., *IEEE Trans Biomed Eng* 26(5): 273–279.

Gross, G. W., Harsch, A., Rhoades, B. K. & Göpel, W. (1997). Odor, drug and toxin analysis with neuronal networks in vitro: extracellular array recording of network responses., *Biosens Bioelectron* 12(5): 373–393.

Gross, G. W., Rhoades, B. K., Azzazy, H. M. & Wu, M. C. (1995). The use of neuronal networks on multielectrode arrays as biosensors., *Biosens Bioelectron* 10(6-7): 553–567.

Heuschkel, M. O., Fejtl, M., Raggenbass, M., Bertrand, D. & Renaud, P. (2002). A three-dimensional multi-electrode array for multi-site stimulation and recording in acute brain slices., *J Neurosci Methods* 114(2): 135–148.

Hirase, H., Nikolenko, V., Goldberg, J. H. & Yuste, R. (2002). Multiphoton stimulation of neurons., *J Neurobiol* 51(3): 237–247.

Hofmann, F., Guenther, E., Hämmerle, H., Leibrock, C., Berezin, V., Bock, E. & Volkmer, H. (2004). Functional re-establishment of the perforant pathway in organotypic co-cultures on microelectrode arrays., *Brain Res* 1017(1-2): 184–196.

Jimbo, Y., Kasai, N., Torimitsu, K., Tateno, T. & Robinson, H. P. C. (2003). A system for mea-based multisite stimulation., *IEEE Trans Biomed Eng* 50(2): 241–248.

Kaech, S. & Banker, G. (2006). Culturing hippocampal neurons., *Nat Protoc* 1(5): 2406–2415.

Kopanitsa, M. V., Afinowi, N. O. & Grant, S. G. N. (2006). Recording long-term potentiation of synaptic transmission by three-dimensional multi-electrode arrays., *BMC Neurosci* 7: 61.

Kraus, T., Verpoorte, E., Linder, V., Franks, W., Hierlemann, A., Heer, F., Hafizovic, S., Fujii, T., de Rooij, N. F. & Koster, S. (2006). Characterization of a microfluidic dispensing system for localised stimulation of cellular networks., *Lab Chip* 6(2): 218–229.

Legrand, J.-C., Darbon, P. & Streit, J. (2004). Contributions of nmda receptors to network recruitment and rhythm generation in spinal cord cultures., *Eur J Neurosci* 19(3): 521–532.

Lutz, C., Otis, T. S., DeSars, V., Charpak, S., DiGregorio, D. A. & Emiliani, V. (2008). Holographic photolysis of caged neurotransmitters., *Nat Methods* 5(9): 821–827.

Marom, S. & Shahaf, G. (2002). Development, learning and memory in large random networks of cortical neurons: lessons beyond anatomy., *Q Rev Biophys* 35(1): 63–87.

Morin, F. O., Takamura, Y. & Tamiya, E. (2005). Investigating neuronal activity with planar microelectrode arrays: achievements and new perspectives., *J Biosci Bioeng* 100(2): 131–143.

Natarajan, A., Molnar, P., Sieverdes, K., Jamshidi, A. & Hickman, J. J. (2006). Microelectrode array recordings of cardiac action potentials as a high throughput method to evaluate pesticide toxicity., *Toxicol In Vitro* 20(3): 375–381.

Nerbonne, J. M. (1996). Caged compounds: tools for illuminating neuronal responses and connections., *Curr Opin Neurobiol* 6(3): 379–386.

Pancrazio, J. J., Gray, S. A., Shubin, Y. S., Kulagina, N., Cuttino, D. S., Shaffer, K. M., Eisemann, K., Curran, A., Zim, B., Gross, G. W. & O'Shaughnessy, T. J. (2003). A portable microelectrode array recording system incorporating cultured neuronal networks for neurotoxin detection., *Biosens Bioelectron* 18(11): 1339–1347.

Pine, J. (1980). Recording action potentials from cultured neurons with extracellular microcircuit electrodes., *J Neurosci Methods* 2(1): 19–31.

Puchalla, J. L., Schneidman, E., Harris, R. A. & Berry, M. J. (2005). Redundancy in the population code of the retina., *Neuron* 46(3): 493–504.

Reppel, M., Igelmund, P., Egert, U., Juchelka, F., Hescheler, J. & Drobinskaya, I. (2007). Effect of cardioactive drugs on action potential generation and propagation in embryonic stem cell-derived cardiomyocytes., *Cell Physiol Biochem* 19(5-6): 213–224.

Segev, R., Goodhouse, J., Puchalla, J. & Berry, M. J. (2004). Recording spikes from a large fraction of the ganglion cells in a retinal patch., *Nat Neurosci* 7(10): 1154–1161.

Shoham, S., O'Connor, D. H., Sarkisov, D. V. & Wang, S. S.-H. (2005). Rapid neurotransmitter uncaging in spatially defined patterns., *Nat Methods* 2(11): 837–843.

Steidl, E.-M., Neveu, E., Bertrand, D. & Buisson, B. (2006). The adult rat hippocampal slice revisited with multi-electrode arrays., *Brain Res* 1096(1): 70–84.

Stett, A., Egert, U., Guenther, E., Hofmann, F., Meyer, T., Nisch, W. & Haemmerle, H. (2003). Biological application of microelectrode arrays in drug discovery and basic research., *Anal Bioanal Chem* 377(3): 486–495.

Takayama, S., Ostuni, E., LeDuc, P., Naruse, K., Ingber, D. E. & Whitesides, G. M. (2003). Selective chemical treatment of cellular microdomains using multiple laminar streams., *Chem Biol* 10(2): 123–130.

Thomas, C. A., Springer, P. A., Loeb, G. E., Berwald-Netter, Y. & Okun, L. M. (1972). A miniature microelectrode array to monitor the bioelectric activity of cultured cells., *Exp Cell Res* 74(1): 61–66.

Tscherter, A., Heuschkel, M. O., Renaud, P. & Streit, J. (2001). Spatiotemporal characterization of rhythmic activity in rat spinal cord slice cultures., *Eur J Neurosci* 14(2): 179–190.

Venkataramani, S., Davitt, K. M., Xu, H., Zhang, J., Song, Y.-K., Connors, B. W. & Nurmikko, A. V. (2007). Semiconductor ultra-violet light-emitting diodes for flash photolysis., *J Neurosci Methods* 160(1): 5–9.

Wagenaar, D. A. & Potter, S. M. (2002). Real-time multi-channel stimulus artifact suppression by local curve fitting., *J Neurosci Methods* 120(2): 113–120.

Wieboldt, R., Gee, K. R., Niu, L., Ramesh, D., Carpenter, B. K. & Hess, G. P. (1994). Photolabile precursors of glutamate: synthesis, photochemical properties, and activation of glutamate receptors on a microsecond time scale., *Proc Natl Acad Sci U S A* 91(19): 8752–8756.

Wirth, C. & Lüscher, H.-R. (2004). Spatiotemporal evolution of excitation and inhibition in the rat barrel cortex investigated with multielectrode arrays., *J Neurophysiol* 91(4): 1635–1647.

Xiang, G., Pan, L., Huang, L., Yu, Z., Song, X., Cheng, J., Xing, W. & Zhou, Y. (2007). Microelectrode array-based system for neuropharmacological applications with cortical neurons cultured in vitro., *Biosens Bioelectron* 22(11): 2478–2484.

Zemelman, B. V., Lee, G. A., Ng, M. & Miesenböck, G. (2002). Selective photostimulation of genetically charged neurons., *Neuron* 33(1): 15–22.

8

Detection of Optical Radiation in NO$_x$ Optoelectronic Sensors Employing Cavity Enhanced Absorption Spectroscopy

Jacek Wojtas
Military University of Technology
Poland

1. Introduction

Currently there are two main reasons for seeking new methods and technologies that aim to develop new and more perfect sensors detecting various chemical compounds. The first reason is man's striving for an ever better understanding of the surrounding world and the universe. Second, sensors are used to ensure safety, e.g. in the vicinity of factories, in an important objects like airports, in environmental protection, health care, etc. These applications have a significant impact on the performance of sensors. This chapter addresses the issue of some nitrogen oxides (NO$_x$) sensor designs using some of the most sensitive methods such as cavity enhanced absorption spectroscopy (CEAS) and cavity ring down spectroscopy (CRDS).

Nitrogen oxides are compounds of nitrogen and oxygen. For example, among them very important are nitric oxide (NO), nitrogen dioxide (NO$_2$) and nitrous oxide (N$_2$O). According to the HITRAN database, in standard atmosphere[1] their concentration is as follows: NO – about 0.3 ppbv[2], NO$_2$ – about 0,023 ppbv, N$_2$O – about 320 ppbv. However, in real ambient air their concentrations are strongly related to meteorological conditions and emission sources (anthropogenic and natural). They are compounds that play a significant role in many different fields. They are important greenhouse gases, and their reactions with H$_2$O (water) lead to acid rains. For example, nitrous oxide is used as an anaesthetic, especially in dentistry and minor surgery. It produces mild hysteria and laughter. Thus it is also known as 'laughing gas'. Atmospheric photochemistry induces a complicated conversion mechanism between nitrogen oxides [Godish, 2004]. Moreover, NO, NO$_2$ and N$_2$O are also characteristic decomposition compounds which are the main products of specific explosives materials. Many of them contain NO$_2$ groups, which can be detected using spectroscopic detection methods [Moore, 2007].

There are many methods for NO$_x$ detection. For example, in the case of gas chromatography and mass spectrometry, a detection limit of a few dozen ppb is reported

---

[1] Based on US Standard Atmosphere 1976 from HITRAN database

[2] In the air pollution monitoring, concentration of the substance is often expressed in units of *ppmv* (parts per one million by volume) or *ppbv* (parts per one billion by volume). It specifies the number of molecules of the absorber for all the molecules present in a given volume.

[Shimadzu Scientific Instruments, Drescher & Brown, 2006]. Detection methods using the photoacoustic phenomenon provide a sensitivity of about 20 ppb [Grossel et al., 2007]. In gas detection applications, a special role is played by optoelectronic methods. CRDS and CEAS methods belong to optoelectronic absorption methods, but there are a lot of other optoelectronic methods for the detection of gases and hazardous substances. In Fig. 1 the most popular are shown. All of them have advantages and disadvantages. However, in respect of the specific properties are all widely used. Due to the theme of this chapter the absorption methods will be discussed in more detail.

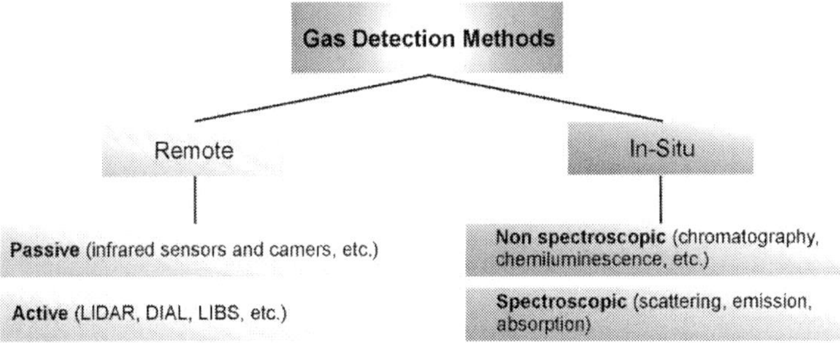

Fig. 1. Popular gas detection methods

In passive methods, an optical radiation emitted by a thermal object is registered. These systems are widely used in infrared cameras and infrared sensors. Such a solution does not require often very costly, radiation sources.

Active methods are more frequently used in remote (standoff) gas detection application. The most commonly are LIDAR's (LIght Detection And Ranging). Typical LIDAR consists of a transmitter that emits laser pulses and an optical radiation receiver. Laser pulses after being scattered in the clouds, aerosols or dust are registered with a photoreceiver. Next, the signal is processed in a digital processing unit. Such a system is used to monitor rainfall, clouds and smoke emerging from chimneys or for the detection of gaseous pollutants of the atmosphere [Mierczyk et al., 2008, Karasinski et al., 2007]. DIAL (Differential Absorption LIDAR) is based on laser radiation measurement at a peak of absorption and at a trough. Thus, a differential signal is received, which is used to determine concentration profiles and mass emissions of various species [Chudzynski et al., 1999]. Laser-induced breakdown spectroscopy (LIBS) uses a highly energetic laser pulses as the excitation source to form plasma, which atomizes and excites samples. The emission from the plasma plume is registered and analyzed with the detection system [Owsik & Janucki, 2004].

The group of examination methods that are used exactly in the place of occurring gas (in-situ) includes non-spectroscopic and spectroscopic methods. These are widely used chemical methods, which belong to the former. They are based on the use of certain chemical reactions, which may indicate the presence of the substance sought [Sigrist, 1994]. The latter are very popular. They can be divided into scattering, emission and absorption methods. In the first, radiation scattered with the sample is examined [Li et al., 2005]. In the second, a probing radiation beam causes the sample excitation. Next, the detection system registers and analyzes the spectrum emitted with the sample. In the third type of spectroscopic method, the absorbed radiation is analyzed. In all spectroscopic methods the properties of the sample are determined on the basis of the measured spectral characteristics of radiation [Lagalante, 1999].

Absorption spectra can be defined as the set of all electron crossings from lower energy levels to higher ones. They cause an increase in molecules energy. In case of the emission spectra there is inverse situation. The spectra correspond to the reduction of molecules energy as a result of electrons transitions from higher energy levels to lower ones. Scattering spectra rely on a change in the frequency spectra diffuse radiation in relation to the frequency of incident radiation, due to the partial change of the photon energy as a result of impact with the molecules. However, in this case there is no effect of radiation absorption or emission [Saleh & Teich, 2007, Sigrist 1994].

2. Principles of absorption spectroscopy

Each gas molecule has a very characteristic arrangement of electron energy levels (vibrational and rotational). As a result of light absorption, particles go to one of the excited states and then in various ways lose energy. Absorption spectroscopy refers to spectroscopic techniques that measure the absorption of radiation, as a function of wavelength, due to its interaction with a sample. The sample absorbs energy, i.e., photons, from the radiating field. The intensity of the absorption varies as a function of wavelength and this variation is the absorption spectrum [Sigrist, 1994]. Absorption spectroscopy is performed across the electromagnetic spectrum. A source of radiation and very sensitive photoreceiver is used which records radiation passing through the absorber sample. During the last several years absorptions methods for gas detection were significantly developed. The simple setup, which shows the idea of absorption method, is presented in Fig. 2.

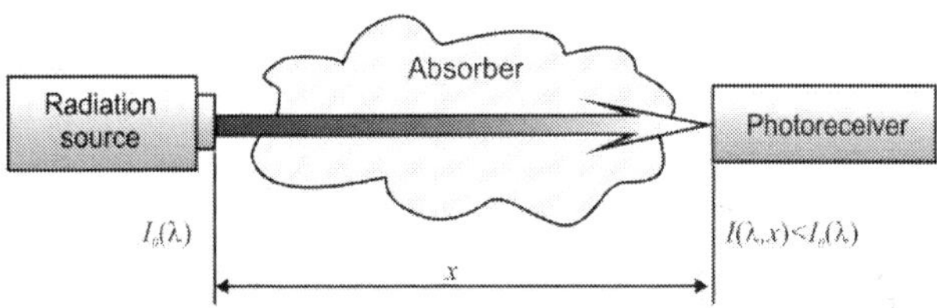

Fig. 2. The absorption method idea.

An arc lamp, LED (Light Emitting Diode) or laser emitting a wavelength matched to the absorption lines of the test gas could be applied as the source of radiation. If an absorber is placed between the source and photoreceiver, the intensity of radiation is weakened. The type and concentration of the test absorber can be inferred on this basis. The intensity of radiation registered with the photoreceiver can be determined using the Lambert-Beer law

$$I(\lambda,x) = I_0(\lambda)\exp(-x\sigma(\lambda)C),\tag{1}$$

where $I_0(\lambda)$ is the intensity of radiation emitted by the source, x is the path of light in the absorber, C - concentration of the investigated gas, while $\sigma(\lambda)$ is the absorption cross section. The cross section is the characteristic parameter of the gas and it can be determined during the laboratory experiment. Knowledge regarding the intensity of radiation emitted from the source, the intensity of received radiation, the absorption cross section and the distance x, provides the possibility of gas concentration calculation from the formula

$$C = \log\left(\frac{I_0(\lambda)}{I(\lambda,x)}\right) \cdot \left(\sigma(\lambda) \cdot x\right)^{-1} . \qquad (2)$$

One of the most common gas detection systems is differential optical absorption spectroscopy (DOAS). The first system was applied by Ulrich Platt in the 1970's. Currently, similar arrangements are applied to the monitoring of atmospheric pollutants, including the detection of NO_x, in terrestrial applications, in air and in the space, e.g. GOME and SCIAMACHY satellite. Sensitivity of the method depends on the distance between the radiation source and the photoreceiver. For systems where this distance is a few kilometres, the sensitivity of the DOAS method is better than 1 ppb in the case of NO_2 detection [Martin et al., 2004, Wang et al., 2005, Noel et al., 1999].

In order to lengthen the optical path and to improve the sensitivity of absorption methods, reflective multipass cells are used, e.g. in tuneable diode laser absorption spectroscopy (TDLAS). This method is characterized by high sensitivity. Applications cells with lengths of a few dozen meters provide the possibility to achieve a sensitivity of 1 ppb and higher [Jean-Franqois et al., 1999, Horii et al., 1999].

There are many differ concepts applied to gas detection and identification. However, optoelectronic methods enable a direct and selective measurement of concentration on the level of a single ppb.

3. Idea of the CRDS and other cavity enhanced methods

Cavity ring down spectroscopy for the first time was applied to determine the reflectivity mirrors by J.M. Herbelin [Herbelin et al., 1980]. CRDS provides a much higher sensitivity than conventional absorption spectroscopy. The idea of the CRDS method is shown in Fig. 3. In this method there is applied an optical cavity with a high quality factor that is made up of two concave mirrors with very high reflectivity R. This results in a long optical path, even up to several kilometres [Busch & Busch, 1999].

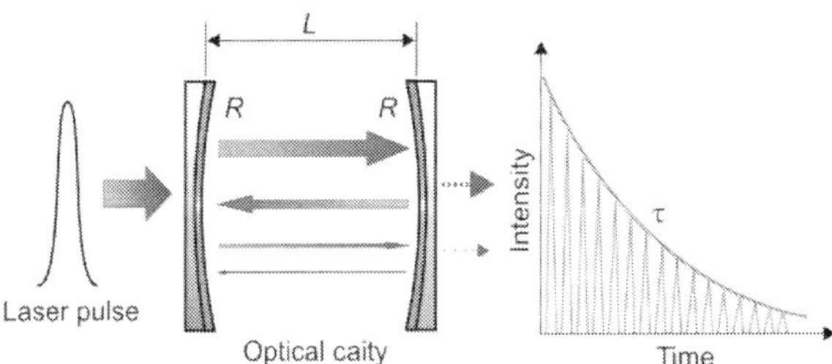

Fig. 3. Cavity ring down spectroscopy idea.

A pulse of optical radiation is injected into the cavity through one of the mirrors. Then inside the cavity multiple reflections occur. After each reflection, part of the radiation exiting from the cavity is registered with the photodetector. The output signal from the photodetector is proportional to the intensity of radiation propagated inside the optical cavity. If the laser wavelength is matched to the absorption spectra of gas filling the cavity, the cavity quality decreases. Thus, parameters of the signal from the photodetector are changed. Thanks to this, the absorption coefficient and concentration of gas can be determined. The methods of their determination will be discussed in a subsequent section.

3.1 Characteristics of common cavity enhanced systems

Currently there are used many types of cavity enhanced systems that are characterized by different technical constructions and properties. The literature shows that most of them use:

- P-CRDS method (called Pulsed), which uses pulsed lasers [O'Keefe & Deacon, 1988],
- CW-CRDS method (called Continuous Wave) applying continuous operation lasers [He & Orr, 2000],
- CEAS and ICOS (Integrated Cavity Output Spectroscopy) methods basis on off-axis arrangement of the radiation beam and optical cavity [Kasyutich et al., 2003a],
- cavity evanescent ring-down spectroscopy (EW-CRDS), which uses the evanescent wave phenomenon [Pipino, 1999],
- fibber-optic CRDS (F-CRDS) [Atherton et al., 2004],
- ring-down spectral photography (RSP) – a broadband spectroscopy of optical losses [Czyzewski et al., 2001, Stelmaszczyk et al., 2009, Scherer et al., 2001].

The greatest sensitivity of the method is characterized by P-CRDS, CW-CRDS and CEAS [Ye et al., 1997, Berden et al., 2000]. For this reason they are often used for detecting and measuring gas concentrations [Kasyutich et al., 2003b]. The P-CRDS method was first used in 1988 to measure the absorption coefficient of gas [O'Keefe & Deacon, 1988]. Typical schematic layout is shown in Fig. 4.

This method involves the use of a pulsed radiation source, characterized by a broad spectrum of the pulse. This leads to the excitation of multiple longitudinal of the resonance cavity, and also reduces the sensitivity. Sensitivity of the P-CRDS usually reaches values corresponding to the absorption coefficients of the order of 10^{-6} - 19^{-10} cm^{-1} [Busch & Busch, 1999].

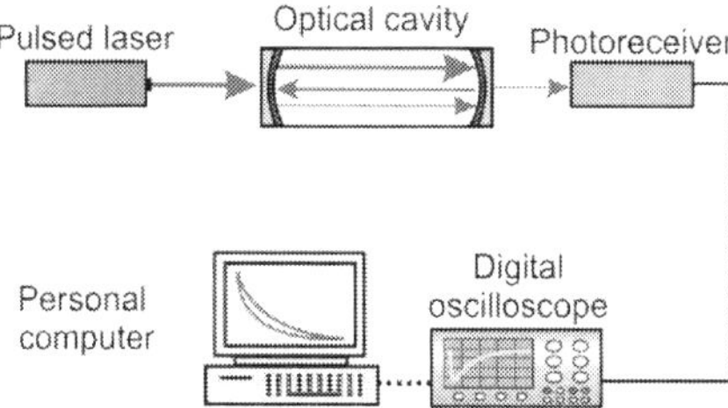

Fig. 4. Diagram of the P-CRDS setup.

CW-CRDS for gas detection has been used since 1997 [Romanini et al., 1997]. A simplified diagram of the experimental setup is shown in Fig. 5. The use of continuous operating lasers in the CRDS technique was possible through the use of different laser beam modulators (e.g. acusto-optic) [Berden et al., 2000]. Due to the narrow spectral lines available with these lasers, operation in a single longitudinal mode is possible in longer optical cavities. Thanks to this CW-CRDS has the highest sensitivity among the cavity enhanced methods. The extreme sensitivity of this method reaches the level of absorption coefficients of up to 10^{-14} cm^{-1}. Due to the high spectral resolution of CW-CRDS, the method is often used in absorption spectra measurements [Busch & Busch, 1999].

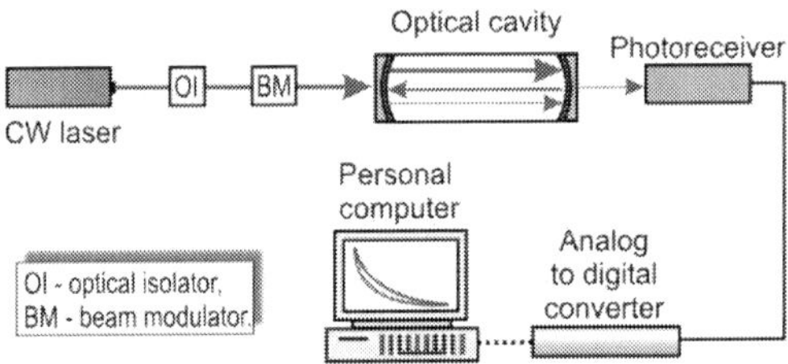

Fig. 5. Experimental CW-CRDS.

The main drawback of this method is the very high sensitivity of the mechanical instability. If the laser frequency is matched to the cavity mode, there is a very efficient storage of light (Fig. 6). However, fluctuations in the frequency of their own cavity, for example due to a change in its length due to mechanical vibrations, cause the optical resonance phenomenon to become impossible and it lead to high volatility of the output signal [Berden et al., 2000].

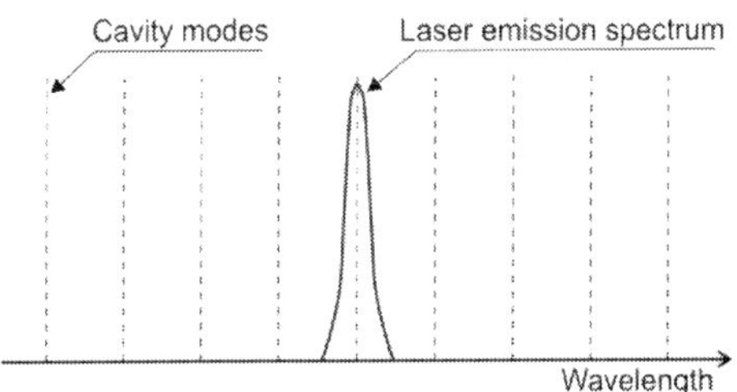

Fig. 6. Coupling of the modes structure of the cavity and cw type laser in the CW-CRDS.

In 1998, R. Engeln proposed a new method – cavity enhanced absorption spectroscopy (also called ICOS), whose principle of operation is very similar to CRDS. The main difference relates to a laser and the optical cavity alignment [Engeln et al., 1998]. In this technique the laser beam is injected at a very small angle in respect to the cavity axis (Fig. 7). As the result, a dense structure of weak modes is obtained or the modes do not occur due to overlapping. Sometimes, in addition to the output mirror, a piezoelectric-driven mount that modulates the cavity length is used in order to prevent the establishment of a constant mode structure within the cavity [Paul et al., 2001]. The weak mode structure causes that the entire system is much less sensitive to instability in the cavity and to instability in laser frequencies. Additionally, due to off-axis illumination of the front mirror, the source interference by the optical feedback from the cavity is eliminated. CEAS sensors attain a detection limit of about 10^{-9} cm^{-1} [Berden et al., 2000, Courtillot et al., 2006]. Therefore, this method creates the best opportunity to develop a portable optoelectronic sensor of nitrogen oxides.

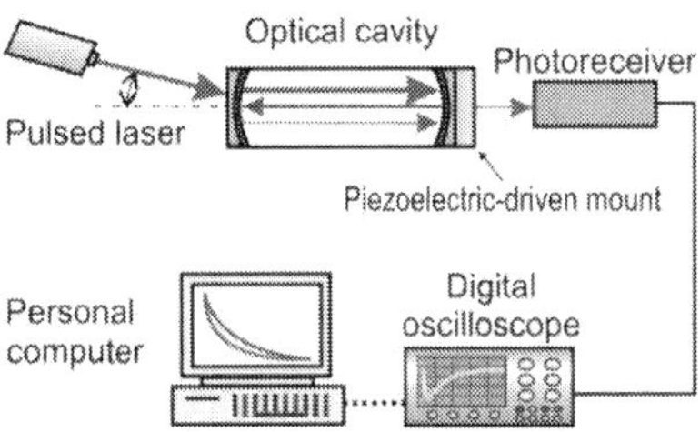

Fig. 7. The scheme of CEAS setup.

3.2 Methods for gas concentration determination used in cavity enhanced spectroscopy

In the methods described in the previous section, several methods are used to determine the gas concentration: by measuring the decay time of the signal, by measuring the phase shift and by measuring the signal amplitude [Busch & Busch, 1999, Berden et al., 2000, Wojtas et al., 2005].

If the laser pulse duration is negligibly short and only the main transverse mode of the cavity is excited, then exponential decay of radiation intensity can be observed

$$I(t) = I_0 \exp\left(-\frac{t}{\tau}\right). \tag{3}$$

If intrinsic cavity losses can be disregarded, the decay time of signal in the cavity (τ) depends on the reflectivity of mirrors R, diffraction losses and the extinction coefficient a, i.e. the scattering and absorption of radiation occurring in the gas filling the cavity

$$\tau = \frac{L}{c(1 - R + \alpha L)}, \tag{4}$$

where L is the length of the resonator, c - speed of light. Determination of the concentration of the examined gas is a two-step process. First, measurement of the signal decay time (τ_0) in the optical cavity not containing the absorber (tested gas) is performed (Fig. 8-A), and then measuring the signal decay time τ in the cavity filled with the tested gas is carried out (Fig. 8-B). Knowing the absorption cross section (σ) of the examined gas, its concentration can be calculated from the formula

$$C = \frac{1}{c\sigma}\left(\frac{1}{\tau} - \frac{1}{\tau_0}\right), \tag{5}$$

where

$$\tau_0 = \frac{L}{c(1 - R)}. \tag{6}$$

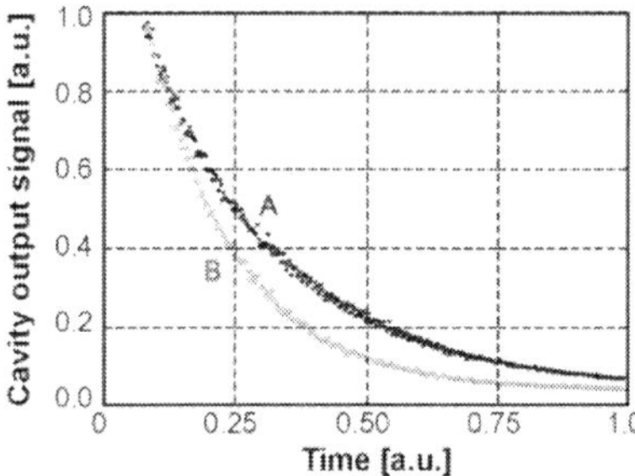

Fig. 8. Examples of signals at the output of the optical cavity without absorber (A) and at the output of the cavity filled with absorber (B)

Based on equation (4) and (5), the lowest concentration (concentration limit) of analyzed gas molecules (C_{lmt}), which causes a measurable change of the output signal, can be determined from the formula

$$C_{lmt} = \frac{1}{c\sigma\tau_0}\delta_\tau = \frac{(1-R)}{\sigma \cdot L}\delta_\tau \; , \tag{7}$$

where δ_τ is the relative precision of the decay time measurement (uncertainty). The relationship between uncertainty δ_τ and τ_0 can be described as

$$\delta_\tau = \frac{\tau_0 - \tau_{lmt}}{\tau_0} \cdot 100\% \; , \tag{8}$$

where τ_{lmt} denotes a decay time for minimal absorber concentration.

In the other hand, C_{lmt} can be treated as the detection limit of the sensor. It is a function of two variables: the decay time for the empty cavity (τ_0) and uncertainty (δ_τ). Furthermore, the decay time τ_0, according to the formula (6), depends on the length of the resonator and the reflectivity mirrors. The longer this time, the longer effective path of absorption, the greater the sensitivity of the sensor and the lower concentrations of the absorber can be measured.

Another way of gas concentration determination is measurements of the phase shift between the respective harmonics of the signal (e.g. the first) at the input and output optical cavity [Herbelin et al. 1980, Engeln et al. 1996]. In these measurements, lock-in amplifiers are frequently used. The phase shift occurs due to cavity ability to the energy (radiation) storage, as in the case of the charging process of the capacitor. The value of $tan(\varphi)$ is associated with the decay of radiation in the cavity dependence

$$tan(\varphi) = 4\pi f \tau \; , \tag{9}$$

where f denotes the modulation frequency. The gas concentration can be calculated by comparing the phase (φ) when the resonator is filled with test gas and the phase shift (φ_0) for the resonator without gas

$$C = \frac{4\pi f}{c\sigma}\left(\frac{1}{tg(\varphi)} - \frac{1}{tg(\varphi_0)} \right). \tag{10}$$

In techniques with an off-axis arrangement light source and optical cavity, the gas concentration is often determined by measuring the amplitude of the signal from the photodetector. Application of the system synchronization of laser and cavity modes is not required. It simplifies the experimental system. Thanks to this, the intensity from individual reflections of radiation from the output mirror can be summed [O'Keefe et al., 1999, O'Keefe, 1998]

$$I_{os} = -I_{in} \frac{(1-R)^2 e^{-\alpha L}}{2\ln(R \cdot e^{-\alpha L})}.$$

(11)

In the case of a single pass, the transmitted light pulse is described by

$$I_{op} = I_{in}(1-R)^2 e^{-\alpha L}.$$

(12)

Comparing expressions (11) and (12) it can be shown that for small absorption coefficients a and high reflectivity mirrors $(R \rightarrow 1)$ ratio of the I_{OS}/I_{OP} can be expressed with the formula

$$\frac{I_{os}}{I_{op}} = -\frac{1}{2[\ln(R)-\alpha L]} \rightarrow \frac{1}{2(1-R+\alpha L)},$$

(13)

thus

$$C = \frac{\ln(R)}{\sigma \cdot L} \frac{I_{os}-I_{op}}{I_{os}}.$$

(14)

An important drawback of this method is the necessity of knowledge of the mirrors reflectivity to determining the gas concentration. In practical realisations it is difficult to ensure.

4. NO$_x$ sensors project

Basic experimental setups of the cavity enhanced methods were described in the third section. All of them consist of pulse laser (or cw laser with modulator), beam directing and shaping system (mirrors, diaphragms, diffraction grating), optical cavity and photoreceiver with signal processing system (e.g. digital oscilloscope in the simplest case). First of all, the sensor project should take into account the appropriate matching cavity parameters and the laser emission wavelength to the test gas absorption spectrum (Fig. 9).

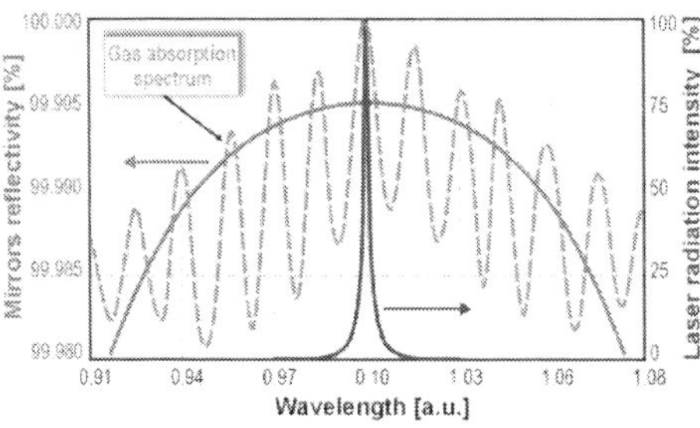

Fig. 9. Illustration of matching the laser emission wavelength and cavity mirrors transmission.

Moreover, it is necessary to apply adequate optical cavity, which provides repeatedly reflection of the laser radiation. To ensure multiple reflections, the cavity must be stable, i.e. the light after reflection from the mirrors must be re-focused (Fig. 10.a). In the case of an unstable cavity, the laser beam after a few reflections leaves the cavity, and thus there are large losses (Fig. 10.b).

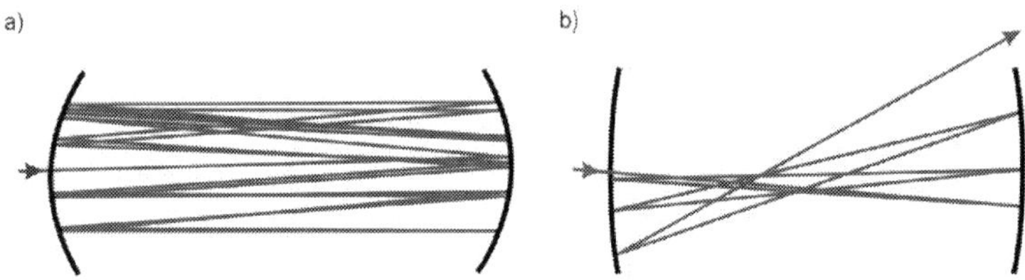

Fig. 10. Schematic illustration of the reflections in stable cavity (a) and in unstable one (b).

For the cavity to be stable, the selected curvature rays of the mirrors (r_1, r_2) and the distance between them (L) should be appropriate. The relation between these parameters describes the so-called stability criterion [Busch & Bush, 1999]

$$0 \leq g_1 \cdot g_2 \leq 1 , \tag{15}$$

where the parameters g_1 and g_2 are respectively

$$g_1 = \left(1 - \frac{L}{r_1}\right) , \tag{16}$$

$$g_2 = \left(1 - \frac{L}{r_2}\right) . \tag{17}$$

The optical signal from the cavity is registered with a photoreceiver, the operating spectrum of which should be matched to the selected absorption line of the gas. It usually is characterized by high gain, high speed and low dark current. In addition to the photodetector, the photoreceiver frequently includes different type of preamplifier which is used to amplify the signal from the photodetector. The preamplifier should have a wide dynamic range, low noises, high gain and an appropriately selected frequency band [Rogalski & Bielecki, 2006]. Next, the signal from the preamplifier is digitized with a high sampling rate (e.g. 100 MS/s). Data from the analogue-to-digital converter (ADC) are transmitted to a computer, for example through a USB interface. Special computer software provides processing of the measuring data and gas concentration determination. A scheme of a signal processing in the cavity enhanced sensor is presented in Fig. 11.

Observation of NO_x molecules can be done at electronic transitions which are characterized by a broad absorption spectra providing a relatively large mean absorption cross section within the range of several nanometres. Therefore the use of broadband multimode lasers is possible. In the case of nitrogen dioxide, the absorption spectrum has a band in the 395 - 430 nm range with a mean cross section of about $6 \cdot 10^{-19}$ cm^2 (Fig. 12a). There are various light sources applied, e.g. blue – violet LED's or diode lasers or even broadband supercontinuum sources [Wojtas et al., 2009, Holc et al., 2010, Stelmaszczyk et al., 2009].

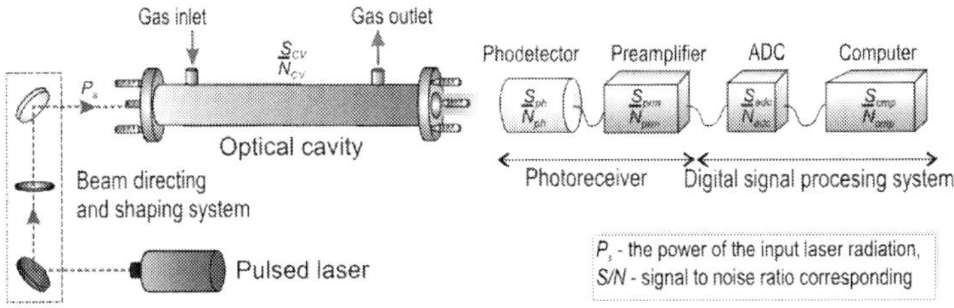

Fig. 11. Block diagram of NO$_x$ sensor.

Assuming that determination of the gas concentration basis on the temporal analysis, the sensor sensitivity (in generally) depends on the mirrors reflectivity, cavity length and uncertainty of decay time measurements (Fig. 12b). The sensitivities of the laboratory NO$_2$ sensors reach 0.1 ppb. Our approaches to the nitrogen dioxide sensor were already described in several papers [Wojtas et al., 2006, Nowakowski et al., 2009].

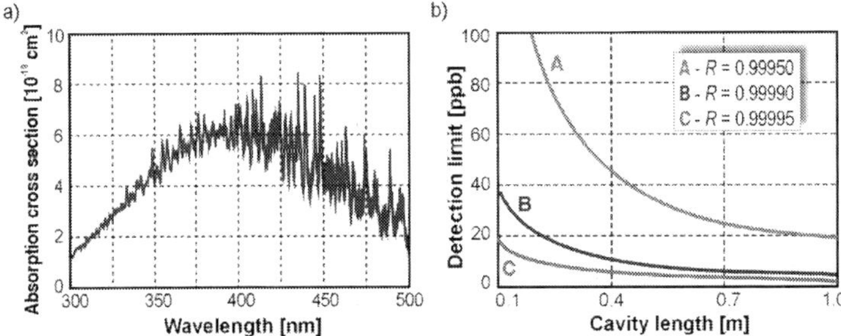

Fig. 12. NO$_2$ absorption spectrum (a) and dependence of the concentration limit on the cavity length and the reflectivity of mirrors R (b).

However, for many other compounds (like N$_2$O and NO) the electronic transitions correspond to the ultraviolet spectral range [HITRAN, 2008], where neither suitable laser sources nor high reflectivity mirrors are available. For example, reflectivities of available UV mirrors do not exceed the value of 90%. Therefore, a higher sensitivity of the NO and N$_2$O sensor can be obtained using IR absorption lines (Fig. 13).

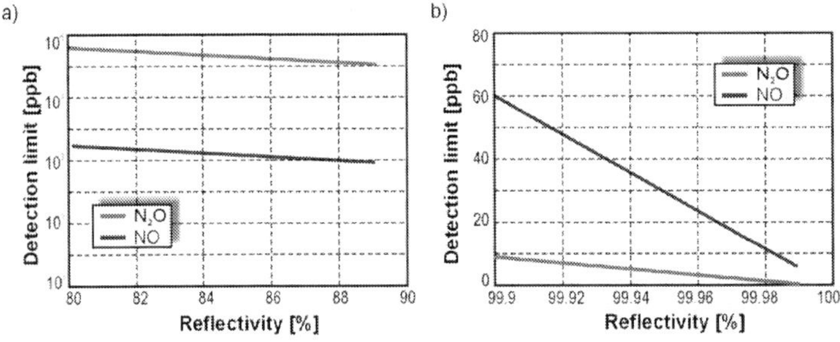

Fig. 13. Detectable concentration limit versus cavity mirrors reflectivity in UV (a) and in IR wavelength ranges (b).

The analyses show that the IR wavelength range provides the possibility to develop NO and N_2O sensor, the sensitivity of which could reach the ppb level (Rutecka, 2010). For instance, at the wavelength ranges of 5.24 µm – 5.28 µm and 4.51 µm – 4.56 µm the absorption cross section reaches the value 3.9×10^{-18} cm^2 for N_2O and 0.7×10^{-18} cm^2 for NO. Additionally, there is no significant interference of absorption lines of other atmosphere gases (e.g. CO, H_2O). There could only be observed a low interference of H_2O, which can be minimized with the use of special particles of a filter or dryer. Both NO and N_2O absorption spectrum are presented in Fig. 14 and in Fig. 15 respectively.

In this spectral range, quantum cascade lasers (QCL) are the most suitable radiation sources for experiments with cavity enhanced methods. Available QCL's provide high power and narrowband pulses of radiation [Namjou et al., 1998, Alpes Lasers SA]. The *FWHM* duration time of their pulses reaches hundreds of microseconds pulses while the repetition rate might be of some kHz. Moreover, their emission wavelength can be easy tuned to the maxima of N_2O and NO absorption cross section.

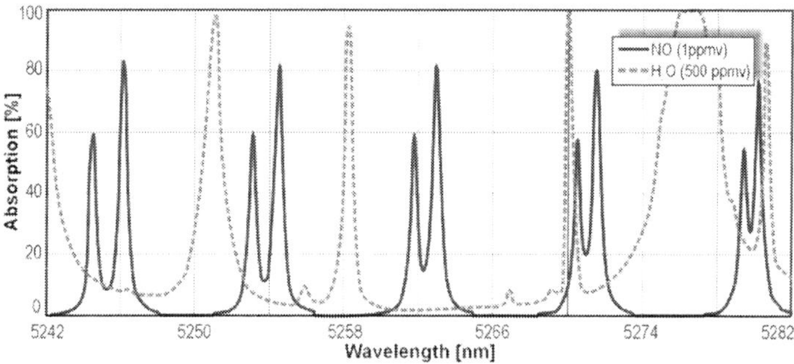

Fig. 14. NO absorption spectrum [Hitran, 2008].

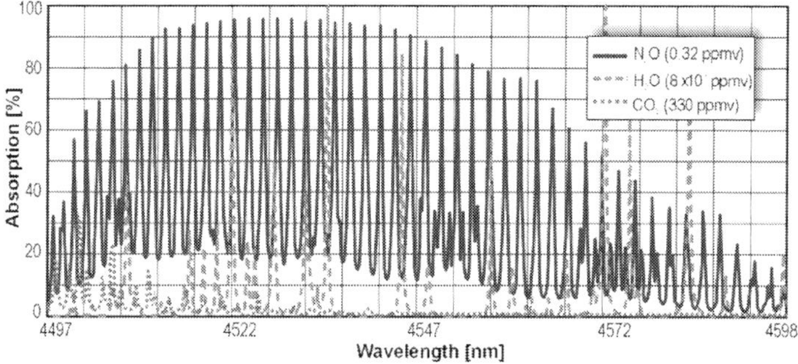

Fig. 15. N_2O absorption spectrum [Hitran, 2008].

5. Signal to noise ratio of the sensor

As we have seen, the reflectivity of the mirrors has a significant impact on the theoretical sensitivity of the sensor. According to the equation (7), the sensor sensitivity is higher when the mirror reflectivity and cavity length are increased (Fig. 12 and Fig. 13). However, then a lower level of optical signal reaches the photodetector. Therefore, the signal-to-noise ratio (*SNR*) of the system is very important.

5.1 Optical cavity parameters

Usually, for the cavities, such parameters like, e.g., the finesse F, the time of a photon life τ_p, the transmission function $T(R,\lambda)$ and signal-to-noise ratio S_{cv}/N_{cv} are determined [Wojtas & Bielecki, 2008].

The finesse F characterizes the quality of the cavity and determines an effective number of a roundtrip of optical radiation in the cavity up to its energy reaching the level of $1/e$. The finesse F can be found from the formula

$$F = \pi \frac{\sqrt{R}}{1-R}. \tag{18}$$

The time of a photon life is described by the equation

$$t_p = \frac{2nLF}{c}, \tag{19}$$

where n is the refractive index. The transmission function of the optical cavity is known as the Airy formula. It has the following form

$$T(R,\phi) = \frac{(1-R)^2}{(1-R)^2 + 4R\sin^2(\frac{\phi}{2})}, \tag{20}$$

where ϕ is the radiation phase shift during one roundtrip inside the cavity

$$\phi = \frac{4\pi nL}{\lambda}, \tag{21}$$

and λ is the optical radiation wavelength. The graphical representation of Eq. (20) is presented in Fig. 16.

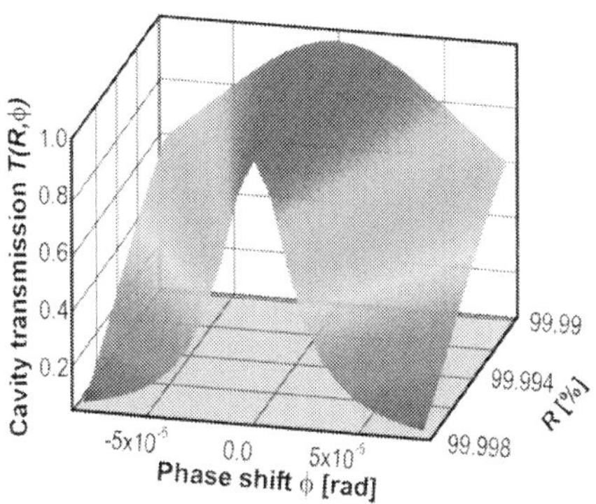

Fig. 16. Graphical representation of the transmission function of an optical cavity.

It shows a strong influence of the mirrors reflectivity on the selectivity of an optical cavity. The transmission of the cavity is maximum wherever ϕ is the integral multiple of 2π.

The optical cavity signal-to-noise ratio (S_{cv}/N_{cv}) is connected with its transmission function. S_{cv}/N_{cv} is directly proportional to the power of radiation matched to the transmission function of a cavity and to an absorption band of the examined gas. However, S_{cv}/N_{cv} is inversely proportional to the power of undesirable radiation transmitted through a cavity because of non-zero values of the mirrors' transmissions. The formula describing a signal-to-noise ratio of the cavity is

$$\frac{S_{cv}(\lambda)}{N_{cv}(\lambda)} = \frac{\left[T(R(\lambda),\phi)\right]^2}{\left[1 - R(\lambda)\right]^2}.$$

(22)

Assuming that a length of optical cavity is 0.5 m and it is consists of two concave mirrors with the reflectivity of 0.999976, then $S_{cv}/N_{cv}=1.7\times10^9$ ($F = 1.3\times10^5$, $\tau_f=5.2\times10^{-4}s$).

5.2 Analysis of detection system parameters

Due to the high value of SNR of the optical cavity, the signal-to-noise ratio of an electronic circuit is the crucial parameter of the cavity enhanced sensor. The signal from the cavities is registered with different types of photodetectors; depending on the spectral range. In the case of ultraviolet (UV), visible (VIS) and near infrared (NIR) region (approximately from form 100 nm up to 1.5 μm) the most popular are photomultiplier tubes (PMT's). They are characterized by high gain, high speed and low dark current. Because of PMT high resistance, transimpedance preamplifiers are usually used to amplify signal from PMT. They are characterized by a wide dynamic range [Rogalski & Bielecki, 2006].

In the medium infrared (MIR) part of the spectrum there are two types of photodetectors: thermal and quantum. Thermal photodetectors use infrared energy as heat, and their responsivity is independent of the wavelength. But they have disadvantages because their response time is slow and detectivity is low. Therefore, quantum photodetectors are used in the practical implementations of cavity enhanced methods. They offer higher responsivity and faster response speed. To achieve higher performance, i.e. a wider frequency band and higher detectivity (D^*), they are cooled. There are several cooling methods: thermoelectric cooling (TEC), cryogenic cooling (e.g. dry ice or liquid nitrogen) and mechanical cooling (e.g. Stirling coolers). The most popular are HgCdTe (mercury-cadmium-telluride, MCT) photoconductive and photovoltaic detectors. There are available MCT photodetectors that use monolithic optical immersion technology and TEC cooling. They offer high detectivity (about 10^{12} cm $\sqrt{Hz}/W$) and high speed (up to 1GHz). To amplify the signal from the MCT photodetector, transimpedance preamplifiers are applied as well [Hamamatsu, 2011, Piotrowski et al., 2004, VIGO System S.A.].

5.2.1 Photoreceiver with photomultiplier tube

To determine the signal-to-noise ratio of the photoreceiver, the PMT equivalent scheme is necessary. The scheme is presented in Fig. 17. The current source I_s represents the current of useful signal, R_p and C_p are the resistance and capacitance of the photomultiplier respectively [Wojtas et al., 2008].

PMT noise sources are as follows: the current source I_{ns} represents the shot noise from useful signal, the current source I_{nd} represents shot noise of anode dark current, I_{nb} is the current sources of noise from background radiation and I_{nRL} is the thermal noise of load resistance.

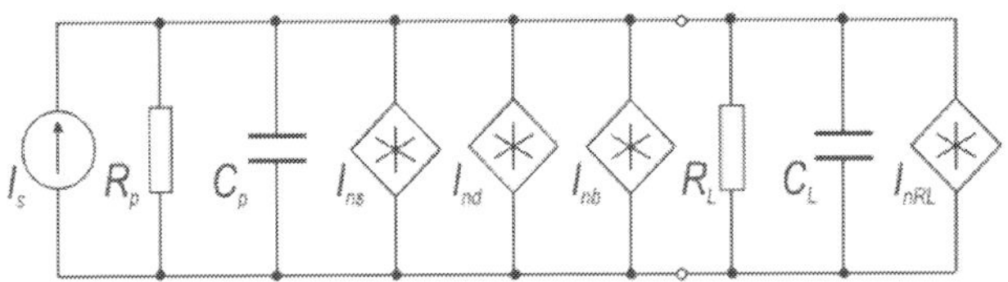

Fig. 17. PMT equivalent scheme.

In the case when all the described noise sources will be taken into consideration, PMT signal-to-noise ratio can be determined by the formula [Wojtas & Bielecki, 2008]

$$\frac{S_{ph}}{N_{ph}} = \frac{I_s^2}{I_{ns}^2 + I_{nd}^2 + I_{nb}^2 + I_{nRL}^2} \cdot \tag{23}$$

Assuming that during cavity enhanced experiments background noise can be eliminated, and a photoemission process is described by the Poisson model, and all stages of PMT will have the same gain, then

$$\frac{S_{ph}}{N_{ph}} = \frac{(P_s \cdot S_p \cdot G_p)^2}{2\,q\,\Delta\,f_n\,(G_p\,S_p\,P_s + I_{da})\dfrac{\delta}{\delta-1} + \dfrac{4\,k\,T_o\,\Delta\,f_n}{R_L}}, \tag{24}$$

where P_s is the power of optical radiation, G_p is the PMT gain, S_p is the photocathode sensitivity, q is the electron charge, Δf_n is the noise bandwidth, I_{da} is the anode dark current, δ is one stage of the PMT gain, k is the Boltzmann constant, and T_0 is the temperature [Flyckt & Marmonier, 2002].
The noise bandwidth can be determined from the formula

$$\Delta f_n = \frac{\pi}{2}\Delta f_{3dB} \approx \frac{1}{4R_L(C_L + C_p)}, \tag{25}$$

where Δf_{3dB} represents 3dB frequency bandwidth.
Because PMT can be treated as a current source the best preamplifier configuration is a transimpedance preamplifier. Moreover, its input circuit does not affect photodetector polarization. The scheme of a transimpedance preamplifier is presented in Fig. 18.
In the case when one photoelectron is emitted by the PMT photocathode, the output voltage signal of the transimpedance preamplifier can be described by the formula

$$V_{prm} = \frac{q \cdot G_p \cdot R_f}{R_f \cdot C_{eq}' - t_i} \cdot \left[\exp\left(\frac{-t}{R_f \cdot C_{eq}'}\right) - \exp\left(\frac{-t}{t_i}\right)\right], \tag{26}$$

where C_{eq}' is PMT and a load circuit equivalent capacitance located in the feedback circuit, and t_i is PMT pulse duration. The Miller theorem states that C_{eq}' is $(G_{OL} + 1)$ times lower then C_{eq} (G_{OL} is the amplifier open-loop gain). In the appropriate developed circuit, the value of C_{eq}' is lower than 0.1 pF.

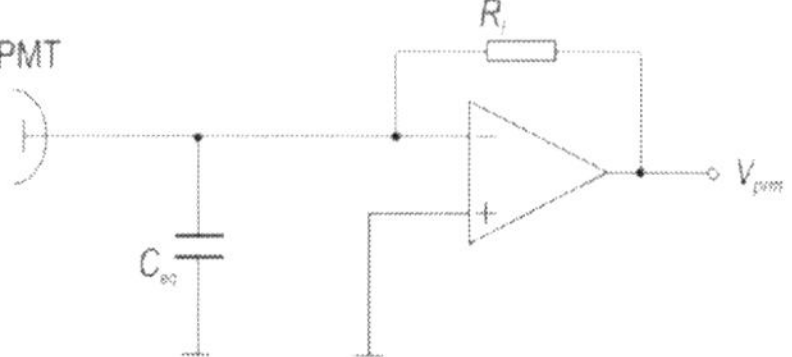

Fig. 18. Scheme of the transimpedance preamplifier.

Analysis showed that an increase in R_f caused that the output pulse duration is longer and longer (Fig. 19). Because of this, to reach a high value of gain and to avoid signal distortion, the next stage of amplifier should be used. Because of the low output resistance of the transimpedance preamplifier (< 50 Ω), a voltage amplifier can be used.

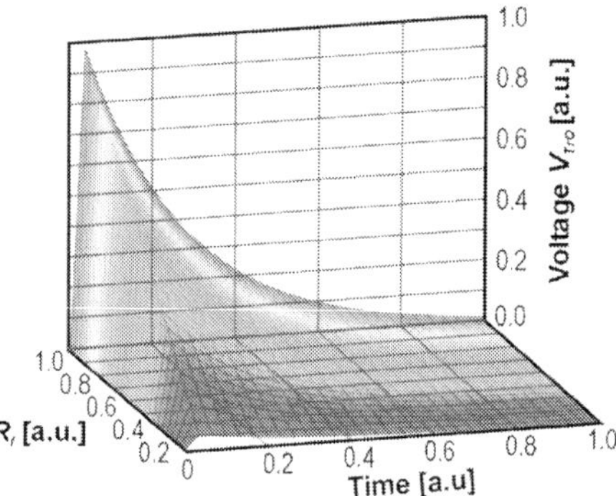

Fig. 19. Example of transimpedance preamplifier output signal.

To determine the SNR of the photoreceiver, an equivalent scheme is necessary (Fig. 20).

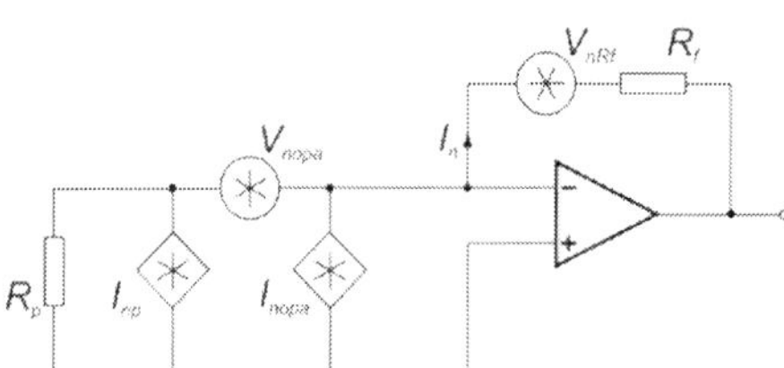

Fig. 20. Equivalent scheme of the first stage of the photoreceiver.

The noise of the operational amplifier is represented by the voltage source V_{nopa} and the current source I_{nopa}. The noise source I_{nph} is equivalent to the PMT noise. In this case, the total current noise I_{nt} is described by the formula

$$I_{nt}^2 = I_{np}^2 + \left(V_{nopa} \frac{R_p + R_f}{R_p R_f} \right)^2 + I_{nopa}^2 + \left(\frac{V_{nRf}}{R_f} \right)^2 , \tag{27}$$

where V_{nRf} is the thermal noise determined by the equation

$$V_{nRf}^2 = 4kT_o R_f \Delta f_n . \tag{28}$$

The output voltage noise of the transimpedance preamplifier can be defined as

$$V_{nprm} = I_{nt} R_f , \tag{29}$$

and of the SNR of the photoreceiver can be described with the formula

$$\left(\frac{S_{prm}}{N_{prm}} \right)_{pmt} = \frac{I_s^2}{I_{nt}^2} . \tag{30}$$

Usually, the amplified signal from the preamplifier is fed to an analogue digital converter (ADC). This circuit also adds its noise. Assuming a 12-bit ADC and the same quantization steps δ_{adc}, its noise can be determined by the formula

$$V_{nadc}^2 = \frac{\delta_{adc}^2}{12} . \tag{31}$$

The analysis showed that the SNR of the detection system consists of PMT, preamplifier and ADC, and can be described by the formula

$$\left(\frac{S_{adc}}{N_{adc}} \right)_{pmt} = \frac{\left(P_s S_p G_p \right)^2 \cdot R_f^2}{\gamma R_f^2 \Delta f_n + V_{nadc}^2} , \tag{32}$$

where

$$\gamma = 2q(G_p S_p P_s + I_{da}) \frac{\delta}{\delta - 1} + \left(V_{nopa} \frac{R_p + R_f}{R_p \cdot R_f} \right)^2 + I_{nopa}^2 + \frac{4kT_o}{R_f} . \tag{33}$$

5.2.2 Photoreceiver with a MCT photodiode

The noise equivalent scheme of the photoreceiver using a MCT photodiode and a transimpedance preamplifier is presented in Fig. 21. The signal current generator I_{ph} represents the detected signal. Noises in a photodiode are represented by three noise generators: I_{nph} - the shot noise associated with photocurrent, I_{nd} - the shot noise of a dark current, while I_{nb} - the shot noise from a background current [Bielecki 2002]. In the scheme, the value of the load resistance of the photodetector depends on the feedback resistance R_f and the preamplifier gain G. The resistor R_f affects both the level of the preamplifier output signal and its noise. The noise current generator I_{nf} is the thermal noise current and excess noise of the feedback resistance. Since the thermal noise of I_{nf} is inversely related to the square root of the resistance, R_f should be of great value. The R_{sh} is the shunt

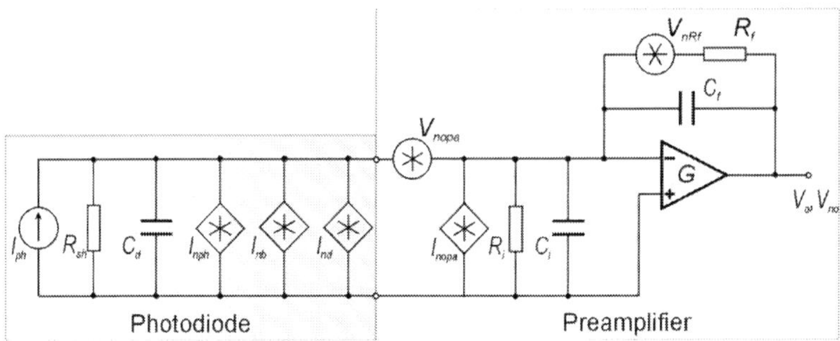

Fig. 21. Scheme of the photoreceiver with a photodiode.

resistance of a photodiode. The equivalent photoreceiver noise is the square root of each component noise squares sum [Bielecki et al., 2009]. Thus, the signal-to-noise ratio can be described with the simplified expression

$$\left(\frac{S_{prm}}{N_{prm}}\right)_{pht} = \frac{I_{ph}^2}{\left(I_{nph}^2 + I_{nd}^2 + I_{nb}^2 + I_{nopa}^2 + \dfrac{4kT\Delta f}{R_f}\right) + \left(V_{nopa}\dfrac{R_{eq}^2}{1+\omega^2\tau_{eq}^2}\right)^2},\tag{34}$$

where

$$R_{eq} = \frac{R_f R_{sh}}{R_f + R_{sh}} \quad \text{and} \quad \tau_{eq} = R_{eq}\left(C_f + C_d\right).\tag{35}$$

Only the modulus of feedback loop impedance and photodetector impedance is included. Furthermore, it could be assumed that in experiments applying cavity enhanced methods, current I_{nb} can be ignored. Moreover, intensity of the radiation reaching the photodiode is rather low, thus shot noise associated with the photocurrent is negligibly. In practical realisations (low frequency and $R_{sh} >> R_f$), the SNR of the system consisting in a photodiode, preamplifier and ADC can be determined from equation

$$\left(\frac{S_{adc}}{N_{adc}}\right)_{pht} = \frac{\left(R_i P_s\right)^2}{\left(\dfrac{R_i(A\Delta f)^{1/2}}{D*}\right)^2 + I_{nopa}^2 + \dfrac{4kT\Delta f}{R_f} + \left(\dfrac{V_{nopa}}{R_f}\right)^2 + V_{nadc}^2},\tag{36}$$

where R_i - photodiode current responsivity, A – detector active area.

5.3 Methods of SNR improving

Analyses in the previous section showed a significant influence of preamplifier feedback resistance (R_f) on the output photoreceiver signal. In an appropriately developed photoreceiver, the preamplifier shouldn't degrade photoreceiver performance. In Fig. 22 ADC noise, preamplifier noise and photodetector noise for different values of R_f were presented.

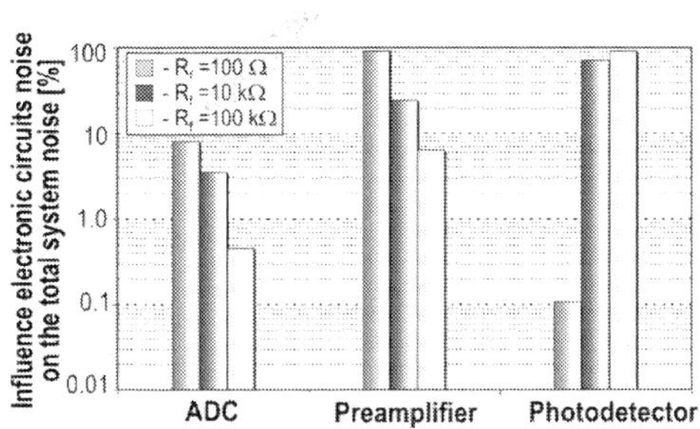

Fig. 22. Comparison noise sources of electronic circuit for different values of R_f.

In the case of $R_f = 100$ Ω, the highest influence on the total electronic system noise was preamplifier noise at ~92%. However, an increase in R_f caused a decrease in influence preamplifier noise on the total signal processing system noise. For $R_f = 100$ kΩ, the noise of photodetector is equal to ~93% of the total signal processing system noise and preamplifier is only ~6%. ADC noise is below 8%. Furthermore, the value of R_f also has a strong influence on the bandwidth of the system. In Fig. 23, the dependence SNR of the signal processing system and a preamplifier output pulse fall time on the R_f is presented.

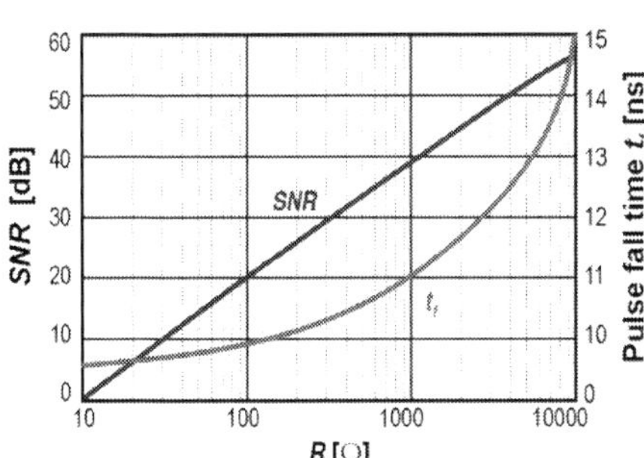

Fig. 23. Dependence of electronic circuit SNR and fall time of output pulse on resistance R_f.

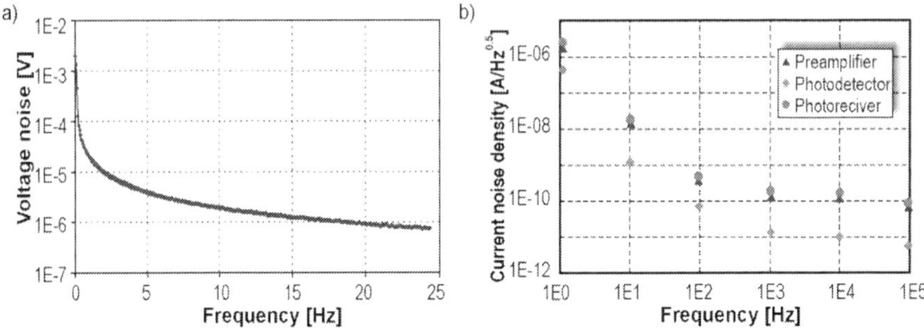

Fig. 24. Voltage noise (a) and current noise density (b) of the photoreceiver.

Experiments have shown that in the low frequency region the $1/f$ noise is dominant (Fig. 24a). Therefore, in order to minimize the adverse impact of such noise on the detectivity of the receiver (and SNR as well), a high pass filter is frequently used which limits the frequency bandwidth by several kilohertz. In the higher frequency region, there is dominant g-r noise by recombination of electrons and holes. Although the density of this noise is less than $1/f$ (Fig. 24b), the upper limit frequency should be suitably matched to the recorded signal bandwidth to avoid SNR degradation.

SNR of the cavity enhanced system can be additionally improved by the use of one of the advanced methods of signal detection, i.e. coherent averaging [Lyons, 2010]. This technique can be implemented in the software of the digital signal processing system. The software is usually installed in a personal computer. Thanks to this, increase in the SNR is directly proportional to the root of a number of the averaging samples n_{smpl},

$$\frac{S_{cmp}}{N_{cmp}} = S_{adc} \left(\frac{N_{adc}}{\sqrt{n_{smpl}}} \right)^{-1} . \tag{37}$$

Thanks to improving SNR, uncertainty of decay time determination is likely to reach values below 0.5% (e.g. in the case of 10 000 averaging samples). Hence, the detection limit can achieve the value of about 2×10^{-9} cm^{-1} (Fig. 25).

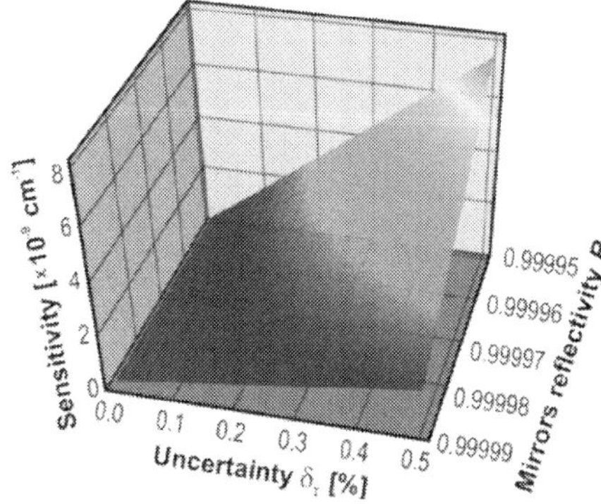

Fig. 25. Dependence cavity enhanced sensor sensitivity on decay time precision determination and cavity mirrors reflectivity.

6. Conclusion

In this chapter, characterisations of absorption spectroscopy methods were shown. The methods provide the possibility of absorption spectra investigations. This kind of spectra can be defined as the set of all electron crossings from lower energy levels to higher ones. They caused an increase in molecules energy. In practical implementations, a source of radiation and very sensitive photoreceiver is used which records radiation passing through the absorber sample. One of the most common gas detection systems is differential optical absorption spectroscopy. Such arrangements are applied to the monitoring of atmospheric pollutants, including the detection of NO_x, in terrestrial applications, in air and in space, e.g. GOME and SCIAMACHY satellite.

Cavity enhanced spectroscopy is the one of the most sensitive absorption methods. The greatest sensitivity is provided by P-CRDS, CW-CRDS and CEAS methods. CRDS was applied to determine the mirrors reflectivity for the first time in the early 1980's. This method provides a much higher sensitivity than conventional absorption spectroscopy. An optical cavity with a high quality is applied that is made up of two concave mirrors with very high reflectance R. This results in a long optical path, even up to several kilometres. To determine the gas concentration several different methods are used: by measuring the decay time of the signal, by measuring the phase shift, and by measuring the signal amplitude. All of them were described in detail.

Furthermore, the basic experimental setups of cavity enhanced methods were described. Generally, they consist of pulse laser (or cw laser with modulator), beam directing and shaping system (mirrors, diaphragms, diffraction grating), optical cavity and photoreceiver with signal processing system (e.g. digital oscilloscope in the simplest case). First of all, the sensor project should take into account the appropriate matching cavity parameters and the laser emission wavelength to the test gas absorption spectrum.

Observation of NO_x molecules can be done at electronic transitions which are characterized by a broad absorption spectra providing a relatively large mean absorption cross section within the range of several nanometres. Therefore, using broadband multimode lasers is possible. However, for many other compounds (like N_2O and NO), the electronic transitions correspond to an ultraviolet spectral range, where neither suitable laser sources nor high reflectivity mirrors are available. Therefore, a higher sensitivity of the NO and N_2O sensor can be obtained using an IR absorption line.

It was shown that reflectivity of the mirrors has a significant impact on the theoretical sensitivity of the sensor. The sensor sensitivity is higher when the mirror reflectivity and cavity length are increased. However, then a lower level of optical signal reaches the photodetector. Therefore, the signal-to-noise ratio of the system is very important. Thus analyses of the main parameters of the optical cavity, photoreceiver and the signal processing system were performed. In the analyses the most popular photodetectors were taken into consideration. In the UV, VIS and NIR spectral regions, the photomultiplier is characterized with high performance. Photodetectors designed for MIR operation require an additional cooling system. Thanks to this they can achieve a higher performance, i.e. a wider frequency band and higher detectivity (D^*). Because of the many advantages, MCT photodetectors are frequently used in cavity enhanced applications.

Analyses showed a significant influence of preamplifier feedback resistance (R_f) on the output photoreceiver signal. In appropriately developed photoreceiver, the preamplifier shouldn't degrade photoreceiver performance. The *SNR* of the cavity enhanced system can be additionally improved by the use of one of the advanced methods of signal detection, i.e. coherent averaging.

Cavity enhanced sensors are able to measure NO_x concentration at ppb level. Their sensitivity is comparable with the sensitivities of instruments based on other methods, e.g. gas chromatography or mass spectrometry. The developed sensor can be applied for monitoring atmosphere quality. Using the sensor, the detection of vapours from some explosive materials is also possible.

7. Acknowledgment

The researchers are supported by the Ministry of Science and High Education of Poland in 2009-2011.

8. References

Alpes Lasers SA, 1-3 Max.-de-Meuron, C.P. 1766, CH-2001 Neuchâtel, http://www.alpeslasers.ch/products.html#products_top

Atherton, K.J., Yu, H., Stewart, G. & Culshaw, B. (2004). Gas detection with fibre amplifiers by intra-cavity and cavity ring-down absorption, *Measurement Science and Technology* Vol. 15, pp. 1621–1628

Berden, G., Peeters, R. & Meijer, G. (2000). Cavity ring-down spectroscopy: Experimental schemes and applications, International Reviews In Physical Chemistry, Vol. 19, No. 4. pp. 565-607

Bielecki, Z. (2002). Maximization of signal to noise ratio in infrared radiation receivers, *Opto-Electron. Rev.*, 10, pp. 209-216

Bielecki, Z., Kolosowski, W., Sedek, E., Wnuk, M. & Wojtas, J. (2009). Multispectral detection circuits in special application, *Transactions on Modelling and Simulations*, WIT Press – WIT, Vol 48, Print ISBN: 1-84564-187-0; On-line ISBN: 1-84564-364-5; Print ISSN: 1746-4064, pp. 217-228

Busch, K.W., Busch, M.A. (1999). *Cavity-Ringdown Spectroscopy*, ACS Symposium series, American Chemical Society, Washington DC

Chudzynski, S., Czyzewski, A., Skubiszak, W., Stacewicz, T., Stelmaszczyk, K., Szymanski, A. & Ernst, K. (1999). Practical solutions for calibration of DIAL system, Optica Applicata 29, pp. 477 – 485

Courtillot I., Morville J., Motto-Ros & Romanini D. (2006). Sub-ppb NO2 detection by optical feedback cavity-enhanced absorption spectroscopy with a blue diode laser. *Appl. Phys. B*, 85, pp. 407–412

Czyzewski, A., Chudzynski, S., Ernst, K., Karasinski, G., Kilianek, L., Pietruczuk, A., Skubiszak, W., Stacewicz, T., Stelmaszczyk, K., Koch, B. & Rairoux, P. (2001). Cavity Ring-Down Spectrography, *Optics Commun*, 191, 271 – 275

Drescher, S.R., Brown, S.D. (2006). Solid phase microextraction-gas chromatographic-mass spectrometric determination of nitrous oxide evolution to measure denitrification in estuarine soils and sediments, *Journal of Chromatography A*, 1133(1–2), pp. 300–304

Engeln, R., Berden, G. & Meijer, G. (1996). Phase shift cavity ring down absorption spectroscopy, *Chemical Physic Letters*, 262, pp. 105 – 109

Engeln, R., Berden, G., Peeters, R. & Meijer, G. (1998). Cavity enhanced absorption and cavity enhanced magnetic rotation spectroscopy, *Review Of Scientific Instruments*, Vol. 69, No. 11, pp. 3763 – 3769

Flyckt, S.O., Marmonier, C. (2002). *Photomultiplier tubes, principles, and applications*, Photonics, Brive, France

Godish, T. (2004). *Air Quality*, Lewis Publishers, ISBN 156670586X, 9781566705868

Grossel, A., Ze´ninari, V., Joly, L., Parvitte, B., Durry, G. & Courtois, D. (2007). Photoacoustic detection of nitric oxide with a Helmholtz resonant quantum cascade laser sensor, *Infrared Physics & Technology*, 51, pp. 95–101

Hamamatsu, Solid State Division (2011). *Technical information SD-12, Characteristics and use of infrared detectors*, Cat. No. KIRD9001E04, Mar. 2011 DN

He, Y., Orr, B.J. (2000). Ringdown and cavity-enhanced absorption spectroscopy using a continuous-wave tunable diode laser and a rapidly swept optical cavity, *Chemical Physics Letters*, 319, pp. 131–137

Herbelin, J.M., McKay, J.A.., Kwok, M.A., Ueunten, R.H., Urevig, D.S., Spencer, D.J. & Benard, D.J. (1980). Sensitive measurement of photon lifetime and true reflectances in optical cavity by a phase-shift method, *Applied Optics*, Vol. 19, Issue 1, pp. 144-147

HITRAN (2008). High-resolution transmission molecular absorption database, http://www.hitran.com

Holc, K., Bielecki, Z., Wojtas, J., Perlin, P., Goss, J., Czyzewski, A., Magryta, P. & Stacewicz, T. (2010). Blue laser diodes for trace matter detection, *Optica Applicata*, Vol. XL, No. 3, 2010, pp. 641-651

Horii, C.V., Zahniser, M.S., Nelson, D.D., McManus, J.B. & Wofsy, S.C. (1999). Nitric Acid and Nitrogen Dioxide Flux Measurements: a New Application of Tunable Diode Laser Absorption Spectroscopy, *Proc.SPIE*, v. 3758, pp. 152-161

Jean-Franqois, D., Ritz, D. & Carlier, P. (1999). Multiple-pass cell for very-long-path infrared spectrometry, Applied Optics, Vol. 38, No. 19, pp. 4145-4151

Kasyutich, V.L., Bale, C.S.E., Canosa-Mas, C.E., Pfrang, C., Vaughan , S. & Wayne, R.P. (2003a). Off-axis continuous-wave cavity-enhanced absorption spectroscopy of narrow-band and broadband absorbers using red diode lasers, *Applied Physics B*, Vol. 75, pp. 755-761

Kasyutich, V.L., Bale, C.S.E., Canosa-Mas, C.E., Pfrang, C., Vaughan, S. & Wayne, R.P. (2003b). Cavity-enhanced absorption: detection of nitrogen dioxide and iodine monoxide using a violet laser diode, *Applied Physics B*, Vol. 76, No. 6, pp. 691-698

Karasinski, G., Kardas, A.E., Markowicz, K., Malinowski, S.P., Stacewicz, T., Stelmaszczyk, K., Chudzynski, S., Skubiszak, W., Posyniak, M., Jagodnicka, A.K., Hochhertz, C. & L. Woeste (2007). LIDAR investigation of properties of atmospheric aerosol, *The European Physical Journal Special Topics*, 144, pp. 129–138

Lagalante, A.F. (1999). Atomic Absorption Spectroscopy: A Tutorial Review, *Applied Spectroscopy Reviews*, Vol. 34(3), pp.173–189

Li, P., Shi, K. & Liu, Z. (2005). Optical scattering spectroscopy by using tightly focused supercontinuum, *Optics Express*, Vol. 13, No. 22, pp. 9039-9044

Lyons, R.G. (2010). *Understanding Digital Signal Processing*, Addison Wesley Pub Co Inc., 3rd Edition, ISBN-13: 978-0-13-702741-5, ISBN-10: 0-13-702741-9

Martin, R.V., Parrish, D.D., Ryerson, T.B., Nicks Jr., D.K., Chance, K., Kurosu, T.P., Jacob, D.J., Sturges, E.D., Fried, A. & Wert, B.P. (2004). Evaluation of GOME satellite measurements of tropospheric NO_2 and HCHO using regional data from aircraft campaigns in the southeastern United States, *Journal of Gophysical Research*, Vol. 109, 11 pp.

Mierczyk, Z., Zygmunt, M., Gawlikowski, A., Gietka, A., Kaszczuk, M., Knysak, P., Mlodzianko, A., Muzal, M., Piotrowski, W. & Wojtanowski, J. (2008). Two-wavelength backscattering lidar for stand off detection of aerosols, Lidar Technologies, Techniques, and Measurements for Atmospheric Remote Sensing IV. Edited by Singh, Upendra N.; Pappalardo, Gelsomina. *Proceedings of the SPIE*, Volume 7111, pp. 71110R-71110R-9

Moore, D.S. (2007). Recent Advances in Trace Explosives Detection Instrumentation, *Sens Imaging*, 8, pp. 9–38

Namjou, K., Cai, S., Whittaker, E.A., Faist, J., Gmachl, C., Capasso, F., Sivco, D.L. & Cho, A.Y. (1998) .Sensitive absorption spectroscopy with a room-temperature distributed-feedback quantum-cascade laser, *Optics Letters*, V23, 3, pp. 219 – 221

Noel, S., Bovensmann, H., Burrows, J.P., Frerick, Chance', J., K.V. & Goede, A.H.P. (1999). Global Atmospheric Monitoring with SCIAMACHY, *Physal. Chemical. Earth*, Vol. 24, No. 5, pp. 427-434

Nowakowski M., Wojtas J., Bielecki Z., & Mikolajczyk J. (2009). Cavity enhanced absorption spectroscopy sensor, Acta Phys. Pol. A, 116, 363–367

O'Keefe, A. (1998). Integrated cavity output analysis of ultra-weak absorption, *Chemical Physics Letters*, 293 (5-6), pp. 331-336

O'Keefe, A., Deacon, D.A.G. (1988). Cavity ring-down optical spectrometer for absorption measurements using pulsed laser sources, *Review of Scientific Instruments*, No. 59, pp. 2544-2554

O'Keefe, A., Scherer, J.J. & Paul, J.B. (1999). CW integrated cavity output spectroscopy, *Chemical Physic Letters*, 307 (5-6), pp. 343-349

Owsik J., Janucki J. (2004). Laser-induced breakdown spectrometer for non-destructive diagnostics", *Proc. SPIE*, 4962, pp. 135–142

Paul, J.B., Lapson, L. & Anderson, J.G. (2001). Ultrasensitive absorption spectroscopy with a high-finesse optical cavity and off-axis alignment, Applied Optics, Vol. 40, No. 27, pp. 4904-4910

Piotrowski A., Madejczyk P., Gawron W., Klos K., Romanis M., Grudzien M., Rogalski A. & Piotrowski J. (2004). MOCVD growth of Hg1xCdxTe heterostructures for uncooled infrared photodetectors. Opto-Electron. Rev., 12, pp. 453–458

Pipino, A.C.R. (1999). Ultrasensitive surface spectroscopy with a miniature optical resonator, *Physical Rewiew Letters*, Vol. 83, No. 15, pp. 3093-3096.

Rogalski, A., Bielecki, Z. (2006). Detection of optical radiation (chapter A1.4), *Handbook of optoelectronics*, Taylor & Francis, New York, London pp. 73-117

Romanini, D., Kachanov, A. A., Sadeghi, N. & Stoeckel, F. (1997). Diode laser cavity ring down spectroscopy, Chemical Physics Letters, No 270, pp. 538-545

Rutecka, B., Wojtas, J., Bielecki, Z., Mikołajczyk, J. & M. Nowakowski. (2010). Application of an optical parametric generator to cavity enhanced experiment. *Proc. of SPIE*, vol. 7745, pp. 77450I-1- 77450I-8

Saleh B.E.A., Teich M.C. (2007). *Fundamentals of Photonics*, John Wiley & Sons, 2nd Edition ISBN: 978-0-471-35832-9

Scherer, J.J., Paul, J.B., Jiao, H. & O'Keefe, A. (2001). Broadband ringdown spectral photography, *Applied Optics*, Vol. 40, No. 36, pp. 6725-6732

Shimadzu Scientific Instruments, 7102 Riverwood Drive, Columbia, MD 21046, Available from: http://www.mandel.ca/application_notes/SSI_GC_Green_Gasses_Lo.pdf

Sigrist, M.W. (1994). *Air monitoring by spectroscopic techniques*, John Wiley & Sons, ISBN-10: 0-471-55875-3. ISBN-13: 978-0-471-55875-0

Stelmaszczyk, K., Fechner, M., Rohwetter, P., Queißer, M., Czyzewski, A., Stacewicz, T. & L. Wöste, (2009). Towards Supercontinuum Cavity Ringdown Spectroscopy, *Appl. Phys. B*, 94, pp. 396–373

Stelmaszczyk, K., Rohwetter, P., Fechner, M., Queißer, M., Czyzewski, A., Stacewicz, T. & Wöste, L. (2009). Cavity Ring-Down Absorption Spectrography Based on Filament-Generated Supercontinuum Light, *Optics Express*, 17(5), 3673 – 3678

VIGO System S.A., http://www.vigo.com.pl/index.php/en/main_menu/strona_glowna

Wang, P., Richter, A., Bruns, M., Burrows, J.P., Junkermann, W., Heue, K.P., Wagner, T., Platt, U. & Pundt, I. (2005). Airborne multi-axis DOAS measurements of tropospheric SO2 plumes in the Po-valley, Italy, *Atmosphric. Chemical. Physics Discussion*, No 5, pp. 2017–2045

Wojtas, J., Bielecki, Z. (2008). Signal processing system in the cavity enhanced spectroscopy. *Opto-Electron. Rev.*, 16(4), pp. 44-51

Wojtas, J., Bielecki, Z., Mikołajczyk, J. & Nowakowski, M. (2008). Signal processing system in portable NO_2 optoelectronic sensor, *Sensor+Test 2008 Proceedings*, Nurnberg, Germany pp. 105-108

Wojtas, J., Bielecki, Z., Stecewicz, T., Czyżewski, A., Mikołajczyk, J., Nowakowski, M. & Rutecka, B. (2009). Ultrasensitive NOx optoelectronic sensor. *Photonics Letters of Poland*, Vol. 1 (2), pp.85-87

Wojtas, J., Czyzewski, A., Stacewicz, T. & Bielecki, Z. (2006). Sensitive detection of NO_2 with Cavity Enhanced Spectroscopy, *Optica Applicata*, Vol. 36, No. 4, pp. 461-467

Wojtas, J., Czyżewski, A., Stacewicz, T., Bielecki, Z. & Mikolajczyk J. (2005). Cavity enhanced spectroscopy for NO2 detection, Proc. SPIE Vol. 5954, pp. 174-178

Ye, J., Ma, L.S. & Hall, J.L. (1997). Ultrasensitive high resolution laser spectroscopy and its application to optical frequency standards, *28th Annual Precise Time and Time Interval (PTTI) Applications and Planning meeting, Proceedings*, L. A. Breakiron, Ed., US Naval Observatory, Washington D.C., p. 289

9

Use of Optoelectronics to Measure Biosignals Concurrently During Functional Magnetic Resonance Imaging of the Brain

Bradley J MacIntosh[1,2], Fred Tam[1] and Simon J Graham[1,2]
*[1]Sunnybrook Research Institute, Sunnybrook Health
Sciences Centre, Toronto, Ontario
[2]Department of Medical Biophysics, University of Toronto,
Toronto, Ontario
Canada*

1. Introduction

In the rapidly advancing world of biomedical imaging and engineering, magnetic resonance imaging (MRI) has become an indispensable technique to visualize normal and diseased anatomy non-invasively in the human body. The 2003 Nobel prize in Physiology and Medicine to biophysicists Lauterbur and Mansfield is a testament to the significant health care advances that have followed from the early development of MRI technology. In the past three decades, MRI technology has not only matured, but has continued to diversify to encompass new applications in science and medicine. The focus of this chapter is one such application, that involves use of optoelectronic devices to measure electrical biosignals during concurrent mapping of brain activity with a technique called functional MRI (fMRI).

Since the development of early MRI systems, there has been an ongoing need to develop ancillary devices adjacent to or within the magnet bore. For example, clinical MRI systems are now equipped with sensors to measure heart rate, respiratory rate, and electrocardiograph signals that provide essential summaries of the physiological status of the patient during scanning. These and other devices (e.g. wheelchairs, incubators, power injectors to deliver drugs and contrast agents, and interventional devices such as catheters) cover a wide range of functions and electromechanical complexity. However, they are all carefully designed with common consideration of several critical factors. The MRI system uses three electromagnetic fields to produce an image. First, radio-frequency (RF) coils are used to transmit and receive RF fields to and from the patient, at a frequency of approximately 100 MHz. Second, a very strong, static, homogeneous main magnetic field is required primarily for signal-to-noise ratio (SNR) considerations. This field typically has a strength of 1.5 or 3.0 Tesla, or approximately 50 000 – 100 000 times the strength of the Earth's magnetic field, with spatial uniformity to approximately less than 1 part per million over a 20 cm diameter spherical volume. Third, time-varying magnetic gradient fields are produced along orthogonal directions by gradient coils to encode MRI signals spatially, with amplitudes of approximately 10 mT/m and slew rates of approximately 100 T/m/s.

The use of electromagnetic fields and their relation with basic MRI theory have implications for ancillary MRI devices in several different ways. The formation of MR images relies on resonant excitation of magnetization primarily related to hydrogen nuclei on water molecules within biological tissues, and subsequent reception of RF signals after spatial encoding and additional manipulations that introduce signal contrast between different normal tissue types and disease states. For 1.5 T and 3.0 T MRI systems, the resonant frequencies of interest are 64 MHz and 128 MHz, respectively. The use of imaging gradients for spatial encoding leads to a bandwidth of interest of several 100 kHz surrounding these resonance frequencies. Within this bandwidth, it is important that devices introduce negligible RF interference with the MRI system, and *vice versa*. In addition, patient safety must also be considered (Schaefers 2008). Substantial attractive forces can easily occur between the very large static magnet field and ancillary devices, if the device contains ferromagnetic components. Such interactions also distort the static magnetic field uniformity and can introduce unacceptable spatial distortions and signal loss in MR images. There is also the potential for the time-varying electromagnetic fields generated by the MRI system to interact with electrically conductive devices to cause unwanted heating (Lemieux, Allen et al. 1997), given that the RF power amplifiers used in MRI systems are in the range of 1–30 kW. Consequently, much attention has been paid to the establishment of safety standards, through bodies such as the American Society for Testing and Materials (ASTM; http://www.astm.org/), and to evaluation of the extent of compatibility of ancillary devices with MRI systems at different magnetic fields (see http://www.mrisafety.com/). According to the new system of terms recently developed by the ASTM, devices are classified as MR-safe, MR-conditional, or MR-unsafe. MR-safe indicates no known hazards in all MR environments, whereas MR-unsafe indicates the converse. MR-conditional indicates devices can be used under specific MRI conditions, determined by specific device testing.

For obvious reasons, the safety aspects related to MRI ancillary devices are subject to established international standards or guidelines that are applicable in different countries. Although similar standards have been developed to characterize the electromagnetic emissions from such devices, associated with regulatory bodies (e.g. see http://www.fcc.gov/oet/), these standards do not evaluate specifically the potential for electromagnetic interference between the device and the MRI system itself. Electromagnetic interference primarily affects SNR and contrast-to-noise ratio in a manner that has become more stringent with time as the fidelity of MRI systems continues to improve with new technological innovation. Notwithstanding this issue, the static and time-varying electromagnetic fields used in MRI systems usually ensure that conventional electrical design approaches are not feasible, and necessitate dedicated ancillary device development.

In this regard, optoelectronics have played and continue to play a very important role in development of devices that generate biosignals measured concurrently with MRI. Optoelectronics provide several major advantages: 1) the use of non-ferromagnetic optical fibres instead of coaxial cables eliminates a major route for unintended electromagnetic emissions within the critical RF bandwidth of the MRI system, while eliminating possible magnetic field distortions and attractive forces; and 2) optical fibres are not susceptible to interference from external RF fields. For these compelling reasons, optoelectronics have been incorporated into the receiver chains of modern MRI systems, with substantial SNR benefits. Ancillary devices are important when it comes to the field of blood oxygenation level-dependent (BOLD) functional MRI (fMRI) neuroscience research (Ogawa, Tank et al. 1992; Hennig, Speck et al. 2003). Functional MRI is a popular imaging tool used in neuroscience because it measures brain activation associated with cognitive, sensory or motor tasks and behaviours. Compared to other functional neuroimaging techniques like positron emission tomography, near infrared diffuse optical tomography, electroencephalography (EEG), and magnetoencephalography, the BOLD fMRI technique has very attractive characteristics. Neither ionizing radiation nor injectable contrast agents are required. Activity of neurons is inferred through coupled changes in blood oxygenation, volume and flow that are reflected in magnetic resonance signals. Extensive research has addressed the biophysical basis for the fMRI signal, and enough is known about the neurovascular relationship at this point (Logothetis, Pauls et al. 2001; Menon 2001; Attwell and Iadecola 2002) to be confident that fMRI signals are strongly coupled with brain activity in healthy individuals and many patient populations (although there are provisos (Cohen, Ugurbil et al. 2002; Roc, Wang et al. 2006)). The spatial and temporal resolution provided by fMRI (millimeters and seconds, respectively) is not presently achievable throughout the brain volume by any other functional neuroimaging method.

During fMRI, brain activity is often measured for a specific behavioural task. The details of the task are determined by the mental processing under study, be it sensory, motor, or cognitive. With many tasks, there remains a need for an additional, concurrent measurement that serves either to quantify behavioural performance, or to provide an additional probe of mental state. Through appropriate signal processing, these concurrent measurements can be used retrospectively during post-processing to improve the detection of brain activity by fMRI analysis methods (Friston, Holmes et al. 1995).

This chapter focuses on biosignals (i.e. electrical signals indicative of physiological function) that can be measured simultaneously during fMRI. The focus herein will be on example biosignals that complement and augment brain activation maps calculated from fMRI data. In each of the cases that we will explore, optoelectronics have played a critical role in technological advancement. Three biosignals are considered: electroencephalography (EEG), electromyography (EMG) and electrodermal activity (EDA). Each technique has a common methodological approach: electrodes placed on the skin, amplification of the signals of interest, followed by digitization and subsequent transmission of signals from the magnet room to a display or recording device. Early research in this field identified the safety considerations associated with using electrodes in the MRI environment (Lemieux, Allen et al. 1997; Allen, Polizzi et al. 1998). EEG, EMG, and EDA devices are now commercially available for fMRI applications.

2. Electroencephalagraphy and functional MRI

Electroencephalography (EEG) is a non-invasive technique used to measure voltages on the scalp that are produced by the electrical current related to neuronal signaling in the brain. Participants wear a cap covered with numerous electrodes (typically ranging between 32 and 128). Neurons in the brain are electrically charged cells by virtue of the membrane proteins that pump ions in and out of the cell to create a resting potential. Signal propagation along neurons occurs by action potentials, which are rapid changes in the membrane potential caused by the opening of ionic channels. The electrodes detect the sum voltage signals, primarily from post-synaptic potentials, that are conducted to the surface of the head. Pyramidal neurons in the brain's cortex are thought to produce the strongest EEG signals because of their alignment and their firing characteristics. The frequency range of EEG signals is from 4 to 100 Hz and can been deconstructed into different frequency bands.

For example, alpha rhythms are found within 8–12 Hz activity and are present when the participant is awake with their eyes closed (Goldman, Stern et al. 2002).

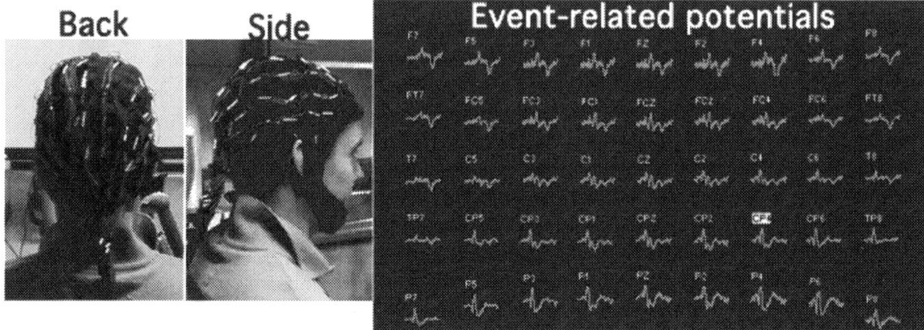

Fig. 1. Left: Back and side views of a participant wearing a 64-channel electroencephalography (EEG) cap. Wires from individual electrodes are collected in a bundle at the back of the head. Right: Event-related potentials (ERP) are shown across each of the different electrode sites, listed according to their position on the head. In this example a motor response (button press) is made in response to a decision task. On inspection, the electrode sites in the region surrounding P1 and P3 have large ERP amplitudes. EEG ERP are typically in the micro-volt range. Data shown in this figure were collected outside of the MRI scanner room and are therefore free of MRI-related artifacts.

EEG voltages are low amplitude (typically microvolts) and are easily contaminated by environmental noise. As a result, EEG can involve lengthy temporal recording over tens of minutes. Signal averaging over multiple trials is commonly employed in the study of event-related potentials (ERP), the term used to describe the EEG signals that are derived from tasks that are separated into single or multiple brief events to resolve different temporal and spatial aspects of neural activity (Figure 1). Further, intensive signal processing is often used to reduce noise levels. Within an MRI system, EEG signals are additionally degraded to the point that conventional EEG hardware cannot be utilized. First, time-varying changes in the magnetic flux through a loop geometry (which can be created through the EEG electrodes and wires) will produce an electromotive force (EMF) by Faraday's law of induction (Allen, Polizzi et al. 1998). These voltages can be produced in EEG time series data either by movement of an electrode in the static magnetic field, or by a stationary electrode in the presence of a time-varying magnetic field. The former case introduces artifacts arising from head motion, which can be postural in nature, or can arise from the pulsatility of blood flow in the scalp, an effect known as the "ballistocardiogram artifact" (Allen, Polizzi et al. 1998). These artifacts are typically approximately the same order of magnitude as EEG signals. Regarding the latter case, the magnetic field changes produced by rapidly varying imaging gradients can produce very large artifacts, several orders of magnitude larger than EEG signals and the primary MRI-induced artifact in EEG data.

Fortunately, gradient artifacts can be greatly attenuated through optoelectronic technology. Today, EEG systems for use with fMRI typically use twisted-pair electrodes with short lead lengths, local amplification, and conversion to optical signals to reduce EMF artifacts substantially. This strategy ensures that if the MRI system is sufficiently stable in the temporal domain, and EEG data are sampled at sufficiently high frequency, characteristic artifacts from specific gradient waveforms can be determined by temporal averaging and then simply subtracting from EEG data, effectively suppressing the problem (Allen, Polizzi et al. 1998).

There are also mechanisms by which EEG hardware influences MRI signals. During EEG-fMRI, the EEG cap (Figure 1) can cause spatial distortions in some types of MRI data acquisition, particularly echo planar imaging   (Jezzard and Balaban 1995) and spiral imaging (Glover and Lai 1998) that are commonly used in BOLD fMRI. These image distortions arise from spatial non-uniformity imposed on the static magnetic field due to the magnetic susceptibility properties of the cap, electrodes, electrode gels, wires, or pre-amplifiers, and due to the magnetic fields associated with the currents induced in EEG hardware according to the mechanisms summarized above (Bonmassar, Hadjikhani et al. 2001). Empirically, water-based electrode gels have been found superior to oil-based gels within the MRI environment, because the water gels have magnetic susceptibility properties more similar to those of the head (Bonmassar, Hadjikhani et al. 2001). Irrespective of the type of electrode gel, the function of the gel is to reduce the scalp resistance to improve the effective electrode contact and improve EEG signal strength. Unfortunately, the gel has potential to localize RF power deposition during the resonant excitation component of MR image formation. Although this effect is not problematic at 1.5 T and 3.0 T for BOLD fMRI, which involves relatively modest RF excitation, substantial heating is possible when EEG is conducted simultaneously with other MRI acquisitions. The effect is more pronounced at 3.0 T, as power deposition is proportional to the square of the static magnetic field. For this reason, certain MRI applications (for example, "fast spin echo imaging" and perfusion imaging of cerebral blood flow using "arterial spin labeling") may be contraindicated during EEG due to concerns of exceeding heating safety thresholds established by regulatory agencies (e.g. the US Food and Drug Administration).

It is evident that there are numerous challenges involved in conducting simultaneous EEG and fMRI measurements. Addressing these challenges is worthwhile for several different motivations. The major clinical motivation is the investigation and evaluation of epilepsy (Warach, Ives et al. 1996; Seeck, Lazeyras et al. 1998; Schomer, Bonmassar et al. 2000; Benar, Gross et al. 2002; Benar, Aghakhani et al. 2003). Very succinctly, EEG can be used to identify the onset of epileptic discharges and fMRI can be used to help localize where the seizure originated. Although the extent to which this idea will work in clinical applications still needs further investigation, such work has potential in the future to assist in selecting for individual patients the appropriate therapeutic intervention, such as drug treatment, or neurosurgery.

It can be said that "the sum of EEG and fMRI measurements is greater than their individual parts". This is because due to the physical principles underlying each modality, fMRI and EEG signals are highly complementary. EEG reflects electrical activity of populations of neurons in the brain (i.e. pyramidal neurons), over millisecond timescales. However, EEG exhibits limited spatial resolution due to the well known, "inverse problem" of detecting multiple current sources within a volume based on the signals observed in an array of remote detectors (Plonsey 1963). The spatial resolution of EEG is typically on the order of 1 cm, and EEG signals attenuate substantially with depth within the brain. In contrast, BOLD-fMRI measures changes in blood oxygenation with much lower temporal resolution, but markedly improved spatial resolution throughout the entire brain volume. In addition to the complementary nature of fMRI and EEG, the highly variable nature of individual behaviour and brain activity as a function of time often argues for the acquisition of EEG and fMRI data simultaneously. For example, subtle memory and learning effects can mean that equivalent behavioural performance in repeated tasks, measured sequentially by fMRI and EEG may not be manifested as the same brain activity. In such circumstances, it is essential to acquire EEG and fMRI data simultaneously as brain activity evolves.

In one illustrative basic neuroscience application, fMRI was used to identify the brain regions that were correlated with the EEG alpha band rhythms while participants lay in the MRI scanner, awake with their eyes closed. This experiment is an example of a "resting-state" study, during which the participant is not required to perform specific behavioural tasks. Experiments that probe resting-state brain activity, and the correlations in resting-state activity that occur in specific networks throughout the brain, have been increasingly popular in recent years. EEG-fMRI is well poised to advance this line of research (Musso, Brinkmeyer et al. 2010). Due to lack of task demands, resting state studies are simpler to conduct on patient populations and therefore attractive from a clinical perspective.

3. Electromyography and functional MRI

Turning attention now to another important biosignal, electromyography (EMG) is the study of electrical signals underlying muscle activity. EMG recordings are often used to evaluate muscle firing patterns, such as during walking (Hof, Elzinga et al. 2002). This is possible because as in the case of EEG, muscle activity is recorded with millisecond temporal resolution. EMG is useful clinically in the study of numerous movement disorders and neurological diseases such as stroke, motor neurone disease and others (Kautz and Brown 1998; Moreland, Thomson et al. 1998).

Briefly summarizing how the EMG signal relates to muscle physiology, the basic cellular unit of skeletal muscle is the muscle fibre. The functional unit within the fibres is called the sarcomere, which contains thick and thin filaments that are arranged alongside a scaffolding of cross-bridges (Gulrajani 1998). A muscle contraction involves the shortening of the muscle fibres, as described by "sliding filament theory". This process involves the sliding movement of the thick and thin filaments within the muscle, over top of and toward each other. As the overlap between the thin and thick filaments increases, the muscle length decreases. Such a muscle contraction is initiated by electrical signals that are like volleys of voltages, known as action potentials, that occur at different fibres within the muscle and originate from the controlling motor neurons in the central nervous system. A single "motor unit" is defined as a single motor neuron and all the muscle fibres that it innervates. The EMG signal is the accumulated voltage from all motor units and muscle fibres as they each act to contract the muscle.

EMG measurements are usually made by placing electrodes on the surface of the skin above the muscle of interest, although subcutaneous placement is also possible. In comparison to EEG signals that although they are noisy they show oscillatory or transient waveform characteristics, EMG signals appear as noise bursts during muscle contractions. The EMG signal is a zero-mean stochastic process and the standard deviation of the EMG signal is proportional to: 1) the number of active motor units and 2) the rate at which the motor units are activated (Clancy, Morin et al. 2002). The EMG signal is typically larger in amplitude compared to the EEG signal, ranging up to approximately 50 mV over a bandwidth of 10 – 200 Hz (Basmajian and De Luca 1985). Signal processing for EMG typically involves rectifying the voltage signal, lowpass filtering, and then evaluating the resulting signal envelope.

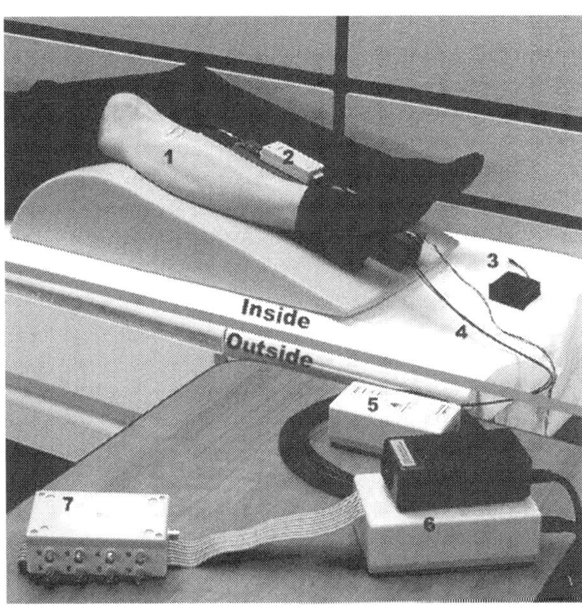

Fig. 2. MRI-compatible electromyography (EMG) system that is capable of measuring muscle biosignals inside the MRI environment and during concurrent fMRI data acquisition. The picture was taken outside of the MRI scanner room. The text "Inside" and "Outside" are used to denote hardware components that are positioned inside and outside of the MRI scanner room during EMG recording sessions. Item 2 is the pre-amplifier device that is capable of measuring 8 EMG channels (i.e. 8 muscles) simultaneously by making use of an optical multiplexer. The pre-amplifier includes optoelectronics to convert voltage signals to fibre optic signals that are sent out of the MRI scanner room via waveguide.

Several studies have involved combined EMG with fMRI (Toma, Honda et al. 1999; Liu, Dai et al. 2000; Dimitrova, Kolb et al. 2003; van Duinen, Zijdewind et al. 2005; MacIntosh, Baker et al. 2007). Similar to the early EEG-fMRI literature, initial EMG-fMRI recordings involved discarding portions of the EMG data collected during fMRI acquisition. This was because the time-varying gradient magnetic fields generated at specific time periods during fMRI saturated EMG preamplifiers. In these early studies, EMG signals were also measured by decreasing the temporal resolution of fMRI to create quiescent periods (Liu, Dai et al. 2000), an approach that was not ideal from an experimental point of view.

Some of the characteristics of a modern fMRI-compatible EMG system are large dynamic range and limited pass frequency bandwidth. With respect to the latter, EMG preamplifiers with 15 to 80 Hz and 33 to 80 Hz bandwidth were found to significantly reduce the amount of variance when EMG data were collected in a static magnetic field (MacIntosh, Baker et al. 2007). EMG signals are converted to optical signals at the preamplifier and transmitted outside the MRI scanner room using fibre optic cables. Figure 2 shows a set-up of an fMRI-compatible EMG system that was developed for use in our laboratory.

The amplitude of fMRI-induced artifacts in EMG signals depends on factors that are analogous to those described above for EEG-fMRI. First, the EMG artifact will depend on

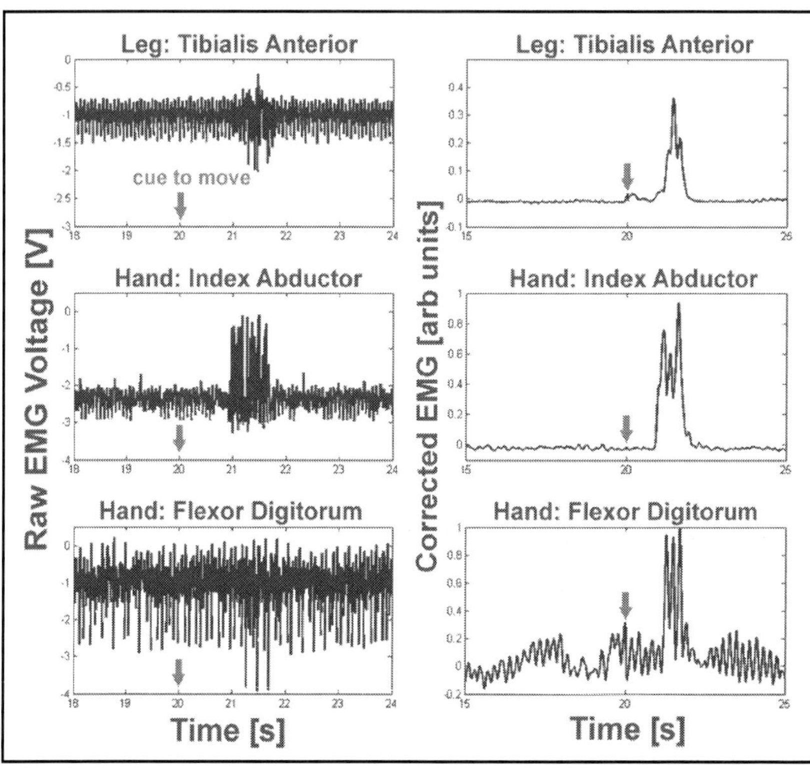

Fig. 3. Electromyography signals from 3 different muscles measured concurrently during an fMRI brain activation scan. On the left the raw amplified EMG signals are influenced by the muscle biosignal of interest as well as the signal artifact associated with the time-varying magnetic fields generated by the imaging gradient hardware active during fMRI. The muscle EMG signal occurs approximately 1 s after the participant is given a cue to move the muscle, which corresponds to t=20 s on the time axes. The three muscles are: 1) tibialis anterior, which is located on the lower leg and used to rotate the ankle; 2) index abductor, the muscle that moves the index finger towards the thumb; 3) flexor digitorum, the muscle on the forearm that flexes the wrist. EMG time series shown on the right have been corrected so as to suppress the artifact contribution. The suppression works well for the tibialis anterior and index abductor, but less so in the hardest case of the flexor digitorum. Details on post-processing approaches are discussed below. These data are based on published work (MacIntosh, Baker et al. 2007).

how much the electrode moves in the main magnetic field as a result of muscle contractions and relaxations. This can be problematic if the task is intended to create large movements. Second, the EMG artifact will depend on the locations of electrodes on the patient, in relation to the time-varying magnetic fields produced by the imaging gradients during fMRI. For example, the imaging gradient along the longitudinal axis of the magnet bore produces small time-varying fields at isocentre (where the head is located for fMRI). These fields increase linearly with distance from isocentre to a maximum approximately at the entrance to the magnet bore, and then reduce substantially in amplitude with increasing distance outside the magnet, toward the end of the patient table. Regarding this point, Figure 3 shows that the fMRI-induced EMG artifact is more substantial for recordings of hand muscles, compared to those of the lower leg, and that fMRI-induced artifact at the forearm is even larger, as observed in EMG signals from a muscle that moves the wrist (MacIntosh, Baker et al. 2007). These effects will depend on the specific MR imaging sequence used, as well as the specific characteristics of the EMG preamplifiers.

EMG-fMRI studies are useful for improving understanding of how the brain controls muscle movements, as reflected in recent literature (van Duinen, Zijdewind et al. 2005). We have shown that the use of EMG to estimate the onset, offset, and duration of brief muscle contractions, and subsequent use of this information in fMRI data processing, leads to improved visualization of motor networks (MacIntosh, Baker et al. 2007). EMG-fMRI has also been used to show differences in brain activation associated with active, passive and electrically stimulated muscle contractions (Francis, Lin et al. 2009).

4. Electrodermal activity and functional MRI

The skin plays an important role in many biological processes such as sensory and motor exploration, immunity, and thermoregulation. In the latter process, eccrine sweat glands in the skin secrete water and salts to regulate body temperature. They are innervated by a structure in the brain known as the hypothalamus. Sympathetic nerves make the connection between the hypothalamus and the sweat gland, as part of the autonomic nervous system (sometimes referred to as the visceral nervous system). Secretion of sweat in skin is therefore a window into the human body's visceral controls in the sense that they occur without being conscious. An example of the autonomic nervous system acting with little willful control is the fight or flight response that occurs when we are in a stressful situation.

The voltages related to electrodermal activity (EDA) are very weak and poorly localized. Therefore, skin conductance is the preferred physiological recording parameter to characterize this aspect of the autonomic nervous system. Changes in EDA are described on two time scales of changing skin conductivity: skin conductance level (SCL), reflecting low frequency changes (i.e. approximate frequencies are < 0.1 Hz); and skin conductance responses (SCRs), reflecting more rapid changes (i.e. approximate frequencies are > 0.5 Hz). Both types of fluctuations obviously occur at much lower frequencies than EEG and EMG measures.

EDA is also a useful surrogate measure of cognitive and emotional processes and therefore has appeal in psychology and neuroscience research. One important aspect of cognition is arousal, the psychological state of heightened activity or attention. Being alert by directing our attention to a specific task allows us to carry out certain actions or behaviours. Arousal plays an important role when our world is altered quickly, such as due to a postural instability (Sibley, Lakhani et al. 2010; Sibley, Mochizuki et al. 2010). In this case, autonomic and cortical responses help to avoid a fall or injury. EDA responses relate to arousal as they are part of the behavioural inhibition system (with sensitivity to punishment and avoidance motivation), whereas heart rate is part of the behavioural activation system (sensitivity to reward and approach motivation) (Fowles 1980). Interestingly, EDA and heart rate are two biosignals that are used in a lie-detector test.

The first EDA-fMRI studies investigated the effects of attention on cognitive tasks (Williams, Brammer et al. 2000). Subsequently, other researchers used EDA and fMRI to monitor brain responses associated with gambling (Patterson, Ungerleider et al. 2002), biofeedback (Critchley, Melmed et al. 2001), motor tasks (MacIntosh, Baker et al. 2007), motor recovery after stroke (MacIntosh, McIlroy et al. 2008) and sense of effort (Mochizuki, Hoque et al. 2009). A simple circuit diagram for measuring skin conductance as part of an fMRI-compatible EDA system is shown in Figure 4. The optoelectronic components are listed as IF-E96 and IF-D91. IF-E96 is a plastic fibre optic red light emitting diode (commercially available at: http://i-fiberoptics.com/). Optical cable can easily be mounted to this component to transmit EDA signals outside of the MRI scanner room via a waveguide and converted back to a voltage at the IF-D91 photodiode detector. The fMRI-compatible EDA system shown in Figure 4 is MRI-safe for use at 1.5 and 3 T scanners. Also shown is the battery box that contains a non-magnetic 9V lithium battery.

The physiological frequency content in EDA is narrow and low frequency, so sharp lowpass filters are advisable. One hardware implementation that is therefore advisable in EDA design is a slew-rate limited preamplifier (Ives, Warach et al. 1993), which yields more effective lowpass filtering. The instrumentation amplifier is shown as INA-120 in Figure 4.

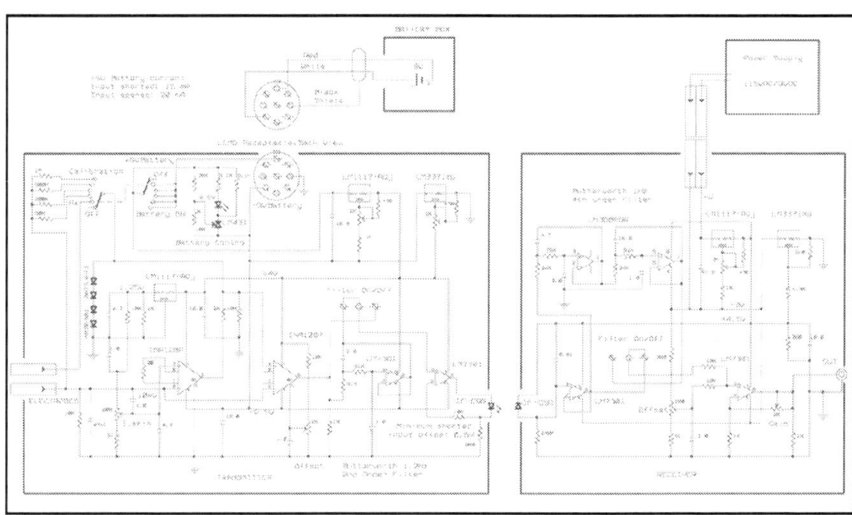

Fig. 4. The circuit diagram for a low-cost MRI-compatible electrodermal activity (EDA) system that was prototyped at Sunnybrook Research Institute. The circuit design is based on the Wheatstone bridge preamplifier with a gain of 1000 and is based on early work described by others (Shastri, Lomarev et al. 2001). Additional details associated with this system can be found elsewhere (MacIntosh, Mraz et al. 2007).

EDA-fMRI can be used to look at how behaviour changes over the course of an experiment. If we use the EDA data shown in Figure 5 as a model in the fMRI analysis then we can identify the brain regions that are correlated with the EDA trial-by-trial trends. Figure 6 shows the EDA trial averages that were used as model predictors in the fMRI analysis.

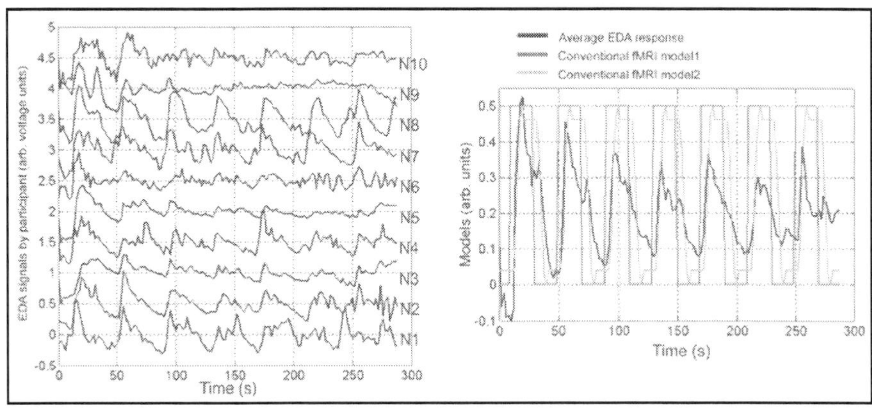

Fig. 5. Left: Electrodermal activity time series data from healthy adult participants who performed a hand motor task that required concentration and arousal to perform effectively. Each trace corresponds to a single participant (i.e. N1 corresponds to participant 1, N2 corresponds to participant 2, for a total of N=10 participants). The task, affectionately known as the "Spock task", involved separating the middle and ring fingers, requiring some concentration and dexterity. Participants alternated between 20 seconds of performing the task and 20 second of rest. Fluctuations in EDA signals across the seven task trials are more evident in some participants compared to others. Right: The average EDA time series is shown (black) from all participants. Two typical fMRI models are shown in gray for comparison and it is clear that the EDA time series data has unique temporal features that are not captured by the fMRI models (which assume constant neural activity). For example, the amplitude of the EDA signals at later trials is decreased compared to early trials. These data are based on published work (MacIntosh, Mraz et al. 2007).

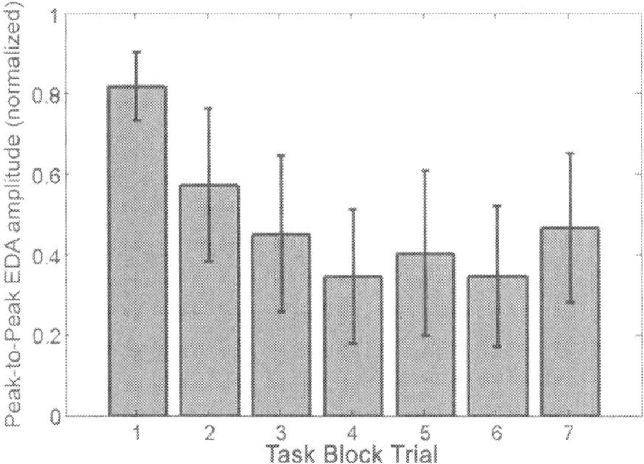

Fig. 6. Normalized peak-to-peak EDA amplitude plotted as a function of the task block trials. After the second trial the amplitude is consistently reduced. Error bars denote standard deviation.

5. Other examples of optoelectronics used in fMRI

A recent example of optoelectronics used in fMRI is a system that is able to measure diastolic blood pressure through an fMRI experiment (Myllyla, Elseoud et al. 2011). This device uses a fibre optic sensor to measure skin movements associated with blood pressure changes in the underlying blood vessels. The system works in the MRI scanner room and is useful for collecting blood pressure data because it is non-invasive (i.e. no intra-arterial line) and it does not involve an inflatable cuff. Fibre optic technology has also been used to monitor movements, or kinematics, during fMRI. This has been done to measure amplitude and velocity in ankle movements, using optical fibres specially configured as bend sensors (MacIntosh, Mraz et al. 2004).

6. Software solutions for reducing fMRI-induced artifacts in EEG, EMG and/or EDA

Having discussed some biosignals that are of interest during fMRI, it is important to now consider post-processing software approaches to minimize or correct for some residual or inevitable fMRI-induced artifacts. For software approaches to be effective, the measured voltages must fall within the dynamic range of the amplifier system. There is little that can be done when an amplifier saturates. Post-processing software approaches that attempt to reduce the fMRI-induced artifacts include: adaptive noise canceling, direct artifact subtraction (Allen, Polizzi et al. 1998; Allen, Josephs et al. 2000), filtering (Hoffmann, Jager et al. 2000), and principal component analysis (PCA) and independent component analysis (ICA) (McKeown, Makeig et al. 1998; McKeown and Radtke 2001). In the direct artifact subtraction approach, the average artifact signal is calculated over a short time-window and then retrospectively subtracted from the original biosignal (Allen, Josephs et al. 2000). The timing of the artifact with respect to the MRI acquisition is critical, therefore high temporal resolution sampling (i.e. 10 microsec) of the biosignal is required.

PCA and ICA are data-driven feature extraction and data reduction techniques that have been used to reduce fMRI-induced signal contamination in EEG data (Negishi, Abildgaard et al. 2004) and EMG (MacIntosh, Baker et al. 2007; van der Meer, Tijssen et al. 2010). In PCA, biosignal data are decomposed into orthogonal principal components that are ranked according to the amount they explain the total variance. In ICA, biosignal data are represented as linearly independent components (ICs), in what is sometimes referred to as blind source separation. Both PCA and ICA rely on redundancy across trials or across sampling sources to separate signals. Although useful, it can sometimes be challenging with these techniques to separate the components of interest (those containing biosignal) from those that contain artifact and noise.

7. Conclusion

Several components of an MRI system can affect the operation of ancillary devices for use during imaging. The RF transmit and receive fields, large static magnetic field, and time-varying gradient magnetic fields must all be considered in device design. Measuring electrical biosignals in an MRI environment poses a significant challenge as a consequence, but over the past decades numerous solutions have been proposed and successfully implemented. The scope of human brain mapping continues to grow with assistance from an increasing number of peripheral devices that can be used during fMRI to make concurrent measurements. Viable products are now available that enable high quality, simultaneous fMRI, EEG, EMG, and EDA recordings.

This chapter illustrates that optoelectronics are a recurring engineering theme when it comes to measuring biosignals in the MRI environment. Conversion to optical signals, which can then be passed outside the magnet room via fibre optic cable and subsequently converted back to electrical signals in the console area with minimal loss or signal degradation, is of substantial benefit. Optoelectronics will, therefore, continue to play a key role in MRI-compatible hardware design.

8. Acknowledgment

The authors declare that they have no conflicts of interest with respect to the content of this work. We would like to acknowledge the Heart and Stroke Foundation Centre for Stroke Recovery as well as Richard Mraz, Nicole Baker and John Ives for their technical expertise and assistance. We would like to acknowledge Sunnybrook Research Institute electronics engineer, Vladimir Verpakhovski, for fabrication of the electrodermal activity system. We thank Dr. Laura Middleton and Mike Sage for EEG-setup photos shown in Figure 1. And we thank David Dongkyung Kim for comments on the chapter.

9. References

Allen, P. J., O. Josephs, et al. (2000). "A method for removing imaging artifact from continuous EEG recorded during functional MRI." *Neuroimage* 12(2): 230-239.

Allen, P. J., G. Polizzi, et al. (1998). "Identification of EEG events in the MR scanner: the problem of pulse artifact and a method for its subtraction." *Neuroimage* 8(3): 229-239.

Attwell, D. and C. Iadecola (2002). "The neural basis of functional brain imaging signals." *Trends Neurosci* 25(12): 621-625.

Basmajian, J. V. and C. J. De Luca (1985). *Muscles Alive. Their Functions Revealed by Electromyography.*

Benar, C., Y. Aghakhani, et al. (2003). "Quality of EEG in simultaneous EEG-fMRI for epilepsy." *Clin Neurophysiol* 114(3): 569-580.

Benar, C. G., D. W. Gross, et al. (2002). "The BOLD response to interictal epileptiform discharges." *Neuroimage* 17(3): 1182-1192.

Bonmassar, G., N. Hadjikhani, et al. (2001). "Influence of EEG electrodes on the BOLD fMRI signal." *Hum Brain Mapp* 14(2): 108-115.

Clancy, E. A., E. L. Morin, et al. (2002). "Sampling, noise-reduction and amplitude estimation issues in surface electromyography." *J Electromyogr Kinesiol* 12(1): 1-16.

Cohen, E. R., K. Ugurbil, et al. (2002). "Effect of basal conditions on the magnitude and dynamics of the blood oxygenation level-dependent fMRI response." *J Cereb Blood Flow Metab* 22(9): 1042-1053.

Critchley, H. D., R. N. Melmed, et al. (2001). "Brain activity during biofeedback relaxation: a functional neuroimaging investigation." *Brain* 124(Pt 5): 1003-1012.

Dimitrova, A., F. P. Kolb, et al. (2003). "Cerebellar responses evoked by nociceptive leg withdrawal reflex as revealed by event-related FMRI." *J Neurophysiol* 90(3): 1877-1886.

Fowles, D. C. (1980). "The three arousal model: implications of gray's two-factor learning theory for heart rate, electrodermal activity, and psychopathy." *Psychophysiology* 17(2): 87-104.

Francis, S., X. Lin, et al. (2009). "fMRI analysis of active, passive and electrically stimulated ankle dorsiflexion." *Neuroimage* 44(2): 469-479.

Friston, K. J., A. P. Holmes, et al. (1995). "Analysis of fMRI time-series revisited." *Neuroimage.* 2(1): 45-53.

Glover, G. H. and S. Lai (1998). "Self-navigated spiral fMRI: Interleaved versus single-shot." *Magnetic Resonance in Medicine* 39: 361-368.

Goldman, R. I., J. M. Stern, et al. (2002). "Simultaneous EEG and fMRI of the alpha rhythm." *Neuroreport* 13(18): 2487-2492.

Gulrajani, R. (1998). *Bioelectricity and Biomagnetism,* John Wiley & Sons Inc.

Hennig, J., O. Speck, et al. (2003). "Functional magnetic resonance imaging: a review of methodological aspects and clinical applications." *J Magn Reson Imaging* 18(1): 1-15.

Hof, A. L., H. Elzinga, et al. (2002). "Speed dependence of averaged EMG profiles in walking." *Gait Posture* 16(1): 78-86.

Hoffmann, A., L. Jager, et al. (2000). "Electroencephalography during functional echo-planar imaging: detection of epileptic spikes using post-processing methods." *Magn Reson Med* 44(5): 791-798.

Ives, J. R., S. Warach, et al. (1993). "Monitoring the patient's EEG during echo planar MRI." *Electroencephalogr Clin Neurophysiol* 87(6): 417-420.

Jezzard, P. and R. S. Balaban (1995). "Correction for geometric distortion in echo planar images from B0 field variations." *Magn Reson Med* 34(1): 65-73.

Kautz, S. A. and D. A. Brown (1998). "Relationships between timing of muscle excitation and impaired motor performance during cyclical lower extremity movement in post-stroke hemiplegia." *Brain* 121 (Pt 3): 515-526.

Lemieux, L., P. J. Allen, et al. (1997). "Recording of EEG during fMRI experiments: patient safety." *Magn Reson Med* 38(6): 943-952.

Liu, J. Z., T. H. Dai, et al. (2000). "Simultaneous measurement of human joint force, surface electromyograms, and functional MRI-measured brain activation." *J Neurosci Methods* 101(1): 49-57.

Logothetis, N. K., J. Pauls, et al. (2001). "Neurophysiological investigation of the basis of the fMRI signal." *Nature.* 412(6843): 150-157.

MacIntosh, B. J., S. N. Baker, et al. (2007). "Improving functional magnetic resonance imaging motor studies through simultaneous electromyography recordings." *Hum Brain Mapp* 28(9): 835-845.

MacIntosh, B. J., W. E. McIlroy, et al. (2008). "Electrodermal recording and fMRI to inform sensorimotor recovery in stroke patients." *Neurorehabil Neural Repair* 22(6): 728-736.

MacIntosh, B. J., R. Mraz, et al. (2004). "Optimizing the experimental design for ankle dorsiflexion fMRI." *Neuroimage* 22(4): 1619-1627.

MacIntosh, B. J., R. Mraz, et al. (2007). "Brain activity during a motor learning task: An fMRI and skin conductance study." *Hum Brain Mapp* 28(12): 1359-1367.

McKeown, M. J., S. Makeig, et al. (1998). "Analysis of fMRI data by blind separation into independent spatial components." *Human Brain Mapping* 6: 160-188.

McKeown, M. J. and R. Radtke (2001). "Phasic and tonic coupling between EEG and EMG demonstrated with independent component analysis." *J Clin Neurophysiol* 18(1): 45-57.

Menon, R. S. (2001). "Imaging function in the working brain with fMRI." *Current Opinion in Neurobiology.* 11(5): 630-636.

Mochizuki, G., T. Hoque, et al. (2009). "Challenging the brain: Exploring the link between effort and cortical activation." *Brain Res* 1301: 9-19.

Moreland, J. D., M. A. Thomson, et al. (1998). "Electromyographic biofeedback to improve lower extremity function after stroke: a meta-analysis." *Arch Phys Med Rehabil* 79(2): 134-140.

Musso, F., J. Brinkmeyer, et al. (2010). "Spontaneous brain activity and EEG microstates. A novel EEG/fMRI analysis approach to explore resting-state networks." *Neuroimage* 52(4): 1149-1161.

Myllyla, T. S., A. A. Elseoud, et al. (2011). "Fibre optic sensor for non-invasive monitoring of blood pressure during MRI scanning." *J Biophotonics* 4(1-2): 98-107.

Negishi, M., M. Abildgaard, et al. (2004). "Removal of time-varying gradient artifacts from EEG data acquired during continuous fMRI." *Clin Neurophysiol* 115(9): 2181-2192.

Ogawa, S., D. W. Tank, et al. (1992). "Intrinsic signal changes accompanying sensory stimulation: functional brain mapping with magnetic resonance imaging." *Proc Natl Acad Sci U S A* 89(13): 5951-5955.

Patterson, J. C., 2nd, L. G. Ungerleider, et al. (2002). "Task-independent functional brain activity correlation with skin conductance changes: an fMRI study." *Neuroimage* 17(4): 1797-1806.

Plonsey, R. (1963). "Reciprocity Applied to Volume Conductors and the Ecg." *IEEE Trans Biomed Eng* 10: 9-12.

Roc, A. C., J. Wang, et al. (2006). "Altered hemodynamics and regional cerebral blood flow in patients with hemodynamically significant stenoses." *Stroke* 37(2): 382-387.

Schaefers, G. (2008). "Testing MR safety and compatibility: an overview of the methods and current standards." *IEEE Eng Med Biol Mag* 27(3): 23-27.

Schomer, D. L., G. Bonmassar, et al. (2000). "EEG-Linked functional magnetic resonance imaging in epilepsy and cognitive neurophysiology." *J Clin Neurophysiol* 17(1): 43-58.

Seeck, M., F. Lazeyras, et al. (1998). "Non-invasive epileptic focus localization using EEG-triggered functional MRI and electromagnetic tomography." *Electroencephalogr Clin Neurophysiol* 106(6): 508-512.

Shastri, A., M. P. Lomarev, et al. (2001). "A low-cost system for monitoring skin conductance during functional MRI." *J Magn Reson Imaging* 14(2): 187-193.

Sibley, K. M., B. Lakhani, et al. (2010). "Perturbation-evoked electrodermal responses are sensitive to stimulus and context-dependent manipulations of task challenge." *Neurosci Lett* 485(3): 217-221.

Sibley, K. M., G. Mochizuki, et al. (2010). "The relationship between physiological arousal and cortical and autonomic responses to postural instability." *Exp Brain Res* 203(3): 533-540.

Toma, K., M. Honda, et al. (1999). "Activities of the primary and supplementary motor areas increase in preparation and execution of voluntary muscle relaxation: an event-related fMRI study." *J Neurosci* 19(9): 3527-3534.

van der Meer, J. N., M. A. Tijssen, et al. (2010). "Robust EMG-fMRI artifact reduction for motion (FARM)." *Clin Neurophysiol* 121(5): 766-776.

van Duinen, H., I. Zijdewind, et al. (2005). "Surface EMG measurements during fMRI at 3T: accurate EMG recordings after artifact correction." *Neuroimage* 27(1): 240-246.

Warach, S., J. R. Ives, et al. (1996). "EEG-triggered echo-planar functional MRI in epilepsy." *Neurology* 47(1): 89-93.

Williams, L. M., M. J. Brammer, et al. (2000). "The neural correlates of orienting: an integration of fMRI and skin conductance orienting." *Neuroreport* 11(13): 3011-3015.

10

Applications and Optoelectronic Methods of Detection of Ammonia

Paul Chambers, William B. Lyons, Tong Sun and Kenneth T.V. Grattan
City University, London
United Kingdom

1. Introduction

This chapter describes applications of ammonia in agriculture, pharmaceutical and environmental industries and optoelectronic methods of its detection.

In Section 2 the discovery, the chemical structure, reactivity and application of ammonia are reviewed. Applications in agriculture, cleaning products, pharmaceutical industries, beauty products, the benefits and dangers of ammonia to human health, environment and industrial processes are also discussed.

Section 3 describes the rotational-vibrational molecular processes that cause the optical absorption of light in the infrared spectrum. The infrared and ultraviolet absorption spectra are also shown. Data relating to the absorption of light by ammonia at ultraviolet wavelengths is also shown.

Existing optical ammonia gas detection methods that utilise lasers and broadband sources and the evanescent field of an optical fibre are described in Section 4. This includes a discussion of optical sources, optical fibers, gas cell designs and detectors that are used in optoelectronic gas sensing systems.

In section 5, the limiting effects of fundamental noise sources, such as photon noise, resistor noise and optical source noise on sensor sensitivity are described. The selective performance of optoelectronic gas sensors is also discussed, i.e. the discrimination of the sensor to different gases.

2. Ammonia: the chemical

The ammonia molecule consists of one nitrogen atom covalently bound to three hydrogen atoms, the pyramidal configuration is shown in Figure 1. The structure of the covalent bond results in the compound being neutral in charge, but there remain two unfilled electron pairs in the valence band. As ammonia has an unfilled valence band, it is a weak base, with a Ph. of approximately 12. Ammonia exists in the gas phase in the in the environment, as the boiling point of ammonia is -33.35°C.

The production of ammonia by the distillation of animal hoofs, horns and hide scraps is recognised as a very old method of the extraction of ammonia. Written references to the use of ammonia date back to the thirteenth century in Catalan literature (Miller (1981)), while Felty (1982) noted that the name ammonia is derived from the salt sal-ammoniac. Sal-ammoniac (salt comprised of ammonium chloride) or salt of Ammon was named after the Egyptian chief

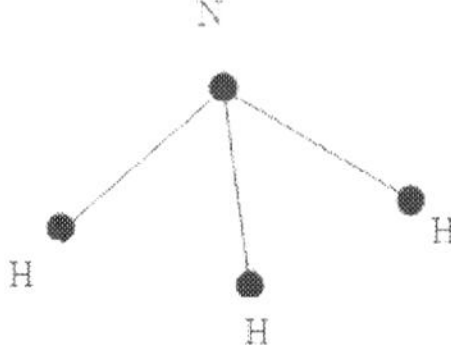

Fig. 1. Diagram of ammonia molecule consisting of 1 nitrogen and 3 hydrogen atoms which are covalently bound.

God Ammon, as it is possible that it was extracted from camel dung near the Temple of Jupiter, in ancient Egypt (in present day Libya) in around 332 B.C..

In Miller (1981), it is also described how Johann Kunkel van Lowenstern noticed that ammonia gas could be produced by the addition of lime to sal-ammoniac. In 1773 Joseph Priestley was the first chemist to identify and extract pure gaseous ammonia by applying heat to aqueous ammonia mixed with sal-ammoniac. Using the Haber-Bosch process of nitrogen fixation(Howard & Rees (1996)), ammonia is now the most widely synthetically produced chemical. Ammonia reacts with acids or neutral substances, such as water, sulfuric acid and nitric acid. These reactions result in the formation of anhydrous ammonia, which is acidic, ammonium sulfate and ammonium nitrate. Compounds in the Amine functional group are derived from the ammonia molecule, where one, or more, of the hydrogen atoms is replaced with an alternate chemical arrangement.

Ammonia currently has a wide range of uses and applications. When used in agriculture, ammonia forms a source of nitrides in fertilisers which promotes plant growth. In cleaning products the action of ammonia hydroxide, as an acid , aids in removing contamination from surfaces. In hair conditioners, ammonia aids the blending of colour into hair. Pharmaceutical processes use ammonia as a buffer to control the P.H. level for solution preparation. Dissociated ammonia atmospheres are employed in steel processing for the annealing of steel to aid corrosion resistance (Levey & van Bennekom (1995),Samide et al. (2004)). Chilled ammonia is used, as a binding agent, to remove carbon dioxide from the exhausts from fossil fuel burning power plants (Darde et al. (2008)).

While ammonia gas is necessary for these processes, it is dangerous to people in excessive concentrations if inhaled, as anhydrous ammonia is corrosive (Close et al. (1980)). Ammonia is also destructive when present in semiconductor fabrication facilities (Sun et al. (2003)). For safety reasons and process monitoring applications, it is therefore important to monitor the concentrations of ammonia and optoelectronic methods can provide an accurate means of achieving this.

3. Optical absorption spectrum of ammonia

The literature relating to the vibrational and electronic optical absorption spectra is reviewed in this section. This includes data relating to the infrared and ultraviolet absorption spectra.

3.1 Infrared absorption spectrum

Incident optical radiation on the ammonia molecule causes vibrations of the inter-atomic distances between the nitrogen and hydrogen atoms in the pyramidal structure. This causes the partial absorption of optical power at characteristic wavelengths, including the original vibration and higher frequency (shorter wave-length) harmonics, resulting in the absorption spectrum. The rotational-vibrational modes, as reviewed by McBride & Nicholls (1972) are shown in Figure 2. The non-degenerate symmetric v_1 and v_2 vibrational modes shown in Figures 2(a) and 2(b), preserve the pyramidal shape, while the degenerate v_3 and v_4 modes, shown in figures 2(c) and 2(d) distort the three dimensional shape.

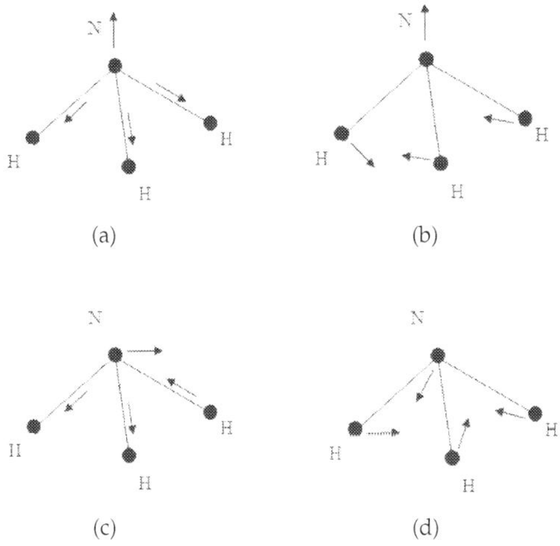

Fig. 2. The rotational-vibrational modes of the ammonia (NH_3) molecule
(a): v_1 vibration
(b): v_2 vibration
(c): v_3 vibration
(d): v_4 vibration

The optical absorbance of a gas is defined in terms of its cross-section. The absorbance cross-section is denoted by $\sigma(\lambda)$. The Beer-Lambert Law, which is shown in Equation 1, describes the relationship between the incident optical power, I_0, and the transmitted power intensity, I_1, at each wavelength, λ, in terms of the absorbance cross-section, $\sigma(\lambda)$, gas path-length, l, and gas concentration, c.

$$\ln\left(\frac{I_0}{I_1}\right) = \sigma(\lambda) \times l \times c \tag{1}$$

The cross-sections of atmospheric gases are contained in the HITRAN database. While the database is updated, with the last update in 2008 (http://www.hitran.com), the last update of the ammonia gas data was in the 1986 edition (Rothman et al. (1987)). The articles documenting the molecular vibrations of ammonia and their characteristic wave-lengths are detailed in Table 1.

The cross-sectional absorbance spectrum up to 8μ m using data from the Hitran database, which was detailed in Table 1, is shown in Figure 3. It can be observed that the maximum

Wavelength	Absorption Band	Reference
1.89–2.09 μm	v_1+v_4 v_3+v_4	Brown & Margolis (1996)
6.00 μm	$2v_2/v_4$	Cottaz et al. (2000)
5–8 μm	$3v_2 - v_2$ $v_2 + v_4 - v_2$	Cottaz et al. (2001)
4.00 μm	$4v_2 - v_2$ $v_1 - v_2$ $v_3 - v_2$	
3 μm	v_1 v_3 $2v_4$	Kleiner et al. (1999) Guelachvili et al. (1989)
4 μm	$3\,v_2/v_2 + v_4$	Kleiner et al. (1995)
1.89–2.09 μm	v_1+v_2 v_2+v_3 $v_2+2\,v_4$	Urban et al. (1989)

Table 1. The rotational-vibrational coupling of ammonia gas that gives the infra-red absorption of ammonia gas

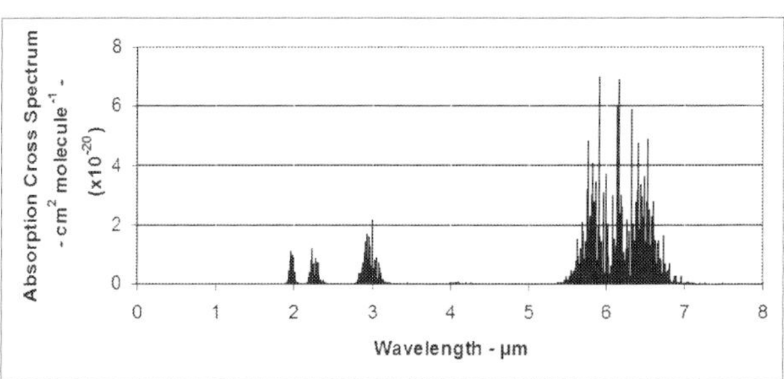

Fig. 3. The infrared absorption cross-section of ammonia gas. The data were selected from the 1986 edition of the HITRAN database (Rothman et al. (1987)), which includes the data described in Table 1

cross-section absorbance in this wavelength range is approximately 7×10^{-20} cm^2 molecule^{-1} at around 6 μ m.

3.2 Ultraviolet absorption spectrum

Ammonia also absorbs optical power at ultraviolet wavelengths. The ultraviolet electronic absorption is caused by the interaction of light with electrons in the valence band of the ammonia molecule (Burton et al. (1993)). The absorption spectrum of ammonia is shown in Figure 4. The data was obtained from the results of Cheng et al. (2006) that are contained in the Mainz UV spectral database (Keller-Rudek & Moortgat (2006)).

The peak absorbance shown in Figure 4 in the ultraviolet absorption spectrum is appoximately 2×10^{-17} cm^2 molecule^{-1}. This is around one thousand times greater than was the case in the

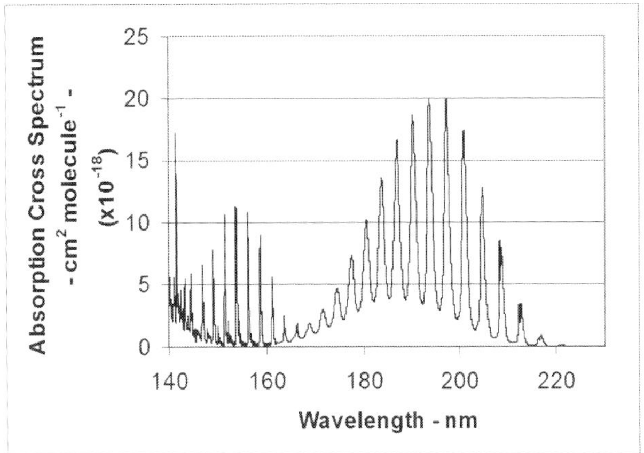

Fig. 4. Ulraviolet absorption cross-section absorption spectrum of NH_3. Absorption spectrum data shown were taken from Cheng et al. (2006)

infrared spectrum. As the absorbance cross section is in the exponent of Equation 1, this leads to a difference in the gas absorption of many orders of magnitude.

4. Optical methods of detection

This Section reviews a range of optoelectronic methods for the detection of gases including how they are applied for the detection of ammonia.

Early optical gas analysers relied upon the photoacoustic properties of gases, at the time this was referred to as the "Tyndall-Röntgen effect". The "Tyndall-Röntgen effect" in gases is analogous to the "Bell effect", which is the development of an audible sound arising from the intermittent exposure of a solid or liquid to radiation. Early gas sensing systems that utilised the photoacoustic effect were developed before, during and since World War II, in Britain, the U.S.S.R. and Germany. An example of an early gas detection method due to Veingerov (1938), which is described in Hill Hill & Powell (1968), is shown in Figure 5. The gas analyser, which was named an "optico-accoustic" analyser, operated by passing intensity modulated optical radiation from a Nernst Glower Source through a highly polished tube to a telephone receiver. The pressure variations induced by the intermittent optical beams resulted in a differing expansion of the gases present in the sample gas cell. This, in turn, induced the generation of acoustic tones that were picked up by the telephone earpiece (microphone). These tones were indicative of the gases present in the sample gas cell. The branch-resonator enabled the pressure fluctuations developed to be amplified, so that the detected signal could be enhanced.

Concurrently with the work by Veingerov, Luft developed a null-balance arrangement (see Hill & Powell (1968)). This, and systems developed from it, were later referred to as LIRA (Luft Infra-Red Analyser, Luft (1947)) type analysers, an example of which is shown in Figure 6.

The systems operate by passing two alternately chopped optical beams through a reference gas cell and a sample or measurement gas cell to a detector. Initially the device was

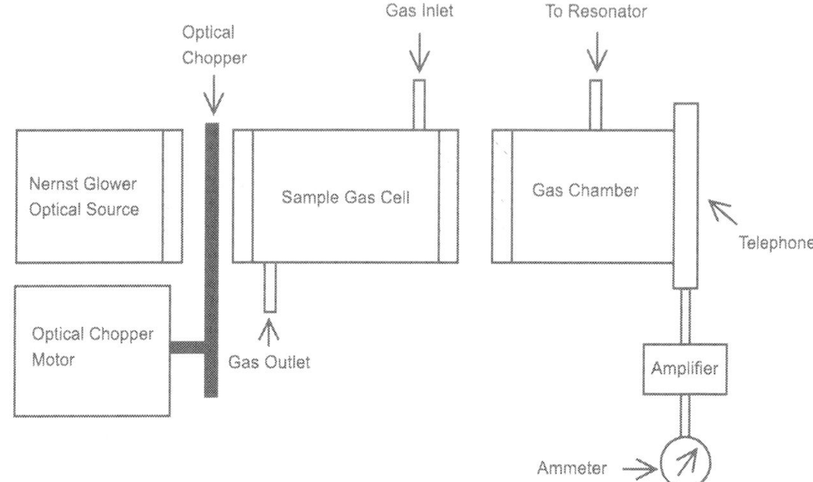

Fig. 5. An early "optico-acoustic" gas detection arrangement due to Vengerov (see references Hill & Powell (1968); Veingerov (1938))

"null-balanced" by filling both the reference and sample cells with a gas that had no absorption lines in the spectrum of interest and equalising the intensity of the optical beams by adjusting a blocking trimmer screw. The insertion of the sample gas into the sample or measurement cell leads to a signal modulation at the detector which is proportional to the target gas concentration in the sample gas. It was found that, by the insertion of a reference gas cell in series with the measurement gas cell, the selectivity of the system to the target gas could be improved. The LIRA system was able to detect CO_2 concentrations of less than 10 ppm (parts per million).

Hill & Powell (1968) also described the development of early gas analysers that were manufactured during the 1950s and 1960s and the development of early infra-red detectors.

Goody (1968) explored the possibility of selectively detecting a specified target gas by a correlation technique. Goody introduced a pressure modulated "cross-correlating spectrometer", a device which involved passing light from an optical source through two sequential gas cells and an optical filter, before impinging on an optical detector, see Figure 7. The first gas cell contained the gas volume to be analysed (the measurement cell) and the second contained only the target gas (the reference cell). By modulation of the target gas pressure within the reference gas cell, a modulation of the detected optical power at the output of the measurement gas cell was observed. This magnitude of the output modulation was related to the concentration of target gas within the measurement gas cell. The method showed high rejection of drifts in source power and had high rejection of contaminant gas. A NH_3 sensor based on this method was found to be 140 times less sensitive to N_2O contaminant gas, even though the spectral absorption of N_2O is significantly stronger than the spectral absorption of NH_3 in the band used.

Taylor et al. (1972) provide details of a similar system, intended to measure remotely (from a satellite) the temperature of the upper atmosphere from the spectral transmission of CO_2. The system gathered light reflected from the earths atmosphere, and passed it though a pressure

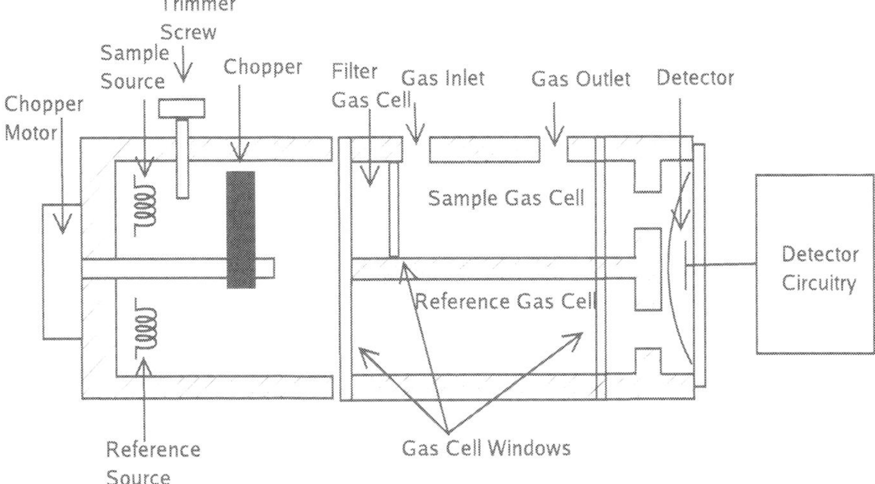

Fig. 6. A Null-Balance Lira (Luft Infra-Red Gas Analyser) gas detection system Hill & Powell (1968)

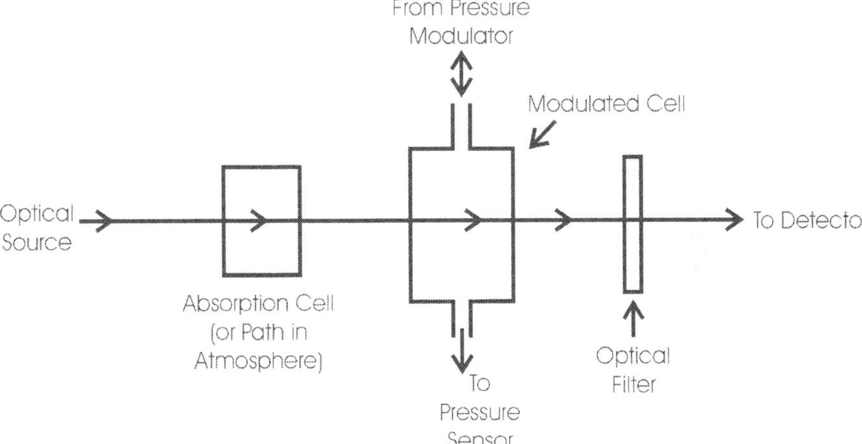

Fig. 7. Pressure Modulation Spectroscopy system (reproduced from Goody Goody (1968))

modulated reference gas cell to a detector. Their system showed a sensitivity of 1°C. This method utilised the spectral emission of CO_2 at 15 μm.

A reported method of modulating the transmission of the reference cell was that of Stark modulation. This is the line-splitting effect that results when a high electric field is applied to a gas. It is only effective on molecules having a significant dipole moment, e.g. H_2O, CH_4 etc. Edwards & Dakin (1993) investigated the use of Stark modulation for the detection of ammonia and water vapour, both of industrial significance, using optical fibre-based systems.

This system demonstrated the use of optical fibres for gas detection, an area which is now discussed in detail.

4.1 Optical fibre implementations

Optical detection systems using optical fibres offer a number of advantages over bulk optical sensing systems. This section develops the review by Dakin and Chambers (2004) to show the application of sensors using an optical fibre for the detection of ammonia. The principal advantage is that a robust passive sensing head may be sited remotely from the monitoring station, which is a useful feature in severe environments. This also allows for the development of multiplexed networked systems, where a single interrogation unit can monitor many low-cost passive sensing heads via a predictable propagation medium (i.e. the optical fibre). Conventional silica fibres have the disadvantage that transmission is restricted to the visible and near-infra-red region (0.6 μm to 2.0 μm). Fluoride and other fibres may be used to extend the operation of these sensors further into the infra-red, allowing accurate detection of gases with infra-red absorption in the mid- and far-infra-red. Unfortunately, these fibres are expensive and less robust. Optical fibre sensors are also generally believed to be safe for use in explosive atmospheres. However, the safety of optical fibre sensors is not unqualified, as it has been established that in the case of very high powers, i.e. of the order of 100 mW, or greater in multi-mode fibre, explosive risks may present themselves (Hills et al. (1993); Zhang et al. (1992)). Conventional optical fibres have a very small acceptance aperture, which severely restricts the amount of light that can be coupled into a fibre. Thus the power launched into optical fibres from high-radiance near-infra-red (NIR, $\sim$0.7 μm-$\sim$1.5 μm) Light Emitting Diodes (LEDs) is rarely above 1mW, even when large core optical fibres are used, and by comparison the spectral radiance of incandescent filament lamps is usually at least an order of magnitude less. Longer wavelength LEDs ($>\sim$1.5 μm) often have a lower spectral radiance. As the launched power is relatively low, sensitive light detection systems are required to produce operational sensors. With laser sources, there is no difficulty in achieving launch efficiencies in excess of 80% into multi-mode fibres. Consequently high powers can be launched, and the detection system constraints are eased substantially.

It was realised that narrow-linewidth diode lasers could readily be used in fibre-optic environmental detection systems. Inaba et al. (1979) suggested the use of a dual-wavelength laser to realise a differential absorption method that could be used over many kilometres of low-loss optical fibre in cases where it was necessary to locate the sensing head remotely from the measuring equipment. This typically involved the comparison of the received powers at two, or more, different wavelengths, each having passed through a remote measurement gas cell, so that the differential absorption of the two wavelengths by the gas sample could be used to infer the concentration of the target gas. The method required that the target gas possessed suitable gas absorption bands within the spectral transmission window of the optical fibre.

Culshaw et al. (1998) have surveyed some of the system topologies that may be used with laser-based optical gas detection systems and quantified the expected system sensitivities, which are of the order of less than 1 ppm. Stewart et al. (2004) and Whitenett et al. (2004) have realised some of these topologies, which included a Distributed FeedBack (DFB) wavelength modulated laser cavity ring-down approach that showed a methane detection sensitivity of 50 ppm.

A laser-based detection system for the detection of NO_2 gas (which is an industrial hazard and common environmental pollutant) was developed by Kobayashi et al. (1981). This was

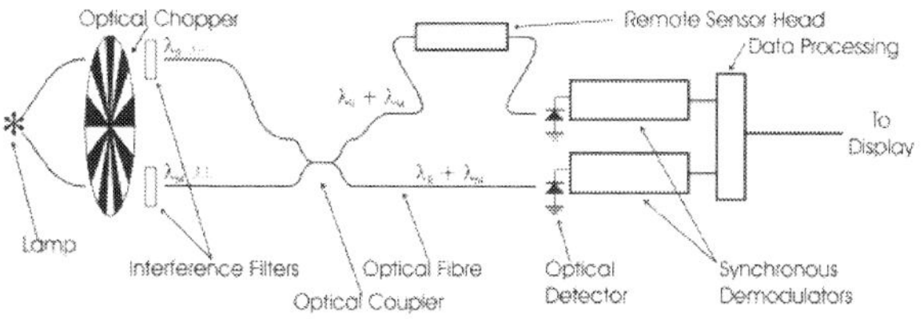

Fig. 8. Schematic of a differential fibre-optic detection system (redrawn from a diagram in Hordvik et al. (1983)).

achieved by splitting light, from an Ar-ion multi-line laser, into two paths, one passing through a measurement gas, and the other being transmitted directly to the measurement unit as a reference signal. The detection unit contained two filters to separate the two chosen laser lines, and these were then detected on separate optical receivers. One of these chosen laser lines coincided with a strong absorption line in the NO_2 absorption spectrum, whilst the other absorption line was somewhat weaker, hence giving a differential absorption method, by which the concentration of NO_2 in the measurement cell could be found. The system had an estimated detection limit of 17 ppm. The advantages of this dual-wavelength system were that the measurement was dependent on neither the optical power spectrum from a single source, which could drift, nor the system transmission, which could be affected by optical alignment, surface contamination, etc. It was realised that the selection of light sources used in this type of detection system was not necessarily limited to lasers, but broad-band sources such as filtered incandescent lamps or LEDs could also be used.

Hordvik et al. (1983) developed a fibre-optic system for the remote detection of methane gas (CH_4), see Figure 8. This system used a halogen lamp light source, which was alternately chopped into two separately filtered paths. One path was passed through a narrow-band interference filter, centred at the same wavelength as a strong absorption band of CH_4 (Q-branch centred at 1.666 μm), whilst the other filter covered a broader spectral range, and consequently had lower average absorption. These two complementary-modulated beams were combined by means of a fibre-coupler, with two output ports. Light from one was passed through a measurement cell to an optical detector (the measurement signal), and light from the other was passed directly to an optical detector (reference signal). By comparison of the optical powers in the narrow-band and broadband beams of the light that had and had not passed through the measurement cell, it was possible to calculate the CH_4 concentration.

A somewhat similar system based on the use of optical fibres and optical fibre couplers, but with the innovative use of compact LED light sources, was developed by Stueflotten et al. (1982). The schematic of the system is shown in Figure 9. Again, two different optical filter wavelengths were used, to give differential attenuation in strong and weak gas absorption regions. This was proposed for remote measurement in hazardous industrial environments, such as off-shore oil platforms. The systems above developed by Hordvik and Stueflotten both had a reported detection limit of approximately 5000 ppm (0.5% vol/vol) of methane.

4.2 Sensing using inelastic processes
Other forms of spectrophotometric processes rely on Raman scattering. A Raman scattering gas detection method is now briefly reviewed.

Raman scattering involves the inelastic scattering of light, i.e. first absorption and then delayed re-emission of light at a different wavelength to that incident on the material. The Raman process represents a form of scattering in which an incident photon may gain energy from (the anti-Stokes Raman process), or donate energy to (the Stokes Raman process) a vibrational or rotational energy level in a material. This produces a re-emitted photon of different energy and, hence, of a different wavelength. A method of detection that exploits Raman spectroscopy was developed by Samson & Stuart (1989) using the detection system shown in Figure 10. Raman scattering in gases is generally very weak, but the emission usually occurs in a well defined spectrum.

In the system developed by Samson and Stuart, the laser excites the gas and a mirror is used to reflect the incident light back through the interaction zone. Another concave mirror reflector doubles the level of Raman light received by the collection lenses. The alternative inelastic scattering process of fluorescence is rare in gases, and consequently is not commonly used for optical gas sensors, but fluorescence cannot be ignored when using Raman sensing, as it can cause crosstalk if it occurs in optical glass components or at mirror surfaces. Fortunately, Raman lines for simple gases are narrow compared to fluorescence emission which is usually relatively broadband. Raman detection systems may be employed to monitor the concentration of ammonia and ammonia based compounds in industrial atmospheres (Schmidt et al. (1999)).

4.3 Comb filter modulator for partially matching several spectral lines

Instead of detecting a gas using a single line of its absorption spectrum, or using a broadband source to cover many absorption lines, there are advantages in using some form of optical "comb" filter, with several periodic narrow transmission windows, in order to match several

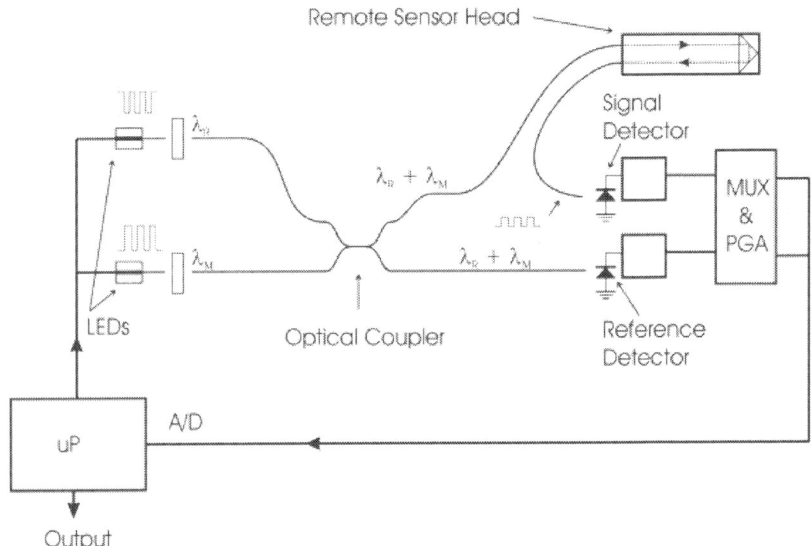

Fig. 9. Schematic of the differential fibre-optic detection system (redrawn from a diagram of Stueflotten et al. (1982))

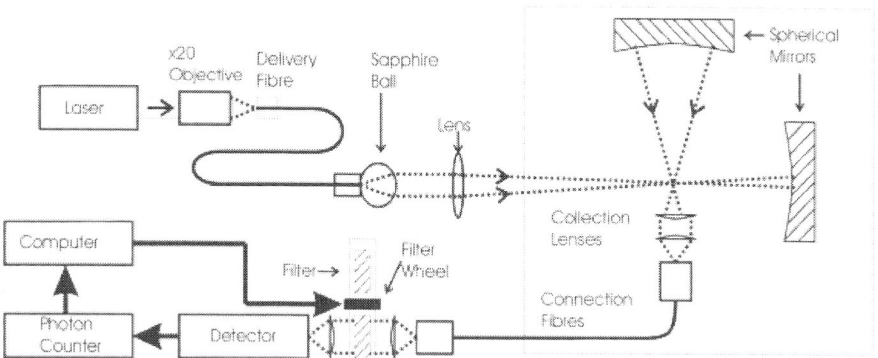

Fig. 10. Schematic of a gas sensor using Raman Scattering
(Redrawn from Samson & Stuart (1989))

spectral lines simultaneously. Such a comb filter can be scanned, correlated or wavelength modulated, through a set of gas lines to give an intensity modulation. The method has the advantage that it may allow improved selectivity, as a synthetic multiple-narrow-line comb-filter spectrum allows simultaneous measurement on several spectral lines. This reduces interference effects, which can cause complications with laser sources. A method of doing this, with a scanned Fabry-Perot comb filter, has been demonstrated (Dakin et al. (1987); Mohebati & King (1988)) with application to methane detection. Dakin *et al.* implemented a system that passed light sequentially from a source through a Fabry-Perot interferometer to a detector. By changing the spacing of the plates of the interferometer the transmission of fringes of the interferometer were tuned to match the absorption spectrum of the methane target gas. Dakin et al. (1987) reported a sensitivity limit of 100 ppm. The disadvantage of the Fabry-Perot filter is that it has a regular frequency spacing, whereas the gas absorption features are not normally equally spaced. A variation of the method is to use the correlation spectroscopy complementary source modulation technique with a filter that replicates the gas transmission spectrum. Recently in Vargas-Rodriguez & Rutt (2009) have demonstrated that, using this approach at 3.3 μ m, a minimum detection level of 0.023 % methane could be detected with a 1 s integration time.

4.4 Photoacoustic ammonia gas detection

A wavelength modulation can also be used for a photo-acoustic optoelectronic gas concentration measurement. Kosterev & Tittel (2004) demonstrated a noise limited detection of 0.65 ppm v. The system operated by the wavelength modulation of light from a 1.53 μm laser source with a quartz tuning fork. The tuning fork vibration frequency was twice that of the modulation of the laser source. The detected current from the optical detector could then be demodulated to find the gas concentration.

4.5 Sol-gel ammonia detection

Gases, including ammonia, may also be detected by the application of a chemical indicator dye to the surface of an optical fibre. The Sol-gel process enables the deposition and immobilisation of the chemical dye on to the surface of the optical fibre. The dye then absorbs light when in the presence of the gas to be sensed. The process can be applied to a wide range of chemical processes, however, the interactions of the dye with contaminant gases and humidity must be carefully considered (Malins et al. (1999)).

4.6 Ultraviolet optical detection of ammonia

The relatively intense ultraviolet absorption spectrum of ammonia, which was shown in Section 3.2, enables precise and selective detection of ammonia gas. Chambers et al. (2007) have demonstrated that ammonia gas can be detected at levels of ppm with low-cost ultraviolet LED light sources and detectors. Manap et al. (2009) has shown that the ultraviolet measurement was highly selective as contamination gases were not identifiable. With the recent development of these systems, it is necessary that the performance ultraviolet optical components are analysed (Eckhardt et al. (2007)).

5. Sources of noise

The accuracy of an optoelectronic sensor is limited by the selectivity and sensitivity of the sensor. These design considerations are now discussed.

In the design of optoelectronic gas sensor, it is important that the sensor measures solely the target gas that it was designed to measure. This is termed the selectivity of the sensor. In a gas absorption sensor selectivity issues can arise from contaminant gases, or the fouling of optical components, with an absorption spectrum that overlaps the gas to be sensed in the wavelength range of the optical source. Usually, the careful design of a sensor can eliminate this issue.

The measurement of an optoelectronic system will always be limited by a form of fundamental noise. These noise sources include resistor noise, photon noise, source noise and, in photo-acoustic systems, acoustic noise. Optical noise sources will now be discussed with their impact on measurement.

With optical absorption gas sensors it is necessary to accurately measure the optical power transmitted from the measurement gas cell. The output from the measurement detector photodiode is an electrical current that is proportional to the incident optical intensity. When broadband optical sources are used, the transmitted spectral power density is usually small, of the order of nW nm^{-1}, making it necessary to use a transimpedance amplifier to transform the detector current into a measurable voltage. This makes it necessary to use a sizeable feedback resistor, which is a significant source of thermal noise. This is usually the dominant source of noise in sensors with a low output optical power level. The thermal voltage noise, $V_{thermal}$, using the thermal noise equation is given by:

$$V_{thermal} = \sqrt{4kTR_{SO}B},\qquad(2)$$

where k is Boltzmann's constant, T is the resistor absolute temperature in kelvins, R_{SO} is the parallel resistance of the photodiode shunt resistance and feedback resistance and B is the post-detection noise bandwidth.

Shot noise (photon noise) describes the random arrival of photons at a detector and is described by Poisson Statistics. Photon noise is expressed by the following equation:

$$I_{Shot\,Noise} = \sqrt{2qI_{Sig}B},$$

where q is the electronic charge, I_{Sig} is the photocurrent generated and B is the post-detection noise bandwidth. With a shot noise limited system has reached the fundamental noise floor. The noise from an optical source is due to source related intensity or phase fluctuations. These variations have been analytically quantified and described by Tur Tur et al. (1990). They derived a method for calculating the relative intensity noise from an optical source.

$$S_{in}(f) = \frac{0.66 I_0^2}{\Delta v} \qquad (3)$$

Tur et al. (1990) showed that the optical source noise may be described by Equation 3, where Δv is the FWHM bandwidth (in Hertz) of the emission from the source, and the optical power from the source is I_0. Source noise tends to be the dominant source of noise in laser coupled gas detection systems.

6. Conclusions

As the globally most produced chemical, with a range of applications in agriculture, cleaning products, pharmaceutical industry, steel processing and carbon dioxide capture processes ammonia is vitally important to modern life. The monitoring of ammonia concentration is essential as, not least the gas has a pungent odour, but is also extremely toxic.

The infrared and ultraviolet molecular absorption mechanisms were discussed and their resulting spectra shown. A range of optoelectronic detection systems were described. The intention is to show how these sensors may be adapted to domestic, agricultural and industrial environments. With the growing awareness of the importance and dangers of ammonia, it is highly likely that optoelectronic sensors will be further researched and developed.

7. References

Brown, L. R. & Margolis, J. S. (1996). Empirical line parameters of NH_3 from 4791 to 5294 cm^{-1}, *JQSRT* 56: 283–294.

Burton, G. R., Chan, W. F., Cooper, G. & Brion, C. E. (1993). The electronic absorption spectrum of nh3 in the valence shell discrete and continuum regions. absolute oscillator strengths for photoabsorption (5Ű200 ev), *Chem. Phys.* 177: 217–231.

Chambers, P., Lyons, W. B., Lewis, E., Sun, T. & Grattan, K. T. V. (2007). The potential for development of an NH_3 optical fibre gas sensor, *Journal of Physics: Conference Series, Third International Conference on Optical and Laser Diagnostics*, Vol. 012015.

Cheng, B.-M., Lu1, H.-C., Chen, H.-K., Bahou, M., Lee, Y.-P., Mebel, A. M., Lee, L. C., Liang, M.-C. & Yung, Y. L. (2006). Absorption cross sections of nh3, nh2d, nhd2, and nd3 in the spectral range 140-220 nm and implications for planetary isotopic fractionation, *The Astrophysical Journal* 647(2): 1535.

Close, L. G., Catlin, F. I. & Cohn, A. M. (1980). Acute and chronic effects of ammonia burns of the respiratory tract, *Arch Otolaryngol.* 106(3): 151–158.

Cottaz, C., Kleiner, I., Tarrago, G., Brown, L.R. (2001). Assignments and intensities of $^{14}NH_3$ hot bands in the 5-8μm ($3v_2 - v_2$, $v_2 + v_4 - v_2$) and 4μm ($4v_2 - v_2$, $v_1 - v_2$, $v_3 - v_2$) regions, *J.Mol.Spectrosc.* 209: 30–49.

Cottaz, C., Kleiner, I., Tarrago, G., Brown, L. R., Margolis, J. S., Poynter, P. L., Pickett, H. M., Fouchet, T. & Drossart, P. (2000). Line positions and intensities in the $2v_2/v_4$ vibrational system of $^{14}NH_3$ near 5-7 μm, *JQSRT* 203: 285–309.

Culshaw, B., Stewart, G., Dong, F., Tandy, C. & Moodie, D. (1998). Fibre optic techniques for remote spectroscopic methane detection from concept to system realisation, *Sensors & Actuators B: Chemical* 51: 25–37.

Dakin J.P. & Chambers P. (2004). Review of methods of optical gas detection by direct optical spectroscopy with emphasis on correlation spectroscopy, *NATO Science Series*, Volume 224, Part 2, 457-477,

Dakin, J. P., Wade, C. A., Pinchbeck, D. & Wykes, J. S. (1987). A novel optical fibre methane system, *SPIE volume: 734 Fibre Optics '87: Fifth Internation Conference on Fibre Optics and Optoelectronics*.

Darde, V., Thomsen, K., van Well, W. J. M. & Stenby, E. H. (2008). Chilled ammonia process for CO_2 capture, *Greenhouse Gas Control Technologies 9, Proceedings of the 9th International Conference on Greenhouse Gas Control Technologies (GHGT-9)*, Vol. 1, pp. 1035–1042.

Eckhardt, H. S., Klein, K.-F., Spangenberg, B., Sun, T. & Grattan, K. T. V. (2007). Fibre-optic uv systems for gas and vapour analysis, *J. Phys.: Conf.* 85: 012018.

Edwards, H. O. & Dakin, J. P. (1993). Gas sensors using correlation spectroscopy compatible with fibre-optic operation, *Sensors & actuators B: Chemical* 11: 9.

Felty, W. L. (1982). From camel dung, *J. Chem. Educ.* 59: 170.

Goody, R. (1968). Cross-correlating spectrometer, *Journal of the Optical Society of America* 58(7): 900–908.

Guelachvili, G., Abdullah, A. H., Tu, N., Rao, K. N., Urban, S. & Papoušek, D. (1989). Analysis of high-resolution fourier transform spectra of $^{14}NH_3$ at 3.0μm, *Journal of Molecular Spectroscopy* 133: 345–364.

Hill, D. W. & Powell, T. (1968). *Non-Dispersive Infra-Red Gas Analysis in Science, Medicine and Industry*, Adam Hilget Ltd.

Hills, P. C., Samson, P. J. & Webster, I. (1993). Optical fibres are intrinsically safe: reviewing the myth, *Australian Journal of Electrical and Electronics Engineering* 10(3): 207–220. ISSN:0725-2986.

Hordvik, A., Berg, A. & Thingbø, D. (1983). A fibre optic gas detection system, *proceedings 9th international conference on Optical Communications, 'ECOC 83'*, p. 317.

Howard, J. B. & Rees, D. (1996). Structural basis of biological nitrogen fixation, *Chem. Rev.*, 96: 2965Ű–2982.

Inaba, H., Kobayasi, T., Hirama, M. & Hamza, M. (1979). Optical-fibre network system for air polution monitoring over a wide area by optical absorption method, *Electronics Letters* 15(23): 749–751.

Keller-Rudek, H. & Moortgat, G. K. (2006). *MPI-Mainz-UV-VIS Spectral Atlas of Gaseous Molecules*, Max-Planck-Gesellschaft zur Förderung der Wissenschaften e.V., chapter Spectral-Atlas-Mainz.

Kleiner, I., Brown, L. R., Tarrago, G., Kou, Q.-L., Picque, N., Guelachvili, G., Dana, V. & Mandin, J.-Y. (1999). Line positions and intensities in the vibrational system v_1, v_3 and $2v_4$ of $^{14}NH_3$ near 3 micron, *J.Mol.Spectrosc.* 193: 46–71.

Kleiner, I., Tarrago, G. & Brown, L. R. (1995). Positions and intensities in the 3 v_2/v_2 + v_4 vibrational system of $^{14}NH_3$ near 4 micron, *Journal of Molecular Spectroscopy* 173: 120–145.

Kobayashi, T., Hirana, M. & Inaba, H. (1981). Remote monitoring of NO_2 molecules by differential absorption using optical fibre link, *Applied Optics* 20(19): 3279.

Kosterev, A. A. & Tittel, F. K. (2004). Ammonia detection by use of quartz-enhanced photoacoustic spectroscopy with a near-ir telecommunication diode laser, *Applied Optics* 43: 6213–6217.

Levey, P. R. & van Bennekom, A. (1995). A mechanistic study of the effects of nitrogen on the corrosion properties of stainless steels, *Corrosion* 51(12): 911–921.

Luft, K. F. (1947). Anwendung des ultraroten spektrums in der chemischen industrie, *Angew. Chem.* 19(B): 2.

Malins, C., Doyle, A., MacCraith, B. D., Kvasnik, F., Landl, M., Simon, P., Kalvoda, L., Lukas, R., Pufler, K. & Babusík, I. (1999). Personal ammonia sensor for industrial environments., *J Environ Monit.* 1(5): 417–422.

Manap, H., Muda, R., O'Keeffe, S. & Lewis, E. (2009). Ammonia sensing and a cross sensitivity evaluation with atmosphere gases using optical fiber sensor, *Procedia Chemistry, Proceedings of the Eurosensors XXIII conference*, Vol. 1, pp. 959–962.

McBride, J. . P. & Nicholls, R. W. (1972). The vibration-rotation spectrum of ammonia gas i.

Miller, G. C. (1981). Chemical of the month: Ammonia, *J. Chem. Educ.* 58: 424–425.

Mohebati, A. & King, T. A. (1988). Remote detection of gases by diode laser spectroscopy, *Journal of Modern Optics* 35(3): 319–324.

Rothman, L. S., Gamache, R. R., Goldman, A., Brown, L. R., Toth, R. A., Pickett, H. M., Poynter, R. L., Flaud, J.-M., Camy-Peyret, C., Barbe, A., Husson, N., Rinsland, C. P. & Smith, M. A. H. (1987). The hitran database: 1986 edition, *Appl.Opt.* 26: 4058–4097.

Samide, A., Bibicu, I., Rogalski, M. S. & Preda, M. (2004). Surface study of the corrosion of carbon steel in solutions of ammonium salts using mössbauer spectrometry, *Journal of Radioanalytical and Nuclear Chemistry* 261(3): 593–596.

Samson, P. J. & Stuart, A. D. (1989). Fibre optic gas sensing using Raman spectroscopy, *Proceedings of 14th Australian conference on optical fibre technology*, pp. 145–148.

Schmidt, K., Michaelian, K. & Loppnow, G. (1999). Identification of major spieces in industrial metal-refining solutions with raman spectroscopy, *Applied Spectroscopy* 53(2): 139–143.

Stewart, G., Shields, P. & Culshaw, B. (2004). Development of fibre laser systems for ring-down and intracavity gas spectroscopy in the near-IR, *Measurement Science Technology* 15: 1621–1628.

Stueflotten, S., Christensen, T., Iversen, S., Hellvik, J. O., Almås, K., Wien, T. & Graav, A. (1982). An infrared fibre optic gas detection system, *Proceedings OFS-2 international conference*, pp. 87–90.

Sun, P., Ayre, C. & Wallace, M. (2003). Characterization of organic contaminants outgassed from materials used in semiconductor fabs/processing, *Characterization And Metrology For Ulsi Technology: 2003 International Conference on Characterization and Metrology for ULSI Technology. AIP Conference Proceedings*, Vol. 683, pp. 245–253.

Taylor, F. W., Haughton, J. T., Peskett, G. D., Rogers, C. D. & Williamson, E. J. (1972). Radiometer for remote sounding of the upper atmosphere, *Applied Optics* 11(1): 135–141.

Tur, M., Shafir, E. & Blotekjaer, K. (1990). Source-induced noise in optical systems driven by low coherence sources, *Journal of Lightwave Technology* 8: 183–189.

Urban, S., Tu, N., Rao, K. N. & Guelachvili, G. (1989). Analysis of high-resolution fourier transform spectra of $^{14}NH_3$ at 2.3 μm, *Journal of Molecular Spectroscopy* 133: 312–330.

Vargas-Rodriguez, E. & Rutt, H. N. (2009). Design of CO, CO_2 and CH_4 gas sensors based on correlation spectroscopy using a fabry-perot interferometer, *Sensors & actuators B: Chemical* 137: 410–419.

Veingerov, M. L. (1938). Eine methode der gasanalyze beruhend auf dem optisch-akustischen Tyndall-Röntgeneffekt, *Dokl. Akad. Nauk SSSR.* 19: 687.

Whitenett, G., Stewart, G., Yu, H. & Culshaw, B. (2004). Investigation of a tuneable mode-locked fiber laser for application to multipoint gas spectroscopy, *Journal of Lightwave Technology* 22(3): 813–819.

Zhang, D. K., Hills, P. C., Zheng, C., Wall, T. F. & Samson, P. (1992). Fibre optic ignition of combustible gas mixtures by the radiative heating of small particles, *Proceedings of the 24th International Symposium on Combustion (code 19626)*, Pittsburgh, PA, USA, pp. 1761–1767. ISSN:0082-0784.

11

Optical-Fiber Measurement Systems for Medical Applications

Sergio Silvestri and Emiliano Schena
University Campus Bio-Medico of Rome
Italy

1. Introduction

After telecommunications, also medicine has been revolutionized by optical fibers. They were firstly used, in the early sixties, to visualize internal anatomical sites by illuminating endoscopes. The essential technological solution to obtain good quality images was the introduction of "cladding" during the fifties. The result was the development of minimally invasive tools that have become essential for medical diagnosis and surgery. But optical fibers offer the potential for much more than illumination or imaging tasks. For example, they can also be utilized to sense physiological parameters.

The subject of present chapter is, therefore, a description of the design and measurement principles utilized in fiber optic sensors (FOSs) with a particular reference to biomedical applications.

FOSs development started in the sixties, but the high component costs and the poor interest of the medical community delayed the industrial expansion. The cost reduction of key optical components allowing to realize even disposable or mono-patient FOSs, the increase of components quality, the development of miniaturization, and the availability of plug and play and easy-to-use devices are the main reasons of the growth that is taking place in the use of FOSs.

Moreover, FOSs are characterized by some crucial advantages respect on the conventional transducers that allow to satisfy requirements for use in medical applications: they are robust, may have good accuracy and sensitivity, low zero- and sensitivity-drift, small size and light weight, are intrinsically safer than conventional sensors by not having electrical connection to the patient, large bandwidth, and show immunity from electromagnetic interference. This last feature allows to monitor parameters of physiological interest also during the use of electrical cauterization tools or in magnetic resonance imaging. At present, FOSs are used to measure physical variables (e.g., pressure, force, strain, and fluid flow) and also chemical variables (oxygen concentration in blood, pH, pO_2, and pCO_2).

The simplest FOSs classification is based on the subdivision in intrinsic and extrinsic sensors. In an intrinsic sensor the sensing element is the optical fiber itself, whereas an extrinsic sensor utilizes the optical fiber as a medium for conveying the light, whose physical parameters are, in turn, related to the measurand.

Due to different requirements for miniaturization and safety, in medical applications, these sensors are usually further divided in: invasive sensors, which are inserted into the body, therefore they must be miniaturized and biocompatible; non-invasive sensors, placed near the body or on the skin surface.

A number of measurement principles can be utilized to realize transducers based on the variation of fiber optic properties with physical or chemical variables, or based on variation of light parameters in the fiber. As the wide variety of techniques developed to design FOS for medical applications, just some of them are here described in detail.

This chapter is divided into subsections where a concise description of the measurement principle of FOSs is presented along with the main medical applications. Particular emphasis is placed on the metrological characteristics of the described FOSs and on the comparison with conventional sensors. Measurement principles include interferometry-based, intensity-based, fiber Bragg grating and laser Doppler velocimetry sensors.

In the following sections, the four abovementioned working principles and their use in specific medical applications to sense variables of physiological interest are investigated. The performances of the sensing methods are also presented with particular reference to the description of commercially available sensors.

2. Interferometry-based and intensity-modulated fiber optic sensors

FOSs can be realized with a working principle based on a large number of interferometric configurations, e.g., Sagnac interferometer, Michelson interferometer, Mach-Zehnder interferometer, and Fabry-Perot interferometer (Yoshino et al., 1982a) (Davis et al., 1982). Typically, these approaches show an extremely high sensitivity although cross-sensitivity represents a significant drawback: first of all the influence of temperature may introduce quite high measurement uncertainties (Grattan & Sun, 2000).

These FOSs can be designed as intrinsic sensors, where the sensing element is the fiber itself, or as extrinsic sensors, where a small size sensing element is attached at the tip of an optical fiber. The most common configuration is the second, where the sensing element, placed at the tip of the optical fiber, causes changes of light parameters in a well-known relation with the measurand. In this case, the optical fiber is employed to transmit the radiation emitted by a light source (e.g., laser or diode) and to transport the radiation, modulated by the measurand, from the sensing element to a photodetector (e.g., an optical spectrum analyzer). Thanks to this solution, the sensor can be used also for invasive measurements, as the largest part of the measurement system (light source and photodetector) can be placed far from the miniaturized sensing element, due to the very limited energy losses of light in the fiber.

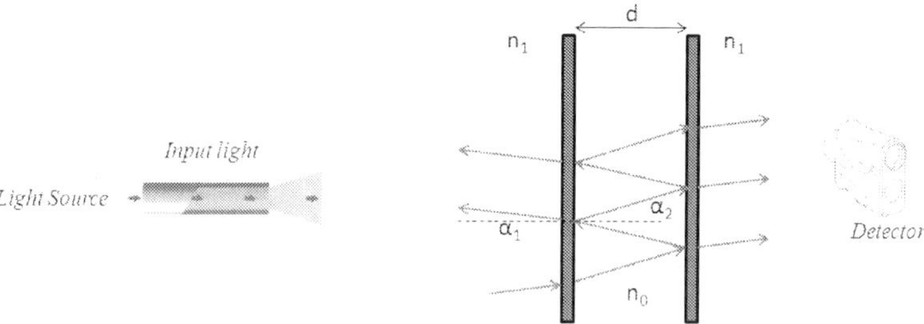

Fig. 1. Schematic representation of a Fabry-Perot interferometer.

In medical applications, mainly dedicated to force and pressure monitoring (Rolfe et al., 2007), the most common design is based on the interferometer configuration proposed by Fabry and Perot (Fabry & Perot, 1898), also known as multi-beam interferometer because many beams interfere in one resonator. A typical realization is composed of two parallel high reflecting mirrors placed at distance d (Figure 1). If d is variable, the instrument is called a Fabry-Perot interferometer. If d is fixed, whereas the incident light angle varies, the instrument is called a Fabry-Perot etalon. The Fabry-Perot interferometer allows to distinguish very close radiation wavelengths.

The Fabry-Perot cavity is usually utilized as secondary element of the sensor. Its output is an electromagnetic radiation with a wavelength that is function of d. In order to have high performances a measurement system based on Fabry-Perot interferometer needs a photodetector discriminating radiations with very close wavelengths. The working principle can be described as follows. When a light beam, emitted by a light source (e.g., a laser), enters between the two mirrors, a multiple reflections phenomenon takes place. The electromagnetic waves in the cavity can interact constructively or destructively, depending on if they are in phase or out of phase respectively. The condition of constructive interference, corresponding to a peak of transmitted light intensity, happens if the difference of optical path length between the interacting beams is an integer multiple of the light wavelength. The phase difference between interacting beams, and therefore the intensity of transmitted light, depends on the distance d between the mirrors. Considering for simplicity the same value for the refractive index upward the first surface and downward the second mirror (n_1), the intensity of transmitted light can be expressed as follows (Peatross & Ware, 2008):

$$I = I_0 \cdot \frac{(1-R)^2}{(1-R)^2 + 4 \cdot R \cdot \sin^2\left(\frac{\delta}{2}\right)} = I_0 \cdot \frac{1}{1 + F \cdot \sin^2\left(\frac{\delta}{2}\right)} \tag{1}$$

Where I_0 is the intensity of the incoming wave, F is the cavity's coefficient of *finesse* that can be expressed by the following equation:

$$F = \frac{4 \cdot R}{(1-R)^2} \tag{2}$$

R is the reflectance of both mirrors, and δ, the phase difference between each succeeding reflection, is a function of the radiation wavelength (λ), the distance between the two mirrors (d), and the angle between the radiation direction and the normal to the mirror surface (α_1):

$$\delta = \frac{4 \cdot \pi \cdot d \cdot n_1}{\lambda} \cdot \cos\alpha_1 \tag{3}$$

In order to increase the sensitivity of the Fabry-Perot interferometer, it is desirable that the intensity (I) varies strongly with δ. Equation 1 shows that sensitivity of I with δ increases when F is increased. Therefore, the sensitivity of the device increases when F, and consequently R, is increased, as shown by equation 2. For the above mentioned reasons, important parameters of a Fabry-Perot interferometer are: the difference between two succeeding transmission peaks (free spectral range) and the value of R. In fact, the difference between the maximal and the minimal peaks of the transmitted radiation increases with R, moreover, the trend of I as a function of d becomes sharper when R increases: this makes easier the determination of d variations. A mirror with a very high reflectance (R) is usually obtained by coating the internal surface of the two mirrors.

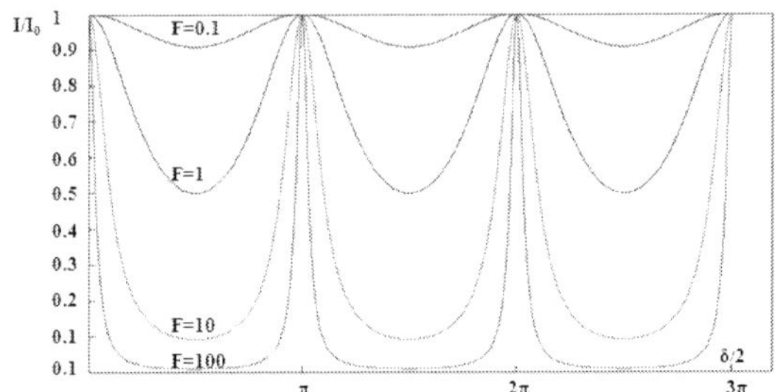

Fig. 2. Ratio between the intensities of transmitted and incident radiation as a function of δ/2 for different values of cavity's finesse coefficient.

Figure 2 shows the ratio between the intensity of transmitted and incident light as a function of δ/2, considering a normal incident radiation ($\cos\alpha_1 \approx 1$), for the following F values: F=0.1 (R≈2.4 %), F=1 (R≈17 %), F=10 (R≈54 %), F=100 (R≈82 %).

Thanks to the use of an optical fiber coupled to a Fabry-Perot cavity, the measurement system can be miniaturized, with the light source and the photodetector separated from the sensing element (the cavity). Moreover, the small size of the sensing element, along with the flexibility of the fiber optic with small outer diameter, allows to directly insert the sensing element into the body for use in clinical applications where an invasive measurement is required. Some sensors, showing the above described working principle, designed for medical applications are reported in Section 2.1.

The intensity-modulated FOSs are characterized by a working principle based on the intensity variation of the reflected light into the fiber related to a displacement induced by the measurand on a secondary element. A basic configuration shows one or more optical fibers with the extremity placed at a known distance from a movable mirror having high reflectance. The radiation, emitted by a source and conveyed into the fiber, is reflected by the mirror: the distance (d) between the fiber tip and the mirror is related to the measurand magnitude. The intensity of the back-reflected light coupled to the fiber is a fraction of the incident light intensity and depends on the distance between the fiber and the reflecting surface, or on a deformation of the surface: an increase of the distance causes a decrease of the back-reflected intensity as shown in figures 3a, 3b, and 3c. This principle, when applied to a secondary transducer, allows to measure several physical variables: temperature, pressure, force, fluid velocity and volumetric flow rate.

More complex configurations have been realized with solutions improving sensor performances (Puangmali et al., 2010).

Other methods applied to the design of intensity-modulated FOSs are based on the light coupling of two fibers (Lee, 2003). In this configuration, schematically reported in figure 4, the radiation emitted by a light source is conveyed within a fiber optic, whose distal extremity is placed in front of another fiber. The intensity of the light transmitted into the second fiber, and measured by a photodetector placed at its distal tip, is related to the distance (d) between the two fiber tips: the transmitted intensity decreases when d increases, as shown in figures 4a, 4b, and 4c. The measurand can be a displacement or a physical variable causing the displacement, such as force, pressure or temperature.

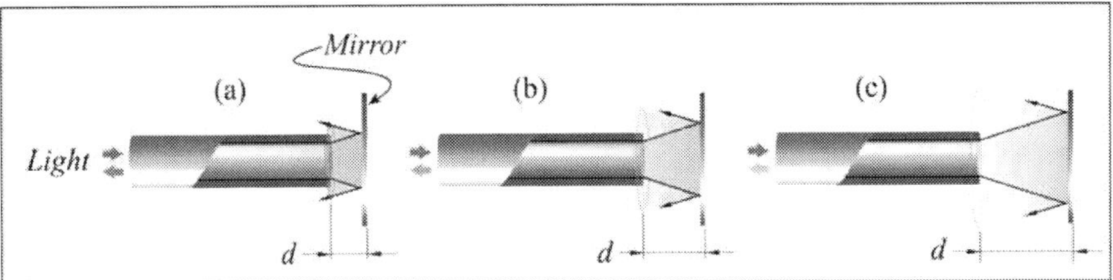

Fig. 3. Schematic representation of common intensity-modulated FOSs realized with a fiber and a reflecting surface. The intensity of the reflected radiation coupled to the fiber at different distances d between the fiber and the mirror (a, b, and c).

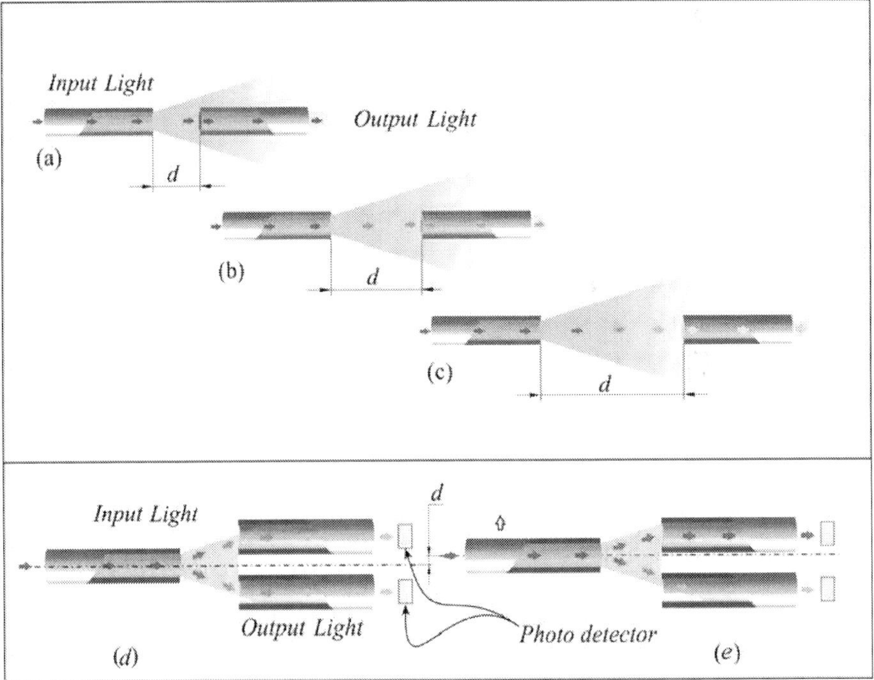

Fig. 4. Schematic representation of intensity-modulated FOSs realized with two fibers or more. The intensity of the coupled radiation of the two fibers is function of their distance d: if d increases the intensity of collected light decreases (a, b, and c). Sensors' performances can be improved using differential configuration (d, and e).

The above described configuration can be improved using three or more fibers in differential configuration for the compensation of changes in the light source intensity or losses in the fiber, as shown in figure 4d and 4e.

An interesting example of sensor designed with this working principle is a flow-meter based on the vortex shedding phenomenon: the light intensity transmitted between the fibers is modulated by the periodical mechanical motion caused by vortex shedding. The light intensity is converted by a photodiode in an electric signal related to the flow rate. This sensor can be used to perform measurements of fluid flow rate also at high temperature (350 °C) (Wroblewski & Skuratovsky, 1985).

A further design to realize intrinsic intensity-modulated FOSs is based on microbending. The bending of an optical fiber, in fact, causes an attenuation of the light intensity conveyed. As it is well known, the ever-present radiation loss into the cladding region causes an attenuation when the light passes through a fiber. The intensity loss into the cladding region can be increased if the fiber is bent (figure 5). Also this working principle allows to sense pressure, force, fluid velocity and volumetric flow rate.

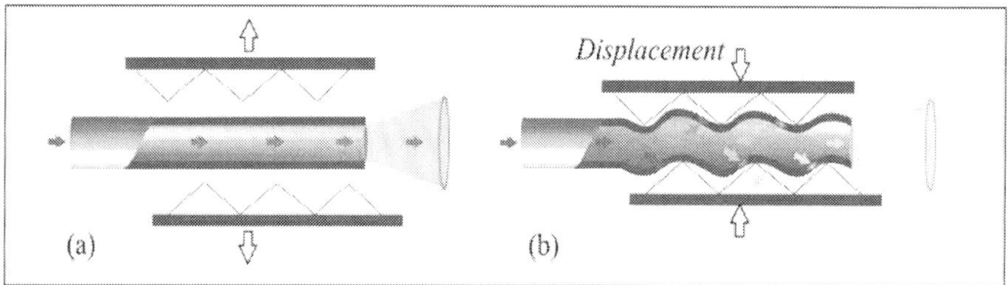

Fig. 5. a and b: principle of work of FOS based on microbending that transduces a displacement causing the bending into an intensity attenuation.

Generally speaking, intensity-modulated FOSs are characterized by the main advantage of requiring a modest amount of electronic interfaces (Udd, 2006).

2.1 Interferometry-based and intensity-modulated fiber optic sensors: medical

Intracranial pressure (ICP) is the cerebrospinal fluid pressure inside the skull. Being the skull a rigid case, any volumetric increase of its content may raise the ICP. The ICP monitoring is probably the most important application of FOSs in the medical field at the moment. During the sixties there was a significant propulsion in the development of pressure FOSs for measurement of intravascular blood pressure (Lekholm & Lindstrom, 1969) and in the early seventies some patents were issued describing FOSs to measure the ICP.

The ICP value is principally due to two main components: cerebrospinal fluid (CSF) volume, that is responsible for ICP baseline and can cause, in pathological conditions, an increase of this parameter; vasogenic components causing small fluctuations of cerebral blood volume, that can increase the ICP value in conditions of hypercapnea or increase of cerebral metabolism. Mass lesions (tumors, pus or hematoma), vascular engorgement (e.g., in case of traumatic brain injury), cerebral oedema or hydrocephalus may also increase the ICP value.

Its monitoring is, therefore, essential in patients with traumatic brain injury, tumors or pus, where the ICP increase is a common cause of ischemia, intracranial hemorrhages or brain herniation. Since the ICP value varies continuously, an uninterrupted record of the ICP should be obtained in order to avoid the loss of diagnostic data.

Normal ICP values depend on age, position and clinical conditions. In supine position it ranges from 7 mmHg to 15 mmHg for adults and from 3 mmHg to 7 mmHg for children; an ICP exceeding 20 mmHg needs therapeutic treatment (Smith, 2008).

Although some attempts have been performed to introduce non-invasive or minimally-invasive methods, e.g., the estimation of ICP by the measurement of tympanic membrane displacement (Shimbles et al., 2005) or by ultrasound-based techniques (Yoshino et al., 1982b), the ICP monitoring usually requires invasive transducers. Transducers can be placed in parenchymal, ventricular, epidural, subdural, or subarachnoid locations, although measurements obtained from the last three sites appear less accurate (Bratton et al., 2007). Also, lumbar puncture can be utilized to estimate the ICP, but this indirect measurement not always correlates with the ICP value. In the clinical practice, the monitoring is performed in several ways: 1) through a catheter placed in ventricular, epidural or subarachnoid spaces and connected to an external strain gauge; 2) through a micro strain gauge typically placed in ventricular or parenchymal catheters; 3) through FOSs guided inside the ventricles, brain parenchyma, subdural or subarachnoid spaces.

The standard proposed by the Association for the Advancement of Medical Instrumentation (AAMI) provides the performances that a device intended for ICP measurement should assure. The device should have a pressure range between 0 mmHg and 100 mmHg, an accuracy better than ±2 mmHg in the range from 0 mmHg to 20 mmHg, and lower than 10 % of the measured value in the range from 20 mmHg to 100 mmHg (Bratton et al., 2007).

Commercially available FOSs, intended to monitor ICP value, are based on two working principles. The former was introduced by T. E. Hansen, who described an interesting micro-tip FOS for medical applications (Hansen, 1983). Successively, A. Lekholm and L. Lindstrom realized a FOS with the same measurement principle for intravascular use (Lekholm & Lindstrom, 1969) and A. Wald dedicated it to monitor the ICP (Wald et al., 1977). The working principle is based on the presence of two groups of fiber bundles connected to a LED and to a photodetector respectively. At the common end of the bundle is placed a thin metal membrane reflecting the light. In this way, the light emitted by LED is conveyed to the fibers connected to the photodetector, as schematically reported in figure 6.

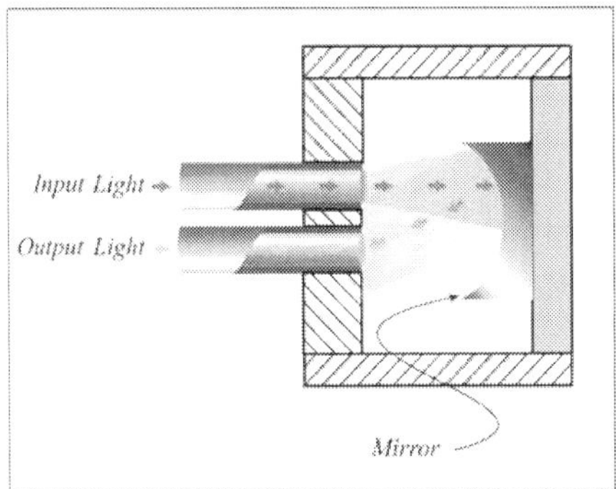

Fig. 6. Schematic representation of the sensing element of the FOS made by Camino Laboratories: the output light variation is related to the pressure that causes a mirror displacement.

The coupling and, therefore, the photodetector signal, depends on the membrane deflection caused by an external pressure. These FOSs have small size, e.g., an outer diameter of 1.5 mm or 0.85 mm for the unshielded case, and a frequency response flat from 0 Hz up to 15 kHz. On the other hand, they show zero drift if a temperature step from 20 °C to 37 °C is applied, recovering the baseline after about 40 s.

The second FOS design is based on Fabry-Perot interferometry (Fabry & Perot, 1898). These sensors were introduced in the late seventies (Mitchell, 1989). The sensing element shows two parallel optical reflecting surfaces, as schematically reported in figure 1, and one of them is a pressure sensitive diaphragm. A variation of the external pressure causes a deflection of the diaphragm, reducing the optical cavity depth. The optical cavity is, therefore, with variable dimension allowing light intensity curve having multiple *maxima* and *minima* that depend on cavity depth.

At present, some fiber optic devices based on micro-optical mechanical systems (MOMS) to monitor the ICP are commercially available.

FISO Technologies, Inc. has developed some pressure FOSs for medical applications constituted by a Fabry-Perot cavity whose optical length changes with the physical parameters to be measured. The FOP-MIV pressure sensor is a miniaturized Fabry-Perot cavity constituted by a micromachined silicon diaphragm membrane, acting as pressure sensing element (Chavko et al., 2007). When pressure increases, the thin membrane is deflected and the Fabry-Perot cavity depth is reduced, in this way the small cavity depth variations are related to pressure variations. Vacuum inside the cavity prevents changes of internal pressure caused by gas thermal expansion that would, otherwise, distort the pressure measurement. A high vacuum is maintained inside the cavity therefore, the FOP-MIV measures absolute pressure. Being one of the smallest pressure sensors commercially available, FOP-MIV is well designed for many medical applications where size is an important issue (Hamel & Pinet, 2006). The optical nature of the FOP-MIV, makes the sensor immune to electromagnetic field or radiofrequency interferences regularly encountered in operating rooms or MRI environment. FOP-MIV is characterized by a measurement range up to 300 mmHg, an accuracy equal to 1.5 % of full scale output (or ±1 mmHg), a resolution better than 0.3 mmHg, a thermal effect sensitivity of 0.1 %/°C; a zero drift thermal effect of 0.4 mmHg/°C.

Innerspace, Inc. produces a device to monitor ICP also based on the Fabry-Perot cavity: the pressure, deflecting the diaphragm, alters the cavity depth and thus the optical cavity reflectance at a given wavelength. If a LED source is used, the spectrally modulated reflected light can be split into two wavebands by a dichroic mirror. The ratio of the two signals provides a pressure estimation immune to the typical light level changes occurring in FOS systems. The measurement range is from -10 mmHg up to +100 mmHg, the linearity and hysteresis is ±2 mmHg from 0 mmHg to 10 mmHg and 10% of reading from -10 mmHg to 125 mmHg (Mignani & Baldini, 1995).

Camino Laboratories realized the ICP monitoring through an intensity-modulated based FOS. A dual-beam reference, using a secondary fiber optic path, is joined to the pressure measuring fiber link, but unaffected by pressure variations. The sensor is based on the intensity modulation technique with dual-beam referencing. The radiation conveyed within a fiber optic is coupled to a fiber by a bellow with a reflecting surface. The intensity of the coupled radiation depends on the position of the bellow's tip, which is function of the pressure, as shown in figure 6. The measurement range is from 0 mmHg up to 100 mmHg, linearity and hysteresis is ±2 mmHg from -10 mmHg to +50 mmHg and 6% of read value from +50 mmHg to +125 mmHg, the frequency response shows an attenuation of -3dB from 33 Hz to 123 Hz, a zero drift <2 mmHg (first 24 hours) and less than 1 mmHg/day (first 5 days).

At present, the "gold standard" technique for ICP monitoring is considered a catheter inserted into the lateral ventricle and connected to an external strain gauge (Smith, 2008). The main advantages include the chance to perform a periodic external calibration and a slightly lower cost than microstrain gauge and fiber optic devices (Bratton et al., 2007). However, also these sensors have their own set of potential complications, including obstruction or disconnecting of the tubing, occasional difficulties to place in the presence of brain swelling and shift (Ostrup et al, 1987), migration of the catheter out of the ventricle (Al-Tamimi et al., 2009), and infections occur in up to 11% of cases (Steiner & Andrews, 2006). Microstrain gauge and FOS show similar metrological characteristics but with a higher cost. The additional advantages of FOS devices are: the immunity to the electromagnetic interferences that allows to monitor the ICP during magnetic resonance or during the use of electrical cauterization tools, no obstruction and electrical hazard, and a quite low zero-drift (Crutchfield et al., 1990).

Arterial pressure invasive measurements can also be performed through FOSs. The most widely used configuration is Fabry-Perot interferometry. Many studies describe miniaturized pressure sensors (Ceyssens et al., 2008; Totsu et al., 2005). FOS allows to obtain a large bandwidth (some kHz), an accuracy better than 4 % of the read value, and a measurement range that covers the physiological pressure values (less than 300 mmHg).

Wolthuis *et al.* developed a Fabry-Perot FOS able to perform a concurrent measurement of pressure (with resolution of 1 mmHg, measurement range from 1 mmHg to 1000 mmHg and flat frequency response up to 1000 Hz) and temperature (with resolution of 0.2 °C, rise time of 20 ms, and measurement range from 10 °C to 60 °C) (Wolthuis et al., 1993). RJC Enterprises, LLC realized further developments of the sensor using the same principle of measurement. For temperature measurement, the outer surface of a thin silicon layer defines the optical reflecting cavity; the refractive index of silicon changes with temperature altering the optical cavity reflectance spectra The transducer (for pressure and temperature) contains a 850 nm LED whose emission reaches the sensor via an optical fiber. In the sensor's optical reflecting cavity, the spectral distribution of the LED light is modified as a function of cavity depth, and this spectrally altered light is reflected back down the fiber to the instrument. Light returning to the instrument is optically split into two spectral components; the photocurrents from these two components form a ratiometric signal which in turn correlates with changes in the measured parameter (Wolthuis et al., 1993). The sensor shows some advantages: small size (the maximum dimension is 300 μm), resolution of 0.02 °C and 0.1 mmHg, accuracy of 0.1 °C and ±1 mmHg (or 2 % of read value), bandwidth up to 500 Hz limited only by supporting instrumentation, measurement range from 15 °C to 55 °C and from 500 mmHg to 1100 mmHg (absolute pressure).

An interesting application of these sensors is related to the intra-aortic balloon pumping (IABP) therapy, which is a therapy of circulatory support often used to help patients recovery from critical heart diseases, cardiac surgery or to wait until a transplant is performed. A catheter, terminated by an inflatable balloon, is introduced through the femoral artery and is positioned into the descending aorta just below the subclavian artery. The inner lumen of the catheter can be used to monitor systemic arterial pressure and the outer lumen is used for the delivery of gas to the balloon. The balloon must be rapidly inflated with the onset of the diastole and deflated when the systole happens. The synchronization between the balloon pumping and the heartbeat can be done either by using electrocardiogram (ECG) signals or aortic-pressure waveform (Trost & Hillis, 2006). The second approach is required when ECG signals are distorted or unavailable (Dehdia et al., 2008). FOSs are used to measure blood pressure, so that to give a trigger to the balloon. The traditional technique to monitor the aortic pressure is performed by an external electrical sensor, which measures pressure applied through a fluid-filled catheter. Many drawbacks are associated with this method such as, among others, the damping effects due to catheter elasticity or the presence of air micro-bubbles and the presence of catheter vibrations. These drawbacks are eliminated by the use of a FOS mounted at the tip of the catheter. It also shows a good accuracy and resolution, and a large bandwidth. Moreover, it is miniaturized (MOMS) decreasing the risk of ischemia, that is the main risk associated with IABP therapy (Pinet, 2008).

FISO Technologies, Inc has developed a pressure FOS for this application constituted by a Fabry-Perot cavity. The working principle has been presented in the previous section. Thanks to the small size (diameter of 550 µm) the sensor shows a large bandwidth limited by the signal conditioner at 250 Hz, the resolution is 0.5 mmHg (Pinet et al., 2005). Coupling this sensor with a faster signal conditioner could result in a better dynamic response of traditional catheter tip pressure sensor and can be used in presence of strong EM fields (e.g., during magnetic resonance investigation).

FOS is also used in IABP therapy by Arrow International, Inc. and by Maquet Getinge Group. It shows an accuracy of 4 mmHg or 4 % of read values whichever is greater, and a pressure range from 0 mmHg to 300 mmHg.

Opsens produces a FOS with the same working principle to monitor blood pressure. Small size (catheter diameter equal to about 250 µm), pressure range from -50 mmHg to +300 mmHg, and good resolution (0.5 mmHg) and accuracy (0.5 mmHg or 1 % of full scale) are the main advantages. It can also be used to monitor ICP and urodynamic pressure.

FOSs based on Fabry-Perot interferometry are also used to perform direct measurements of intra-tracheal pressure. Direct intra-tracheal pressure measurements show some advantages to monitor respiratory mechanics. This measurement is usually performed by introducing a catheter into the endotracheal tube (ETT) and connecting it to a conventional pressure transducer. In pediatric respiratory monitoring this approach is usually not performed because it implies an excessive occlusion of the narrow pediatric tubes (Guttmann et al., 2000). Samba Sensors designs a FOS for intra-tracheal pressure monitoring. It consists of a membrane that changes the depth of a microcavity with pressure (Fabry-Perot interferometry). It shows some advantages such as: small size (the silicon sensor chip diameter is equal to 420 µm, the fiber optic diameter is 250 µm or 400 µm), measurement range between -50 cmH_2O and +350 cmH_2O, low temperature drift (< 0.2 $cmH_2O/°C$), accuracy of 0.5 cmH_2O or 2.5 % of reading between -50 cmH_2O to +250 cmH_2O and 4 % of reading between +250 cmH_2O to +350 cmH_2O, short response time (1.3 ms), moreover the introduction of the sensor into the ETT does not significantly increase the fluid-dynamic resistance, therefore the pressure drop across the ETT (Sondergaard et al., 2002). This company produces another FOS, with similar metrological characteristics of the above described sensor, used in urology to monitor bladder pressure and the pressure in the lower urinary tract, especially important in paraplegic patients for the high risk of urinary tract infections.

An interesting application of intensity-based FOSs was presented by Babchenko *et al*. They performed the measurement of the respiratory chest circumference changes (RCCC) through a bending-based FOS composed of a bent optic fiber connected to the chest so that its curvature radius changes during respiration due to RCCC. The light intensity transmitted through the fiber depends on the bent, so that provides an indirect measurement of the RCCC (Babchenko et al., 1999).

3. Fiber Bragg grating sensors

About thirty years have passed since the introduction of fiber Bragg grating (FBG) sensors, when Hill *et al.* discovered the phenomenon of photosensitivity (Hill et al., 1978). They discovered that an optical fiber with a germanium-doped core can show a light-induced permanent change in the core's refractive index thanks to an electromagnetic wave with high intensity and a particular wavelength. An optical fiber core characterized by periodic refractive index changes constitutes a FBG. Only eleven years later, a milestone study describing a FBG-based sensor was published (Meltz et al., 1989). The development of such sensors was mainly delayed due to the high cost and the realization difficulties, that only during the nineties showed a considerable reduction. Nowadays, there are two main techniques to realize fiber grating: interferometric and phase mask method. These techniques allow for the manufacturing of some different types of grating classified as: FBG; long-period fiber grating; chirped fiber grating; tilted fiber grating; and sampled fiber grating (Lee, 2003). The gratings are called fiber Bragg gratings (FBGs) if the grating spatial period has the order of magnitude of hundreds of nanometers, or long period gratings (LPGs) if the spatial period has the order of magnitude of hundreds of micrometers.
FBG sensors are characterized by a working principle based on the coupling of the forward propagating core mode to the backward propagating core mode. When light propagates through a fiber with a Bragg grating, a phenomenon of radiation reflection happens only for a narrow range of wavelengths, other wavelengths are transmitted. The wavelength placed in the middle of the reflected range is called Bragg wavelength λ_B, and can be expressed by the following equation:

$$\lambda_B = 2 \cdot n_{eff} \cdot \Lambda \qquad (4)$$

where Λ is the spatial period of the grating, and n_{eff} is the effective refractive index of the fiber core.
The working principle, commonly used in FBG sensors, is based on λ_B shift due to a variation of the spatial period of the grating, or on λ_B shift due to a refractive index variation of the core, as shown in equation 4. The former is the technique typically implemented to perform strain measurement; the second is mainly used to monitor changes of temperature, as a consequence of the temperature influence on n_{eff}. As the measurand directly modulates the signal into the fiber, FBGs are examples of fiber optic intrinsic sensors.
Figure 7 shows the λ_B shift from λ_{B1} to λ_{B2} caused by a strain.
Therefore, the relative variation of the Bragg wavelength can be expressed as a function of the fractional changes of the spatial period and of the effective refractive index:

$$\frac{\Delta \lambda_B}{\lambda_B} = \frac{\Delta \Lambda}{\Lambda} + \frac{\Delta n_{eff}}{n_{eff}} \qquad (5)$$

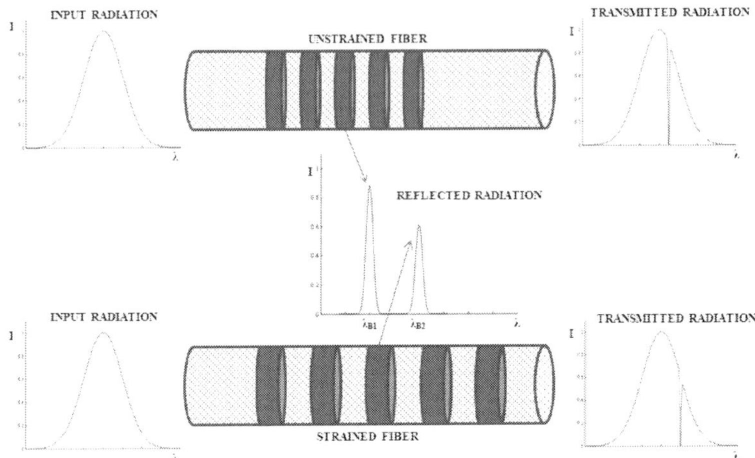

Fig. 7. Shift of Bragg wavelength due to fiber optic strain.

The relative λ_B change can be also expressed showing its dependence on the fiber strain, ε, and on the temperature variation, ΔT:

$$\frac{\Delta \lambda_B}{\lambda_B} = P_e \cdot \varepsilon + \left[P_e \left(\alpha_s - \alpha_f \right) + \varsigma \right] \cdot \Delta T \tag{6}$$

where P_e is the strain-optic coefficient, α_s and α_f are the thermal expansion coefficients of the fiber bonding material and of the fiber respectively, and ς is the thermo-optic coefficient (You & Yin, 2002). In the environmental temperature range, for a silica fiber, the thermal effect is dominated by the effect of temperature changes on n_{eff} (dn_{eff}/dT) that is one order of magnitude greater than the effect of thermal expansion. In fact, about 95% of the wavelength shift is caused by the effect of temperature changes on n_{eff}.

FBG response (shift of λ_B) for strain input and temperature changes depends on λ_B and FBG types (Shu et al., 2002). Typical strain sensitivities are: 0.64 pm/$\mu\varepsilon$ at λ_B near to 830 nm; 1 pm/$\mu\varepsilon$ at λ_B near to 1300 nm; 1.2 pm/$\mu\varepsilon$ at λ_B near to 1550 nm. Typical temperature sensitivities are: 6.8 pm/°C at λ_B near to 830 nm; 10 pm/°C at λ_B near to 1300 nm; 13 pm/°C at λ_B near to 1550 nm (Rao, 1998).

Although the effect of one of the two components on λ_B shift is predominant, both components can be separated through the implementation of specific configurations, e.g., a reference FBG added to the main sensor (Xu et al., 1994), in order to attenuate the influence of the undesired effect and to improve the repeatability of the measurement system.

As these sensors can detect physical variables such as strain, temperature, force, pressure, vibrations, they have been widely utilized for clinical applications in some medical fields, described in detail in the following section.

3.1 Fiber Bragg grating sensors: medical applications

Although this technology is characterized by a high potential for the monitoring of diagnostic variables thanks to the immunity of electromagnetic field, the non-toxicity, the possibility of realizing miniaturized sensors, the biocompatibility, and it has been introduced in researches for medical application for more than 20 years, it is not yet widely used in clinical practice (Mishra et al., 2011).

In the late nineties Rao *et al.* described a FBG-based system for medical applications (Rao et al., 1998). In particular, they developed two temperature sensors: the former was used for cardiac monitoring estimating the stroke volume with thermodilution technique, the second one was realized for *in vivo* blood temperature monitoring. In both cardiac output and blood temperature monitoring the use of FBG-based sensors allows to substitute traditional sensing elements that are electrically active or powered, therefore poorly appropriate where high electro-magnetic fields are present or in some other medical applications. These sensors were also used to monitor heart muscle activity: the sound generated by the heartbeats causes vibrations of a membrane. The stretches and contractions of the FBGs, mounted on the membrane, can perform vibrations measurement (Gurkan et al., 2005).

FBG was also used for other interesting medical applications. One of them is related to the invasive ablation treatment of Atrial Fibrillation (AF). AF is a cardiac arrhythmia that typically causes poor heart blood pumping. There are two treatments for AF: pharmaceutical drugs and electrical defibrillation (cardioversion), and invasive ablation surgery. Surgical ablation is performed by, among many methods, radiofrequency (RF) waves conveyed in an invasive probe. Some studies show that lesion size, caused by RF ablation, is related to electrode-tissue contact force (Haines, 1991). The strong influence of contact force on the amount of the ablated tissue shows the importance of its monitoring. A FBG-based sensor has been developed to perform a real time monitoring of the contact force (Yokoyama et al., 2008). The sensor (TactiCath, Endosense SA), incorporated in the ablation catheter, is composed of three optical fibers to monitor the deformation of the catheter tip. Three FBGs, mounted on the deformable body, allow to relate the body deformation with the applied force representing the above mentioned contact force between the ablation catheter and the tissue. An infrared laser radiation (wavelength ranging from 1520 nm to 1570 nm) conveyed in three fibers is partially reflected at a particular wavelength (λ_B) related to the body deformation, so that to the contact force. This system allows to measure contact force along three different directions (parallel, perpendicular, and at 45° with the tissue) at frequency of 10 Hz, the discrimination threshold is better than 10^{-3} kgf ($9.8 \cdot 10^{-3}$ N), a further advantage is the small size (catheter tip diameter equal to 3.5 mm).

FBG-based force transducers could play an important role in minimally invasive surgery (e.g., laparoscopic surgery widely used in appendectomy, stomach surgery, surgery of the colon and of the rectum) or in minimally invasive robotic surgery (e.g., Da Vinci), to provide a feedback of the forces applied to the tissue during the surgery. The force monitoring allows to minimize the tissue damage (Song et al., 2009). Microsurgery is characterized by a wide number of applications for FBG-based sensors. Sun *et al.* realized a FBG-based sensor to monitor force between tool and tissue during retinal microsurgery (Iordachita et al., 2009). In this application, the force monitoring can help the surgeon, since the intensity of interaction forces is usually lower than human perception. The sensor is characterized by: good resolution (0.25 mN) that allows to monitor very low forces, and a temperature compensation through the use of a further reference FBG that improves the sensor repeatability.

During alternative surgical technique for tumors removal, as laser-induced thermotherapy (LITT), FBG sensors could result useful. LITT destroys the neoplastic tissues inducing hyperthermia through laser light delivered through optical fiber probe placed in correspondence of the tumor. During the LITT treatment, the monitoring of tissue temperature could avoid inefficacy in the treatment caused by excessive temperature values of the tissue, that can cause the damage of the healthy tissue close to the neoplastic one, or to low temperature values that can cause an incomplete ablation of the neoplastic tissue. FBG sensor with spatial resolution of 0.25 mm has been designed to monitor the tissue temperature during LITT (Li et al., 2009) and to use its value to control the temperature in the boundary region at 35 °C to improve the safety of the LITT treatment (Ding et al., 2010), *in vivo* measurements during this procedure are also performed (Webb et al., 2000).

4. Fiber optic sensors for laser Doppler velocimetry

Generally speaking, the design of these sensors is based on the "Doppler shift" described by Christian Doppler in 1842 in an article entitled "On the colored light of double stars and some other heavenly bodies". If an electromagnetic or acoustic wave is reflected back by a moving object, it undergoes the Doppler shift phenomenon, i.e., the frequency of the reflected wave is different from the frequency of the incident one. This frequency difference is related to the speed of the moving object and can be expressed by the following relation (Beckwith et al., 1995):

$$\Delta f = \left(\frac{2V}{\lambda} \right) \cdot \cos \beta \cdot \sin \left(\frac{\alpha}{2} \right)$$

$$(7)$$

being V the moving object speed, λ the wavelength of the incident wave, α the angle between the axis of the incoming wave and the observer, and β the angle between the moving object velocity direction and the bisector of the angle between the axis of the incoming wave and the segment connecting the object and the observer, as shown in figure 8.

If the object moves within a fluid flow, its velocity can be approximately equal to the average velocity of the fluid flow, therefore the frequency shift measurement becomes an indirect measurement of the flow rate.

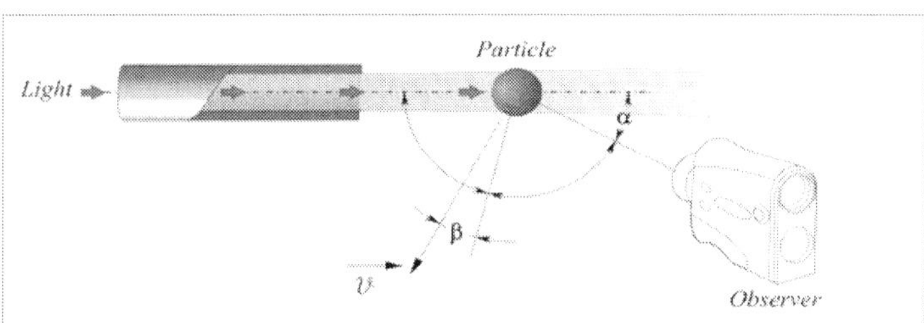

Fig. 8. By illuminating a moving particle with the light transported through a fiber, a frequency variation of the light reaching the observer is obtained.

The performances of these sensors have been improved through the dual-beam approach, a schematic representation is depicted in figure 9.

The radiation is split into two beams, through a beam splitter, successively crossed into an intersection region through a mirror. This region defines a sampling volume, placed into the fluid flow, where the particle velocity is measured. In the intersection region, an interference pattern, characterized by alternating interference fringes, is produced. When a particle passes through the fringes, it causes periodic variations of the intensity of the scattered light.

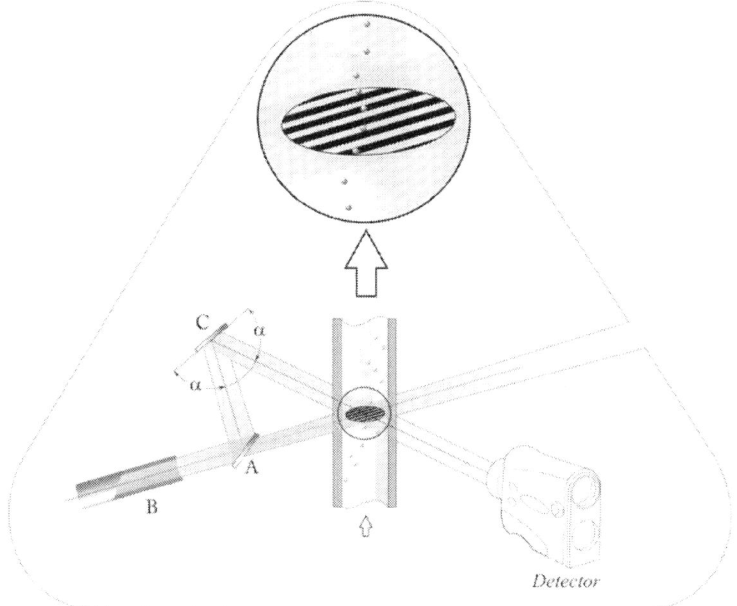

Fig. 9. Schematic representation of the working principle of FOS based on laser Doppler differential approach and its components: the beam splitter (A), the fiber optic (B), the mirror (C).

The light intensity time plot, monitored by a photodetector, reports these periodic variations and allows to estimate the flow rate.
The particle speed (V) is related to the frequency fluctuations (fr) of the scattered intensity and to the distance between two contiguous fringes (γ) as follows:

$$V = \gamma \cdot f_r = \frac{\lambda}{2 \cdot \sin\left(\frac{\theta}{2}\right)} \cdot f_r \tag{8}$$

where θ is the angle between the two beams.
The use of fiber optics to transmit the incident and the reflected light enhances the chances to perform both invasive and non-contact measurements allowing for monitoring very low flow rates, such as blood flow in capillaries: the use of light waves instead of sound waves allows to estimate the speed of red blood cells when they moves into the capillaries. A detailed description of the medical applications is reported in the following section.

4.1 Laser Doppler velocimetry: medical applications

Laser Doppler velocimetry (LDV) sensors in medical arena are used to perform measurements of blood flow rate in some tissues, i.e., perfusion. As the perfusion is a primary parameter in the local transport of oxygen, nutrients, heat, *et cetera* its monitoring is useful in several diagnoses. The initial applications, used for measuring the average speed in single vessels, were determined through the estimation of the frequency shift of the light backscattered by the red blood cells in the flowing stream (Tanaka & Benedek, 1974). Their development and commercialization started with the prototype designed by Stern *et al.* to monitor skin blood flow rate (Stern et al., 1977) and with the study regarding the application of LDV in microvascular tissue by Bonner & Nossal (Bonner & Nossal, 1981). The light penetrating the tissue is backscattered by red blood cells moving in different directions and with various velocities. Therefore, the backscattered light shows a frequency shift, measurements are based on the changes of the power spectrum of the backscattered light. In the years during and following the above reported two researches, LDV has been used for several medical applications in the following briefly described.

The *in vivo* monitoring of blood flow in the optical nerve head, the sub-foveal choroid, and the iris is performed with the following aims: scientific (research about the vascularization of the optical nerve), clinical (allowing to evaluate alterations in blood flow), and to quantify the effect of medical therapies in pathologic conditions (Riva et al., 2010). Riva *et al.* introduced this technique to measure blood flow in the optical nerve (Riva et al., 1982). An important innovation that allows to simplify the alignment between the measurement systems and the patient's eye was introduced by Geiser *et al.* (Geiser et al., 1999).

This technique was introduced by Shepherd and Riedel in the measurement of intestinal mucosal blood flow (Shepherd & Riedel, 1982). The measurement system consisted on a probe in contact with the tissue placed at the distal extremity of two optical fibers. The former conveyed the light, emitted by a helium-neon laser, investing a small area of intestinal tissue, the second carried the backscattered light from the tissue to a photodetector. The backscattered light produces a spectrum of frequencies (from 0 Hz to 20 kHz) analyzed in real time that provides the output of the system. Further *in vivo* trials were realized by Kiel *et al.* with valuable results (Kiel et al., 1985).

LDV was also applied to monitor blood flow in some other anatomical sites such as cerebral blood flow (Skapherdinsson et al., 1988), hepatic blood flow (Arvidsson et al., 1988), renal blood flow, gingival blood flow, and bone blood flow.

5. Conclusion

In this chapter we wanted to give an idea of the potentialities of FOSs in medical applications without pretending to be exhaustive. Intentionally we avoided to mention FOSs dedicated to monitor physiological parameters such as blood or gastric pH, blood and respiratory oxygen and carbon dioxide, bilirubin concentration and other chemical variables. Although, also in these fields, the diffusion of FOSs is far to be negligible.

The design of new fiber-based sensors is a rather active area of research with a number of possibilities to be discovered in the near future.

As professionals working in the biomedical engineering field we would emphasize that medical practitioners do not have the experience, neither the time, to carefully handle an optical-fibre sensor, above all in emergency situations. One of the greatest challenge for the optical-fibre-sensor industry is therefore to offer 'plug and play' devices that are easy to use and tolerant to rough handling. Moreover, also the cost of the necessary instrumentation should be strongly reduced.

Application-oriented customizations, reduced costs and robust designs are important for the commercial success of such sensing technology, especially for the medical market.

6. Acknowledgment

This work has been carried out under the financial support of Regione Lazio in the framework of the ITINERIS2 project (CUP code F87G10000120009).
Authors gratefully acknowledge ITAL GM s.r.l. for the precious support provided.
Authors would like to thank Ms. Sonia Rubegni for her precious collaboration in the realization of figures.

7. References

Al-Tamimi, Y. Z.; Helmy, A.; Bavetta, S. & Price, S. J. (2009). Assessment of zero drift in the Codman intracranial pressure monitor: a study from 2 neurointensive care units. *Neurosurgery*, Vol. 64, No. 1, pp. 94-98

Arvidsson, D.; Svensson, H. & Haglund U. (1988). Laser-Doppler flowmetry for estimating liver blood flow, *American Journal of Physiology, Gastrointestinal and Liver Physiology*, Vol. 254, No. 4, pp. G471-G476.

Babchenko, A.; Khanokh, B.; Shomer Y. & Nitzan N. (1999). Fiber optic sensor for the measurement of respiratory chest circumference changes. *Journal of Biomedical Optics*, Vol. 4, No. 2, pp. 224-229.

Beckwith, T. G.; Marangoni, R. D. & Lienhard, J. H. (1995). *Mechanical Measurements* (Fifth Edition), Addison-Wesley publishing Company, Inc., ISBN: 0-201-56947-7, United States of America.

Bonner, R. & Nossal, R. (1981). Model for laser Doppler measurements of blood flow in tissue, Applied Optics, Vol. 20, no. 12, pp. 2097-2107.

Bratton, S. L. et al. (2007). Intracranial pressure monitoring technology. *Journal of Neurotrauma*, Vol. 24, No. 1, pp. S45- S54.

Ceyssens, F.; Driesen, M.; Wouters, K.; Puers, R. & Leuven, K. U. (2008). A low-cost and highly integrated fiber optical pressure sensor system. *Sensors Actuators A: Physical*, Vol. 145-146, pp. 81-86.

Chavko, M.; Koller, W. A.; Prusaczyk, W. K. & McCarron, R. M. (2007). Measurement of blast wave by a miniature fiber optic pressure transducer in the rat brain. *Journal of Neuroscience Methods*, Vol. 159, No. 2, pp. 277-281.

Crutchfield, J. S.; Narayan, R. K; Robertson, C. S. & Michael, L. H. (1990). Evaluation of fiber optic intracranial pressure monitoring. *Journal of Neurosurgery*, Vol. 72, No. 3, pp. 482-487.

Davis, C.; Carome, E. F.; Weik, M. H.; Rzekiel, S. & Einfig, R. E. (1982). *Fiberoptic sensor technology handbook*, Optical Technologies Inc., Herndon.

Dehdia, J. D.; Chakravarthy K. R. N. & Ahmed A. A. (2008). Intra aortic Balloon Pumping (IABP): past, present and future. *Indian Journal of Anaesthesia*, Vol. 52, No. 4, pp. 387-391.

Ding, Y; Chen, N.; Chen, Z.; Pang, F.; Zeng, X. & Wang, T. (2010). Dynamic temperature monitoring and control with fully distributed fiber Bragg grating sensor, *Proceedings of SPIE, Optics in Health Care and Biomedical Optics IV*, Vol. 7845, Beijing, China.

Fabry, C. & Perot. A. Measure de petites epaisseurs en valeur absolue. *Academie des sciences*, Paris, Comptes Rendus, 1898.

Geiser, M. H.; Dierman, U. & Riva, C. E. (1999). Compact laser Doppler choroidal flowmeter, *Journal of Biomedical Optics*, Vol. 4, No. 4.

Grattan, K. T. V. & Sun, T. (2000). Fiber optic sensor technology: an overview. *Sensors and Actuators A: Physical*, Vol. 82, No 1-3, pp. 40-61.

Gurkan, D., Starodubov, D. & Yuan, X. (2005). Monitoring of the heartbeat sounds using an optical fiber bragg grating sensor, *Proceedings of 4th IEEE Conference Sensors*, Irvine, Ca.

Guttmann, J.; Kessler, V.; Mols, J.; Hentschel, R.; Hoberthur, C. & Geiger, K. (2000). Continuous calculation of intratracheal pressure in the presence of pediatric endotracheal tubes. *Critical Care Medicine*, Vol. 28, No. 4, pp. 1018-1026.

Haines, D.E. (1991). Determinants of lesion size during radiofrequency catheter ablation: the role of electrode-tissue contact pressure and duration of energy delivery. *Journal of Cardiovascular Electrophysiology*, Vol. 2, No. 6, pp. 509-515.

Hamel, C. & Pinet, E. (2006). Temperature and pressure fiber-optic sensors applied to minimally invasive diagnostics and therapies. *Progress in Biomedical Optics and Imaging*, Vol. 7, No. 6, ISSN: 1605-7422.

Hansen, T. E. (1983). A fiber optic micro-tip pressure transducer for medical applications. *Sensors and Actuators*, Vol. 4, pp. 545-554.

Hill, K. O.; Fujii, Y.; Johnson, D.C. & Kawasaki, B. S. (1978). Photosensitivity in optical fiber waveguides: application to reflection filter fabrication, *Applied Physics Letters*, Vol. 32, No. 10, pp. 647–649, 0003-6951.

Iordachita, I.; Sun, Z.; Balicki, M.; Kang, J.U.; Phee, S.J.; Handa, J.; Gehlbach, P. & Taylor R. (2009). A sub-millimetric, 0.25 mN resolution fully integrated fiber-optic force-sensing tool for retinal microsurgery. *International Journal of Computer Assisted Radiology and Surgery*, Vol. 4, No. 4, pp. 383-390.

Kiel, G. W.; Riedel, G. L.; Di Resta, G. R. & Shepherd, A. P. (1985). Gastric mucosal blood flow measured by laser-Doppler velocimetry, *American Journal of Physiology, Gastrointestinal and Liver Physiology*, Vol. 249, No. 4, pp. G539-G545.

Lee, B. (2003). Review of the present status of the optical fiber sensors. *Optical Fiber Technology*, Vol. 9, No. 2, pp. 57-79.

Lekholm, A. & Lindstrom, L. (1969). Optoelectronic transducer for intravascular measurements of pressure variations. *Medical & Biological Engineering*, Vol. 7, No. 3, pp. 333-335.

Li, C.; Chen, N.; Chen, Z. & Wang, T. (2009). Fully distributed chirped FBG sensor and application in laser-induced interstitial thermotherapy, *Proceedings of SPIE, Communications and Photonics Conference and Exhibition*, Vol. 7634, ISBN: 978-1-55752-877-3, Shanghai, China.

Meltz, G.; Morey, W. W. & Glenn, W. H. (1989). Formation of Bragg gratings in optical fibre by a transverse holographic method, *Optics Letters*, Vol. 14, No. 15, pp. 823–825.

Mignani, A. G. & Baldini, F. (1995). In-vivo biomedical monitoring by fiber-optic systems. *Journal of Lightwave Technology*, Vol. 13, No. 7, pp. 1396-1406.

Mishra, V.; Singh, N.; Tiwari, U. & Kapur, P. (2011). Fiber grating sensors in medicine: current and emerging applications. *Sensors and Actuators A: Physical*, Vol. 167, No. 2, pp. 279-290.

Mitchell, G. L. A review of Fabry-Perot interferometric sensors. Optical Fiber Sensors. *Proceedings of the 6th International Conference*, Paris, 1989.

Ostrup, R. C.; Luerssen, T. G.; Marshall, L. F. & Zornow, M. H. (1987). Continuous monitoring of intracranial pressure with miniaturized fiber optic device. *Journal of Neurosurgery*, Vol. 67, No. 2, pp. 206-209.

Peatross, J. & Ware, M. (2008). Multiple parallel interfaces, In: *Physics of Light and Optics*, pp. 141-155, http://optics.byu.edu/BYUOpticsBook_Fall2008.pdf

Petkus, V.; Ragauskas, A. & Jurkonis, R. (2002). Investigation of intracranial media ultrasonic monitoring model. *Ultrasonics*, Vol. 40, No. 1-8, pp. 829-833.

Pinet, E. Medical applications: saving lives. (2008). *Nature Photonics*, Vol. 2, pp. 150-152.

Pinet, E.; Pham, A. & Rioux, S. (2005). Miniature fiber optic pressure sensor for medical application: an opportunity for intra-aortic balloon pumping (IABP) therapy. *Proceedings of SPIE*, 17th International Conference on optical Sensors , Vol. 5855, pp. 234-237.

Puangmali, P.; Althoefer, K. & Seneviratne L. D. (2010). Mathematical model of intensity modulated bent-tip optical fiber displacement sensors. *IEEE Transactions on Instrumentation and Measurement*, Vol. 59, No. 2, pp. 283- 291.

Rao, Y. J. (1998). Fiber Bragg grating sensors: principles and applications, In: *Grattan, K.T.V. & Meggitt, B. T. (Eds.), Optical Fiber Sensor Technology*, Vol. 2, pp. 355–380, Chapman & Hall, ISBN 0 412 78290 1, London, U.K.

Rao, Y. J.; Webb, D. J.; Jackson, D.A.; Zhang, L. & Bennion, I. (1998). Optical in-fiber bragg grating sensor systems for medical applications. *Journal of Biomedical Optics*, Vol. 3, No. 1, pp. 38-44.

Riva, E. C.; Geiser, M. & Petrig, B. L. (2010). Ocular blood flow assessment using continuous laser-Doppler flowmetry, *Acta Ophthalmologica*, Vol. 88, No. 6, pp. 622-629

Riva, E. C.; Grunwald, G. E. & Sinclair, S. H. (1982). Laser Doppler measurement of relative blood velocity in the human optic nerve head, *Investigative Ophthalmology & Visual Science*, Vol. 22, No. 2, pp. 241-248.

Rolfe, P.; Scopesi, F. & Serra, G. (2007). Advances in fiber-optic sensing in medicine and biology. *Measurement Science and Technology*, Vol. 18, No. 6, pp. 1683-1688.

Shepherd, A. P. & Riedel, G. L. (1982). Continuous measurement of intestinal mucosal blood flow by laser-Doppler velocimetry, *American Journal of Physiology, Gastrointestinal and liver Physiology*, Vol. 242, No. 6, pp. G668-G672.

Shimbles, S.; Dodd, C.; Banister, K.; Mendelow, A. D. & Chambers, I. R. (2005) Clinical comparison of tympanic membrane displacement with invasive ICP measurements. *Physiological Measurement*, Vol. 26, No. 6, pp. 197-199, 2005.

Shu, X.; Liu, Y.; Zhao, D.; Gwandu, B.; Floreani, F.; Zhang, L. & Bennion, I. (2002) Dependence of temperature and strain coefficients on fiber grating type and its application to simultaneous temperature and strain measurement. *Optics Letters*, Vol. 27, No. 9, pp. 701–703.

Skarphedinsson, J. O.; Harding, H. & Thoren, P. (1988). Repeated measurements of cerebral blood flow in rats. Comparisons between the hydrogen clearance method and laser Doppler flowmetry, *Acta Physiologica*, Vol. 134, No. 1, pp. 133-142.

Smith, M. (2008). Monitoring intracranial pressure in traumatic brain injury. *Anesthesia & Analgesia*, Vol. 106, No. 1, pp. 240-248.

Sondergaard, S.; Karason, S.; Hanson, A.; Nilsson, K.; Hojer, S.; Lundin, S. & Stenqvist, A. O. (2002). Direct measurements of intratracheal pressure in pediatric respiratory monitoring. *Pediatric Research*, Vol. 51, No. 3, pp. 339-345.

Song, H.; Kim, K.; Suh, J. & Lee. (2009). Development of optical FBG force measurement system for the medical application, *Proceedings of SPIE*, Vol. 7522, Singapore, November, 2009.

Steiner, L. A. & Andrews, P. J. D. (2006). Monitoring the injured brain: ICP and CBF. *British Journal of Anaesthesia*, Vol. 97, No. 1, pp. 26-38.

Stern, M. D. et al. (1977). Continuous measurement of tissue blood flow by laser-Doppler spectroscopy, *American Journal of Physiology, Heart and Circulatory Physiology*, Vol. 232, No. 4, pp. H441-448.

Tanaka, T. & Benedek, G. (1974). Measurement of velocity of blood flow (in vivo) using a fiber optic catheter and optical mixing spectroscopy, *Applied Optics*, Vol. 14, No. 1, pp. 186-196.

Totsu, K.; Haga, Y. & Esashi M. (2005). Ultra-miniature fiber-optic pressure sensor using white light interferometry. *Journal of Micromechanics and Microengineering*, Vol. 15, No. 1.

Trost, G. C. & Hillis, L. D. (2006). Intra-aortic balloon counterpulsation. *The American Journal of Cardiology*, Vol. 97, No. 9, pp. 1391-1398.

Udd, E. (2006). *Fiber optic sensors*, John Wiley & Sons, Inc., ISBN-13 978-0-470-06810-6, Hoboken, New Jersey.

Wald, A.; Post, K.; Ransohoff, J., Hass, W. & Epstein, F. (1977). A new technique for monitoring epidural intracranial pressure. *Medical Instrumentation*, Vol. 11, No. 6, pp. 352-354.

Webb, D. J.; Hathaway, M. W. & Jackson, D. A. (2000). First in-vivo trials of a fiber Bragg grating based temperature profiling system. *Journal of Biomedical Optics*, Vol. 5, No. 1, pp. 45-50.

Wolthuis, R. A.; Mitchell, G. L.; Saaski, E.; Hartl, J. C. & Afromowitz, M. A. (1991). Development of medical pressure and temperature sensors employing optical spectrum modulation. *IEEE Transactions on Biomedical Engineering*, Vol. 38, No. 10, pp. 974-981, ISSN: 0018-9294.

Wolthuis, R. A.; Mitchell, G. L.; Hartl, J. C. & Saaski, E. (1993). Development of a dual function sensor system for measuring pressure and temperature at the tip of a single optical fiber. *IEEE Transactions on Biomedical Engineering*, Vol. 40, No. 3, pp. 298-302, ISSN: 0018-9294.

Wroblewski, D. J. & Skuratovsky, E. (1985). Vortex shedding flowmeter with fiber optic sensor. In: *31st International Instrumentation Symposium*, San Diego, CA, pp. 653-656.

Xu, M. G.; Archambault, J. L.; Reekie, L. & Dakin, J. P. (1994). Thermally-compensated bending gauge using surface mounted fiber gratings. *International Journal of Optoelectronics*, Vol. 9, pp. 281–283.

Yokoyama, K.; Nakagawa, H.; Shah, D. C.; Lambert, H.; Leo, G.; Aeby, N.; Ikeda, A.; Pitha, J. V.; Sharma, T.; Lazzara, R. & Jackman, W.M. (2008). Novel contact force sensor incorporated in irrigated radiofrequency ablation catheter predicts lesion size and incidence of steam pop and thrombus. *Circulation: Arrhythmia and Electrophysiology*, Vol. 1, pp. 354-362.

Yoshino, T.; Kurosawa, K.; Itoh, K. & Ose, T. (1982a). Fiber- optic Fabry-Perot interferometer and its sensor applications. *IEEE Journal of Quantum Electronics*, Vol. 18, No. 10, pp. 1624- 1633, 0018-9480.

Yoshino, T.; Kurosawa, K.; Itoh, K. & Ose, T. (1982b). Fiber- optic Fabry-Perot interferometer and its sensor applications. *IEEE Transactions on Microwave Theory and Techniques*, vol. 30, No. 10, pp. 1612-1621, ISSN: 0018-9480.

You, F. T. S & Yin, S. (2002). *Fiber Optic Sensors*, Marcel Dekker, Inc., ISBN: 0-8247-0732-X, New York, NY.

Part 3

Lasers in Optoelectronics

12

The Vertical-Cavity Surface Emitting Laser (VCSEL) and Electrical Access Contribution

Angelique Rissons and Jean-Claude Mollier
Institut Superieur de l'Aeronautique et de l'Espace (ISAE),
Université de Toulouse
France

1. Introduction

This chapter aims at highlighting the influence of the electrical access of different kinds of Vertical-Cavity Surface-Emitting Laser (VCSEL) (Rissons & Mollier, 2009).

By presenting an overview of electrical equivalent circuit modeling and characterization for the VCSEL chip, based on the Scattering parameters, an overview of the VCSEL performance is given in order to provide a good knowledge of this device behavior in various operation modes and avoid its inadequate utilization (Rissons et al., 2003), (Perchoux et al., 2008), (Bacou et al., 2009).

Due to the emergence of the short distance optical communication, the VCSELs have become, in the past ten years, a key component of the optical interconnections. The current progress in the VCSEL technology makes this laser diode promising for different application fields and competitive with respect to the Edge-Emitting laser (EEL) and the Light Emitting Diode (LED) (providing the laser performance while keeping the cost effectiveness of the LED)(Koyama, 2006),(Suematsu & Iga, 2008).

Above all the VCSELs have been designed to fulfill the need of optoelectronic circuit planarization. Since its invention in 1977 by Prof. K. Iga (Iga et al., 1988) and its first commercialization at the end of the 90's, the VCSEL structure finds itself in a state of constant progress. Today, it covers a wide wavelength emission range (from the blue-green band up to the infrared) that enables the usage of these components in various application fields, not only in the short distance digital communications (LAN, Avionic network (Ly et al., 2008), ...) but also in consumer applications (laser printers, laser mice, display systems, etc.).

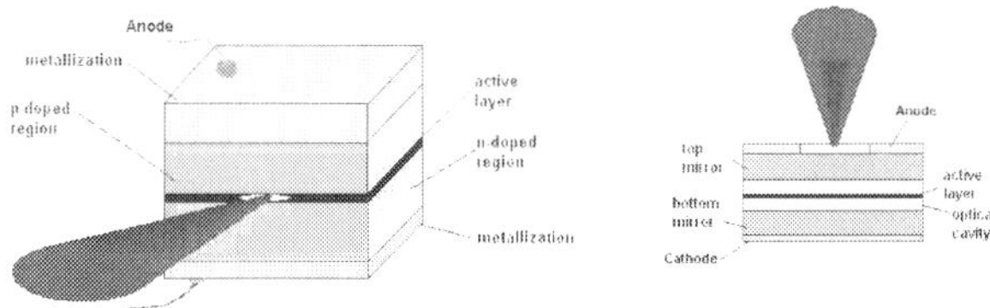

Fig. 1. comparison between Edge Emitting laser (on the left) and VCSEL (on the right)

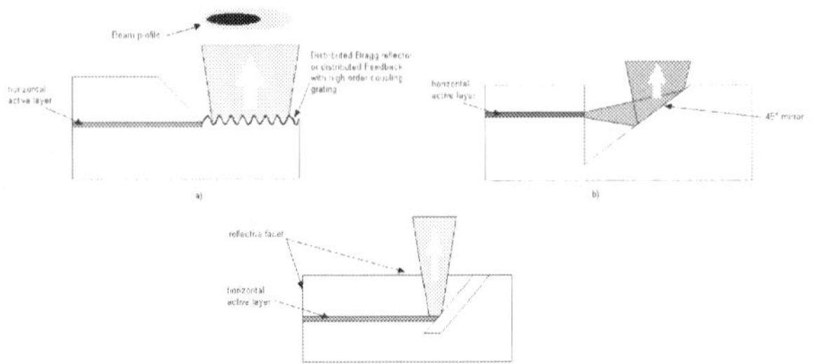

Fig. 2. First structure proposal of Surface Emitting laser

All these applications can be made possible thanks to:

- The vertical emission perpendicular to the semiconductor layers makes easier the integration (1D and 2D Array) according to the electrical packaging constraints.

- By combining the small size and the perpendicular emission, this component responds to the criterion of planarization providing a high integration level.

- The low volume of the active layer (AL) due to the quantum wells (QW) involves a sub-milliamps threshold current and a low electrical power consumption.

- The VCSEL presents a low thermal dependence close to the room temperature.

- The serial fabrication reduces the cost and allows on-wafer testing.

- Due to its cylindrical geometry, the light-beam cross section is circular.

All these reasons have led to the growth of the VCSEL market in a wide application range especially in the integrated optical sub-assemblies of transceivers for short distance optical links. Despite of the numerous advantages of the VCSEL, the available performance has to be improved to face the Distributed-Feedback (DFB) lasers. This improvement could be achieved thanks to the progress in the VCSEL technology, in particular for the structures of 1.3 and 1.55μm (850nm VCSEL technology having matured in the past ten years). Another way to enhance the VCSEL performance and avoid its inadequate utilization, is to have a perfect knowledge of the component behavior in various operating modes. It can be completed by modeling and by measurements, though special attention must be paid to the electrical contribution which hugely modifies the frequency response (O'Brien et al., 2007) (Rissons et al., 2003). Moreover, the current advance in VCSEL technology (intrinsic structure and chip submount) requires a constant review of the model and the characterization before the VCSEL utilization (Bacou et al., 2010). Hence we suggest to dedicate a chapter to the VCSEL technology, that aims at giving an overview of the advances, the physical behavior, and its various structures including the electrical access regarding VCSELs.

2. VCSEL technology

2.1 VCSEL emergence

The vertical light emitting laser idea emerged between 1975-1977 to satisfy the planarization constraints of the integrated photonics according to the microelectronic technology available then. At first, laser diode researchers proposed structured based on beam deflection (see Fig. 2). After two years of proposals, Prof. K. Iga (Iga et al., 1988) suggested the best solution of

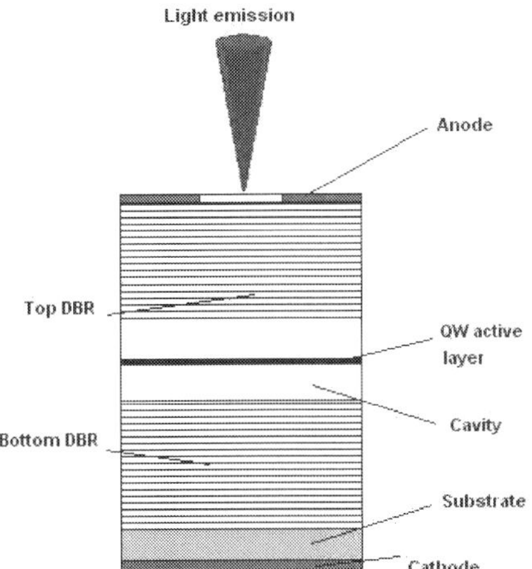

Fig. 3. Vertical-Cavity Surface-Emitting Laser

Vertical-Cavity Surface-Emitting Laser with GaInAsP/InP and AlGaAs/GaAs active region for optical fiber communications, for the optical disks, optical sensing and optical processing. The first goal of Prof. Iga was to grow a monolithic structure in a wafer and test the component before separation. In 1979, the first lasing surface emitting laser (SEL) was obtained with a GaInAsP/InP structure at 77K under pulsed regime. The threshold current was about 900mA within 1.3 or 1.55μm wavelength. In 1983, the first lasing at room temperature (RT) under pulsed operation with a GaAs active region was achieved but the threshold current remained higher than the Edge Emitting Laser (EEL)). In spite of the poor VCSEL performance in those days, the progress of the microelectronic technology gave the opportunity to the researcher to improve the VCSEL structure in view of threshold reduction at RT. After a decade of improvement attempts, the first continuous wave (CW) operation at RT was obtained by Iga with a GaAs structure. At the same time, Ibariki (Ibaraki et al., 1989) introduced, into the VCSEL structure, doped Distributed Bragg Reflector (DBR) as mirrors as well for the current injection. Jewell (Jewell et al., 1989) presented the first characterisation of Quantum Wells (QW) GaAs Based Vertical-Cavity Surface Emitting Laser where the DBR and QW introduction is an important breakthrough for the VCSEL technology advance: DBR involves the increase of the reflection coefficient and the QW strongly reduces the threshold current up to few milliamps.

Furthermore, the growth of the VCSEL structure by Molecular Beam Epitaxy (MBE) was a crucial advance toward its performance enhancement. MBE led to a broad-based production (mainly for the AlGaAs/GaAs structure) involving cost effectiveness. Thus, at the beginning of the 90's, we could find the 850nm VCSEL structure presented on Fig. 3, there were still two major drawbacks: the high electrical resistivity of the DBR and the optical confinement through the top DBR. Finding a solution for these problems represented a new challenge in the VCSEL technology.

During 90's, the VCSEL technology research was divided into two branches: on one hand, improving the 850nm VCSEL performance and, on the other hand, designing a 1.3 and a 1.55 μm VCSELs.

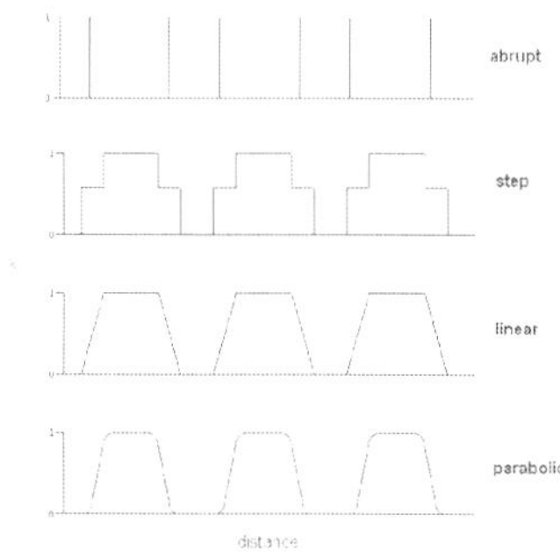

Fig. 4. Various doping profile of $Al_xGa_{1-x}As$

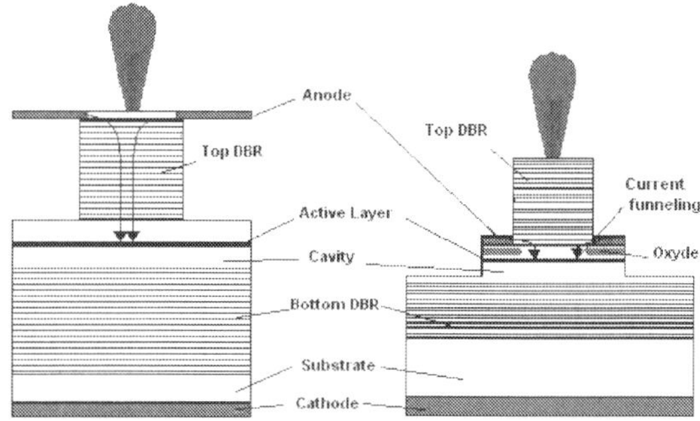

Fig. 5. Mesa structure

2.2 850nm VCSEL

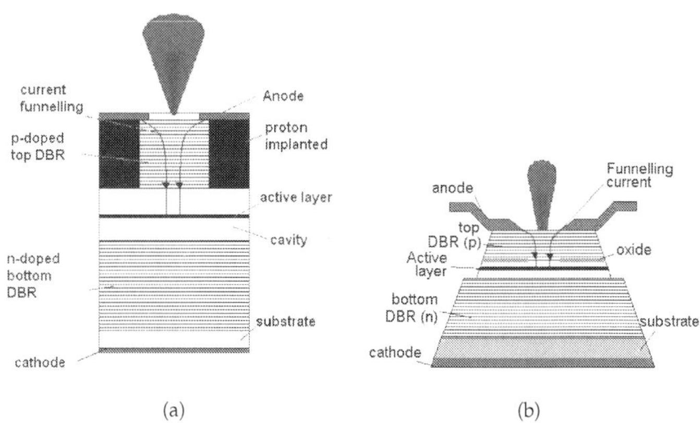

Fig. 6. (a) Proton implanted VCSEL, (b) Oxide confined VCSEL

Before attaining the maturity of the 850nm VCSEL technology numerous research works have been carried-out. Firstly, the resistivity of the DBR has been reduced by modifying the doping profile of $Al_xGa_{1-x}As$ (Kopf et al., 1992). Fig. 4 shows different doping profiles. The first VCSEL generations had an abrupt doping profile providing a good reflectivity but a high resistivity ($> 100\Omega$). By modifying the doping profile at each AlGaAs/GaAs junction, the best compromise between high reflectivity and low resistance has been obtained. The parabolic profile (see fig.4) finally gave the best performance.

The current and photon confinements were other technological bottlenecks. Up to the day, many GaAs-VCSEL structures were proposed. Structures presented on Fig.5 and 6 are the most common. By using a MESA DBR, the optical confinement has been improved. This technology allowed two possibilities for the top electrodes: on the top of the DBR (left side of Fig.5) and closer to the cavity (right side of Fig.5). These structures provided good performance but the technology is in disagreement to the broad-based production. The proton implanted structure presented on the right side of Fig.6 is the first serial produced VCSEL. The top DBR contains an insulating proton (H^+) layer to limit the current spreading below the top electrode. Nevertheless, this method doesn't reduce enough the injection area to avoid a transverse carrier spreading into the active layer (Zhang & Petermann, 1994). The main consequence is a multimode transverse emission. Indeed, the coexistence of the optical field and the current funnelling in the same area degrades the VCSEL operation. The oxide confined structure Fig.6 provides a good compromise between the beam profile and high optical power. Indeed, the diameter of the oxide aperture has an influence on the multimode transverse behavior and the output power. If the oxide aperture diameter is smaller than $5\mu m$, the VCSEL has a singlemode transverse emission nonetheless the optical power is lower than 1mW. To obtain a high power VCSEL (about 40mW), the diameter of the oxide aperture has to be wider ($25\mu m$) but the beam profile is strongly multimode transverse.

Another point to be emphasized for the use of the 850nm VCSEL is the thermal behavior. As in any semiconductor, the carrier number is strongly dependent on the temperature, while involving fluctuations of the optical power, the wavelength and threshold current (Scott et al., 1993). The earmark of the VCSEL is the parabolic threshold current (I_{th}) evolution close to a temperature characteristic. If this characteristic of temperature is close to the ambiant, it has the advantage of avoiding a thermal control for its applications. However the thermal behavior degrades the carrier confinement due to the Joule effect through the DBRs and modifies the refractive index of the DBR. These phenomena are responsible for the multimode transverse emission and strong spatial hole burning.

By knowing these drawbacks, it is possible to consider the VCSEL as a median component between good laser diodes and LED. Its low cost had allowed its emergence into the short distance communication applications to increase the bit rate while keeping cost effectiveness.

2.3 1.3 and $1.55\mu m$ **VCSEL**

The emergence of the 1.3 and $1.55\mu m$ VCSELs was quite different than the 850nm ones. In fact, the telecom wavelength laser market was widely filled by the DFB lasers whose performances are well adapted to the telecom market. Bringing into the market, the LW-VCSEL, the following assets have to be kept versus the DFB: high integration level and cost reduction with relatively good performance. By considering the numerous bottlenecks of the LW-VCSEL technology, it takes up a challenge. The first $1.3\mu m$ CW operation was demonstrated by Iga (Baba et al., 1993),(Soda et al., 1983) in 1993 with an InGaAs/InP based active layer at 77K. The upper mirror was constituted by 8.5 pairs of p-doped MgO-Si material with Au/Ni/Au layers at the top and 6 pairs of n-doped SiO/Si materials at the bottom. The materials given a 1.3 and 1.5 μm wavelength are not compatible with a monolithic growth. To provide a wavelength emission within 1.1 - 1.6 μm range, the most suitable semiconductor compound is InGaAsP/InP. Even if the wavelength's range is easy to reach, the InGaAsP/InP are not well optimized for the DBR (Shau et al., 2004). Only 12-15 AlAs/AlGaAs pairs are needed to fabricate a DBR with 99% reflectivity. By taking into account a low refractive index difference (0.3) between InP/InGaAsP layer pairs, more than 40 pairs are required to have 99% reflectivity. The thickness of DBRs has strong consequences on the VCSEL interest, not only in terms of integration but also in terms of heat sinking. In other hand, AlAs/AlGaAs DBR couldn't be grown on InP substrate due to a lattice mismatch. The problem encountered with the DBR utilization has a strong impact into the LW VCSEL technology. In 1997, Salet et Al. (Salet et al., 1997) demonstrated a pulsed RT operation of single-mode InGaAs/InP VCSEL at 1277nm. The structure was composed by a bottom n-doped InGaAsP/InP DBR (50 pairs) with 99.5% reflectivity and a top p-doped $SiO_2 : Si$ reflector. The threshold current at 300K was 500mA. For each kind of VCSEL, the vertical common path of carrier and photon

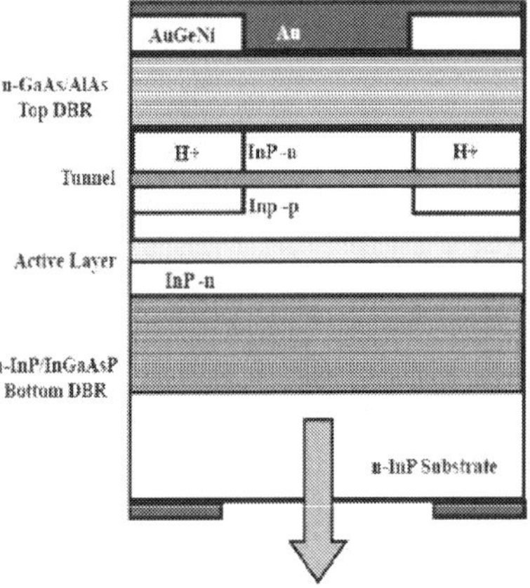

Fig. 7. 1.55 μm VCSEL with tunnel junction ((Boucart et al., 1999)

flow has a strong influence on the multimode transverse emission, this unwanted behavior is linked with thermal problems. One of the solutions to segregate the carrier and photon paths was brought by the tunnel junction introduction into the structure. The tunnel junction was discoverd by L. Esaki in 1951 (Esaki, 1974). This junction is composed by two highly doped layers: $n^{++} = p^{++} = 1 - 2 \cdot 10^{19} cm^{-3}$. In the case of LW-VCSEL, the tunnel junction acts as a hole generator. With a reverse bias, the electron tunnelling between the valence and the conduction band involves a wide hole population. The tunnel junction has to be localised just above the active layer. Moreover it presents numerous advantages such as the reduction of the intra valence band absorption due to P doping, the threshold current reduction by improving the carrier mobility, the optical confinement. So the tunnel junction is an important technological breakthrough in the LW VCSEL technology. Today, all LW VCSEL include a tunnel junction. In 1999,Boucart et al.(Boucart et al., 1999) demonstrated a RT CW operation of a 1.55μm VCSEL consisting in a tunnel junction and a metamorphic mirror (Fig. 7). A metamorphic mirror is a GaAs DBR directly grown on the InP active layer. The threshold current of this structure was 11mA.

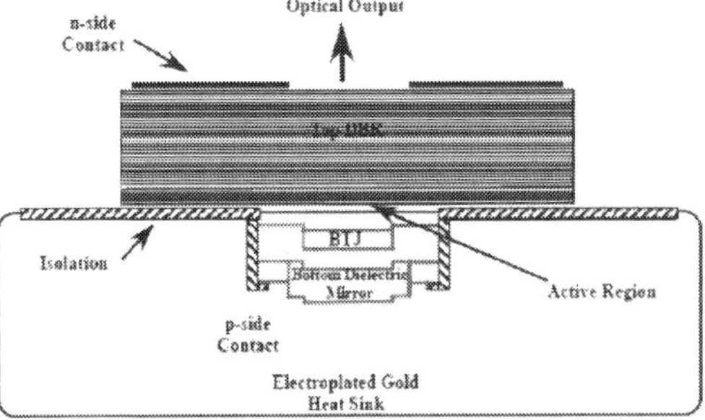

Fig. 8. 1.55 μm Vertilas structure

At the same time, Vertilas (Ortsiefer et al., 2000) presented a variation of the Boucart's structure with a bottom dielectric mirror as shown by Fig. 8. The dielectric mirror provides a 99.75% reflectivity with only 2.5 pairs of $CaF_2/a - Si$. Today this kind of structure are commercialised by Vertilas. The performance of these VCSELs, that are in current progress, make them very competitive in the 1.55μm VCSEL market.

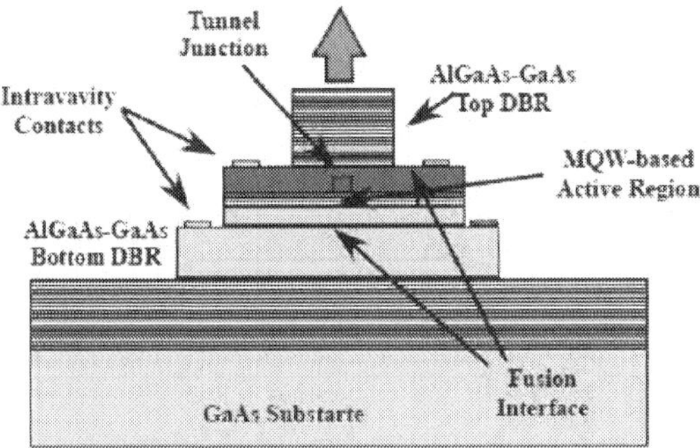

Fig. 9. Wafer fused BeamExpress VCSEL

Another technological breakthrough was the wafer bonding (or wafer fusion) technique. Wafer fusion has been developed by University of California Santa Barbara in 1995 (Babic et al., 1995). Chemical bonds are directly achieved between two materials without an intermediate layer at the heterointerface. A variant of the wafer fusion technique has been demonstrated by Kapon et al. (Syrbu et al., 2004) in order to apply the "localised wafer fusion" to a serial production. This process was developed and patented at Ecole Polytechnique Fédérale de Lausanne (EPFL) where the BeamExpress spin-off emerged. Fig.9 shows the 1.55μm BeamExpress structure. Besides the originality of the localised wafer fusion technique, the carrier injection is also improved by using a double intracavity contact avoiding a current flow through the DBR. Thus a singlemode transverse emission is reached. Today, BeamExpress leads the market in terms of optical power: 6.5mW at 1.3μm and 4.5 mW at $1.5\mu m$ (Kapon et al., 2009).

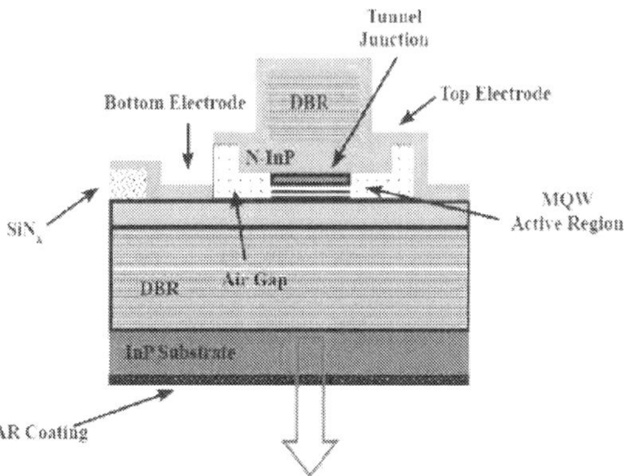

Fig. 10. Monilithic structure of Raycan VCSEL

In 2002, Raycan, a spin-off supported by the Korean Government, launched a project of monolithic long-wavelength VCSEL. They attempted to monolithically grow InAlGaAs DBR and InGaAs-based quantum well active layer on an InP substrate. This technique was unconsidered before because 99% reflectivity of an InAlGaAs-based DBR required more than 40 pairs. Raycan employed the metal-Organic Chemical Vapour Deposition (MOCVD) technique to fabricate the longwavelength VCSEL. For $1.55\mu m$ VCSELs, the top and bottom DBR were grown as 28 and 38 pairs of undoped InAlGaAs/InAlAs layers. And for the $1.3\mu m$ VCSELs, the top and bottom DBRs consisted of 33 and 50 layers respectively. The 0.5λ thick active region consists of seven pairs of strain-compensated InAlGaAs QW. The lower pair number of the top DBR was compensated by using an InAlGaAs phase-matching layer and a Au metal layer. Fig. 10 presents the structure of $1.55\mu m$ Raycan VCSEL. Reliable structure (Rhew et al., 2009) are being commercialised since 2004.

2.4 Electrical access topology

Up to this point, we have presented the main VCSEL structures without taking into account the electrical access topology. Knowing the VCSEL structure facilitates the understanding of the mecanism of electron-photons but it is insufficient to foresee the VCSEL behavior under modulation. As for the edge emitting laser (Tucker & Pope, 1983), the VCSEL modulation response is affected by parasitic elements due to the connection with the input electrical source. The electrical access is the most influential in the VCSEL array configuration. Despite of its high integration level the VCSEL technology, the electrical connection ensuring the driving is not immediate and requires an optimization in order to match the VCSEL with its driving circuit. Up to the day, the VCSEL are shipped into various packages. Each package is available for an associated frequency application range. The increases in frequency involve a specific electrical access to limit parasitics effects. But as it will be shown, even for the VCSEL chip, the electrical access modifies the VCSEL frequency response. Before continuing, let us dwell on the different chip types and the packages.

The VCSEL chip topology presents top and bottom electrodes. According to the intrinsic structure, we could have two kind of VCSELs: the "top-emitting VCSELs" where the signal is brought through the top electrode and the ground linked to the bottom electrode, on the other hand, the "bottom-emitting VCSELs" have the ground contact on the top and the signal contact on the bottom. Thus, the topology of the chip will depend on the top and bottom emission.

- **Microstrip electrical access**
  A great deal of VCSEL arrays are manufactured with a signal access on the top and a bottom common ground as we can see on Fig.11.

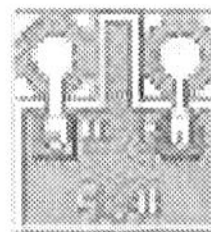

Fig. 11. Microstrip access

This topology is perfectly adapted to the top emitting VCSEL. It allows to share the ground face of each VCSEL of an array. The signal, on the top, is achieved by a microstrip line matched to each VCSEL of the array. Such a structure has the advantage of reducing the spacing between each VCSEL of an array. In order to test the VCSEL, it is necessary to mount the array on a TO package or on a submount with etched strip lines.

Due to its technological simplicity, TO package is a common packaging for a single

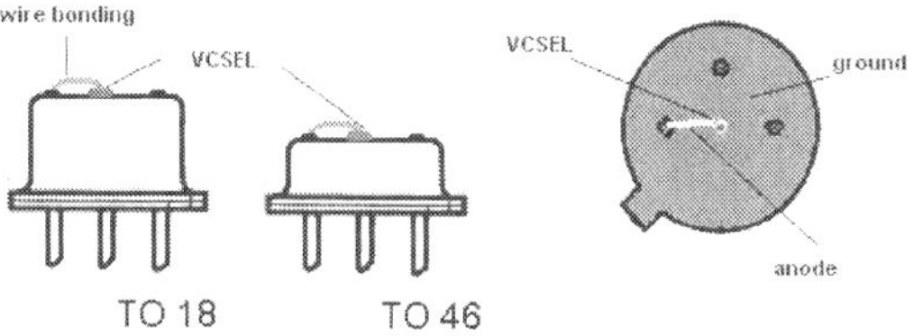

Fig. 12. TO package

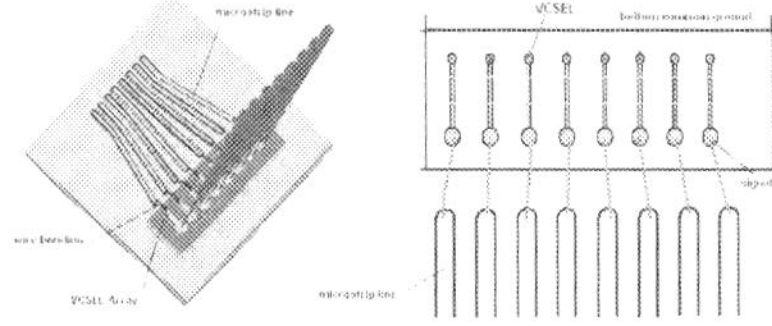

Fig. 13. VCSEL array submount

VCSEL and sometimes for VCSEL array. As shown in Fig.12, the VCSEL is fused on the top of the TO package. The ground contact, on the bottom, is carried out through the welding, that is to say, the ground is linked to the metal can. The signal contact is provided by a wire bonding between the VCSEL strip line and a pin isolated to the metal can. Many VCSELs are available on TO package with connector, lens caps or pigtailed. The utilisation of a TO packaged VCSEL is easy and allows to do many characterisations such as optical power versus bias current, optical spectrum, linewidth and Relative Intensity Noise (RIN). Unfortunately it is not well adapted for the high frequency application. Actually, the TO packaging presents a frequency limitation between 2 an 4 GHz, often below VCSEL cut-off frequency. That is why the utilisation of a TO packaged VCSEL is inadvisable for high frequency modulation. Reliable mathemathical extraction procedures are available for the frequency response study (Cartledge & Srinivasan, 1997) but, in a goal of integration in an optical sub-assembly, the modulation frequency or the bit rate would not be optimized.

In the case of a VCSEL array, the TO package is not well adapted. Thus it is necessary to set the array on a submount with etched strip lines. As it is presented by Fig. 13, the electrical connection with the array is realized by using wire bondings.
The common ground of the VCSEL array is linked with the ground of the etched strip lines. However, this submount involves parasitic effects clearly visible under modulation operation (Rissons & Mollier, 2009). A coupling between adjacent VCSEL is observable: when one VCSEL is modulated, the neighbouring VCSEL lase without any injection (we will return to this point in a further section). This coupling increase with the frequency but according to these drawback, the microstrip line electrical access is not the best configuration for the frequency modulation.

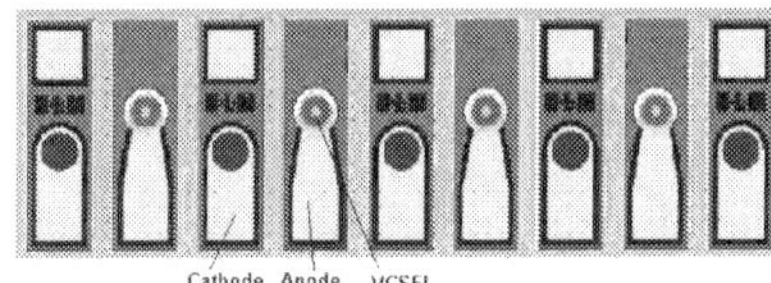

Fig. 14. Coplanar electrical access VCSEL array

- **Coplanar electrical access**
  Another available VCSEL array chip presents a coplanar access. This topology is in a good agreement with the planarization. As Fig.14 shown, not only the anode but also the cathode (which is rised by via-hole) are on the top of the chip. This topology have the advantage to minimize the length of the electrical access and reduce the parasitics phenomena. Moreover, the coplanar access allows an impedance matching to limit the electrical reflection on the VCSEL input. This configuration is ideal for the RF test because the RF probe could be placed closer to the chip. Regarding to the VCSEL array, no coupling phenomema between adjacent VCSEL have been observed. Finally the integration is easyer than the microstrip access due to the ground on the top. Nevertheless, wire bondings are required to connect the VCSEL array with its driver.

- **Bottom-emitting VCSEL chip**
  The electrical access toplogy previously presented is not adapted to the bottom-emitting VCSEL. The flip-chip bonding is required for the electrical contact. This technique has the advantage to be suitable for the integration on a CMOS circuit. Several VCSEL manufacturers provide this kind of chip. Fig.15 shows the topology of a Raycan VCSEL chip. In counterpart, the RF testing is difficult because the bottom emission implies the impossibility of optical power collection.

Fig. 15. Bottom-emitting Raycan VCSEL chip

3. Optoelectronic model: rate-equations and equivalent circuit model

This section aims at presenting a complete model of VCSEL in order to be able to simulate the VCSEL behavior before its implementation in an optical sub-assembly. Firstly, the rate equations are defined according to the VCSEL structure and simplified in compliance with the operating mode. The steady state model and characterization through the light current model is developed. Secondly, we will be interested in the dynamic behavior of the VCSEL. This approach is based on the comparison between the rate equations and an electrical equivalent circuit to obtain the relationships between intrinsic parameters and equivalent circuit elements (Tucker & Pope, 1983),(Bacou et al., 2010). The electrical equivalent circuit approach consists in describing the physical phenomena occurring into the VCSEL structure by resistive, inductive and capacitive elements. The behavioral electrical equivalent circuit is cascaded with the electrical access circuit according to each submount.

3.1 VCSEL rate equations

As for each laser diodes, the electron-photon exchanges into the VCSEL are modeled by a set of coupling rate equations. These equations relate the physical mecanisme inside the VCSEL structure, thus each approximation has to take into account the variant of each VCSEL.

The carrier rate equation is the difference between the carrier injection and the carrier recombinations. The photon rate equation is the difference between the generated photons participated to the stimulated emission and the lost photons. These equations can be written as the following form:

$$\frac{dN}{dt} = \frac{\eta_i \cdot I}{q \cdot N_w} - \left(A + B \cdot N + C \cdot N^2 \right) N - G \cdot S + F_N(t) \tag{1}$$

$$\frac{dS}{dt} = \Gamma \cdot \beta \cdot B \cdot N^2 + N_w \cdot G \cdot S - \frac{S}{\tau_S} + F_S(t) \tag{2}$$

Where:

- N is the carrier number in one QW, S is the photon number in the cavity.

- N_w is the QW number. η_i is the internal quantum efficiency. I is the injected current. So $\frac{\eta_i \cdot I}{q \cdot N_w}$ represents the population injection into each QW.

- A is the non-radiative recombinations (by recombinant center), B is the bimolecular recombination (representing the random spontaneous emission), C is the Auger recombination coefficient which can be neglected for the sub-micron emitting wavelength. We can consider $A + B \cdot N + C \cdot N^2 = \tau_n^{-1}$ where τ_n is the carrier lifetime which could be taken as a constant according to the laser operation mode.

- G is the modal gain. It depends to the carrier and photon number through the relationship $G(N,S) = g_0 \cdot \frac{N-N_{tr}}{1+\epsilon S}$ where N_{tr} is the transparency electron number, ϵ is the gain compression factor. The modal gain coefficient g_0 is expressed as $g_0 = v_{gr} \cdot \Gamma \cdot \frac{a}{V_{act}}$ where a is the differential gain coefficient, V_{act} is the active layer volume, v_{gr} the group velocity and Γ is the confinement factor.

- The term $\Gamma \cdot \beta \cdot B \cdot N^2$ corresponds to the spontaneous emission contributing to the lasing mode. β, the spontaneous emission coefficient, relates the portion of the spontaneous emission which will be amplified.

- τ_S is the photon lifetime into the cavity. It is linked to the loss by the relationship $\tau_S^{-1} = v_{gr} \cdot (\alpha_i + \alpha_m)$, α_i represents the internal losses and α_m, the mirror losses.

These equations are adapted to a QW laser through the η_i value and the presence of N_w. The confinement factor takes into account the vertical light emission and the DBR contribution. Moreover the values of each intrinsic parameters depend on the VCSEL structure.

The two last term $F_N(t)$ and $F_S(t)$ have to be taken in part. In fact, $F_N(t)$ and $F_S(t)$ are the carrier and photon Langevin functions respectively, representing the carrier and electron fluctuations. These fluctuations are due to the stochastic evolution of N and S associated to the noise generation. Indeed, the operation of the laser diode is affected by several noise sources whose influence varies according to the different regimes. For targeted applications, the preponderant noise source is the spontaneous emission. The randomness of the spontaneous emission generates amplitude and phase fluctuations of the total optical field. Moreover, these photons which are produced in the laser cavity follow the feedback of the stimulated photons and interact with them. By taking into account the wave-corpuscule duality of the light, a quantum approach is well suited to describe the emission noise generation including the photon-electron interaction: each state of photon or electron is associated to a noise pulse. For the purposes of noise generation quantification, recombination and absorption rates in the cavity allow the utilization of the electron and photon Langevin forces to give a mathematical representation of the optical emission noise.

To complete the VCSEL modeling, rate equations have to be solved according to each operation mode.

3.1.1 Steady state resolution

The first step of the rate equation resolution considers the case of the steady state. This resolution aims at to extract the relationship of the threshold current, threshold carrier number, and current/photons relations above threshold. It also allows to valid which approximation degrees are reliable.

When the steady state is reached, the rate equations are equal to 0 such as:

$$0 = \frac{\eta_i \cdot I}{q \cdot N_w} - \left(A + B \cdot N + C \cdot N^2\right) N - G \cdot S \tag{3}$$

$$0 = \Gamma \cdot \beta \cdot B \cdot N^2 + N_w \cdot G \cdot S - \frac{S}{\tau_S} \tag{4}$$

Fig.16 represents the carrier and photon evolution versus the bias current where the red straight line corresponds to an asymptotic representation and the dotted line correponds to a physical representation. So we will study both cases begining by an asymptotic resolution involving that the spontaneaous emission $\Gamma \cdot \beta \cdot B \cdot N^2$ and the gain compression $\epsilon \cdot S$ are neglected.

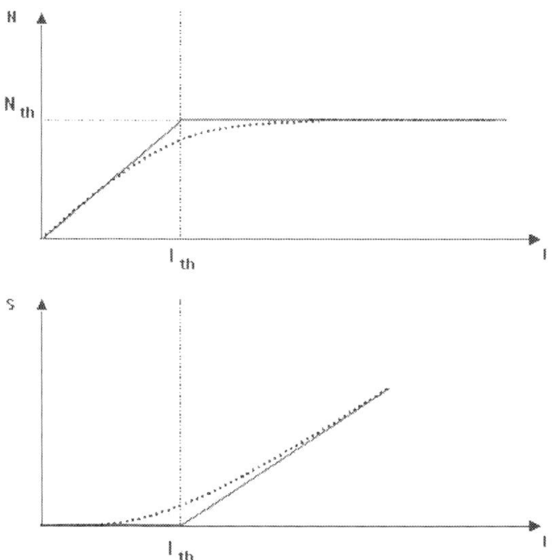

Fig. 16. Evolution of carrier number (N) and photon number S versus the bias current I

- Below threshold: $I < I_{th}$, $N < N_{th}$ et $S \approx 0$, according to Fig.16
  From the equation 3, we can extract:

$$N = \frac{\tau_n \cdot \eta_i \cdot I}{q \cdot N_w}, \forall N \leq N_{th} \tag{5}$$

- Above threshold: $S > 0$ and $I > I_{th}$
  From the equation 4, we obtain:

$$N_w \cdot g_0 \left(N - N_{tr}\right) = \frac{1}{\tau_S} \tag{6}$$

Involving:

$$N = N_{tr} + \frac{1}{N_w \cdot \tau_S \cdot g_0} = cte = N_S \tag{7}$$

Which is equivalent to say : while the simulated emission occures, the carrier number stops to increase linearly with the current because the carrier consumption is compensated by the injected current. Thus, we can consider that the carrier number is constant above the threshold.

From the equations 5 and 7, the following relationship could be extracted:

$$I_{th} = \frac{q \cdot N_w \cdot N_{th}}{\eta_i \cdot \tau_N} \tag{8}$$

Thus from the equations 3, 6 and 8, we obtain:

$$S = \frac{\tau_S \cdot \eta_i}{q \cdot N_w} \left(I - I_{th}\right) \tag{9}$$

Then for $I > I_{th}$, the photon number linearly increases with the current which verifies the well-known relationship versus the optical power such as:

$$P_{opt} = v_{gr} \cdot \alpha_m \cdot h \cdot v \cdot S = \eta_d \cdot \frac{h \cdot v}{q \cdot N_w} \left(I - I_{th}\right) \tag{10}$$

where η_d is the slope efficiency such as: $\eta_d = \eta_i \frac{\alpha_m}{\alpha_i + \alpha_m}$.

By considering the spontaneaous emission contribution, we can express the real N(I) and S(I) curves close to the threshold according to the dotted line on Fig.16. Henceforth in equation 4, the term $\Gamma \cdot \beta \cdot B \cdot N$ is different from zero. As the following equation is valid:

$$N_w \cdot g_0 \left(N_{th} - N_{tr}\right) = \frac{1}{\tau_S} \tag{11}$$

The photon rate equation becomes:

$$0 = \Gamma \cdot \beta \cdot B \cdot N^2 + g_0 \cdot N_w \cdot \left(N - N_{th}\right) \cdot S \tag{12}$$

So for each injected current, the photon number is not negligible because of the spontaneous emission. Thus, the amplified spontaneous emission is expressed in the following relationship:

$$S = \frac{\Gamma \cdot \beta \cdot B \cdot N^2}{g_0 \cdot N_w \cdot \left(N - N_{th}\right)} \tag{13}$$

By injecting Equation 13 in Equation 3, the relationship between N and I is written as:

$$\frac{\eta_i \cdot I}{q \cdot N_w} - \frac{N}{\tau_N} - \left(N - N_{tr}\right) \frac{\Gamma \cdot \beta \cdot B \cdot N^2}{N_w \cdot \left(N - N_{th}\right)} = 0 \tag{14}$$

The solution of the 3rd equation corresponds to the evolution of the carrier number close to the threshold. According to Fig.16 and while I is close to its threshold value, the spontaneaous emission slows the increase of the carrier number to reach the steady-state.

Now, considering the gain compression, that is to say for high value of photon number involving: $\epsilon \cdot S > 1$. The rate equation becomes:

$$0 = \frac{\eta_i \cdot I}{q \cdot N_w} - \frac{N}{\tau_N} - g_0 \frac{N - N_{tr}}{1 + \epsilon \cdot S} \cdot S \tag{15}$$

$$0 = N_w \cdot g_0 \left[\frac{N - N_{tr}}{1 + \epsilon \cdot S} - \left(N_{th} - N_{tr}\right)\right] \cdot S \tag{16}$$

Thus:

$$\frac{N_{th} - N_{tr}}{N - N_{tr}} = \frac{1}{1 + \epsilon \cdot S} \tag{17}$$

and

$$S = \frac{N_w \cdot g_0 \cdot \tau_S}{\epsilon} \left(N - N_{th}\right) \tag{18}$$

By injecting Equations 17 and 18 in Equation 15, we obtain:

$$\frac{\eta_i \cdot I}{q \cdot N_w} - \frac{N}{\tau_N} - \frac{g_0}{\epsilon} \left(N - N_{th}\right) = 0 \tag{19}$$

In fact the overflow of the carrier population versus the steady-state value causes an overflow of photon responsible to the gain compression. We can note δN the carrier overflow, so that: $N = N_{th} + \Delta N$.

From the equation 19, we obtain:

$$\delta N = \frac{\eta_i}{q \cdot N_w} \frac{\tau_N \cdot \epsilon}{\tau_N \cdot g_0 + \epsilon} \left(I - I_{th} \right) = 0 \tag{20}$$

If $\tau_N = 10^{-9}s$, $\epsilon = 10^{-7}$, $g_0 = 10^4 s^{-1}$, $I - I_{th} = 10^{-3}A$ and $\eta_i = 10^{-2}$, δN will be approximatively equal to 10^3 which is negligible against N_{th}. Hence, we can admit the validity of the asymptotic approximation.

3.1.2 Small-signal linearization

Now, the VCSEL is modulated by the superimposition of the bias current I_0 and a sinusoidal current $\Delta I(t)$ so that $\Delta I(t) << I_0$. Due to the small-signal approximation, $N(t)$ and $S(t)$ follow $I(t)$. Let us now linearizing the rate equations according to these conditions. As $X(t) = X_0 + \Delta X(t)$ (where $X(t) = I(t), N(t)$ or $S(t)$), $\frac{dX_0}{dt} = 0$ and $\frac{dX(t)}{dt} = \frac{d\Delta X(t)}{dt}$. By putting $\dot{X} = \frac{dX}{dt}$, we have $\frac{d\delta X}{dt} = \delta \dot{X}$. As $\dot{N}$ depends on I, N, S and $\dot{S}$ on N and S, we have:

$$\Delta \dot{N} = \frac{\partial \dot{N}}{\partial I} \Delta I + \frac{\partial \dot{N}}{\partial N} \Delta N + \frac{\partial \dot{N}}{\partial S} \Delta S \tag{21}$$

$$\Delta \dot{S} = \frac{\partial \dot{S}}{\partial N} \Delta N + \frac{\partial \dot{S}}{\partial S} \Delta S \tag{22}$$

After linearization, we obtain:

$$\Delta \dot{N} = \frac{\eta_i}{q \cdot N_w} \Delta I - \left(g_0 \cdot \frac{S_0}{1 + \epsilon \cdot S_0} + A + 2 \cdot B \cdot N_0 + 3 \cdot N_0^2 \right) \Delta N + g_0 \frac{N_0 - N_{tr}}{(1 + \epsilon \cdot S_0)^2} \Delta S \tag{23}$$

$$\Delta \dot{S} = \left(2 \cdot \Gamma \cdot \beta \cdot B \cdot N_0 + N_w \cdot g_0 \frac{S_0}{1 + \epsilon \cdot S_0} \right) \Delta N + \left(N_w \cdot g_0 \frac{N_0 - N_{tr}}{(1 + \epsilon \cdot S_0)^2} - \frac{1}{\tau_S} \right) \Delta S \tag{24}$$

The equation set can be expressed as follows:

$$\begin{bmatrix} \Delta \dot{N} \\ \Delta \dot{S} \end{bmatrix} = [M] \begin{bmatrix} \Delta N \\ \Delta S \end{bmatrix} + \frac{\eta_i}{q \cdot N_w} \begin{bmatrix} \Delta I \\ 0 \end{bmatrix} \tag{25}$$

M is a matrix such as:

$$[M] = \begin{bmatrix} -\gamma_{NN} & -\gamma_{NS} \\ \gamma_{SN} & -\gamma_{SS} \end{bmatrix} \tag{26}$$

where:

$$\gamma_{NN} = A + 2BN_0 + 3CN_0^2 + \frac{g_0 \cdot S_0}{1 + \epsilon \cdot S_0} \tag{27}$$

$$\gamma_{NS} = \frac{g_0 (N_0 - N_{tr})}{(1 + \epsilon S_0)^2} \tag{28}$$

$$\gamma_{SN} = 2\Gamma \beta B N_0 + \frac{N_w \cdot g_0 \cdot S_0}{1 + \epsilon S_0} \tag{29}$$

$$\gamma_{SS} = \frac{1}{\tau_S} - \frac{N_w \cdot g_0 (N_0 - N_{tr})}{(1 + \epsilon S_0)^2} \tag{30}$$

By considering the sinusoidal modulation as $\Delta I = I_m \cdot e^{j\omega t}$, $\Delta N = N_m \cdot e^{j\omega t}$ and $\Delta S = S_m \cdot e^{j\omega t}$, and with $\frac{d}{dt} = j\omega$, the Equation 25 becomes:

$$\begin{bmatrix} \gamma_{NN} + j\omega & \gamma_{NS} \\ -\gamma_{SN} & \gamma_{SS} + j\omega \end{bmatrix} \begin{bmatrix} \Delta N \\ \Delta S \end{bmatrix} = \frac{\eta_i}{q} \begin{bmatrix} \Delta I \\ 0 \end{bmatrix} \tag{31}$$

By using the Cramer rule, we obtain:

$$N_m = \frac{\eta_i I_m}{q} \frac{1}{\Delta} \begin{vmatrix} 1 & \gamma_{NS} \\ 0 & \gamma_{SS} + j\omega \end{vmatrix} \tag{32}$$

$$S_m = \frac{\eta_i I_m}{q} \frac{1}{\Delta} \begin{vmatrix} \gamma_{NN} + j\omega & 1 \\ -\gamma_{SN} & 0 \end{vmatrix} \tag{33}$$

where Δ is the matrix determinant such as:

$$\dot{\Delta} = \gamma_{NN}\gamma_{SS} + \gamma_{NS}\gamma_{SN} - \omega^2 + j\omega(\gamma_{NN} + \gamma_{SS}) = \omega_R^2 - \omega^2 + j\omega\gamma_R \tag{34}$$

ω_R is the resonance frequency such as: $\omega_R^2 = \gamma_{NN}\gamma_{SS} + \gamma_{NS}\gamma_{SN}$
et γ_R is the damping factor: $\gamma_R = \gamma_{NN} + \gamma_{PP}$

Which is conform to a transfer function of a second order system. The resonance frequency is an important parameters in the determination of the VCSEL frequency bandwidth. We will make some approximation according to the small signal regime. By considering a bias current I_0 above threshold ($I > 2 \cdot I_{th}$), in this context, the spontaneous emission and the gain compression can be neglected. Moreover as we are above threshold, the non-radiative and bimolecular recombination are negligible against the stimulated emission. So γ_{NN}, γ_{NS}, γ_{SN} and γ_{SS} become:

$$\gamma_{NN} = g_0 \cdot S_0 \tag{35}$$

$$\gamma_{NS} = g_0 (N_0 - N_{tr}) \tag{36}$$

$$\gamma_{SN} = N_w \cdot g_0 \cdot S_0 \tag{37}$$

$$\gamma_{SS} = \frac{1}{\tau_S} - N_w \cdot g_0 (N_0 - N_{tr}) \tag{38}$$

Which allow us to determine the resonance frequency as:

$$f_R = \frac{\omega_R}{2\pi} = \frac{1}{2 \cdot \pi} \cdot \sqrt{\frac{g_0 \cdot S}{\tau_S}} = \sqrt{\frac{g_0 \cdot \eta_i}{q \cdot N_w} (I - I_{th})} \tag{39}$$

This relationship is only reliable above threshold. If the bias current had been close to the threshold, the approximation couldn't be suitable. Moreover, this resonance frequency doesn't take into account the electrical access. We will see how to take into account the electrical access in a model.

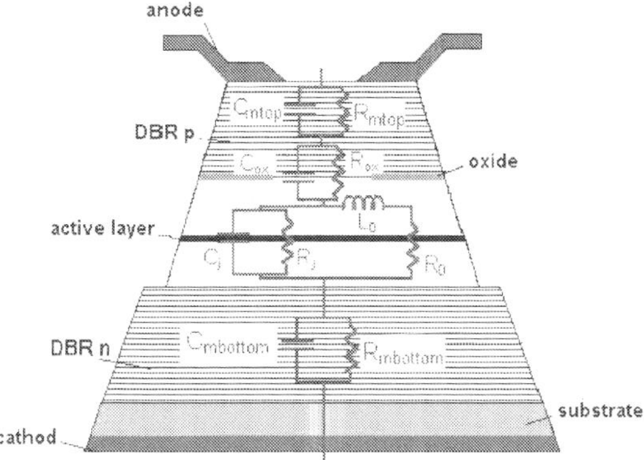

Fig. 17. Behavioral equivalent circuit of a 850nm VCSEL

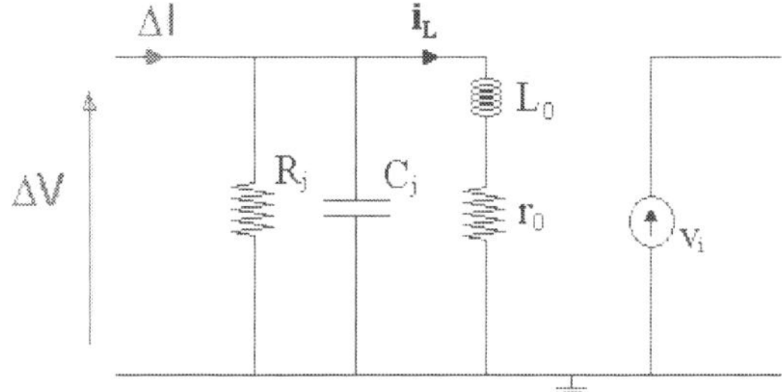

Fig. 18. Small-signal driving of a VCSEL cavity equivelent circuit

3.2 Behavioral equivalent circuit

This approach consists in describing the physical phenomena occuring into the VCSEL structure by an equivalent circuit as shown in Fig.17 for an oxide-confined VCSEL emitting at 850nm. This concept is frequently used in EEL since it has been proposed by R.S Tucker (Tucker & Pope, 1983)in the 80's. As the electronic funneling through the VCSEL structure is different than the EEL one, the electrical equivalent circuit needs to be adapted by including the multi-quantum well (QW), as well as the active layer represented by a RLC resonant circuit (17): R_j and C_j are associated to the carrier exchanges, L_0 and R_0 are the photon storage and resonance damping respectively. For the 850nm VCSEL, the doped DBR, which is a stack of doped heterojunctions, is equivalent to distributed RC cells: R_{mtop} and C_{mtop} for the top DBR, $R_{mbottom}$ and $C_{mbottom}$ for the bottom one. As the current confinement is performed by an oxide aperture, a $R_{ox}C_{ox}$ cell is added to the circuit (Brusenbach et al., 1993). This behavioral electrical equivalent circuit can be directly cascaded with the electrical access according to each submount.

By using the linearized rate equation and the Kirchhoff equation of the circuit, relationships between intrinsic parameters and circuit element can be achieved.

As the rate equation considers the carrier number in each QW and the photon number into the cavity, the Kirchhoff equations are limited to the equivalent circuit of the active region. The equation of the cavity equivalent circuit are expressed according to the convention given by Fig.18 where ΔV and ΔI are the input voltage and current respectively, and i_L is the photonic current related to the photon flow variation.

Then, we obtain the following equations:

$$\frac{d\Delta V}{dt} = \frac{\Delta I}{C_j} - \frac{\Delta V}{R_j \cdot C_j} - \frac{i_L}{C_j} \tag{40}$$

$$\frac{di_L}{dt} = \frac{\Delta V}{L_0} - \frac{R_0 \cdot i_L}{L_0} \tag{41}$$

With these equations the relationship with the instrinsic parameters can be etablished.

3.3 Relationship between rate equation and equivalent circuit

To write the relationships between VCSEL intrinsic parameters and the circuit elements, Equations 23,24,40, 41 have to be compared by using the well-known relationship derived from the voltage-current characteristic of a junction diode:

$$\frac{\Delta V}{V_T} = \frac{\Delta I}{I_0} = \frac{\Delta N}{N_0} \tag{42}$$

Where V_T is assumed to be a constant according to the semiconductor material: $V_T = \frac{m \cdot k \cdot T}{q}$. Thus:

$$\frac{\Delta N}{dt} = \frac{N_0}{V_T} \cdot \frac{d\Delta V}{dt} \tag{43}$$

That allow us to obtain directly:

$$C_j = \frac{N_0 \cdot q \cdot N_w}{V_T \cdot \eta_i} \tag{44}$$

$$R_j = \left[\frac{N_0 \cdot q \cdot N_w}{V_T \cdot \eta_i} \cdot \left(A + 2 \cdot B \cdot N_0 + 3 \cdot C \cdot N_0^2 + \frac{g_0 \cdot S_0}{1 + \epsilon \cdot S} \right) \right]^{-1} \tag{45}$$

The relationship between the current i_L and the photon fluctuations ΔS is:

$$g_0 \cdot \frac{N_0 - N_{tr}}{(1 + \epsilon \cdot S_0)^2} = \frac{N_0}{V_T \cdot C_j} i_L \tag{46}$$

By injecting Equation 46 into Equation 24, we obtain:

$$L_0 = \frac{1}{C_j} \cdot \frac{(1 + \epsilon \cdot S)^2}{g_0 \left(N_0 - N_{tr} \right) \cdot \left(2 \cdot \Gamma \cdot \beta \cdot B \cdot N_0 + \frac{g_0 \cdot S_0}{1 + \epsilon \cdot S_0} \right)} \tag{47}$$

$$\frac{R_0}{L_0} = \frac{1}{\tau_S} - g_0 \frac{N_0 - N_{tr}}{(1 + \epsilon \cdot S0)^2} \tag{48}$$

From the convention of Fig. 18, the resonance frequency can be expressed as follow:

$$f_R = \frac{\omega_R}{2 \cdot \pi} = \frac{1}{2 \cdot \pi} \cdot \sqrt{\frac{1}{L_0 \cdot C_j} + \frac{R_0}{L_0} \cdot \frac{1}{R_j \cdot C_j}} \tag{49}$$

So by modulating the VCSEL in small signal conditions, some intrinsic parameters can be extracted. The task would be easy if the contribution of the electrical access were to be negligeable. In practice, the strong contribution of the electrical access in the frequency response hide the VCSEL response. That's why the equivalent circuit approach is an excellent tool to cascade the electrical access and the laser diode. We will see how to extract the instrinsic resonance frequency and some intrinsic parameters.

4. VCSEL chip characterisation

As the frequency response of a laser diode is in the microwave domain, the modulation required a particular care. Indeed, the driving signal could be consider as a simple current-voltage couple but as an electromagnetic field that propagate as a standing-wave. thus, the connection between the VCSEL and the transmission line being achieved by a transmission line, an impedance matching between the driver and the VCSEL is required yet it is not the case of many available VCSEL. This impedance mismatch involves a high reflection on the VCSEL input, that is to say the modulation signal is not totally transmitted to the VCSEL, and the energy of the electromagnetic field radiates nearby the transmission line. To characterize the VCSEL chip by taking into account the electrical access, the scattering (S) parameters measurement with a vector network analyzer is the most suitable. This method allows not only to measure the frequency response of the VCSEL but also to extract the electrical access effect.

4.1 S-parameters measurement

In microwave circuit, the voltage, current and impedance cannot be measured in a direct manner. The measurable quantities are the amplitude and phase of the transmitted wave as compared with those incident wave. These quantities are defined as the Scaterring Matrix (S-Matrix). For a two port system, the S-matrix is written as follow:

$$[S] = \begin{bmatrix} S_{11} & S_{12} \\ S_{21} & S_{22} \end{bmatrix} \tag{50}$$

Where:

- S_{11} is the reflection coefficient on the input (1-port)
- S_{22} is the reflection coefficient on the output (2-port)
- S_{21} is the transmission coefficient through the system from the 1-port to the 2-port
- S_{12} is the transmission through the system from the 2-port to the 1-port

For a laser diode, the S matrix becomes:

$$[S] = \begin{bmatrix} S_{11} & 0 \\ S_{21} & 1 \end{bmatrix} \tag{51}$$

$S_{12} = 0$, $S_{22} = 1$ if no optical feed-back into laser cavity is considered. So that for the microwave-photonic S-Matrix measurement, we have to use a device which is able to measure a microwave reflection coefficient (S_{11}) and a microwave-photonic transmission coefficient (S_{21}). To reach this measurement, a Vector Network Analyser (VNA) is required. The experimental setup depends on the VCSEL structure (three different emitting wavelengths: 850nm, 1310nm and 1550nm) and the electrical access. For the LW-VCSEL, the VNA contains an optical rack able to measure directly the microwave-photonic S_{21} coefficient. For the 850nm VCSEL, an optical fiber linked to a calibrated photodetector allows the S_{21} measurement according to the experimental setup of Fig.20.

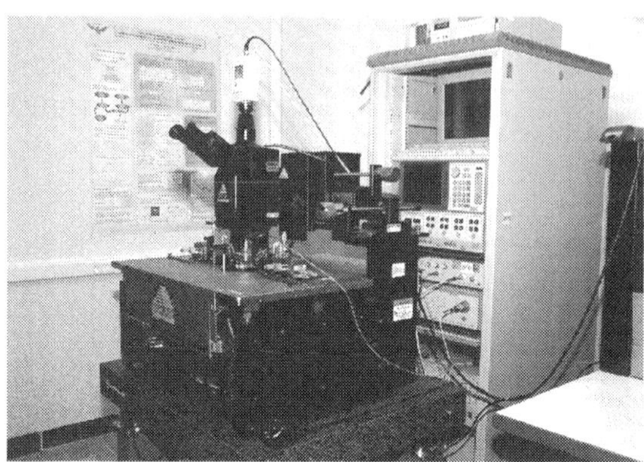

Fig. 19. RF probe station connected with a vector network analyzer

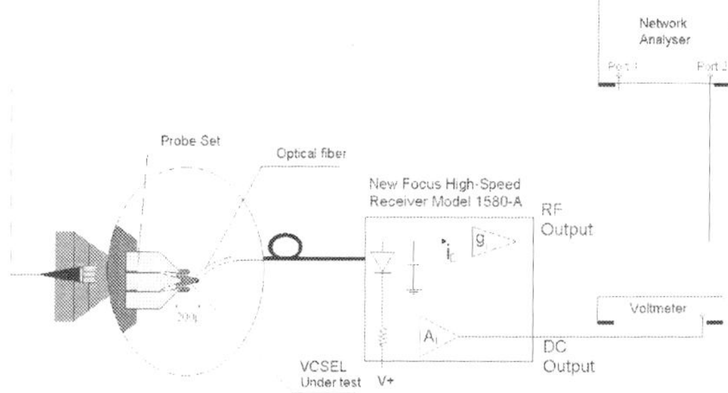

Fig. 20. Measurement of VCSEL response in opto-microwave domain

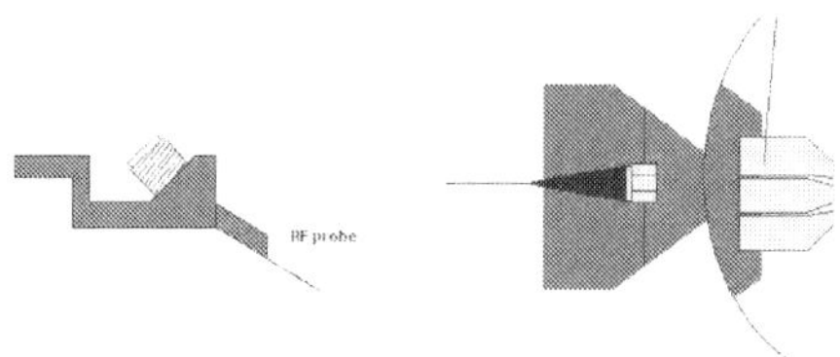

Fig. 21. RF probe

For the chip measurement, the VNA is coupled with a probe station as shown by Fig.19. The driving signal is brought by an RF probe as presented in Fig.21.

4.2 Experimental results

The experimental results presented in this section are achieved with the same experimental setup presented in the previous section. By thismeasurement technique, the influence of the electrical access is highlighted for two kind of VCSEL array topology. The TO packaging is excluded for two reasons: it doesn't allow the driving of only one VCSEL from an array and other extraction techniques have already been presented such as relative intensity noise measurement (Majewski & Novak, 1991), subtraction procedure (Cartledge & Srinivasan, 1997). So, we will focus on a VCSEL array with microstrip line electrical access and a VCSEL array with coplanar access. The model is validated by comparing the measurement and S-parameters simulation by implementing the equivalent circuit in the ADS^{TM} software (RF simulation tool).

4.2.1 Microstrip line electrical access

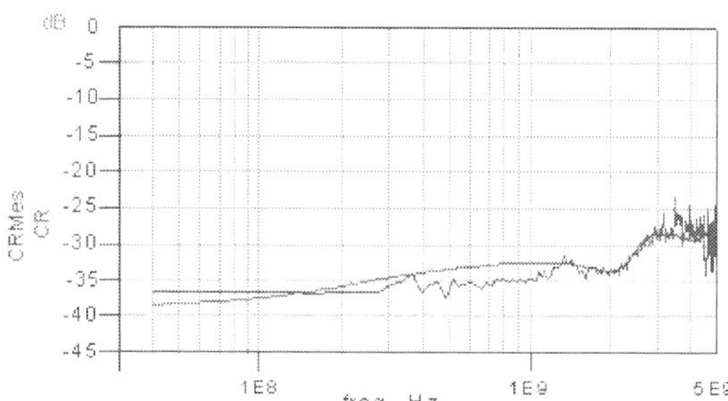

Fig. 22. Crosstalk measurement (blue curve) and simulation (red curve) of a VCSEL array with microstip access

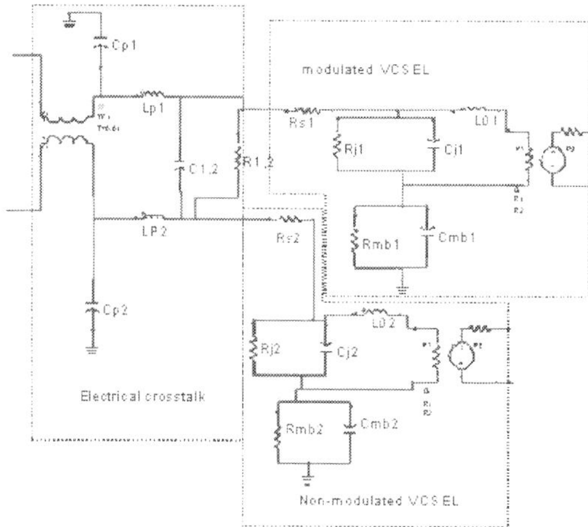

Fig. 23. Electrical equivalent circuit of two VCSELs including the crosstalk contribution

The measurement of S-parameters have been achieved on 850nm VCSEL arrays with microstrip line access. The microstrip line access requires the submount presented on Fig.13 to complete the RF tests and the integration in an optical subassembly. However this circuit involves parasitic effects clearly visible on the S_{11} and S_{21} measurement and a coupling between adjacent VCSEL. As we assume a negligible optical crosstalk, this coupling is the result of an impedance mismatch due to the wire bonding (Nakagawa et al., 2000). The RF modulation can circulate on the neighbouring wire by capacitive and inductive coupling. Through these couplings, carriers are injected in the adjacent laser involving a parasitic light emission. From the S_{21} measurement, it is thus possible to get the crosstalk versus frequency. Fig.22, printing the crosstalk up to 5GHz, is obtained by the difference between the S_{21} of the non-driven VCSEL and the modulated VCSEL.

By using the VCSEL electrical equivalent circuit and by the available crosstalk model (represented by a mutual inductance, a parasitic inductance and capacitance), an electrical equivalent circuit of two VCSELs can be developed as shown in Fig.23. The simulation results are given by the red curve of Fig.22. The characterization and the modeling of this crosstalk is not significant below 1GHz (less than -35dB) but it increases quickly above this frequency. The degradation of the S_{21} measurement due to this coupling makes impossible the extraction of reliable value of the VCSEL intrinsic parameters.

4.2.2 Coplanar access

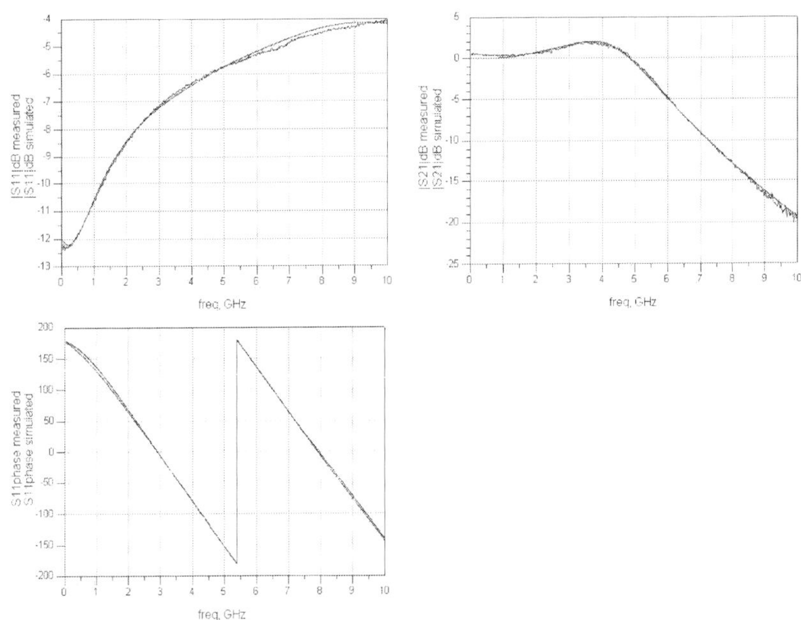

Fig. 24. Comparison between simulated and measured S_{11} module and phase and S_{21} module of a 850nm VCSEL

The second kind of measurements is achieved on oxide confined 850nm VCSEL arrays with a coplanar access such as given by Fig.14. The measured scattering parameters of the VCSEL chip including the electrical access are so smouth (without numerical averaging of the VNA) that the extraction of the VCSEL cavity S_{21} becomes available for the extraction of the equivalent circuit elements. According to the range of values of intrinsic parameters and by using the Equations 44,45, 47 and 48, a value of each intrinsic parameters can be fixed with a good agreement as it is shown in the comparison between the simulated S_{21} (Fig.24, red curve) and the measured S_{21} (Fig.24, blue curve). As the VCSEL tested are

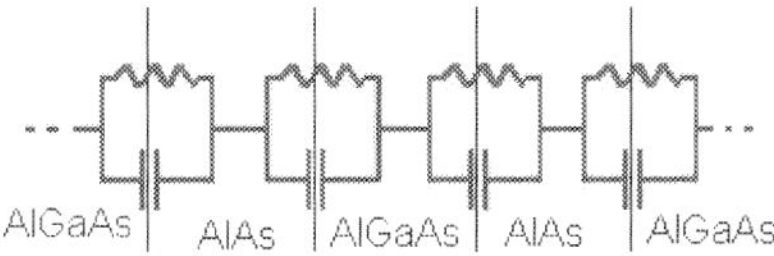

Fig. 25. RC distributed of DBR heterojunctions

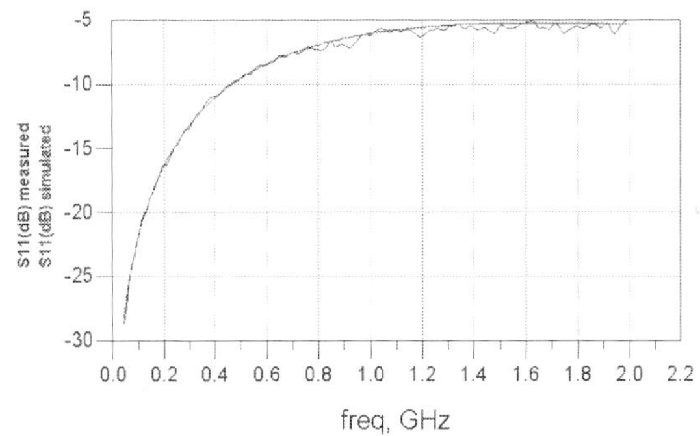

Fig. 26. Measured and simulated S_{11} DBR with $10\mu m$ ring contact

850nm oxide confined VCSEL, with a doped DBR, additional measurement can be achieved to extract the DBR equivalent circuit. As aforementioned, DBRs are constituted by a stack of heterojunctions. Each heterojunction implies RC parallel cell. Consequently, if we consider the whole DBR, we have as much resistances (R_i) and capacitances (C_j) as interfaces (Brusenbach et al., 1993). By considering that each interface of a DBR is identical, the electrical effects in a mirror is represented by an equivalent resistance R_m and capacitance C_m such as:

$$C_m = \frac{C_i}{2n - 1} \tag{52}$$

$$R_m = (2n - 1) \cdot R_i \tag{53}$$

The top and bottom DBR having different doping and layer pair numbers, the equivalent

Φ (μm)	R_m (Ω)	C_m (pF)	$R_{i calculated}$ (Ω)	$C_{i calculated}$ (pF)	$R_{i,simulated}$ (Ω)	$C_{i,simulated}$ (pF)
10	36	1.43	10.18	5.05	10.21	5.05
12	20.8	3	5.88	10.6	5.88	10.6
14	14.11	2.93	3.99	10.35	3.99	10.35

Table 1. Comparison of the DBR resistances and capacitances

resistances and capacitances are thus different for the two mirrors.

Indeed, DBR stacks have been tested to check the values of the capacitances and resistances. The S_{11} of 8-p type DBR layer pairs grown by ULM Photonics GmbH (which provided VCSEL arrays). The test wafer contains different mesa with a ring contact on the top. Thus, the measurement is achieved by putting the signal pin of the coplanar probe on the ring contact and the ground pin on the substrate. As the geometry of the wafer hasn't a coplanar access, the measurement frequency range is limited to 2 GHz (above this frequency the signal is to degraded to be exploited). By using the same approach than the preceeding one, the small signal equivalent circuit is implemented in ADS^{TM} software and compared with the measurements. The characterization having been performed for different mesa diameters, the

Parameters	Units	Values
A	s^{-1}	$[1.10^8; 1, 3.10^8]$
B	s^{-1}	$[0, 7.10^{-16}; 1, 8.10^{-16}] * V_{act}$
N_{tr}	–	$[0, 83.10^{24}; 4, 4.10^{24}] * V_{act}$
τ_p	ps	$[1; 6]$
a	m^2	$[0, 2.10^{-20}; 3, 7.10^{-20}]$
V_{gr}	$m.s^{-1}$	$[8, 33.10^7; 8, 6.10^7]$
Γ	–	$[0, 045; 0, 06]$
β	–	$[10^{-5}; 10^{-4}]$
η_i	–	$[0, 6; 0, 86]$

Table 2. Range of 850 nm VCSEL intrinsic parameters

resistances and capacitances values can be calculated according to Equations 52 and 53 and the values for each diameters is presented in table 1. Fig.26 gives the comparison between the measurement and the simulation of S_{11} module of $10\mu m$ diameter DBR. The agreement between the S_{11} simulation and measurement is quite good and allows us to implement the R_m and C_m values into the VCSEL equivalent circuit.

From Equations 44, 45, 47, 48 and the values given by the table 2 for 850nm VCSEL, the intrinsic parameter values are extracted. A, B, β, N_{tr}, τ_S, a, v_{gr}, η_i and Γ are obtained by optimization according to R_j, C_j, L_0, R_0 and the known VCSEL parameters V_T, q, N_w, I_0, I_{th}, V_{act}. Thus for a $25\mu m$ oxide aperture VCSEL, we get the values:

$V_T = 0.063V$, $N_w = 3$, $I_0 = 2mA$, $I_{th} = 0.9mA$, $\tau_p = 6ps$, $v_{gr} = 8.5 \cdot 10^7 m/s$, $a = 1.76 \cdot 10^{-20} m^2$, $\beta = 10^4$, $\Gamma = 0.049$, $\eta_i = 0.6$, $A = 10^8 \cdot s^{-1}$, $B = 88 \cdot s^{-1}$, $N_{tr} = 2.41$.

By the way of the Equation, the extracted values are:

- The threshold carrier number: $N_{th} = 3.04 \cdot 10^6$

- The photon number: $S = 2.47 \cdot 10^4$

- The modal gain: $G = 5.55 \cdot 10^{10}$

- The compression factor: $\epsilon = 2.11 \cdot 10^{-6}$

The model gives then realistic results under certain conditions of VCSEL.

For the $1.55\mu m$ VCSEL with a coplanar access, the same approache is used. The intracavity contacts avoid the contribution of the DBR into the electrical circuit. These wavelengths allow to extend the frequency range ut to 20GHz by using the calibrated optical rack with the VNA HP 8510C. However, this frequency increase shows a S_{21} response with a -60dB/decade slope in spite of the expected -40dB/decade. By knowing that a VCSEL has a -40dB/decade slope, the -60dB/decade slope is assumed to be induced by the electrical access topology (Fig.27).

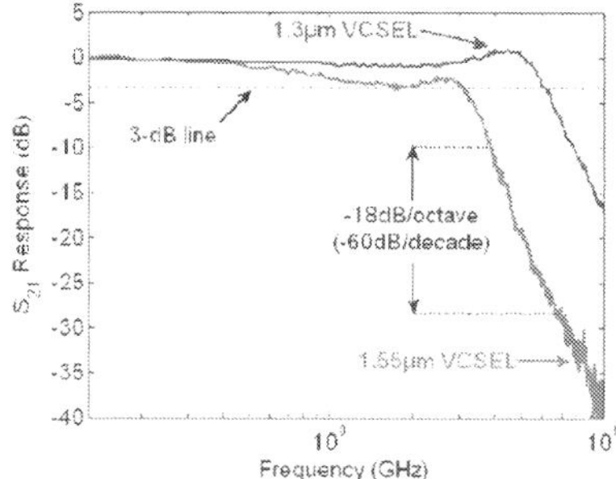

Fig. 27. Measured and extracted S_{21} responses for a 1.3μm VCSEL at a fixed bias current. The measurement takes into account the response of the electrical access as well as the cavity while the extracted curve shows the VCSEL cavity response only

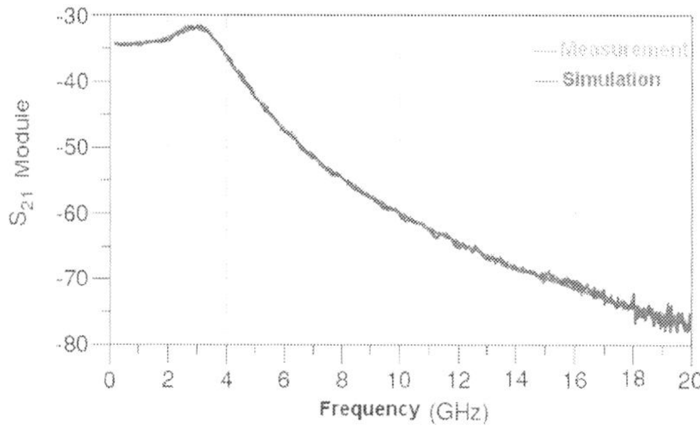

Fig. 28. Comparison between simulated and measured S_{21} module of a 1.55μm VCSEL

This effect has been modeled by using the tranfer matrix (T-matrix) formalism as presented by A. Bacou in (Bacou et al., 2009) and (Bacou et al., 2010). The method using the T-matrix consists in the splitting of the VCSEL chip into two cascaded subsystems representing the electrical access and the optical cavity, respectively. The equivalent electrical circuit defining the behavior of the electrical access is combined with the T-Matrix formalims to remove the parasitic contribution with the S_{21} measurement response. The results allow to extract the S_{21} of the VCSEL as shown by Fig.28 given a -40dB/Decade slope.

5. Conclusion

As it has been presented, the VCSEL technology is in current progress due to its wide advantages. The vertical emission allows a high integration level offering the possibility of 1D and 2D array. Due to the small size and the closeness of each VCSEL in an array, the possibility of electrical coupling on the electrical access is increased particularly for microwave frequency modulations. That's why the electrical access has a higher importance than it has for edge emitting laser diodes. A sight of the consequences of a mismatched electrical access has been presented here through a microwave-photonic study based on an electrical equivalent circuit model and the S_{11} and S_{21} parameters measurements. The results presented for the VCSEL array with a microstip-line electrical access shows that influence. Indeed, a coupling between the access lines involves an electrical crosstalk having repercussions on the emitted light by the VCSEL. The measurements achieved on a VCSEL array with a coplanar access bring the proof that a matched electrical access really improves the frequency response of the VCSEL array. Nevertheless, the matching is reached only for a certain frequency range. For wider frequency band of measurement, an increase of the S_{21} slope has been observed. Even if the response of the VCSEL has been extracted, the electrical access design has to be improved for an impedance matching according to the frequency utilization. Thus this modeling an characterisation could further help in the VCSEL design and in the VCSEL integration.

6. References

Baba, T., Yogo, Y., Suzuki, K., Koyama, F. & Iga, K. (1993). Near room temperature continuous wave lasing characteristics of gainasp/inp surface emitting laser, *Electronics Letters* 29(10): 913 –914.

Babic, D., Streubel, K., Mirin, R., Margalit, N., Bowers, J., Hu, E., Mars, D., Yang, L. & Carey, K. (1995). Room-temperature continuous-wave operation of 1.54- mu;m vertical-cavity lasers, *Photonics Technology Letters, IEEE* 7(11): 1225 –1227.

Bacou, A., Hayat, A., Iakovlev, V., Syrbu, A., Rissons, A., Mollier, J.-C. & Kapon, E. (2010). Electrical modeling of long-wavelength vcsels for intrinsic parameters extraction, *Quantum Electronics, IEEE Journal of* 46(3): 313 –322.

Bacou, A., Hayat, A., Rissons, A., Iakovlev, V., Syrbu, A., Mollier, J.-C. & Kapon, E. (2009). Vcsel intrinsic response extraction using t -matrix formalism, *Photonics Technology Letters, IEEE* .

Boucart, J., Starck, C., Gaborit, F., Plais, A., Bouche, N., Derouin, E., Goldstein, L., Fortin, C., Carpentier, D., Salet, P., Brillouet, F. & Jacquet, J. (1999). 1-mW CW-RT monolithic VCSEL at 1.55 μm, *IEEE Photonics Technology Letters* 11(6): 629–631.

Brusenbach, P., Swirhun, S., Uchida, T., Kim, M. & Parsons, C. (1993). Equivalent circuit for vertical cavity top surface emitting lasers, *Electronics Letters* 29(23): 2037 –2038.

Cartledge, J. & Srinivasan, R. (1997). Extraction of dfb laser rate equation parameters for system simulation purposes, *Lightwave Technology, Journal of* 15(5): 852 –860.

Esaki, L. (1974). Long journey into tunneling, *Proceedings of the IEEE* 62(6): 825 – 831.

Ibaraki, A., Kawashima, K., Furusawa, K., Ishikawa, T. & Yamaguchi, T. (1989). Buried heterostructure GaAs/GaAlAs distributed Bragg reflector surface emitting laser with very low threshold (5.2 mA) under room temperature CW conditions, *Japanese Journal of Applied Physics* 28: L667+.

Iga, K., Koyama, F. & Kinoshita, S. (1988). Surface emitting semiconductor lasers, *Quantum Electronics, IEEE Journal of* 24(9): 1845 –1855.

Jewell, J., Scherer, A., McCall, S., Lee, Y., Walker, S., Harbison, J. & Florez, L. (1989). Low-threshold electrically pumped vertical-cavity surface-emitting microlasers, *Electronics Letters* 25(17): 1123 –1124.

Kapon, E., Sirbu, A., Iakovlev, V., Mereuta, A., Caliman, A. & Suruceanu, G. (2009). Recent developments in long wavelength vcsels based on localized wafer fusion, *Transparent Optical Networks, 2009. ICTON '09. 11th International Conference on.*

Kopf, R. F., Schubert, E. F., Downey, S. W. & Emerson, A. B. (1992). N- and p-type dopant profiles in distributed bragg reflector structures and their effect on resistance, *Applied Physics Letters* 61(15): 1820 –1822.

Koyama, F. (2006). Recent advances of vcsel photonics, *Journal Of Lightwave Technology* Vol.24(No.12): 4502–4513.

Ly, K., Rissons, A., Quentel, F., Pez, M., Gambardella, E. & Mollier, J.-C. (2008). Bidirectional link mock-up for avionic applications, *IEEE Conference on Avionic FiberOptic (AVFOP)*, Publisher, San Diego, pp. xx–yy.

Majewski, M. & Novak, D. (1991). Method for characterization of intrinsic and extrinsic components of semiconductor laser diode circuit model, *Microwave and Guided Wave Letters, IEEE* 1(9): 246–248.

Nakagawa, S., Hu, S.-Y., Louderback, D. & Coldren, L. (2000). Rf crosstalk in multiple-wavelength vertical-cavity surface-emitting laser arrays, *Photonics Technology Letters, IEEE* 12(6): 612–614.

O'Brien, C., Majewski, M. & Rakic, A. (2007). A critical comparison of high-speed vcsel characterization techniques, *Lightwave Technology, Journal of* 25(2): 597–605.

Ortsiefer, M., Shau, R., Böhm, G., Köhler, F., Abstreiter, G. & Amann, M.-C. (2000). Low-resistance inga(al)as tunnel junctions for long wavelength vertical-cavity surface-emitting lasers, *Japanese Journal of Applied Physics* 39(Part 1, No. 4A): 1727–1729.
URL: *http://jjap.jsap.jp/link?JJAP/39/1727/*

Perchoux, J., Rissons, A. & Mollier, J.-C. (2008). Multimode vcsel model for wide frequency-range rin simulation, *Optics Communications* 281(1): 162–169.

Rhew, K. H., Jeon, S. C., Lee, D. H., Yoo, B.-S. & Yun, I. (2009). Reliability assessment of 1.55-μm vertical cavity surface emitting lasers with tunnel junction using high-temperature aging tests, *Microelectronics Reliability* 49(1): 42–50.

Rissons, A. & Mollier, J.-C. (2009). Critical study of the vertical-cavity surface emitting laser electrical access for integrated optical sub-assembly, *EMC Europ Workshop*, Publisher, Athens, pp. xx–yy.

Rissons, A., Mollier, J., Toffano, Z., Destrez, A. & Pez, M. M. (2003). Thermal and optoelectronic model of VCSEL arrays for short-range communication, *in* C. Lei & S. P. Kilcoyne (ed.), *Society of Photo-Optical Instrumentation Engineers (SPIE) Conference Series*, Vol. 4994 of *Society of Photo-Optical Instrumentation Engineers (SPIE) Conference Series*, pp. 100–111.

Salet, P., Gaborit, F., Pagnod-Rossiaux, P., Plais, A., Derouin, E., Pasquier, J. & Jacquet, J. (1997). Room-temperature pulsed operation of 1.3 mu;m vertical-cavity lasers including bottom ingaasp/inp multilayer bragg mirrors, *Electronics Letters* 33(24): 2048–2049.

Scott, J., Geels, R., Corzine, S. & Coldren, L. (1993). Modeling temperature effects and spatial hole burning to optimize vertical-cavity surface-emitting laser performance, *Quantum Electronics, IEEE Journal of* 29(5): 1295–1308.

Shau, R., Ortsiefer, M., Rosskopf, J., Boehm, G., Lauer, C., Maute, M. & Amann, M.-C. (2004). Long-wavelength inp-based vcsels with buried tunnel junction: properties and applications, Vol. 5364, SPIE, pp. 1–15.
URL: *http://link.aip.org/link/?PSI/5364/1/1*

Soda, H., Motegi, Y. & Iga, K. (1983). Gainasp/inp surface emitting injection lasers with short cavity length, *Quantum Electronics, IEEE Journal of* 19(6): 1035–1041.

Suematsu, Y. & Iga, K. (2008). Semiconductor lasers in photonics, *Lightwave Technology, Journal of*.

Syrbu, A., Iakovlev, V., Suruceanu, G., Caliman, A., Rudra, A., Mircea, A., Mereuta, A., Tadeoni, S., Berseth, C.-A., Achtenhagen, M., Boucart, J. & Kapon, E. (2004).

1.55- mu;m optically pumped wafer-fused tunable vcsels with 32-nm tuning range, *Photonics Technology Letters, IEEE* 16(9): 1991 –1993.

Tucker, R. S. & Pope, D. J. (1983). Microwave circuit models of semiconductor injection lasers, *IEEE Transactions On Microwave Theory And Techniques* Vol. 31(No. 3): 597–605.

Zhang, J.-P. & Petermann, K. (1994). Beam propagation model for vertical-cavity surface-emitting lasers: threshold properties, *Quantum Electronics, IEEE Journal of* 30(7): 1529 –1536.

13

Effects of Quantum-Well Base Geometry on Optoelectronic Characteristics of Transistor Laser

Iman Taghavi and Hassan Kaatuzian
Amirkabir University of Technology (Tehran Polytechnic)
Iran

1. Introduction

Since the discovery of the transistor, the base current (minority carrier recombination) has been the key to the device operation. Among all developments through which Bipolar Junction Transistors (BJT) have progressed, the most innovating modification may be the replacement of the homojunction emitter material by a larger-energy-gap material, thus forming a Heterojunction Bipolar Transistor (HBT). In a silicon bipolar junction transistor, both homo and hetero structures, the base current is dissipated as heat, i.e. through nonradiative recombination, but can yield substantial radiative recombination for a direct-bandgap semiconductor like GaAs. Thus we can reinvent the base region and its mechanism (i.e. carrier recombination and transport function) to decrease current gain significantly and achieve stimulated recombination.

Researches on III–V high-current-density high-speed HBT, revealed the ability of the transistor to operate electrically and optically in the same time. The modified transistor, called in literature as light emitting transistor (LET), works as a three port device with an electrical and an optical outputs (Feng et al. 2004a). Further improvements including wavelength tunability obtained by incorporating a place for better carrier confinement, called a Quantum-Well (QW) (Feng et al. 2004b). Carrier recombination in quantum well (QW) can be modified with a reflecting cavity changing the optoelectronic properties of QWLET (Walter et al. 2004). Room temperature continuous wave operation of such a device at GHz develops a novel three-terminal device, called a HBT Transistor Laser (HBTL) or briefly TL, in which laser emission produces by stimulated recombination (Feng et al. 2005).

In the TL, the usual transistor electrical collector is accompanied with an optical collector, i.e. the above mentioned QW, inserted in the base region of the HBT. Stimulated recombination, unique in the TL, causes "compression" in the collector I-V characteristics and decrease in gain. Combined functionality of an HBT, i.e. amplification of a weak electrical signal, and that of a diode laser, i.e. generating laser emission, is observed in the TL. In other words, a modulating base current leads to modulated signals of the laser output power and collector current. It raises the possibility of replacing some metal wiring between components on a circuit board or wafer chip with optical interconnections, thus providing more flexibility and capability in optoelectronic integrated circuits (OEIC) (Feng et al. 2006a). It has been planned that TL is appropriate for telecommunication and other applications because of its capability of achieving a large optical bandwidth (BW) and a frequency response without sharp resonance. Processing speed can increase when the TL-equipped microprocessors are commercialized in the future.

Utilizing the I-V characteristics of a TL (Then et al., 2007a) together with common transistor charge analysis, dynamic properties and collector current map are extracted where the above mentioned "compression" is observed in the common emitter gain ($\beta \equiv I_C / I_B$)(Chan et al. 2006). A Transistor Laser with increased (external) mirror reflection demonstrates empirically (Walter et al. 2006) lower threshold current and higher collector breakdown voltage while increased breakdown at lower currents is observed on the collector I-V characteristics if the base region cavity Q is spoiled. Electrical properties of TL will therefore become similar to those of normal transistor (βstim→βspon) (Chan et al. 2006). Trade-off between collector current gain and the differential optical gain of a heterojunction bipolar transistor laser TL has been demonstrated analytically before (Then et al. 2009). The electrical-optical gain relationship shows that a reduction in the transistor current gain is accompanied by an increase in the differential optical gain of the TL and, as a consequence, results in a larger optical modulation bandwidth. Another trade-off, i.e. electrical gain-optical bandwidth, has also been reported for which one can utilize in order to extract the maximum possible optical bandwidth of TL (Taghavi & Kaatuzain, 2010).

We proceed in section 2 by introducing the TL crystal structure and fabrication issues, while details are left with proper references for an interested reader. In section 3 carrier dynamics and charge control model in TL will be discussed which is necessary for interpretation of device physics. This section determines a simplified method for analytical modelling of TL and their modifications through the years TL has been invented. TL optical bandwidth and current gain are formulized in the section 4. The method will be utilized in section 5 to simulate QW-base geometry effects on optoelectronic characteristics of the TL. The investigated transistor laser has an electrical bandwidth of more than 100GHz. Thus the structure can be modified, utilizing the methods mentioned in the section 5, to equalize optical and electrical cut-off frequencies as much as possible. Indeed, any improvement in optoelectronic characteristics of TL, i.e. optical bandwidth and current gain, is currently necessary in order to incorporate the TL in OEIC`s. The most urgent work needed in the future is an investigation of other quantum-well parameters effect on TL characteristics. Among these parameters are width and number of quantum wells. Section 6 is devoted to discussions about new researches in the field of TL including MQWTL. The chapter will be completed with conclusion in this section.

2. TL structure

The device studied here is based on n-p-n heterojunction bipolar transistor (n-InGaP/ p-GaAs +InGaAs quantum well/n-GaAs). The epitaxial structure of the crystal used for the HBTL, demonstrated in Fig. 1, consist of a 5000 Å n-type heavily doped GaAs buffer layer, followed by three layers of n-type forming the bottom cladding layers. These layers are followed by an n-type subcollector layer, a 600 Å undoped GaAs collector layer, and an 880 Å p-type GaAs base layer (the active layer), which includes (in the base region) a 120 Å InGaAs QW (designed for $\lambda \approx 1000$ nm). The base area is $\approx 4000\mu m^2$ and we can treat the QW base region as three distinct subregions: Two 10^{19} cm^{-3} doped GaAs regions and one 160Å undoped InGaAs QW within them. The epitaxial HBTL structure is completed with the growth of the n-type emitter and upper cladding layers. Finally, the HBTL structure is capped with a 1000 Å heavily doped n-type GaAs contact layer. The HBTL fabrication process utilizes eight mask layers for three wet chemical etching steps, three dry (plasma) etching steps, and three metallization steps for top-side contacts. The HBTL sample is cleaved front-to-back using conventional etching methods to a length of 150μm between Fabry-Perot facets. Detailed fabrication process of HBTL has been described in (Chan et al., 2006; Feng et al., 2005; Feng et al., 2006a).

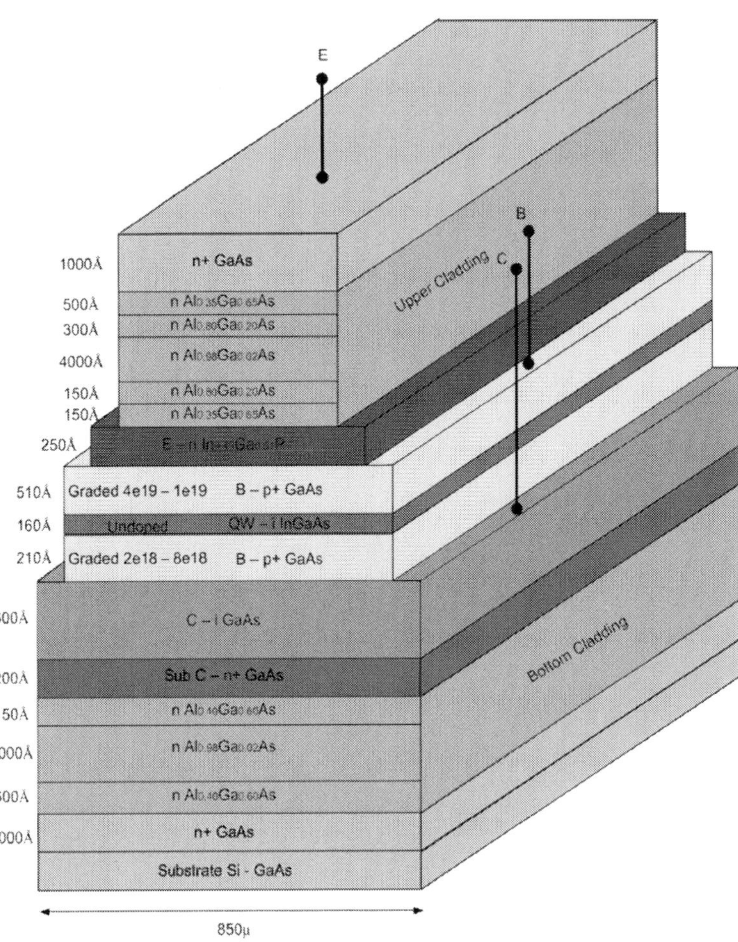

Fig. 1. Epitaxial structure of heterojunction bipolar Transistor Laser (HBTL). A QW is incorporated within the base to improve the recombination efficiency.

Although other device structures like Tunnel Junction Transistor Laser (TJTL)(Feng et al., 2009), VSCEL (Shi et al., 2008), Heterojunction Field effect (HFETL)(Suzuki et al., 1990) and Distributed Feedback Transistor Laser (DFBTL)(Dixon, 2010) have been proposed in the literature, we focus solely on Heterostructure Bipolar Junction transistors.

3. Charge analysis

The three-terminal TL is fundamentally different from a diode laser (single p-n junction), i.e. a two-terminal device, where the injection current is carried totally by recombination. There is no third terminal and collector current "competition". In TL the collector current I_C ($I_E=I_B+I_C$) is the fraction of emitter current "collected" at the reverse-biased base collector junction, with the remainder I_B supplying base recombination radiation. We show here that the carrier lifetime can be extracted from the HBT I-V characteristics, thus allowing the intrinsic optical frequency response of the TL to be derived by modifying the coupled carrier-photon rate equations.

3.1 Carrier dynamics

Being an essential feature of transistor operation, QW-base recombination of electron-hole supported by the base current I_B causes photon generation while carrier injection into the base region is provided by the emitter current I_E. Solving the continuity equation for the above mentioned three regions of QW-base, expressed in equations 1 and 2 for base bulk and QW region respectively, can shape distribution of base injected electrons from emitter.

$$\partial n/\partial t = D\,\partial^2 n/\partial x^2 - n/\tau_{bulk} \quad (bulk\ base) \tag{1}$$

$$\partial n/\partial t = D\,\partial^2 n/\partial x^2 - n/\tau_{qw} \quad (QW) \tag{2}$$

Where n=n(x,t) is the base electron distribution and the quantities τ_{bulk} and τ_{qw} are the recombination lifetimes in the GaAs and in the InGaAs QW, respectively. Sudden decrease in τ_{qw} at lasing threshold is a result of faster recombination rate in the QW thus base carrier tilt at emitter is intensified compared to spontaneous emission. The term n/τ_{qw} in equation 2 represents for spontaneous and stimulated photon generation and coupling of the optical field into the device operation. The bulk lifetime, τ_{bulk}, in the GaAs layer with 10^{19} doping is approximated as 193 ps. Uniform diffusion constant, $D{\approx}26\ cm^2/s$, is assumed throughout the base region and diffusion current, $I=qAD\partial n/\partial x$, is supposed continuous across the GaAs/InGaAs-QW/GaAs interface. The position of QW, W_{EQW}, is assessed here by its central location distance from the emitter-base interface and is the variable parameter during analysis of section 5 where the effect of QW position will be investigated.

Zero charge density at collector-base junction and $\partial n/\partial t=0$ form the initial conditions for the continuity equations 1 and 2. The calculated charge distribution for increasing base current (I_B) corresponding to the I-V characteristics of HBTL are shown plotted in Fig.2. The device currents are then calculated from common formula for diffusion current. Deviation from triangular approximation is observed in Fig. 2 and the profile is inclined more steeply upward near the emitter-base interface due to faster recombination in QW compared with the bulk base and reduced base width ($W_{QW}<W_{EC}$). At the lasing threshold, τ_{qw} decreases abruptly agreeing with $\beta_{stim}< \beta_{spon}$ as stimulated recombination speeds up the overall rate of recombination in the QW. Consequently, the emitter-end "tilt" of the carrier population is more noticeable beyond the threshold.

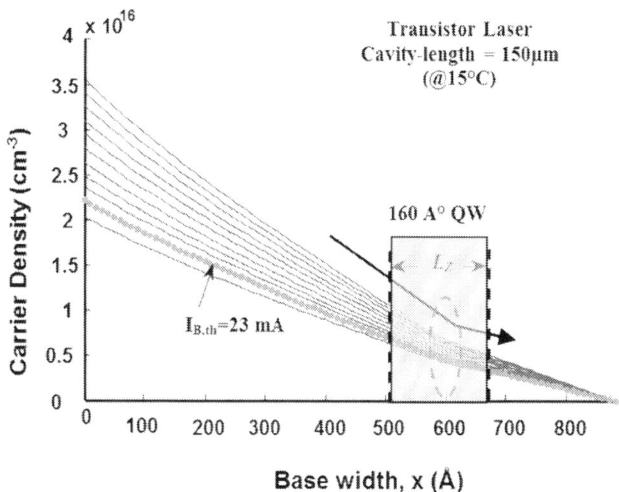

Base width, x (Å)

Fig. 2. Calculated minority carrier distribution for increasing I_B corresponding to the I-V characteristics. The shape tilts inside the QW due to faster recombination and reduced base width.

3.2 Charge control model

Charge control model for TL consists of superposition of two different triangular charge populations named as Q_1 and Q_2 (Feng et al., 2007). As TL integrates two distinguish operations, i.e. HBT electrical and semiconductor diode laser optical outputs, it might be supposed as a two-collector device. One is the base-collector junction and the other is QW called correspondingly electrical and optical collector. In the model of charge control Q_1 and Q_2 are responsible for transporting minority carriers to the QW (for optical operation) and keeping transport of carriers to the electrical collector (for electrical operation) respectively. Q_1 and Q_2 are obtained from carrier density profile, Fig. 3 and calculated so that

$$Q_1 = q\Delta n_1 A W_{EQW}/2 \tag{3}$$

$$Q_2 = q\Delta n_2 A W_B/2 \tag{4}$$

Where Δn_1 and Δn_2 are extracted from bias data, i.e. emitter base bias voltage (v_{BE}) and I_C, so that (Δn_1+ Δn_2) exhibits total carrier density at the emitter-base interface. Deriving I_C and I_B from I-V characteristics of TL (W_{EQW}=590 Å), we can estimate Q_1 and Q_2 as

$$I_C = Q_2/\tau_{t,2} \tag{5}$$

$$I_B = (Q_1 + Q_2)/\tau_{bulk} + Q_1/\tau_{t,1} \tag{6}$$

Where $\tau_{t,1}$ is the transient time from emitter to QW and $\tau_{t,2}$ is the transient time across the entire base (W_B=880Å) and calculated as below.

$$\tau_{t,1} = W_{EQW}{}^2/2D \approx 0.67ps \tag{7}$$

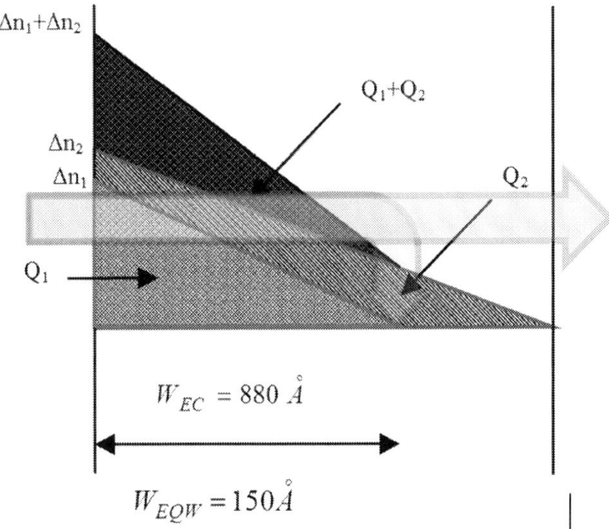

Fig. 3. Charge Control Model illustrates the role of Q_1 and Q_2 triangles in the entire base carrier distribution. Large amount of carriers enter the base while a little of them survive to the collector.

$$\tau_{t,2} = W_B{}^2/2D \approx 1.5ps \tag{8}$$

Two base carrier populations, Q_1 and Q_2, are combined to form the overall effective base recombination lifetime, τ_B, of the entire base population which is expressed as

$$1/\tau_B = 1/\tau_{bulk} + \tau_{t,2}\,Q_1/(Q_1 + Q_2) \tag{9}$$

Since $\tau_{bulk} >> \tau_{t,1}$, $\tau_B \approx (1+Q_2/Q_1).\tau_{t,1}$. Beyond the lasing threshold, τ_B decreases when stimulated emission accelerates the overall rate of recombination in the base of the TL. This is obvious in the steeper slope of the Q_1 population, transporting a larger proportion of the base carriers to a faster ($\tau_{B,stim}$) optical collector (QW), leading to a shorter overall base recombination lifetime, τ_B, and decreased gain β. τ_B values for charge density distributions near the threshold and beyond, plotted in Fig. 4 of (Feng et al., 2007).

4. Device optoelectronic characteristics

Among several characteristics, either optical or electrical, of the TL, small signal optical frequency response and current gain are most interesting. The reason arises from the fact that the TL needs suitable optoelectronic characteristics to be incorporated in an Optoelectronic Integrated Circuit (OEIC). The intrinsic frequency response of the TL can be improved towards 100 GHz owing to carrier lifetime determined by a thin base and the ability of the TL to inject and withdraw stored charge within picoseconds (forcing recombination to compete with E-C transport). Methods are required to improve the optical bandwidth to equalize both bandwidths as much as possible. The following describes a brief analysis of these characteristics.

4.1 Small-signal analysis of optical frequency response

When integrated over the entire base width, continuity equations (1) and (2) can result in an approximation for optical frequency response. The result for the TL is a modification of the coupled carrier-photon equations, formulated by Statz & deMars (Feng et al., 2007) which is expressed as

$$dN/dt = I_E/q - (I_C/q) + N/\tau_{B,spon} + \upsilon g \Gamma N_p \tag{10}$$

$$dN_p/dt = \upsilon g \Gamma N_p - N_p/\tau_p \tag{11}$$

Where $N \approx (Q_1+Q_2)/q$ is the total base minority carrier population, υ is the photon group velocity, Γ is the optical confinement factor of active medium (i.e. QW), τ_P is the photon lifetime, g is the gain per unit length of the active medium, and N_p is the total generated photon number. The small signal optical frequency response is obtained as below

$$\frac{\Delta P_m(\omega)}{\Delta I_B(\omega)} = \frac{\Gamma_b \tau_p/(qAW_B)}{[1-(\omega/\omega_n)^2]+j2(\omega/\omega_n)\xi} \tag{12}$$

Where $\Delta P_m(\omega)$ is the modulated photon density, $\Delta I_B(\omega)$ is the modulated base current and Γ_b is the optical confinement factor of the waveguide. The natural frequency and damping ratio are:

$$\omega_n^2 \approx \left(1/\tau_p \tau_{B,spon}\right)(I_B/I_{B,th} - 1) \tag{13}$$

$$\xi = (2\omega_n \tau_{B,spon})^{-1} + 0.5(\omega_n \tau_p) \tag{14}$$

For $\zeta \leq 0.7$ the resonance frequency and bandwidth are:

$$\omega_r = \omega_n\sqrt{(1 - 2\xi^2)} \tag{15}$$

$$\omega_{-3db} = \omega_n\sqrt{(1 - 2\xi^2) + \sqrt{4\xi^4 - 4\xi^2 + 4}} \tag{16}$$

While these values for $\zeta \geq 0.7$ are expressed as:

$$\omega_r = 0 \tag{17}$$

$$\omega_{-3db} = \omega_n\sqrt{(1 - 2\xi^2) + \sqrt{4\xi^4 - 4\xi^2 + 2}} \tag{18}$$

We can obtain τ_P from the common equation:

$$^1/_{\tau_p} = \left(\frac{c}{n}\right)\{\alpha_i + (1/2L)\ln(1/R_1 R_2)\} \tag{19}$$

For cleaved cavity laser of lengths L=150 and 400(µm) with $R_1=R_2=0.32$ and a photon absorption factor (α) of 5cm-1, the approximate photon lifetime are τ_P=1.5 and 3.6(ps) respectively. It should be noticed that we neglect the capture and escape lifetimes in this small signal analysis. Optical gain is also assumed independent on photon and carrier populations. Although the difference in final results is negligible, an interested reader can refer to (Faraji et al., 2009) for more information.

In equations (13) and (14) the parameter $\tau_{B,spon}$ is a specific value of overall base recombination lifetime, τ_B, evaluated in spontaneous emission region at lasing threshold. According to Fig. 4

of (Feng et al.,2007) (not shown here) this value for a TL with W_{EQW}=590Å is about 2.5ps. Under bias condition such that I_B=33mA ($I_B/I_{B,th}$=1.5), small signal optical frequency response for two TL with above mentioned cavity length is sketched in Fig. 4. For comparison, the curve also contains optical response of an ordinary diode laser (DL) with a large base recombination lifetime of $\tau_{B,spon}$=100ps. When compared with traditional diode laser, fast modulation response of TL is due to this difference in their optical frequency response which suffers a resonance peak in low frequencies in the case of diode laser.

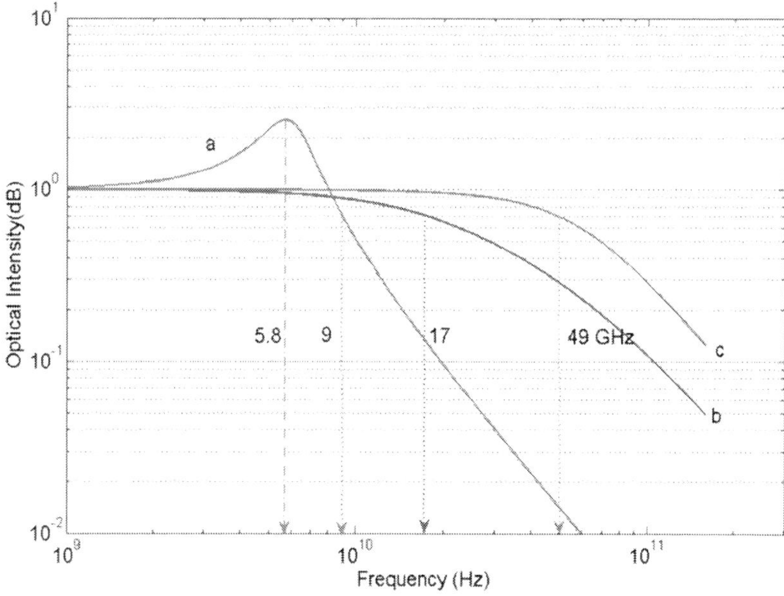

Fig. 4. Small signal optical frequency response of (a) Diode laser with L=400μm ($\tau_{B,spon}$=100ps), (b) TL with L=400μm ($\tau_{B,spon}$=2.5ps), (c) TL with L=150μm ($\tau_{B,spon}$=2.5ps)

The very short effective carrier lifetime of a TL, i.e. the same order of magnitude as the cavity photon lifetime, results in a near unity resonance peak and provides in effect a significantly larger bandwidth , which is typically not available when the bandwidth is limited by carrier-photon density resonance, as in a diode with a large carrier lifetime (0.1–1ns). We see that by charge analysis, if a QW base HBT is capable of laser operation, its "speed" of operation is not affected by common recombination, and the limitation of spatial charge "pileup" as in a diode, but by the base dynamics of charge transport described in this section. Indeed, only fast recombination is able to compete with fast collection of carriers.

4.2 Current gain

Thanks to special features of QW incorporated in its base region, the most recent form of laser, the Transistor Laser, govern carrier recombination and hence electrical current gain , $\beta=\Delta I_C/\Delta I_B$. This gain therefore becomes a crucial parameter in both transistor and laser operations of TL. As previously mentioned, stimulated emission, an extra feature of transistor, makes the collector I-V characteristics "compressed" and decreases β. The gain value for both the transistor laser and an ordinary HBT (i.e. a Q-spoiled TL) was demonstrated in Fig. 6 of (Walter et al., 2006). Charge control analysis can be utilized in conjunction with these values of current gain to determine the TL terminal currents and estimate base transit and recombination lifetimes. Conversely, one can calculate theoretically β for a TL as described below.

The base and collector currents and the static common-emitter current gain are given for an HBT as

$$I_E = I_B + I_C = Q_B/\tau_B + Q_B/\tau_t \tag{20}$$

$$\beta \equiv \Delta I_C/\Delta I_B = I_C/I_B = \tau_B/\tau_t \tag{21}$$

where Q_B is the charge stored in the base, τ_B is the base recombination lifetime, and τ_t is the base transit time. For the same HBT with a quantum well inserted in the base, a QWHBT, the base current and current gain can be modified and expressed as

$$I_B \equiv I_{BQW} = \left(Q_B/\tau_B + Q_B/\tau_{QW}\right) = \left(Q_B/\tau_{BQW}\right) \tag{22}$$

$$\beta_{QW} = \tau_{BQW}/\tau_t \tag{23}$$

where τ_{QW} is the recombination lifetime of the quantum well while τ_{BQW} is the effective recombination lifetime of the base and quantum well. For an QWHBT transistor laser with stimulated base recombination, a QWHBTL, the base current and current gain can be further modified and expressed as

$$I_B \equiv I_{BTL} = \left(Q_B/\tau_B + Q_B/\tau_{QW} + Q_B/\tau_{st}\right) = \left(Q_B/\tau_{TL}\right) \tag{24}$$

$$\beta_{TL} = \tau_{TL}/\tau_t \tag{25}$$

where τ_{TL} is the effective base recombination lifetime of the transistor laser including τ_{st}, the stimulated recombination lifetime.

5. QW-base geometry effects

Since the first report of successful operation of TL, a number of publications reported different material combinations (Dixon et al., 2006), incorporation of Quantum Well(s) (QW) in the base, modulation characteristics (Then et al., 2009)(Feng et al.,2006)(Then et al., 2007b), use of tunnel junctions (Feng et al., 2009), electrical-optical signal mixing and multiplication (Feng et al., 2006a), etc. Various schemes have also been employed to increase the modulation bandwidth of the TL (Then et al., 2008)(Taghavi & Kaatuzian, 2010). A proposal has also been put forward to use the gain medium as an optical amplifier. Several models have appeared in the literature for terminal currents, gain, modulation bandwidth, etc. of TL (Feng et al., 2007)(Faraji et al., 2008)(Basu et al., 2009)(Then et al., 2010)(Zhang & Leburton, 2009). However, these models based on rate equations, give values for optoelectronic characteristics of TL. In addition, the TL has been simulated both numerically (Shi et al., 2008)(Kaatuzian & Taghavi, 2009) and by CAD, i.e. software packages, (Shi et al., 2008)(Duan et al., 2010).

The electron recombination plays an important role in determining the base current due to spontaneous and stimulated emission in the QW, so that the electronic characteristics of the

TL have a strong dependence on its laser operation. This is proved by the decrease in the electrical gain $\beta=\Delta I_C/\Delta I_B$ due to enhanced carrier recombination in the base region when stimulated emission occurs. According to the model described in section 4, effects of geometrical parameters of QW-base on small signal optical frequency response and current gain should be significant. In other words optoelectronic characteristics of TL are directly affected by parameters like QW position in the base (W_{EQW}), well width (W_{QW}) (and barrier width in the case of multiple quantum-wells TL), number of QWs, base width (W_B), etc. For instance, TL characteristics depend on the relative position of QW because of the quasi-linear density profile of the base minority carriers (Fig. 2). Although, the mentioned first order model of charge control does not distinguish between bulk and QW carriers, TL characteristics dependency on QW and base width are still acceptable. Herein, we utilize analysis described in section before to model the effect of QW position on optical bandwidth and current gain of TL. As a goal, we look for an enhanced performance for the TL by "restructuring" it.

5.1 QW location effect

Being integrated inside the base region of the TL, the quantum-well structural parameters have certainly significant influence on both optical and electrical properties of TL, e.g. gain and bandwidth for optical and electrical outputs. This subsection is dedicated to the mentioned effects while other parameters will be studied later.

5.1.1 Carrier profile

According to the model, constant base-emitter voltage and collector current makes ($\Delta n_1 + \Delta n_2$), Δn_1, Q_2 unchanged. Altering the position of QW within the base can therefore change Q_1, i.e. portion of carriers responsible for lasing. The new values for Q_1 and $\tau_{t,1}$ (base transit time) when QW moves are as below

$$Q_{1,new} = Q_1 * (W_{EQW,new}/W_{EQW,old}) \tag{26}$$

$$\tau_{t,1,new} = \tau_{t,1,old} * (W_{EQW,new}/W_{EQW,old})^2 \tag{27}$$

For the TL described in previous sections we may set $W_{EQW,old}$=590Å. As a result, minority carrier profile of base region solved analytically for W_{EQW}=150 Å is sketched in Fig. 5. Also Fig. 6 demonstrates the modified charge control model. Displacement of QW position, W_{EQW}, towards emitter can cause two noteworthy effects simultaneously. First, τ_B falls due to reduction of Q_1 while Q_2 is constant. Second, in accordance to the diffusion formula for I_E and Fig. 6, I_E would increase significantly as a result of steeper charge density profile at the base emitter interface.

These conditions make base current (I_B) to rise abruptly beyond the threshold current (22 mA). Due to (Kaatuzian, 2005) the threshold current is constant during this movement of QW. In other words, moving the QW in this direction decreases the electrons which are "trapped" and recombined with holes within it. Thus more carriers are allocated to electrical operation and fewer involved in light generation, i.e. fewer photons created within the well. This phenomenon, exhibited by animation in Fig. 7, means that carriers arrive sooner at QW. As a result of "faster rich time" of minority carriers to the well, it is predicted that both τ_B and β decreases drastically (by about 4 or 5 times for τ_B).

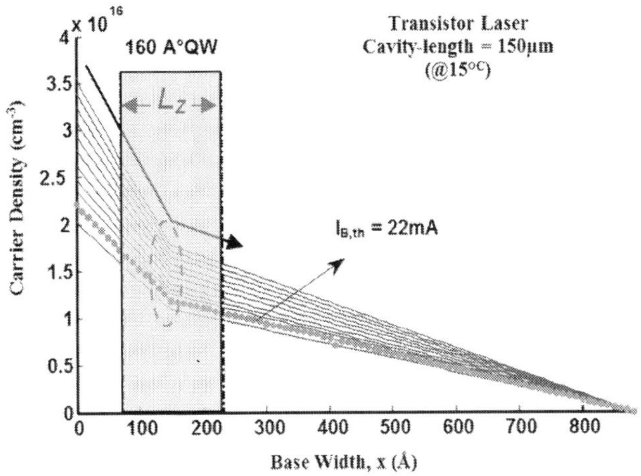

Fig. 5. Calculated minority carrier distribution for QW moved toward emitter; the profile deviates more from triangular form of regular HBT.

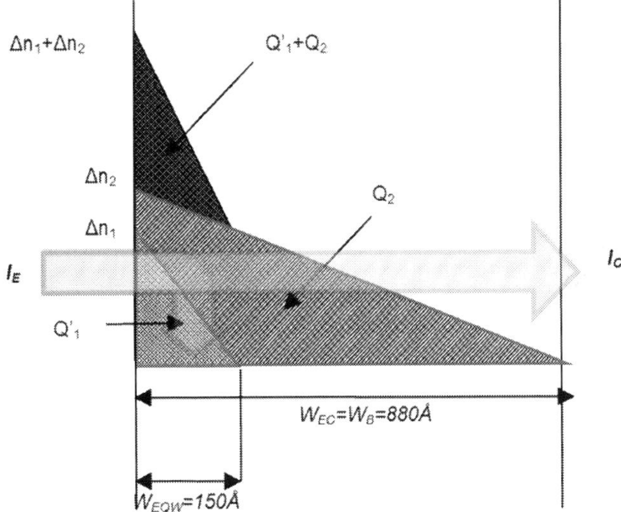

Fig. 6. Modified charge control model for W_{EQW}=150Å. Q_2 is constant while Q_1 decreased to Q_1'.

5.1.2 Base recombination lifetime

Overall effective base recombination lifetime of the entire base population, τ_B, can be calculated using the concept of charge control model and is changed for different QW locations, i.e. W_{EQW}. The special calculation method of τ_B value for different base currents is based on a reverse estimation of them from experimental data for W_{EQW}=590Å and the charge relations (physical model). The method involves a set of calculations which generates new values of τ_B using old values of τ_B (i.e. Fig. 4 in (Feng et al., 2007)) and charge relations according to the charge control model of section 3. The result for new τ_B values at I_B=22mA, i.e. $\tau_{B,spon}$ which equals τ_B evaluated at I_{Bth}, for QW located in different places through the bulk base is sketched in Fig. 8. To show the reliance of base recombination lifetime on I_B and W_{EQW} simultaneously, τ_B has been drawn in three dimensions (3D) in Fig. 9. Assessing these two figures, it is obvious that the closer to the collector the QW, the less variant with I_B the τ_B. Spontaneous base recombination lifetime (τ_{Bspon}) is obtained by cross section of 3-D scheme of τ_B and $I_B=I_{th}$=22mA plane as well.

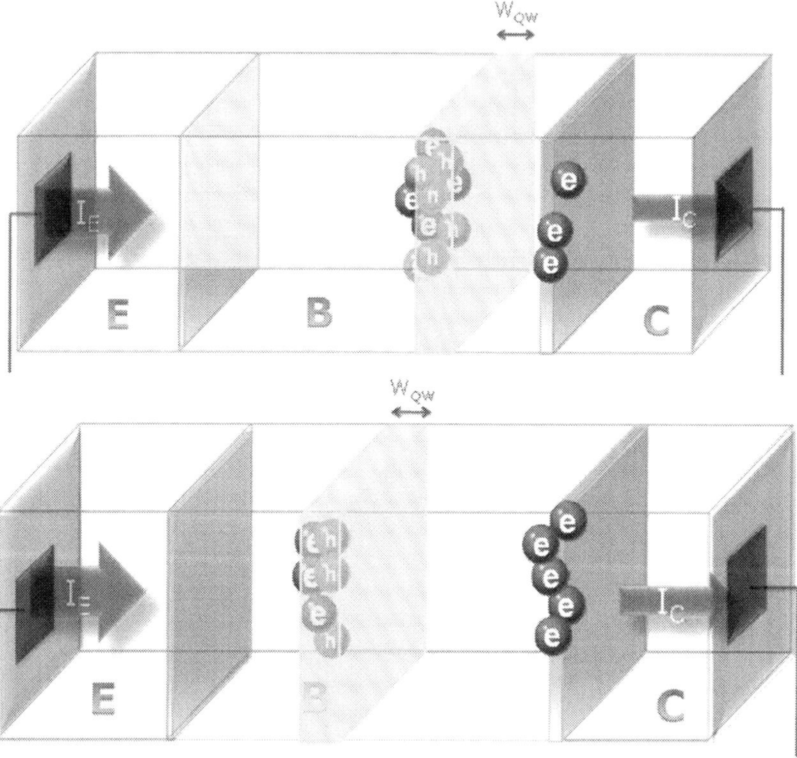

Fig. 7. Animated diagram illustrates a decrease in "trapped" electrons recombined with holes when QW moves from a location near collector (top) towards emitter (down). Carrier "reach time" to QW is thus decreased.

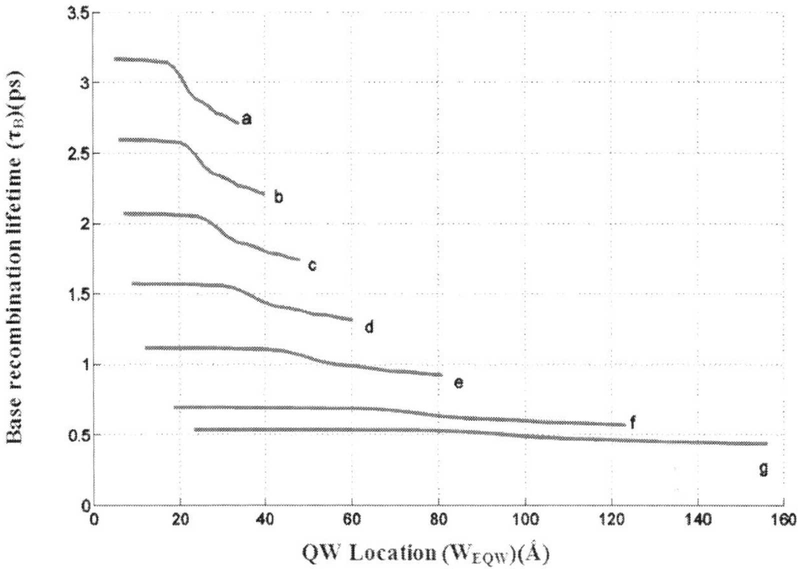

Fig. 8. Calculated base recombination lifetime (τ_B) versus W_{EQW} (a)690 (b)590 (c)490 (d)390 (e)290 (f)190 (g)150(Å).

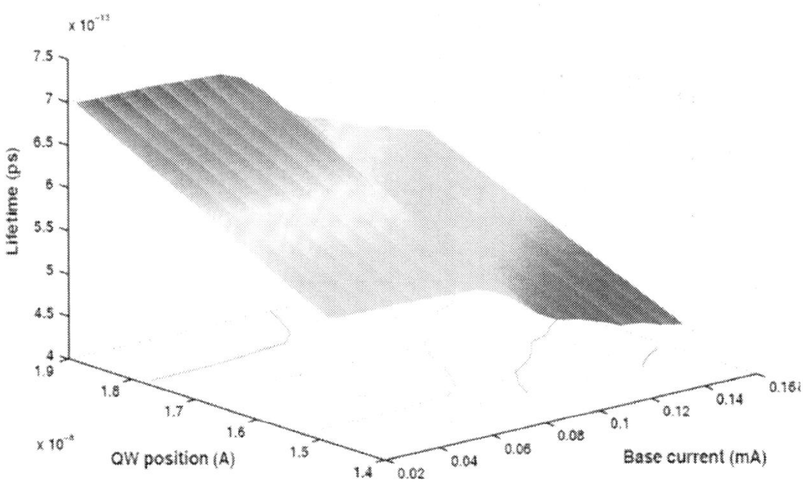

Fig. 9. 3-D scheme of base recombination lifetime (τ_B) versus I_B and W_{EQW} placed in different positions through the base.

5.1.3 Small-signal optical frequency response

Calculating the optical frequency response of transistor laser in 4.1, one can sketch the response for different QW locations. Table 1 shows the simulation results of device parameters for different QW locations through the base region while Fig. 10 and Fig. 11 show optical response and bandwidth dependence on QW location, respectively. As it is obvious from the values for simulated ζ, all of them exceed 0.7 and no resonance peak, the limiting factor for diode lasers, occurs due to QW dislocation through such a thin base layer. Non unity resonance peak in diode lasers with a large carrier lifetime (0.1 – 1 ns), sets limit on f_{-3db} and restricts performance of diode laser.

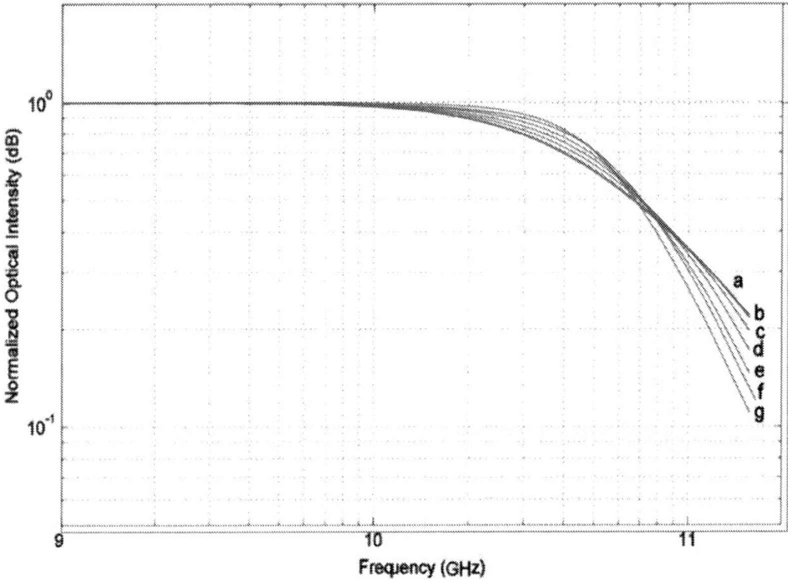

Fig. 10. Optical frequency response for TL with W_{EQW} (a)150 (b)190 (c)290 (d)390 (e)490 (f)590 (g)690(Å).

Increase in optical bandwidth (f_{-3db}) is a direct result of sliding the QW toward collector. At the first glance opposite direction of increase for f_{-3db} and $\tau_{B,spon}$ is confusing and seems mistaken. However, it would be clear if we sketch the f_{-3db} versus $\tau_{B,spon}$ (using approximation of f_{-3db} by equation 18). As Fig. 12 demonstrates, the bandwidth rises abnormally with $\tau_{B,spon}$ increases while reach a peak at about $\tau_{B,spon}$=3.37 ps and then decreases. This fact shines a novel idea that there may be an optimum point for QW to be located in order to maximize the optical bandwidth. Utilizing this figure in conjunction with calculated $\tau_{B,spon}$, one can move the QW toward collector to W_{EQW}=730Å to achieve maximum optical bandwidth (~54 GHz, taking into account the error due to approximate equation). As a result, the QW should be placed closed to collector as much as possible in order to achieve the maximum bandwidth.

Quantum-Well Position (Å)	$\tau_{B,spon}$ (ps)	fn (undamed natural ferquency)(GHz)	ξ (damping ratio)	Simulated -3dB Bandwidth(GHz)
150	0.54	125.8	1.7675	38.7
190	0.7	109.7	1.5595	39.3
290	1.11	87.5	1.2288	42.3
390	1.57	74	1.0365	45.2
490	2.1	64	0.9063	47.4
590	2.57	57	0.8103	48.9
690	3.03	53.5	0.7424	50.9

Table 1. Simulated device parameters for different W_{EQW}; No resonance peak due to QW movement when $\xi \geq 0.7$.

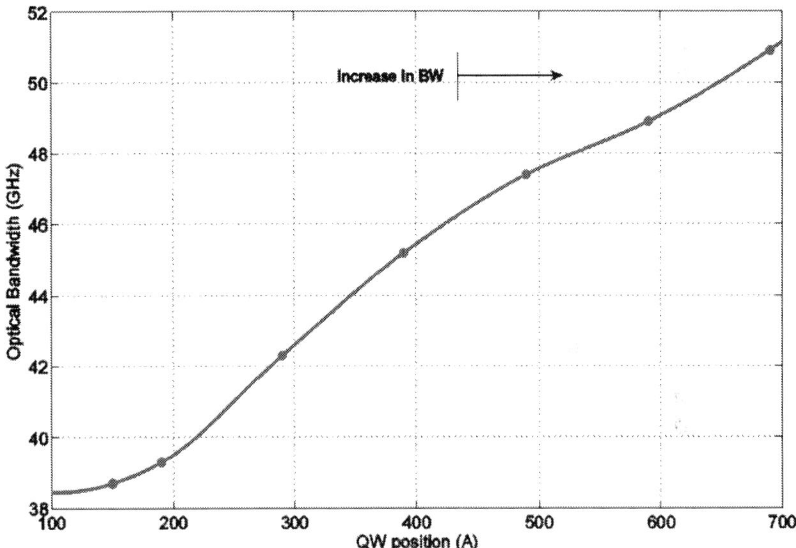

Fig. 11. QW location effect on optical bandwidth; an HBTL with well near its collector has larger optical bandwidth while electrical bandwidth is the same.

5.1.4 Current gain

We predicted previously that optical bandwidth increment would be at cost of sacrificing the current gain (β). Further investigation of QW effect on TL optoelectronic characteristics, this time electrical, leads in an interesting result that matures the previous founding. Collector current gain for an HBTL was calculated in subsection 4.2 in equations (24) and (25). As a result of QW movement towards collector, simulation at constant bias currents shows that τ_{TL} and consequently β declines. Fig. 13 shows this change in optical bandwidth and current gain versus displacement of QW while bias voltage, i.e. v_{be}, forces base current to be constant at I_B=33mA for W_{EQW}=590 Å. This bias enforcement does not disturb generality of the simulation results.

The opposite dependence of BW and β to W_{EQW}, is a trade-off between TL optoelectronic characteristics. Other experimental and theoretical works proved the described "trade-off" between β and f_{-3db} as well (Then et al.,2009),(faraji et al., 2009). They also predict analytically the above mentioned direct dependence of f_{-3db} on $\tau_{B,spon}$. In (Then et al., 2008) the authors utilized an auxiliary base signal to enhance the optical bandwidth. As a merge of their work and the present analysis we can find the optimum place for QW that leads to better results for both β and f_{-3db} of a TL. It means we can use both method, i.e. auxiliary base signal and QW dislocation method, simultaneously. A suggestion for finding an "optimum" QW location consists of two steps . First we focus on β and make it larger by locating the QW close to emitter, e.g. W_{EQW}<300 Å, which results in BW<43 GHz. Then we use auxiliary AC bias signal to trade some gain for BW. It should be noted that β less than unity is not generally accepted if TL is supposed to work as an electrical amplifier.

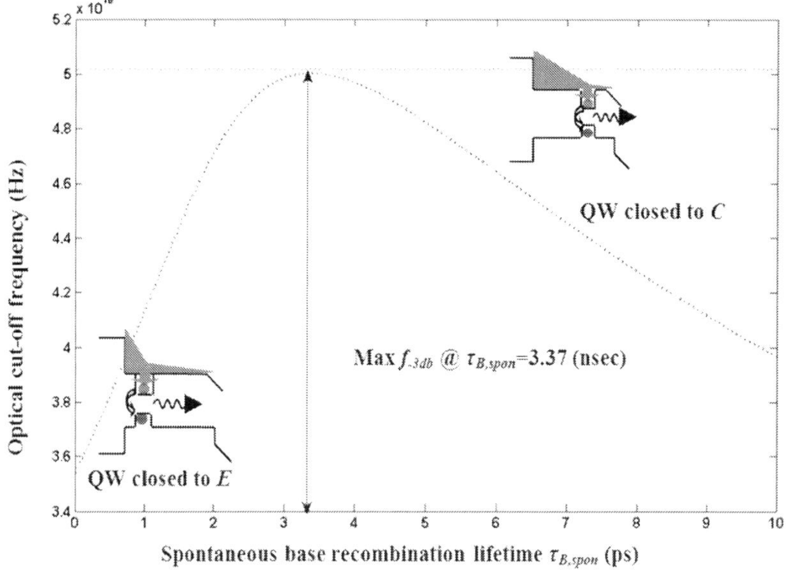

Fig. 12. Calculated optical cut-off frequency (f_{-3db}) versus $\tau_{B,spon}$. Bandwidth is maximum for $\tau_{B,spon}$ corresponding to W_{EQW}≈730Å.

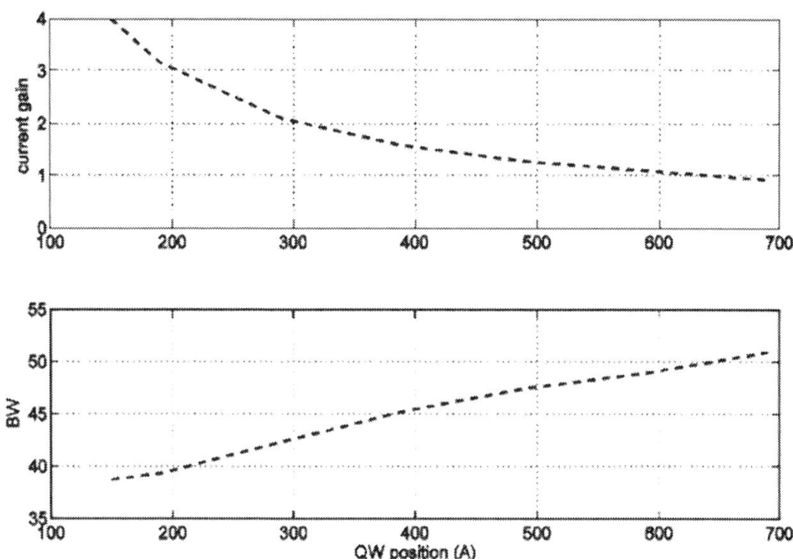

Fig. 13. Electrical gain-optical bandwidth trade-off in an HBTL.

5.2 Quantum-well width

Being an optical collector, the QW plays a significant role as it governs both optical and electrical characteristics of TL. Among QW-base geometry parameters, other than QW location, one can investigate the width of QW incorporated within the base region. For instance, we can change the well width while other geometrical parameters, like QW location, are left unchanged in order to examine how the optical frequency response and current gain alter. Base minority carrier recombination lifetime was calculated before as in equation (9) which was an independent function of well width. Charge control model and charge analysis based on this model, described in previous sections, should be completed in order to do this analysis. The base carrier lifetime, τ_B, can be written as below (Then et al., 2007c)

$$1/\tau_B = \sigma \upsilon_{th} N_r \tag{28}$$

Where υ_{th} is the thermal velocity of carriers, N_r is the density of possible recombination sites and σ is the cross section of carrier capture. σ is a measure of the region that an electron has the possibility to capture and recombine with a hole and is proportional to well width (W_{QW}). In the other hand, N_r depends on the hole concentration, i.e. N_A of the base region. So we can evaluate τ_B as

$$1/\tau_B = G W_{QW} N_A \tag{29}$$

where G is a proportionality factor defined by other geometrical properties of the base. Using this equation one can extract base recombination lifetime of base minority carriers for different base doping densities. Calculations exhibit an indirect relation between τ_B and well width, agreeing with a larger QW width enhancing the capture cross section for electrons. Moreover, the larger N_A, the greater the recombination and hence the smaller τ_B.

The results for optical frequency response based on Statz-deMars equations of section 4 can be utilized to evaluate the optical properties of a TL for varying well width. Indeed, equations (13) and (14) require $\tau_{B,spon}$ not τ_B, as described above, therefore threshold current should be calculated for different QW widths. An expression for base threshold current of TL is as below

$$J_{th} = qn_0\tau_{cap}/v\tau_{qw}\tau_{rb} = \frac{qn_0}{\tau_{qw}}\left[1 + \left(\frac{1-v}{v}\right)\frac{\tau_{cap}}{\tau_{rb0}}\right] \tag{30}$$

where n_0 is minority carrier density in steady-state (under dc base current density of J_0), τ_{cap} is the electron capture time by QW (not included in charge control model for simplicity), τ_{qw} is the QW recombination lifetime of electron and τ_{rb0} is the bulk lifetime (or direct recombination lifetime outside the well, also ignored in our model). The base geometry factor, v, gives the fraction of the base charge captured in the QW and defines as (Zhang & Leburton, 2009)

$$v = (W_{qw}/W_b)(1 - x_{qw}/W_b) \tag{31}$$

where W_{qw} is the QW width, the factor we investigate here, W_b is the base width and x_{qw} is the QW location, similar but not equal to previously defined parameter of W_{EQW}. By setting all the constants, one can calculate $\tau_{B,spon}$ and then small-signal optical frequency response and bandwidth of TL for a range of QW widths. Optimization is also possible like what we did for QW location.

6. Conclusion and future prospects

An analytical simulation was performed to predict dependence of TL optoelectronic characteristics on QW position in order to find a possible optimum place for QW. Simulated base recombination lifetime of HBTL for different QW positions exhibited an increase in optical bandwidth QW moved towards the collector within the base. Further investigations of optical response prove the possibility of a maximum optical bandwidth of about 54GHz in WEQW≈730 Å. Since no resonance peak occurred in optical frequency response, the bandwidth is not limited in this method. In addition, the current gain decreased when QW moved in the direction of collector. The above mentioned gain-bandwidth trade-off between optoelectronic parameters of TL was utilized together with other experimental methods reported previously to find a QW position for more appropriate performance. The investigated transistor laser has an electrical bandwidth of more than 100GHz. Thus the structure can be modified, utilizing the displacement method reported in this paper, to equalize optical and electrical cut-off frequencies as much as possible.

In previous sections we consider the analysis of a single quantum well (SQW) where there is just one QW incorporated within the base region. This simplifies the modelling and math-related processes. In practice, SQWTL has not sufficient optical gain and may suffer thermal heating which requires additional heat sink. Modifications needed to model a multiple QW transistor laser (MQWTL). First one should rewrite the rate equations of coupled carrier and photon for separate regions between wells. Solving these equations and link them by applying initial conditions, i.e. continuity of current and carrier concentrations, is the next step. In addition to multiple capture and escape lifetime of carriers, tunnelling of the 2-dimensional carriers to the adjacent wells should be considered. For wide barriers one may use carrier transport across the barriers instead the mentioned tunnelling. Simulation results for diode laser (Duan et al., 2010), as one of the transistor laser parents, demonstrate considerable enhancement in optical bandwidth and gain of the device when increasing the number of quantum wells (Nagarajan et al., 1992), (Bahrami and Kaatuzian, 2010). Like the well location modelled here in this chapter, there may be an optimum number of quantum wells to be incorporated within the base region. Due to its high electrical bandwidth (≥ 100 GHz), it is needed to increase the optical modulation bandwidth of the TL. Base region plays the key role in all BJT transistors, especially in Transistor Lasers.

Like Quantum-Well, base structural parameters have significant effects on optoelectronic characteristics of TL which can be modelled like what performed before during this chapter. Among these parameters are base width (Zhang et al., 2009), material, doping (Chu-Kung et al., 2006), etc. For example, a graded base region can cause an internal field which accelerates the carrier transport across the base thus alters both the optical bandwidth and the current gain considerably.

7. References

Basu, R., Mukhopadhyay, B. & Basu, P.K. (2009). Gain Spectra and Characteristics of a Transistor Laser with InGaAs Quantum Well in the Base. *Proceeding of International Conference on Computers and Devices for Communication,* India, Dec. 2009

Bahrami Yekta, V. & Kaatuzian, H. (2010). Design considerations to improve high temperature characteristics of 1.3µm AlGaInAs-InP uncooled multiple quantum well lasers: Strain in barriers. *Optik, Elsevier,* doi:10.1016/j.ijileo.2010.03.016

Chan, R., Feng, M., Holonyak, N. Jr., James, A. & Walter, G. (2006). Collector current map of gain and stimulated recombination on the base quantum well transitions of a transistor laser. *Appl. Phys. Lett.,* Vol. 88, No. 143508

Chu-kung, B.F., Feng, M., Walter, G., Holonyak, N. Jr., Chung, T., Ryou, J.-H., Limb, J., Yoo, D., Shen, S.C. & Dupuis, R.D. (2006). Graded-base InGaN/GaN heterojunction bipolar light-emitting transistors. *Appl. Phys. Lett.* Vol. 89, No. 082108

Dixon, F., Feng, M. & Holonyak, N. Jr. (2010). Distributed feedback transistor laser. *Appl. Phys. Lett.* Vol. 96, No. 241103

Dixon, F., Chan, R., Walter, G., Holonyak, N. Jr. & Feng, M. (2006). Visible spectrum light-emitting transistors. *Appl. Phys. Lett.* Vol. 88, No. 012108

Duan, Z., Shi, W., Chrostowski, L., Huang, X., Zhou, N., & Chai, G. (2010). Design and epitaxy of 1.5 µm InGaAsP-InP MQW material for a transistor laser. *Optics Express,* Vol. 18, Issue 2, pp. (1501-1509), doi:10.1364/OE.18.001501

Faraji, B., Pulfrey, D.L. & Chrostowski, L. (2008). Small-signal modeling of the transistor laser including the quantum capture and escape lifetimes. *Appl. Phys. Lett.,* Vol. 93, No. 103509

Faraji, B., Shi, W., Pulfrey, D.L. & Chrostowski, L. (2009). Analytical modeling of the transistor laser. *IEEE Journal of Quantum Electron.,* Vol. 15, No. 3, pp (594-603)

Feng, M., Holonyak, N. Jr. & Hafez, W. (2004a). Light-emitting transistor: light emission from InGaP/GaAs heterojunction bipolar transistors. *Appl. Phys. Lett.* Vol. 84, No. 151

Feng, M., Holonyak, N. Jr. & Chan, R. (2004b). Quantum-well-base heterojunction bipolar light-emitting transistor. *Appl. Phys. Lett.,* Vol. 84, No. 11

Feng, M., Holonyak, N. Jr., Walter, G. & Chan, R. (2005). Room temperature continuous wave operation of a heterojunction bipolar transistor laser. *Appl. Phys. Lett.* Vol. 87, No. 131103

Feng, M., Holonyak, N. Jr., Chan, R., James, A. & Walter, G. (2006a). Signal mixing in a multiple input transistor laser near threshold. *Appl. Phys. Lett.* Vol. 88, No. 063509

Feng, M., Holonyak, N. Jr., James, A., Cimino, K., Walter, G. & Chan, R. (2006b). Carrier lifetime and modulation bandwidth of a quantum well AlGaAs/InGaP/GaAs/InGaAs transistor laser. *Appl. Phys. Lett.* Vol. 89, No. 113504

Feng, M., Holonyak, N. Jr., Then, H.W. & Walter, G. (2007). Charge control analysis of transistor laser operation. *Appl. Phys. Lett.* Vol. 91, No. 053501

Feng, M., Holonyak, N. Jr., Then, H.W., Wu, C.H. & Walter, G. (2009). Tunnel junction transistor laser. *Appl. Phys. Lett.* Vol. 94, No. 041118

Kaatuzian, H. (2005). *Photonics, Vol. 1*, AmirKabir University of Technology press, , Tehran, Iran

Kaatuzian, H. & Taghavi,I. (2009). Simulation of quantum-well slipping effect on optical bandwidth in transistor laser. *Chinese optics letters*. doi:10.3788/COL20090705.0435, pp. 435–436

Nagarajan, R., Ishikawa, M., Fukushima, T., Geels, R. & Bowers, E. (1992). High speed quantum-well lasers and carrier transport effects. *IEEE Journal of Quantum Electron*, Vol. 28, No. 10, pp (1990-2008)

Shi, W., Chrostowski, L & Faraji, B. (2008). Numerical Study of the Optical Saturation and Voltage Control of a Transistor Vertical-Cavity Surface-Emitting Laser. *IEEE Photonics Technology Letters*, Vol. 20, No. 24

Suzuki, Y., Yajima, H., Shimoyama, K., Inoue, Y., Katoh, M. & Gotoh, H. (1990). (Heterojunction field effect transistor laser). *Ellectronics Letters*, Vol. 26, No. 19

Taghavi,I & Kaatuzian, H. (2010). Gain-Bandwidth trade-off in a transistor laser : quantum well dislocation effect. *Springer, Opt Quant Electron*. doi: 10.1007/s11082-010-9384-0, pp. 481–488

Then, H.W., Walter, G., Feng, M. & Holonyak, N. Jr. (2007a). Collector characteristics and the differential optical gain of a quantum-well transistor laser. *Appl. Phys. Lett* Vol. 91, No. 243508

Then, H.W., Feng, M. & Holonyak, N. Jr. (2007b). Optical bandwidth enhancement by operation and modulation of the first excited state of a transistor laser. *Appl. Phys. Lett.* Vol. 91, No. 183505

Then, H.W., Feng, M., Holonyak, N. Jr. & Wu, C.H. (2007c). Experimental determination of the effective minority carrier lifetime in the operation of a quantum-well n-p-n heterojunction bipolar light-emitting transistor of varying base quantum-well design and doping. *Appl. Phys. Lett.* Vol. 91, No. 033505

Then, H.W., Walter, G., Feng, M. & Holonyak, N. Jr. (2008). Optical bandwidth enhancement of heterojunction bipolar transistor laser operation with an auxiliary base signal. *Appl. Phys. Lett.* Vol. 93, No. 163504

Then, H.W., Feng, M. & Holonyak, N. Jr. (2009). Bandwidth extension by trade-off of electrical and optical gain in a transistor laser: three-terminal control. *Appl. Phys. Lett.* Vol. 94, No. 013509

Then, H.W., Feng, M. & Holonyak, N. Jr. (2010). Microwave circuit model of the three-port transistor laser. *Appl. Phys. Lett.* Vol. 107, No. 094509

Walter, G., Holonyak, N. Jr., Feng, M. & Chan, M. (2004). Laser operation of a heterojunction bipolar light-emitting transistor. *Appl. Phys. Lett.* Vol. 85, No. 4768

Walter, G., James, A., Holonyak, N. Jr., Feng, M. & Chan, R. (2006). Collector breakdown in the heterojunction bipolar transistor laser. *Appl. Phys. Lett.* Vol. 88, No. 232105

Zhang, L. & Leburton, J.-P. (2009). Modeling of the transient characteristics of heterojunction bipolar transistor lasers. *IEEE Journal of Quantum Electron*. doi:10.1109/JQE.2009.2013215, pp. (359–366)

14

Intersubband and Interband Absorptions in Near-Surface Quantum Wells Under Intense Laser Field

Nicoleta Eseanu

Physics Department, "Politehnica" University of Bucharest,
Bucharest
Romania

1. Introduction

The intersubband transitions in quantum wells have attracted much interest due to their unique characteristics: a large dipole moment, an ultra-fast relaxation time, and an outstanding tunability of the transition wavelengths (Asano et al., 1998; Elsaesser, 2006; Helm, 2000). These phenomena are not only important by the fundamental physics point of view, but novel technological applications are expected to be designed.

Many important devices based on intersubband transitions in quantum well heterostructures have been reported. For example: far- and near-infrared photodetectors (Alves et al., 2007; Levine, 1993; Li, S.S. 2002; Liu, 2000; Schneider & Liu, 2007; West & Eglash, 1985), ultrafast all-optical modulators (Ahn & Chuang, 1987; Carter et al., 2004; Li, Y. et al., 2007), all optical switches (Iizuka et al., 2006; Noda et al., 1990), and quantum cascade lasers (Belkin et al., 2008; Chakraborty & Apalkov, 2003; Faist et al., 1994).

It is well-known that the optical properties of the quantum wells mainly depend on the asymmetry of the confining potential experienced by the carriers. Such an asymmetry in potential profile can be obtained either by applying an electric/laser field to a symmetric quantum well (QW) or by compositionally grading the QW. In these structures the changes in the absorption coefficients were theoretically predicted and experimentally confirmed to be larger than those occurred in conventional square QW (Karabulut et al., 2007; Miller, D.A.B. et al., 1986; Ozturk, 2010; Ozturk & Sökmen, 2010).

In recent years, with the availability of intense THz laser sources, a large number of strongly laser-driven semiconductor heterostructures were investigated (Brandi et al., 2001; Diniz Neto & Qu, 2004; Duque, 2011; Eseanu et al., 2009; Eseanu, 2010; Kasapoglu & Sokmen, 2008; Lima, F. M. S. et al., 2009; Niculescu & Burileanu, 2010b; Ozturk et al., 2004; Ozturk et al., 2005; Sari et al., 2003; Xie, 2010).

These works have revealed important laser-induced effects:

i) the confinement potential is dramatically modified; ii) the energy levels of the electrons and, to a lesser extent, those of the holes are enhanced; iii) the linear and nonlinear absorption coefficients can be easily controlled by the confinement parameters in competition with the laser field intensity.

The intersubband transitions (ISBTs) have been observed in many different material systems. In recent years, apart from GaAs/AlGaAs, the InGaAs/GaAs QW structures have attracted much interest because of their promising applications in optoelectronic and microelectronic devices: multiple-quantum-well modulators and switches (Stohr et al., 1993), broadband photodetectors (Gunapala et al., 1994; Li J. et al., 2010, Passmore, et al., 2007), superluminescent diodes used for optical coherence tomography (Li Z. et al., 2010) and metal-oxide-semiconductor field-effect-transistors, i. e. MOSFETs (Zhao et al., 2010). The optical absorption associated with the excitons in semiconductor QWs have been the subject of a considerable amount of work for the reason that the exciton binding energy and oscillator strength in QWs are considerably enhanced due to quantum confinement effect (Andreani & Pasquarello, 1990; Jho et al., 2010; Miller, R. C. et al., 1981; Turner et al., 2009; Zheng & Matsuura, 1998).

As a distinctive type of dielectric quantum wells, the near-surface quantum wells (n-sQWs) have involved increasing attention due to their potential to sustain electro-optic operations under a wide range of applied electric fields. In these heterostructures the QW is located close to vacuum and, as a consequence, the semiconductor-vacuum interface which is parallel to the well plane introduces a remarkable discrepancy in the dielectric constant (Chang & Peeters, 2000). This dielectric mismatch leads to a significant enhancement of the exciton binding energy (Gippius et al., 1998; Kulik et al., 1996; Niculescu & Eseanu, 2010a) and, consequently, it changes the exciton absorption spectra as some experimental (Gippius et al., 1998; Kulik et al., 1996; Li, Z. et al., 2010; Yablonskii et al., 1996) and theoretical (Niculescu & Eseanu, 2011a; Yu et al., 2004) studies have demonstrated.

The rapid advances in modern growth techniques and researches for InGaAs/GaAs QWs (Schowalter et al., 2006; Wu, S. et al., 2009) create the possibility to fabricate such heterostructures with well-controlled dimensions and compositions. Therefore, the differently shaped InGaAs/GaAs near-surface QWs become interesting and worth studying systems.

We expect that the capped layer of these n-sQWs induces considerable modifications on the intersubband absorption as it did on the interband excitonic transitions (Niculescu & Eseanu, 2011). To the best of our knowledge this is the first research concerning the intense laser field effect on the ISBTs in InGaAs/GaAs differently shaped near-surface QWs.

In this chapter we are concerned about the intersubband and interband optical transitions in differently shaped n-sQWs with symmetrical/asymmetrical barriers subjected to intense high-frequency laser fields. We took into account an accurate form for the laser-dressing confinement potential as well as the occurrence of the image-charges. Within the framework of a simple two-band model the consequences of the laser field intensity and carriers-surface interaction on the absorption spectra have been investigated.

The organization of this work is as follows. In Section 2 the theoretical model for the intense laser field (ILF) effect on the intersubband absorption in differently shaped n-sQWs is described together with numerical results for the electronic energy levels and absorption coefficients (linear and nonlinear). In Section 3 we explain the ILF effect on the exciton ground energy and interband transitions in the same QWs, taking into account the repulsive interaction between carriers and their image-charges. Also, numerical results for the 1S-exciton binding energy and interband linear absorption coefficient are discussed. Finally, our conclusions are summarized in Section 4.

2. Intersubband transitions in near-surface QWs under intense laser field

The intersubband transitions (ISBTs) are optical transitions between quasi-two-dimensional electronic states ("subbands") in semiconductors which are formed due to the confinement of the electron wave function in one dimension. The conceptually simplest band-structure engineered system that can be fabricated is a quantum well (QW), which consists of a thin semiconductor layer embedded in another semiconductor with a larger bandgap. Depending on the relative band offsets of the two semiconductor materials, both electrons and holes can be confined in one direction in the conduction band (CB) and the valence band (VB), respectively. Thus, allowed energy levels which are quantized along the growth direction of the heterostructure appear (Yang, 1995). These levels can be tailored by changing the QW geometry (shape, width, barrier heights) or by applying external perturbations (hydrostatic pressure, electric, magnetic and laser fields). Whereas, of course, optical transitions can take place between VB and CB states, in this Section we are concerning only with ISBTs between quantized levels within the CB.

2.1 Theory

Let us consider an InGaAs n-sQW embedded between symmetrical/asymmetrical GaAs barriers. It is convention to define the QW growth as the z-axis. According to the effective mass approximation, in the absence of the laser field, the time-independent Schrödinger equation is

$$-\frac{\hbar^2}{2m^*}\frac{\partial^2}{\partial z^2}\varphi(z)+\left[V(z)+V_{self}(z)\right]\varphi(z)=E\varphi(z). \tag{1}$$

where m^* is the electron effective mass, $V(z)$ is the confinement potential in the QW growth direction and $V_{self}(z)$ describes the repulsive interaction in the system consisting of an electron and its image-charge.

$$V_{self}(z)=\frac{e_0^2}{2\varepsilon}\left(\frac{\varepsilon-1}{\varepsilon+1}\right)\frac{1}{2d_0}. \tag{2}$$

Here ε is the semiconductor dielectric constant, $e_0^2=e^2/4\pi$, and d_0 is the distance between the electron and its image-charge without laser field. For the three differently shaped n-sQWs studied in this work the potential $V(z)$ reads as follows. For a square n-sQW, $V(z)$ has the well-known form

$$V^{SQW}(z)=\begin{cases}\infty, & z<-L_c \\ V_0, & -L_c<z<0 \ and \ z>L_w \\ 0, & 0\le z\le L_w\end{cases} \tag{3a}$$

For a graded n-sQW, the confinement potential is

$$V^{GQW}(z)=\begin{cases}\infty, & z<-L_c \\ V_0, & -L_c<z<0 \\ V_r\dfrac{z}{2L_w}, & 0\le z\le L_w \\ V_r=\sigma V_0, & z>L_w\end{cases} \tag{3b}$$

For a semiparabolic n-sQW, $V(z)$ is given by

$$V^{sPQW}(z) = \begin{cases} \infty, & z < -L_c \\ V_0, & -L_c < z < 0 \\ V_r\left(\dfrac{z}{L_w}\right)^2, & 0 \leq z \leq L_w \\ V_r = \sigma V_0, & z > L_w \end{cases} \tag{3c}$$

The quantities L_w and L_c are the well width and capped layer thicknesses, respectively; V_0 is the GaAs barrier height in the QW left side (with cap layer); V_r is the barrier height in the QW right side and σ is the barrier asymmetry parameter.

Under the action of a non-resonant intense laser field (ILF) represented by a monochromatic plane wave of frequency ω_{LF} having the vector potential $\vec{A}(t) = \vec{u} A_0 \cos(\omega_{LF} t)$, the Schrödinger equation to be solved becomes a time-dependent one due to the time-dependent nature of the radiation field. Here $\vec{u}$ is the unit vector of the polarization direction (chosen as z-axis). By applying the translation $\vec{r} \to \vec{r} + \vec{\alpha}(t)$ the equation describing the electron-field interaction dynamics was transformed by Kramers (Kramers, 1956) as

$$-\frac{\hbar^2}{2m^*}\Delta\Psi(\vec{r},t) + V(\vec{r} + \vec{\alpha}(t))\Psi(\vec{r},t) = i\hbar\frac{\partial}{\partial t}\Psi(\vec{r},t). \tag{4}$$

The expression

$$\vec{\alpha}(t) = \vec{u}\alpha_0 \sin(\omega_{LF}\, t) \tag{5}$$

describes the quiver motion of the electron under laser field action. α_0 is known as the laser-dressing parameter, i. e. a laser-dependent quantity which contains both the laser frequency and intensity,

$$\alpha_0 = \frac{e\,A_0}{m^*\omega_{LF}}. \tag{6}$$

Thus, in the presence of the laser field linearly polarized along the z-axis, the confinement potential $V(z + \alpha(t))$ is a time-periodic function for a given z and it can be expanded in a Fourier series,

$$V(z + \alpha(t)) = \sum_{k}\sum_{v=-\infty}^{\infty}(-1)^v V_k J_v(k\alpha_0)\exp(ikz)\exp(-iv\omega_{LF}t) \tag{7}$$

where V_k is the k-th Fourier component of $V(z)$ and J_v is the Bessel function of order v. In the high-frequency limit, i.e. $\omega_{LF}\tau \gg 1$, with τ being the transit time of the electron in the QW region (Marinescu & Gavrila, 1995) the electron "sees" a laser-dressed potential which is obtained by averaging the potential $V(z + \alpha(t))$ over a laser field period,

$$\tilde{V}(z,\alpha_0) = \frac{\omega_{LF}}{2\pi} \int_0^{2\pi/\omega_{LF}} V(z+\alpha(t))dt = \sum_k V_k J_0(k\alpha_0)\exp(ikz). \tag{8}$$

Note that this approximation remains valid provided that the laser is tuned far from any resonance so that only photon absorption was taken into account and photon emission was disregarded.

Following some basic works (Gavrila & Kaminski, 1984; Lima, C.A.S. & Miranda, 1981; Marinescu & Gavrila, 1995) the laser-dressed electronic eigenstates in the QW are the solutions of a time-independent Schrödinger equation,

$$\left[-\frac{\hbar^2}{2}\frac{d}{dz}\left(\frac{1}{m^*(z)}\frac{d}{dz} \right) + \tilde{V}(z,\alpha_0) \right]\tilde{\varphi}(z) = E\,\tilde{\varphi}(z). \tag{9}$$

Here $\tilde{\varphi}(z)$ is the laser-dressed wave function of the electron. The envelope wave functions and subband energies in this modified potential can be obtained by using a transfer matrix method (Ando & Ytoh, 1987; Tsu & Esaki, 1973).

In order to characterize the intersubband transitions in laser-dressed n-sQWs the $E_1 \rightarrow E_2$ transition energy, $E_{tr} = E_2 - E_1$, and square of the optical matrix element, M_{21}^2, are worth being calculated. Our results for n-sQWs revealed that these quantities depend on the QW width and shape as well as on laser field parameter and cap layer thickness.

The dipole matrix element of the $E_i \rightarrow E_f$ transition is defined by

$$M_{fi} = \int \varphi_f^* |e| z \varphi_i dz \tag{10}$$

Also, the first-order absorption coefficient, $\beta^{(1)}(\omega_{ex})$, and the third-order absorption coefficient, $\beta^{(3)}(\omega_{ex},I)$, for an optical transition between two subbands can be calculated by using the compact density matrix method (Bedoya & Camacho, 2005; Rosencher & Bois, 1991) and a typical iterative procedure (Ahn & Chuang, 1987). The linear and nonlinear absorption coefficients are written (Unlu et al., 2006; Ozturk, 2010) as:

$$\beta^{(1)}(\omega_{ex}) = \sqrt{\frac{\mu}{\varepsilon}} |M_{21}|^2 \frac{m^* k_B T}{L_{eff}\pi\hbar^2}$$

$$\times \ln\left\{ \frac{1+\exp\left[(E_F - E_1)/k_B T\right]}{1+\exp\left[(E_F - E_2)/k_B T\right]} \right\} \times \frac{\omega_{ex}\hbar/\tau_{in}}{(E_2 - E_1 - \hbar\omega_{ex})^2 + (\hbar/\tau_{in})^2} \tag{11}$$

$$\beta^{(3)}(\omega_{ex},I) = -\sqrt{\frac{\mu}{\varepsilon}}\left(\frac{I}{2\varepsilon_0 n_r c} \right)|M_{21}|^2 \frac{m^* k_B T}{L_{eff}\pi\hbar^2}$$

$$\times \ln\left\{ \frac{1+\exp\left[(E_F - E_1)/k_B T\right]}{1+\exp\left[(E_F - E_2)/k_B T\right]} \right\} \times \frac{\omega_{ex}\hbar/\tau_{in}}{\left[(E_2 - E_1 - \hbar\omega_{ex})^2 + (\hbar/\tau_{in})^2\right]^2}$$

$$\times \left\{ 4|M_{21}|^2 - \frac{|M_{22} - M_{11}|^2}{\left(E_2 - E_1\right)^2 + \left(\hbar / \tau_{in}\right)^2} \times \left[\left(E_2 - E_1 - \hbar\omega_{ex}\right)^2 - \left(\hbar / \tau_{in}\right)^2 + 2\left(E_2 - E_1\right)\left(E_2 - E_1 - \hbar\omega_{ex}\right) \right] \right\} \tag{12}$$

Here I is the optical intensity of the incident electromagnetic wave (with the angular frequency ω_{ex}) that excites the semiconductor nanostructure and leads to the intersubband optical transition, μ is the permeability, $\varepsilon = \varepsilon_0 n_r^2$ with n_r being the refractive index, $\hbar\omega_{ex}$ is the pump photon energy produced by a tunable laser source, E_F represents the Fermi energy, E_1 and E_2 denote the quantized energy levels for the initial and final states, respectively; k_B and T are the Boltzmann constant and temperature, respectively, c is the speed of light in free space, L_{eff} is the effective spatial extent of the electrons and τ_{in} is the intersubband relaxation time.

The total absorption coefficient is given by

$$\beta\left(\omega_{ex}, I\right) = \beta^{(1)}\left(\omega_{ex}\right) + \beta^{(3)}\left(\omega_{ex}, I\right) \tag{13}$$

The absorbtion coefficients in n-sQWs under laser field depend on both laser-dressing parameter and QW geometry (well shape, barrier asymmetry, cap layer thickness).

2.2 Electronic properties
2.2.1 Laser-dressed confinement potential and energy levels

Within the framework of effective-mass approximation the ground and the first excited energetic levels for an electron confined in differently shaped $In_{0.18}Ga_{0.82}As/GaAs$ near-surface QWs: square (SQW), graded (GQW), and semiparabolic (sPQW) under high-frequency laser field were calculated. We used various QW widths L = 100 Å, 150 Å, and 200 Å, different cap layer thicknesses, L_c, between 5 Å and 200 Å, for n-sQWs with symmetrical (σ = 1) or asymmetrical (σ = 0.6 and 0.8) barriers. The small In atoms concentration in the QW layer allowed us to take the same value of the electron effective mass in all regions (barriers, QW), i. e. $m^* = 0.0665\, m_0$. Also, we used E_F = 6.49 meV which corresponds to about 1.6×10^{17} electrons/cm³ in GaAs, and τ_{in} = 0.14 ps (Ahn & Chuang, 1987).

Fig. 1 displays the "dressed" potential profiles of the conduction band (CB) for three differently shaped n-sQWs (SQW, GQW and sPQW) with QW width L_w = 150 Å, identical barriers (σ = 1) and cap layer thickness L_c = 20 Å, under various laser intensities described by the laser parameter values α_0 = 0; 40; 80 and 100 Å. Only two energy levels have been taken into account for all the n-sQWs investigated in this work: E_1 (ground state) and E_2 (first excited state). They are are plotted in Fig. 1, too.

We see that for all studied n-sQWs the increasing of the laser parameter dramatically modifies the potential shape which is responsible for quantum confinement of the electrons. Up to $\alpha_0 \leq L_w / 2$ two effects are noticeable: i) while the effective "dressed" well width (i. e. the lower part of the confinement potential) decreases with the laser intensity, the width of the upper part of this "dressed" QW increases; ii) a reduction of the effective well height at the interface between the capped layer and the QW ($z = 0$).

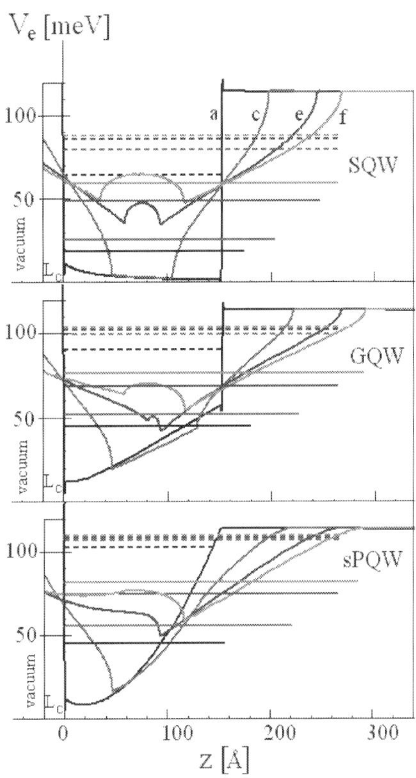

Fig. 1. (Color online) Laser-dressed confinement potentials of three differently shaped
In$_{0.18}$Ga$_{0.82}$As/GaAs near-surface QWs with identical barriers and the corresponding energy
levels: ground state E_1 (solid lines) and first excited state E_2 (dashed lines). Notations a, c,
e, and f stand for various laser parameter values, α_0 = 0 (black), 40 Å (olive), 80 Å (purple),
and 100 Å (orange), respectively. QW width and cap layer thickness are L_w = 150 Å and L_c
= 20 Å, respectively.

Therefore, under an intense laser field a distinctive blue-shift of the electronic energy levels occurs, as expected (Brandi et al., 2001; Diniz Neto & Qu, 2004; Eseanu, 2010; Kasapoglu & Sökmen, 2008; Lima, F. M. S. et al., 2009; Niculescu & Burileanu, 2010b; Ozturk et al., 2004; Ozturk et al., 2005). This laser-induced push-up effect is more pronounced in the semiparabolic QW due to the stronger geometric confinement.

For $\alpha_0 > L_w/2$ a supplementary barrier having a "hill"-form appears into the well region and, as a consequence, the formation of a double well potential in the InGaAs layer is predicted. Similar laser-induced phenomena were reported for a GaAs/AlGaAs square QW (Lima, F. M. S. et al., 2009) and for coaxial quantum wires (Niculescu & Radu, 2010c).

The two energy levels E_1 (ground state) and E_2 (first excited state) depend on both laser field characteristics (frequency and intensity, combined in the laser parameter) and QW confining geometry (shape, width, cap layer thickness and barrier asymmetry).

In Figs. 2 A-C the energy levels E_1 and E_2 are plotted as functions of laser parameter in three differently shaped n-sQWs (SQW, GQW and sPQW) with L_w = 200 Å for various cap layer thicknesses.

We observed that:

i.   the laser-induced increasing of the ground level energy E_1 is more pronounced for $\alpha_0 > 40$ Å in all three differently shaped n-sQWs with the same L_w ;

ii.   the first excited level energy E_2 is almost unaltered by the laser field in the GQW and sPQW for $\alpha_0 > 40$ Å; instead, in the SQW, E_2 has a significant rising.

The reasons are as follows: i) the ground level E_1 is localized in the lower part of the laser-dressed QW and it is significantly moved up only by an intense laser field; ii) in the GQW and sPQW the stronger geometric confinement competes with the laser-induced push-up of the energy levels. As a consequence, the excited level which is localized in the upper part becomes less sensitive to the laser action.

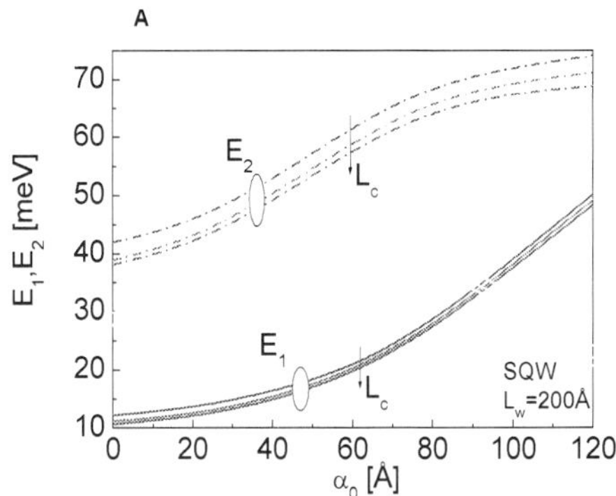

Fig. 2. A

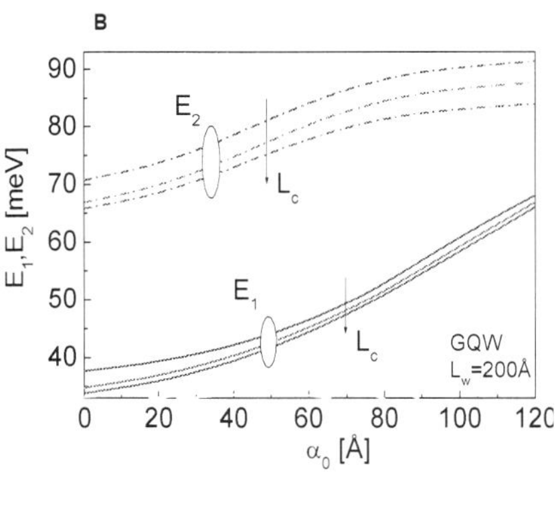

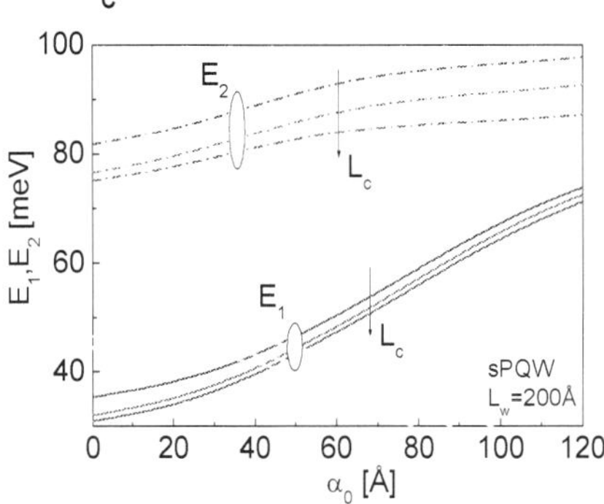

Fig. 2. Energy levels (ground state and first excited state) vs. laser parameter in three differently shaped n-sQWs: a) SQW, b) GQW and c) sPQW with L_w = 200 Å. Each arrow indicates the rising of the cap layer thickness.

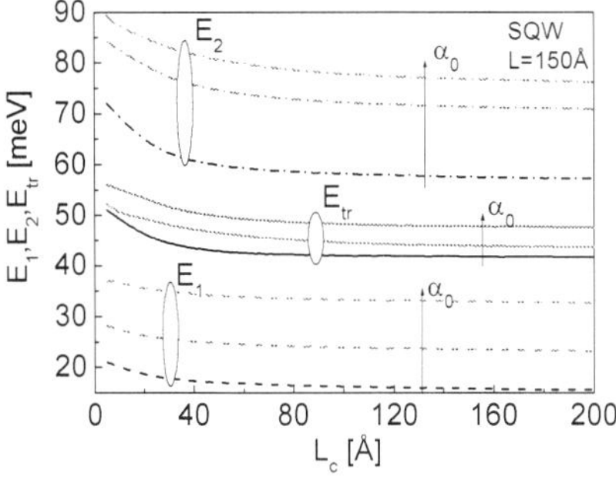

Fig. 3. Energy levels and transition energy vs. cap layer thickness in a near-surface SQW with L_w = 150 Å under various laser intensities. Each arrow indicates the rising of the laser parameter.

Also, in the range of thin cap layers, the energy levels E_1 and E_2 depend on the capped layer thickness (L_c). Fig. 3 displays the energies E_1, E_2 and transition energy, $E_{tr} = E_2 - E_1$, vs. L_c in a near-surface SQW with L_w = 150 Å for several values of the laser parameter. This variation is similar in GQW and sPQW.

We note that, up to $L_c \cong 40$ Å, the energies E_1, E_2 and, consequently, the transition energy first rapidly decrease with cap layer thickness and, for further large L_c values, these three quantities turn out to be insensitive to the dielectric effect afforded by the cap layer. The reason for this behavior is that the effect of the image-charge is reduced by a thicker cap layer.

2.2.2 The $E_1 \rightarrow E_2$ transition energy

As the intense laser field induces obvious changes in the electronic levels a noticeable dependence of the $E_1 \rightarrow E_2$ transition energy, $E_{tr} = E_2 - E_1$, on the laser parameter, α_0, is expected. Also, this energy is modified by the QW confining properties. Figs. 4 A-C present the grouped data plot of the E_{tr} and square of the dipole matrix element (in units of e^2), M_{21}^2, as functions of the laser parameter, for differently shaped n-sQWs with various cap layer, L_c = 20, 50, 100, 200 Å and the same width, L_w = 150 Å. For SQW and GQW (Figs. 4 A, B) the transition energy, E_{tr}, increases up to a certain value of the laser parameter, α_{0M}, and then it begin to decrease. The critical laser parameter, α_{0M}, increases for larger QWs (Table 1). A similar behavior have been reported for regular (i.e. uncapped) GaAs/AlGaAs SQW and PQW (Eseanu, 2010; Ozturk et al., 2004). As suggested by Ozturk et al. (2004), for $\alpha_0 < \alpha_{0M}$ the transition energy E_{tr} increases due to reduction of the effective "dressed" well width. Instead, for $\alpha_0 > \alpha_{0M}$ the subbands levels E_1 and E_2 tends to localize in the upper part of the laser-dressed well (which has a larger width) and thus they become closer to each other. Our calculations show a difference between the increasing rates of E_1 and E_2 values in SQW and GQW under laser field action. Therefore, E_{tr} may have a maximum for a certain value α_{0M}. This critical value of the laser parameter seems to have a weak dependence on the cap layer thickness (Table 1).

L [Å]	α_{0M} [Å] (L_c = 20 Å)	α_{0M} [Å] (L_c = 50 Å)	α_{0M} [Å] (L_c = 100 Å)	α_{0m} [Å] (L_c = 20 Å)	α_{0m} [Å] (L_c = 50 Å)	α_{0m} [Å] (L_c = 100 Å)
SQW;150	39.7	40.4	40.0	45.0	46.3	42.4
SQW;200	67.0	68.3	66.6	71.0	72.0	69.7
GQW;150	32.5	32.0	31.3	41.0	38.9	-
GQW;200	61.9	59.6	56.8	62.0	-	-

Table 1. Laser parameter critical values α_{0M} (for the maximum of E_{tr}) and α_{0m} (for the minimum of M_{21}^2).

For sPQW we note a different behavior, i.e. E_{tr} reduces monotonically as the laser parameter increases (Fig. 4C). The supplementary quantum confinement of the electron localization in sPQW comparing with SQW and GQW could be the explanation.

For the three studied QW structures and for all the widths under our investigation the values of E_{tr} are diminuted by increasing L_c (vertical arrows indicate the L_c rising). This effect can be explained by the stronger reduction of the ground level energy values E_2 for larger L_c (Fig. 3).

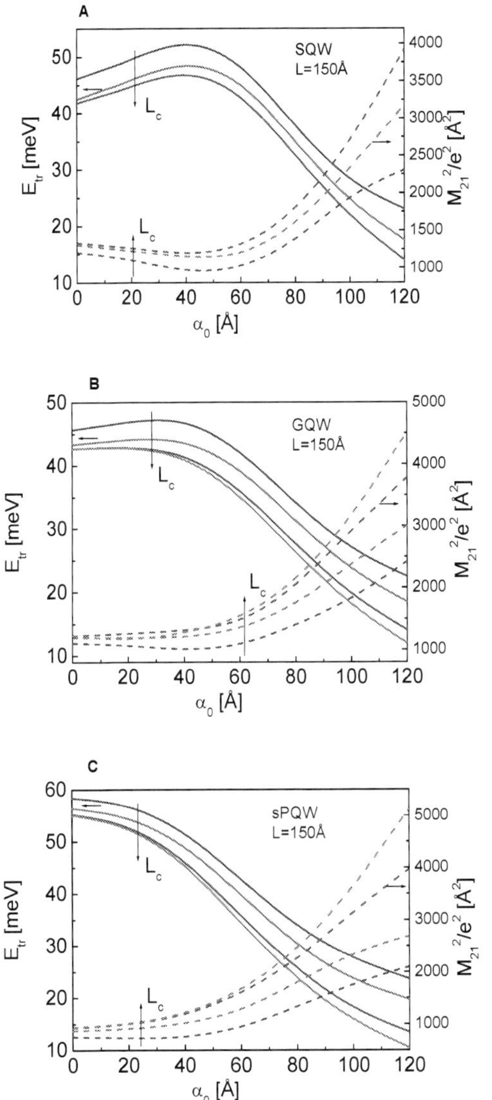

Fig. 4. Grouped data plot of the $E_1 \rightarrow E_2$ transition energy and the square of the matrix element (in units of e^2) vs. laser parameter, for differently shaped n-sQWs: A) square, B) graded, and C) semiparabolic, with various L_c.

Also, the transition energy depends on the QW width as Fig. 5 shows for a near-surface SQW with two values of the cap layer thickness, under several laser intensities.

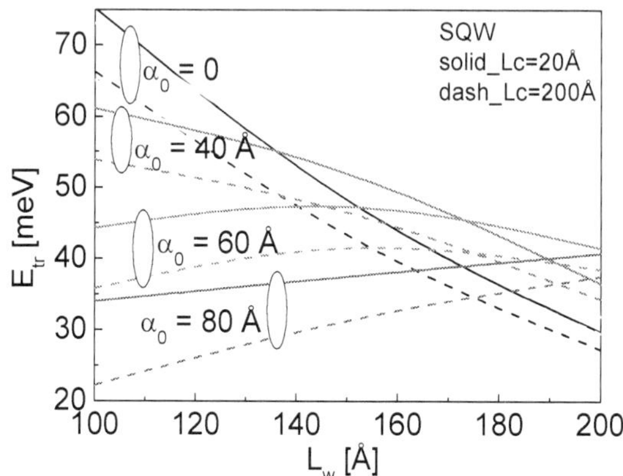

Fig. 5. Transition energy vs. QW width in a near-surface SQW for several laser parameters and two values of the cap layer thickness.

The variation of $E_{tr} = f(L_w)$ is modulated by the laser field. In the absence of radiation ($\alpha_0 = 0$) or in a low laser field ($\alpha_0 < 40$ Å) E_{tr} diminishes in large QWs, as expected, because the levels E_1 and E_2 move down and become closer to each other. For uncapped QW-heterostructures this result is a customary one, for example: in a GaAs/Al$_{0.3}$Ga$_{0.7}$As SQW with L_w = 50 - 65 Å (Helm, 2000), in GaAs/Al$_{0.3}$Ga$_{0.7}$As SQW and sPQW with L_w = 100 - 200 Å (Eseanu, 2010) and in a In$_{0.5}$Ga$_{0.5}$As/AlAs SQW with L_w = 20 - 100 Å (Chui, 1994).

By applying an intense laser field the reduction of E_{tr} (as function of L_w) becomes weaker, but for high laser parameter values (α_0 = 80 Å), E_{tr} turn to rise with L_w, especially in the presence of a thick cap layer. The reason for this behavior is the competition between the geometric quantum confinement and laser-induced increasing of the energy levels.

In the asymmetrical n-sQWs (GQW and sPQW) another factor modifiying the transition energy appears. This is the asymmetry parameter of the QW barriers, σ, (see Eqs. 2b and 2c). In Fig. 6 the transition energy, E_{tr}, as a function of the barrier asymmetry is plotted for GQW and sPQW with L_w = 200 Å and a thin cap layer (20 Å). As seen in this figure E_{tr} is enhanced by the increasing σ. The dependence $E_{tr} = f(\sigma)$ is also modulated by the laser field and, to a lesser extent, by the QW shape.

In the GQW the increasing of E_{tr} is almost linear for all laser parameter values, but the rising slope is higher under intense laser field (α_0 = 80 Å) due to the laser-induced push-up effect on the energy levels. Instead, the stronger geometric confinement leads to a deviation from linearity of the curve $E_{tr} = f(\sigma)$ in the sPQW.

Now we should emphasize an important issue. As a consequence of the laser-induced shift in the subband transition energy it clearly appears the possibility that E_{tr} can be tuned by the joint action of the laser field and a supplementary external perturbation such as: electric field (Ozturk et al., 2004; Ozturk et al., 2005) or hydrostatic pressure (Eseanu, 2010), these two cases refering to uncapped QWs. The last case was named "laser- and pressure-driven optical absorption" (LPDOA). The variable cap layer thickness, especially in the range of thin films, in simultaneous action with an intense laser field could be a new method (suggested by the present study) to adjust the transition energy, E_{tr} .

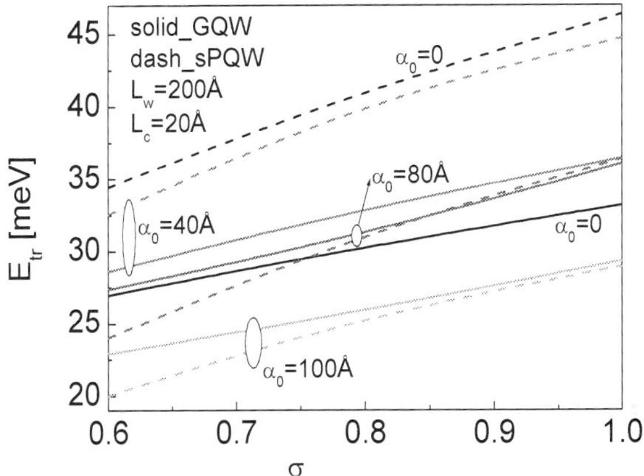

Fig. 6. Transition energy vs. asymmetry parameter of the QW barriers in GQW (solid lines) and sPQW (dashed lines) with L_w = 200 Å and L_c = 20 Å under various laser intensities.

2.2.3 The square dipole matrix element

The intense laser field strongly modifies the dipole matrix element of the $E_1 \rightarrow E_2$ transition, M_{21}^2, for all the three n-sQWs presented in this work, but in a different manner (Figs. 4 A, B, C).

In the symmetrical structure SQW (Fig. 4 A) M_{21}^2 has a relative minimum at a critical value α_{0m} of the laser parameter. This value is very close to that for which the transition energy E_{tr} has its maximum. A similar behavior have been reported for regular (i.e. uncapped) GaAs/AlGaAs SQWs (Eseanu, 2010). The explanation for the minimum of M_{21}^2 can be connected with the laser intensity dependent overlap of the two wave functions $\Psi_1(z)$ and $\Psi_2(z)$.

In the presence of laser field, the SQW shape is dramatically modified: as α_0 increases the lower part of the confinement potential becomes more and more narrow while the upper part becomes wider. Therefore, the energy subbands will be pushed up to the top of the well. We may identify two regimes:

i.   for $\alpha_0 < \alpha_{0m}$ the second subband E_2 is localized in the upper part of the "dressed" well and the ground state E_1 is still localized in the lower part. As α_0 increases, the ground state wave function $\Psi_1(z)$ becomes more compressed in the vicinity of $z = 0$ and its overlap with the first excited wave function $\Psi_2(z)$ (which has a minimum in $z = 0$) reduces.

ii.   by further increasing α_0, for $\alpha_0 > \alpha_{0m}$, the E_1 level is also pushed up to the wider upper part of the QW. Thus the carrier confinement decreases and the two wave functions $\Psi_1(z)$ and $\Psi_2(z)$ are spread out (or delocalized) in the potential barrier regions. As a consequence their overlapping is enhanced.

In the asymmetrical structures GQW and sPQW, M_{21}^2 monotonically increases with laser parameter (Figs. 4 B, C). This effect is determined by the strong electron localization in the graded barriers. Therefore, the two wave functions $\Psi_1(z)$, $\Psi_2(z)$ are localized in the upper part of the laser-dressed QW.

For the three studied n-sQWs and for all the widths under our investigation the values of M_{21}^2 are enhanced by growing L_c (vertical arrows indicate the L_c rising in Figs. 4 A, B, C). This effect can be explained by the broadening of the effective n-sQW width for thicker cap layers. Thus, the ground state wave function $\Psi_1(z)$ becomes more extended in the heterostructure and its overlap with the first excited wave function $\Psi_2(z)$ is enhanced.

2.3 Linear and nonlinear optical absorption

The intersubband linear and nonlinear absorption coefficients given by the Eqs. (11)-(12) depend on the transition energy, $E_{tr} = E_2 - E_1$, and on square of the dipole matrix element, M_{21}^2. In the n-sQWs under our investigation these quantities are strongly modified by the laser field parameter, but in a different manner (Fig. 4). As a consequence, the absorption coefficients $\beta^{(1)}(\omega_{ex})$ and $\beta^{(3)}(\omega_{ex}, I)$ are significantly changed by the laser intensity (included in the laser-dressing parameter α_0). However, there are still four variables to take into account: the pump photon energy, $\hbar\omega_{ex}$, cap layer thickness, L_c, asymmetry parameter of the QW barriers, σ, and QW potential shape.

In Fig. 7 the dependence of the linear absorption coefficient $\beta^{(1)}$ on the pump photon energy in differently shaped n-sQWs with L_w = 150 Å and symmetrical barriers, under various laser intensities α_0 = 0, 40 Å, and 80 Å, is plotted. The values of L_c are 20 Å and 200 Å

Fig. 8 displays the variation of $\beta^{(1)}$ on the pump photon energy in a near-surface SQW with L_c = 50 Å under various laser intensities α_0 = 0, 40 Å, and 80 Å. The values of L_w are 100 Å and 200 Å.

One can see from Figs. 7 and 8 that the increasing laser intensity generates a noticeable shift of the absorption peak position toward higher/lower photon energies and an obvious reduction of the $\beta^{(1)}$ magnitude. The trend of this shift significantly depends on the QW shape and width, also on the laser parameter. Several regimes occur:

1.   in SQW and GQW with L_w = 150 Å the absorption peaks are blue-shifted for $\alpha_0 < L_w / 2$, but red-shifted for $\alpha_0 > L_w / 2$;

2.   in GQW with L_w = 200 Å the absorption peaks (not plotted here) are blue-shifted for all the studied α_0 values, but the shift induced by a strong laser field (α_0 = 80 Å) is smaller;

3.   in sPQW with L_w = 150 Å the linear absorption peaks are red-shifted for all the studied values of the laser parameter; a similar variation (not plotted here) was observed for L_w = 200 Å.

4.   in SQW the absorption peaks are red-shifted for L_w = 100 Å, but for L_w = 200 Å they are blue-shifted (Fig. 8).

For SQW and GQW these effects can be connected with the different laser-dependence of the two energy levels E_1 and E_2 (Fig. 2). For sPQW the stronger geometric confinement competes with the laser-induced change of the transition energy. The explanation used for the results plotted in Fig. 4 remains valid for the absorption peak positions too. In all n-sQWs under our study the peak magnitude of the $\beta^{(1)}$ is diminuted by the laser field. The reason for this reduction is the laser-induced spreading of the carriers wave functions in the laser-dressed QW barriers.

Concerning the effect of the cap layer on the absorption peak position an obvious red-shift occurs for thicker cap layers, especially in sPQW (Fig. 7). The broadening of the effective n-sQW width for large cap layer thickness leading to a reduction of E_{tr} is the reason for this "dielectric effect".

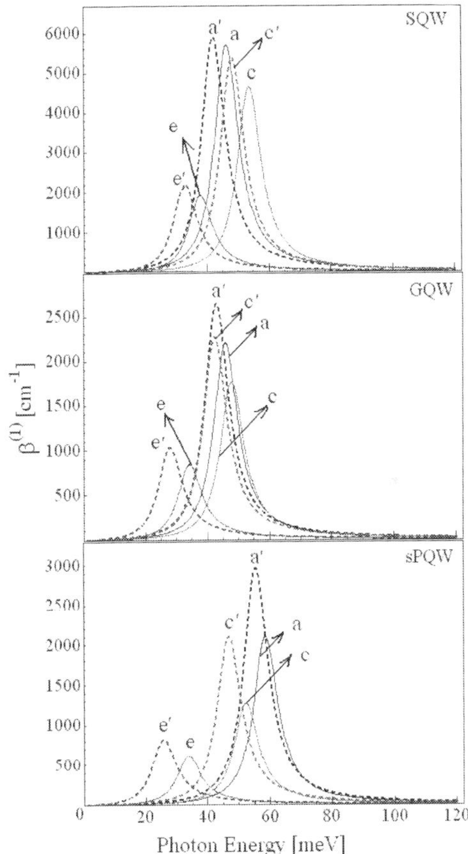

Fig. 7. Linear absorption coefficient, $\beta^{(1)}$, vs. pump photon energy in differently shaped n-sQWs with L_w = 150 Å under three laser intensities (same notations as in Fig. 1). L_c = 20 Å (solid lines) and 200 Å (dashed lines).

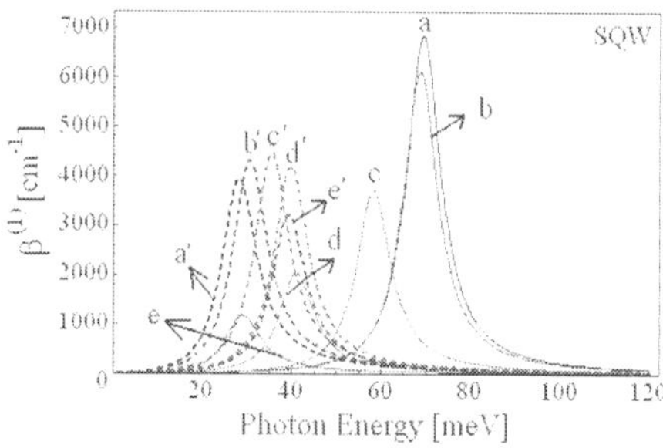

Fig. 8. Linear absorption coefficient, $\beta^{(1)}$, vs. pump photon energy in a n-sSQW with L_c = 50 Å under various laser dressing parameters. Notations a, b, c, d, and e stand for: $\alpha_0 = 0$ (black), 20 Å (blue), 40 Å (olive), 60 Å (red), and 80 Å (purple), respectively. The QW widths are 100 Å (solid lines) and 200 Å (dashed lines).

In Fig. 9 the total absorption coefficient, $\beta(\omega_{ex}, I) = \beta^{(1)}(\omega_{ex}) + \beta^{(3)}(\omega_{ex}, I)$, as a function of the photon energy for different values of the incident optical intensity I, with and without applied laser field, in differently shaped n-sQWs having L_w = 150 Å and symmetrical barriers, is plotted. The cap layer thickness is L_c = 20 Å.

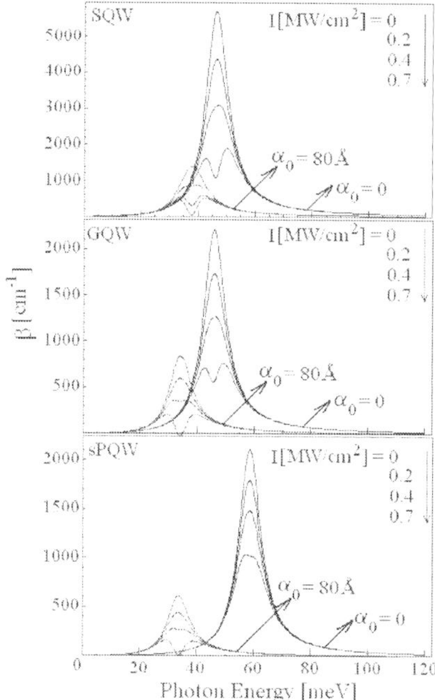

Fig. 9. Total absorption coefficient vs. pump photon energy in differently shaped n-sQWs with L_w = 150 Å, L_c = 20 Å and symmetrical barriers for two laser parameters $\alpha_0 = 0$ (black) and 80 Å (purple). The vertical arrow indicates the rising of the exciting intensity.

The main findings for $\beta(\omega_{ex}, I)$ are:

i.   there is no shift of the resonant peak positions with incident optical intensity, as expected;

ii.   the total absorption coefficient β is significantly diminuted by the increasing optical intensity due to the negative nonlinear term, $\beta^{(3)}(\omega_{ex}, I)$. Therefore, both linear and nonlinear contributions should be taken into account in the calculated absorption coefficient β near the resonance frequency ($E_{tr} \cong \hbar\omega_{ex}$), especially for high exciting intensities.

iii. the magnitude peak of the total absorption coefficient is dramatically reduced by the intense laser field;

iv.   the resonant peak of β can be bleached at sufficiently high incident optical intensities; the bleaching effect generates a split up of the resonant peak into two peaks. For our n-sQW structures the bleaching intensities depend on the confining potential shape, laser-dressing parameter, and cap layer thickness.

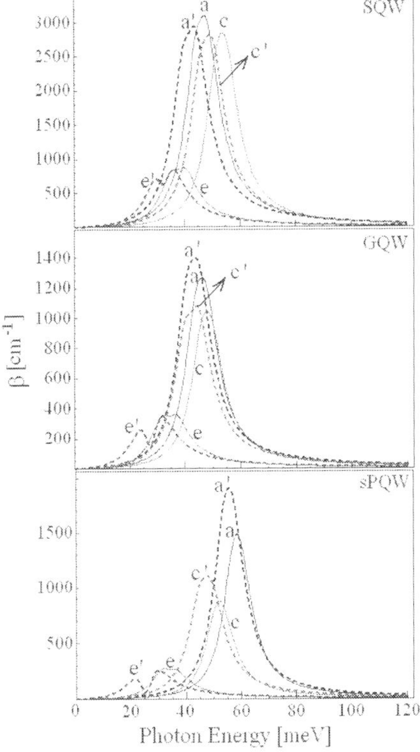

Fig. 10. Total absorption coefficient, $\beta = \beta^{(1)} + \beta^{(3)}$, vs. pump photon energy in differently shaped n-sQWs under the same conditions as in Fig. 7, and supplementary, with exciting intensity $I = 0.4$ MW/cm².

Similar results were reported for GaAs/AlGaAs uncapped QWs having different shapes: inverse V-type (Niculescu & Burileanu, 2010b) or graded-type (Ozturk, 2010).

In the particular case of a relatively low exciting intensity, $I = 0.4$ MW/cm², the dependence of the total absorption coefficient β on the laser-dressing parameter (Fig. 10) is similar with that for $\beta^{(1)}(\omega_{ex})$ (see Fig. 7). Therefore, the exciting intensity may be used to modulate the intersubband absorption.

Concluding, we note that the optical properties of the differently shaped n-sQWs could be tuned by proper tailoring of the heterostructure parameters (well shape and width, cap layer thickness, barrier asymmetry) and/or by varying the laser field intensity. The switch between the strong absorption and induced laser transparency regimes can be suitable for good performance optical modulators.

3. Interband transitions in near-surface QWs under intense laser field

Excitons play an important role in the optical properties of quantum wells. Compared to those in three-dimensional bulk semiconductors, the exciton binding energy and oscillator strength in QWs are considerably enhanced due to quantum confinement effect, so excitons in QWs are more stable.

The electronic and optical properties of the semiconductor QWs are significantly modified by applying intense high-frequency laser fields and this effect provides new degrees of freedom for technological applications.

3.1 Theory

Let us consider again the n-sQW structure described in the Section 2.1. According to the effective mass approximation, in the absence of the laser field, the exciton Hamiltonian is

$$H = H_{ez} + H_{hz} + \frac{P_\perp^2}{2M} + \frac{p_\perp^2}{2\mu} + U_{eh} \, . \tag{14}$$

Here

$$H_{jz} = -\frac{\hbar^2}{2m_{jz}^*}\frac{\partial^2}{\partial z_j^2} + V_j(z_j) + V_{self}(z_j) \tag{15}$$

are the single-particle 1D Hamiltonians. The symbol $j = e,h$ denotes the electron (hole) and V_j is the confinement potential in the growth direction (z-axis). The forms of V_j for the three n-sQWs are given by the Eqs. (3 a, b, and c), but V_0 must be replaced by V_e (for electron) and V_h (for hole). The last term in Eq. (15) represents the electrostatic energy of the repulsive interaction between the carriers and their image-charges

$$V_{self}(z_j) = \frac{e_0^2}{2\varepsilon}\left(\frac{\varepsilon - 1}{\varepsilon + 1}\right)\frac{1}{2d_{0j}} \, . \tag{16}$$

Here ε is the semiconductor dielectric constant, $e_0^2 = e^2/4\pi$, and d_{0j} is the distance between the electron (hole) and its image-charge. The third and fourth terms of Eq. (14) are the kinetic operators of the mass-center motion and relative motion of an exciton, respectively, in the (x-y) plane.

The last term in Eq. (14) describes the total electron-hole Coulomb interaction (Chang & Peeters, 2000) as

$$U_{eh}\left(\rho,z_e,z_h\right) = -\frac{e_0^2}{\varepsilon}\left[\frac{1}{\sqrt{\rho^2+\left(z_e-z_h\right)^2}}+\left(\frac{\varepsilon-1}{\varepsilon+1}\right)\frac{1}{\sqrt{\rho^2+d_{e-h}^2}}\right],\tag{17}$$

where ρ denotes the in-plane relative coordinate $\rho=\left|\boldsymbol{\rho}_e-\boldsymbol{\rho}_h\right|$; d_{e-h} is the distance between the electron (hole) and hole-image (electron-image) along the z-axis. For the sake of simplicity, in Eqs. (14–17) we have assumed that the electron and hole effective masses as well as the dielectric constant do not vary inside the whole heterostructure.

By applying a non-resonant intense laser field (ILF) which has the polarization direction parallel to the QW growth direction (z-axis), in the high-frequency limit (Marinescu & Gavrila, 1995), the electron (hole) "sees" a laser-dressed potential which is obtained by averaging the confining potential $V_j\left(z_j+\alpha_j(t)\right)$ over a period

$$\tilde{V}\left(z,\alpha_{0j}\right)=\frac{\omega_{LF}}{2\pi}\int_0^{2\pi/\omega_{LF}} V_j\left(z_j+\alpha_j(t)\right)dt.\tag{18}$$

The quantity

$$\alpha_j(t)=\alpha_{0j}\sin\left(\omega_{LF}t\right)\tag{19}$$

describes the motion of the electron (hole) in the laser field of frequency ω_{LF}; α_0 is the laser-dressing parameter, similar with that given by the Eq. (6), i.e. $\alpha_{0j}=eA_0/\left(m_{jz}^*\omega_{LF}\right)$. The laser-dressed repulsive interaction between the carriers and their self-image charges can be approximated by a soft-core potential (Lima, C.A.S. & Miranda, 1981) as

$$\tilde{V}_{self}\left(z,\alpha_{0j}\right)=\frac{e_0^2}{4\varepsilon}\left(\frac{\varepsilon-1}{\varepsilon+1}\right)\frac{1}{\sqrt{d_{0j}^2+\alpha_{0j}^2}}\tag{20}$$

Following some important studies (Gavrila & Kaminski, 1984; Lima, C.A.S. & Miranda, 1981; Marinescu & Gavrila, 1995) the dressed single-particle eigenstates are the solutions of time-independent Schrödinger equations

$$\left[-\frac{\hbar^2}{2m_{jz}^*}\frac{\partial^2}{\partial z_j^2}+\tilde{V}_j\left(z_j,\alpha_{0j}\right)+\tilde{V}_{self}\left(z_j,\alpha_{0j}\right)\right]\tilde{\varphi}_j\left(z_j\right)=E_j\tilde{\varphi}_j\left(z_j\right),\quad j=e,h\,.\tag{21}$$

In order to obtain the envelope wave functions and subband energies of both electron and hole in these modified potentials we have used a transfer matrix method (Ando & Ytoh, 1987; Tsu & Esaki, 1973).

Within the same approximation of the soft-core potential the dressed exciton binding energy was calculated by using the replacement

$$U_{eh}\left(\rho,z_e,z_h\right)\rightarrow-\frac{e_0^2}{\varepsilon}\left[\frac{1}{\sqrt{\rho^2+\left(z_e-z_h\right)^2+\alpha_0^2}}+\left(\frac{\varepsilon-1}{\varepsilon+1}\right)\frac{1}{\sqrt{\rho^2+d_{e-h}^2+\alpha_0^2}}\right]\tag{22}$$

in the Hamiltonian (14). In the Eq. (23) $\alpha_0 = eA_0 / \left(\mu\omega_{LF} \right)$ is the laser parameter for the *exciton*. Assuming a strong quantization in the QW potentials the separate trial wave function for the ground exciton state (1S) is chosen (Miller, R.C. et al., 1981) as

$$\Psi_{ex} = N\tilde{\varphi}_e\left(z_e\right)\tilde{\varphi}_h\left(z_h\right)\chi\left(\rho,\lambda\right) \tag{23}$$

Here N is a normalization constant, while $\varphi_e\left(z_e\right)$ and $\varphi_h\left(z_h\right)$ are the ground state eigenfunctions of the laser-dressed Hamiltonians $\tilde{H}_{ez}$ and $\tilde{H}_{hz}$, respectively. These operators are obtained by replacing the Eqs. (18) and (21) in Eq. (15). $\chi\left(\rho,\lambda\right) = \exp\left(-\rho/\lambda\right)$ is a 2D hydrogenic function with λ as variational parameter. Then, the exciton binding energy is calculated as

$$E_b = E_e + E_h - \min_{\lambda}\frac{\left\langle\Psi_{ex}\left|H\right|\Psi_{ex}\right\rangle}{\left\langle\Psi_{ex}\left|\Psi_{ex}\right.\right\rangle}, \tag{24}$$

where E_e and E_h are the single-particle energies of the electron and hole, respectively.

In order to systematically investigate the ILF effects on the optical n-sQW properties, the linear absorption coefficient of the excitonic transitions induced by an exciting light which is polarized along the z-direction have been calculated. We denote by ω_{ex} the angular frequency of the light from a tunable source. Assuming the dipole approximation, the absorption coefficient (Bassani & Parravicini, 1975) depends on the oscillator strength per unit area, f, as

$$\beta\left(\omega_{ex}\right) = \frac{4\pi^2 e^2 \hbar f}{n_r\, c m_0 L}\frac{\Gamma/2\pi}{\left(\hbar\omega_{ex} - E_{ex}\right)^2 + \left(\Gamma/2\right)^2} \tag{25}$$

where the following notations were used: m_0 the free-electron mass, n_r the refractive index of the QW material, L the total thickness of the structure, $E_{ex} = E_{g\,well} + E_e + E_h - E_b$ the excitonic transition energy and Γ the linewidth.

On the other hand, the oscillator strength per unit area is connected to the excitonic wave function (Bassani & Parravicini, 1975) by

$$f = \frac{2M^2}{m_0 E_{ex}}\left|\left\langle\tilde{\varphi}_e\left(z_e\right)\left|\tilde{\varphi}_h\left(z_h\right)\right.\right\rangle\right|^2\left|\chi(0)\right|^2, \tag{26}$$

where M is the optical transition matrix element between the valence and conduction bands, $\varphi_j\left(z_j\right)$ are the single-particle wave functions under applied laser field, and $\left|\chi(0)\right|^2$ denotes the probability of finding the electron and hole at the same position. For the allowed optical transitions between the hh1 and e1 subbands we obtain

$$\beta\left(\omega_{ex}\right) = B\left|\left\langle\tilde{\varphi}_e\left(z_e\right)\left|\tilde{\varphi}_h\left(z_h\right)\right.\right\rangle\right|^2\left|\chi(0)\right|^2\frac{\Gamma/2}{\left(\hbar\omega_{ex} - E_{ex}\right)^2 + \left(\Gamma/2\right)^2} \tag{27}$$

where the quantity B is almost frequency-independent in the frequency range of the interband transitions (Bastard, 1981). So, B was taken as a constant.

3.2 Electronic properties

The numerical calculations were carried out for $In_{0.18}Ga_{0.82}As$/GaAs differently shaped n-sQWs with the well width L_w = 50 Å or 80 Å and various cap layer thicknesses, L_c = 40 Å, 80 Å, 120 Å, by using the material parameters (Gippius et al., 1998) listed in Table 2 and a broadening parameter Γ = 1 meV (Sari et al., 2003). The chosen values for L_w and L_c allowed us to validate the calculation results by the reported experimental data (Gippius et al., 1998; Kulik et al., 1996; Li, Z. et al., 2010; Yablonskii et al., 1996).

Material	GaAs	$In_{0.18}Ga_{0.82}As$
V_e (meV)	113	0
V_h (meV)	75	0
m_e/m_0	0.067	0.067
m_{hz}/m_0	0.35	0.35
$m_{h\perp}/m_0$	0.112 *	0.112 *
E_g^Γ (meV)	1519	1331
ε_s	12.5	12.5
ε_∞	10.9	10.9

Table 2. Material parameters used in calculations. (*Maialle & Degani, 2001)

For two nanostructures, GQW and sPQW, one of the GaAs barriers (the cap layer) has an unchanged height, V_e (electron) or V_h (hole), and the other barrier is variable, $V_r = \sigma V_{e/h}$. The following values for the asymmetry parameter have been used: σ = 0.6, 0.8, and 1. The SQW structure is a symmetrical one (σ = 1). The laser parameter was chosen in the range of ILF, i. e. $\alpha_0 \in [0, 120$ Å$]$.

3.2.1 Single-particle energies

The strong laser effect on the potential profile and on energy levels of the conduction band (CB) for SQW, GQW and sPQW with L_w = 150 Å was discussed in the Section 2.2.1. For narrower n-sQWs with L_w = 50 Å or 80 Å the main laser-induced properties of the CB are very similar (Eseanu, 2011; Niculescu & Eseanu, 2011). In contrast, the laser parameter augmentation has little effect on the heavy-hole potential (not plotted here) because $\alpha_0 \propto 1/m^*$ (see Eq. (20) and Table 2).

In Figs. 11A and 11B the single-particle electron and hole energies, respectively, are plotted as functions of the laser intensity, α_0, in three differently shaped n-sQWs with L_c = 40 Å, for several values of the asymmetry parameter: σ = 0.6, 0.8, and 1.

We can see that the ground state energy in the conduction band first rapidly raises with the laser parameter up to $\alpha_0 \cong 40$ Å, i. e. $\alpha_0 = L_w/2$, and for further large laser intensities the E_e-level becomes less sensitive to the laser field (Fig. 11A). This is a typical behavior for low-dimensional systems under ILFs, as mentioned in Section 2.2.1. Instead, the valence band states (Fig. 11B) are less affected by the radiation field because of the weaker change of the laser-dressed confinement potential, in comparison with the electronic potential.

The values of the single-particle energies are greater in sPQW than in GQW due to the stronger geometric confinement. Also, the increasing of the barrier asymmetry, σ, generates an enhancement of the energy levels of both electrons and holes in GQW and sPQW (Figs.

11 A, B), as expected. For example, the electronic energies increase almost linearly with σ, as the inset of Fig. 11A shows.

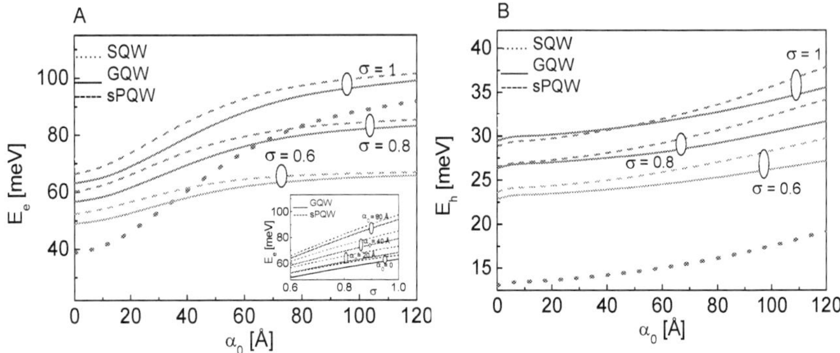

Fig. 11. The single-particle energy of the electron (A) and hole (B) as functions of the laser intensity, α_0 , in three differently shaped n-sQWs with L_c = 40 Å, for various asymmetry parameter values: σ = 0.6 (orange), 0.8 (magenta), and 1 (dark cyan). Inset A: the electronic energy vs. the asymmetry parameter for GQW and sPQW under different laser intensities.

The single-particle electron and hole energies are significantly modified by the capped layer thickness as Figs. 12 A, B prove for two types of n-sQWs with L_w = 50 Å and identical barriers, under different laser intensities.

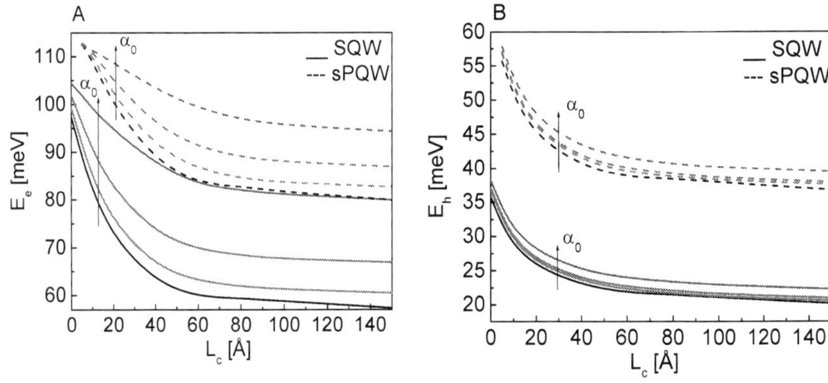

Fig. 12. The single-particle energy for electron (A) and hole (B) vs. capped layer thickness under different laser field intensities: α_0 = 0 (black), 10 Å (blue), 20 Å (olive), and 40 Å (red) for two types of n-sQWs with L_w = 50 Å and identical barriers.

It can be seen that the ground state energy in the conduction band first rapidly decreases with cap layer up to $L_c \cong 50$ Å, and for further thick cap layers it becomes insensitive to the dielectric enhancement. For thin cap layers, the stronger geometric confinement of the electron in the sPQW comparing with SQW leads to an enhanced shift of the E_e-level toward higher energies. Also, the electron energy presents a significant laser-induced augmentation which is more pronounced for sPQWs with thicker cap layer. There are two reasons for this behavior:

i. as the cap layer thickness increases the contribution of the image-charge is weakened and the ILF effect becomes dominant;
ii. under laser field action, the potential profile of SQW changes faster than in sPQW, so that the ground state of square structure are faster moved up in the upper part of the dressed QW.

In contrast, the valence band states are less affected by the radiation field. With the decrease of L_c the contribution of dielectric confinement to the hole energy becomes dominant leading to a pronounced blueshift. This result are in good agreement with the previous theoretical and experimental data in the absence of the laser field (Gippius, 1998; Tran Thoai, 1990).

3.2.2 Exciton binding energy

A noticeable dependence of the electron-hole interaction on the laser parameter, α_0, is expected as a consequence of the obvious laser-induced modification in the electronic structure of the CB. Fig. 13 displays the 1S-exciton binding energy as a function of the laser parameter for SQW, GQW and sPQW with L_c = 40 Å and different barrier asymmetries. We observe that for all the heterostructures under our study the exciton binding energy, E_b, first increases with α_0, reaches a maximum value at $\alpha_0 \cong 6$ Å, and then it experiences a pronounced diminution. Finally, for large enough laser amplitudes E_b drops to the $In_{0.18}Ga_{0.82}As$ bulk characteristic value, 4.45 meV (Chang & Peeters, 2000). In the range of our calculations this value is reached for the sPQW with σ = 0.6 under intense laser field, α_0 = 120 Å.

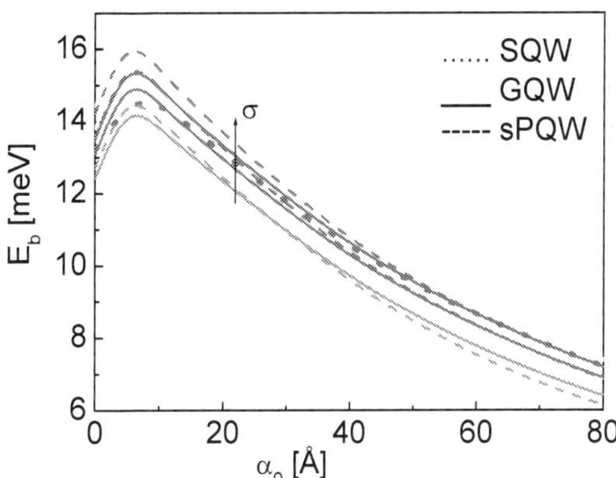

Fig. 13. The exciton binding energy vs. the laser-field intensity in three differently shaped n-sQWs: SQW (doted line), GQW (solid lines) and sPQW (dashed lines) with L_c = 40 Å and various asymmetry parameter values (the same as in Fig. 11).

The explanation of this behavior may be the following: under a low laser field ($\alpha_0 < 10$ Å) the electronic ground level is localized in the lower part of the dressed QW which has a narrow width; therefore, the small distance between the electron and hole leads to a considerable exciton binding energy. By increasing the laser field intensity the geometric confinement of the carriers is reduced and thus, they can easily penetrate into the QW barriers. This "travel" results in a pronounced diminution of E_b .

As the barrier asymmetry parameter, σ, increases the exciton binding energy is enhanced due to the higher right barrier ($V_r = \sigma V_{e/h}$) which moderates the spreading of the carriers wavefunctions in the QW barriers. Therefore, under intense laser field a competition between the laser intensity and barrier asymmetry in "tailoring" the exciton binding energy occurs.

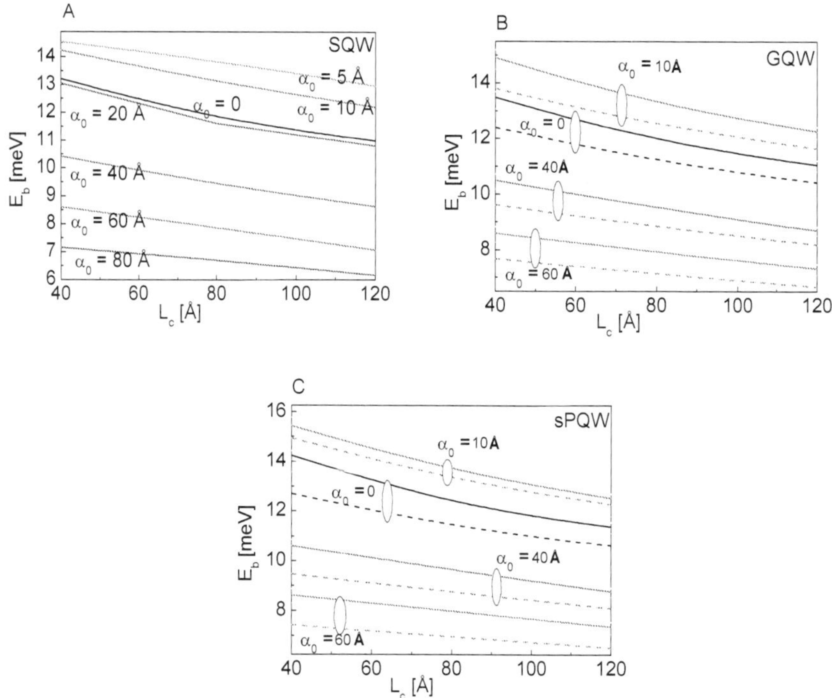

Fig. 14. The exciton binding energy vs. the cap layer thickness in three differently shaped n-sQWs with L_c = 40 Å, for two asymmetry parameter values: σ = 0.6 (dashed lines) and 1 (solid lines) under different laser field intensities α_0 : 0 (black), 5 Å (light magenta), 10 Å (magenta), 20 Å (blue), 40 Å (olive), 60 Å (red) and 80 Å (purple); A) SQW, B) GQW and C) sPQW.

In Figs. 14 A-C the exciton binding energy is plotted as a function of the cap layer thickness, L_c , in three differently shaped n-sQWs under various laser field intensities. For the asymmetrical structures GQW and sPQW two values of the asymmetry parameter, σ = 0.6 and 1, were used.

We observe that for $L_c \geq 40$ Å and for all laser parameter values used in this work, the exciton binding energy in the three differently shaped n-sQWs is reduced as L_c increases. A similar behavior has been reported for a square n-sQW with L_w = 50 Å in two cases: i) under various laser intensities (Niculescu & Eseanu, 2010a) and ii) in the absence of the laser field (Kulik et al., 1996). This dielectric enhancement of E_b obtained by using narrow capped layers, which are experimentally available, could be exploited in new optoelectronic devices.

The exciton binding energy properties (presented above) reveal the transition from two-dimensional confinement effect produced by the wave function squeezing in a symmetrical/asymmetrical InGaAs quantum well to the three-dimensional behavior of the exciton, as α_0 increases. A similar variation of the ground state binding energy has been reported for excitons in quantum well wires (Sari et al., 2003) and in n-sQWs with symmetrical barriers (Niculescu & Eseanu, 2011), under intense laser field.

3.3 Exciton absorption spectra

The exciton absorption coefficient given by the Eq. (27) depends on the exciton transition energy, E_{ex}, whose components have been discussed in Section 3.2 and on the overlap integral $\left|\left\langle \tilde{\varphi}_e(z_e) \middle| \tilde{\varphi}_h(z_h) \right\rangle\right|^2$. Figs. 15 A, B display the simultaneous dependence of this overlap on the cap layer thickness and laser field parameter in two n-sQWs (SQW and sPQW) with L_w = 50 Å and identical barriers. As expected, this quantity decreases with α_0, especially in SQW, due to laser-induced spreading of the carriers wave functions in the dressed n-sQW barriers.

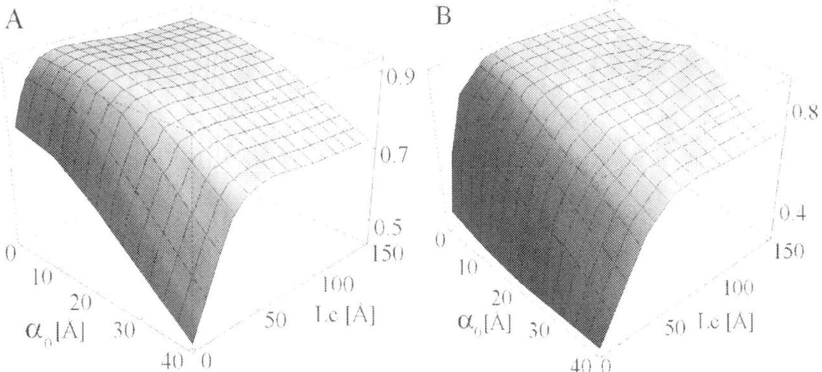

Fig. 15. The dependence of the overlap integral on the cap layer thickness, L_c , and laser field intensity, α_0 in two n-sQWs: A) SQW and B) sPQW with L_w = 50 Å and identical barriers.

We can also observe that, if α_0 is kept unchanged, there is an optimal range of the cap layer thickness corresponding to a broad maximum of the overlap integral. As α_0 increases this optimal range is shifted to larger L_c values in SQW, whereas it remains practically unchanged in the stronger confining structure sPQW. Moreover, since in the sPQW the particles are mostly localized in the QW region, the average electron-hole distance is less sensitive to the laser amplitude leading to an overlap integral nearly independent on the α_0 values.

Fig. 16 displays the overlap dependence on the barrier asymmetry parameter and laser field intensity in an asymmetrical n-sQW, namely GQW with L_c = 80 Å. The intense laser field reduces the overlap, as discussed before. Instead, the overlap integral is enhanced by the increasing barrier asymmetry (σ), particularly in low laser fields ($\alpha_0 \leq$ 40 Å). A similar variation was obtained for sPQW having the same L_c = 80 Å (not plotted here).

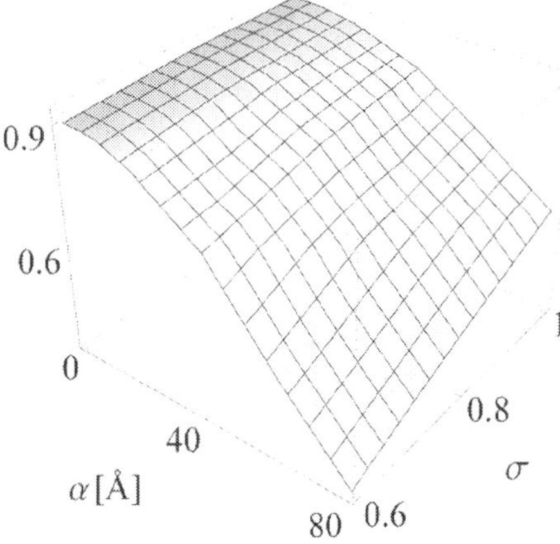

Fig. 16. The dependence of the overlap integral on the barrier asymmetry parameter, σ, and laser field intensity, α_0 in an asymmetrical n-sQW, namely GQW with L_c = 80 Å.

This asymmetry effect can be explained by the spreading effect of the wave functions in a QW with reduced barriers. The increasing of the height barrier in the QW right side ($V_r = \sigma V_{e/h}$) in joint action with a low laser field lead to a stronger localization of the wave functions inside the well. Again, the intense laser field action competes with the asymmetry parameter, in the present case on the overlap integral. This effect could be used for particular applications.

The exciton absorption coefficient (EAC) is modified by several parameters: pump photon energy, capped layer thickness L_c , laser intensity α_0 , and barrier asymmetry σ. In Figs. 17 A, B the dependence of this quantity vs. the pump photon energy in differently shaped n-sQWs with L_w = 80 Å, under various laser intensities α_0 = 0, 20 Å , 40 Å, and 60 Å, is plotted as follows: a) for SQW with identical barrier heigths and different L_c (Fig. 17A); b) for GQW and sPQW having two asymmetry parameter values (Fig. 17 B).

Firstly, one can see that the increasing laser intensity generates a significant shift of the absorption peak position to the high photon energies (blue-shift) and a simultaneous reduction of the exciton absorption coefficient magnitude, for all the cases under our study.

A similar effect was observed for square and semiparabolic n-sQWs of the same width L_w = 50 Å, having identical barrier heights and variable cap layer thickness (Niculescu & Eseanu, 2011). As expected, the shifts of the EAC peaks are more pronounced in the SQW (Fig. 17A) due to the weaker quantum confinement afforded by this structure in comparison with GQW and sPQW.

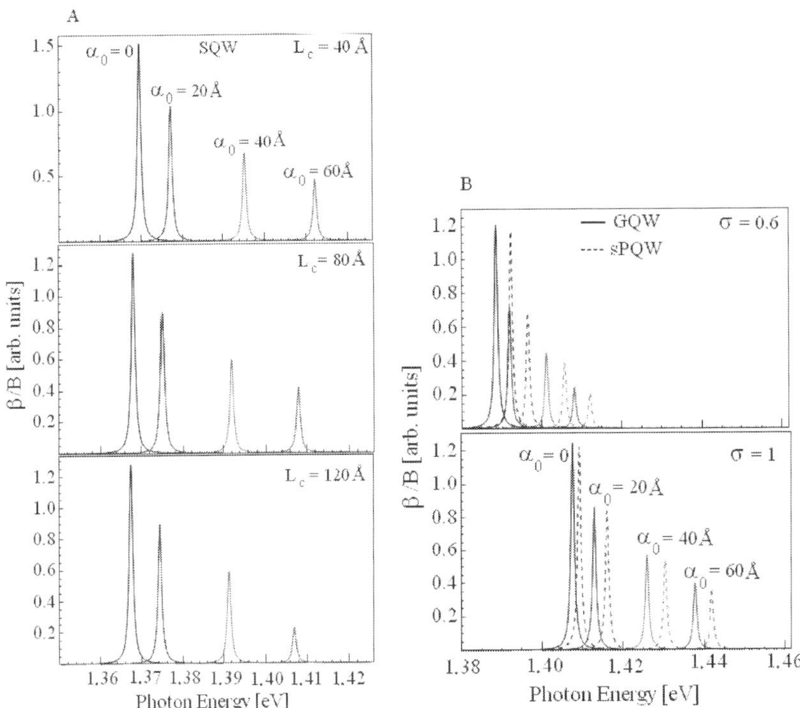

Fig. 17. The exciton absorption coefficient as a function of the pump photon energy in: A) SQW for three values of the cap layer thickness; B) GQW (solid lines) and sPQW (dashed lines) for two barrier asymmetries, under various laser intensities α_0 = 0 (black), 20 Å (blue), 40 Å (olive), 60 Å (red).

The reason for the EAC laser-induced reduction is two-fold:
i.   the overlap integral diminution under intense laser fields (Figs. 15, 16);
ii.   the typical laser-induced blue-shift of the E_e-levels.
Secondly, we observe that the EAC peak positions show a red-shift for thicker cap layers (in the range 40 Å - 80 Å) whereas for $L_c > 80$ Å these positions are insensitive to the cap layer variation.

Thirdly, we note that in the asymmetrical n-sQWs (GQW and sPQW) the blue-shift of the exciton absorption peak positions for σ = 1 (identical barriers) is more pronounced comparing with the case σ = 0.6.

For the $In_{0.18}Ga_{0.82}As/GaAs$ near-surface SQW with L_w = 50 Å and variable L_c, in the absence of the laser field, our results (Niculescu & Eseanu, 2010a) agree with the photoluminescence excitation measurements (Gippius et al., 1998; Kulik et al., 1996; Li, Z. et al., 2010; Yablonskii et al., 1996) and with theoretical normal-incidence reflectivity spectra (Yu et al., 2004).

It is important to emphasize that the red-shift induced by the increasing L_c or by the decreasing σ can be effectively compensated using the blue-shift caused by the enhanced laser parameter. The changes induced by the joint action of the laser intensity and capped layer thickness or barrier asymmetry parameter in differently shaped QWs can be exploited to design new optoelectronic devices.

4. Conclusion

Summarizing, we note that the optical properties of the $In_{0.18}Ga_{0.82}As/GaAs$ differently shaped near-surface quantum wells under intense laser fields could be tuned by proper tailoring of the heterostructure parameters (well shape and width, cap layer thickness, barrier asymmetry) and/or by varying the laser field intensity. This characteristics holds for both intersubband and interband absorptions. For example, in the exciton spectra of narrow $In_{0.18}Ga_{0.82}As/GaAs$ n-sQWs, our calculations show that the red-shift of the absorption peaks induced by the rising cap layer thickness or by barrier asymmetry diminution can be effectively compensated using the blue-shift caused by enhancing laser parameter. These phenomena could be exploited for particular applications. Also, for both intersubband and interband transitions, the switch between the strong absorption and induced laser transparency regimes can be suitable for good performance optical modulators.

To the best of our knowledge this is the first investigation on the intersubband absorption in symmetrical/asymmetrical InGaAs/GaAs n-sQWs under intense laser fields. We expect that these results will be of help for experimental works focused on innovative optoelectronic devices.

5. Acknowledgement

The author would like to express her special thanks to Professor E. C. Niculescu for helpful discussions and constant support.

6. References

Ahn, D. & Chuang, S.L. (1987). Intersubband optical absorption in a quantum well with an applied electric field. *Phys. Rev. B*, vol. 35, No. 8, (1987), 4149-4151, ISSN 1098-0121.

Alves, F.D.P.; Karunasiri, G.; Hanson, N.; Byloos, M.; Liu, H.C.; Bezinger, A. & Buchanan, M. (2007). NIR, MWIR and LWIR quantum well infrared photodetector using interband and intersubband transitions. *Infrared Phys. & Technol.* vol. 50, No. 2-3, (April 2007), 182-186, ISSN 1350-4495.

Ando, Y. & Ytoh, T. (1987). Calculation of transmission tunneling current across arbitrary potential barriers . *J. Appl. Phys.* vol. 61, No. 4, (1987),1497-1502, ISSN 1089-7550.

Andreani, L.C. & Pasquarello, A. (1990). Accurate theory of excitons in $GaAs$-$Ga_{1-x}Al_xAs$ quantum wells. *Phys. Rev. B*, vol. 42, No. 14, (1990), 8928-8938, ISSN 1098-0121.

Asano, T.; Noda, S. & Sasaki, A. (1998). Absorption magnitude and phase relaxation time in short wavelength intersubband transitions in InGaAs/AlAs quantum wells on

GaAs substrates. *Physica E: Low-dimensional Systems and Nanostructures*, vol. 2, No. 1-4, (July 1998), 111-115, ISSN 1386-9477.

Bassani, F. & Parravicini, G. P. (1975). *Electronic States and Optical Transitions in Solids*; Pergamon: Oxford, 1975, Ch.5, ISBN 0080168469.

Bastard, G. (1981). Hydrogenic impurity states in a quantum well: A simple model. *Phys. Rev. B*, vol. 24, No. (1981), 4714-4722, , ISSN 1098-0121.

Bedoya, M. & Camacho, A.S. (2005). Nonlinear intersubband terahertz absorption in asymmetric quantum well structures. *Phys. Rev. B*, vol. 72, No. 15, (2005), 155318, ISSN 1098-0121.

Belkin, M.A.; Capasso, F.; Xie, F.; A. Balyanin, A.; Fischer, M.; Whitmann, A. & Faist, J. (2008). Room temperature terahertz quantum cascade laser source based on intracavity difference-frequency generation. *Appl. Phys. Lett.*, vol. 92, No. 20, (2008), 201101, ISSN 1077-3118.

Brandi, H. S.; Latge, A. & Oliveira, L. E. (2001) Dressed-band approach to laser-field effects in semiconductors and quantum-confined heterostructures. *Phys. Rev. B*, vol. 64, No. (2001), 035323... ISSN 1098-0121.

Carter, S.G.; Ciulin, V.; Sherwin, M.S.; Hanson, M.; Huntington, A.; Coldren, L.A. & Gossard, A.C. (2004). Terahertz electro-optic wavelength conversion in GaAs quantum wells: Improved efficiency and room-temperature operation. *Appl. Phys. Lett.* vol. 84, No. 6, (2004), 840-842, ISSN 1077-3118.

Chakraborty, T. & Apalkov, V. M. (2003). Quantum cascade transitions in nanostructures. *Adv. Phys.*, vol. 52 No. 5, (2003), 455-521, ISSN 1460-6976.

Chang, K. & Peeters, F. M. (2000) Asymmetric Stark shifts in $In_{0.18}Ga_{0.82}As/GaAs$ near-surface quantum wells: The image-charge effect. *J. Appl. Phys.* vol. 88, No. 9, (2000), 5246-5251, ISSN 1089-7550.

Chui, H.C.; Martinet, E.L.; Fejer, M.M. & Harris, J.S. (1994). Short wavelength intersubband transitions in InGaAs/AIGaAs quantum wells grown on GaAs. *Appl. Phys. Lett.*, vol. 64, No. 6, (Febr. 1994), 736-, ISSN 1077-3118.

Diniz Neto, O. O. & Qu, F. (2004). Effects of an intense laser field radiation on the optical properties of semiconductor quantum wells. *Superlatt. Microstruct.*, vol. 35, No. 1-2, (Jan-Febr. 2004), 1-8, ISSN 0749-6063.

Duque, C.A.; Kasapoglu, E.; Sakiroglu, S.; Sari H. & Sökmen, I. (2011). Intense laser effects on nonlinear optical absorption and optical rectification in single quantum wells under applied electric and magnetic field. *Appl. Surf. Sci.*, vol. 157, No. 6, (Jan. 2011), 2313-2319, ISSN 0169-4332.

Elsaesser, T. (2006). Ultrafast Dynamics of Intersubband Excitations in Quantum Wells and Quantum Cascade Structures, in Paiella, R. *Intersubband Transitions in Quantum Structures*, McGraw-Hill Companies, ISBN 0-07-145792-5, pp. 137-140.

Eseanu, N.; Niculescu, E. C. & Burileanu, L. M. (2009). Simultaneous effects of pressure and laser field on donors in $GaAs/Ga^{1-x}Al^xAs$ quantum wells. *Physica E: Low-dimens. Syst. Nanostruct.*, vol. 41, No. 8, (Aug. 2009), 1386-1392, ISSN:1386-9477.

Eseanu, N. (2010). Simultaneous effects of laser field and hydrostatic pressure on the intersubband transitions in square and parabolic quantum wells. *Phys. Lett. A*, vol. 374, No. 10, (Febr. 2010), 1278- 1285, ISSN 0375-9601.

Eseanu, N. (2011). Intense laser field effect on the interband absorption in differently shaped near-surface quantum wells, *Phys. Lett. A*, vol. 375, No. 6 , (Febr. 2011), 1036-1042, ISSN 0375-9601.

Faist, J.; Capasso, F.; Sivco, D.L.; Sirtori, C.; Hutchinson, A.L. & Cho, A.Y. (1994). Quantum Cascade Laser. *Science*, vol. 264, No. 5158, (Apr. 1994), 553-556, ISSN 1095-9203.

Gavrila, M. & Kaminski, J. Z. (1984). Free-free transitions in intense high-frequency laser fields. *Phys. Rev. Lett.*, vol. 52, No. 8, (1984), 613-616, , ISSN 0031-9007.

Gippius, N. A.; Yablonskii, A. L.; Dzyubenko, A. B.; Tikhodeev, S. G.; Kulik, L. V.; Kulakovskii, V. D. & Forchel, A. (1998). Excitons in near-surface quantum wells in magnetic fields: Experiment and theory. *J. Appl. Phys.* vol. 83, No. 10, (1998), 5410-5417, ISSN 1089-7550.

Gunapala, S.D.; Bandara, K.S.M.V.; Levine, B.F.; Sarusi, G.; Park, J.S.; Lin,T.L.; Pike, W.T. & Liu, J.K. (1994). High performance InGaAs/GaAs quantum well infrared photodetectors. *Appl. Phys. Lett.* vol. 64, No. 25, (1994), 3431-3433, ISSN 1077-3118.

Helm, M. (2000). The Basic Physics of Intersubband Transitions, in *Intersubband Transitions in Quantum Wells:*

Physics and Device Applications I, Capasso, F. and Liu, H. C. (eds.), *Semiconductors and Semimetals,* vol. 62, 1-32 and 73-80, Academic Press, ISBN 0-12-752171-2, San Diego, USA.

Iizuka, N.; Kaneko, K. & Suzuki, N. (2006). All-optical switch utilizing intersubband transition in GaN quantum wells. *IEEE J. Quantum Electr.,* vol. 42, No. 8, (Aug. 2006), 765-771, ISSN 0018-9197.

Jho, Y. D.; Wang, X.; Reitze, D. H.; Kono, J.; Belyanin, A. A.; Kocharovsky, V. V.; Kocharovsky, Vl. V. & Solomon, G. S. (2010). Cooperative recombination of electron-hole pairs in semiconductor quantum wells under quantizing magnetic fields. *Phys. Rev. B,* vol. 81, No. 15 (2010), 155314, ISSN 1098-0121.

Karabulut, I.; Atav, U.; Safak, H. & Tomak, M. (2007). Linear and nonlinear intersubband optical absorptions in an asymmetric rectangular quantum well. *Eur. Phys. J. B,* vol. 55, No. 3, (Febr. I, 2007), 283-288 ISSN1434-6036.

Kasapoglu E. & Sökmen, I. (2008). The effects of intense laser field and electric field on intersubband absorption in a double-graded quantum well. *Physica B: Condensed Matter,* vol. 403, No. 19-20, (Oct. 2008), 3746-3750, ISSN 0921-4526.

Kramers, H. (1956). *Collected Scientific Papers,* North-Holland, Amsterdam, 1956, p. 866.

Kulik, L. V.; Kulakovskii, V. D.; Bayer, M.; Forchel, A.; Gippius, N. A. & Tikhodeev, S. G. (1996). Dielectric enhancement of excitons in near-surface quantum wells. *Phys. Rev. B* vol. 54, No. 4, (1996) R2335- R2338, ISSN 1098-0121.

Levine, B. F. (1993). Quantum-well infrared photodetectors. *J. Appl. Phys.,* vol. 74, No. 8, (1993), R1-R81, ISSN 1089-7550.

Li, J.; Choi, K.K.; Klem, J.F.; Reno, J.L. & Tsui, D.C. (2006). High gain, broadband InGaAs/InGaAsP quantum well infrared photodetector. *Appl. Phys. Lett.* vol. 89, No. 8, (2006), 081128, ISSN 1077-3118.

Li, S.S. (2002). Multi-color, broadband quantum well infrared photodetectors for mid-, long-, and very long wavelength infrared applications. *International Journal of High Speed Electronics and Systems,* vol. 12, No. 3, (2002), 761-801, ISSN 0129-1564.

Li, Y.; Bhattacharya, A.; Thomodis, C.; Moustakas, T.D. & Paiella, R. (2007). Ultrafast all-optical switching with low saturation energy via intersubband transitions in GaN/AlN quantum-well waveguides. *Opt. Express,* vol.15, No. 26, (2007), 17922-17927, ISSN 1094-4087.

Li, Z. ; Wu, J.; Wang, Z. M.; Fan, D.; Guo A.; Li, S.; Yu, S. Q.; Manasreh O.; & Salamo, G. J. (2010). InGaAs Quantum Well Grown on High-Index Surfaces for Superluminescent Diode Applications, *Nanoscale Res. Lett.* vol. 5, No. 6, (2010),1079-1084, ISSN 1556-276X.

Lima, C. A. S. & Miranda, L. C. M. (1981). Atoms in superintense laser fields. *Phys. Rev. A,* vol. 23, No. 6, (1981), 3335-3337, ISSN 1050-2947.

Lima, F. M. S.; Amato, M. A.; Nunes, O. A. C.; Fonseca, A. L. A.; Enders, B. G. & Da Silva Jr., E. F. (2009). Unexpected transition from single to double quantum well potential induced by intense laser fields in a semiconductor quantum well. *J. Appl. Phys.*, vol. 105, No. (2009), 123111, ISSN 1089-7550.

Liu, H.C. (2000). Quantum well infrared photodetector physics and novel devices, in *Intersubband Transitions in Quantum Wells: Physics and Device Applications I*, Capasso, F. and Liu, H. C. (eds.), *Semiconductors and Semimetals*, vol. 62, 129-196, Academic Press, ISBN 0-12-752171-2, San Diego, USA

Maialle, M. Z. & Degani, M. H. (2001). Transverse magnetic field effects upon the exciton exchange interaction in quantum wells. *Semicond. Sci. Technol.*, vol. 16, No. (2001), 982-985, ISSN .

Marinescu, M. & Gavrila, M. (1995). First iteration within the high-frequency Floquet theory of laser-atom interactions. *Phys. Rev. A*, vol. 53, No. 4, (1995), 2513-2521, ISSN 1094-1622.

Miller, D.A.B.; Chemla, D.S. & Schmitt-Rink, S. (1986). Relation between electroabsorption in bulk semiconductors and in quantum wells: The quantum-confined Franz-Keldysh effect. *Phys. Rev. B*, vol. 33, No. 10, (1986), 6976-6982 , ISSN 1098-0121.

Miller, R. C.; Kleinman, D. A.; Tsang, W. T. & Gossard, A. C. (1981). Observation of the excited level of the excitons in GaAs quantum wells. *Phys. Rev. B*, vol. 24, No. 2, (1981), 1134-1136, ISSN 1098-0121.

Niculescu, E. C. & Eseanu N. (2010a) Dielectric enhancement of the exciton energies in laser dressed near-surface quantum wells. *Superlatt. Microstruct.*, vol. 48, No. 4, (Oct. 2010), 416-425, ISSN 0749-6063.

Niculescu, E. C. & Burileanu, L.M. (2010b). Nonlinear optical absorption in inverse V-shaped quantum wells modulated by high-frequency laser field. *Eur. Phys. J. B*, vol. 74, No. 1, (March I, 2010), 117-122, ISSN 1434-6036.

Niculescu, E. C. & Radu, A. (2010c). Laser-induced diamagnetic anisotropy of coaxial nanowires. *Curr. Appl. Phys.*, vol. 10, No. 5, (2010), 1354-1359, ISSN 1567-1739.

Niculescu, E. C. & Eseanu N. (2011). Interband absorption in square and semiparabolic near-surface quantum wells under intense laser field. *Eur. Phys. J. B*, vol. 79, No. 3, (Febr. I, 2011), 313-319, ISSN 1434-6036.

Noda, S.; Uemura, T.; Yamashita, T. & Sasaki, A. (1990). All-optical modulation using an *n*-doped quantum-well structure. *J. Appl. Phys.* vol. 68, No. 12, (1990), 6529-6531, ISSN 1089-7550.

Ozturk, E.; Sari H. & Sokmen I. (2004). The dependence of the intersubband transitions in square and graded QWs on intense laser fields. *Solid State Comm.*, vol. 132, No. 7, (Nov. 2004), 497-502, ISSN: 0038-1098.

Ozturk, E.; Sari H. & Sokmen I. (2005). Electric field and intense laser field effects on the intersubband optical absorption in a graded quantum well. *J. Phys. D: Appl. Phys.*, vol. 38, No. 6, (2005), 935-941, ISSN 1361-6463.

Ozturk, E. (2010). Nonlinear optical absorption in graded quantum wells modulated by electric field and intense laser field. *Eur. Phys. J. B*, vol. 75, No. 2, (2010), 197-203, ISSN 1434-6036.

Passmore, B. S.; Wu, J.; Manasreh, M. O. & Salamo, G. J. (2007). Dual broadband photodetector based on interband and intersubband transitions in InAs quantum dots embedded in graded InGaAs quantum wells. *Appl. Phys. Lett.*, vol. 91, No. 23, (2007), 233508, ISSN 1077-3118.

Rosencher, E. & Bois, Ph. (1991). Model system for optical nonlinearities: Asymmetric quantum wells. *Phys. Rev. B*, vol. 44, No. 20, (1991), 11315-11327, ISSN 1098-0121.

Sari, H.; Kasapoglu, E.; Sokmen, I. & Gunes, M. (2003). Optical transitions in quantum well wires under intense laser radiation. *Phys. Lett. A*, vol. 319, No. 1-2, (Dec. 2003), 211-216, ISSN 0375-9601.

Schneider, H. & Liu, H.C. (2007). *QuantumWell Infrared Photodetectors. Physics and Applications*, Springer Berlin Heidelberg New York, ISBN-10 3-540-36323-8.

Schöwalter, M.; Rosenauer, A. & Gerthsen, D. (2006). Influence of surface segregation on the optical properties of semiconductor quantum wells. *Appl. Phys. Lett.*, vol. 88, No. 11, (2006), 111906, ISSN 1077-3118.

Stöhr, A.; Humbach, O.; Zumkley, S.; Wingen, G.; David, G.; Jäger, D.; Bollig, B.; Larkins E. C. & Ralston, J. D. (1993). InGaAs/GaAs multiple-quantum-well modulators and switches. *Opt. Quant. Electronics*, vol. 25, No. 12, (Dec. 1993), S865-S883, ISSN 1572-817X.

Tran Thoai, D. B.; Zimmermann, R.; Grundmann, M. & Bimberg, D. (1990). Image charges in semiconductor quantum wells: Effect on exciton binding energy. *Phys. Rev. B*, vol. 42, No. 9, (1990) 5906-5909, ISSN 1098-0121.

Turner, D.B.; Stone, K.W.; Gundogdu, K. & Nelson, K.A. (2009). Three-dimensional electronic spectroscopy of excitons in GaAs quantum wells. *J. Chem. Phys.*, vo.l. 131, (2009), 144510, ISSN 0021-9606.

Tsu, R. & Esaki, L. (1973). Tunneling in a finit superlattice. *Appl. Phys. Lett.*, vol. 22, No. 11 (1973), 562-564, ISSN 1077-3118.

Unlu, S.; Karabulut, I. & Safak H. (2006). Linear and nonlinear intersubband optical absorption coefficients and refractive index changes in a quantum box with finite confining potential. *Physica E: Low-dimensional Systems and Nanostructures*, vol. 33, No. 2, (2006), 319–324, ISSN 1386-9477.

West, L.C. & Eglash, S.J. (1985). First observation of an extremely large-dipole infrared transition within the conduction band of a GaAs quantum well. *Appl. Phys. Lett.*, vol. 46, No. 12, (1985), 1156-1158, ISSN 1077-3118.

Wu, S.; Huang, Z.; Liu, Y.; Huang, Q.; Guo, W. & Cao, Y. (2009). The effects of indium segregation on exciton binding energy and oscillator strength in GaInAs/GaAs quantum wells. *Superlatt. Microstruct.*, vol. 46, No. 4, (Oct. 2009), 618-626, ISSN 0749-6063.

Xie, W. (2010). Laser radiation effects on optical absorptions and refractive index in a quantum dot. *Opt. Commun.* vol. 283, No. 19 (Oct. 2010), 3703-3706, ISSN 0030-4018.

Yablonskii, A. L.; Dzyubenko, A. B.; Tikhodeev, S. G.; Kulik, L. V. & Kulakovskii, V. D. (1996). Dielectric enhancement of magnetoexcitons in surface quantum wells. *JETP Lett.*, vol. 64, No. 1, (July 1996), 51-56, ISSN 1090-6487.

Yang R. Q. (1995). Optical intersubband transitions in conduction-band quantum wells. *Phys. Rev. B* vol. 33, No. 16, (Oct. 1995-II), 11958-11968, ISSN 1098-0121.

Zhao, H.; Chen, Y.T.; Yum, J. H.; Wang, Y.; Zhou, F.; Xue, F; & Lee, J.C. (2010). Effects of barrier layers on device performance of high mobility $In_{0.7}Ga_{0.3}As$ metal-oxide-semiconductor field-effect-transistors. *Appl. Phys. Lett.* Vol. 96, No. 10, (2010), 102101, ISSN 1077-3118.

Zheng, R.S. & Matsuura, M. (1998). Exciton binding energies in polar quantum wells with finite potential barriers. *Phys. Rev. B*, vol. 58, No. 16, (1998), 10769-10777, ISSN 1098-0121.

Xie, W. (2010). Laser radiation effects on optical absorptions and refractive index in a quantum dot. *Opt. Commun.* vol. 283, No. 19, (Oct. 2010), 3703-3706, ISSN 0030-4018.

15

Using the Liquid Crystal Spatial Light Modulators for Control of Coherence and Polarization of Optical Beams

Andrey S. Ostrovsky, Carolina Rickenstorff-Parrao
and Miguel Á. Olvera-Santamaría
Universidad Autónoma de Puebla, Facultad de Ciencias Físico Matemáticas
Mexico

1. Introduction

Beginning from the mid-1980's the liquid crystal (LC) spatial light modulators (SLMs) have been used in many optical applications, such as optical data processing, beam shaping, optical communication, adaptive optics, real-time holography, etc. In these applications LC-SLMs are used for both amplitude and phase modulation of optical field. The modulation characteristics of LC- SLMs have been widely studied and reported (Lu & Saleh, 1990). A special attention has been given to the optimization of LC-SLM parameters in order to provide the amplitude-only and phase-only modes of modulation (Yamauchi & Eiju, 1995).

Recently, in connection with the heightened interest in the vector coherence theory of electromagnetic fields (Ostrovsky et al, 2009a), a new possible application of LC-SLMs has been found. It has been shown (Shirai & Wolf, 2004) that LC-SLM of certain configuration can realize the controlled changes of statistical properties of an electromagnetic beam, namely the degree of coherence and the degree of polarization. This fact can be successfully used for generating a secondary partially coherent and partially polarized optical source with the desired characteristics. Somewhat later the technique proposed by Sahirai and Wolf has been improved using the systems of two LC-SLMs coupled in series (Ostrovsky et al., 2009b) or in parallel (Shirai et al., 2005). Unfortunately, the mentioned techniques have not been yet realized in practice because of the lack of commercial LC-SLMs with proper characteristics. Here we propose an alternative technique that uses widely available commercial LC-SLMs and, hence, can be easily realized in practice.

In this chapter, we present a comparative analysis of the techniques for controlling coherence and polarization by LC SLMs. When doing this we consider both the known, properly referenced, techniques and the original ones proposed by the authors. We illustrate the efficiency of the proposed technique with the results of numerical simulation and physical experiments. To facilitate the perception of material, we anticipate the main matter by the backgrounds concerning the fundamentals of vector coherence theory and the elements of theory and design of LC-SLMs. We hope that this chapter will help the specialists and postgraduate students in optics and optoelectronics to be well guided in the subject dispersed in numerous publications.

2. Fundamentals of the unified theory of coherence and polarization

As well known (Wolf, 2007), the second-order statistical properties of a random planar (primary or secondary) electromagnetic source can be completely described by the so-called cross-spectral density matrix

$$\mathbf{W}(\mathbf{x}_1,\mathbf{x}_2) = \begin{pmatrix} \langle E_x^*(\mathbf{x}_1)E_x(\mathbf{x}_2)\rangle & \langle E_x^*(\mathbf{x}_1)E_y(\mathbf{x}_2)\rangle \\ \langle E_y^*(\mathbf{x}_1)E_x(\mathbf{x}_2)\rangle & \langle E_y^*(\mathbf{x}_1)E_y(\mathbf{x}_2)\rangle \end{pmatrix}, \tag{1}$$

where $E_x(\mathbf{x})$ and $E_y(\mathbf{x})$ are the orthogonal components of the electric field vector $\mathbf{E}(\mathbf{x})$, asterisk denotes the complex conjugate and the angle brackets denote the average over the statistical ensemble (here and further on, for the sake of simplicity, we omit the explicit dependence of the considered quantities on frequency ν). Using this matrix, the following three fundamental statistical characteristics of the source can be defined:
the power spectrum

$$S(\mathbf{x}) = \mathrm{Tr}\mathbf{W}(\mathbf{x},\mathbf{x}), \tag{2}$$

the spectral degree of coherence

$$\mu(\mathbf{x}_1,\mathbf{x}_2) = \frac{\mathrm{Tr}\mathbf{W}(\mathbf{x}_1,\mathbf{x}_2)}{[\mathrm{Tr}\mathbf{W}(\mathbf{x}_1,\mathbf{x}_1)\mathrm{Tr}\mathbf{W}(\mathbf{x}_2,\mathbf{x}_2)]^{1/2}}, \tag{3}$$

and the spectral degree of polarization

$$P(\mathbf{x}) = \left(1 - \frac{4\mathrm{Det}\mathbf{W}(\mathbf{x},\mathbf{x})}{[\mathrm{Tr}\mathbf{W}(\mathbf{x},\mathbf{x})]^2}\right)^{1/2}. \tag{4}$$

In Eqs. (2) - (4) Tr stands for the trace and Det denotes the determinant.
It may be shown that $0 \le |\mu(\mathbf{x}_1,\mathbf{x}_2)| \le 1$ and $0 \le P(\mathbf{x}) \le 1$. It is said that an electromagnetic source is completely coherent and completely polarized if $|\mu(\mathbf{x}_1,\mathbf{x}_2)| = 1$ and $P(\mathbf{x}) = 1$, whereas, when $|\mu(\mathbf{x}_1,\mathbf{x}_2)| = 0$ and $P(\mathbf{x}) = 0$, a source is referred as completely incoherent and completely unpolarized, respectively. In the case of intermediate values of μ and P a source is referred as partially coherent and partially polarized.
As a typical example of such a source, which will be used bellow, we mention the so-called Gaussian Schell-model source (Wolf, 2007) which, in particular case, is given by the diagonal cross-spectral density matrix

$$\mathbf{W}(\mathbf{x}_1,\mathbf{x}_2) = \begin{pmatrix} W_{xx}(\mathbf{x}_1,\mathbf{x}_2) & 0 \\ 0 & W_{yy}(\mathbf{x}_1,\mathbf{x}_2) \end{pmatrix} \tag{5}$$

with elements

$$W_{ii}(\mathbf{x}_1,\mathbf{x}_2) = S_{0i}\exp\left(-\frac{\mathbf{x}_1^2+\mathbf{x}_2^2}{4\sigma^2}\right)\exp\left(-\frac{(\mathbf{x}_1-\mathbf{x}_2)^2}{2\sigma_i^2}\right), \quad i=x,y, \tag{6}$$

where S_{0i}, σ_i and σ are positive constants.. It should be noted that the degree of coherence and the degree of polarization of this source take a rather simple form, i.e.

$$\mu(\mathbf{x}_1,\mathbf{x}_2) = \frac{1}{S_{0x} + S_{0y}} \left[S_{0x} \exp\left(-\frac{\xi^2}{2\sigma_x^2}\right) + S_{0y} \exp\left(-\frac{\xi^2}{2\sigma_y^2}\right) \right], \tag{7}$$

$$P(\mathbf{x}) = \frac{\left|S_{0x} - S_{0y}\right|}{S_{0x} + S_{0y}}, \tag{8}$$

where $\xi = \left|\mathbf{x}_1 - \mathbf{x}_2\right|$.

Now we consider the propagation of the electromagnetic beam characterized by the cross-spectral density matrix $\mathbf{W}(\mathbf{x}_1,\mathbf{x}_2)$ through a thin polarization-dependent screen whose amplitude transmittance is given by the so-called Jones matrix (Yariv & Yeh, 1984)

$$\mathbf{T}(\mathbf{x}) = \begin{pmatrix} t_{xx}(\mathbf{x}) & t_{xy}(\mathbf{x}) \\ t_{yx}(\mathbf{x}) & t_{yy}(\mathbf{x}) \end{pmatrix}, \tag{9}$$

with elements $t_{ij}(\mathbf{x})$ being the random (generally complex) functions of time. It can be readily shown (Shirai & Wolf, 2004) that the cross-spectral density matrix of the beam just behind the screen is given by the expression

$$\mathbf{W}'(\mathbf{x}_1,\mathbf{x}_2) = \left\langle \mathbf{T}^{\dagger}(\mathbf{x}_1)\mathbf{W}(\mathbf{x}_1,\mathbf{x}_2)\mathbf{T}(\mathbf{x}_2)\right\rangle, \tag{10}$$

where the dagger denotes the Hermitian conjugation and the angle brackets again denote the average over the statistical ensemble. Taking into account Eq. (10) with due regard for definitions given by Eqs. (3) and (4), it becomes obvious that, in general, $\mu'(\mathbf{x}_1,\mathbf{x}_2) \neq \mu(\mathbf{x}_1,\mathbf{x}_2)$ and $P'(\mathbf{x}) \neq P(\mathbf{x})$. This fact can be used to realize the modulation of coherence and polarization by means of a random polarization-dependent screen. To minimize the light loss, the elements of the matrix $\mathbf{T}(\mathbf{x})$ must be of the form

$$t_{ij}(\mathbf{x}) = \exp\left[i\varphi(\mathbf{x})\right], \tag{11}$$

where $\varphi(\mathbf{x})$ is some real random function. To provide the desired statistical characteristics of modulation, the function $\varphi(\mathbf{x})$ has to be generated by computer. The most appropriate candidate for physical realization of such a computer controled random phase screen is the LC-SLM.

3. Elements of the theory and design of LC-SLMs

The LC represents an optically transparent material that has physical properties of both solids and liquids. The molecules of such a material have an ellipsoidal form with a long axis about which there is circular symmetry in any transverse plane. The spatial organization of these molecules defines the type of LC (Goodman, 1996). From the practical point of view, the most interesting type is so-called nematic LC, for which the molecules have a parallel orientation with randomly located centres within entire volume of the material. Further we will consider the LCs exclusively of this type.

Because of its geometrical structure the nematic LC exhibits anisotropic optical behaviour, possessing different refractive indices for light polarized in different directions. From the optical point of view the nematic LC can be considered as an uniaxial crystal with ordinary refraction index n_o along the short molecular axis and extraordinary refraction index n_e along the long molecular axis, so that it can be characterized by the so-called birefringence parameter

$$\beta = \frac{\pi d}{\lambda}(n_e - n_o),\tag{12}$$

where λ is the wavelength of light and d is the thickness of LC layer.

When LC material is placed in a container with two glass walls it receives the name of LC cell. The glass walls of the LC cell are linearly polished to provide the selected directions in which the LC molecules are aligned at the boundary layers. If the glass walls are polished in different directions, then LC molecules inside the cell gradually rotate to match the boundary conditions at the alignment layers, as illustrated in Fig. 1. Such a LC cell received the name of twisted LC cell. The angle ϕ between the directions of polishing is refered as the twist angle.

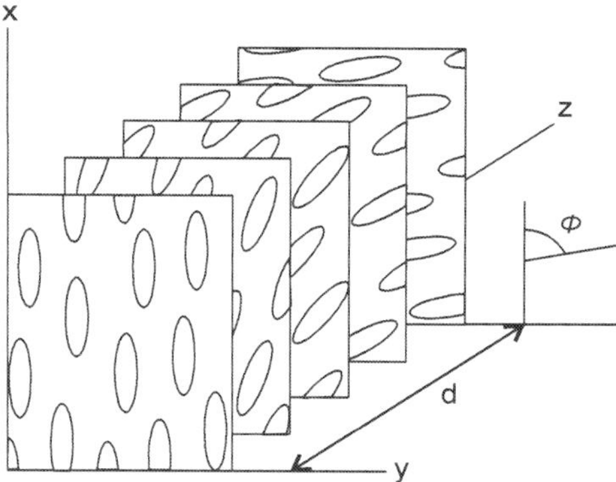

Fig. 1. Twist of LC molecules due to the boundary conditions at the alignment layers.

According to (Yariv & Yeh, 1984), the amplitude transmittance of twisted LC cell with front molecules aligned along x-axis is given by the Jones matrix

$$\mathbf{J}_{LC} = \mathbf{R}(-\phi)\exp(-i\beta)\begin{pmatrix} \cos\gamma - i\dfrac{\beta}{\gamma}\sin\gamma & \dfrac{\phi}{\gamma}\sin\gamma \\[2mm] -\dfrac{\phi}{\gamma}\sin\gamma & \cos\gamma + i\dfrac{\beta}{\gamma}\sin\gamma \end{pmatrix},\tag{13}$$

where

$$\mathbf{R}(\phi) = \begin{pmatrix} \cos(\phi) & \sin(\phi) \\ -\sin(\phi) & \cos(\phi) \end{pmatrix},\tag{14}$$

is the coordinate rotation matrix and parameter γ is defined as

$$\gamma = \sqrt{\beta^2 + \phi^2}\;.\tag{15}$$

Bellow we consider two important particular cases, namely when $\phi = 0°$ and $\phi = 90°$. In the first case we will reffer to the LC cell as 0°-twist LC cell and in the second case we will refer to it as 90°-twist LC cell.
For 0°-twist LC cell Eq. (13) takes the form

$$\mathbf{J}_{LC} = \begin{pmatrix} \exp(-i2\beta) & 0 \\ 0 & 1 \end{pmatrix}. \tag{16}$$

As can be seen from Eq. (16), the element j_{xx} of matrix $\mathbf{J}_{LC}$ describes the phase-only modulation, and hence, under certain conditions, the 0°-twist LC cell can be used to provide the modulation of coherence and polarization discussed in the previous section.
For 90°-twist LC cell Eq. (13) takes the form

$$\mathbf{J}_{LC} = \exp(-i\beta) \begin{pmatrix} \dfrac{\pi}{2\gamma}\sin\gamma & -\cos\gamma - i\dfrac{\beta}{\gamma}\sin\gamma \\ \cos\gamma - i\dfrac{\beta}{\gamma}\sin\gamma & \dfrac{\pi}{2\gamma}\sin\gamma \end{pmatrix}, \tag{17}$$

with

$$\gamma = \sqrt{\beta^2 + (\pi/2)^2} . \tag{18}$$

As can be seen from Eq. (17), the general element j_{ij} of matrix $\mathbf{J}_{LC}$ this time describes the joint amplitude and phase modulation, and hence, the 90°-twist LC cell can not be directly used to realize the modulation of coherence and polarization. Nevertheless, placing the 90°-twist LC cell between a pair of polarizers whose main axes make angles ψ_1 and ψ_2 with x direction, as is shown in Fig. 2, it is possible to achieve the phase-only modulation (Lu & Saleh, 1990).

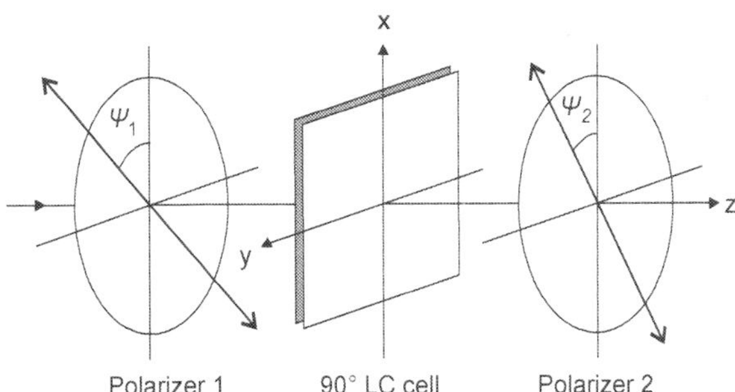

Fig. 2. 90°-twist LC cell sandwiched between two polarizers.

The Jones matrix for the system shown in Fig. 2 is given by

$$\mathbf{T} = \mathbf{J}_P(\psi_2)\mathbf{J}_{LC}\mathbf{J}_P(\psi_1) , \tag{19}$$

where $\mathbf{J}_{LC}$ is the matrix given by Eq. (17) and

$$\mathbf{J}_P(\psi) = \begin{pmatrix} \cos^2 \psi & \cos\psi \sin\psi \\ \cos\psi \sin\psi & \sin^2 \psi \end{pmatrix} \tag{20}$$

is the Jones matrix of polarizer. On substituting from Eqs. (17) and (20) into Eq. (19) with values $\psi_1 = 0°$ and $\psi_2 = 90°$, we obtain

$$\mathbf{T} = \exp(-i\beta) \begin{pmatrix} 0 & 0 \\ \cos\gamma - i\dfrac{\beta}{\gamma}\sin\gamma & 0 \end{pmatrix} . \tag{21}$$

The matrix given by Eq. (21) contains the only non-zero element

$$t_{yx} = \exp(-i\beta)\left[\cos\gamma - i\frac{\beta}{\gamma}\sin\gamma \right], \tag{22}$$

which can be expressed in the complex exponential form

$$t_{yx} = \left| t_{yx} \right| \exp\left[-i \arg(t_{yx}) \right], \tag{23}$$

with modulus

$$\left| t_{yx} \right| = \left[1 - \left(\frac{\pi}{2\gamma} \right)^2 \sin^2\gamma \right]^{1/2} , \tag{24}$$

and argument

$$\arg(t_{yx}) = \beta + \tan^{-1}\left(\frac{\beta}{\gamma}\tan\gamma \right). \tag{25}$$

The quantities given by Eqs. (24) and (25), as functions of the birefringence parameter β with due regard for Eq. (18), are plotted in Fig. 3.

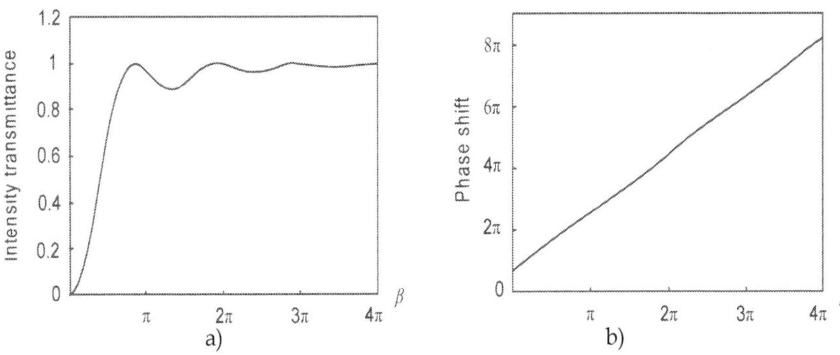

a)   b)

Fig. 3. Complex transmittance given by Eq. (23): a) amplitude transmittance, Eq. (24); b) phase transmittance, Eq. (25).

As can be seen from this figure, starting from some critical value of the birefringence parameter β, approximately equal to $3\pi/2$, the amplitude transmittance $\left| t_{yx} \right|$ approachs to unity while the phase transmittance rises linearly having a slope of approximately 2β. Thus, when the birefringence parameter satisfies the condition $\beta \geq 3\pi/2$, the matrix given by Eq. (21) can be well approximated as

$$\mathbf{T} = \begin{pmatrix} 0 & 0 \\ \exp(-i2\beta) & 0 \end{pmatrix}, \tag{26}$$

i.e. the 90°-twist LC cell sandwiched between two crossed polarizers can be considered as the phase-only modulator.

Till now we assumed that parameter β characterizing the LC cell has a fixed value. Nevertheless, as well known (Lu & Saleh, 1990) , applying to the LC cell an electric field normal to its surface, the birefringence parameter is no longer constant and changes in accordance with

$$\beta(\delta) = \frac{\pi}{\lambda}[n_e(\delta) - n_o]d, \tag{27}$$

where δ is the tilt of the LC molecules with respect to the z axis caused by the electric field. Besides, the relationship between extraordinary refraction index and the molecular tilt can be approximated as

$$n_e(\delta) \approx (n_e - n_o)\cos^2 \delta + n_o. \tag{28}$$

It has been shown (Lu & Saleh, 1990) that the dependence between tilt angle and applied voltage has the form

$$\theta(V_{\mathrm{rms}}) = \begin{cases} 0 & , \ V_{\mathrm{rms}} \leq V_c \\ \dfrac{\pi}{2} - 2\arctan\left\{\exp\left[-\left(\dfrac{V_{\mathrm{rms}} - V_c}{V_0}\right)\right]\right\} & , \ V_{\mathrm{rms}} > V_c \end{cases}, \tag{29}$$

where V_c is the threshold voltage, V_0 is the saturation voltage, and V_{rms} is the effective voltage. Combining Eqs. (27) – (29), it is possible to show that the birefringence parameter β finds to be approximately proportional to the inverse value of the applied voltage.

Finally, we are ready to define a LC-SLM as an electro-optical device composed by a large number of LC cells (pixels) whose birefringence indices are controlled by the electrical signals generated by computer and applied individually to each cell by means of an array of electrodes. The amplitude transmittance of 0°-twist LC-SLM or 90°-twist LC-SLM can be described by Eqs. (16) and (26), respectively, replacing parameter β by spatial function $\beta(\mathbf{x})$.

4. Techniques for control of coherence and polarization by means of LC-SLMs

4.1 Single 0°-twist LC-SLM

We begin with the technique based on the use of a 0°-twist LC-SLM (Shirai & Wolf, 2004). It is assumed that the incident light represents a linear polarized laser beam characterized by the cross-spectral density matrix

$$\mathbf{W}(\mathbf{x}_1,\mathbf{x}_2) = E_0^2 \exp\left(-\frac{\mathbf{x}_1^2 + \mathbf{x}_2^2}{4\varepsilon^2}\right)\begin{pmatrix} \cos^2\theta & \cos\theta\sin\theta \\ \cos\theta\sin\theta & \sin^2\theta \end{pmatrix}, \tag{30}$$

where E_0 is the value of power spectrum at the beam centre, ε is the effective (rms) size of the source, and θ is the angle that the direction of polarization makes with the x axis. It can be readily verified that for such a beam $\mu(\mathbf{x}_1,\mathbf{x}_2)=1$ and $P(\mathbf{x})=1$, i.e. the beam described by Eq. (30) is completely coherent and completely (linearly) polarized.

If the extraordinary axis of the LC is aligned along the y direction the transmittance of $0°$-twist LC-SLM, in accordance with the previous section, is given by matrix

$$\mathbf{T}_{LC1}(\mathbf{x}) = \begin{pmatrix} 1 & 0 \\ 0 & \exp[-i2\beta_1(\mathbf{x})] \end{pmatrix} \tag{31}$$

(here the subscript "1" is used for the sake of simplicity of posterior consideration). It is assumed that the birefringence $\beta_1(\mathbf{x})$ has the form

$$\beta_1(\mathbf{x}) = \frac{1}{2}[\varphi_0 + \varphi(\mathbf{x})], \tag{32}$$

where φ_0 is a constant and $\varphi(\mathbf{x})$ is a computer generated zero mean random variable which is characterized by the Gaussian probability density

$$p[\varphi(\mathbf{x})] = \frac{1}{\sqrt{2\pi}\,\sigma_\varphi}\exp\left(-\frac{\varphi^2(\mathbf{x})}{2\sigma_\varphi^2}\right), \tag{33}$$

with variance $\langle\varphi^2(\mathbf{x})\rangle = \sigma_\varphi^2$ and cross correlation defined at two different points as

$$\langle\varphi(\mathbf{x}_1)\varphi(\mathbf{x}_2)\rangle = \sigma_\varphi^2\exp\left(-\frac{\xi^2}{2\alpha_\varphi^2}\right), \tag{34}$$

where $\xi = |\mathbf{x}_1 - \mathbf{x}_2|$ and α_φ is a positive constant characterizing correlation width of $\varphi(\mathbf{x})$. On substituting from Eqs. (30) - (32) into Eq. (10), one obtains

$$\mathbf{W}'(\mathbf{x}_1,\mathbf{x}_2) = E_0^2 \exp\left(-\frac{\mathbf{x}_1^2 + \mathbf{x}_1^2}{4\varepsilon^2}\right)$$

$$\times\begin{pmatrix} \cos^2\theta & \exp(-i\varphi_0)\langle\exp[-i\varphi(\mathbf{x}_2)]\rangle\cos\theta\sin\theta \\ \exp(i\varphi_0)\langle\exp[i\varphi(\mathbf{x}_1)]\rangle\cos\theta\sin\theta & \langle\exp\{i[\varphi(\mathbf{x}_1)-\varphi(\mathbf{x}_2)]\}\rangle\sin^2\theta \end{pmatrix}. \tag{35}$$

On making use of Eqs. (33) and (34), it can be shown that (Ostrovsky et al, 2009b, 2010)

$$\langle\exp[\pm i\varphi(\mathbf{x})]\rangle = \exp\left(-\frac{\sigma_\varphi^2}{2}\right), \tag{36}$$

$$\langle\exp\{+i[\varphi(\mathbf{x}_1)\pm\varphi(\mathbf{x}_2)]\}\rangle = \langle\exp\{-i[\varphi(\mathbf{x}_1)\pm\varphi(\mathbf{x}_2)]\}\rangle = \exp\left\{-\sigma_\varphi^2\left[1\pm\exp\left(-\frac{\xi^2}{2\alpha_\varphi^2}\right)\right]\right\}. \tag{37}$$

Then, Eq. (35) can be rewritten as

$$\mathbf{W}'(\mathbf{x}_1,\mathbf{x}_2) = E_0^2 \exp\left(-\frac{\mathbf{x}_1^2 + \mathbf{x}_1^2}{4\varepsilon^2}\right)$$

$$\times \begin{pmatrix} \cos^2\theta & \exp(-i\varphi_0)\exp\left(-\frac{\sigma_\varphi^2}{2}\right)\cos\theta\sin\theta \\ \exp(i\varphi_0)\exp\left(-\frac{\sigma_\varphi^2}{2}\right)\cos\theta\sin\theta & \exp\left\{-\sigma_\varphi^2\left[1-\exp\left(-\frac{\xi^2}{2\alpha_\varphi^2}\right)\right]\right\}\sin^2\theta \end{pmatrix}. \tag{38}$$

To simplify the consequent analysis, we assume that σ_φ is large enough to accept the following approximations (Shirai & Wolf, 2004):

$$\exp\left(-\frac{\sigma_\varphi^2}{2}\right) \approx 0, \tag{39}$$

$$\exp\left\{-\sigma_\varphi^2\left[1-\exp\left(-\frac{\xi^2}{2\alpha_\varphi^2}\right)\right]\right\} \approx \exp\left(-\frac{\xi^2}{2\eta_\varphi^2}\right), \tag{40}$$

where $\eta_\varphi = \alpha_\varphi/\sigma_\varphi$. In this case Eq. (38) can be rewritten approximately as

$$\mathbf{W}'(\mathbf{x}_1,\mathbf{x}_2) = E_0^2 \exp\left(-\frac{\mathbf{x}_1^2 + \mathbf{x}_1^2}{4\varepsilon^2}\right)\begin{pmatrix} \cos^2\theta & 0 \\ 0 & \exp\left(-\frac{\xi^2}{2\eta_\varphi^2}\right)\sin^2\theta \end{pmatrix}. \tag{41}$$

According to definitions given by Eqs. (3) and (4), we find

$$\mu'(\xi) = 1 - \left[1-\exp\left(-\frac{\xi^2}{2\eta_\varphi^2}\right)\right]\sin^2\theta, \tag{42}$$

$$P'(\mathbf{x}) = \left|\cos 2\theta\right|. \tag{43}$$

Equations (42) and (43) show that the modulated beam is, in general, partially coherent and partially polarized. The degree of polarization changes in the range from 1 to 0 with a proper choice of polarization angle θ of the incident beam. The degree of coherence, for a fixed value of θ can be varied by a proper choice of parameters α_φ and σ_φ of the control signal $\varphi(\mathbf{x})$, as it is shown in Fig. 4.

We would like to point out the following two shortcomings of the described technique. Firstly, as can be seen from Eqs. (42) and (43) this technique does not provide the independent modulation of the degree of coherence and the degree of polarization since both of them depend at the same time on the polarization angle θ. Secondly, as can be seen from Fig. 4, this technique does not allow to obtain the values of the degree of coherence in a whole desired range from 1 to 0.

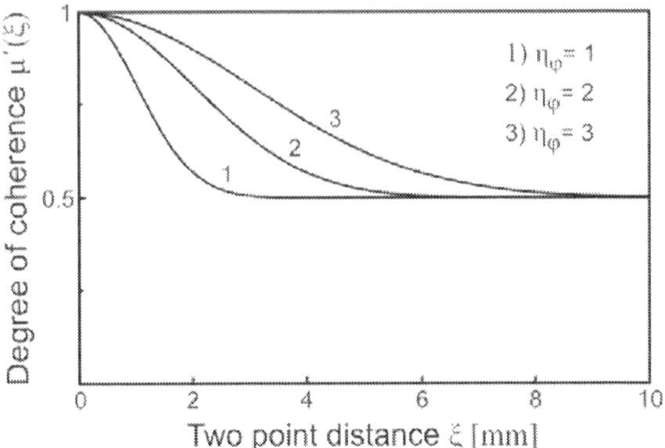

Fig. 4. Degree of coherence given by Eq. (42) for $\theta = \pi/4$ and different values of η_φ.

4.2 Two 0°-twist LC-SLMs coupled in series

To avoid the shortcomings mentioned above, the authors (Ostrovsky et al, 2009) proposed to use instead of a single 0°-twist LC-SLM the system of two 0°-twist LC-SLMs coupled in series as shown in Fig. 5.

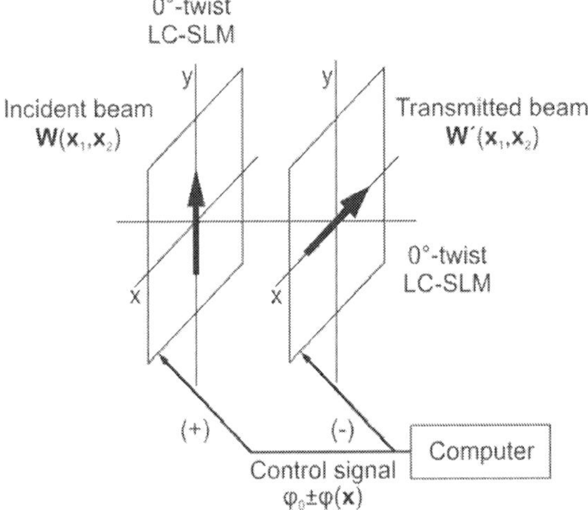

Fig. 5. System of two crossed 0°-twist LC-SLMs coupled in series. The bold-faced arrows denote the extraordinary axis of liquid LC.

The transmittance of the first SLM is just the same as in previous technique, while the transmittance of the second one, whose extraordinary axis is aligned in the x direction, is given by matrix

$$\mathbf{T}_{LC2}(\mathbf{x}) = \begin{pmatrix} \exp[-i2\beta_2(\mathbf{x})] & 0 \\ 0 & 1 \end{pmatrix}, \tag{44}$$

where birefringence $\beta_2(\mathbf{x})$ has the form

$$\beta_2(\mathbf{x}) = \frac{1}{2}\left[\varphi_0 - \varphi(\mathbf{x})\right], \tag{45}$$

with φ_0 and $\varphi(\mathbf{x})$ of the same meaning as stated in the context of Eq. (32). The transmittance of the system composed by two crossed $0°$-twist LC-SLMs is given by matrix

$$\mathbf{T}(\mathbf{x}) = \mathbf{T}_{LC2}(\mathbf{x})\mathbf{T}_{LC1}(\mathbf{x}) = \exp(-i\varphi_0)\begin{pmatrix} \exp[i\varphi(\mathbf{x})] & 0 \\ 0 & \exp[-i\varphi(\mathbf{x})] \end{pmatrix}. \tag{46}$$

On substituting from Eqs. (30) and (46) into Eq. (10) with due regard for relation (37), one obtains

$$\mathbf{W}'(\mathbf{x}_1, \mathbf{x}_2) = E_0^2 \exp\left(-\frac{\mathbf{x}_1^2 + \mathbf{x}_1^2}{4\varepsilon^2}\right)$$

$$\times \begin{pmatrix} \exp\left\{-\sigma_\varphi^2\left[1 - \exp\left(-\frac{\xi^2}{2\alpha_\varphi^2}\right)\right]\right\}\cos^2\theta & \exp\left\{-\sigma_\varphi^2\left[1 + \exp\left(-\frac{\xi^2}{2\alpha_\varphi^2}\right)\right]\right\}\cos\theta\sin\theta \\ \exp\left\{-\sigma_\varphi^2\left[1 + \exp\left(-\frac{\xi^2}{2\alpha_\varphi^2}\right)\right]\right\}\cos\theta\sin\theta & \exp\left\{-\sigma_\varphi^2\left[1 - \exp\left(-\frac{\xi^2}{2\alpha_\varphi^2}\right)\right]\right\}\sin^2\theta \end{pmatrix} \tag{47}$$

and then, using approximations (39) and (40),

$$\mathbf{W}'(\mathbf{x}_1, \mathbf{x}_2) = E_0^2 \exp\left(-\frac{\mathbf{x}_1^2 + \mathbf{x}_1^2}{4\varepsilon^2}\right)\exp\left(-\frac{\xi^2}{2\eta_\varphi^2}\right)\begin{pmatrix} \cos^2\theta & 0 \\ 0 & \sin^2\theta \end{pmatrix}. \tag{48}$$

According to definitions given by Eqs. (3) and (4), we find

$$\mu'(\xi) = \exp\left(-\frac{\xi^2}{2\eta_\varphi^2}\right), \tag{49}$$

$$P'(\mathbf{x}) = \left|\cos 2\theta\right|. \tag{50}$$

As can be seen from Eqs. (49) and (50), the output degree of coherence in this case does not depend on direction of the input polarization and changes in the whole desired range from 1 to 0.

4.3 Two 0°-twist LC-SLMs coupled in parallel

The result resembling the one given above can be also obtained using the system of two $0°$-twist LC-SLMs coupled in parallel. Such a system has been described in (Shirai et al, 2005). Here we propose a somewhat modified version of this technique.

The technique is based on the use of two $0°$-twist LC-SLMs with orthogonal orientations of their extraordinary axes placed in the opposite arms of a Mach-Zehnder interferometer as it is shown in Fig. 7. The polarizing beam splitter at the interferometer input separates the orthogonal beam components $E_x(\mathbf{x})$ and $E_y(\mathbf{x})$ so that each of them can be independently

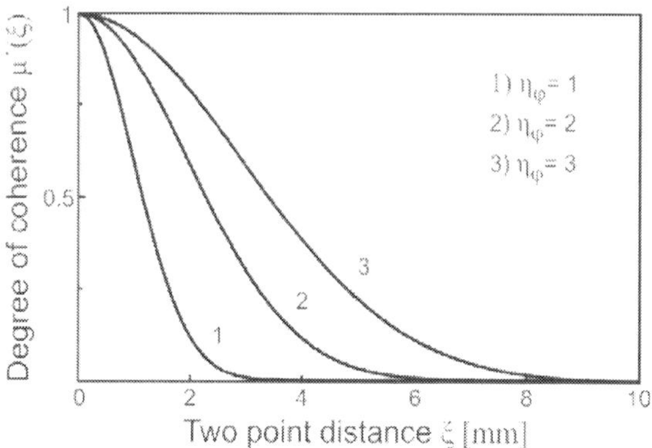

Fig. 6. Degree of coherence given by Eq. (49) for $\theta = \pi/4$ and different values of η_φ.

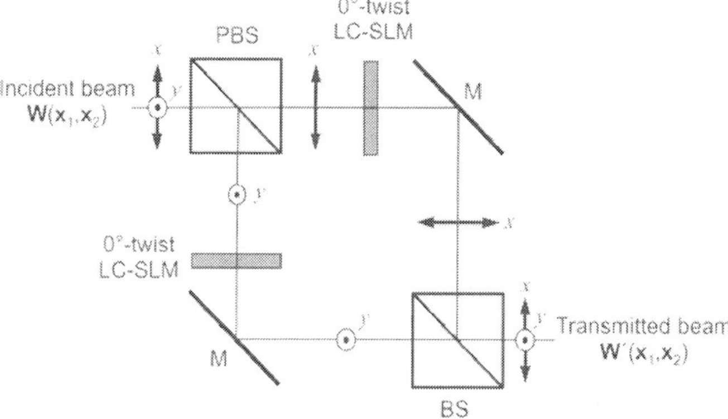

Fig. 7. System of two crossed 0°-twist LC-SLMs coupled in parallel: PBS, polarizing beam splitter; M mirror; BS beam splitter. The bold-faced arrows and circled dots denote polarization directions.

modified by different LC-SLMs. The modified beam components are superimposed in the conventional beam splitter at the interferometer output. Disregarding the negligible changes of coherence and polarization properties of the electromagnetic field induced by the free space propagation within the interferometer, one can represent the considering system as a thin polarization-dependent screen with the transmittance given by matrix

$$\mathbf{T}(\mathbf{x}) = \begin{pmatrix} \exp[-i2\beta_1(\mathbf{x})] & 0 \\ 0 & \exp[-i2\beta_2(\mathbf{x})] \end{pmatrix}.$$

(51)

As before, we assume that parameter β of each 0°-twist LC-SLM has the form

$$\beta_{1(2)}(\mathbf{x}) = \frac{1}{2}\left[\varphi_0 + \varphi_{1(2)}(\mathbf{x})\right],$$

(52)

with φ_0 and $\varphi_{1(2)}(\mathbf{x})$ of the same meaning as stated in the context of Eq. (32). It is assumed also that the variables $\varphi_1(\mathbf{x})$ and $\varphi_2(\mathbf{x})$ are generated by two different computers so that they can be considered as statistically independent with the separable joint probability density

$$p[\varphi_1(\mathbf{x})\varphi_2(\mathbf{x})] = p[\varphi_1(\mathbf{x})]p[\varphi_2(\mathbf{x})]. \tag{53}$$

Following Subsection 4.1 and using in addition relation (Ostrovsky et al, 2010)

$$\left\langle \exp\{i[\varphi_{1(2)}(\mathbf{x}_2) \pm \varphi_{2(1)}(\mathbf{x}_1)]\}\right\rangle = \exp\left(-\sigma_\varphi^2\right), \tag{54}$$

it may be readily shown that the cross-spectral density matrix $\mathbf{W}'(\mathbf{x}_1,\mathbf{x}_2)$, the degree of coherence $\mu'(\mathbf{x}_1,\mathbf{x}_2)$ and the degree of polarization $P'(\mathbf{x})$, in this case, are just the same as ones given by Eqs. (48) – (50).

4.4 Two 90°-twist LC-SLMs coupled in parallel

The considered above techniques for modulation of coherence and polarization are based on the use of 0°-twist LC-SLMs. The consistence of these techniques is well grounded in theory but no relevant experimental results have been yet reported. This situation can be explained by the lack at present of commercial 0°-twist LC-SLMs with the required characteristics. Here we propose an alternative technique which uses widely available commercial 90°-twist LC-SLMs and, hence, can be easily realized in practice. The proposed technique is sketched schematically in Fig. 8.

In spite of its outward resemblance with the technique described in the previous subsection, this technique has two essential distinctions. Firstly, instead of 0°-twist LC-SLMs here the orthogonally aligned 90°-twist LC-SLMs are used in the opposite arms of the Mach-Zehnder interferometer. Secondly, the conventional beam splitter at the output of interferometer is replaced by the polarizing one. As a result, taking into account that two polarizing beam splitters coupled in series act as crossed polarizers, each arm of interferometer can be considered as the system shown in Fig. 2 with $\psi_1 = 0°$ and $\psi_2 = 90°$. In accordance with Section 3 such a system realizes the phase-only modulation of the correspondent orthogonal component of the incident beam.

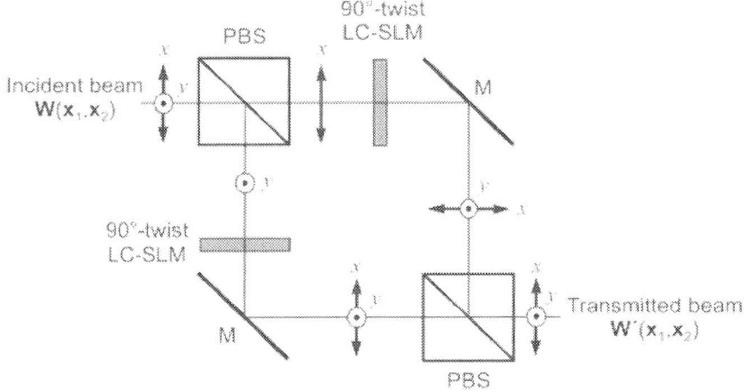

Fig. 8. System of two crossed 90°-twist LC-SLMs coupled in parallel: PBS, polarizing beam splitter; M, mirror. The bold-faced arrows and circled dots denote polarization directions.

Again disregarding the negligible changes of coherence and polarization properties of the electric field induced by the free space propagation within the interferometer, the system shown in Fig. 8 can be considered as a thin polarization-dependent screen with transmittance given by matrix

$$\mathbf{T}(\mathbf{x}) = \begin{pmatrix} 0 & \exp[-i2\beta_2(\mathbf{x})] \\ \exp[-i2\beta_1(\mathbf{x})] & 0 \end{pmatrix}, \tag{55}$$

where $\beta_{1(2)}(\mathbf{x})$ has the same meaning as in Eq. (51). It can be readily shown that for this technique the cross-spectral density matrix has a slightly different form, namely

$$\mathbf{W}'(\mathbf{x}_1,\mathbf{x}_2) = E_0^2 \exp\left(-\frac{\mathbf{x}_1^2 + \mathbf{x}_1^2}{4\varepsilon^2}\right)\exp\left(-\frac{\xi^2}{2\eta_\varphi^2}\right)\begin{pmatrix} \sin^2\theta & 0 \\ 0 & \cos^2\theta \end{pmatrix}, \tag{56}$$

but the degree of coherence $\mu'(\mathbf{x}_1,\mathbf{x}_2)$ and the degree of polarization $P'(\mathbf{x})$ are given again by Eqs. (49) and (50).

Concluding, it is appropriate to mention here that the proposed technique, as well the one presented before, provides generating the beam of a Gaussian Schell-model type given by Eq. (6) with parameters $S_{0x} = E_0^2 \sin^2\theta$, $S_{0y} = E_0^2 \cos^2\theta$ and $\sigma_x = \sigma_y = \eta_\varphi$.

5. Measurements of coherence and polarization

In practice the efficiency of the techniques described in previous section can be verified by measuring the degree of coherence and the degree of polarization of modulated electromagnetic beam. The main idea of such a measuring is well known (Wolf, 2007) and consists in the implementation of four two-pinhole Young's experiments with different states of polarization of the radiation emerged from each pinhole. Nevertheless, the practical realization of such an idea proves to be difficult in consequence of physical impossibility to insert the needed optical elements just behind the pinholes. Here we present the technique for measuring the degree of coherence and the degree of polarization proposed by the authors (Ostrovsky et al, 2010), which permits to avoid this difficultness.

The technique consists in applying the Mach-Zehnder interferometer shown in Fig. 9. This allows the physical insertion of the appropriate optical elements for simultaneous and independent transforming the orthogonal beam components. The polarizers P_1 and P_2 serve to cut off only one of the orthogonal field components, while the removable rotators R_1 and R_2 serve to produce the rotation of one of the transmitted field component through 90° (for such a purpose a suitably oriented half-wave plate can be used). The operation description of the technique is given bellow.

The determination of the elements W'_{ij} of the matrix $\mathbf{W}'(\mathbf{x}_1,\mathbf{x}_2)$ is realized by means of the following four experiments. In the first experiment the polarizers P_1 and P_2 are aligned to transmit only x components of the incident field without any subsequent rotation of the plane of polarization. In the second experiment P_1 and P_2 are aligned to transmit only y components of the incident field again without any subsequent rotation of the plane of polarization. In the third and the fourth experiments the polarizers P_1 and P_2 cut off the different orthogonal components of the incident field and the corresponding polarization rotator R_1 or R_2 serves to allow the interference of these components.

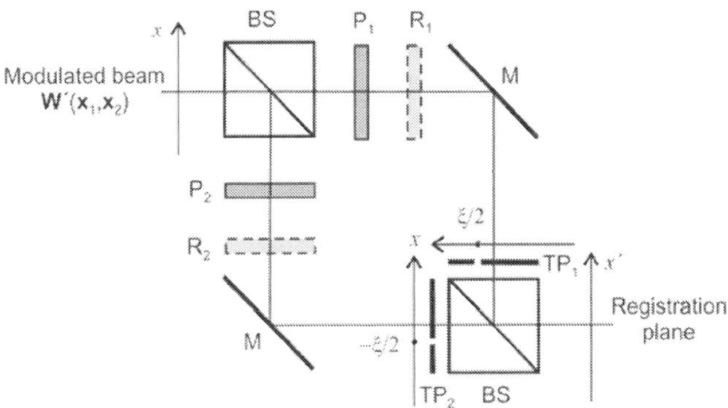

Fig. 9. System for measuring the statistical properties of modulated beam: BS, beam splitter; M, mirror; TP, translating pinhole; P_1, P_2, polarizers; R_1, R_2; polarization rotators.

The spectral density or power spectrum of the field observed at the output of the interference system in each experiment can be described by the spectral interference law, which under certain conditions can be written as (Wolf, 2007)

$$S_{ij}(x') = S_i'\left(\frac{\xi}{2}\right) + S_j'\left(\frac{\xi}{2}\right) + 2\left|W_{ij}'\left(\frac{\xi}{2}, -\frac{\xi}{2}\right)\right|\cos\left[\frac{k\xi}{z_0}x' + \alpha_{ij}\left(\frac{\xi}{2}, -\frac{\xi}{2}\right)\right] \quad (i, j = x, y), \tag{57}$$

where S_i' and S_j' are the power spectra of the field components in the pinhole position $\xi/2$, k is the wave number, z_0 is the geometrical path between the pinhole plane and the observation plane, and $\alpha_{ij} = \arg W_{ij}'$. From the physical point of view, Eq. (57) describes an image with periodic structure, known as the interference fringe pattern. The measure of the contrast of the interference fringes is the so-called visibility coefficient defined as

$$V_{ij}(\xi) = \frac{S_{ij}'^{\max}(x') - S_{ij}'^{\min}(x')}{S_{ij}'^{\max}(x') + S_{ij}'^{\min}(x')}. \tag{58}$$

On substituting from Eq. (57) with $\cos(.) = \pm 1$ into Eq. (58), we readily find that

$$\left|W_{ij}'(\xi)\right| = \frac{1}{2}\left[S_i'\left(\frac{\xi}{2}\right) + S_j'\left(\frac{\xi}{2}\right)\right]V_{ij}(\xi). \tag{59}$$

The spectra S_i' and S_j' can be easily measured when one of the pinholes is covered by an opaque screen. The phase α_{ij} can be measured by determining the location of maxima in the interference pattern. Hence, measuring in each experiment the visibility V_{ij}, power spectra $S_{i(j)}'$, and phase α_{ij}, one can determine all the elements W_{ij}' of the matrix $\mathbf{W}'(\mathbf{x}_1, \mathbf{x}_2)$. The degree of coherence and the degree of polarization of the modulated beam can be then calculated using definitions given by Eqs. (3) and (4).

6. Experiments and results

To verfy the proposed technique in practice, we conducted some physical experiments. The experimental setup used in experiments is sketched in Fig. 10. The principal part of the experimental setup was composed of two Mach-Zehnder interferometers coupled by the common beam splitter. The first interferometer realized the modulation of the incident beam as it has been described in Subsection 4.4, while the second one served for measuring the degree of coherence and the degree of polarization of the modulated beam as it has been described in Section 5.

As the primary source we used a highly coherent linearly polarized beam generated by He-Ne laser (Spectra-Physics model 117A, λ=633 nm, output power 4.5 mW) which can be well described by the model given by Eq. (30). The laser was mounted in a rotary stage that allowed changing the polarization angle θ without any loss of light energy. As the 90°-twist nematic LC-SLMs we used the computer controlled HoloEye LC2002 electro-optical modulators which have resolution of 800 × 600 pixels (32 µm square in size) and can display the control signal with 8 bit accuracy (256 gray levels). The control of LC-SLMs was realized independently by two computers using a specially designed program for generating the random signals which obey the desired Gaussian statistics. To realize the measurements of the degree of coherence we used two pinholes with diameter of 200 µm mounted on motorized linear translation stages.

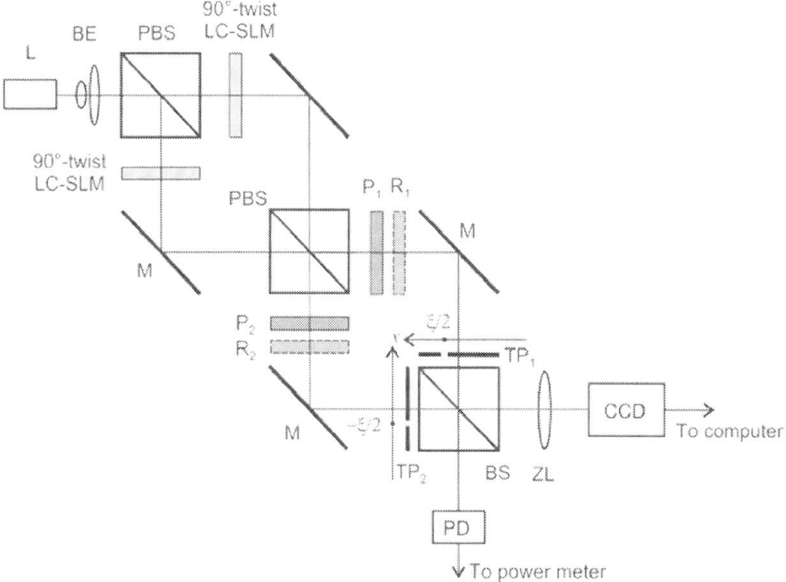

Fig. 10. Experimental setup: L, laser; BE, beam expander; ZL, zoom-lens; PD, photodiode; the other abbreviations are just the same as in Figs. 8 and 9.

We realized two sets of experiments. In the first set we measured the degree of polarization for different values of the polarization angle of incident beam and for the fixed value (≈ 1)

of parameter η_φ of the control signal. The results of these experiments are shown in Fig. 11. In the second set we chosed $\theta = \pi/4$ and measured the degree of coherence varying the parameters α_φ and σ_φ of the control signal to ensure the chosen values of parameter η_φ. The results of these experiments are shown in Fig. 12. As a whole, the results obtained in both sets of experiments are in a good accordance with theoretical predictions.

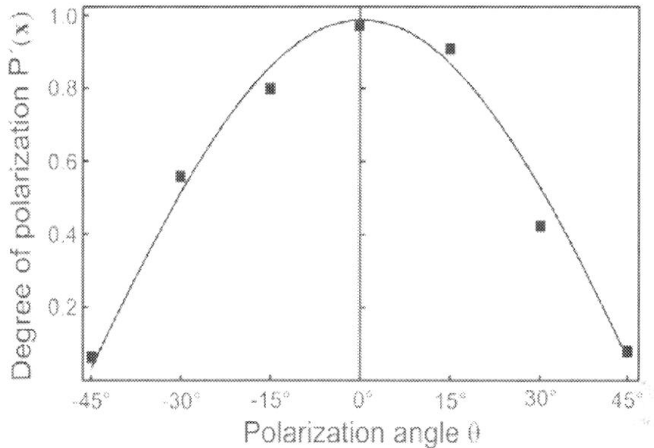

Fig. 11. Results of measuring the degree of polarization for $\eta_\varphi \approx 1$.

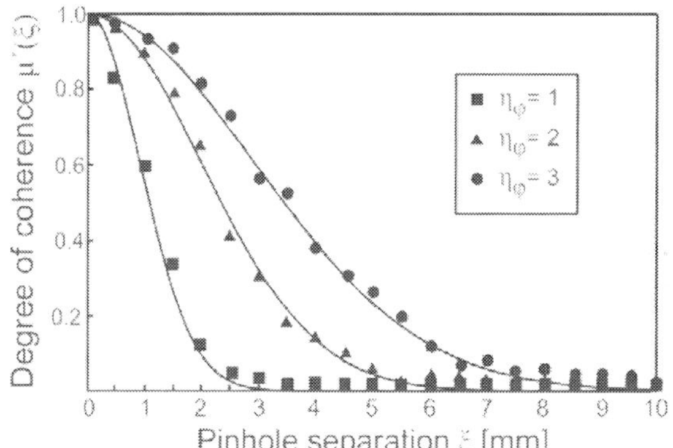

Fig. 12. Results of measuring the degree of coherence for $\theta = \pi/4$ and different values of parameter η_φ.

7. Conclusion

In this chapter the problem of modulating the coherence and polarization of optical beams has been considered. It has been shown that the LC-SLM represents an ideal tool for practical realizing such a modulation. We have analized the known techniques of optical modulation based on the use of 0°-twist LC-SLM and have proposed a new technique based on the use of two 90°-twist LC-SLMs. Because of the wide commercial availability of 90°-

twist LC-SLMs the proposed technique proves to be the most attractive one. The justifiability of this technique has been corroborated by the results of physical experiments.

8. Acknowledgment

The authors gratefully acknowledge the financial support from the Benemérita Universidad Autónoma de Puebla under project VIEP: OSA-EXT-11-G.

9. References

Goodman, J. W. (1996). *Introduction to Fourier Optics* (2nd Edition), McGraw-Hill, ISBN 0-07-024254-2, USA.

Lu, K. & Saleh B. E. A. (1990). Theory and design of the liquid crystal TV as an optical spatial phase modulator. *Optical Engineering*, Vol.29, No.3, (March 1990), pp. (240-246). ISSN 0091-3286.

Ostrovsky A. S. (2006). *Coherent Mode Representations in Optics*, SPIE Press, ISBN 0-8194-6350-7, Bellingham WA, USA.

Ostrovsky A. S.; Martínez-Vara P.; Olvera-Santamaría M. Á. & Martínez-Niconoff G. (2009a). Vector coherence theory : An overview of basic concepts and definitions, In: *Recent Research Developments in Optics*, S. G. Pandalai, (113-132), Research Singpost, ISBN 978-81-308-0370-8, Kerala, India

Ostrovsky A. S.; Martínez-Niconoff G.; Arrizón V.; Martínez-Vara P.; Olvera-Santamaría M. Á. & Rickenstorff C. (2009b). Modulation of coherence and polarization using liquid crystal spatial light modulators. *Optics Express*, Vol.17, No.7, (March 2009), pp. (5257-5264). ISSN 1094-4087.

Ostrovsky A. S.; Rodríguez-Zurita G.; Meneses-Fabián C.; Olvera-Santamaría M. Á. & Rickenstorff C. (2010). Experimental generating the partially coherent and partially polarized electromagnetic source. *Optics Express*, Vol.18, No.12, (June 2010), pp.(12864-12871). ISSN 1094-4087.

Shirai T. & Wolf E. (2004). Coherence and polarization of electromagnetic beams modulated by random phase screens and their changes on propagation in free space. *Journal of the Optical Society of America A*, Vol.21, No.10, (October 2004), pp. (1907-1916). ISSN 1084-7529.

Shirai T.; Korotkova O. & Wolf E. (2005). A method of generating electromagnetic Gaussian Schell-model beams. *Journal of Optics A: Pure and Applied Optics*, Vol.7, No.5, (March 2005), pp. (232-237). ISSN 1464-4258.

Wolf E. (2007). *Introduction to the Theory of Coherence and Polarization of Light*, Cambridge University Press, ISBN 9780521822114, Cambridge, UK.

Yamauchi M. & Eiju T. (1995). Optimization of twisted nematic liquid crystal panels for spatial light phase modulation. *Optical Communications*, Vol.115, No.1, (March 1995), pp. (19-25). ISSN 0030-4018.

Yariv, A. & Pochi, Y. (1984). *Optical Waves in Crystals*, Wiley, ISBN 0-471-09142-1, USA.

16

Recent Developments in High Power Semiconductor Diode Lasers

Li Zhong and Xiaoyu Ma
*National Engineering Research Center for Optoelectronic Devices,
Institute of Semiconductors, Chinese Academy of Sciences
Beijing
China*

1. Introduction

Due to a number of advantages of diode lasers, such as small size, light weight, high efficiency etc., it has been the focus of the laser field from the beginning of the birth and has been widely used in industrial, military, medical, communications and other fields. Especially, to a great extent, a tremendous growth in the technology of solid-state lasers has been complemented by laser diode array designs for pumping such solid-state lasers. Significant applications continue to exist at common solid state laser systems such as yttrium aluminum garnet doped with neodymium or ytterbium (Nd:YAG or Yb:YAG, respectively) requiring pump light in the 780 nm to 1000 nm range. Driven by the increasing demands of high-performance high-power laser pumping source and direct industrial processing applications, tremendous breakthrough have been realized in the main optical-electronic performances of high power semiconductor diode lasers, such as ultra-high peak power, super-high electro-optical conversion efficiency, low beam divergence, high brightness, narrow spectrum linewidth, high operation temperature, high reliability, wavelength stabilization and fundamental transverse mode operation etc. These achievements are attributed to a combination of the maturity of semiconductor material epitaxy, the optimization of the laser waveguide structure, the cavity surface-passivation technology as well as the high effective cooling and packaging technologies. The Occident and Japan keep ahead in this field with several large corporations actively engaged in this market, for example, Coherent, IMC, SDL, OPC, HPD, Spectrum-Physics of the U.S., OSRAM, JOLD, Frauhorf of Germany, THALES of France, SANYO, SONY of Japan, and ATC of Russia etc. The wavelength of these industrial products ranges from 630 nm to 1550 nm, and optical output power levels from several W to 10 kW class. In China, prominent progresses have also been made at a rapid rate. Advances in the design and manufacture of the bars, together with effective means of stacking and imaging monolithic semiconductor laser arrays (bars), have enabled the production of robust sources at market-competitive costs. In particular, the diode lasers for these systems have to meet high demands in relation to efficiency, power, reliability and manufacturability, which following the desire for reducing the cost per watt and the cost per hour's lifetime for the customer. However, with the enhancement of the power and beam quality, a series of new practical problems arise in the aspect of engineering, such as high cost of the high-current and low-voltage power supply and short life span of micro-channel heat sink cooling etc.. Gradually, single-emitter semiconductor laser devices and mini-bars with high power and high beam quality are becoming the mainstream research trend and replacing the traditional cm-bars. On the other hand, the reduced divergence angle accelerates the improvement of beam quality, which is directly reflected in the decrease of the fiber diameter and the increase of the output power for fiber-coupled diode laser module. Here we review and discuss the state of the art of high power semiconductor diode lasers, including single emitters, bars, horizontal bar arrays and vertical bar stacks, with the typical data presented. Several key technological problems concerning the improvements of diode lasers performance, the optimization of packaging architectures and the developments of high beam quality of diode lasers will be discussed in section 2~5, respectively. In section 6, we conclude with some thoughts on the future study directions and the developing tendency for high power diode lasers.

2. Status of high-power diode laser technology and characteristics

2.1 Laser diode chip technology

Over the recent years, high power diode lasers have seen a tremendous evolution in material epitaxial growth technology, epi-structure optimization technique, cavity surface-passivation technology etc.. Epitaxial structure is designed for a specific range of operation to optimize a combination of optical, electrical and thermal performance, generally minimizing both operating voltage and internal loss to achieve high efficiency with long cavities for high-average-power and high-brightness applications. The details of these structures, such as material compositions, layer thicknesses, asymmetric or symmetric waveguide structure design, and doping profiles are selected to ensure that manufacturability and reliability are not compromised. Important developments in epitaxial growth technology include the reporting of low loss materials (about 1 cm^{-1} for AlGaAs for example), the development of the strained materials with attendant benefits on gain and bulk defect pinning, and the development of aluminum-free materials such as InGaAs and InGaAsP with the latter material having been reported to wavelengths below 800 nm. A number of careful studies are being reported on filament formation and current crowding in semiconductor lasers and methods for avoiding their deleterious effects. With the improvement of the high-quality, low defect density epitaxial growth technology of semiconductor materials, the resonator cavity length of the existing cm bar has been increased from 0.6 ~ 1.0 mm to 2.0 ~ 5.0 mm, making a significant increase of the output power. The large cavity length ensured low thermal and electrical resistivities of the devices by increasing their active area. The cavity length is selected mainly depending on desired operation power and is optimized for best power conversion efficiency (PCE) at the given condition.

In continuous wave (CW) operation the output power from high power laser bars usually is limited by the thermal load that the assembly may dissipate. Failure modes, like wear out of the output power or bulk failures are critical in CW operation. For quasi continuous wave (qCW) applications the reliable output power in general is not thermally limited. The robustness of the output facet of the devices and the degradation of the assembly under the cyclic thermal load become the critical matter. Special methods of facet design and treatment have been employed to increase the COMD power threshold and suppress its degradation over the operating life of the device, such as facet passivation with a dielectric layer, regrown non-absorbing mirror (ReNAM), intermixed non-absorbing mirror (iNAM), ultra-high vacuum (UHV) cleaved facets etc..(Yanson et al., 2011) NAM-based techniques (ReNAM and iNAM) require further development to achieve required reliability figures. Dielectric passivation and cleave-in-a-vacuum techniques are found to be the two best performing facet engineering solutions. At 980nm, dielectric facet passivation can be employed with a pre-clean cycle to deliver a device lifetime in excess of 3,000 hours at increasing current steps. Vacuum cleaved emitters have delivered excellent reliability at 915nm, and can be expected to perform just as well at 925 and 980nm. By preventing exposure of a freshly cleaved facet to oxygen, the formation of surface oxides and shallow levels is avoided without the need for ion plasma cleaning.(Tu et al., 1996) A capping layer, also deposited in a vacuum, seals the facet and stops the penetration of oxygen. Single emitters fabricated with these two techniques are packaged into fiber-coupled modules with 10W output and 47% efficiency. (Tu et al., 1996)

2.2 Far field divergence angle control

As the basic unit of the integration of semiconductor laser system, the performance of different structure and different types of semiconductor laser device directly contributes to the development of semiconductor laser systems, one of the most important developments is the reduction of the beam divergence and the increase of the output power. According to the definition of the beam quality, the beam divergence angle is proportional to the beam-parameter product (Q or BPP), which is a measure of the beam quality. Therefore the beam quality is under the direct control of the far field divergence angle. Overall, the waveguide structure of semiconductor lasers leads to a serious asymmetry far-field beam quality. In the fast axis direction, the output beam can be considered to be fundamental mode, but the divergence angle is large. The compression of the fast axis divergence angle can effectively reduce the requirements for the fast axis collimator aperture. While in the slow axis direction, the output beam is multi-mode and the beam quality is poor. The beam quality can be directly improved by reducing the divergence angle in the slow axis direction, which is the research focus in the field of the high-beam quality semiconductor laser.

The research focus in the control of the fast axis divergence angle is how to balance the fast axis divergence angle and the electro-optical conversion efficiency. Although a number of research institutions had press release of the continued access to fast axis divergence angle of only 3° and even 1°, but based on the consideration of the power, the electro-optical conversion efficiency and the cost, it is difficult to promote practical applications in the short term. In the early year of 2010, the P. Crump etc. of German Ferdinand-Braun Institute has reported the fast axis divergence angle of 30° (95% of the energy range) obtained through the use of large optical cavity and low-limiting factors, meanwhile the electro-optical conversion efficiency of the device is 55%, which is the basic standards to practical devices. The fast axis divergence angle of the current commercial high-power semiconductor laser devices are also dropped from the original of about 80° (95% of the energy range) to below 50°, which substantially lower the requirements for the numerical aperture of the collimator.

In the slow-axis divergence angle control, recent studies have shown that, in addition to the device's own structure, the combination of the drive current density and the thermal effects of semiconductor lasers affect the slow axis divergence angle. The slow axis divergence of a single emitter with long cavity length is of the most easy to control, whereas in the array device, with the increase of the fill factor, the intensification of thermal cross-talk between the emitting elements will lead to the increase of the slow-axis divergence angle. In the year of 2009, the Bookham company of Switzerland has successfully reduced the slow-axis divergence (95% of the energy range) of 9xx-nm 10 W commercial devices with 5 mm cavity length from 10º ~ 12º to about 7º. In the same year, the Osram Company of German and the Coherent Company of the United States has reduced the slow-axis divergence of the array (95% of the energy range) to 7º level.

2.3 High-temperature performance of laser bars and arrays

Since the performance of a diode laser is operating temperature dependent, high-power diode laser pump modules usually need a cooling system to control their operating temperature. However, some diode laser applications require that high-power diode laser pump modules operate in a high temperature environment without any cooling. In addition, the diode laser pump modules have to provide both high peak power and a nice pulse shape because certain energy in each pulse is required. At such a high temperature, semiconductor quantum well gain drops significantly, and the carrier leakage and the Auger recombination rate increases. Thus, the laser bar has a high threshold and low slope efficiency, resulting in very low power efficiency. To reach certain power level at high temperature, the pump current has to be much higher than that at room temperature. More waste heat is generated in the active region of the diode laser bar. In addition, tens of milliseconds pulse width with a few percent duty cycle forces the laser bar to operate in a "CW" mode.(Ziegler et al., 2006; Puchert et al., 2000; Voss et al., 1996) The Lasertel Company has presented the development of high-temperature 8xx-nm diode laser bars for diode laser long-pulse (>10 milliseconds) pumping within a high-temperature (130 °C) environment without any cooling.(Fan et al., 2011) The epi-structure is based on a large optical cavity separate confinement heterostructure with Al-free active region. By adjusting Aluminum concentration in the AlGaInP barrier, introducing strain in quantum well (QW) and adjusting the width of QW, optimizing the strain and the width of quantum well, the gain is maximized, the loss and carrier leakage especially at high temperature is minimized and the optical confinement of the waveguide is also be improved. Under the operation condition (130 °C, 15 ms pulse width, 5 Hz frequency and 100-A current pulse), the high-temperature laser bars show robust and consistent performance, reaching 60 W (peak) power and having good pulse shape, as shown in Figure 1. The laser bars do not show any degradation after 310,000 15-millisecond current pulse shots. They demonstrated over 40-millisecond long-pulse operation of the 8xx-nm CS bars at 130 °C and 100 A. Regardless of the pulse shape, this laser bar can lase at extremely high temperature and output pulse can last for 8 ms/2ms at 170 °C/180 °C respectively, both driven by 60 A current pulses with 5-Hz frequency, 10 millisecond pulse width. This is the highest operating temperature for a long-pulse 8xx-nm laser bar. Figure 2 shows the high-temperature performance of the 3-bar stack array and its pulse shape at 130 °C. The peak power of the 3-bar array reaches 165 kW at 100A and 130 °C, but the pulse shape is very sensitive to the current and the power of the array drops much faster than that of the CS bar, which may be attributed to the package difference between the CS bar and stack array.

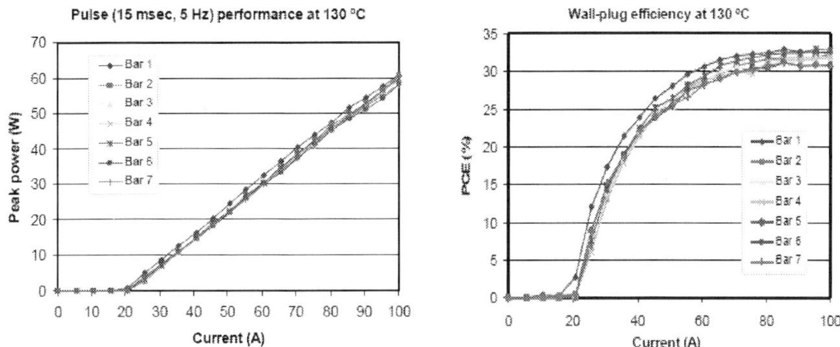

Fig. 1. Pulsed (15 msec PW and 5 Hz frequency) L-I performance and wall-plug efficiency at 130°C.

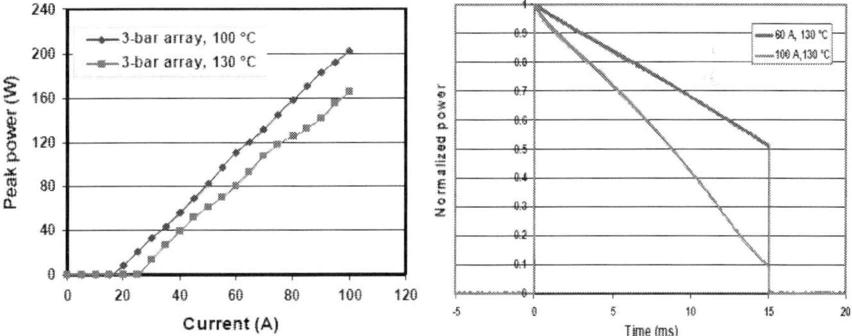

Fig. 2. High temperature performance of 3-bar array (left) and its pulse shape at 130 °C (right).

3. The latest high-power semiconductor laser devices

3.1 Developments of the standard cm laser diode bar

The increasing of reliable power level of diode lasers will enable wider deployment, higher power systems and new market which were previously unavailable, such as direct materials processing etc. Also, it will lead to lower cost to the user, per watt of useable output. The standard length of the laser bars of 1 cm array has been established to obtain high power output, i.e. many emitters has been lateral parallel monolithic integrated in the scale of the slow axis direction into one diode laser bar, Laser bars provide a magnitude of output power of single emitter by integration the single emitters at the wafer level, which has long been a most commonly used form for high-power semiconductor lasers. According to the application, whether high brightness or high power operation is intended, the filling factor is varied between 10% to 80%.

In the early year of 2008, the Spectra-Physics Company of the United States reported the access of 800 W/bar, 1010 W/bar, 950 W/bar maximum output power respectively from the center wavelength of 808 nm, 940 nm, 980 nm cm bar, with corresponding maximum PCE values of 54.3%, 67.4%, and 70%. (Li et al., 2008) Figure 3 summarizes the P-I-PCE curves of bars.

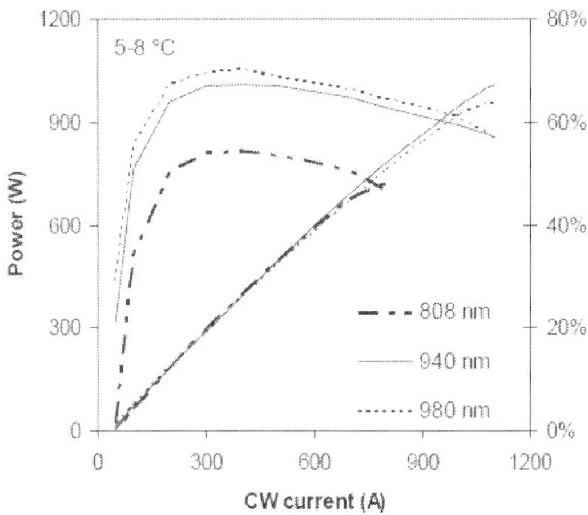

Fig. 3. P-I-PCE curves of bars with double-side cooling from the Spectra-Physics Lasers Company.

All of the bars had a FF of 83% (65 emitters, each 125-µm wide and 5.0-mm long, with a separation of 150 µm). All bars were bonded using the double-sided micro-channel heat sink cooling technology with the circulating water at 5-8°C (Li et al., 2008). These output power are the maximum continuous output power level of the current laboratory. In addition, many other semiconductor laser company, such as the German company JENOPTIK, Switzerland company Oclaro also continued to prepare the kilowatt diode laser array. J. Müller ect. of the Oclaro has made it clear that the access to 1.5 kW/bar array devices is not a problem based on the existing technologies. At the same time, the output power of high beam quality, low fill factor cm bar is also increasing. Table 1 gives the BPP value of cm bars with different fill factors obtained by the Limo Company of German. From the results of Table 1, it can be concluded that for a certain horizontal dimensions of semiconductor laser array device, in the case of the same divergence angle, the BPP value is proportional to the fill factor, i.e. the lower the fill factor, the smaller of the BPP value, and the better the beam quality. Currently, In the 9xx-nm wavelength range 150 W or higher CW output power levels have become the standard for high filling factor bars.(Lichtenstein et al., 2005; Krejci et al., 2009; Crump et al., 2006) In the 80x range devices with 100 W and more output power were demonstrated (Tu et al., 1996; Ziegler et al., 2006). The output characteristic of an 808 nm device optimized for 140-160 W CW output power is shown in Figure 4 (Müller et al., 2010).An output power of 185 W is achieved for CW 200A drive current. For currents above 180 A the thermal limitation of the device is clearly visible in a pronounced rolling of the P-I curve limiting the achievable brightness in this operation mode. The output power of 9xx-nm cm bar with 20% fill factor is up to 180 W/bar in CW condition, the BPP value is down to 5.9 mm·mrad after the symmetry of the fast and slow axis of the beam, and the commercial devices can be work above the level of 80 W/bar with long-term stability; the

Emitter Width (μm)	100	200	150	100	5
Pitch (μm)	200	400	500	500	200
Number of Emitters	49	24	19	19	49
Filling Factor (%)	50	50	30	20	2.5
Beam Divergence FA (FW 90%) (°)	80	80	80	80	80
Beam Divergence SA (FW 90%) (°)	12	12	12	12	12
BPP FA (mm·mrad)	0.35	0.35	0.35	0.35	0.35
BPP SA (mm·mrad)	257	251	149	100	13
Ratio BPP SA/BPP FA	735	720	428	285	37
Symmetrized BPP (mm·mrad)	9.5	9.4	7.2	5.9	2.1
Beam Diameter at NA=0.22 (μm)	121	120	92	75	27

Table 1. the BPP of cm bars with different structures.

output power of 2.5 % fill factor cm bar is up to 50 W/bar in CW condition, the BPP value is down to 2.1 mm·mrad after the symmetry of the fast and slow axis of the beam, the current device is still under developing, and need to improve the stability further. The reliable output power of CW operated bars is in first place limited by the cooling capability of the assembly. Bars operated in a qCW mode deliver a significantly higher reliable output power, because the thermal load is reduced by a factor which is inverse proportional to the duty cycle of the operation mode. Figure 5 and 6 give the typical performance of our lab's 808nm laser diode bar packaged by conductive heatsink (qCW 300 W/barand 200 W/bar with 60% PCE) and 980nm laser diode bar packaged by micro-channel heatsink (CW 200W/bar), respectively.

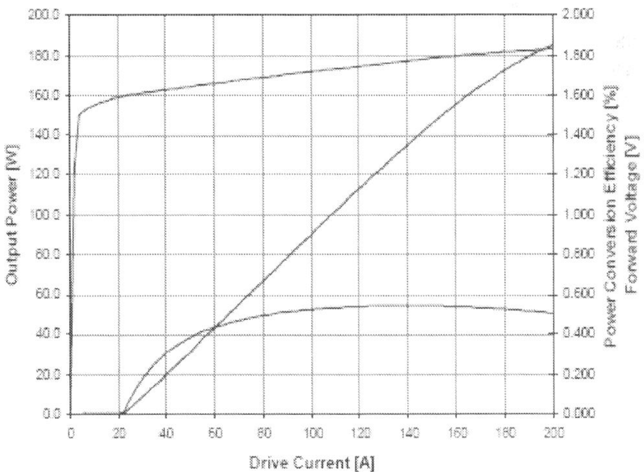

Fig. 4. PIV-PCE characteristic of a 50% filling factor 808 nm bar on micro channel cooler at 25 °C water temperature from the Oclaro Company.

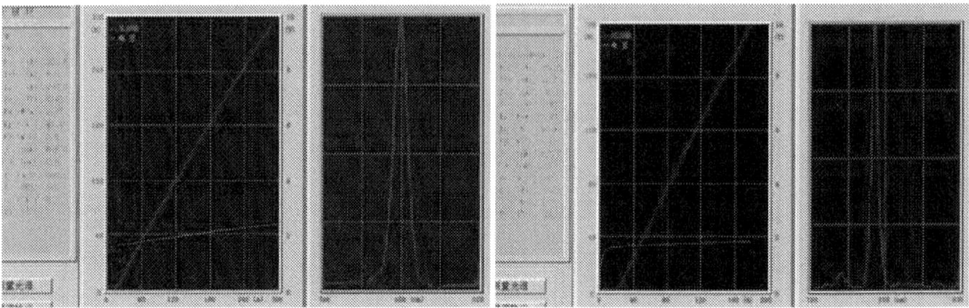

Fig. 5. The typical P-I-V curve and spectrum of 808-nm bar packaged by conductive heatsink in our lab. (a) qCW 300 W/barand (b) qCW 200 W/bar with 60% PCE

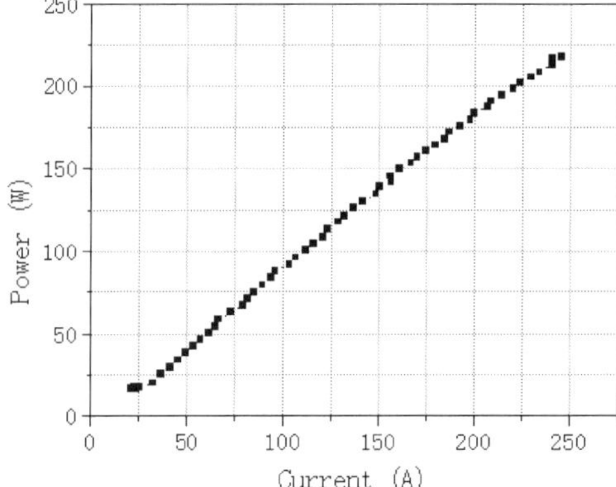

Fig. 6. The typical P-I curve of CW 200 W/bar 980-nm bar packaged by microchannel heatsink in our lab.

However, with the requirement for the decrease of the fill factor to increase the output power and increase the beam quality, a series of new problems arise gradually, especially high cost problem of its accompanying high-current and low-voltage power supply and short life span problem of the micro-channel heat sink cooling etc. Most of the power test of cm bar is subject to the limit of maximum current supply, instead of its own output power limit. In engineering, a combination of several volts with several hundreds of current will produce many practical problems. On the other hand, the micro-channel heat sink must be used to dissipate the high heat flux generated by the cm bar with ultra-high power and high beam quality, but the utmost ability of heat dissipation for the existing water-cooled micro-channel heat sink is no doubt become the biggest obstacle for the further improvements of power and beam quality. The recent developed double-sided micro-channel cooling technology has limited effect on reducing the thermal resistance, and it doesn't seem to satisfy the sustainable expansibility for the power enhancement of the cm bar. In addition, it also can not be ignored that the relatively short life span of the micro-channel heat sink has been the current bottleneck for high-power semiconductor lasers. Although there are some other new efficient technologies arise, such as phase change thermal cooling, spray cooling and micro-heat-pipe technology etc., it is difficult to satisfy the practical applications in the short term because of the performance characteristics, the cost and the compatibility problems. Due to the two constraints mentioned above, in recent years, the major research institutions and high-power semiconductor laser suppliers no longer blindly pursue the improvement of the output power of cm bars, but gradually shift the focus to develop single emitter and mini bars with high power and good beam quality.

3.2 Developments of the single emitter

Compared with the cm bar, the single emitter, which possesses the independent electrical and thermal operation environment, can avoid the thermal crosstalk between the emitting units, so that to have the obvious advantages in life span and beam quality. In addition, owing to the low drive current of the single emitter, the requirement for the drive power can be reduced significantly even in the serial operation. Meanwhile, the heat from a single emitter is relatively low, so the conduction heat sink can be used directly for heat dissipation to avoid the short life span problem brought by the micro-channel heat sink. And the independent thermal operation environment can make it operate under high power density. Currently, optical power linear density of a single emitter can be up to above 200 mW/µm with narrow spectral width, while that of a cm bar is only about 50~85 mW/µm. Especially, the independent thermal and electrical operation environment can make a significant reduction in the risk of device failure. Under the support of the high stable gold tin solder packaging technology, average life of the commercial high-power single emitters can reach to above 100 thousands hours, which is much higher than that of cm bars and hence reduces the use-cost of devices. Based on the advantages listed above, the single emitter exhibits the trend that it will replace the cm bar gradually to become the mainstream semiconductor laser device with high power and high beam quality.

In recent years, single emitters have been developed rapidly. Especially driven by the demand of high power fiber lasers for high-brightness fiber-coupled pumping modules, the single emitters with 90~100 µm strip width, which is matched to the 105 µm/125 µm multimode fiber tail, are improved significantly in output power and beam quality. Currently, the continuous output power of 9xx-nm single emitters can reach 20~25 W/emitter; while that for 8xx-nm can be also beyond 12 W/emitter. In terms of the commercial devices, the 9xx-nm single emitters with the stripe width of 90~100 µm, prepared by IPG, Oclaro, JDSU and other high-power semiconductor laser device suppliers, can operate continuously and steadily over 10 W/emitter, and above 100 W output power can be achieved from the synthesis of the fiber-coupled multiple single emitters.

3.3 Developments of the mini-bar devices

Although the output power of the single emitter is improved rapidly, there is still a wide gap compared to the cm bar. To meet the needs of different applications on power, mini-bar, a new type of high-power semiconductor laser devices, appears and is being developed rapidly. The mini-bar is obtained by integrating several single emitters on a substrate, which is actually the compromise and optimization in the structure of the cm bar and the single emitter. It combines the advantages of the cm bar and the single emitter, and its drive current, lifetime, output power density and spectral width are between those of the cm bar and the single emitter. Also taking into account the high beam quality and the demand for fiber laser pumping source, the development of the mini-bar mainly focuses on the low fill factor devices with the strip width of 100 µm. In 2009, the Osram Company and the DILAS Company collaborated to fabricate the mini-bar with the filling factor of 10%, which contained five 980 nm single emitters with the stripe width of 100 µm and the cavity length of 4 mm. It can achieve the continuous output power of higher than 80 W and electro-optical conversion efficiency of above 60%. The output power of its internal single unit is 16 W/emitter, close to the level of a single emitter. It is worth mentioning that the device shows the similar characteristics of a single emitter in the life test. When the failure happened in the single unit within the mini bar, the entire device was not destroyed but only showed the decay of the output power. Due to the excellent power and life features, mini-bar is now being popularized rapidly to apply in the fiber-coupled pumping modules. Currently, the 9xx-nm commercial devices based on 100 µm single emitters can operate at 8 W/emitter with long-term stability, while the 808 nm devices can also be up to 5 W/emitter level.

4. Advances of package technology

The major characteristics of high power lasers, such as maximum useful output power, wavelength, lifetime are not only limited by the diode or semiconductor structure itself, but also strongly by the quality of the package, such as heat transfer from the junction and the cooling mechanisms used to remove the heat. The type of packaging technology applied to semiconductor laser diode arrays is key to enabling the high average power performance of these devices. Needless to say, packaging techniques, including mounting methods, novel cooling mechanisms and array cooler design are currently a very active area of research and development. Due to the higher standards that have to be met in terms of alignment precision, subcomponent preparation, and soldering-process control, highly specialized equipment and well-experienced operators are needed. Therefore, the packaging process accounts for more than 50% of the total production costs of a packaged diode laser bar. Not only must the laser diode package be capable of efficiently shedding the large heat intensities generated at the laser diode array with only a small temperature rise at the device, it must also be low cost to implement if it is to be attractive to commercial users. Expansion matched, non corrosive, non erosive, low thermal resistance and high thermal conductivity are some of the keywords for the packaging in the near future.

4.1 Mounting of diode laser bars

It is known from mounting other high power semiconductors to heat sinks, the widely used mounting technology is soldering. Diverse soldering technologies have been established according to certain operational conditions. Today, indium and AuSn technology are the two main soldering techniques used for the manufacturing of commercial diode laser bars.

Since indium is very soft and ductile, it allows to compensate for the thermal expansion mismatch between heat sink materials and the GaAs material of the laser bar during the mounting process as well as during operation. Also the ductility of indium is very high, so that indium can equalize the displacement between bar and heat sink due to thermal expansion without forming cracks and voids between bar and heat sink. But the operation temperature is limited if mechanical or thermo-mechanical cycling with high plastical deformation rates for the indium interface is applied and long-term reliability problems presented for the packages. Industrial laser sources typically require a mean time to failure

(MTTF) of at least 20,000 hrs under rated conditions. In long-pulse operation of diode lasers, this reliability is difficult to achieve using a conventional bar assembly (i.e., Cu heatsink with indium solder) due to solder migration. In contrast, hard AuSn solder is comparably stiff. D. Lorenzen et al made comparisons between In and AuSn soldered laser diode bars, and presented that AuSn packaged diode lasers turned out to have clearly higher destruction currents in hard-pulse mode. (Lammert et al., 2005) The reliability of AuSn solder is very good. It can eliminate a number of packaging related failure modes, notably stability during on-off cycling operation. 50 W hard-pulse operation at 8xx-nm has demonstrated a reliability of MTTF > 27 khrs, which is an order of magnitude improvement over traditional packaging. And at 9xx nm a reliability of MTTF >17 khrs at 75 W has been realized (Schleuning et al., 2007) AuSn is well suited for high-temperature applications as generally requested for military lasers. But all forces that result from a mismatch between heat sink material (usually CTE-matched submounts, made, e.g. , of copper/tungsten) and the GaAs laser bar are transferred directly to the laser bar. Because of no stress and strain compensation in this system, hard solders can cause the bar to bow or smile. A package with smile will reduce the brightness or optical efficiency of the bar as a whole and can produce spectral broadening and reliability issues due to uneven heat removal with very high power diode bars. A trade-off must be considered between superior thermal cycling characteristics on one side and the increase of the thermal resistance on the other side, which is caused by the lower thermal conductivity of the CTE-matched submount materials and an additional joint between submount and heat sink.

4.2 Cooling technology

The cooling techniques of semiconductor laser arrays and stacks will be directly related to lifetime of lasers, resulting in rapid temperature increase in active parts of lasers, and therefore leading to catastrophic optical damage, and sometimes burning out of semiconductor lasers. Thermal management and thermal stresses are critical high-power laser diode packaging problems. Depending on the thermal power density, two different types of heat sinks are used: active and conductive, as shown in Figure 7. Depending on the specific requirements, different cooling techniques may be used. For QCW operational mode, conductive cooling using massive copper heat sinks mounted on thermo-electric coolers are preferred. For CW and long-pulse operational mode, active cooling is necessary. The active heat sinks can further be subdivided into liquid-cooled micro- or macro-channel heat sinks, liquid-impingement jets, and evaporative sprays.

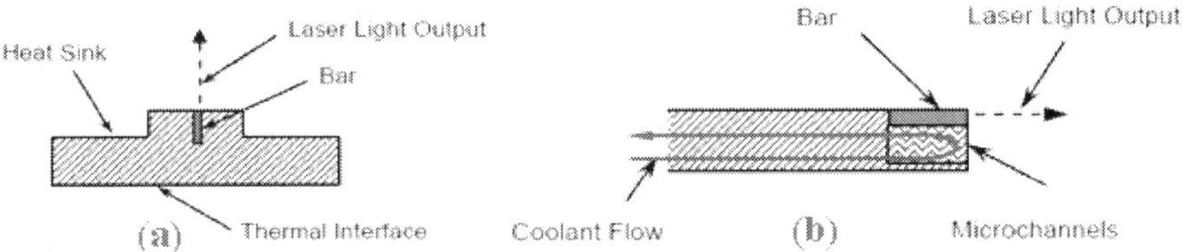

Fig. 7. (a) A passive heat sink cooled laser bar (b) A microchannel heat sinks cooled laser bar.

4.2.1 Expansion-matched packages

The standard heat sink material in nearly all commercially available packages is copper, because of its excellent thermal conductivity (approx. $380Wm^{-1}K^{-1}$), its good mechanical machining properties, and its comparatively low price. But the coefficient of thermal expansion (CTE) of copper (16.8 ppm/K) is much larger than those of ceramics and laser diodes (6.7 ppm/K), requiring significant compromises to reduce thermal stresses and warping, such as the use of compliant solders or adhesives. Thick adhesives reduce thermal performance. Compliant solders can reduce thermal stresses, but they are susceptible to fatigue failure under thermal cycling. Alternative materials and material combinations are investigated mainly to achieve thermal expansion-matching between the heat sink and the laser bar, and to improve the cooling capabilities of the package and to isolate the cooling water circuit electrically from the laser bar, reducing the requirements on the cooling liquid quality and preventing voltage-driven corrosion in the cooling circuit. To ensure the packaging manufacture technology is low cost in terms of its implementation and reduce both CTE and thermal resistance simultaneously, an increasing number of new packaging materials has been developed and is continuing to be developed. The packaging of semiconductor diode lasers has moved increasingly from the traditional indium soldering on copper heat-sinks to the use of hard-solders on CTE matched materials. (Schleuning et al., 2007; McNulty et al., 2008; Du et al., 2008) It is known that diamond, being a non-metallic substance, would have presented itself as the obvious material of choice for heat sinks if it were it not for the cost and the difficulty of manufacturing suitable shapes. To ensure the packaging manufacture technology is low cost in terms of its implementation and reduce both CTE and thermal resistance simultaneously, an increasing number of new packaging materials has been developed and is continuing to be developed. The key design parameters for CTE matching are material selection and layer thickness. CTE values on the mounting surface are fine tuned by varying both of these while remaining mindful of the impact on thermal resistance. When bonded layers of materials with different CTE values are heated, the material with lower CTE will restrict the expansion of the higher CTE material (e.g., Cu). This is the basis for engineering the effective CTE of the heatsink mounting surface, as shown in Figure 8. (Srinivasan et al., 2007) The concept can be applied to a variety of material combinations, including CuW on Cu heatsinks, ceramic on Cu, multi-layer assemblies, and more exotic composite material designs. Other advanced materials fall into six categories: monolithic carbonaceous materials, metal matrix composites, carbon/carbon composites, ceramic matrix composites, polymer matrix composites, and advanced metallic alloys. (Zweben, 2004) Fraunhofer ILT used one of metal matrix composites materials - copper heat sinks with inserted molybdenum layers shown in Figure 9, and realized a coefficient of thermal expansion (CTE) of <9 ppm/K and a thermal resistance of <1 K/W. (Leers, 2007) But this kind of materials have relatively high densities, which is an issue in applications that are portable or subjected to shock loads in shipping or service. Another research was about diamond containing composite materials, such as Silver Diamond, Copper Diamond and Silicon Carbide Diamond (ScD). By adding properly dimensioned top and bottom copper layers a CTE of 7-8 ppm/K and a thermal resistance of 0.7 K/W have been achieved (Schleuning et al., 2007). As an example, heat sink of ScD core with copper brazed on the top and bottom sides was shown in Figure 10. This heat sink provides an expansion-matched and thermally equivalent or even better alternative to standard pure copper heat sinks. The micro photoluminescence spectroscopy (μ-PL) was used to analyze

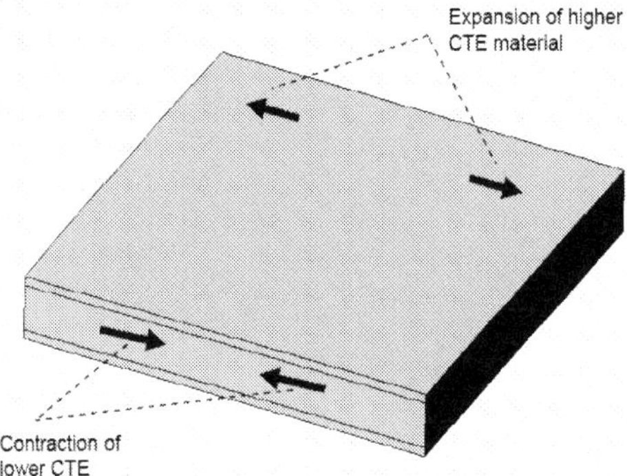

Fig. 8. CET matching concept.

Fig. 9. Diode laser mounted on a conductively expansion-matched heat sink.

Fig. 10. High performance heat sink with ScD core and copper top and bottom layer.

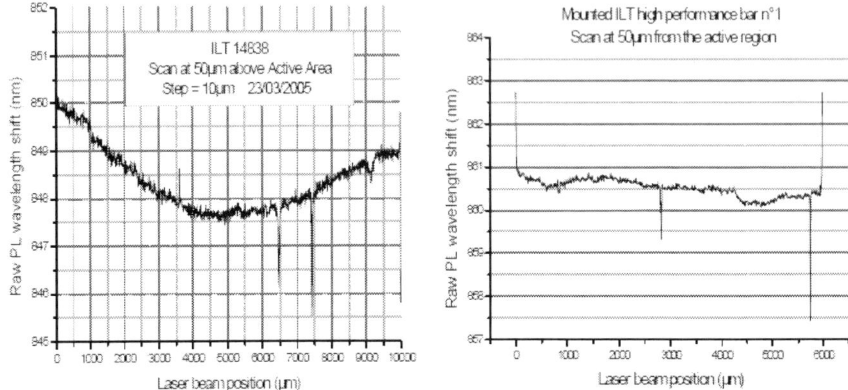

Fig. 11. Comparison of μ-PL measurement of a standard copper heat sink (left) and an expansion matched heat sink (right) with an indium mounted laser.

the induced stress in the laser bar. The precise wavelength of the PL signal depends on the lattice strain and which in turn depends on the mechanical stress the laser bar is exposed to. μ-PL line scans at a standard copper heat sink with an indium mounted laser bar show the increased stress at the edges of the laser. The wavelength shift is around 2 nm compared to the middle. Compared to this, an indium mounted laser bar on a ScD heat sink shows an almost flat line. The edges of the μ-PL scan show no significant changes in the wavelength across the bar as shown in Figure 11. The wavelength shift is around 0.5 nm only, thus the stress induced in the laser bar is reduced by a factor of four.

4.2.2 Active heatsinks

Active heatsinks can be effective platforms for heat removal and are commonly used in high-power applications with either single-bar or multi-bar stacks. Next generation active heatsinks have been designed with the goals of (1) matching the CTE of the GaAs-based bar to reduce the die bonding stress and to allow the use of gold-tin solder, improving the reliability of the overall assembly in long-pulse operation, and (2) eliminating the electro-chemical corrosion and erosion of the mini-channels to improve the long-term reliability of the heatsink itself. In active heatsinks the circulating fluid is typically 200-400 μm from the bar mounting surface. The short heat diffusion length and the large surface area of the mini-channels together support very effective heat removal from laser-diode bars in operation at high average power. Raman Srinivasan et al. reported a water-cooled, mini-channel heatsink with a CTE of 6.8 ppm/°C (near to the nominal 6.5 ppm/°C CTE of GaAs) and a thermal resistance of 0.43 K/W (Du et al., 2008). This mini-channel heatsink is build up with a multi-layer heatsink consisting of both ceramic and Cu layers. The relatively low CTE of the ceramic combined with the high CTE of Cu enables CTE matching of the Cu mounting surface to the GaAs diodelaser bar. The ceramic layers enable electrical isolation of the fluid, eliminating the possibility of electrolytic corrosion. While smaller channels with high flow velocities offer superior thermal performance, these compromise both the manufacturability and operational reliability of the mini-channel heatsink. Taking into consideration the latter requirements, the minimum channel width is best kept above 200 μm, smaller than that of the predecessor product 300 μm. John Vetrovec et al. reported a novel active heat sink high-power laser diodes offering unparalleled capacity in high-heat flux handling and temperature control.(Vetrovec et al., 2010) The heat sink receives diode waste heat at high flux and transfers it at reduced flux to environment, coolant fluid, heat pipe, or structure. The thermal resistance can be < 0.1 K/W. In addition, thermal conductance of the heat sink is electronically adjustable, allowing for precise control of diode temperature and the output light wavelength. When pumping solid-state lasers, diode wavelength can be precisely tuned to the absorption features of the laser gain medium. Figure 12 shows the specially designed micro-channel heat sink structure of our lab. The entire heat sink is divided into five layers, and each layer is made by the lithography graphic arrangement technology forming special micro-channel alignment. The cooling water is not designed to flow along the length direction of the bar, but the direction of the cavity length, so that the cooling effect for each emitters of the bar is uniform. The heat is taken away through the second, third and fourth layer of the heatsink, and then the coolant tumbles down to the bottom of the bar, and flow out. The first layer where to contact the cooling water is made of small fine serration to increase the surface contact area of heat exchange with the water. Thermal simulation of the whole heatsink structure is given in Figure 13. The thermal resistance of the heat sink is down to 0.34 K/W or less with excellent heat dissipation.

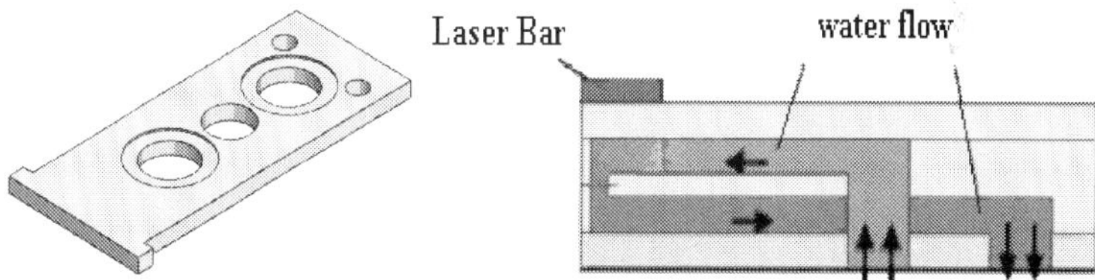

Fig. 12. The appearance and the internal structure of micro-channel heat sink of our lab.

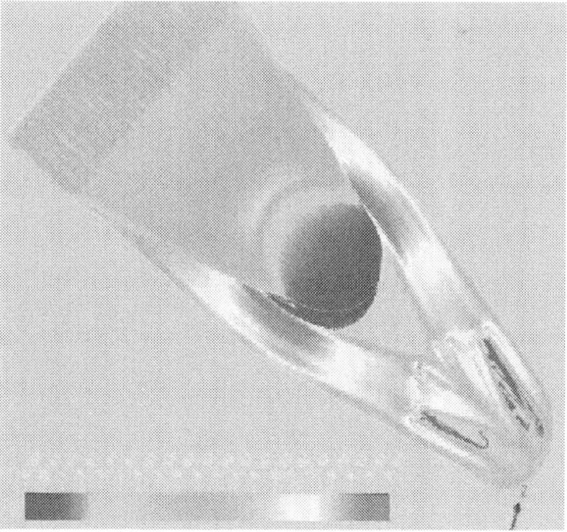

Fig. 13. Thermal simulation of micro-channel heat sink of our lab.

4.2.3 Passive heatsinks

The use of high-thermal-conductivity materials in passive heatsinks is critical to limiting bar operating temperature as required to increase output powers and device lifetimes. The most promising passive heatsink for bars are a solid Cu heatsink with a diamond-based composite submount and a Cu-composite material with no submount. Given the relatively low thermal resistance, improvements in the manufacturability of the former design are of particular interest. Issues associated with the surface roughness and edge quality of the submount have been addressed by a combination of plating and lapping processes and the submount material itself has been engineered to closely match the CTE of the GaAs diode-laser bar.(Du et al., 2008) Simulated results of CTE-matched passive heatsinks for bars (either a Cu-composite heatsink or a Cu heatsink with a diamond-based composite sub-mount) offer a reduction in thermal resistance of 16-19%. Initial CW testing of standard bars at room temperature on Cu-composite heatsinks has demonstrated almost 10 W/bar more than that obtained with the standard Cu CS package. The lead candidate designs for the next-generation heatsinks for single-emitter devices are multi-layer Cu/ceramic heatsink and diamond-based composite heatsink. Modeling of next-generation passive heatsinks for single-emitter devices also shows the opportunity to improve device performance at high power and high temperature. A multi-layer Cu/ceramic design promises a 25% reduction in thermal resistance compared to the standard CuW CT while a diamond-based composite heatsink promises a 50% lower thermal resistance than CuW and it is CTE matched to GaAs. CW COMD testing of devices on the multi-layer heatsink at room temperature show an increase of >20% in the damage threshold.

4.3 Assembly technology

In addition to the requirements of aggressive heat sinking and low cost, the package was also required to be modular. It means to minimize the number of external hydraulic and electrical connections to the large two dimensional arrays that were to be constructed using many of the modular packages.

At the present time, there are almost as many diode array pump architectures as there are solid-state laser designs and manufacturers. For certain applications, such as side pumping of a solid state laser, when higher optical power are required, and the laser beams are not needed to be focused, an array of laser bars can be packaged horizontally. Figure 14 gives the example of this horizontally package structure bar arrays of our lab. For these packages, the output power can range from tens of Watts to hundreds of Watts to even thousand of Watts depending on how many bars are packaged together and the power of each bar. However, the number of bars can be packaged is limited and the output power is limited because these horizontal packaged modules are thermally in series. With the demand of higher output power, vertical bar stacks becomes the choice. Several different diode array vertical packaging concepts have to date been more commonly used than most for the purpose of packaging high power diode laser bars. These packaging methods can be described as (a) rackand-stack in which bars on individual submounts are grouped together and back plane cooled, (b) diodes-ingrooves in which a single submount, grooved with sawn slots to accompany the diode bars is used to package a group of bars which are again back plane cooled, and (c) microchannel cooled diode bars in which coolers are matched one for one to diode bars and the cooler-bar combinations are then grouped to form diode stacks. Each packaging method has its own advantages and disadvantages which depend upon application requirements, optoelectronic performance of the diode bars being mounted, and the cost centers attendant with each packaging technology. (Solarz et al., 1998) Three high-power-bar vertical packaging approaches are shown in Figure 15.

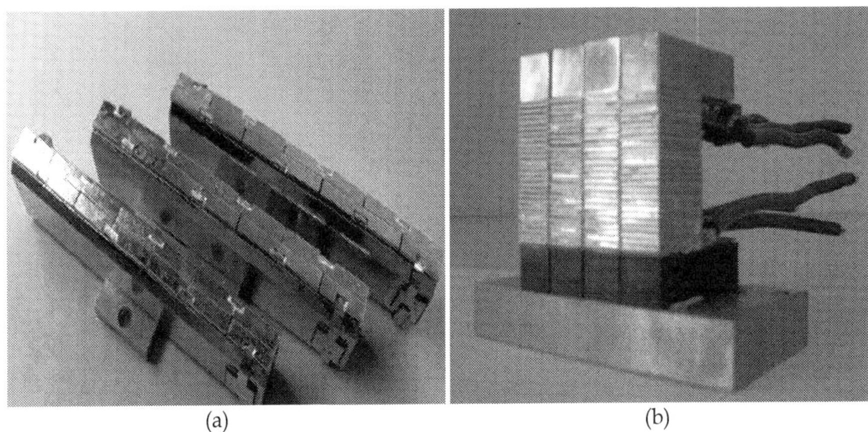

(a)   (b)

Fig. 14. Examples of horizontal bar arrays (a) and vertical bar stacks (b) from our lab.

Since the modularity of the package was required to ensure maintainability of large two-dimensional arrays, it was desired to hold the level of diode array integration to several linear centimeters of diode array bar length per package so that if it became necessary to service a large array, small sections could be replaced cheaply and easily. In many of the military systems, weight and volume envelopes are often the deciding factor as to a system's viability for its intended mission. In commercial systems, it is attractive that the simplicity, improved maintainability, and cost savings can be realized when the hydraulic circuits are simplified. As in Figure 16, a 20-bar, 0.4 mm spaced array was mounted on a simple copper block, which can in turn be mounted on any secondary cooler.(Endriz et al., 1992) Array wavelength spread remained under 3 nm to 2% duty cycle for 1200W operating power (1500 W/cm²). For high duty cycle arrays, an approach manufacturing the simple "linear sub assembly" as the basic building block of the laser array allowed spanning the broad range of bar duty cycles while keeping the volume manufactured laser array unchanged, as shown in Figure 17. (Solarz et al., 1998) Our lab's products of high power semiconductor laser bars and arrays can provide full range of wavelength requirements from 780 to 980nm. Modular devices for QCW (2% or 20% duty factor) operation reached 10-100 kW class output power with MTTF>1×10⁹. For stable CW operation module, kW class output power with MTTF>5000 hours can be provided. Our lab can also supply diversified packaging form devices designed by customer.

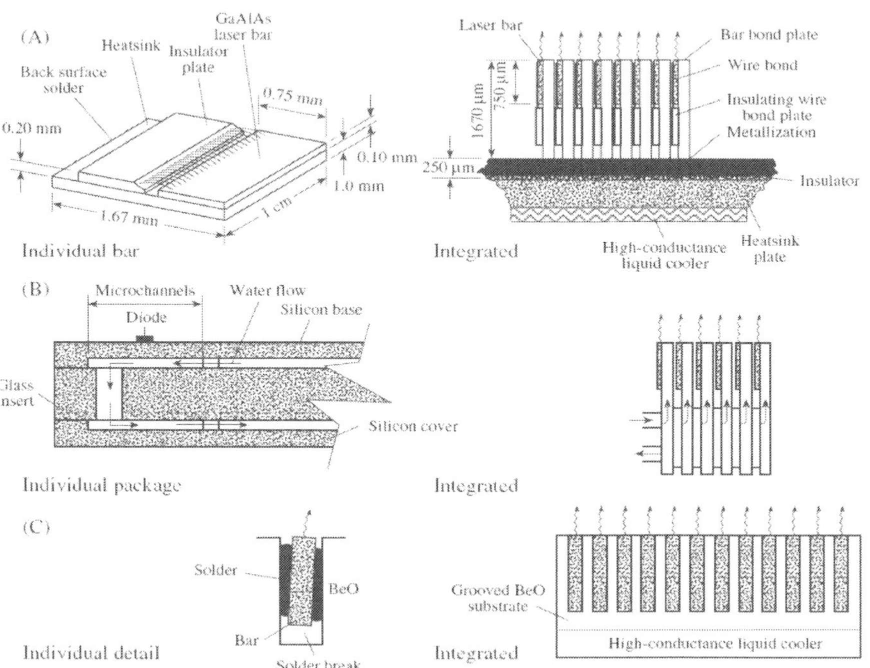

Fig. 15. Three high-power-bar vertical packaging approaches: (A) individual bar mounted on submount and back-plane cooled, (B) individual bars mounted on individual micro channel coolers and (C) diodes-in-grooves mounting into single BeO substrate.

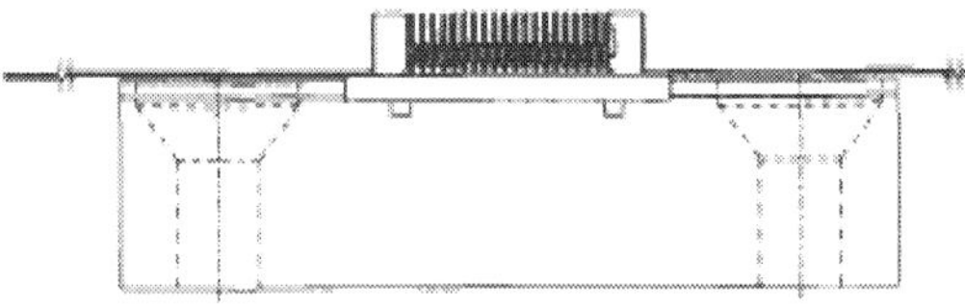

Fig. 16. 20 bar, 0.8 cm² array (1500 W/cm2) mounted on a copper block. This design minimizes overall packaging costs.

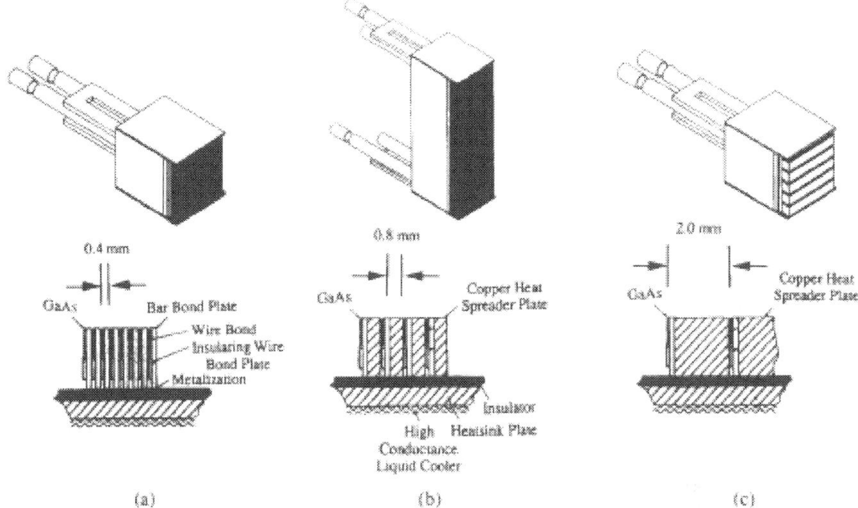

Fig. 17. High duty cycle liquid cooled arrays for slab pumping. (a) 1500W/cm2 peak power array on 1cm x 1cm cooler (>3% duty cycle, 0.4 mm spacing) (b) 800 W/cm^2 peak power array on 1cm x 3cm cooler (> 6% duty cycle, 0.8 mm spacing) (c) 300 W/cm^2 peak power array on 1cm x 1cm cooler (20% duty cycle, 2.0 mm spacing).

5. Developments of high-power semiconductor lasers with high beam quality

Limited by the quantum well waveguide structure, the output beam quality of the semiconductor is poor when compared with the solid-state lasers, CO_2 lasers and other traditional lasers, which impeded the expansion of its applications. In recent years, the beam quality has been improved rapidly by the increase of the output power and the reduction of the divergence angle, which is directly reflected in the decrease of the fiber diameter and the increase of the output power for fiber-coupled diode laser module. The type of single-wavelength fiber-coupled diode laser modules can be divided into several specific forms according to its internal semiconductor laser devices and different types of packages.

5.1 Fiber-coupled integration of single-emitters

In the case of less demand for the output power, a single emitter can be directly coupled into the fiber, as shown in Figure 18. This structure has the advantages of small size, low cost, long life span and mature technology etc., and the output power of up to 8 ~ 10 W/module level can be afforded by numbers of semiconductor laser suppliers currently. In the case of high optical power requirements, the multiple collimated beams emitted from single emitters can be used by fast axis collimators (FAC), in which the fast axis direction is packed closely, and then the polarized beam is combined and coupled into the fiber. In 2009, the Nlight Company integrated 14 single emitter units into a module using this structure and obtained 100 W output power from 105 μm core diameter optical fiber (N.A. =0.15) (Figure

2), with the coupling efficiency of 71%. This type of module structures has the advantages of small size, high brightness, long life etc., but it is difficult and costly to alignment the internal optical components.

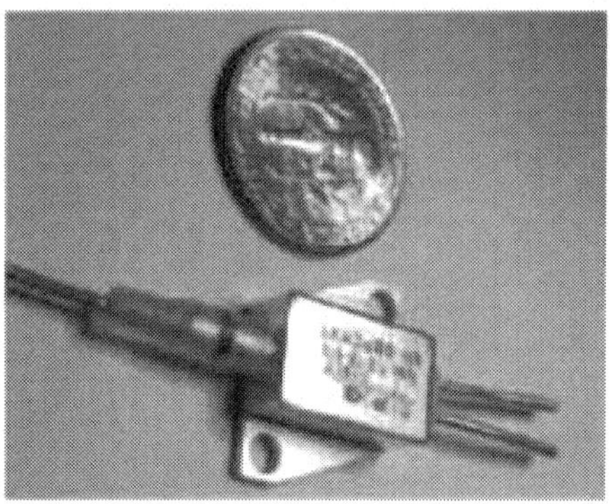

Fig. 18. The fiber-coupled output module from a single emitter.

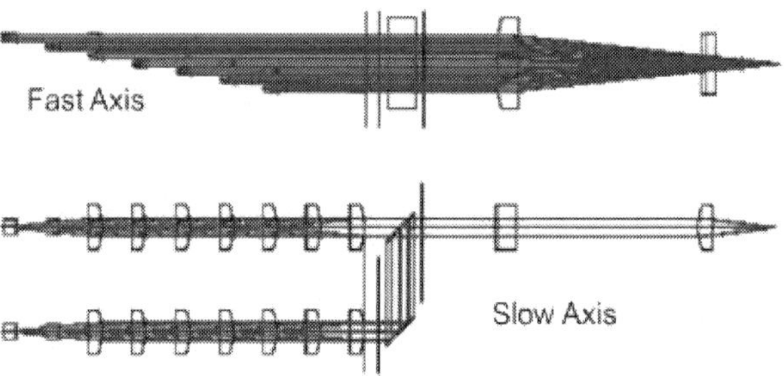

Fig. 19. The optical structure figure of the integrated fiber-coupled output module from multi-single emitters.

5.2 The integrated fiber-coupled output of the mini bars

The integrated fiber-coupled output of mini bars is to make the FAC collimated beams from multiple mini bars arrange closely along the fast axis direction, and then focus and couple the beam into the fiber after combining the polarization beam. In 2007, the DILAS Company achieved the output power of 500 W from 200 μm core diameter optical fiber module (NA = 0.22), with the coupling efficiency of 83%, as shown in Figure 20. This module possesses high brightness and long life advantages, but its integration and alignment is difficult and the cost is high due to large numbers of the optical components in the module.

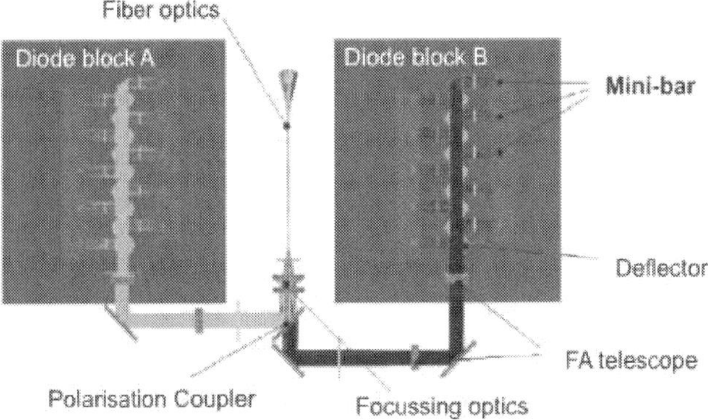

Fig. 20. The integrated fiber-coupled output module from multi-mini bars.

5.3 The fiber-coupled output of the semiconductor laser array stacks packaged with micro-channel heat sink

The fast and slow axis collimated beams from the multiple micro-channel heat sink packaged laser array stack are integrated spatially followed by the beam uniformization, and then focus and couple the beams into fiber, as shown in Figure 21. Currently, the output power of this module is up to 400 W with 200 μm core diameter optical fiber (N.A.= 0.22). The module with this structure has the higher brightness, less optical components and simple structure, but the cost is higher. In addition, the deionized water must be used as cooling medium, so the maintenance demand is high. Meanwhile, since the lifetime of the micro-channel heat sink is short due to the erosion of the deionized water, the lifetime of this module would be only about 20,000 hours if the elaborate cooling water management is unavailable.

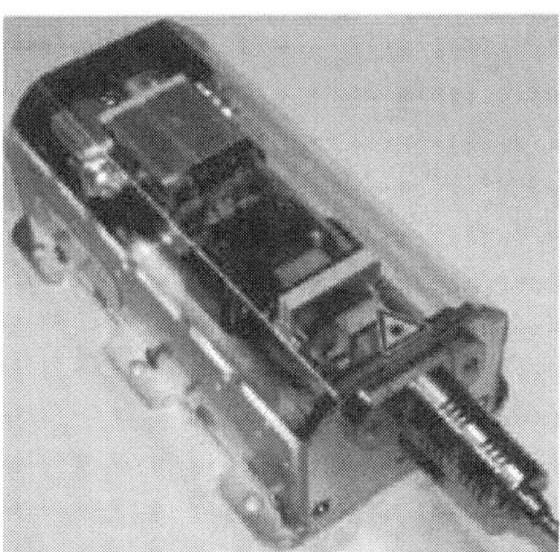

Fig. 21. The integrated fiber-coupled output module from laser array stacks packaged with micro-channel heat sink.

Fig. 22. The integrated fiber-coupled output module from laser array laser arrays packaged by conduction heat sink.

5.4 The fiber-coupled output of the semiconductor laser arrays packaged by conduction heat sink

The beams from the multiple conductive heat sink packaged laser array are integrated spatially after the fast and slow axis collimation, and then go into the optical fiber directly through focusing and coupling system. Currently, the DILAS Company used this idea and achieved the output power of 200 W and 500 W respectively from 200 µm core diameter optical fiber module (NA = 0.22) and 400 µm core diameter optical fiber module, with the coupling efficiency of about 80%. Despite of the relatively lower brightness, compared with other structures, this type of module structures, as shown in Figure 22, has a lot of advantages, such as less optical components, simple structure, long lifetime, maintenance-free and low cost.

As for the direct industrial applications using the high power and high beam quality semiconductor lasers, besides the methods mentioned above, the wavelength-beam-combining technology and the polarization-beam-combining technology can be used to obtain high power and high beam quality if the material processing is insensitive to the wavelength. It can also multiply the output power according to the number of the combining wavelengths and remain the output beam quality unchanged. In this field, the Laserline Company has the leading technologies. They integrated the micro-channel packaged cm bar stacks and achieved high output power from the level of hundreds of watts to the level of megawatts with high beam quality laser processing systems: 2000 W (BPP: 20 mm·mrad), 4000 W (BPP: 30 mm·mrad), 10000 W (BPP: 100 mm·mrad).

6. Conclusion

High power semiconductor laser diode technology is rapidly maturing technically and commercially. Laser diode prices are now low enough to intersect a broad range of commercial applications which has resulted in increased production quantities and competition. In recent years, with the rapid development of semiconductor material epitaxial growth technology, waveguide structure optimization technique, cavity surface-passivation technology, high stability packaging technology, high efficiency cooling technology, particularly driven by the demand of applications in the direct semiconductor laser industrial processing and high-power fiber laser pumping, high power, high operational temperature, high beam quality semiconductor lasers has been developed rapidly, which provide a light foundation for high-quality, high performance solid state laser pumping system, direct semiconductor laser processing equipment and high-performance high-power fiber laser pump source.

Along with diode bar improvements, enhancements in the component and system design must be developed in order to handle the increased current and heat loads at these high powers. The improvement in power limits for laser diode bars will require design advancements for all of the parts of the laser diode system from a multidisciplinary team. Modularity configuration for special application requirements will certainly be a main trend of the developing high power diode lasers. Since the high power laser diode stacks of tomorrow must be looked at the diodes as a part of a large integral system, considerations such as the dimension and weight of a wire or the delivery of water must be weighed by the future designers. When these new elements are combined to form high power stacks coupled with high performance packaging and systems, they will open the doors to new market opportunities.

7. References

Crump P., Wang J., Patterson S., Wise D., Basauri A., DeFranza M., Elim S., Dong W., Zhang S., Bougher M., Patterson J., Das S., Grimshaw M., Farmer J., DeVito M. & Martinsen R. (2006). Diode Laser Efficiency Increases Enable >400-W Peak Power from 1-cm Bars and Show a Clear Path to Peak Powers in Excess of 1-kW. *Proceedings of SPIE*, Vol. 6104, pp.610409-1-10, ISBN, 9780819461469, San Jose, CA, USA, January 23, 2006

Du J.H., Zhou H.L., Schleuning D., Agrawal V., Morales J., Hasenberg T. & Reed M. (2008). 8xx nm kW Conduction Cooled QCW Diode Arrays with both Electrically Conductive and Insulating Submounts. *Proceedings of SPIE*, Vol.6876, pp.687605-1-11, ISBN 9780819470515, San Jose, CA, USA, 21 January 2008

Endriz J. G., Vakili M., Browder G. S., Devito M., Haden J. M., Harnagel G. L., Plano W. E., Sakamoto M., Welch D. F., Willing S., Worland D.P. & Yao H. C. (1992). High Power Diode Laser Arrays. *IEEE J. of Quantum Electronics*, Vol.28, No.4, (April 2009), pp, 952-965, ISSN 0018-9197

Fan L., Cao C. S., Thaler G., Nonnemacher D., Lapinski F., Ai I., Caliva B., Das S., Walker R., Zeng L. F., McElhinney M. & Thiagarajan P. (2011). Reliable High-Power Long-Pulse 8XX-nm Diode Laser Bars and Arrays Operating at High Temperature. *Proceedings of SPIE*, Vol. 7918 pp. 791805-1-7, ISBN 9780819484550, San Francisco, California, USA, January 23, 2011

Krejci M., Gilbert Y., Müller J., Todt R., Weiss S. & Lichtenstein N. (2009). Power Scaling of Bars towards 85mW per 1µm Stripe Width Reliable Output Power. *Proceedings of SPIE*, Vol. 7198, pp.719804-1-12, ISBN 9780819484550, San Jose, CA, USA, January 26, 2009

Lammert R.M., Oh S.W., Osowski M.L., Panja C., Qian D., Rudy P.T., Stakelon T. & Ungar J.E. (2005). Advances in Semiconductor Laser bars and Arrays. *Proceedings of SPIE*, Vol.5887 pp. 58870B-1-11, ISBN 9780819458926, San Diego, CA, USA August 2, 2005

Leers M., Boucke K., Scholz C. & Westphalen T. (2007). Next generation of Cooling Approaches for Diode Laser Bars. *Proceedings of SPIE*, Vol. 6456, pp.64561A-1-10, ISBN 9780819465696, San Jose, CA, USA, January 22-24, 2007

Li H.X., Reinhardt F., Chyr I., Jin X., Kuppuswamy K., Towe T., Brown D., Romero O., Liu D., Miller R., Nguyen T., Crum T., Truchan T., Wolak E., Mott J. & Harri J. High-

Efficiency, (2008). High-Power Diode Laser Chips, Bars, and Stacks. *Proceedings of SPIE*, Vol. 6876, pp.68760G-1-8, ISBN 9780819470515, San Jose, CA, USA, January 21, 2008

Lichtenstein N., Manz Y., Mauron P., Fily A., Schmidt B., Müller J., Arlt S., Weiß S., Thies A., Troger J. & Harder C. (2005). 325 Watt from 1-cm Wide 9xx Laser Bars for DPSSL- and FL-Applications. *Proceedings of SPIE*, Vol. 5711, pp. 1-11, ISBN 9780819456854, San Jose, CA, USA, January 25, 2005

McNulty J.(2008). Processing and Reliability Issues for Eutectic AuSn Solder Joints. *International Microelectronics and Packaging Society. Permission granted from the 41- International Symposium on Microelectronics (IMAPS) Proceedings*, pp.909-916, ISBN 0-930815-86-6, Providence, Rhode Island, November 2-6, 2008

Müller J., Todt R., Krejci M, Manz-Gilbert Y., Valk B., Brunner R., Bättig R. & ichtenstein N. (2010). CW to QCW Power Scaling of High Power Laser Bars. *Proceedings of SPIE*, Vol. 7583, pp.758318, ISBN 9780819479792, San Francisco, California, USA, January 25, 2010

Puchert R., Bärwolff A., Voß M., Menzel U., Tomm J. W. & Luft J. (2000). Transient Thermal Behavior of High-Power Diode-Laser Arrays, *IEEE Transaction on Components, Packaging and Manufacturing Technology*, Part A, Vol.23, No. 1, (March 2000), pp. 95-100, ISSN 1521-3331

Schleuning D., Griffin M., James P., McNulty J., Mendoza D., Morales J., Nabors D., Peters M., Zhou H. L. & Reed M. (2007). Robust Hard-Solder Packaging of Conduction Cooled Laser Diode Bars. *Proceedings of SPIE*, Vol.6456, pp.645604-1-11, (2007) ISBN 9780819465696, San Jose, CA, USA, January 22-24, 2007

Solarz R. W., Emanuel M. A., Skidmore J. A., Freitas B. L. & Krupke W. F. (1998). Trends in Packaging of High Power Semiconductor Laser Bars. *Laser Physics*, Vol. 8, No. 3, (March 1998), pp. 737-740, ISSN 1729-8806

Srinivasan R., Miller R. & Kuppuswamy K. (2007). Next-Generation Active and Passive Heatsink Design for Diode Lasers. *Proceedings of SPIE*, Vol.6456, pp.64561D-1-10, ISBN 9780819465696, San Jose, CA, USA, January 22-24, 2007

Tu L.W., Schubert E.F., Hong M. & Zydzik G.J. (1996). In-Vacuum Cleaving and Coating of Semiconductor Laser Facets Using Thin Silicon and a Dielectric. *Journal of Applied Physics*, Vol. 80, No.11, (December 1996), pp. 6448-6451, ISSN 0021-8979

Vetrovec J., Feeler R. & Bonham S. (2010). Progress in the Development of Active Heat Sink for High-Power Laser Diodes. *Proceedings of SPIE*, Vol.7583, pp.75830K-1-8, ISBN 9780819479792, San Francisco, California, USA, January 25, 2010

Voss M., Lier C., Menzel U., Bärwolff A. & Elsaesser T. Time-Resolved Emission Studies of GaAs/AlGaAs Laser Diode Arrays on Different Heat Sinks. *Journal of Applied Physics*, Vol.79, No.2, (January 1996), pp.1170-1172. ISSN 1089-7550

Yanson, D., Levi, M., Shamay M., Teslera R., Rappaporta, N., Dona Y., Karnia Y., Schnitzera, I., Sicronb N. & Shustermanb S. (2011). Facet Engineering of High Power Single Emitters. *Proceedings of SPIE*, Vol.7918, pp.79180Z-1-12, ISBN 9780819484550, San Francisco, California, USA, January 23, 2011

Ziegler M., Weik F., Tomm J. W., Elsaesser T., Nakwaski W., Sarzala R. P., Lorenzen D., Meusel J. & Kozlowska A. Transient Thermal Properties of High-Power Diode Laser Bars. *Applied Physics Letters*, Vol. 89, No. 26, (December 2006), pp.3506-1-3, ISSN 1077-3118

Zweben C., (2004). New Material Options for High-Power Diode Laser Packaging. *Proceedings of SPIE*, Vol. 5336, pp166-175, ISBN 9780819452443, Bellingham, WA, USA, June 1, 2004

Part 4

Optical Switching Devices

17

Energy Efficient Semiconductor Optical Switch

Liping Sun and Michel Savoie
Communications Research Centre
Canada

1. Introduction

Energy-saving technology that reduces power consumption is of increasing importance due to the ever-increasing demand for Internet services. To prevent the traffic growth from being strangled by energy bottlenecks, novel architectural and technological solutions are indispensable. The most obvious way to cope with the issue is to reduce the energy consumed by the network elements.

Fast optical switching is an important enabler of advanced optical networks, in particular such functions as routing burst and packet optical signals, optical path provisioning and fault restoration. Semiconductor Digital Optical Switches (DOSs) can fulfill such high speed applications due to their nanosecond switching times, step-like switching responses, and immunity to variations in temperature, wavelength, polarization, refractive index and device fabrication tolerances. Moreover, semiconductor DOSs offer the potential for integration with other semiconductor optoelectronic components and thus promise considerable reductions in the size, complexity and cost of an overall optical system.

For optical waveguide switches, fast optical switching may be achieved by a refractive index change, induced either by carrier injection (Zegaoui et al., 2009; Bennett et al., 1990) or by the electro-optic effect (Cao et al., 2009; Agrawal et al., 1995), within III-V semiconductors, such as GaAs-based and InP-based. Compared to carrier-injection switches, electro-optic switches have faster switching speeds but larger switching voltages since the refractive index change induced by an electro-optic effect is about two orders of magnitude smaller than that by carrier injection.

Therefore, until now, most of the commercially available semiconductor DOS products have been based on carrier-injection (Ikezawa et al., 2008). These devices typically utilize carrier-induced Total Internal Reflection (TIR) at a waveguide branching or crossing point to switch the light path from one waveguide to another. Such TIR-based semiconductor switches typically require a large index modulation, e.g. in the order of 0.01, with the region of changed index having a well-defined boundary. Accordingly, efforts have been made to restrict current spreading and to confine the injected carriers to the desired region. Typically, achieving such carrier confinement involves using relatively complex semiconductor device technologies, such as ion implantation (Abdalla et al., 2004; Zhuang et al., 1996), electron-beam lithography (Shimomura et al., 1992), Zn diffusion (Yanagawa et al., 1990), and epitaxial regrowth (Thomson et al., 2008).

A plane view of a typical conventional semiconductor DOS is illustrated in Fig. 1(a), wherein two waveguides intersect at an angle θ forming an X-like waveguide structure. An electrode is provided over a common waveguide region where the waveguides intersect for injecting carriers into a portion of the common region to decrease its refractive index n by an amount Δn sufficient to induce TIR for the input beam and to cause it to turn by the angle θ to form the switched light. In the absence of the carrier injection, i.e. when there is no current flowing through the electrode, the input light continues its propagation along the waveguide past the common region to form the transmitted light. In order to switch the input light, the electrode is positioned so that its edge, which faces the input light, crosses the input waveguide at an angle θ/2. One drawback of the design in Fig. 1 is that, for optimal performance, the electrode should be positioned with its 'receiving' edge centered in the common waveguide region. This makes the structure asymmetrical with respect to the two input waveguides, so that the switch is, essentially, a 'Y' switch that can only function as a 1×2 switch and not as a 2×2 switch.

A typical 2×2 semiconductor TIR DOS with the conventional electrode design is shown in Fig. 1(b). In order to reduce an optical axis misalignment of the reflected light for both input ports, the electrode has to be narrow; disadvantageously, a narrow electrode will generally result in poor reflectivity induced by the current passing through the electrode, leading to a degradation of the switch extinction ratio, i.e. the optical power ratio for the output ports.

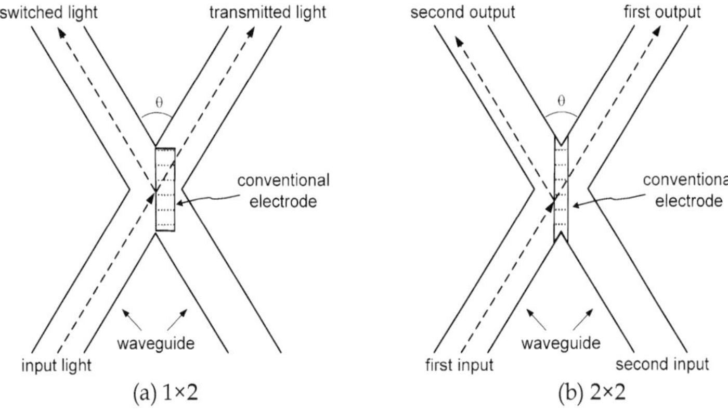

Fig. 1. Schematic diagrams of TIR optical switches with conventional electrodes.

Fig. 2 is a schematic diagram of a 2×2 switch which has a similar waveguide topology as Fig. 1(b), but utilizes the bow-tie electrode in place of the straight electrode of Fig. 1(b) (Li et al., 2001). The bow-tie electrode has a narrow part located at the central point of the waveguide intersection, and the electrode is symmetrical with respect to all four waveguide ends. The switch has a switching angle θ = 2°, and an electrode bow-tie angle of 1.5 degrees. The switch utilizes ion implantation about the electrode to decrease current spreading and confine the injected carriers. One disadvantage of the prior art bow-tie electrode design is that its electrode is still narrow at the waveguide crossing point, potentially enabling the light to leak therethrough under the carrier injection condition. Furthermore, a part of the incoming light that incidents on the second electrode segment may be refracted rather than reflected because of the increased grazing angle. All these limit the waveguide crossing angle θ of the bow-tie type 2×2 switches to about 2° in real device applications. The small switching angle increases the size of the device and limits the integration density.

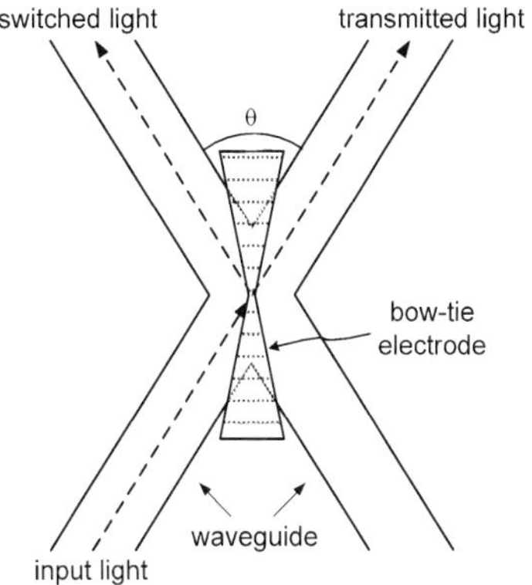

Fig. 2. Schematic diagram of a 2×2 TIR optical switch with bow-tie electrode.

Another disadvantage of the prior art TIR semiconductor switches is their relatively large power consumption, especially for devices with the switching angle at the higher end of the range, as such devices require larger electrical currents. Reconfigurable waveguide DOS devices with small switching current and low crosstalk have been reported recently (Zegaoui et al., 2009; Ng et al., 2007), but their branch angle is only 0.9°. It is therefore desirable to develop TIR switches with greater deflection angles and reduced power consumption.

This chapter presents our approaches to improve the energy efficiency of semiconductor DOSs for applications in next generation green optical networks. In the following sections, we first propose and demonstrate a novel double-reflection switch design that can reduce the power consumption and increase the switch deflection angle. Then, in the next section, an ultra-low-power 2×2 carrier-injection optical switch based on a Mach-Zehnder Interferometer (MZI) formed by Multi-Mode Interference (MMI) couplers is introduced and investigated. Conclusions and future applications are provided in the last section.

2. Double-reflection TIR DOS

This section discusses the design, fabrication and characterization of a novel double-reflection technique to fabricate 1×2 and 2×2 wide-angle AlGaAs/GaAs carrier-injection TIR optical switches. This new development uses only conventional semiconductor fabrication technologies to reduce the switching current and relax the tight restriction on the width of the electrodes located at the switch centre. To compensate for the increased switching current, compositionally graded interfaces are added at the heterojunctions to reduce the switching voltage and thus the power consumption. Switching performance is further improved by applying curved electrodes and carrier-restriction gaps.

2.1 Double-reflection structure
Our approach to improve the performance and reliability of 1×2 and 2×2 TIR optical switches with a double-reflection electrode geometry is illustrated in the two pictures shown

in Fig. 3 below. The electrode in Fig. 3(a), which utilizes a well known 'Y'-shaped waveguide structure, is shaped to induce double reflection of the input light. Its first edge faces the input waveguide and is positioned for turning, in the presence of the carrier injection, the input light by a first deflection angle θ_1, so as to form the first reflected light that propagates generally towards the second edge. The second edge is positioned for turning the first reflected light by a second angle θ_2 towards the switched output waveguide for coupling thereinto as the switched light. This electrode is also shaped and positioned so that, in the absence of the carrier injection, substantially all or at least most of the input light passes under the first edge of the electrode. This design also results in that in the presence of the carrier injection through the electrode, substantially all or at least most of the input light is reflected at the first edge. Note that this feature differentiates the electrode of Fig. 3(a) from the prior art bow-tie electrode of Fig. 2, wherein the input light impinges upon two inclined edges of the bow-tie in substantially equal portions, so that only about half of the input light passes under each bow-tie edge, resulting in a loss of input light since for a large switch branching angle the part of input light that directly incidents on the second edge of the electrode will not be reflected under TIR due to their increased grazing angle.

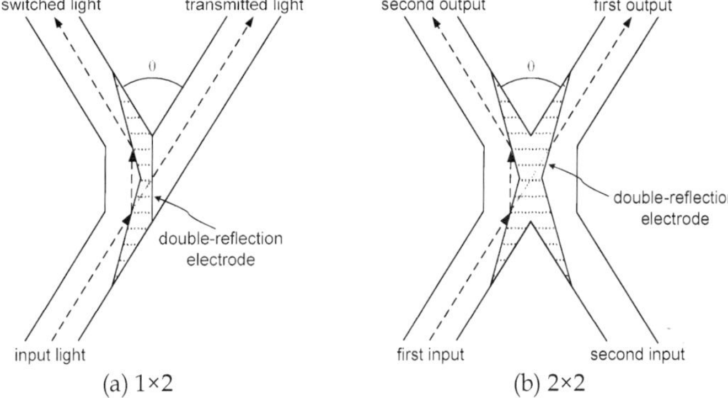

Fig. 3. Schematic diagrams of TIR optical switches with double-reflection electrode.

Continuing to refer to Fig. 3(a), in a further advantage each of the first and second deflection angles is less than the light switching angle θ, and therefore the switching of the input light's direction by the switching angle θ may be accomplished by a smaller refractive index change Δn under the electrode, and thus using a smaller injection current than would have been required for the conventional single-reflection electrode design such as that of Fig. 1. In one currently preferred electrode geometry design, the first and second angles θ_1, θ_2 are both equal to one half of the branching angle θ, i.e. $\theta_1 = \theta_2 = \theta/2$.

Advantageously, this double-reflection structure enables to have the switching angle θ twice as large as in the conventional single-reflection design for the same refractive index change Δn in the index change region, or, alternatively, to turn the light by the same angle θ using only half of the electrical current. Indeed, the reflective index change Δn that is required to induce the TIR at the first and second edges may be estimated from an approximate TIR condition,

$$\theta_i \le \Delta n/n, \ i = 1, 2, \tag{1}$$

which approximately holds for $\Delta n/n \ll 1$ and the angles θ_i measured in radians, so that reducing the turning angle by a factor of 2 reduces the required index change by the same factor.

The double-reflection structure that has been described above for 1×2 optical switches, can be easily adopted for 2×2 switches, such as schematically shown in Fig. 3(b), by adding a second input waveguide and by symmetrically extending the double-reflection electrode so as to provide a second pair of electrode edges for switching the light from the second input waveguide into the first output waveguide by means of double reflection at the third and fourth electrode edges. In the 2×2 switch, the second input waveguide is optically aligned with the second output waveguide so that, in the absence of the carrier injection, input light from the second input waveguide will be transmitted by the branching section into the second output waveguide. In the presence of the carrier injection, light that enters from the second input waveguide experiences two consecutive reflections at the third and fourth edges and is directed into the first output waveguide, while light that enters from the first input waveguide also experiences two consecutive reflections at the first and second edges as described before, and is directed into the second output waveguide. The waveguide and electrode edge structure of the 2×2 switch is substantially symmetrical with respect to an axis of symmetry which bisects the waveguide branching angle. In Fig. 3(b), the electrode is shaped as an 'X', with two cut-outs around the symmetry line in the input and output portions of the structure, so as to reduce the scattering optical loss, the total electrical current required for the switching, and therefore to reduce the power consumption and device heating.

The aforedescribed double-reflection switching addresses important drawbacks of the prior art TIR-based optical switches. It eliminates the optical misalignment problem and electrode width problems of the straight and bow-tie electrodes in the 2×2 switching configurations. As a result, at the same switch crossing angle, the index change, and hence the injection current density, that is required by the TIR condition needs to be only half of that of the prior art "single reflection" design, thereby reducing the injection current density by half for the same switch crossing angle θ. On the other hand, if the same switching current density is to be used as in the prior art TIR-based switches, the switch crossing angle can be doubled, making the overall device more compact and thus allowing more devices to be integrated on a wafer. This newly developed technique enables optical switching at greater waveguide crossing angles θ and/or smaller induced refractive index changes, and thus smaller power consumption.

One potential disadvantage of the aforedescribed multi-reflection switch configuration is that, for the same switch crossing angle θ, the overall area of the switch electrode may be as much as two times larger than that of the prior art single-reflection electrodes, which may partially negate the effect of the reduced current density upon the total power consumption of the device; the larger electrode area and a relatively more complex waveguide branching area may also lead to an increase in the scattering optical loss in the waveguide branching region. Advantageously, these two potentially deleterious effects decrease dramatically by increasing the switch crossing angle θ, and may be reduced to an acceptable level at least for θ greater than about 3-4°.

2.2 Curved electrodes

A schematic layout of our proposed 2×2 switch with double-reflection structure is shown in Fig. 4. Each of the four segments of the switch electrode is curved in a logarithmic spiral shape instead of the conventional straight electrode. This electrode-curvature has been theoretically studied and reported that it could provide high power reflectivity, high extinction ratio, and low scattering loss (Nayyer et al., 2000).

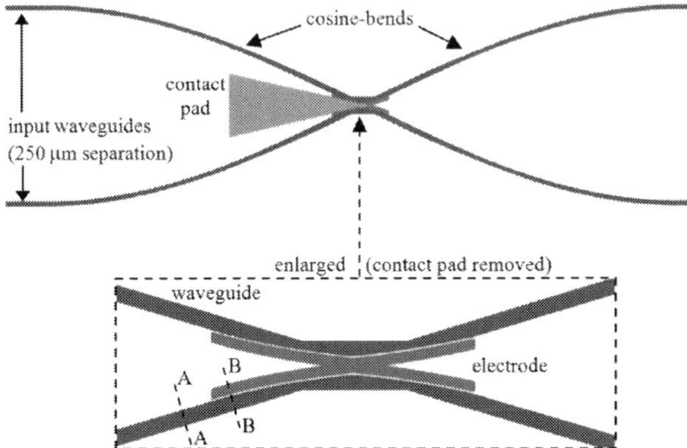

Fig. 4. Schematic layout of a proposed 2×2 double-reflection TIR optical switch.

The fabricated optical switch has the two input waveguides and two output waveguides extended with curved waveguide sections embodied as cosine bends, with a functional form $[h/2-(h/2)\cos(\pi x/l)]$, where h is the height of the bend, l is the length of the bend, and x is the horizontal propagation length. The device further included a top contact pad to provide an electrical contact to the switching electrode for delivering current to the switch electrode. To reduce the crosstalk at the waveguide crossing, adiabatic tapers are used in the waveguide branching region. This waveguide widening can also provide more fabrication freedom for the electrode. As shown in the figure, the double-reflection design enables us to set the electrode width at several times larger than the light penetrating depth and hence provide a sufficient index change for TIR and large extinction ratio.

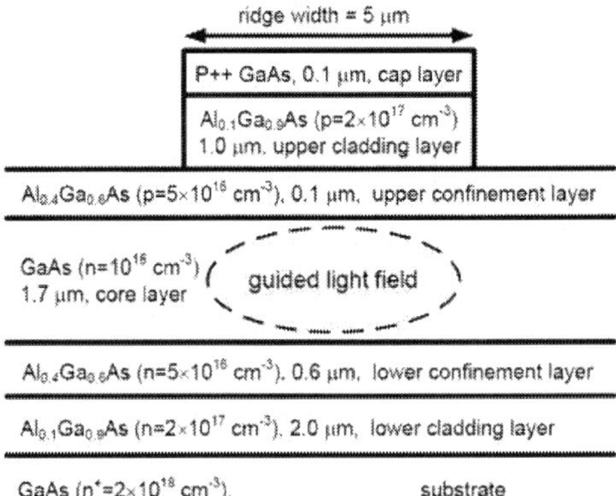

Fig. 5. The waveguide layer structure which specifies the composition ($Al_x Ga_{1-x} As$), doping profile, and layer thickness for each layer.

Fig. 5 is a schematic diagram of the vertical cross-section of the waveguide structure that we used for the 2×2 switch along the line 'A-A' shown in Fig. 4. The switch's strip-loaded waveguide design has a multilayer heterostructure with a W-shaped index profile. The waveguide core layer of GaAs is 1.7 µm thick, and is, to the best of our knowledge, the largest ever reported for a single-mode semiconductor DOS. It provides high coupling efficiency to a single mode fibre.

2.3 Current restriction gap

At the TIR interface inside our switch design, a small isolation gap is introduced between the electrode and waveguide to restrict the current spread into the waveguide regions not covered by the electrode, and to obtain a sharp carrier gradient profile for improved switching efficiency. This current restriction effect of the isolation gap is schematically illustrateed in Fig. 6. It also schematically shows an effective reflection interface wherein the reflection of the guided light occurs in the presence of the carrier injection. When the switching voltage V_b is applied between the switching electrode and a bottom contact, the switching electrical current flows through the current flow region indicated by cross-hatching. Although the gap substantially reduces the current spread into the regions not directly under the switching electrode, it does not eliminate the spreading completely, resulting in the possibility of a small lateral offset of the effective reflection interface from the reflection edge of the electrode. This offset however may be rather small, for example on the order of less than 1 µm, and is smaller than the size of the waveguide branching region.

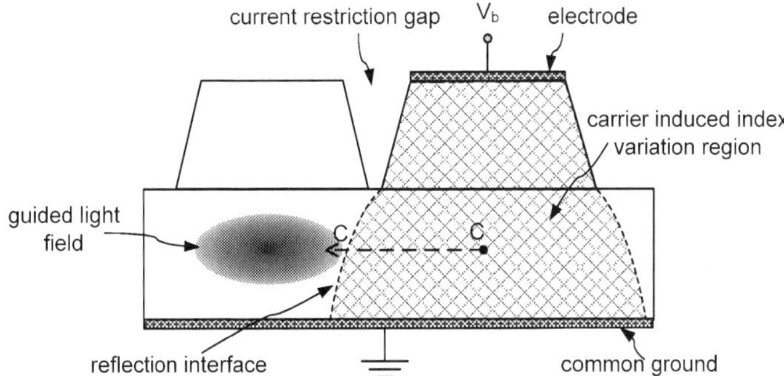

Fig. 6. Schematic diagram of the optical mode, the current flow region and the reflective interface induced by the carrier injection in the cross-sectional view of Fig. 4 taken along the line 'B-B'.

The carrier-induced index variation at the reflection interface is gradual due to the spatial spreading of injection current flow. If an abrupt spatial variation in the carrier concentration or current flow can be formed along the reflection interface, more light can be reflected to the other port. If the waveguide core layer is very thick, it is important to check how the current restriction gap affects the spreading carrier-induced index change profile at the reflection interface.

Fig. 7 is the calculated carrier concentration along line 'C-C', the middle of the core layer, in the cross-section as shown in Fig. 6 with or without a separation gap inside the splitting region, respectively. Fig. 7 was generated based on a two-dimensional simulation of Δn by using ApSys, a 2D/3D FEM optoelectronic simulation software from Crosslight Software Inc., and is based on a top electrode that is 200 μm long and with a bias current of 200 mA. Only the structure from the centerline of the biased waveguide is shown in Fig. 7. The dotted line in Fig. 7 is the edge of the top electrode. As illustrated in Fig. 7, an isolation gap in the splitting region is needed to prevent carriers from spreading beyond the electrode region. The self-heating effect was also included in the 2D FEM simulation. In the simulation, the room temperature is set at 300 K and the heat generated from contact resistance was not included. Calculation shows that the guiding of light to the destination port via modal evolution is unchanged with or without the self-heating temperature change. Fig. 8 is the corresponding carrier-induced refractive index change Δn along the middle of the core layer with or without a separation gap in the splitting region, respectively. The calculation involves a combination of bandfilling, band-gap shrinkage, and plasma effect (Bennett et al., 1990). The maximum index change is calculated to be -0.037 at the midpoint of the biasing electrode. The TIR interface located under the electrode edge is 2.5 μm apart from the biased waveguide centre. By introducing a current restriction gap inside the switch structure, the index change profile at the reflection interface is much more sharp than a switch without a gap inside. Clearly, more light can be steered into the higher-index waveguide by introducing a small gap in the splitting region.

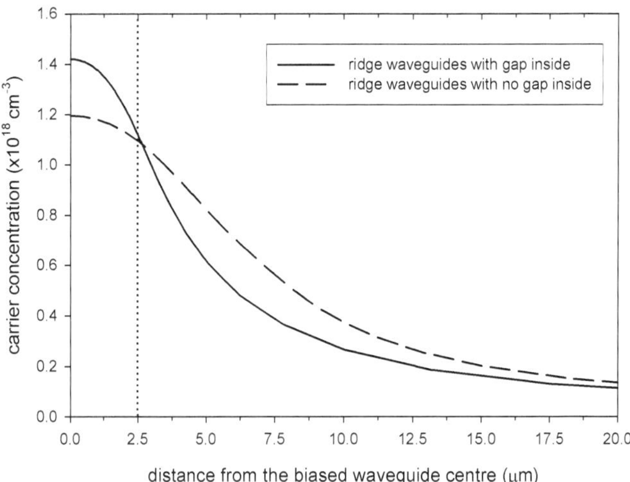

Fig. 7. Carrier concentration along the middle of core layer in the cross section at the junction branching point. The dotted line is the edge of the top electrode.

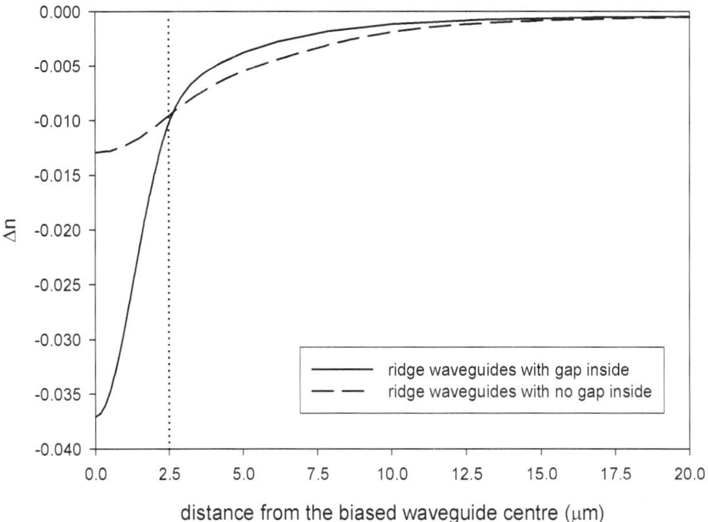

Fig. 8. Carrier-induced refractive index change Δn along the middle of core layer in the cross section at the junction branching point. The dotted line is the edge of the top electrode.

2.4 Graded heterojunction interfaces

The up-to-date growth technology of Metal-Organic Chemical Vapor Deposition (MOCVD) enables us to build semiconductor materials with precise control of thickness and composition at the atomic level. If no grading of the chemical composition of the semiconductor is applied in the vicinity of the heterostructure, all the heterojunction interfaces of the fabricated waveguide using the layer structure shown in Fig. 5 are sharply defined at atomic-level precision. One of the problems introduced by abrupt heterostructure is the band gap discontinuities caused by free charge transferring at the heterojunction between two semiconductors with different bandgap energy (Capasso et al., 1987). Carriers in the large-bandgap material will diffuse over to the small-bandgap material where they occupy band states of lower energy. As a result of the carrier transfer, and electrostatic dipole forms and leads to the band bending. For example, most published results for the $Al_xGa_{1-x}As/GaAs$ heterojunction indicate that the conduction band discontinuity $\Delta E_C{\approx}0.24$ eV for x=0.30 (Wang, 1986).

Fig. 9 is the calculated band diagram of the zero-biased waveguide using the abrupt layer and doping structure shown in Fig. 5. The numerical simulation software (the APSYS of Crosslight Software) uses the FEM method to solve a 2D Poisson's equation for potential voltage, continuity equation for carriers, and thermionic emission theory for a bandgap barrier at the heterojunction interface. The adaptive mesh points were set to be very dense near the heterojunction in order to calculate the bandgap barrier correctly. The band spikes shown in Fig. 9 at the double-heterostructure interfaces near the lower confinement layer will suppress the flow of charge carriers across the junction. This band–barrier rectification effect reduces the carrier injection efficiency of the switch. It is therefore desirable to eliminate such spikes in the conduction or valence band.

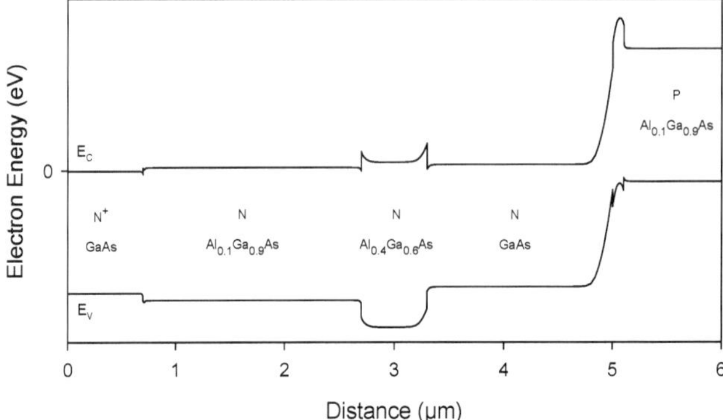

Fig. 9. Band diagram of the zero-biased waveguide using the abrupt layer and doping structure shown in Fig. 5.

Composition grading is a common technique in the design of quantum-well lasers, Heterojunction Bipolar Transistors (HBTs) and modulation-doped transistors. The compositional grading at the junction interface can smooth out a large part of the band discontinuity so that the carrier injection is increased. Usually a graded barrier with a grading length of a few hundred angstroms can effectively minimize the inference of the band spike at the abrupt interface. The smaller the carrier concentration on either side of the heterojunction, the wider the graded region has to be in order to reduce the spike height. The composition grading effects on band spikes are clearly demonstrated in Fig. 10, which shows the calculated conduction band profiles for the double-heterostructure interfaces at the lower confinement layer. Three cases were studied, varying the length of grading width L. The three simulations are for L=0, L=200 Å and L=400 Å using the layer structure shown in Fig. 5. The metallurgical grading of $Al_xGa_{1-x}As$ was set linear in the grading region.

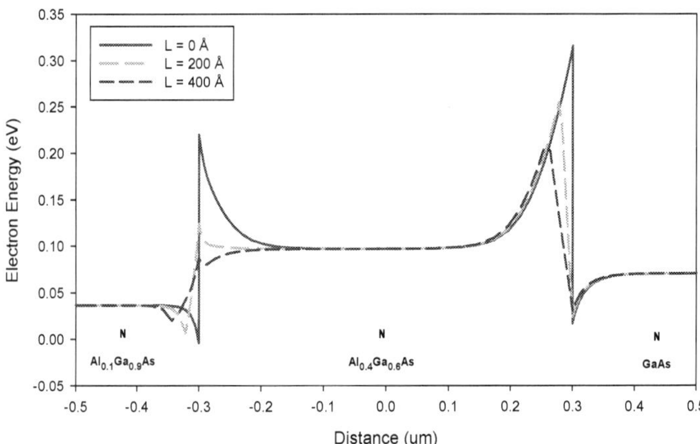

Fig. 10. Conduction band profiles for the double-heterostructure near the lower confinement layer at different grading width L.

The most important advantage of applying graded heterojunctions in the switches is that the reverse bias problem in a multilayer switch structure can be significantly reduced. The n-n heterojunction between $Al_{0.4}Ga_{0.6}As$ and $Al_{0.1}Ga_{0.9}As$ at the bottom of the lower confinement layer (see Fig. 5) would be reverse biased when a voltage is applied to operate the switch. From our simulations, over 80% of the switch driving voltage is applied across the reverse biased n-n heterojunction when all heterojunctions are abrupt. As shown in Fig. 10, this abrupt n-n heterojunction barrier can be effectively reduced or completely eliminated by over 400 Å-length grading. As a result, the switching voltage is further reduced.

2.5 Exemplary demonstrations

A prototype 2×2 carrier-injection optical switch with double-reflection electrode has been developed and fabricated. The waveguide structure, as shown in Fig. 5, has six AlGaAs/GaAs layers grown by MOCVD on an n+ GaAs substrate. For a single-mode waveguide at a wavelength of the input light of 1.55 µm, the ridge width of 5.0 µm was used, with the ridge height of about 1 µm. The GaAs waveguide core layer is lightly doped, $n=10^{16}$ cm^{-3}, to provide an optimized trade-off between low propagation loss and low switching voltage. In order to reduce energy band discontinuities at the heterojunctions that would impede the current flow therethrough, and to reduce the switching voltage, compositionally graded interfaces 20 nm - 40 nm thick were used at all the layer interfaces. The width of the current restriction gap, as illustrated in Fig. 6, is about 1 µm. These gaps may be formed using a double mask technique (van der Tol et al., 1994) to overcome photolithography problems when forming small gaps. With this technique, two masking stripes, which are independently defined in separate lithographic steps, can also be used as masks in a dry etch process to obtain ridge waveguides with small isolation gaps.

Fig. 11 is a Scanning Electron Microscope (SEM) image of the fabricated switch with the input waveguides on the left, viewed from an angle. It should be noted that the fabrication processes only employ conventional technologies, such as UV photolithography, e-beam evaporation and plasma etching. Independently formed electrode and photoresist layers were used as a double-mask in a single dry etch process to form the ridge waveguides and gaps. Uniform gap spacing was obtained using vernier alignment marks with ± 0.1 µm alignment accuracy. Fig. 12 is a SEM picture of the fabricated carrier-restriction gap, with the sample cleaved at a location close to the branching vertex of output waveguides.

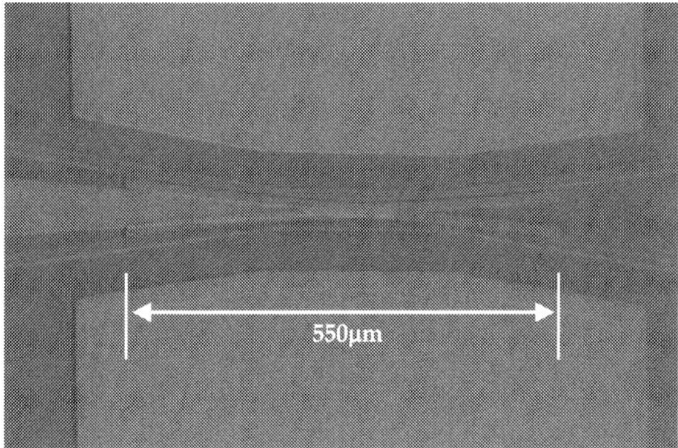

Fig. 11. SEM image of a fabricated 2×2 double-reflection TIR optical switch.

The fabricated switch has an electrode length of 550 μm, a waveguide crossing angle of 7°, and a cosine-bend section of 6.4 mm. The measured propagation losses of straight waveguides with undoped and lightly doped core layers are 0.3 dB/cm and 1.5 dB/cm, respectively, for a wavelength of 1.55 μm. Fig. 13 is the measured switching curves for light coupled into the top left input port. The switched state was obtained for an injection current of 220 mA. At this state, the driving voltage was 2.89 V and the measured device series resistance was 9.8 Ω. So the total power dissipation was at a manageable level of 0.6 W. With the area of the fabricated electrode being 3400 μm², the corresponding switching current density was 6.5 kA/cm². By using this current density and assuming a 10 ns carrier lifetime, the estimated carrier density and Δn are about 2×10^{18} cm^{-3} and -0.02, respectively. Such a high index change may be the result of two facts: one is that the carrier lifetime is quite long in the thick core layer; the other is that the current spreading decreases at very high current densities which results in current "bunching" in the core below the isolation gap and hence yields better power reflectivity. The measured switching time is 15 ns. On-off extinction ratio is 13 dB for the through port and 14 dB for the switched port. The switch is wavelength independent for a wavelength range from 1540 nm to 1570 nm. A very small polarization dependence of TE and TM modes has been observed (<0.5 dB). The measured extra insertion loss of the 2×2 switch compared with the straight waveguide fabricated on the same chip is 3 dB.

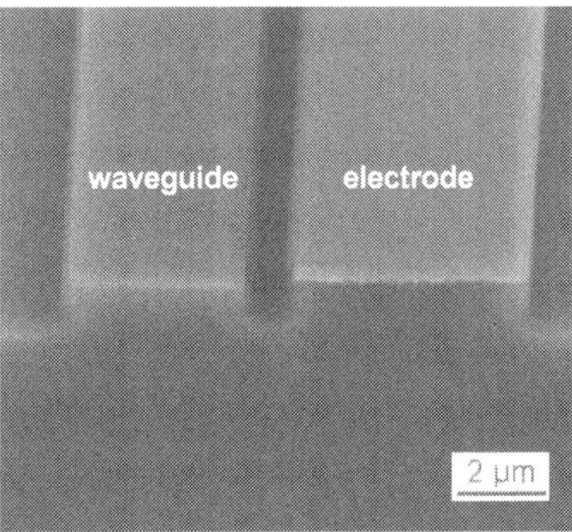

Fig. 12. SEM image of the carrier-restriction gap in an angled view of Fig. 4 taken along the line B-B.

Similarly, we also developed a 1×2 double-reflection optical switch which has a measured total power dissipation of 0.24 W with no extra cooling required to keep the switched state. As shown in Fig. 14, the fabricated switch has an electrode length of 480 μm, and a waveguide crossing angle of 6°. Based on the measured switching curve shown in Fig. 15, the switched state was obtained for an injection current of 105 mA at a bias of 2.30 V. The on-off extinction ratio is 15 dB for the through port and 14 dB for the switched port. The measured extra insertion loss of the 1×2 switch is 2 dB compared with the straight waveguide fabricated on the same chip. The switch is wavelength independent for the measured C-band. The polarization dependence of TE and TM modes input lights has been found to be less than 0.5 dB. The measured switching time is 11 ns which is close to the carrier lifetime in bulk materials. Reducing the guiding layer thickness would increase the switching speed but at the cost of decreasing the fibre coupling efficiency.

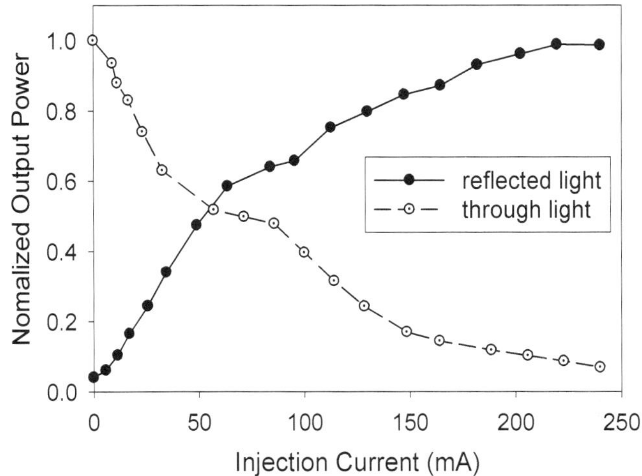

Fig. 13. Measured switching characteristics of a fabricated 2×2 double-reflection TIR optical switch.

(a) Layout of the curved double-reflection electrode design

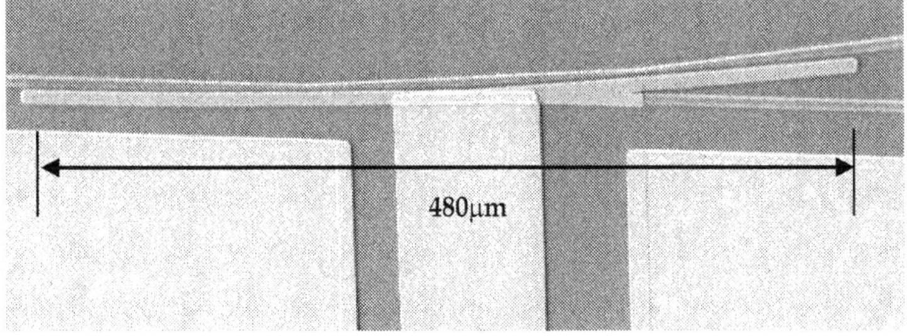

(b) SEM image of a fabricated 1×2 double-reflection TIR switch

Fig. 14. 1×2 TIR optical switch with curved double-reflection electrode.

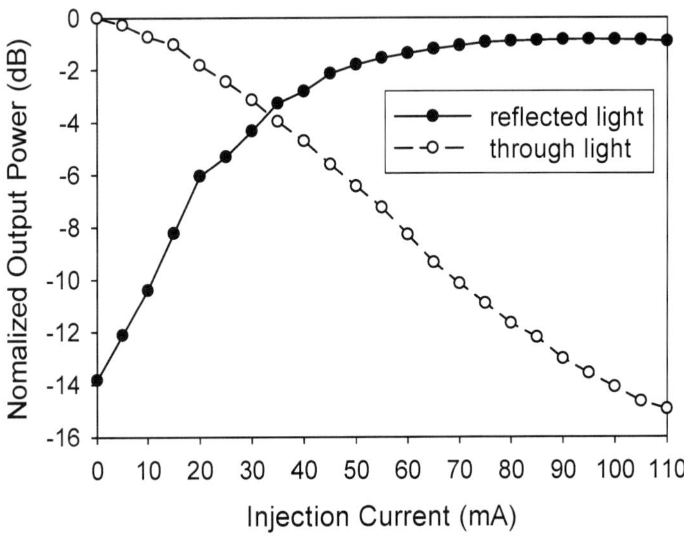

Fig. 15. Measured switching curve of a fabricated 1×2 double-reflection TIR optical switch.

3. 2×2 MMI-MZI DOS

To further reduce the switch power consumption, we have developed a 2×2 GaAs-GaAlAs carrier-injection switch based on a Mach-Zehnder Interferometer (MZI) formed by two Multi-Mode Interference (MMI) couplers. In an MZI-based switch (Cao et al., 2009; Wong et al., 2005), the two MZI arms are formed by two identical single-mode waveguides with electrodes, fabricated on the top of the waveguides, which allow the variation of the phase difference between them, based on the carrier-injection effect. Compared to the TIR switches, MZI-based switches can offer a high on-off extinction ratio and avoid the electrode fabrication complexity. More importantly, the injected current required to achieve π phase difference between the MZI arms is much smaller than that to achieve TIR, and therefore the power consumption of the MZI switches is much lower than that of the TIR devices.

A schematic of the fabricated 2×2 MMI-MZI carrier-injection switch is illustrated in Fig. 16, the device consists of an MZI configuration in which two 2×2 identical MMI couplers are connected by two identical parallel single-mode waveguides, that act as the two MZI arms. Metal electrodes are fabricated on the top of each MZI arm. Each electrode is 250 μm long and 5 μm wide. The MMI couplers are used as the 3-dB optical power splitter and combiner because of their polarization insensitive, fabrication and wavelength-tolerant properties (Besse et al., 1994). The MMI width is 20 μm and the MMI length is 1983 μm. The access (input/output) waveguides are single-mode waveguides with a width of 5 μm. Each access waveguide has two short straight sections sandwiching a long s-bend section to facilitate mode relaxation. The s-bend is designed in such a way that the additional loss caused by bending in this section is negligible.

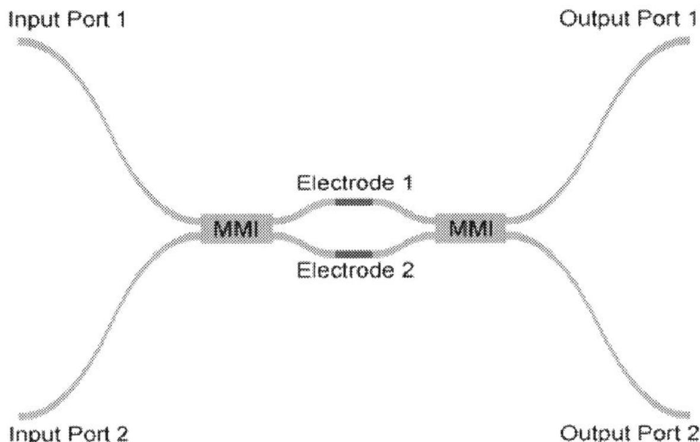

Fig. 16. Schematic diagram of a 2×2 MMI-MZI carrier-injection optical switch.

The separation between the input (output) ports is 250 μm and that between two electrodes is 35 μm. The total length of the device is about 1.4 cm. The operational principle of the 2×2 MMI-MZI switch is as follows. If two identical 2×2 MMI couplers are connected together, the two light beams, equally split from the input light by the first MMI coupler, will merge again after passing through the second MMI coupler but in the cross state. When sufficient carriers are injected into one of the electrodes to produce a π phase difference between the MZI arms, the device changes to the bar state.

The 2×2 MMI-MZI carrier-injection switch is fabricated on the epitaxial layer structure similar to the ridge waveguide structure shown in Fig. 5, with the core layer and two nearby confinement layers undoped. Compositionally graded interfaces 20 nm - 40 nm thick were used at all the layer interfaces to reduce energy band discontinuities at the heterojunctions. The calculated propagation loss of this single-mode waveguide is 0.19 dB/cm for the TE mode and 0.21 dB/cm for the TM mode.

The measured normalized output power at Output Ports 1 and 2 as a function of the injected current of Electrode 1 is shown in Fig. 17. The input light at wavelength 1.55 μm is coupled into the device at Input Port 2. As shown in Fig. 17, the first cross state (Output Ports 1 and 2 are on and off, respectively) occurs at 15.8 mA, and the first bar state (Output Ports 1 and 2 are off and on, respectively) occurs at 44.3 mA. The measured on-off extinction ratio is 20.6 dB for the cross state and 21.5 dB for the bar state. The measured voltage applied to the electrode is 1.35 V for the cross state and 1.48 V for the bar state. Therefore, the corresponding power consumption is 21.3 mW for the cross state and 65.6 mW for the bar state. Because of this very low power consumption, no extra cooling is required to maintain stable switch operation, which has been observed in our measurements. On the other hand, such low power consumption is very important for building the future optical Green Internet. From Fig. 17, we can obtain that the injected current required to achieve π phase difference between the MZI arms is 28.5 mA, which is very close to the theoretically estimated value of 29.4 mA by considering the guiding layer as an intrinsic semiconductor bulk material. Compared to the straight waveguides, the fabricated MMI-MZI devices have almost no measurable additional loss due to the MMI couplers and bend waveguides. The measured propagation loss of straight waveguides is typically around 0.5 dB/cm for a wavelength of 1.55 μm and has very small polarization dependence (< 0.5 dB). Based on these observations, the estimated total on-chip insertion loss would be about 2 dB. The measured switching time of 15 ns is close to the carrier lifetime in the GaAs bulk material. Reducing the guiding layer thickness below 0.5 μm would increase the switching speed to a few ns but at the cost of decreasing the fibre coupling efficiency. It is noted that the switch should be at the cross state when no current is injected but this is not the case as shown in Fig. 17 due to the asymmetrical structure of the device because of fabrication errors.

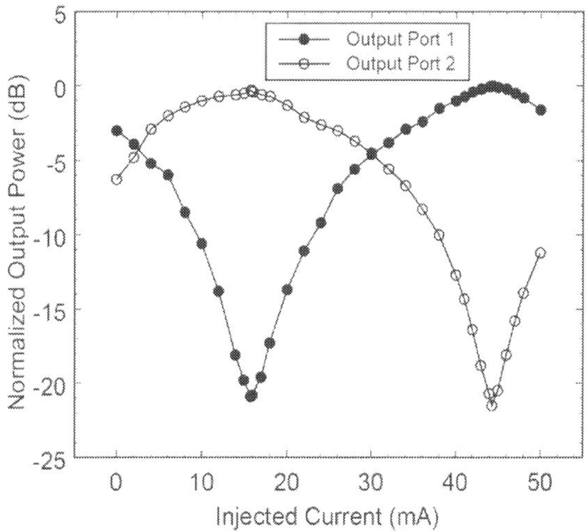

Fig. 17. Switching curves of the 2×2 MMI-MZI optical switch as a function of the injected current of Electrode 1. A light at wavelength 1.55 μm is coupled into the device at Input Port 2. The output optical power is measured at Output Ports 1 and 2.

4. Conclusions

Wide-angle 1×2 and 2×2 AlGaAs/GaAs TIR DOSs were developed by using only conventional semiconductor device fabrication technologies, suitable for monolithic circuit integration due to their small size. The developed 2×2 switch has a 7° waveguide crossing angle and a manageable maximum 0.6W dc power consumption. The 1×2 switch has a 6° branching angle and maximum quarter Watt dc power consumption. They can serve as the basic building blocks for high dense switching matrix. The techniques that we used, such as carrier restriction gap, curved electrodes, double-reflection interfaces and graded heterojunctions, can be easily applied to other switches made with other materials such as InP based, SiGe heterostructures and alike.

An ultra-low-power 2×2 MMI-MZI DOS with maximum 66 mW power consumption has also been demonstrated. Such an optical switch has a low insertion loss, high on-off extinction ratio, and very low power consumption. This device, together with the above TIR DOSs, will have significant importance in energy savings for future optical Green Internet applications, when the traffic moves down from the packet switching/routing layer to the circuit switching layer. The use of dynamic optical path-switching networks will help to accommodate users with massive video-based service demands in the future.

5. References

Abdalla, S., Ng, S., Barrios, P., Celo, D., Delage, A., El-Mougy, S., Golub, I., He, J-J, Janz, S., McKinnon, R., Poole, P., Raymond, S., Smy, T.J. & Syrett, B. (2004). Carrier Injection-Based Digital Optical Switch With Reconfigurable Output Waveguide Arms, *IEEE Photonics Technology Letters*, Vol.16, pp. 1038-1040, ISSN 1041-1135.

Agrawal, N., Weinert, C. M., Ehrke, H.J., Mekonnen, G.G., Franke, D., Bornholdt, C. & Langenhorst, R. (1995). Fast 2x2 Mach-Zehnder Optical Space Switches Using InGaAsP-InP Multiquantum-Well Structures, *IEEE Photonics Technology Letters*, Vol.7, pp. 644-645, ISSN 1041-1135.

Bennett, B. R., Soref, R.A. & Alamo, J.A.D. (1990). Carrier-Induced Change in Refractive Index of InP, GaAs, and InGaAsP, *IEEE Journal of Quantum Electronics*, Vol.26, pp. 113–122, ISSN 0018-9197.

Besse, P.A., Bachmann, M., Melchior, H., Soldano, L.B. & Smit, M.K. (1994). Optical Bandwidth and Fabrication Tolerances of Multimode Interference Couplers, *IEEE Journal of Lightwave Technology*, Vol.12, pp. 1004-1009, ISSN 0733-8724.

Cao, S., Noad, J., Sun, L., James, R., Coulas, D., Lovell, G. & Higgins, E. (2009). Small AC Driving Voltage for MZI-Based GaAs-GaAlAs Electrooptic Modulators/Switches with Coplanar Electrodes," *IEEE Photonics Technology Letters*, Vol.21, pp. 584-586, ISSN 1041-1135.

Capasso F. & Margaritondo, G. (Eds.) (1987). *Heterojunction Band Discontinuities*, Elsevier Science Ltd, ISBN 978-0444870605, Amsterdam.

Ikezawa, K., Iio, S., Suehiro, M., Suzuki, T. & Uneme, S. (2008). Demonstration of Modulation Format Free and Bit Rate Free Characteristics of 2 ns Optical Switch for Optical Routers, *Proceedings of Optical Fiber Communication Conference and Exposition*, no. JWA31, ISBN: 978-1-55752-856-8, San Diego, CA, USA, Feb 2008.

Li, B. & Chua, S.J. (2001). 2x2 Optical Waveguide Switch with Bow-Tie Electrode Based on Carrier-Injection Total Internal Reflection in SiGe Alloy, *IEEE Photonics Technology Letters*, Vol.13, pp. 206-208, ISSN 1041-1135.

Nayyer, J., Niayesh, K. & Yamada, M. (2000). Dynamic Characteristics of Optical Intersecting-Waveguide Modulators/Switches with Curved Electrodes, *IEEE Journal of Lightwave Technology*, Vol.18, pp. 693-699, ISSN 0733-8724.

Ng, S., Janz, S., Mckinnon, W.R., Barrios, P., Delage, A. & Syrett, B.A. (2007). Performance Optimization of a Reconfigurable Waveguide Digital Optical Switch on InGaAsP–InP, *IEEE Journal of Quantum Electronics*, Vol.43, pp. 1147-1157, ISSN 0018-9197.

Shimomura, K., Tanaka, N., Aizawa, T. & Arai, S. (1992). 2V Drive-Voltage Switching Operation in 1.55μm GaInAs/InP MQW Intersectional Waveguide Optical Switch, *Electronics Letters*, Vol.28, pp. 955-956, ISSN 0013-5194.

Thomson, D., Gardes, F.Y., Mashanovich, G.Z., Knights, A.P. & Reed, G.T. (2008). Using SiO_2 Carrier Confinement in Total Internal Reflection Optical Switches to Restrict Carrier Diffusion in the Guiding Layer, *IEEE Journal of Lightwave Technology*, Vol.26, pp. 1288-1294, ISSN 0733-8724.

van der Tol, J.J.G.M., Pedersen, J.W., Metaal, E.G., Ose, Y.S., Groen, F.H. & Demeester, P. (1994). Sharp Vertices in Asymmetric Y-junctions by Double Masking, *IEEE Photonics Technology Letters*, Vol.6, pp. 249-251, ISSN 1041-1135.

Wang, W.I. (1986). On the Band Offsets of AlGaAs/GaAs and Beyond, *Solid-State Electronics*, Vol.29, pp. 133–139, ISSN 0038-1101.

Wong, H. Y., Sorel, M., Bryc, A.C., Marsh, J.H. & Arnold, J.M. (2005). Monolithically Integrated InGaAs-AlGaInAs Mach-Zehnder Interferometer Optical Switch Using Quantum-Well Intermixing, *IEEE Photonics Technology Letters*, Vol.17, pp. 783-785, ISSN 1041-1135.

Yanagawa, H., Ueki, K. & Kamata, Y. (1990). Polarization- and Wavelength-Insensitive Guided-Wave Optical Switch with Semiconductor Y Junction, *IEEE Journal of Lightwave Technology*, Vol.8, pp. 1192-1197, ISSN: 0733-8724.

Zegaoui, M., Choueib, N., Harari, J., Decoster, D., Magnin, V., Wallart, X. & Chazelas, J. (2009). 2x2 InP Optical Switching Matrix Based on Carrier-Induced Effects for 1.55-μm Applications, *IEEE Photonics Technology Letters*, Vol.21, pp. 1357-1359, ISSN 1041-1135.

Zhuang, W., Duan, J., Zou, Z., Yang, P., Shi, Z., Sun, F. & Gao, J. (1996). Total Internal Reflection Optical Switch with Injection Region Isolated by Oxygen Ion Implantation," *Fiber and Integrated Optics*, Vol.15, pp. 27-36, ISSN 0146-8030.

18

On Fault-Tolerance and Bandwidth Consumption Within Fiber-Optic Media Networks

Roman Messmer and Jörg Keller
FernUniversität Hagen
Germany

1. Introduction

Transmission of real-time video signals has long been a domain of copper-based video switchers with or without synchronization or equalization of analog or digital signals. With the beginning era of high definition signals and the associated limitation of maximum cable lengths the need for a change to fiber-optic equipment was essential. Intrafacility transport of video data carrying high definition signals can reach lengths up to kilometers. Using fiber-optic connections for such networks, a new set of reliability requirements arises. Optical fibers feature a lower physical reliability than copper lines. The latter also can be repaired more easily, mostly by the usage of soldering irons. The transformation from circuit-switched lines to packet-switched networks in an application field where the bandwidth of a connection might range over 6 magnitudes from kilobits per second to gigabits per second, thus almost reaching the capacity of the underlying physical link, also influences the planning of such networks.

The goal of this article is to gather the state of the art in two closely related research areas: fault-tolerance and capacity planning for optical media networks. Fault-tolerance can only be achieved by redundancy, i.e. additional capacity has to be provided. Vice versa, sensible capacity planning is only possible when the type of fault-tolerance mechanism applied is known.

Both areas also represent special requirements in optical media networks. Optical links exhibit failure characteristics different from electrical links. Fault-tolerance, i.e. the establishment of an alternate route in the presence of a fault, is time-critical in media networks that transport video and audio streams which have to meet real-time requirements. Capacity planning in media networks is complicated by the fact that the high bandwidth requirements of video connections only allow a few such connections to share a line, while there are other types of signals with much lower bandwidth consumption so that hundreds of those can share a line. The remainder of this article is structured as follows. Section 2 presents the necessary technical information about optical networks and the media data they transport. Section 3 summarizes fault-tolerance concepts for optical media networks. Section 4 then reports on issues of bandwidth consumption and capacity planning in such networks. Finally, Section 5 summarizes the article.

2. Optical networks and transported data

2.1 Network characteristics

All-optical networks (AO) represent connection hardware where not only the transmission lines consist of optical fibers, but where also the switching elements are purely optical. AO networks differ fundamentally from Optical-Electrical-Optical (OEO) networks. The main difference is the active conversion part within OEO networks which converts optical signals into the electrical domain for switching and forwarding using standard protocols, and after data manipulation back into the optical domain. This process takes time and therefore generates a certain switch-intrinsical latency time. Switches with high bandwidth capability consume a lot of electrical power which has to be taken into account during the planning process in more extensive installations. The protocol chosen also limits the usable bandwidth in OEO networks, while AO networks can use 100 percent of the available fiber bandwidth, theoretically. Table 1 compares the main properties of AO and OEO networks.

All Optical	OEO
Protocol independent	Protocol dependent
No conversion delays	Technology immanent latency
Little power consumption	Power consumption rises by network size
Bandwidth as high as carrier allows	Protocol dependent bandwidth
Circuit switched (per color)	Circuit and packet switched

Table 1. Differences between AO and OEO switched networks.

2.2 Networks as graphs

From the perspective of routing, an optical network is a *directed graph* or digraph $G = (V, E)$, where the set of nodes V comprises all sources of signals such as cameras and microphones, all switches and routers, and all sinks of signals such as audio mixers and monitors. The set of arcs $E \subseteq V \times V$ represents the optical links that transport the data. Each link $e \in E$ will be assigned a capacity $c(e)$. A network is *symmetric* if for each edge $(u, w) \in E$, also $(w, u) \in E$ and $c(u, w) = c(w, u)$, i.e. that edges are bidirectional.

A sequence $v_0, v_1, \ldots, v_n$ where $(v_i, v_{i+1}) \in E$ for all i is called a *path* of length n. Typically we assume paths to be simple, i.e. the nodes v_i on the path are all different. A directed graph is *strongly connected* if there is a path from u to w for any pair of nodes $u, w \in V$. In a strongly connected digraph, let us denote the length of the shortest path from u to w by $sp(u, w)$. Then, we call

$$D = \max_{u,w \in V} sp(u, w)$$

the *diameter* of the graph.

A strongly connected digraph is called *doubly connected* if it is still connected after the removal of an edge, or a node with its adjacent edges. If we partition the nodes into two sets V_1 and V_2 of almost equal size, i.e. $-1 \leq |V_1| - |V_2| \leq 1$, then we call this a *bisection* and

$$C(V_1, V_2) = \min \left\{ \sum_{u \in V_1, w \in V_2} c(u, w), \quad \sum_{u \in V_1, w \in V_2} c(w, u) \right\}$$

is the capacity of the cut. The *bisection bandwidth* of a graph is the minimum capacity over all possible bisections.

For a communication network, the connection property ensures reachability, while the doubly-connected property means reachability even in the presence of faults. The diameter bounds the number of hops, and the bisection bandwidth represents the total bandwidth available in a graph when there are many connections from sources to sinks.

Unfortunately, there is no single network that is doubly connected, has low diameter and a high bisection bandwidth, and is cost-efficient in the sense that the number of edges is still linear, i.e. that the average node degree is constant. Thus, a multitude of networks have been developed for different purposes, and with different properties.

A *star* is a symmetric network with node set $V = \{0, \ldots, n-1\}$ and edges $(0, i)$ for $i \geq 1$; node 0 is called the center. A *ring* is a symmetric network with $V = \{0, \ldots, n-1\}$ and edges $(i, i+1 \bmod n)$ for all i. A 2-dimensional *torus* is a symmetric network with node set $V = \{0, \ldots, n-1\}^2$ and edges $((i,j), (i+1 \bmod n, j))$ and $((i,j), (i, j+1 \bmod n))$. A 2-dimensional *grid* or *mesh* is similar to the torus, only the "wrap-around" edges between columns and rows 0 and $n-1$ are amiss. Examples of those networks are given in Figure 1.

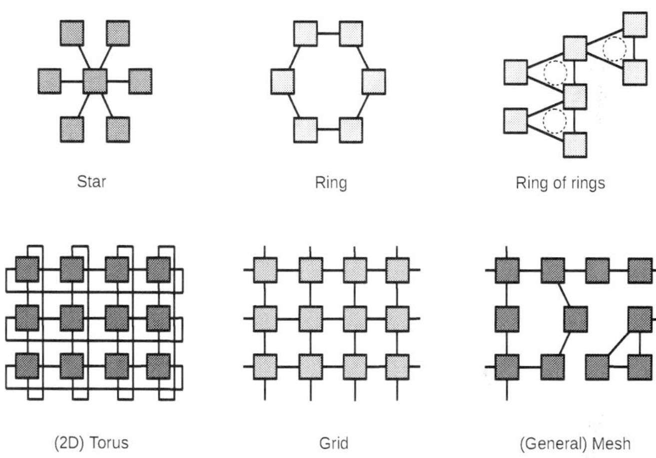

Fig. 1. Examples of interconnection networks.

Large networks typically are constructed as a hierarchy of smaller networks, e.g. one might connect k stars of n nodes each by connecting the center nodes as a k-node ring. Also, networks evolving over time tend to develop some irregularity even if they started from some of the regular networks above. Thus, in a general mesh some edges of the grid can be amiss.

2.3 Media signals

By the term *media* within the broadcast branch we mainly mean audiovisual signals originally generated by microphones and video cameras. Those signals are converted in a way that they easily can be transmitted over electrical or optical networks. Auxiliary and control represent relatively small-bandwidth signals but they occur at a higher number than the media signals. They also can be routed along with the audiovisual signals. In the following we describe the relevant signals in more detail.

2.3.1 Audio signals

We mainly distinguish between *single stereo* audio channels bearing two mono channels *left* and *right* within one data stream and *multichannel* audio, where several audio channels, typically 8 or 64 mono-equivalent channels, are bonded together as one broadband stream. For all concurrently transported signals that originate from a single audio source, synchronicity between left and right channels in a stereo transmission and all up to eight channels within a multichannel transmission is very important. For audio-video connections also the so-called *lip-sync* synchronization exists that specifies a maximum limit of time shift between a video signal and the corresponding audio signals. Even a little time shift between both channels of a stereo signal would destroy the stereophony and the signal would not longer be useable. The audio standard AES3 (Audio Engineering Society) in IEC 60958 standard defines a copper or optical interface for a stereo signal representing two mono audio channels. The consumer version called Sony Philips Digital Interface (S/PDIF) also supports a fiber optic version. With 48 kHz and two channels on 24 bit resolution this results in a net bandwidth of around 2.3 Mbit/s. Packing AES3 frames into ATM cells is defined within standard AES47 (IEC 62365). This allows grouping of several single audio channels into streams of greater bandwidth. Alesis Digital Audio Tape (ADAT) in form of ADAT Lightpipe[1] is designed to carry 8 mono audio channels at 24 bits resulting in about 10 Mbit/s effective data rate. Besides ADAT there is the nowadays most popular Multichannel Audio Digital Interface (MADI) or AES10 interface, supporting up to 64 mono audio channels. MADI is available in two versions, a copper-based (per coax-cable) and an optical-based layout. The AES standard suggests a fixed bandwidth of about 100 Mbit/s.

2.3.2 Video signals

In the standard ITU-R BT 601 of the International Telecommunication Union the coding of digital video signals at *standard resolution* of 525 or 625 lines and 60 or 50 Hz respectively at 8 or 10 bit video resolution has been defined. Video aspect ratios (ratio of image width to height) of standard 4:3 and the newer one 16:9 are possible. The interfaces have been standardized in SMPTE 259M[2], where the effective data rate is specified at 270 Mbit/s.

High resolution video signals carry 720 or 1080 video lines at a great number of different image rates ranging from 24 to 60 Hz. The high definition serial digital interface (HD-SDI) is standardized within SMPTE Standard 292M carrying around 1.5 Gbit/s and within SMPTE 424M carrying around 3 Gbit/s. SMPTE 372M explains another, more rarely used dual link interface whose combined bandwidth also reaches about 3 Gbit/s. In both formats the integration of metadata was standardized. For both versions the integration of 16 synchronized mono audio signals plus adding of user bits for control purposes is provided. The high definition version is committed to a video aspect rate of 16:9 only.

Bandwidth consumptions using *video streams* concern more or less a fixed rate. Within SDTV signals a stream is represented at a data rate of around 50 Mbit/s. The same rate applies to HDTV signals, when the modern XDCAM-HD-4:2:2 (50) codec is used. Currently, besides the older GXF-versions (above 100 Mbit/s), within Central Europe this is the most utilized HDTV streaming format. Designed to be transferred over an old 100 Mbit/s Ethernet link this kind of video transmission perfectly fits into a hierarchy of fixed bandwidth channels.

Though this selection shows only an extract of the large number of different video formats, it covers the widely used ones.

---

[1] As an alternative, ADAT optical interface, US-patent 5297181, can be used.

[2] Society of Motion Picture and Television Engineers, http://www.smpte.org

2.3.3 Control signals

Control signals in the broadcast environment differ in terms of their physical nature. They span from contacts which signal an ON-AIR condition with very little information content up to devices using control protocols for video mixers with a demand of a very short reaction time. Those signals have to be embedded efficiently within the available time slots considering bandwidth and latency.

2.3.4 Overview of bandwidth demand

Table 2 summarizes the different signals and their approximate net bandwidth. Note that there is only a very limited variety of signal bandwidth within the analyzed media signals. Video signals, especially high definition ones consume the majority of the available bandwidth. This fact can be utilized for an online routing heuristic for optimizing fault-tolerance, bandwidth and usage of fiber links (cf. Section 3).

Media bandwidth		
Signal type	Detail	Data rate [Mbit/s]
Video	SDTV	270 ($\sim$ 300)
	HDTV720p	1500
	HDTV1080p	3000
Audio	AES3	2.3 ($\sim$ 3)
	ADAT	10
	MADI	100
Control	RS-types	(upto) 1
Ethernet	100 Mbit/s	100
	1 Gbit/s	1000

Table 2. Overview of media bandwidth demand.

3. Network fault tolerance

Every fiber-optic based network is vulnerable to failures. Because of the huge number of signals a fiber can carry today, a single failure could cause an interruption of thousands of signals. To overcome this problem two main methodologies have been developed. The first is called *protection* which generally means use of redundancies in terms of additional links and physical replication and distribution of equipment in a mere static, predetermined manner, while the second approach *restoration* comprises diverse dynamic methods to recover from failures in very short time. Both approaches must guarantee that a failure of links does not lead to any perceptible or at least interfering interruption of services.

Both methodologies are suitable for solving most of the usually occurring problems. Decisions between them are made concerning a compromise between reliability and efficiency. While additional fiber-links and switches can be an important cost factor, restoration algorithms consume some time until bypass ways can be determined and switched, which could lead to unwanted short interruptions of services. Protection measures in general local-range ring topologies(Goralski, 2002) deal with delays that are composed of several parts. The signal speed in fibers is about 5 μs per kilometer (which can be ignored in local installations). The time for validation of protecting switching commands lies within 0.3 ms. Finally, the latency for bypassing a number of switches also has a significant influence in the protection time depending on the network topology. An ETSI demonstration ring listed in the APS specification 300 746 carrying only 16 nodes reaches an effective restoration time in the range of 20 seconds.

Fault tolerance methods in the computing domain like multiprocessing in Grids (Patrikar, 2009, pp. 531–546) differ from their pendants in the area of network resiliency and are not an issue of this chapter.

Protection in the area of networks means switching one or more source signals from a failed path to a working redundant backup path. When employing *1+1 protection*, there is a redundant path for every connection, and the signal is always sent over both paths to the destination. In the fault-free case, the receiver accepts the signal from one path based on the signal quality. In case of a link failure on one of the paths, the other path is still available. When employing *1:1 protection*, the redundant path is only used upon a failure on the original path. There are variants and generalizations of these basic scenarios. For example, low priority traffic can either be sent on the original or the redundant path. In *1:n protection*, n original paths share a redundant path at least partially. A compromise between 1:1 and 1:n protection is *m:n* protection, where $m < n$ redundant paths are available for n original paths.

These measures are also implemented within the RFC 3031 Multi Protocol Layer Switched Routing (MPLS) protocol. A functionality called *MPLS fast reroute* is implemented. In RFC 4090 an extension to the reservation protocol for Label Switched Paths (LSP) tunnels (RSVP-TE) is proposed. Backup tunnels are established, which work as bypass routes for one or several links. They are all computed in advance of a possible failure. Data traffic is redirected around a failed node or link as close as possible. There are two general protection methods used in MPLS. The *one-to-one-backup* method establishes an LSP that intersects the original path after the failed node, where after is to interpreted as downstream. For each protected LSP there will be a separate backup path. A *facility backup* creates a single LSP that serves several original LSPs. In MPLS language this path is called *bypass tunnel*. In most cases global recovery will require less bandwidth than local recovery. However, calculation of the optimal path in case of global recovery will need more resources and require more time.

3.1 Dynamic network restoration

To restrict the recovery time when engaging network management to calculate a restoration path, media networks mostly rely on heuristics. We will present several of them in the following paragraphs.

General network resilience methods for AO fiber networks are surveyed in (Wosinska et al., 2009). A fast distributed network restoration algorithm is presented in (Chow et al., 1993). In (Ramasubramanian & Chandak, 2008) the authors investigate how to cope with simultaneous failures on two links. They propose to refine the advance calculation of backup paths for point-to-point communications in the presence of single link failures. For every pair of links that may fail concurrently, the backup path for one of those links need not comprise the other (possibly failed) link. They formulate the problem with the help of linear programming and also provide a heuristic. In (Komine et al., 1990) the authors propose a distributed restoration algorithm for broadband networks. Instead of advance path reservation, they use a refined version of flooding to find alternate paths on the occurrence of failures. In (Yang & Hasegawa, 1988) the authors propose a distributed restoration algorithm for broadband networks that uses intelligent routers to restrict the number of messages when flooding in the case of failures. In (Zheng, 1992) the authors propose the use of backup channels for multiple point-to-point communications. Their focus is on guaranteeing real-time properties, notably bounds on end-to-end delay.

3.2 ZirkumFlex

ZirkumFlex (Messmer & Keller, 2010) is an example for a high-speed dynamic restoration heuristic. Therefore it does not rely on any pre-reserved path redundancy but calculates a failure bypass path around one or even several failed nodes or links immediately after failure signalling. If the restoration time, which consists of failure detection, failure signalling, signal detection, calculation of bypass routing, transmitting the routing commands, receiving and switching the signal paths on the router sites, can be held well below 80 ms, then only two video frames are lost and the signal interruption of the video signal is barely visible. Not only several node or link failures can be recovered in a very short time, also point-to-multipoint connections can be handled in addition to point-to-point connections. The tree structure of point-to-multipoint connections even contributes to reduce the time needed to calculate the bypass route.

The algorithm assumes the existence of sufficient bandwidth[3] available to support the bypass paths. Fail-stop as well as byzantine failures can be handled by this method. Failure handling works as following: After reception of failure information, which origins from information of lower protocols like loss-of-light information or CRC errors on signals, the (first) predecessor of the failed node(s) calculates the bypass path on its own topological information. After establishing the new path the links that are not needed anymore are removed by routing commands. The algorithm consists of three parts,

- Relaxation process
- Bypass breadth first search (BFS)
- Inverse link removal

which are briefly detailed now.

The *relaxation process* removes all links which can no longer be used, i.e. the links adjacent to a failed node or link and all further links which are reached only via nodes with indegree and outdegree 1. The predecessor and the successor nodes of the failed node, which represent the starting nodes for the next step (BFS) are moved as well. In case of the successor-BFS the removal process ends at a destination node or at a branch, a predecessor-BFS as well stops at a detected branch or a source node. The predecessor-BFS stops deleting links but still moves the BFS start node until a branch or source node is detected.

The *bypass BFS* run starts from the moved predecessor node and from the moved successor node(s) of the failed node in parallel. As long as new nodes are detected they get the current path information and their neighbors are visited. When the processes find an already visited node, both path representations are concatenated and automatically represent the bypass path between predecessor and/or successors. The corresponding successor BFS processes are stopped. If all successor BFS processes are stopped then also the predecessor BFS is stopped. A DFS run over the resulting bypass paths ensures that the bypass graph is loop-free.

In some cases the bypass routing leads to links in direction opposite to the former original path. The final *inverse link removal* adjusts in that situation. Knowing the distance to the source for each node, an opposite routing direction can easily be detected by decreasing distance information. If the bypass path would route a link from a node with larger distance to a node with smaller distance, this link would represent an unnecessary redundancy and has to be removed.

The resulting bypass path information then can be sent into the network to other routers and switches to establish the necessary connections.

---

[3] In case of optic fiber a sufficient number of fiber links and/or colors is assumed.

4. Bandwidth consumption

Several models exist that consider the difference in bandwidth usage between the two paradigms AO and OEO networks. While pure AO networks route the signal only in the optical domain by means of optical switches and optical links which are still limited to single fibers and a relatively small set of laser colors as routing channels, more classical OEO networks operate in the electrical domain by switching paths, separating channels, shuffling signal streams in the electrical and subsequent optical stream and therefore handling datastream segments in more fine-grained amounts.

Considering both AO and OEO networks, media signals can be embedded into and removed from the network via so-called *add-drop multiplexers* (ADM). They generally integrate several different media streams into a single transport stream which is then sent via a fiber-optic interface to the neighboring switch(es). Figure 2 presents an example of several embedded signal streams in a 5 Gbit/s (5G) transport stream providing the available bandwidth.

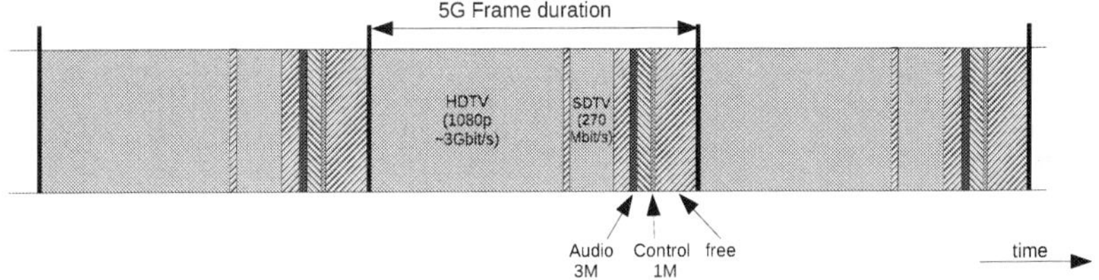

Fig. 2. Bandwidth allocation model for both OEO and AO networks.

Another network functionality within OEO network nodes is the capability to switch signals or groups of signals within the transport stream to other interfaces and to embed the media stream into other transport streams. This means removing (dropping) the media stream to a destination or leaving the media stream within the transport stream, copying and adding the stream to a new one known as multicasting or point-to-multipoint connections.

Concerning media switching there are some considerations to be made: One very important demand which has an impact to bandwidth calculations is the limitation of route-calculation time. Within medium-sized networks flow-based algorithms are too slow, so heuristics are to be chosen.

Load balancing as shown in (Chiu et al., 2002) also plays a minor role in media switching, see Section 4.2.1 for an example. To be considered is the huge bandwidth consumption of the HDTV-signal which e.g. in a 5 Gbit/s network consumes more than half of the available link bandwidth. This means that while switching a new link, the system should take care of keeping at least the bandwidth of one HDTV channel available. Demands with smaller bandwidth requirements should then be satisfied over different links if possible. We introduce a corresponding heuristic in Section 4.2.1.

4.1 Fault-tolerant interconnection methodologies and their bandwidth demand

Within fiber optic networks most of the failures will occur at the weakest points in the network: the optical fiber itself. Experience from practice indicates that *in-house* failures mostly concern single broken fibers, while *remote* line failures often occur during construction works, thereby the whole duct or even several ones concurrently are damaged. When designing a reliable fiber-optic media network this property has to be taken into account.

We now analyze the commonly known network topologies with respect to fault-tolerance and bandwidth consumption with exception of the *bus topology* which is not relevant for fiber optic networks. Layouts for chains, meshes, rings and trees are analyzed in (Gerstel & Zaks, 1994). Restore-management and the referring signal communication within OEO networks is extensively analyzed in the ETSI Automatic Protection Switching (APS) specification 300 746.

4.1.1 Star, double star, star-ring

A *star* topology allows simple adding of new nodes to the network. Failure of one terminal node does not affect the rest of the network. Also cable layouts can easily be modified. But the single star topology generally lacks redundancy and protection capability. A single point of failure is immanent: the central switch. Its failure would totally disrupt any communication on the network. This kind of network topology — combined with a strict redundancy of the central switch — still is very common within OEO media networks in practice. Besides it is the main choice for the emerging AO network topologies.

There are several ways to add fault-tolerance. An easy but expensive method called *double star* simply doubles the links from the central switch to the attached clients. One can choose between 1:1 or 1+1 protection, where at least for external connections like campus to campus one has to ensure that original and redundant links from one terminal to the center are placed in different ducts to really achieve redundancy. This doubles the necessary number of ducts. Using 1+1 protection, nearly transparent switching between failed and spare link is possible. Delays occur only by failure detection, signalling and backup switching which lie in the millisecond range, depending on technology. The disadvantage compared to 1:1 protection is that main path and spare path are busy and cannot be used for any other path reservation or transport of signals with lower priority.

Alternatively, the modification of a star by redesigning the network to a so-called *star-ring* brings a certain amount of redundancy and potential for protection measures. One or more Multiple Access Units represent the central switch base. Additional nodes can more easily be added by inserting additional central switches into the star-ring network without interruption of traffic. With this technique a bigger area than with star-only topology limited by its maximum fiber length can be covered. For this kind of networks some heuristic algorithms already exist. A distributed dynamic bandwidth allocation scheduling mechanism for star-ring protected networks is proposed in (Hwang et al., 2010).

4.1.2 Ring, double ring, multiring

For covering bigger network areas as with star topology as well as realizing backbone structures, *ring* topology efficiently can be used. The performance of the network is fair between all users. On the other hand, a single bidirectional ring can only deal with a single failure. Two failures disrupt all communication between the two isolated parts. Network reconfiguration, i.e. expanding or shrinking the system is problematic, because the ring has to be opened and several bypass and protection measures get obsolete during this step.

A well-known protection technology often used within multiplexing protocols for fiber-optic systems like Synchronous Digital Hierarchy (SDH) is the Multiplex Section-Shared Protection Ring (*MS-SPRing*) (Goralski, 2002). It exists in two variants, called 2-fiber MS-SPRing and 4-fiber MS-SPRing. The former version uses half of the available bandwidth for backup and/or lower priority communication. In case of a link failure this bandwidth is used to re-establish the ring communication. Lower prioritized communication can no longer be supported and is discarded. The latter version uses a double-ring topology with main and cold-spare rings. On failure of the main ring the protection system immediately switches from the main to the spare ring. Future installations as mentioned in (Reading, 2003) probably will make extensive use of the *Resilient Packet Ring* which is described in IEEE standard 802.17. It relies on the so-called self-healing capability of SONET/SDH networks. In contrary to protection measures like MS-SPRing this technology bases on packet-oriented connections which are used for Ethernet protocols for instance. It engages advanced bandwidth recycling mechanisms, a detailed discussion is out of scope.

4.1.3 Mesh and grid

A *mesh* is the most common form of network topologies within OEO networks. It can be tailored to the necessary number of network switches and/or links. Within a mesh or grid each network node is connected to at least one other node. A mesh network can be realized in a self-healing manner. This kind of network is appropriate to network recovery methods protection and restoration. The very common protection method p-cycle is analyzed by (Liu & Ruan, 2006) as well as by (Drid et al., 2009). Within meshes and grids path recovery methods are well surveyed. (Goralski, 2002) deals with the most common protection methods in general. Online restoration methods for example are presented in (Chow et al., 1993), (Komine et al., 1990) or in (Wosinska et al., 2009). The last reference presents software and hardware solutions for resiliency within all-optical networks. Another example for a high-speed algorithm for network-recovery called ZirkumFlex (Messmer & Keller, 2010) has been presented in Section 3.2.

4.2 Bandwidth-based classification of media streams

In Section 2.3 we described the most common media signals and their corresponding bandwidth. Remarkably there is a very limited variety of bandwidth. Grouping together signals with similar bandwidths results in Table 3. Because of the need to split the available bandwidth into discrete channels there is a certain lower limit for low-speed signals like contacts and signal lights (here expressed at 1 Mbit/s).

Description	BW [Mbit/s]	Group
Control lines	1	1
Stereo audio, communication channels	3	2
100 Mbit/s Ethernet, MADI, GXF HD Video	100	3
SDTV Video (audio incl.)	270	4
HD1080i Video (audio incl.)	1500	5
HD1080p Video (audio incl.)	3000	6

Table 3. Signals grouped by bandwidth.

We analyze three different scenarios concerning fiber-optic linking of media signals within broadcast stations. They differ in terms of reliability characteristics and physical realization. First we present the connection of a small studio or an in-house resource which is not intended to be highly transmission-substantial[4]. Though fault-tolerance is an issue, the connection is realized by several redundant fibers of only one single duct.

Second, for contrast, we discuss a local interconnection for a highly transmission-substantial area. In that case fault-tolerance is realized by way-disjoint ducts. The main cost factor is determined by the necessary number of routers and/or interfaces to be used.

---

[4] A *transmission-substantial* facility must work properly in order to maintain the transmission signal path.

The third example represents the transmission-substantial connection from campus to campus. Such fiber links often are bought or rented by the single fiber, so the link itself is the cost factor to consider. This leads to a different kind of fault-tolerance and consequently to a difference in bandwidth usage.

4.2.1 Small studio interconnection

Figure 3 depicts a connection scheme between a small studio and a technical room as a general example of non-redundant wirings in terms of non-disjoint ducts within TV-studios. The majority of failures in this scenario happen by physical damage of single fibers while plugging, or on degradation of links because of dust or similar optical impairments, which in both cases lead to single-link failures.

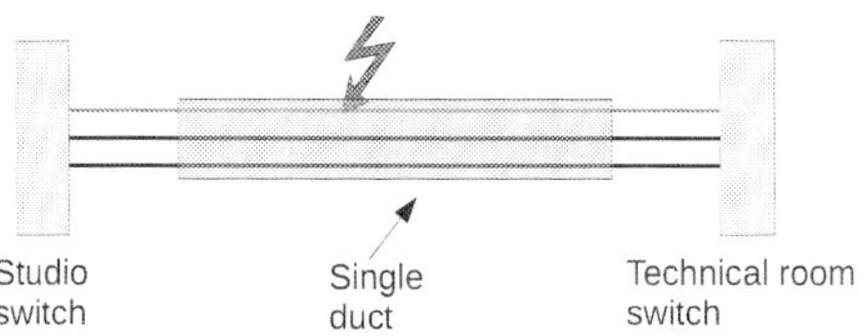

Fig. 3. Small studio interconnection.

We introduce a heuristic that represents an extended variant of the first-fit heuristic used to approximate the online bin-packing problem. To be considered are *fault-tolerance, optimal usage of fiber links* and *load balancing* in a special form that considers the huge differences in bandwidth between audio or control channels and HDTV-video links as shown in Table 3. Using a single duct, there is no reasonable protection method possible like in both other cases using disjoint ducts, so we had to find another scenario. We rely on a kind of load balancing for signals with bandwidth lower than HDTV in combination with the already mentioned ZirkumFlex algorithm for providing fault-tolerance for dynamic path restoration after network failure.

Our proposal guarantees that there will be bandwidth available for HDTV signals as long as possible. It conserves integral HDTV- or at least SDTV[5] bandwidths while paths of signals with small bandwidth can be switched one by one on a first-fit basis.

The proposed online algorithm takes a switch command as input, extracts the requested bandwidth information and checks for all available fibers sequentially if the available bandwidth is greater or equal than the requested one. Unless there is enough bandwidth available in fiber i, fiber $i + 1$ is checked. If the last fiber has been reached and the requested bandwidth is not available, an error is returned. If the switch request can be served by fiber i there are three cases to be distinguished:

- There is enough available bandwidth to serve without reducing the number of HDTV and SDTV channels. Then the switch command immediately is executed on fiber i and the algorithm terminates.

- The number of available HDTV channels on fiber i must be reduced. We denote this condition by $HR(i)$.

---

[5] In this proposal we do not consider the formerly listed 1080i signal. The extension can be implemented in a straight forward way.

```
FOR i=1 to n DO
   if (bw <= cap(i) AND (!HR(i) AND !SR(i))    # no. of HD and SD
   then SWITCH(i), exit                        # channels remain
OD

FOR j=1 to n DO
   if (bw <= cap(j) AND !HR(j))                # no. of HD
   then SWITCH(j), exit                        # channels remain
OD

FOR k=1 to n DO
   if (bw <= cap(k))                           # else
   then SWITCH(k), exit                        # first fit strategy
OD

signal ERROR, exit
```

Fig. 4. Mapping of requests to fibers.

- The number of available SDTV channels must be reduced and the requested signal bandwidth is lower than SDTV. We denote this condition by SR(i).

If $cap(i)$ is the available bandwidth on link i, bw the bandwidth of the current request, and $bw(HDTV)$ the bandwidth consumed by one HDTV signal, i.e. 3000 Mbit/s, then condition HR(i) can be formulated as

$$\left\lfloor \frac{cap(i)}{bw(HDTV)} \right\rfloor - \left\lfloor \frac{cap(i) - bw}{bw(HDTV)} \right\rfloor > 0 \,.$$

Similarly, condition SR(i) can be formulated by substituting $bw(HDTV)$ with $bw(SDTV)$, i.e. 300 MBit/s.

The algorithm is formulated as pseudo-code in Figure 4. The runtime analysis shows that the three concatenated FOR statements question each of n fibers a constant number of times, giving $O(n)$ as runtime performance. This algorithm gives a high priority to HDTV signals which is substantial because typically the number of SDTV signals strongly outweighs the number of HDTV signals and so bandwidth for HDTV signals has to be conserved in the first place.

4.2.2 Intra-campus connection

Connections between broadcast-relevant facilities within a campus suffer of the same problems and failure types as already mentioned in the small studio case. The difference lies on the one hand in the kind of protection the links are equipped with, and on the other hand there are sufficient links available which can be combined in more than one duct.

In this scenario we suggest to consider a kind of signal priority. Figure 5 shows a two-duct combination. In-house connections usually suffer of single fiber failures as mentioned before. Highly transmission-substantial signals should be protected, i.e. for each signal-carrying fiber there should be provided one redundant fiber in the second duct giving a $1 + 1$ protection. Then a failure of several fibers up to one complete duct has no immediate visible (or audible) consequence. Lower prioritized signals can be restored by new calculation of paths or using the mentioned ZirkumFlex algorithm.

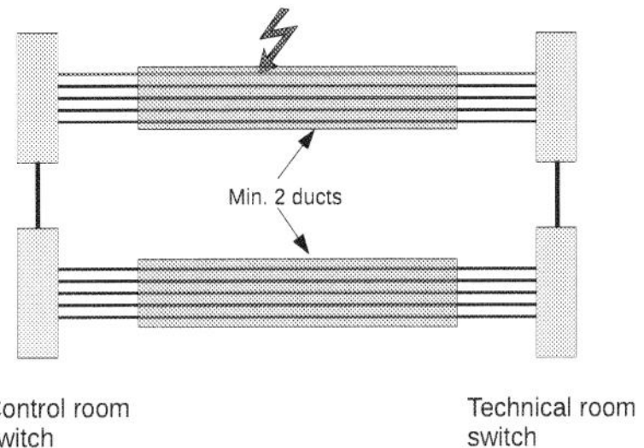

Fig. 5. Intra-campus connection.

4.2.3 Inter-campus connection

Remote studios must be connected via a multitude of links, mostly in dark fiber technology. It is important that the links are laid out geographically diverse, see Figure 6. Depending on the distance and purpose it may be favorable to rent links. In that case it is very important to use protection. CWDM or even DWDM technology is used to exploit the available fiber capacity. Therefore a failure of one link brings a total loss of the transferred signals and the protection link has to get active as quick as possible. This scenario is not suited for any known alternative to $1 + 1$ protection when transmission-substantial lines are in question.

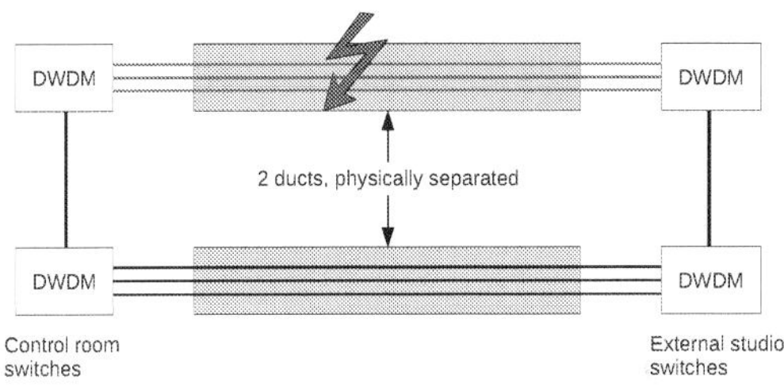

Fig. 6. Inter-campus connection.

5. Conclusion

We analyzed appropriate fault-tolerance concepts for high-speed optical broadcast networks and their influence on the bandwidth of currently used or planned networks of this kind. Besides mentioning the classic protection methodology and some heuristics for restoration methods we presented a novel approach for quick, dynamic restoration of network paths and introduced a heuristic for improved signal occupancy of fiber links in a high-definition video environment.

6. References

Chiu, J.-F., Miao, Y.-B., Lo, C.-W., Huang, Z.-P., Hwang, W.-S. & Shieh, C.-K. (2002). A new preemption avoidance and load balancing policy for traffic engineering in MPLS networks, *Proc. IASTED International Conference on Communications and Computer Networks (CCN 2002)*, pp. 234–239.

Chow, E., Bicknell, J., McCaughey, S. & Syed, S. (1993). A fast distributed network restoration algorithm, *Proc. 12th International Phoenix Conference on Computers and Communications*, pp. 261–267.

Drid, H., Lahoud, S., Cousin, B. & Molnár, M. (2009). A topology aggregation model for survivability in multi-domain optical networks using p-cycles, *Proc. 6th IFIP International Conference on Network and Parallel Computing (NPC '09)*, pp. 211–218.

Gerstel, O. & Zaks, S. (1994). The virtual path layout problem in fast networks, *Proc. 13th ACM Symposium on Principles of Distributed Computing (PODC '94)*, pp. 235–243.

Goralski, W. (2002). *SONET/SDH*, 3rd edn, McGraw-Hill/Osborne.

Hwang, I.-S., Lee, J.-Y., Lee, C.-H. & Shyu, Z.-D. (2010). A novel distributed QoS-based DBA scheduling mechanism for star-ring architecture on ethernet passive optical access networks, *Proc. International Conference on Network-Based Information Systems*, pp. 103–110.

Komine, H., Chujo, T., Ogura, T., Miyazaki, K. & Soejima, T. (1990). A distributed restoration algorithm for multiple-link and node failures of transport networks, *Proc. GlobalCom '90*, pp. 459–463.

Liu, C. & Ruan, L. (2006). p-cycle design in survivable WDM networks with shared risk link groups (SRLGs), *Photonic Network Communications* 11(3): 301–311.

Messmer, R. & Keller, J. (2010). Real-time fault-tolerant routing in high-availability multicast-aware video networks, *Proc. Sicherheit 2010*, pp. 49–60.

Patrikar, M. (2009). *Advanced Technologies*, InTech.

Ramasubramanian, S. & Chandak, A. (2008). Dual-link failure resiliency through backup link mutual exclusion, *IEEE/ACM Trans. Netw.* 16(1): 157–169.

Reading, H. (2003). The future of 'sonet/sdh', *HEAVY READING* 1(6).

Wosinska, L., Colle, D., Demeester, P., Katrinis, K., Lackovic, M., Lapcevic, O., Lievens, I., Markidis, G., Mikac, B., Pickavet, M., Puype, B., Skorin-Kapov, N., Staessens, D. & Tzanakaki, A. (2009). Network resilience in future optical networks, *in* I. Tomkos, M. Spyropoulou, K. Ennser, M. Köhn & B. Mikac (eds), *Towards Digital Optical Networks, COST Action 291 Final Report*, Vol. 5412 of *Lecture Notes in Computer Science*, Springer, pp. 253–284.

Yang, C. H. & Hasegawa, S. (1988). FITNESS: failure immunization technology for network services survivability, *Proc. IEEE Global Telecommunications Conference (GLOBECOM '88)*, pp. 1549–1554.

Zheng, Q. (1992). Fault-tolerant real-time communication in distributed computing systems, *Proc. IEEE Fault-Tolerant Computing Symposium*, pp. 86–93.

19

Integrated ASIC System and CMOS-MEMS Thermally Actuated Optoelectronic Switch Array for Communication Network

Jian-Chiun Liou
Industrial Technology Research Institute
Taiwan

1. Introduction

The advancement of communication technology and growth of internet traffic have continuously driven the fast evolution of networks. Compared to the traditional optoelectronic switch, all-optical switch provides high throughput, rich routing functionalities, and excellent flexibility for rapid signal exchange in fiber optical network. Among various all-optical switches, thermal actuated ring switch provides the advantages of high accuracy, easy actuation, and reasonable switching speed. However, when scale up, thermal ring switch may encounter issues related to fabrication error, non-accurate wavelength response, and large terminal numbers in the control circuit.

Planar-lightwave-circuit switch (PLC-SW), employing thermo-optic (TO) effect of silica glass for light switch, is a very promising technique for communication applications because of low insertion loss, high extinction ratio, long-term stability, and high reliability. There have been many matrix switches designed based on the TO effect with low-loss, polarization insensitive operation, and good fabrication repeatability. For example, 8×8 matrix switches were fabricated by using a single Mach-Zehnder (MZ) switching unit and demonstrated well performance in transmission systems, so as an 8×8 matrix switch and a 16×16 matrix switch by the similar MZ switching unit. 32×four-channel client reconfigurable optical add/drop multiplexer on planar lightwave circuit. However, when scaled up, thermal ring switch may encounter issues related to large terminal numbers in the control circuit, fabrication error and non-accurate wavelength response. For example, if conventional driving circuits are employed for a 16×28 or even larger switch array, 448 or more terminals will be required for control. Such a large number of terminals would complicate the module structure and occupy a large area. On the other hand, when a DC-current is applied for balancing wavelength offset from fabrication error, the input power will result in a temperature elevation of the neighboring switches, thus changes the related refractive indexes and therefore deviate wavelengths.

To solve the aforementioned issues, this chapter proposes a ring resonator with silicon nitride as the core layer and silicon dioxide as the cladding layer was designed and fabricated on silicon substrate, a novel architecture of high selection speed three dimensional (3D) data registration for driving large-array optoelectronic packet switches. The 3D driving architecture can successfully reduce the total numbers of control pads into

31 for 448 switches as well as the scanning time up to 67 % reduction with a higher signal rising speed and smaller circuit area. All the sub-circuits, including power control, digital I/O, analog-to-digital converter, power drivers were integrated into a single IC. On the other hand, instead of DC current control, wavelength lock is realized by amplitude and frequency modulated heating pulses for stabilization of temperature and fine-tune of wavelength from fabrication imperfection and environment fluctuation. This planar-lightwave-circuit has been designed, fabricated, and characterized. It is demonstrated not only the functionality in optical packet switches but also the consistency between simulation and experiment results.

In this research, we design the tunable micro ring resonators and propose the employment of an integrated ASIC system CMOS technology control circuit to compensate the fabrication error and tune as well as lock the wavelength in a thermal-actuated ring-type optical switch through a frequency modulation scheme. The use of a standard and commercial CMOS technology for designing micro resonators entails a set of limitations, such as layer thicknesses, and available materials in an inalterable process sequence. From the MEMS design point of view, those restrictions will limit the electrical properties of fabricated micro-resonators. On the other hand, MEMS integration into a CMOS technology presents unique advantages, like reduction of the parasitic capacitance due to the possibility to monolithically integrate the circuitry, in addition to an expected reduction of the overall production costs.

Additional functionalities can also be added in this circuit by tailoring externally the roundtrip loss or coupling constants of the ring. The design concept can be easily scaled up for large array optical switch system without much change in the terminal numbers thanks to the three dimensional hierarchy of control circuit design, which effectively reduces the terminal numbers into the cubic root of the total control unit numbers. The integrated circuit has been designed, simulated, as well as fabricated, and demonstrated a decent performance with Free Spectral Range (FSR) equal to 1.5nm at 1534 nm and very accurate wavelength modulation to 0.3 nm within 0.01 nm fluctuation for thermal actuated ring type optoelectronic switch.

Thermal actuated optoelectronic ring switch provides the advantages of high accuracy, easy actuation, and reasonable switching speed. However, when scale up, thermal ring switch may encounter issues related to fabrication error, non-accurate wavelength response, and large terminal numbers in the control circuit. Planar-lightwave-circuit switch (PLC-SW), employing thermo-optic (TO) effect of silica glass for light switch, is a very promising technique for communication applications because of low insertion loss, high extinction ratio, long-term stability, and high reliability[1-4]. There have been many matrix switches designed based on the TO effect with low-loss, polarization insensitive operation, and good fabrication repeatability. For example, 8×8 matrix switches were fabricated by using a single Mach-Zehnder(MZ) switching unit and demonstrated well performance in transmission systems [5, 6], so as an 8×8 matrix switch [7] and a 16×16 matrix switch [8] by the similar MZ switching unit. 32×four-channel client reconfigurable optical add/drop multiplexer on planar lightwave circuit [9]. However, when scaled up, thermal ring switch may encounter issues related to large terminal numbers in the control circuit, fabrication error and non-accurate wavelength response. For example, if conventional driving circuits are employed for a 16×28 or even larger switch array, 448 or more terminals will be required for control. Such a large number of terminals would complicate the module structure and occupy a large area. On the other hand, when a DC-current is applied for balancing wavelength offset from fabrication error, the input power will result in a temperature elevation of the neighbouring switches, thus changes the related refractive indexes and therefore deviate wavelengths.

To solve the aforementioned issues, this research proposes a ring resonator with silicon nitride as the core layer and silicon dioxide as the cladding layer was designed and fabricated on silicon substrate, a novel architecture of high selection speed three dimensional data registration for driving large-array optical packet switches. The 3D driving architecture can successfully reduce the total numbers of control pads into 31 for 448 switches as well as the scanning time up to 67 % reduction with a higher signal rising speed and smaller circuit area. All the sub-circuits, including power control, digital I/O, analog-to-digital converter, power drivers were integrated into a single IC. On the other hand, instead of DC current control, wavelength lock is realized by amplitude and frequency modulated heating pulses for stabilization of temperature and fine-tune of wavelength from fabrication imperfection and environment fluctuation. This planar-lightwave-circuit has been designed, fabricated, and characterized. It is demonstrated not only the functionality in optical packet switches but also the consistency between simulation and experiment results.

2. Planar-lightwave-circuit design and principle

2.1 Design of wavelength modulation and lock

In this research, PLC filter will usher as a model system for demonstration. A PLC filter is consisted of arrays of micro-rings (MRs) with the ring radii in the order of tens of micrometers, possessing extremely small-area and providing the characteristic of high selectivity. Based on the design of this device, a four-channel PLC filter is depicted with four MRs, as schematically shown in Fig. 1. The operation principle of this device is as the first column of the optical circuit is on each MR drops an incoming signal λx into one of the output channels Ioutx when the MR resonance frequencies matching that of the incoming signal [10-11]. In this research, the design, fabrication, and measurements of a four-channel thermally tuneable MR-based filter PLC controller is presented. Silicon dioxide and silicon nitride were selected as an optical thin film system to compose the micro-rings on top of a silicon substrate. As shown in Fig. 2, the width of straight and ring waveguide was chosen as 2.0 µm. To ensure single mode operation at TE polarization, the height of waveguides was determined to be 0.45 µm by a 3D finite-difference time-domain simulation using RSoft-FullWave(a finite-difference time-domain solver of Poisson's equations)[12-13]. The simulation flow is electro-thermal transport and optics, then according to follow electrostatic potential equation (1) and carrier continuity equations (2), electrons, holes relative derived among temperature, core index, and radius. The effective index method was employed to determine the radius of the ring as 9.965 µm.

1. Poisson's Equation (Electrostatic Potential)

$$\nabla \cdot \varepsilon \nabla \varphi + q(N_D^+ - N_A^- + p - n) = 0 \tag{1}$$

2. Carrier Continuity Equations (Electrons, Holes)

$$\frac{\partial n}{\partial t} + \nabla \bullet j_n + U = 0 \quad \text{And} \quad \frac{\partial p}{\partial t} + \nabla \bullet j_p + U = 0 \tag{2}$$

3. Lattice Heat Equation (Temperature)

$$\frac{\partial}{\partial t}(c_L + \frac{3}{2}(n+p)k)T = \nabla(K_L \nabla T - \vec{S_n} - \vec{S_p}) + \vec{j_n}\vec{E} + \vec{j_p}\vec{E} + R_{dark}E_g \tag{3}$$

4. Photon Rate Equation (Modal Photons)

$$\frac{\partial S_{m,w}}{\partial t} = (G_{m,w} - \frac{1}{\tau_{m,w}})S_{m,w} + R_{m,w}^{spon} \tag{4}$$

5. Derived Helmholtz Equation (Mode Profile)

$$\nabla^2 \phi + \varepsilon k_o^2 \phi = 0 \tag{5}$$

Where,
n, p: Electron/Hole Density
V: Electrostatic Potential
T: Lattice Temperature
G, R: Optical Gain/Recombination
f: Lasing Frequency
A: Optical Wave Amplitude
S: PhotonNumber
K: Quantum Mechanical Wavefunction
The condition for single mode operation of a Si3N4 optical rib waveguide with a refractive index 2.06 operating at 1.55µm was analyzed using beam propagation method (BPM). The full vectorial BPM was used to analyze the waveguide structure in Fig. 3 for single mode propagation and polarization independence.

Height\width(µm)	0.2-0.4	0.5-0.8	0.9-2.0
0.3	single	multi	multi
0.4	single	single	multi
0.5	single	single	single
0.6	single	single	single
0.7	single	single	single
0.8	single	single	single
0.9	single	single	single

Table 1. Summarizes the waveguide mode.

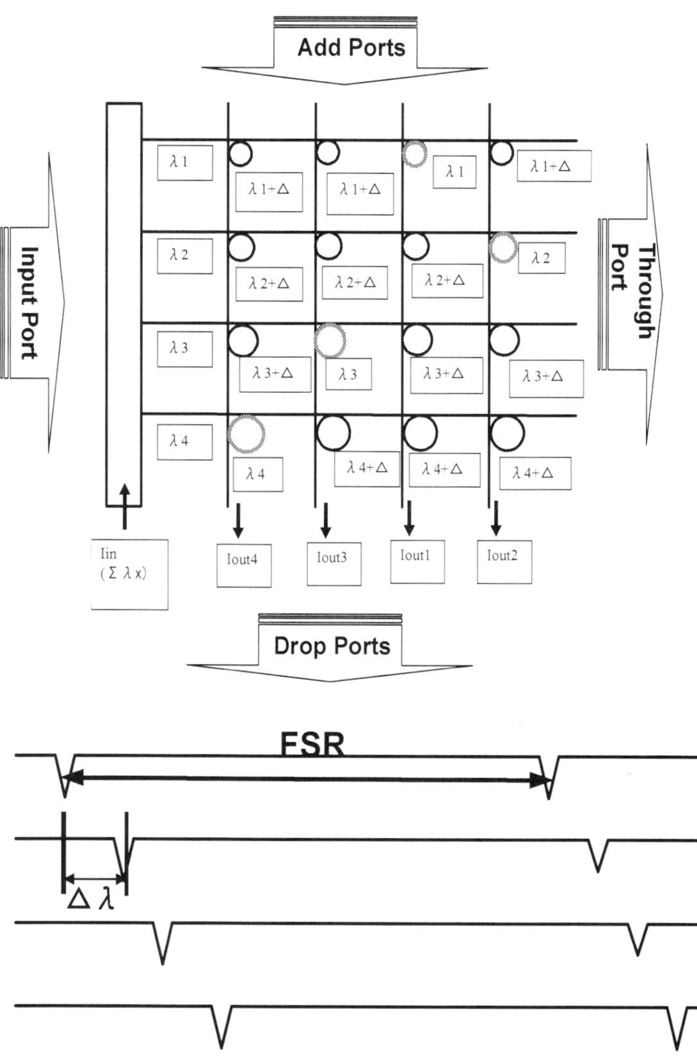

Fig. 1. Schematic of a four-channel PLC.

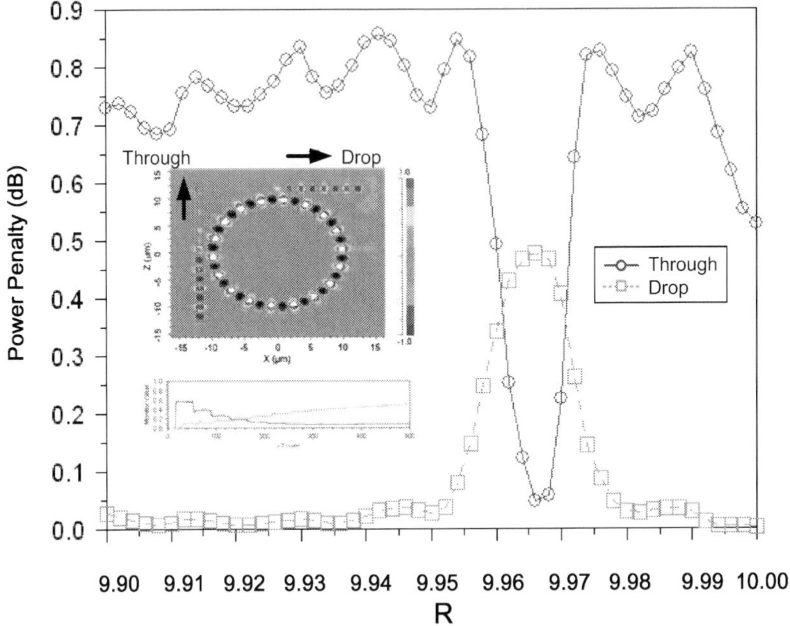

Fig. 2. Transmittance for the radius of micro ring.

According to the BPM calculations, for height = 0.4~0.5µm, width = 0.9~2.0µm enable single mode in the TE polarization. Table 1 summarizes the calculated number of waveguide modes for our studied height and width ratios. Similar single mode output intensity patterns were imaged from other fabricated control waveguide of various height and width. Fig.4(a) 、 (b) and (c) depicts the imaged mode profiles have only one field maximum near the center for both the horizontal and vertical directions. Insets show the BPM calculated fundamental mode profiles from the same waveguide geometries, depicting similar Gaussian-like mode profiles.

Rib waveguide

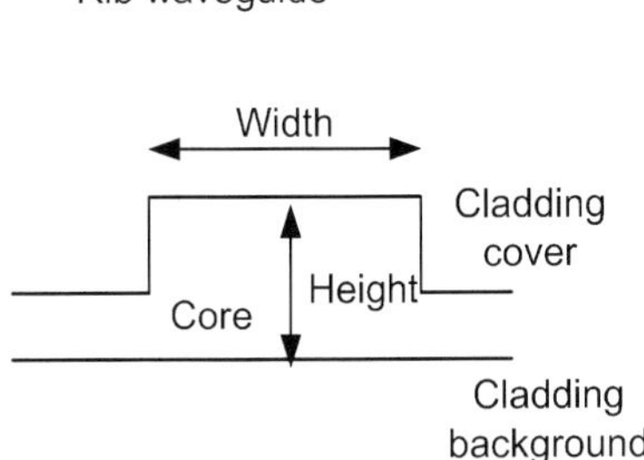

Fig. 3. Schematic cross section of a rib waveguide.

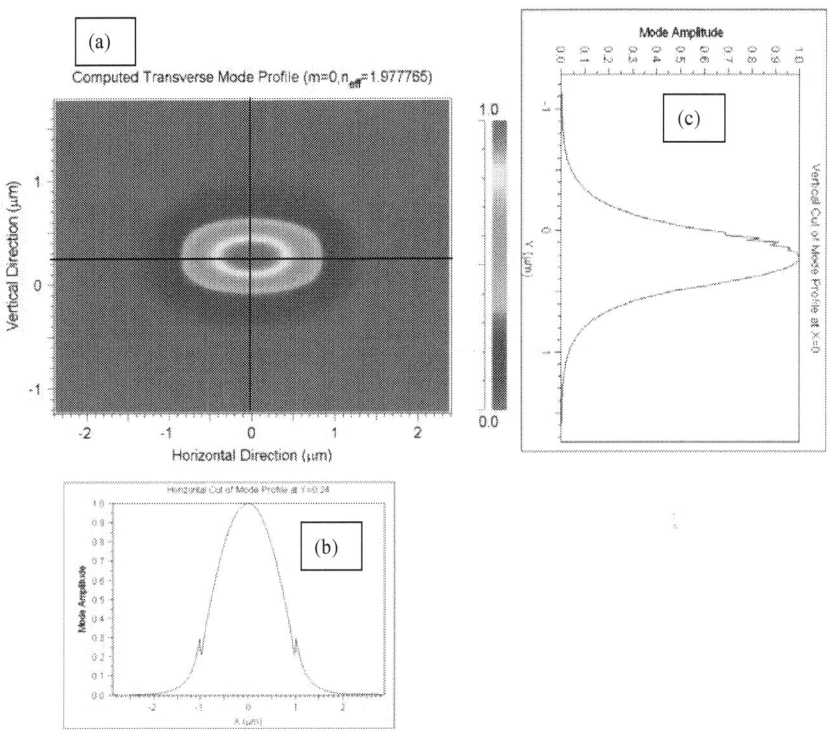

Fig. 4. Simulated fundamental mode profile from BPM-based calculations.

In the process of making the rib waveguide as shown in Table 2, we also examined four different process parameters to see how they affect the final waveguide's dimensions. The first step was to investigate the polarization characteristics of the waveguide due to its geometry. Various waveguide heights ranging from 0.3 to 1.0um and width ranging from 0.9 to 2.0um are considered. While keeping waveguide height constant during the computation, the difference in effective indices of the fundamental TE waveguide modes has been evaluated as the etch depth and waveguide width were varied.

Process number	Process step
1	Thermally grown SiO2 on Si wafer
2	LPCVD deposition of Si3N4 layer
3	Spinning of resist, patterned by photolithography(E-Beam) and structure by RIE
4	Deposition of PECVD SiO2 cladding layer and annealing of layer stack
5	Sputtering a Platinum (Pt) thin film
6	Patterned by photolithography and Pt wet-etch.

Table 2. Process flow for SiO2/Si3N4 coupled microring resonators.

Single mode propagation is an important requirement for optical waveguide devices for use with single-mode fiber; it can reduce the coupling loss. In this research, a technique is used for calculating the field distribution of the Si3N4 rib waveguide. The waveguide was modeled using the three-dimensional full-vectorial beam propagation method (BPM) to calculate the effective refractive indices and modal field profiles for various waveguide widths, heights and etch depths. The scanning electron micrograph (SEM) of Fig. 5 shows such a fabricated rib waveguide.

We experimentally verify the practically single mode nature of our deeply etched rib waveguides by imaging control straight waveguides output intensity patterns. Fig.6 shows the representative imaged waveguide output intensity profile with waveguide of the order of 1550nm.

Fig. 5. SEM image of the waveguide cross section.

Fig. 6. Imaged waveguide output intensity profile.

In the wavelength modulation, thermal optic effect was employed to shift the resonance wavelength at an amount of $\triangle\lambda$ by the tuning of effective index at different temperatures. This wavelength shift can be used to tune the passband to the desired wavelength. The principle of wavelength modulation is shown in Fig. 1, illustrating that the elevation of temperature on one MR switch shifts the center wavelength by $\triangle\lambda$ but remain the same Free Spectral Range (FSR). The term FSR is borrowed from Fabry Perot interferometry, and describes the maximum spectral range one can arbitrarily resolve without the interference from the neighbouring signals. On the other hand, a high extinction ratio can be obtained through the filtering effect from the MRs with a steep wavelength response. A relationship between the radius of the ring R, the effective group index n_g, and the FSR is given by equation (6):

$$FSR = \frac{\lambda^2}{2\pi R n_g} \tag{6}$$

where λ is the wavelength [14-15].

In temperature control, frequency modulation was employed instead of voltage level modulation due to the simplicity of implementation by digital signals. Through frequency modulation, the temperature in the thermally tuneable PLC modules can be maintained almost constant and this will result in a more accurate center wavelength for the optical communication channel. It also ensures rapid response of the PLC module as the heater has been modulated on and off in a high frequency (~MHz). As a result, the PLC module at room temperature was able to achieve a very small temperature fluctuation within 0.1°C which can not be achieved by using traditional DC controls.

In order to compensate the fabrication error of the thermal ring switch, a simple and practical phase-trimming technology was employed to avoid the need of electrical biasing. The phase-trimming technique employs a local heating technology by the employment of a thin-film heater embedded under the optical ring in a feedback loop for the fine tune of the optical phase. However, if DC bias is employed in the phase control, the temperature of the neighbouring switch may encounter drift (cross talk) as well as slow response for temperature compensation. To lower down the cross talk effect, provide more accurate temperature control, and speed up thermal response of the optical ring, a frequency modulated heating scheme is employed by dynamic feedback of the frequency of heating pulses.

To achieve the above goal by frequency modulation for accurate temperature control, this study employs a selection algorithm to select a proper waveform pre-stored in the lookup table in an ASIC(Application specific integrated circuits) chip, in which all waveforms have been simulated and optimized for different temperature situations. Each drop and filter channel is assigned a temperature for the desired wavelength shift. The temperature is maintained by a corresponding waveform from the result of the sum of three signals, including data (address), select, and power.

2.2 Design of three dimensional controller

In traditional control circuit design for thermal optical type switch array, each optical ring requires one heater for wavelength adjustment. As a result, when the optical switches scale up into a large array, the numbers of input/output ports will increase enormously. To handle large array of driving circuits for such a heater array, two dimensional (2D) circuit architecture was employed by traditional driving circuits to reduce the IO number from n*n into 2n+1. However, this reduction still can not meet the requirement for high speed signal scanning with low data accessing points when switch numbers greater than 1000.

To achieve this, in this study, a three dimensional data registration scheme to reduce the number of data accessing points as well as scanning lines for large array optical packet switching chip with switch number more than 1000 is proposed. The total numbers of data accessing points will be N=3× $\sqrt[3]{Y}$ +1, which is 31 for 1000 switches by the 3D novel design, the scanning time is reduced down to 33% (The scanning speed is also increased by 3 times) thanks to the great reduction of lines for 3D scanning, instead of 2D scanning. The property comparison among 1D, 2D, and 3D architectures is listed in Table 3. As the optical switch number increases, a higher order control circuit can effectively reduce the pad number. In addition, the shape and amplitude of the driving signal can be optimized to increase the speed of the response with low driving powers [16].

X :Pads、Y :Switches	X~Y +1	X=2×$\sqrt[2]{Y}$+1	X=3×$\sqrt[3]{Y}$+1 (X : Connection lines ,Y :Switches)
Switches	1000	1024	1000
Ring resistors	1000	1024	1000
Interconnect Pad	1001	65	31

Table 3. Performance comparison among 1D, 2D, and 3D driving schemes.

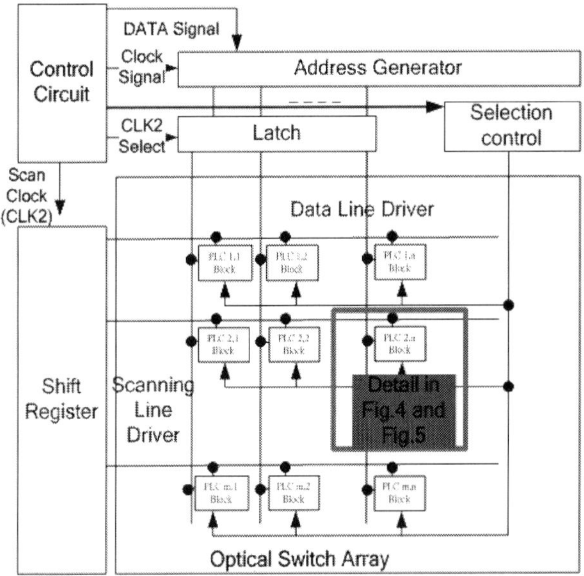

Fig. 7. Block diagram of control algorithm for micro-ring switches.

In the proposed novel 3D design, different from the 2D one, as shown in Fig. 3, the digital driver includes a clock-control circuit, a serial/parallel-conversion circuit, a latch circuit, a level shifter, a D/A converter comprised of a decoder, and an output buffer comprised of an operation amplifier. The D/A converter receive a gray-scale reference voltage from an external source [17-21]. The clock-control circuit receives control signals from an external control circuit. Based on the received control signals, the clock-control circuit attends to control of the latch circuit, the D/A converter [22-24], the output buffer by using a latch-control signal.

The general strategy that we employ is to integrate all relatively small-signal electronic functions into one ASIC to minimize the total number of the components. This strategy demonstrates that both the cost is lowered and the amount of the printed circuit board area is reduced. Based on this concept, a smart 3D multiplexed driver for optical packet switching chip with more than 1000 rings are proposed and the circuit architecture is shown in Fig. 7. Three lines are employed to control the heating of one micro-ring, including voltage, shift register, and data line. The relationship among the waveforms is shown in Fig. 8. Each heater resistor requires a voltage line for the driving current flow and shares the same ground with the other resistors. The resistors are individually addressable to provide unconstraint signal permutations by a serial data stream fed from the controller. The shift register is employed to shift a token bit from one group to another through AND gates to power the switch of a micro-ring group. The selection of a ring is thus a combined selection of the shift register for the group and the data for the specific ring. Such an arrangement allows encoding one data line from the controller to provide data to all of the rings, permitting high-speed printing by shortening the ring selection path and low IC fabrication cost from the greater reduction of circuit component numbers.

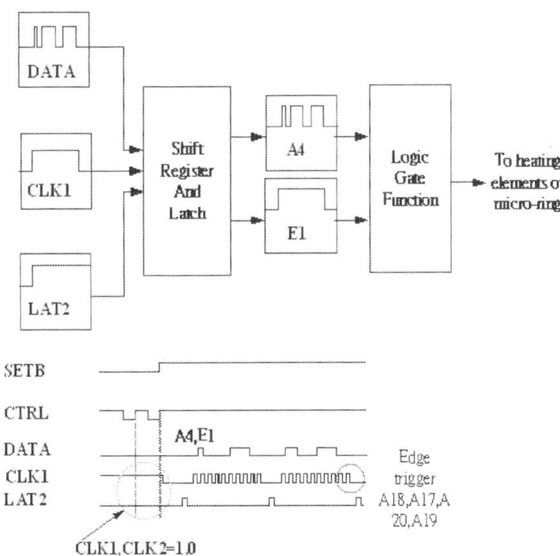

Fig. 8. Example of input waveform for controller from FPGA.

The received optical data information has to be converted into data at an optimal transfer rate (frequency) in order to conform to the ring characteristics. To the end, the clock-control circuit divides the 8-bit optical packet switching signals supplied to the data driver, as shown in Fig. 7, with an aim at lowering the operation frequency. The serial/parallel-conversion circuit converts serial signals of a plurality of channels into parallel signals, and supplies the parallel signals to the latch circuit. The latch circuit temporarily stores the received parallel signals, and supplies them to the level shifter and the D/A converter at predetermined time [25-27]. The level shifter converts a logic level ranging approximately from 3.5 V to 5 V into a ring array voltage level that ranges from 5.5V to 8.5V for various heater resistors as a result of variation processing conditions.

In the signal flow design, optical switches are usually scanned over one by one without jumping on un-activity switches. As a result, for the optical packet switching chip with 448 optical switches, a 1, 2 or 3 dimensional circuit architect will needs 448, 36, and 5 unit times for scanning over all of the switches. Therefore, the scanning time of the 3D multiplexing circuit from the first address line to the 16th, as an example, takes only 5 units of clock time from the simulation result, much faster than that of the 2D configuration with 16 units of clock time. Thus the maximum scanning time for the 3D circuit will be reduced to 30% of that in the 2D case.

To simultaneously write signals into the driving circuit, multiplexing data latches and shift registers are employed by the application of commercial available CMOS ICs. Small numbers of shift registers, control logics, and driving circuits can be electrically connected and integrated with optical packet switching using standard CMOS processes. Fig. 9 shows the driving circuit of the three-dimensional architect. The desired signal for "S" selections and "A" selections can be pre-registered and latched in the circuit for one time writing.

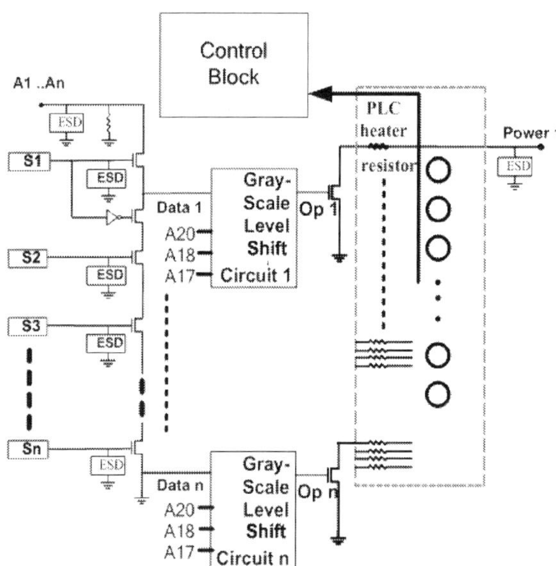

Fig. 9. Architect of three-dimensional driving circuit for micro-ring switches.

The SPICE simulation results on the relationship of input and output signal at 5µs clock time. Fig. 10 demonstrates that not only the switch speed is higher by the level shift device than that of one without level shift circuit, but also the voltage has been enhanced to 5V. An adjustable voltage pulse from 7.98V to 8.02V amplitude modulation is applied to the various heater resistors thanks to the processing condition.

The cooling down of the structure is equally important, though enhancing the speed of the cooling down process might be done by active cooling, but this would require major adaptations to the device and the low cost low power principle would not hold anymore. A much easier way to do this is biasing. In biasing, a DC-current is applied, that will result in a relatively small change in temperature, refractive index and therefore resonance wavelength. To heat the device, the wide pulse width signal or high gray scale level voltage is applied; to cool it down, a narrow pulse width signal or low-level voltage is applied. See Fig. 10 for the simulated behavior of high and low bias driving. The maximum current that can be applied is limited, due to the destruction of the heaters at high powers. The use of a bias will therefore cause a smaller modulation depth, but the modulation will be faster, since the time needed for cooling down is reduced.

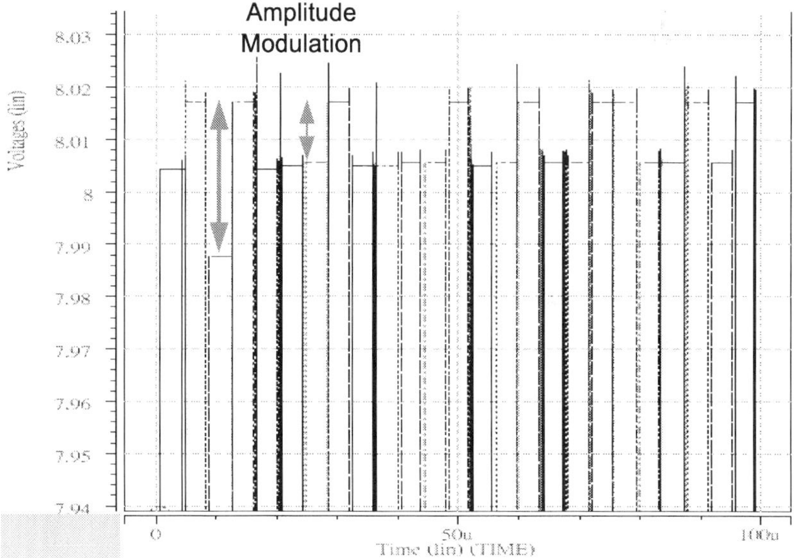

Fig. 10. Transient simulation of the input and output signals of the level shift device.

Power consumption of a narrow pulse width signal or low-level voltage applied for the thermally actuated optical switch array is very small comparing a DC-current applied. The main advancement of this new concept is that the drive signal opening the switch tracks the serial/parallel- conversion circuit, which converts serial signals of a plurality of channels into parallel signals, and supplies the parallel signals to the latch circuit.

If we analyze the total wire power, we calculate that the intermediate interconnection power is the dominant part of the total wire power. The total wire power is scaled down of the three dimensional hierarchy of high gray scale control circuit design, which effectively reduces the terminal numbers into the cubic root of the total control unit numbers.

3. Experimental and results

3.1 Wavelength modulation and lock

By using a Commercial Finite Difference Solver (CFD-RC, USA) for thermo-optical problems, the temperature profile of the MR and the relative changes of refractive indexes can be simulated, as shown in Fig.11(a) and Fig.11(b). Electro-thermal changed temperature by ASIC multiplexing data signal applying to coupled-ring-resonator for adjustment core index. Optimal tunable center wavelength 1511nm conform the shifted core indexes from 2.000 to 2.008. Although the temperature distribution on the ring is about $1\,^\circ C$, the average temperature of the ring is employed as a reference for the temperature control and the tolerance is within $0.1\,^\circ C$. To reduce overshooting and obtain rapid set up of ring temperature, heating pulses with amplitude modulation were employed. Through simulation, optimized driving signal can be obtained to maintain stable wavelength in 0.1 ms by accurate temperature modulation [28]. The temperature fluctuation can be controlled within 0.1°C, with a wavelength variation locked in 0.01 nm, as the measured result shown in Fig. 12.

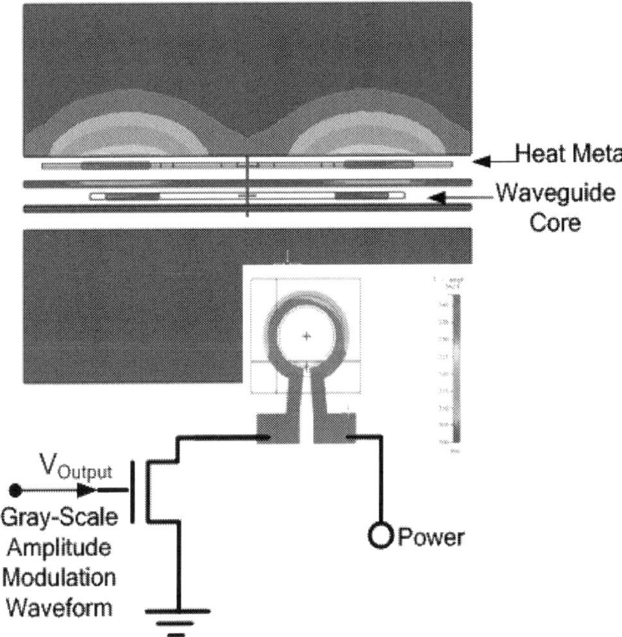

Fig. 11(a). Driving architecture of wavelength lock and simulation profile.

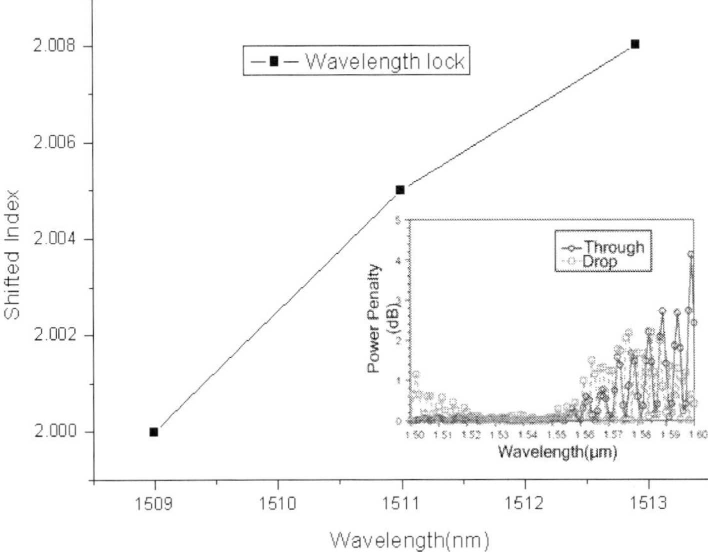

Fig. 11(b). The relative of shifted index and wavelength.

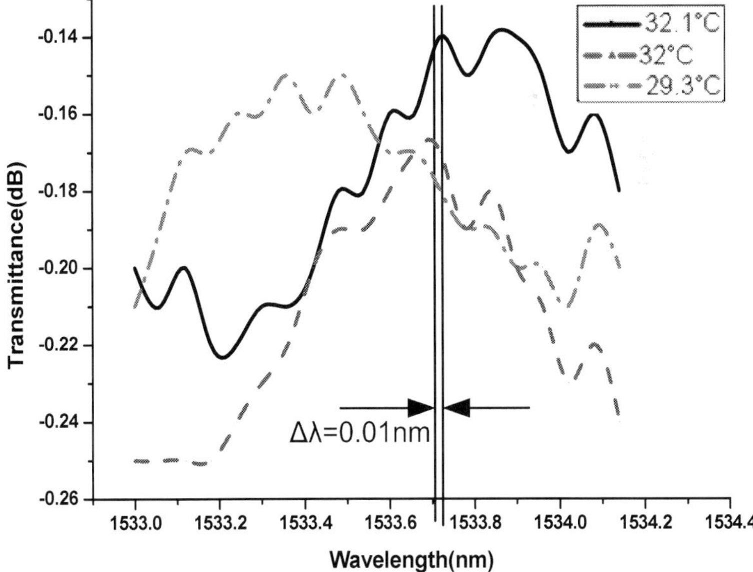

Fig. 12. Wavelength lock.

3.2 Controller and wavelength modulation result

To demonstrate basic functions of the 3D controller, as a result, we were able to reduce the number of electrical terminals to 5 control terminals and 1 power supply terminal. The controller was designed for a 0.35 μm CMOS process with a total circuit area of 2500×2500 μm^2, which is 80% of the circuit area by 2D configuration for 448 switches.

In the Logic Analysis, the relationship between the ASIC input and output is shown in Fig. 13. The input signals include DATA (signal for selected switch action), CLK1(signal to scan DATA signal), CLK2(signal to latch DATA signal or select), CTRL(signal to select enable type), as well as SETB(the time sequence to set up CTRL or power), and the output signals match the designed ASIC signals very well. Fig. 14. shows the image of a fabricated 16×28 matrix switch controller module. In this module, the chip area of 2.5×2.5 mm and was fabricated by a two-poly four-metal (2P4M) 0.35μm twin-well CMOS technology (TSMC, Taiwan Semiconductor Manufacturing Company Ltd). Each transistor is surrounded by full guard ring for preventing electrostatic shock.

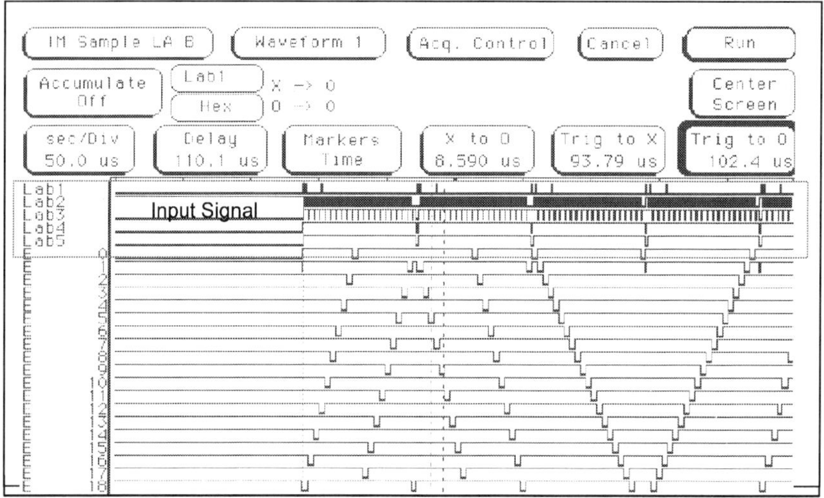

Fig. 13. FPGA verification result.

The testing result of the IC demonstrated the scanning of 448 ring switches takes 60.5 μs for 2D circuit architect while 20.5 μs for the 3D one, representing a time saving of 40 μs or a 67% time reduction. The measurement results of serial output signals for four channels, as shown in Fig. 15, demonstrated a simultaneous operation of four different temperature /wavelength modulations in each channel. By using the optimized driving signals, modulation frequencies up to 10 kHz were measured, resulting in thermal switching speeds in the order of 0.1 ms.

The micro-rings are made with the use of standard clean room fabrication technology. The fabrication of silicon nitride waveguides starts with a six inch diameter polished <100> silicon wafer. First a planar waveguide structure with a SiN(n=2.06@ λ =1550nm) core and SiO2(n=1.452@ λ =1550nm) cladding is formed. Finally the heater layer is deposited by sputtering a Platinum (Pt) thin film and patterned by photolithography and Pt wet-etch. Some results of temperature coefficient of resistance (TCR) measurements on platinum thin films. The shift in center wavelength of the ring λc is a function of the difference in effective index induced by heating the device that is given by equation (7):

$$\delta\lambda c = \frac{\lambda\Delta Neff}{Neff}$$

(7)

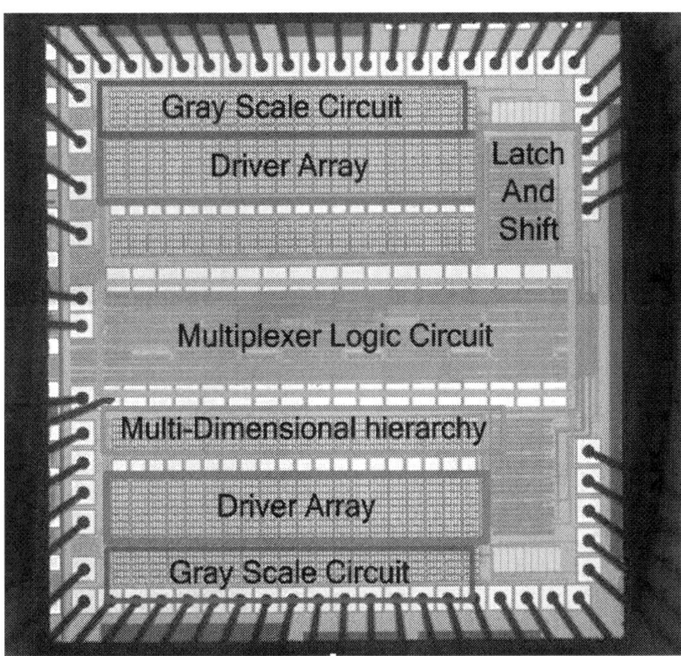

Fig. 14. Photograph of fabricated control IC chip.

In the wavelength modulation, temperature variation induced spectrum shift was measured, and the result is shown in Fig. 16, for the temperature changed from 29°C to 32°C, for which the thermal resonance shift is determined to be 0.1nm/°C. The temperature fluctuation can be controlled within 0.1°C, with a wavelength variation locked in 0.01 nm. The measured values are FSR=1.5nm and center wavelength shift λc=0.3nm at a center wavelength of λ=1532nm.

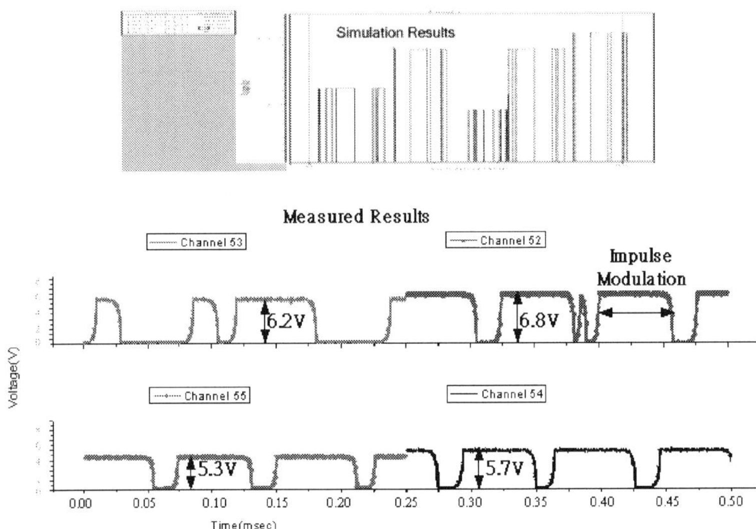

Fig. 15. Measurement result of serial outputs.

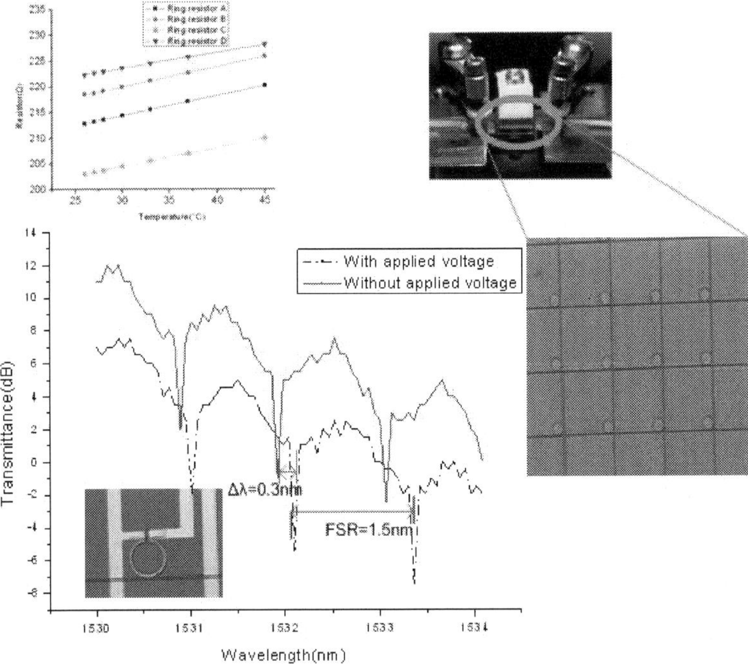

Fig. 16. Wavelength shift of transmission spectrum in coupled-ring-resonator.

4. Conclusion

The next generation of optical networking requires optical switches with complex functionality, small size and low cost. In this research, we have successfully designed and fabricated a silica-based 16×28 PLC-SW controller module in which we incorporated a switch chip based on PLC technology and new driving circuits with a serial-to-parallel signal conversion function. The new driving circuits significantly reduced the number of control terminals, and enabled us to realize a simple module structure suitable for use in a large-scale switch. It has been demonstrated that the scanning of 448 ring switches takes 20.5 µs by the novel 3D architect, representing a 67% time reduction.

On the other hand, thermal-optical effect was employed for wavelength modulation in this optical switch. To reduce overshooting and obtain rapid set up of ring temperature, heating pulses with amplitude modulation were employed. A temperature variation within 0.1°C can be maintained by this design, which can provide a very accurate wavelength modulation to 0.3 nm within 0.01 nm variation.

5. References

C. A. Brackett, et. al.(1993) "A Scalable Multi-wavelength Multihop Optical Network: A Proposal for Research on All-Optical Networks", J.Lightwave Technol., vol. 11, 736-753.

K. Okamoto, K. Takiguchi & Y. Ohmori,(1995) "16-channel optical add/drop multiplexer using silica-based arrayed-waveguide gratings", Electron. Lett., vol. 31, pp. 723-724.

H. Li, C. Lee, W. Lin, L.S. Didde, Y. J. Chen & D. Stone(1997) "8- wavelength photonics integrated 2x2 WDM cross-connect switch using 2xN phased-array waveguide grating (PAWG) multi/demultiplexers", Electron. Lett., vol. 33, pp. 592-594.

C. R. Doerr, et. al.,(1999) "40-wavelength add-drop filter", IEEE photon. Technol. Lett., vol. 11, pp. 1437-1439.

M. Okuno, A. Sugita, T. Matsunaga, M. Kawachi, Y. Ohmori, & K. Katoh,(1993) "8×8 optical matrix switch using silica-based planar lightwave circuits," IEICE Trans. Electron., vol. E76-C, no. 7, pp. 1215–1223.

M. Okuno, K. Kato, Y. Ohmori, M. Kawachi, & T. Matsunaga,(1994) "Improved 8×8 integrated optical matrix switch using silica-based planar lightwave circuits," J. Lightwave Technol., vol. 12, pp. 1597–1606.

T. Goh, A. Himeno, M. Okuno, H. Takahashi, & K. Hattori,(1998) "Highextinction ratio and low-loss silica-based 8×8 thermooptic matrix switch," IEEE Photon. Technol. Lett., vol. 10, pp. 358–360.

T. Goh, M. Yasu, K. Hattori, A. Himeno, M. Okuno, & Y. Ohmori,(1998) "Low-loss and high-extinction-ratio silica-based strictly nonblocking 16×16 thermo-optic matrix switch," IEEE Photon. Technol. Lett., vol. 10, pp. 810–812.

Wenlu Chen, Zhonghua Zhu, Yung Jui(Ray) Chen, Jacob Sun, Boris Grek, & Kevin Schmidt,(2003) "Monolithically integrated 32×four-channel client reconfigurable optical add/drop multiplexer on planar lightwave circuit", IEEE photonics technology letters,vol.15,no10,pp.1413-1415.

J.E. Ford, V. Aksyun, D. Bishop & J. Walker,(1999) "Wavelength add-drop switching using tilting mirrors", J. Lightwave Technol., vol. 17, pp. 904-911.

C. Pu, L. Lin, E. Goldstein & R. Tkach,(2000) "Client-configurable eight channel optical add/drop multiplexer using micromachining technology", IEEE photon. Technol. Lett., vol. 12, pp. 1665-1667.

William M. J. Green, Hendrik F. Hamann, Lidija Sekaric, Michael J. Rooks, & Yurii A. Vlasov,(2006) "Ultra-compact reconfigurable silicon optical devices using micron-scale localized thermal heating", Optical Society of America, pp.1-3.

Andreas Witzig, Matthias Streiff, Wolfgang Fichtner ,"Eigen-mode Analysis of Vertical-Cavity Lasers",pp.1-34.

N. A. Riza & S. Yuan,(1999) "Reconfigurable wavelength add-drop filtering based on a Banyan network topology and ferroelectric liquid crystal fiber-optic switches", J. Lightwave Technol., vol. 17, pp. 1575-1584.

Wenhua Lin, Haifeng Li, Y. J. Chen, M Dagenais & D. Stone,(1996) "Novel Dual-Channel-Spacing WDM Multi/Demultiplexers Based on Phased- Array Waveguide Grating", Photonics Tech. Lett. 8, 1501.

Jian-Chiun Liou, & Fan-Gang Tseng,(2008)" Integrated Control Circuit For Thermally Actuated Optical Packet Switch",Wireless And Optical Communications (WOC) pp.213-218.

Young-Kai Chen, Andreas Leven, Ting Hu, Nils Weimann, Rose Kopf, Al Tate,(2008)" Integrated Photonic Digital-to-Analog Converter for Arbitrary Waveform Generation", Photonics in Switching (PS), Optical Switches and Routing Devices, D-08-2, pp.1-2.

WANG Meng-Yao, WEI PAN, BIN LUO, ZHANG Wei-Li, & ZOU Xi-Hua (2007) "Optimization of gray-scale performance in pixellated-metal-mirror FLC-OASLM by equivalent circuit model" Microelectronics journal, vol. 38, no2, pp. 203-209.

Christopher M Waits, Alireza Modafe & Reza Ghodssi,(2003)"Investigation of gray-scale technology for large area 3D silicon MEMS structures", J. Micromech. Microeng. 13 170-177.

S. Mias, L.G. Manolis, N. Collings, T.D. Wikinson, W.A. Crossland,(2005) Phase-modulating bistable optically addressed spatial light modulators using wide-switching-angle ferroelectric liquid crystal, Opt. Eng. 44 (1) 014003 – 014017.

C.A.T.H. Tee, W.A. Crossland, T.D. Wilkinson, A.B. Davey,(2000) Binary phase modulation using electrically addressed transmissive and silicon backplane spatial light modulators, Opt. Eng. 39 (9) 2527-2534.

S. Mias, N. Collings, T.D. Wilkinson, S. Coomber, M. Stanley, W.A. Crossland,(2003) Spatial sampling in pixelated-metal-mirror ferroelectricliquid- crystal optically addressed spatial-light-modulator, Opt. Eng. 42 (7) 2075 – 2081.

Y. Kuno, Y. Taura, M. Yagura, S. Hamada, N.S. Takashashi, S. Kurita,(1991) Amorphous-silicon photosensor-based ferroelectric-liquidcrystal light valve, Proc. SID 32 (3), 187-190.

W. Li, R.A. Rice, G. Moddel, L.A. Pagano-Stauffer, M.A.Handschy,(1989) Hydrogenated amorphous-silicon photosensor for opticallyaddressed high-speed spatial light modulator, IEEE Trans.Electron Dev. 36 (12) 2959 – 2964.

M.Y. Wang, W. Pan, B. Luo, X.H. Zou, W.L. Zhang, (2007) The t – V min addressing mode and an improved equivalent circuit model for ferroelectric liquid crystal displays, IEE Proc.- Opto- electron , vol. 1, no1, pp. 16-22.

M.Y. Wang, W. Pan, B. Luo, (2004) Study of the respondent characteristicsin ferroelectric liquid crystal spatial light modulator, J. Funct. Mater.Dev. 10 (4) 489 – 492.

Meng-yao Wang, Wei Pan, Bin Luo, Wei-li Zhang and Xi-hua Zou,(2007)"Optimization of gray-scale performance in pixellated-metal-mirror FLC-OASLM by equivalent circuit model "Microelectronics Journal (38) 203-209.

Jian-Chiun Liou & Fan-Gang Tseng (2008) "An Intelligent Multiplexing Control Thermal Actuated Optical Packet Switch", Photonics in Switching (PS), Optical Switches and Routing Devices, D-06-2, pp.1-2.

Part 5

Signals and Fields in Optoelectronic Devices

20

Low Frequency Noise as a Tool for OCDs Reliability Screening

Qiuzhan Zhou, Jian Gao and Dan'e Wu
Jilin University
China

1. Introduction

With the rapid development of the information and science, more and more newly semiconductor devices are used in the electronic equipments or systems, and so is the Optoelectronic Coupled Devices (OCDs). Because of excellent characteristics of it such as small size, long life, non-contact mode and strong anti-interference, OCDs can replace many kinds of devices e.g. relays, transformers, choppers when used in switching circuits, A/D conversion, remote transmission, over-current protection and so on. The reliability of OCDs is very important in numerous applications. It has draw great attention in switching circuit, isolation circuit, analog-digital converter, logic circuit, etc.

However in some high reliability fields, such as navigation and communication of the satellite, it is necessary to make sure of the reliability of the OCDs. In the past, the reliability screening of the OCDs contained ageing experiments; physical analysis at high and low temperature as well as static testing which are either expensive, time-consuming or cannot separate the good ones from the bad ones. So some researchers proposed that using low frequency noise as a reliability indicator.

From the ninety of the last century, we do the research of using noise as reliability screening of the OCDs and improve it continually. So in this paper, we will introduce how to use low frequency noise as a tool for OCDs reliability screening, and summarize what all we had done as well as the latest research.

2. Analysis of noise types in OCDs

Noise as a diagnostic tool for quality control and reliability estimation of semiconductor devices has been widely accepted and used, and there are many papers published in this area. It is very useful to describe the judging rules, which enable us to predict the individual quality of electronic components, based on measurements of their noise.

It is known that an OCD is made of two parts: LED and Photo detector, both of which are p-n junction devices. So it can be concluded that the noise in OCDs below 1 MHz mainly consists of shot noise, 1/f noise, generation-recombination noise and burst noise. Among them, shot noise and 1/f noise are fundamental. It should be noted that the noise that we are interested in here has strong relation to some typical defects in a device. For this reason, it is necessary that the generation mechanisms of 1/f noise, g-r noise and burst noise in OCDs are all briefly discussed, especially on what kinds of defects can lead to these three kinds of noises. And the relation between them should be discussed as well.

2.1 1/f noise

1/f noise spectrum is inversely proportional to frequency in a very wide range. In homogeneous semiconductors, its spectrum can be characterized by a parameter α according to

$$\frac{S_R}{R^2} = \frac{a}{f_N} \tag{1}$$

where S_R is the power spectral density of the noise in the resistance R, N is the total number of free charge carriers, and f is the measurement spot frequency. The parameter α then is the contribution of one electron to the relative noise at 1 Hz, assuming that the N electrons are uncorrelated noise sources.

In addition, it has been found that α is not a constant, whose value is between 10^{-6} and 10^{-3}, but that α depends on the prevailing type of scattering of the electrons and perfection of the crystal lattice. In recent years much progress has been made and found that it is mainly caused by lattice scattering.

Vandamme has shown that the 1/f noise parameter a increases with the concentration of dislocations and its noise spectrum is proportional to a and inversely proportional to the carrier lifetime. Konczakowska research has indicated that there is a strong relation between bipolar device lifetime and 1/f noise.

Usually 1/f noise in a semiconductor device usually can be divided into fundamental 1/f noise and non-fundamental (or excess) 1/f noise. The fundamental 1/f noise is connected with phenomena which are included in the process of the operation of the electronic component. It is believed that this 1/f noise has no relation to the semiconductor surface and the defects in the bulk.

The 1/f noise which is related to device defects is called non-fundamental 1/f noise, which means that this kind of 1/f noise is caused by device surface or bulk defects in most cases. Thus, it is possible for us to evaluate the device quality and reliability according to its magnitude. From this point of view, non-fundamental 1/f noise is of great value to device quality evaluation and reliability prediction. Most of the evidence suggests that in some types of device it is a surface effect, as in the case of a MOSFET where the semiconductor/oxide interface plays an important role, but in other devices, such as a homogeneous resistor, 1/f noise is thought to be a bulk effect associated with a random modulation of the resistance, implying a fluctuation in either the number or the mobility of the charge carriers. For example, M. Mihaila et al have shown that 1/f noise in a specimen with more dislocations is at least one order of magnitude larger than that of the specimen with fewer dislocations.

Different causes for 1/f noise generation have been reported as follows: (1) the fluctuation of surface recombination velocity in the p-n junction, (2) the fluctuation of trapping in the oxide layer in BJTs or in MOSFETs, (3) dislocation 1/f noise, (4) quantum 1/f noise (in dispute). It is obvious that 1/f noise intensity is related to the generation-recombination center (surface defect) numbers in device oxide layer. Thus, the establishment of a relationship between device surface quality and reliability can help us judge and screen devices according to excess noise intensity.

Therefore it is possible for us to evaluate the device quality and reliability according to its magnitude. From this point of view, 1/f noise is of great value to device quality evaluation and reliability prediction. It is verified that crystal defects cause 1/f noise to increase. The experimental results have proved that $1/f$ noise in the specimen with more dislocation is at least one order of magnitude larger than that of the specimen with fewer dislocations.

At present, a major cause of $1/f$ noise in semiconductor devices is traceable to properties of the surface of the material. The generation and recombination of carriers in surface energy states and density of surface states are important factors, but even the interfaces between silicon surfaces and grown oxide passivation are centers of noise generation. It is obvious that $1/f$ noise intensity has a relation to the generation-recombination center (surface defect) numbers in device oxide layer. Thus, the relation between device surface quality and $1/f$ noise is closely related and can be used to screen poor quality devices according to their intensity of excess noise.

2.2 g-r noise

Generation-recombination (g-r) noise distribution is Gaussian and its signal spectrum can be expressed as Lorentzian,

$$\frac{S_R}{R^2} = \frac{S_V}{V^2} = \frac{S_I}{I^2} = \frac{A\tau_0}{1+\left(\omega\tau_0\right)^2} \tag{2}$$

Here, $\tau_0 = 1/\omega_0$ is the characteristic time corresponding to a characteristic (or corner) frequency f_0 or ω_0, and $\omega_0 = 2\pi f$ is the angular frequency of measurement. g-r noise has a Gaussian amplitude distribution function because it is actually made up from the superposition of a very large number of independent random telegraph signal processes with the same characteristic time. The coefficient, A, is a measure of the number of such individual processes. It depends on g-r center density, current and device structures.

It has been found experimentally that g-r noise is often absent in high quality silicon devices, but not yet in heterostructures, where lattice defects are often a problem. In poor quality devices, the g-r noise is generated at the contacts or at the surface. In better samples, the dominant conductance noise source is in the bulk. Hence, g-r noise used as a useful diagnostic tool to study trap centers in compound semiconductors, is indispensable.

By experimentation it has been shown that the defects (dislocation, deep-level impurities) in the emitter junction and surface are the main g-r noise sources of transistor, especially as a p-n junction is in a forward biased state. Jevtic and Lazovic have shown that excess g-r noise in reverse biased p-n junctions can be caused by g-r centers near the metallurgical junction. These centers may be the impurity in metal clusters associated with dislocations.

Thus, it can be seen that g-r noise in a device has a direct relation to semiconductor defects (impurities, damage etc.). Therefore, it has become an effective method of analyzing bulk defects and reliability screening by means of measuring g-r noise in devices.

2.3 Burst noise

A random telegraph signal (RTS) known as burst noise, is often observed in p-n junction devices such as diodes, transistors and detectors operating under forward biased conditions. All authors have attributed the phenomenon to defects located in the neighborhood of the emitter base junction.

Hsu et al. first presented a physical model of explaining burst noise. In this model, it is thought that heavy metal impurities deposited in the charge region of p-n junction is the major cause of this noise.

But Blasquez has found that so-called ``pure'' lattice dislocation can also cause burst noise even when heavy metal impurity deposits have been removed. Therefore it seems that metal impurity precipitates are not indispensable to produce burst noise. Dai et al. has proposed a new burst model, which emphasizes the built-in electric field in the p-n junction and the variation of the potential barrier near the defects. This model not only is consistent with the experimental results given by Blasquez, but also can explain various burst noise waveforms. In addition, Jevtic has also presented a new physical model of noise sources, which is based on the assumptions that a conduction channel (p-inversion layer) exists in degraded p-n junctions and that the current flow through the defects is modulated by traps adjacent to the defects. The model explains the appearance of two polarity and multi-level pulse noise. Although burst noise spectrum is not Gaussian as are the other types of noise, its current noise spectrum has the shape of Lorentzian,

$$S_I(f) = \frac{A_b \tau_b^2}{1 + \omega^2 \tau_b^2} \tag{3}$$

Where A_b is a constant depending on the nature of the defects and τ_b is defined as $1/\tau_b$ $=1/\tau_1+1/\tau_2$. According to the random switch mode, during the time interval dt, an open switch has the probability dt/τ_1 of closing, and a closed switch a probability dt/τ_2 of opening. The information on the defects is contained in the parameter A_b and τ_b.

Besides, burst (or RTS) noise is a problem typical for submicron MOST's or bipolar devices with crystallographic damage in sensitive areas and this noise is also temperature and bias dependent. Many experiments have already shown that lattice dislocation, a serious defect, is the major source of burst noise for both bipolar transistor and integrated circuit. Therefore, devices with burst noise often degrade faster and at least show a poor noise behavior. Fig. 1 shows a noise waveform of time-domain in an OCD with excess noise.

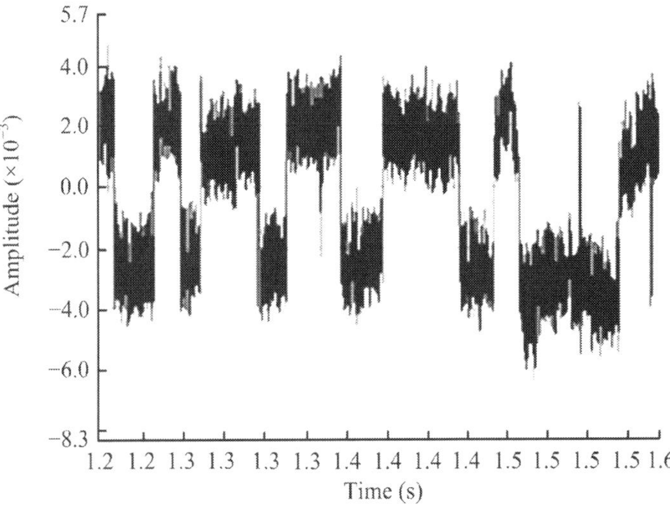

Fig. 1. Noise waveform of time-domain in an OCD with excess noise.

2.4 Brief summary

It can be seen that $1/f$ noise, g-r noise and burst noise are closely related with some defects such as surface defects, impurities and dislocations, etc. Dislocations and electromigration in metallization affect device reliability, and has been identified as the main source of device failure. Thus, it can be said that $1/f$ noise is closely related to the surface states of the semiconductor device, g-r noise related to device bulk defects such as impurities, dislocation, etc., and burst noise related to lattice dislocation as well as heavy metal impurity deposits. Besides, emitter region edge dislocation makes both $1/f$ noise and burst noise increase at the same time in most cases. Strasilla and Struut demonstrated that experimentally observed burst noise consist of a random telegraph signal superimposed on $1/f$ noise, but the two processes were statistically independent.

Hence in order to exclude these defects and meet high reliability, we can use the three independent noise, $1/f$, g-r and burst noise, as reliability indicator for quality estimation of OCDs.

3. The noise measurement and analysis of OCDs

Harder C et al have presented that the noise equivalent circuit of a semiconductor laser diode from the rate equations including Langevin noise sources. This equivalent circuit allows a straightforward calculation of the noise performance of a laser diode combined with extrinsic elements, such as the driving source and the parasitic elements. Recently, using this rate equation, these intrinsic intensity fluctuations in semiconductor laser diode (LD), optoelectronic integrated device (OEID) made by heterojunction phototransistor (HPT) and laser diodes have been analyzed, then the relative intensity noise (RIN) and the correlation between the terminal electrical noise and output optical photocurrent noise have been investigated.

At present, the key to design of low-frequency low noise devices and circuits lies in reducing level of white noise and corner frequency of $1/f$ noise, which has been realized gradually and Whether voltage noise or current noise takes these two parameters as its characteristics. But the present noise measuring apparatus, such as QuanTech2173c/2181 and HP-4470 and so on, only can give out noise of several frequency points or of several fixed frequency, no more give out two pat meters. Thus it can be seen that if one want to understand all-sidely the low-frequency noise performance of a semiconductor device, to make researches on semiconductor noise mechanism and to apply low-frequency noise to analyzing the inherent defect of a device and its reliability and so forth, one must make study of the measurement of low-frequency noise spectrum and the computation of its parameters. This important work is absolutely necessary to understand noise performance of a device or a circuit and specially to develop low-noise devices.

In this part, the noise equivalent circuit of OCDs have been analyzed; then noise spectrum measurement systems of OCDs based on FFT analyzer (CF-920, made in Japan) and virtual instrument are presented, their measuring range is 0. 25Hz- 100kHz and accuracy is higher than 4%. Moreover, the white noise level and corner frequency are computed by applying weighed least square method.

3.1 The noise equivalent circuit of OCDs

Fig. 2 shows the schematic diagram of the OCDs circuit. The OCDs measured and analyzed in this paper are GD315A, made in China. It is well known that if the input current I_0 of a laser diode is less than the threshold current I_{ph}, the noise equivalent circuit of a semiconductor laser diode can also be used to explain the noise performance of an LED. Fig. 3 shows the noise equivalent circuit of OCDs. It is composed of LED and optotransistor shown in the left portion and the right one, respectively. Where R_s is the source resistance, i_{Rs} is the thermal noise of R_s. The quantities in and v_n are the intrinsic noise sources of LED, the sources in and v_n are partially correlated due to the coupled rates. Their noise spectral densities are shown as follows:

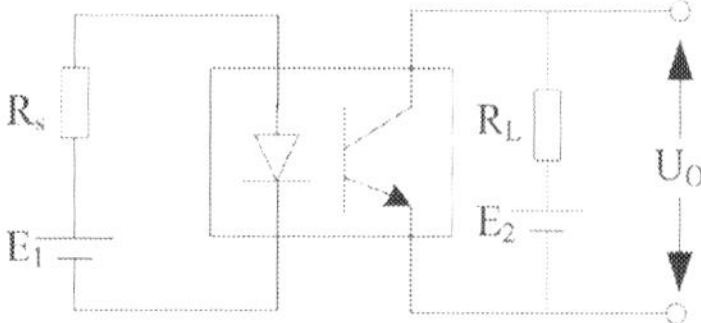

Fig. 2. The schematic diagram of OCDs.

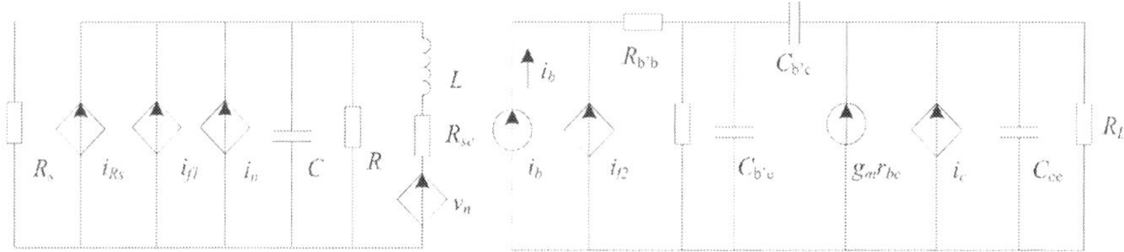

Fig. 3. The noise equivalent circuit of OCDs.

$$S_{i_n}(f) = 2qi_0 + 4q^2 E_{vc} S_0 \tag{4}$$

$$S_{v_n}(f) = \frac{4(mV_T)^2}{\left(n_0\left(AS_0 + (\beta/\tau_s)\right)\right)^2} \frac{1}{\tau_{ph}} + E_{vc} S_0 \tag{5}$$

$$S_{v_n i_n}(f) = 2\frac{mV_T q}{n_0\left(AS_0 + \beta/\tau_s\right)} \frac{1}{\tau_{ph} + 2E_{vc}} S_0 \tag{6}$$

The current i_{fl} denotes the low frequency noise source in LED. The definition of all symbols in Eqs. (4) - (6) and circuit parameters R, C, L, R_{se} in the LED equivalent circuit in Fig. 3 can be found in the references and another publication. In the noise equivalent circuit of the phototransistor, i_b is the base noise current, it is caused by the noise current i_L in LED, hence i_b can be written as γi_L, where γ can be calculated by the current transfer ratio (CTR) of OCDs. Since CTR is defined as I_c/I_0, in the low frequency region all capacitors in the OCDs are omitted; from Fig. 3 we have

$$CTR = \frac{I_c}{I_0} = \frac{\gamma g_m R_L R_{se}\left(r_{b'b} + r_{be}\right)}{\left(R + R_{se}\right)\left(R_L + R_{ce}\right)} \tag{7}$$

Then γ can be calculated by Eq. (7). In the high frequency region, CTR will decrease due to the influence of circuit capacitance. The CTR of OCDs can be obtained by measurement of OCDs (GD315A) as is shown in Fig. 4, where curve 1 is measured in the condition R_L=1kΩ, and curve 2 is R_L=500Ω.

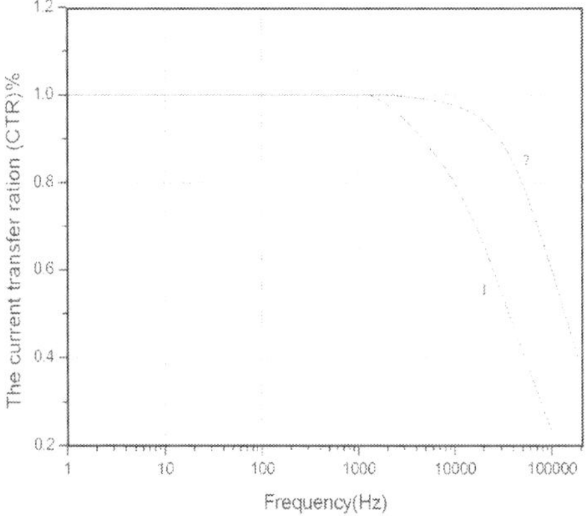

Fig. 4. The CTR of OCDs versus frequency.

In the equivalent circuit of the phototransistor, i_{j2} denotes the low frequency noise source in it. $r_{b'b}$, r_{be}, g_m, $C_{b'e}$, C_{bc}, C_{ce} are the circuit parameters of phototransistor shown in Fig. 3.

3.2 Noise spectrum measurement systems
3.2.1 Noise measurement system based on FFT analyzer

Fig. 5 is the measurement system block scheme of low-frequency noise spectrum testing system we have developed. In our experiments, a dual-channel low frequency amplifier (LNA) chain and the CF-920 cross-spectrum estimator have been used to reduce the background noise of measurement system; hence the noise in the amplifiers will not contribute to the cross-spectrum measured by this system. Therefore, the output voltage noise spectrum of OCDs can be written as

$$S_0(\omega) = \frac{S(\omega)}{K_1 K_2} \tag{8}$$

Where S(ω) is the cross-spectrum measured by CF-920, K_1 and K_2 are the gains of the dual-channel amplifiers LNA I and LNA II, respectively.

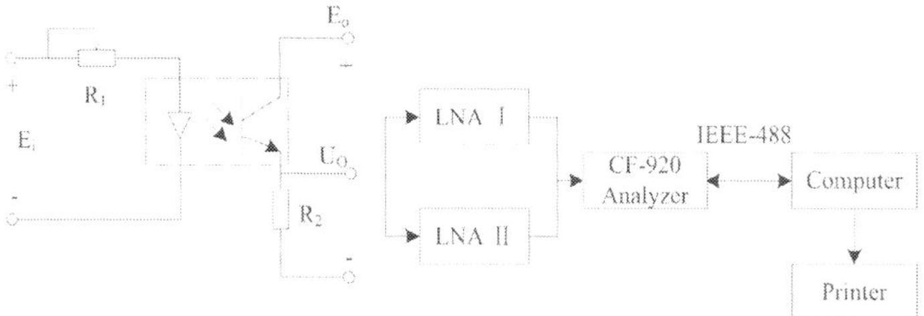

Fig. 5. The noise measurement system of OCDs using CF-920.

In the testing system, the cross-spectrum density estimation method was used to reduce the noise contribution of two preamplifiers. The reason was that two sets of batteries were used as the power supplies for the preamplifiers so that the noises in the two preamplifiers themselves were uncorrelated. So, the measuring system can be used to measure a much smaller signal than usual.

Thus it can be seen that amplifier's self-measurement error has been eliminated basically according to the measuring method in this system. And the measurement accuracy of noise spectrum is mainly decided by CF-920. The measuring range of this system is 0. 25 Hz-100 KHz frequency wide and accuracy is higher than 4%. In the whole measurement process, the measurement and the output of measuring results are controlled automatically by computer.

3.2.2 Noise measurement system based on virtual instrument

Considering the large volume of CF-920 in the above system, which is inconvenient to carry, we design a new measurement system with virtual instrument made by the company of National Instrument and the system block diagram is shown in Fig. 6. It is well known that the virtual instrument platform is widely used in the fields of measurement, auto-control, signal processing and so on, and what the most important is the high precise and small volume.

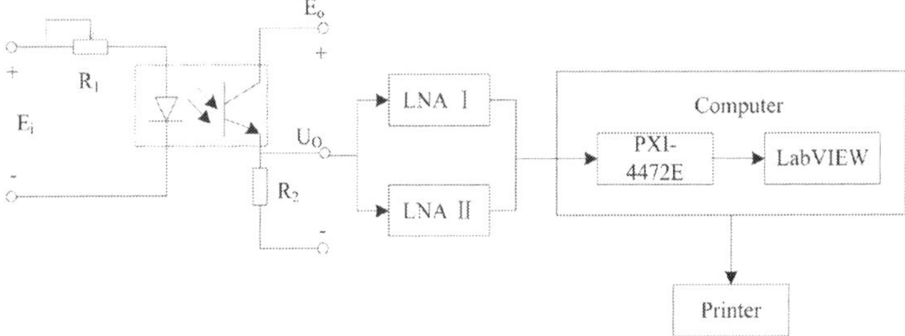

Fig. 6. The noise measurement system of OCDs built on virtual instrument platform.

Where, PXI-4472E is a high-precision 24-bit data acquisition card which acquires the signal from the preamplifiers. The noise signals could be processed for cross-spectral transform, components of noise spectrum estimation and the noise spectrum analysis algorithm in a software developed in LabVIEW. The equivalent input noise power spectrum as Eq. (9) can be tested through low-frequency noise spectrum measurement system.

$$S(f) = A + \frac{B}{f^\alpha} + \sum_{i=1}^{N} \frac{\frac{C_i}{f_{oi}}}{1 + (\frac{f}{f_{oi}})^2} \tag{9}$$

Where A, B and C_i are amplitudes of white noise, 1/f noise and G-R noise respectively; α is a constant 1; f_{oi} is the corner frequency of excess G-R noise; N is the number of G-R noise sources in devices. As long as parameters A, B and C_i are estimated, the magnitude of each type of noise in the device can be determined. A genetic algorithm which would be discussed below was used to fit the parameters, so the coefficient of white noise, 1/f noise and G-R noise were obtained quickly and accurately across the entire spectrum.

3.2.3 Noise analysis of OCDs

According to the noise equivalent circuit given in Fig.3, the noise curves are analyzed in various frequency ranges.

(1) *Low frequency range (1 Hz < f <1 kHz)*: In this frequency range, the $1/f$ noise and generation-recombination g-r noise are dominant. We measured the noise spectrum of 205 OCDs (GD315A made in China), the noise spectrum for four typical devices are shown in Fig. 7. These devices exhibit various low frequency noises. Using the curve fitting method, the analysis results of noise spectrum for four typical devices shown in Fig. 7 are given as follows:

$$\text{NO.4} \quad S_0(f) = 5 \times 10^{-13} + \frac{0.85 \times 10^{-10}}{f} + \frac{0.014 \times 10^{-10}}{1 + (f/15)^2} V^2/Hz$$

$$\text{NO.45} \quad S_0(f) = 2 \times 10^{-13} + \frac{4 \times 10^{-10}}{f} + \frac{0.07 \times 10^{-10}}{1 + (f/15)^2} V^2/Hz$$

$$\text{NO.102} \quad S_0(f) = 1.05 \times 10^{-13} + \frac{1 \times 10^{-10}}{f} + \frac{2.13 \times 10^{-10}}{1 + (f/15)^2} V^2/Hz$$

$$\text{NO.15} \quad S_0(f) = 8 \times 10^{-13} + \frac{6.28 \times 10^{-10}}{f} + \frac{0.64 \times 10^{-7}}{1 + (f/2500)^2} V^2/Hz$$

The noise analysis results show that OCD No. 4 exhibits a normal low frequency noise, No. 45 has excess $1/f$ noise, No. 102 has excess g-r noise (the corner frequency is 15 Hz), and No. 15 has excess $1/f$ noise and g-r noise with burst noise waveform. It means that OCDs No. 4 is a reliable device, No. 45, 102 and 15 coincide with various defects in OCDs, and i.e. their quality is poor.

(2) *Medium frequency range (1 kHz < f <10 kHz)*: In this frequency range, the low frequency noise current i_{f1} and i_{f2} in the OCDs can be omitted; hence, the OCDs output noise is caused by shot noise $i_c(t)$, thermal noise $i_{Rs}(t)$ and the spontaneous emission noise i_n and v_n. Omitting all capacitors in the OCDs, the output noise spectrum can be obtained:

$$S_0(\omega) = R_L^2 (CTR)^2 \left(S_{i_{Rs}}(\omega) + S_{i_n}(\omega) + \frac{S_{v_n}(\omega)}{R_{se}^2} + \frac{2S_{i_n v_n}(\omega)}{R_{se}} \right) + R_L^2 S_{ic}(\omega) \tag{10}$$

Where $S_{iRs} = 4kT/R_s$, $S_{ic}(\omega) = 2qI_c$. According to the Reference written by Harder C et al, the noise current source i_n in parallel with the junction represents mainly the fluctuation in electron population. The first term on the right-hand side of Eq. (4) is the usual shot noise term $2qI_0$ of OCDs, the second term results from the fact that the noise is determined by the sum of emission and absorption. The noise voltage source v_n results from the fluctuation of the photon population.

Since the input current I_0 of LED is less than the threshold current I_{ph}, if we omit the second term of in and v_n, Eq. (11) can be written as follows:

$$S_0(\omega) = 2qI_0 CTR^2 R_L^2 + 2qI_0 R_L^2 \tag{11}$$

Let I_0=10mA, CTR=1, R_L=390Ω, R_s=360Ω, I_c=10mA, from Eq. (11) we obtain the output noise voltage spectrum $S_0(\omega)$=9.7×10^{-16}V^2/Hz. The effective value of $S_0(\omega)$ equals $31.2\,nV/\sqrt{Hz}$ which is smaller than the measurement result (287-$488\,nV/\sqrt{Hz}$). It means that the second term of i_n and v_n caused by the spontaneous emission in LED cannot be omitted.

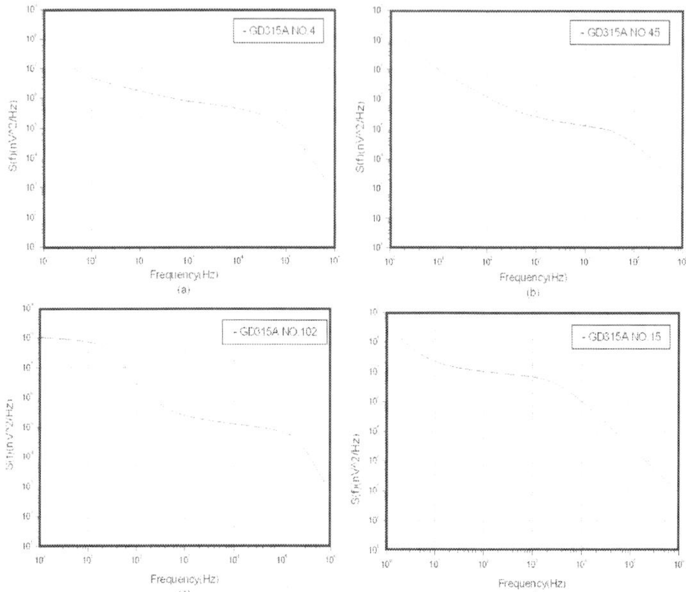

Fig. 7. The noise measurement system of OCDs built on virtual instrument platform.

(3) *High frequency range (10 kHz < f <100 kHz)*: In this frequency range, the white noise is dominant, so that the output noise spectrum of the OCDs is a flat form. However, the noise measurement for 205 OCDs (DG315A) shows that the noise spectrum of all OCDs decreases from 10 to 100 kHz, as shown in Fig. 7. It can be explained as follows: Since $S_0(\omega)$ is directly proportional to CTR2 (Eq. (11)) and CTR decreases from 10 to 100 kHz (Fig. 4), as a consequence, the decrease in noise spectrums of OCDs is caused by the decrease of CTR. Because all capacitors in the OCDs will influence the output noise in this frequency range, the decrease of CTR seems to obey the circuit response.

4. Discussion of the optical screening criterion for OCDs

Generally, it has been already accepted that 1/f noise is closely related to the surface states of the semiconductor device, g–r noise is related to device bulk defects such as impurities, dislocation, and burst noise is related to lattice dislocation as well as heavy metal impurity deposits. From the generation mechanisms of 1/f, g–r and burst noise, it can be seen that the probability to generate these three types of noise by the same defect is quite small although some defects may cause more than one of them simultaneously in some cases.

Hence, in order to exclude these defects and meet high reliability, we can use the three independent noises, 1/f, g–r and burst noise, as reliability indicators for quality estimation of OCDs. In this way even if some defects can cause two or three of them at the same time, such as emitter region edge dislocation which makes both 1/f noise and burst noise increase at the same time in most cases, can also be rejected.

Therefore, because an excess noise is closely associated with some defects in the devices and/or imperfections of technology, noise measurement amplitudes can be used to indicate the defects. In practice, we found that the device with burst noise can be found from its instantaneous waveform in the time domain. The device with g-r noise can be found through noise component analysis or ratio of noise value at 10 Hz to noise value at 1 Hz (which will be explained and proved in later part), which is used to judge whether there is g-r noise or not. And the device with 1/f noise can be judged by the amplitude of voltage noise value at 1 Hz

Therefore, it is necessary that there be three independent screening conditions to meet the requirement of high reliability to reject the devices with excess 1/f, g-r or burst noise. We take the OCDs of GD315A as an example to present the conditions which are presented as follows.

(1) *The value of the noise spectrum at 1 Hz*: Since 1/f noise depends on the surface and bulk defects in OCDs, and it is dominant in the low frequency region, we select the effective value U_0 of noise spectrum at 1 Hz as the reliability indicator. Fig. 8 shows the histogram of the effective value U_0 of 205 OCDs. We can see that the U_0 of most of OCDs obeys the normal distribution; it can be explained by the discretion of OCDs parameters. However, a few of OCDs exhibit excess 1/f noise, it means that these OCDs may contain surface or bulk defects (such as crystal defects and contamination, emitter edge dislocations, electromigration, imperfection of chip bonding, etc.); hence, they should be recognized as poor quality devices.

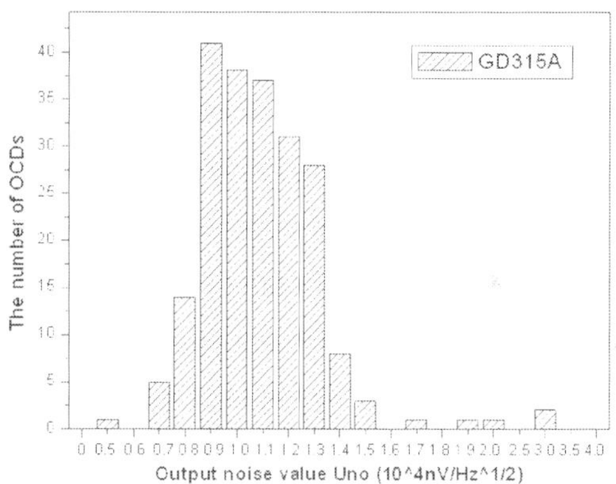

Fig. 8. The histogram of the output noise value $S_0(f)$ of 205 OCDs.

Konczakowska suggested a classification algorithm which has been verified by reliability experiments. On the basis of noise measurement results, we can classify quality groups according to this rule. At first, the mean value and variance of U_0 for 205 OCDs have been calculated, and then the border values for quality groups (i.e., the threshold value) are

$$U_{01} = \bar{U}_0 - \alpha\sigma$$

$$U_{02} = \bar{U}_0 + \alpha\sigma$$

where $\bar{U}_0$ denotes the mean value of U_0, r denotes the variance of U0 and a equals 0.67. In our experiment, the statistical results are $\bar{U}_0 = 11467.5nV/\sqrt{Hz}$, $\sigma = 3194nV/\sqrt{Hz}$, then we obtain $\bar{U}_{01} = 9327.4nV / \sqrt{Hz}$, $\bar{U}_{02} = 13607.7nV / \sqrt{Hz}$. The value of classifying OCDs into three groups according to the 1/f noise is as follows:

- first group –$U_0 \leq 9327.4nV / \sqrt{Hz}$, high quality is expected;
- second group –$9327.4nV/\sqrt{Hz} \leq U_0 \leq 13607.7nV / \sqrt{Hz}$, good quality is expected;
- third group –$U_0 \geq 13607.7nV / \sqrt{Hz}$, poor quality is expected.

(2) *The ratio of noise spectrum at 10 and 1 Hz*: The g-r noise is caused by the defects in the crystal structure of devices and the deep-level impurity in the p-n junction; hence the g-r noise is used as one of the noise reliability indicators. The g-r noise in OCDs can be separated accurately from the output noise spectrum using the curve fitting method; however, it must take a long time and is not suitable for practical application. For the convenience of direct industrial application, we suggest the g-r noise indicator as follows:

$$r = \frac{U_0(f_0)}{U_0(1Hz)} \tag{12}$$

where U_0(1 Hz) is the effective value of noise spectrum at 1 Hz, $U_0(f_0)$ is one at the corner frequency f_0 of g-r noise. From the noise measurement of OCDs, we found that the corner frequency of most OCDs is about 10 Hz, hence we can select f_0=10 Hz.

If the output noise spectrum consists of only white noise and 1/f noise, in the low frequency region, we have

$$S_0(f) = A + \frac{B}{f} \tag{13}$$

If the 1/f noise is dominant, then $S_0(f)$=B/f, the value of r equals $U_0(10Hz) / U_0(1Hz) = 1/\sqrt{10} = 0.33$. If the white noise is dominant, r=0, so that the value of r is in the range of 0-0.33. However, if an OCD exhibits g-r noise, the noise spectrum is shown in a platform below the corner frequency f_0. As a consequence, the value of r is greater than 0.33. In Fig. 9, the histogram of the value r for 205 OCDs is shown. We can see that the value of r for most of the OCDs obeys the normal distribution. The statistical results from 205 GD315A are $\bar{r} = 0.36$ and $\sigma_r = 0.16$. If we select the value of β equal 2, then the threshold value of r is

$$r = \bar{r} + \beta\sigma_r = 0.68$$

The rule of classifying OCDs into two groups according to the g-r noise is as follows:
- $r < 0.68$, first group, high quality is expected;
- r > 0.68, second group, poor quality is expected.

(3) *Burst noise in OCDs*: A number of experiments have already shown that heavy metal impurity deposits and lattice dislocations are the major sources of burst noise for a transistor or IC, so that OCDs with burst noise should always be rejected in any case, because it cannot only effect device reliability, but also hinder the normal operation, especially in digital circuit.

The rule of classifying OCDs into two groups according to the burst noise is as follows:
- no burst noise: first group, high quality is expected;
- with burst noise: second group, poor quality is expected.

As a result, the reliability screening conditions of OCDs are

1. $U_0(1Hz) \geq 13607.7nV / \sqrt{Hz}$

2. $r \geq 0.68$

3. with burst noise

The device will be rejected if it meets any condition of the three. The result of the noise measurement for 205 OCDs is that 23 OCDs are rejected by our reliability screening conditions. Among these OCDs, 19 OCDs are rejected by reliability screening condition 1 (here two OCDs exhibit burst noise). Four OCDs are rejected by condition 2 (here one OCD exhibits burst noise).

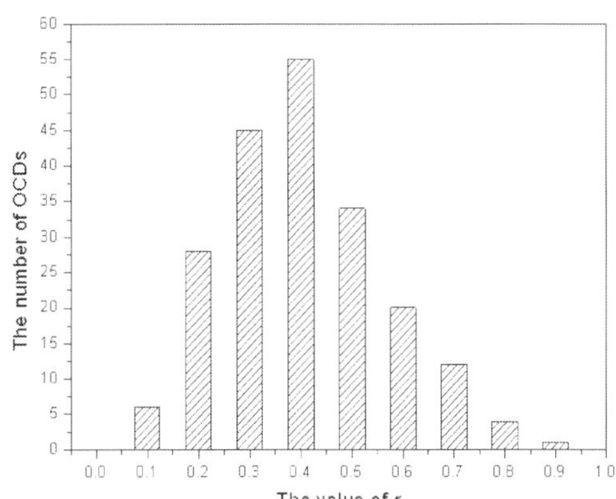

Fig. 9. The histogram of value r of 205 OCDs.

5. Conclusion

In this paper, we analyzed the noise performance of OCDs using the noise equivalent circuit of LED and bipolar transistor.

(1) In the low frequency region (f <1 Hz), the excess 1/f noise and g-r noise can be used as noise reliability indicators; then three reliability indicators have been suggested for quality and reliability screening of OCDs, which have the advantages of general and convenience for direct industrial application.

(2) In the medium frequency region (1 kHz < f <10 kHz), the output noise measurement results show that the fluctuation of the photon population (emission and absorption) in OCDs must be considered. Hence, the value of output noise is larger than usual shot noise in transistor.

(3) In the high frequency region (f > 10 kHz), the output noise will decrease due to the decrease of the current transfer ratio (CTR) of OCDs in this frequency region.

Then the 205 OCDs have been measured by the low noise measurement systems which have high reliability and the satisfactory accuracy. The screening thresholds and experimental results are given. Based on the results, it can be obtained as follows:

(1) It can be found that 1/f noise, g-r noise and burst noise must be used as three independent noise criteria for high reliability estimation.

(2) It is necessary that the estimated error and the maximum value of failure ratio r should be considered together in order to obtain optimal noise thresholds.

(3) In this paper the ratio of Vn(10 Hz) to Vn(1 Hz) was chosen as a screening threshold instead of g–r noise component analysis of noise spectrum, for this reliability indicator is more simple and convenient than the calculation of g–r noise component, especially during the practical screening of a large number of devices.

6. Acknowledgment

Dr. Qiuzhan Zhou would pay the highest respect to his advisor, Prof. Dai Yisong and would like to thank for Prof. Dai's patient guide since 1998. The authors would also thank Prof. Zhang Xinfa for his help on noise parameters measurement. This work is part of Project 60906034 supported by National Natural Science Foundation of China and Project 201115029 supported by Provincial Natural Science Foundation of Jilin Province.

7. References

A. Konczakowska(1995), Quality and 1/f noise of electronic components, *Quality and Reliability Engineering International*, vol. 11, pp. 165-169, ISSN 1099-1638

Bernard O & Philippe S & Jean-Marie P et al(1994), Correlation between electrical laser diodes, *IEEE Transactions on Electron Devices*, vol. 41, pp. 2151-2161, ISSN 0018-9383

Dai.Y& Xu. J(2000). The noise analysis and noise reliability indicators of optoelectron coupled devices, *Solid-State Electronics*, vol. 44, pp. 1495-1500, ISSN 0038-1101

Dai.Y(1991), Optimal low frequency noise criteria used as a reliability test for BJTs and their experimental results, *Microelectronics Reliability*, vol. 31, pp. 75-78, ISSN 0026-2714

Dai.Y(1997), A precision noise measurement and analysis method used to estimate reliability of semiconductor devices, *Microelectronics Reliability*, vol. 37, No. 6, pp. 893-899, ISSN 0026-2714/97

Dai.Y(1991), The performance analysis of cross-spectral density estimator and its applications, *International Journal of Electronics*, vol. 71, pp. 45-53, ISSN 0020-7217

Doru U, Jones BK(1996), Low frequency noise used as a lifetime test of LEDs, *Semiconductor Science and Technology*, vol. 11, pp. 1133-1136, ISSN 0268-1242

Harder C & Katz J & Margalit S et al(1982), Noise equivalent circuit of a semiconductor laser diode, *IEEE J Quant Electron*, vol. 18, pp. 333-337, ISSN 0018-9197

Lei. N,& Du.J & Zhou. Q Z (2009), Fitting Noise Power Spectrum Parameters by Squared Distance Minimization, *Computational Intelligence and Software Engineering, 2009*. pp. 1-4. ISBN: 978-1-4244-4507-3

Levinzon, F. A.(2005), Measurement of low-frequency noise of modern low-noise junction field effect transistors, *Instrumentation and Measurement, IEEE Transactions on*,Vol.54,pp.2427-2432,ISSN 0018-9456

M.M. Jevtic(1995), Noise as a diagnostic and prediction tool in reliability physics, *Microelectronics Reliability*, vol.35, pp. 455-477, ISSN 0026-2714/97

Vandamme L. K. J (1994), Noise as a diagnostic tool for quality and reliability of electronic devices, *Electron Devices, IEEE Transactions on*, vol. 41, pp. 2176-2187, ISSN 0018-9383

Wu,D E &Zhou Q Z &Liu P P(2009) Study of OCDs Reliability Estimation System Based on Bio-Immunology, *Journal of Bionic Engineering*, vol. 6, pp. 306-310, ISSN 1672-6529

Xu.J & Abbott.D & Dai. Y (2000), 1/f, g-r and burst noise used as a screening threshold for reliability estimation of optoelectronic coupled devices, *Microelectronics Reliability*, vol. 40, pp. 171-178, ISSN 0026-2714

21

Electromechanical Fields in Quantum Heterostructures and Superlattices

Lars Duggen and Morten Willatzen
Mads Clausen Institute, University of Southern Denmark
Denmark

1. Introduction

The study of quantum semiconductor structures has received much attention in recent years. Especially group III nitrides have been of considerable interest due to their potential applications as LEDs and laser diodes in the UV region (Kneissl et al., 2003; Nakamura et al., 1994; Yoshida et al., 2008a;b). But also other materials, as the group II oxides have been investigated for their potential as highly efficient laser diodes (Fan et al., 2006; Fujita et al., 2004). Typically, these materials are in the wurtzite crystal phase. Nevertheless some of these materials also exist in the zincblende crystal phase. Furthermore there are the more classical materials as AlGaAs that are in zincblende phase. For this reason we chose to focus on these two crystal structures throughout this chapter.

As opposed to other semiconductor materials as silicone, the above mentioned materials generate electric fields when deformed and vice versa. This so called piezoelectric effect plays a significant role for both band structure and optical gain (Park & Chuang, 1998), which is why it is important to investigate.

In this chapter we start by covering the properties of the wurtzite and zincblende crystal structures and proceed by presenting definitions of stress and strain followed by a derivation of the constitutive relations for the piezoelectric effect from thermodynamic considerations. After establishing the governing equations and justifying for the electrostatic approximation we apply these to a single quantum well structure. Then we present the according calculations for a cylindrical quantum dot along with results from current research papers. We give a brief discussion about the validity of these models and possibilities for improving these. To give a better overview, we briefly present the VFF model as an alternative method for determining strains in quantum wells.

Finally we give a simplified mathematical description of the connection between the computed electromechanical fields and the optical and electrical properties of the quantum structures. Also we provide tables containing various material parameters needed to compute the electromechanical fields as described in this chapter.

2. Crystal structures

As mentioned, there are two predominant crystal structures in semiconductor technology that have piezoelectric properties: zincblende and wurtzite. Silicon based semiconductors are not piezoelectric and will not be covered in this chapter.

A sketch of the zincblende unit cell is shown in Figure 1. zincblende is a combination of two face-centered cubic (fcc) crystals, where one of them is offsetted by $a/4, a/4, a/4$, a being the lattice constant which in this case is the length the cubic edges. It looks quite similar to the diamond structure, but as opposed to the latter, the zincblende structure consists of ionic bonds, causing a nonhomogeneous charge distribution. Macroscopically however, all dipole moments cancel each other and the nonhomogeneous charge distribution is not visible as long as it is not deformed. Notable semiconductor materials that can be fabricated in zincblende form are arsenides such as GaAs, AlAs, or InAs; nitrides such as GaN, AlN, InN, as well as oxides such as ZnO (Auld, 1990; Davydov, 2002; Fonoberov & Balandin, 2003; J.I.Izpura et al., 1999; Vurgaftman et al., 2001).

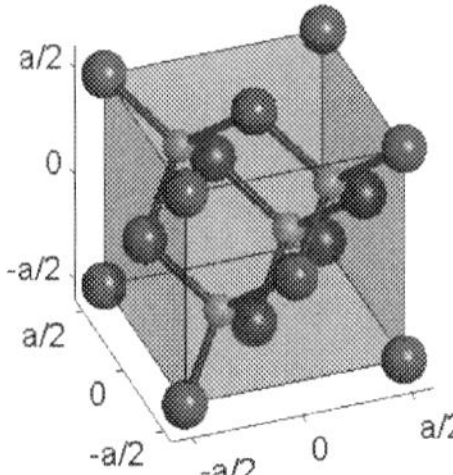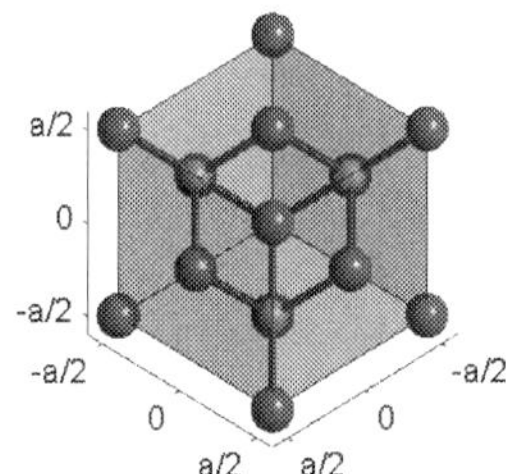

Fig. 1. The zincblende unit cell observed from two different angles. The colors represent the respective atom as e.g. Ga and As for GaAs. Due to ionic bonding they possess different charges. Note that the unit cell to the right has an appearance resembling a hexagonal structure. Indeed we will show that this crystal orientation exhibits piezoelectric properties resembling those of wurtzite (albeit there still are differences)

The other important crystal structure is wurtzite, which is sketched in Figures 2-3. Wurtzite is a hexagonal crystal and, opposed to zincblende, possesses spontaneous polarization, a built-in net polarization in the c-axis direction. Important semiconductor materials grown in wurtzite form include nitrides such as GaN, AlN, and InN as well as oxides such as ZnO and MgO (Fonoberov & Balandin, 2003; Gopal & Spaldin, 2006; Vurgaftman et al., 2001).

2.1 Miller indices

As we will cover the effect of crystal orientation in this chapter, we will briefly introduce the use of Miller indices. They are written in square brackets and consist of three integer digits, i.e. the form is [klm]. The integers refer to directions with respect to the crystal basis. For cubic crystal classes as zincblende the basis vectors are simply the three orthogonal vectors coinciding with the cube edges as can be seen in Figure 4. Note that the indices also describe the orientation of crystal planes by denoting their normal vector. This can be used to construct the indices since one can use intersection points with the basis vectors and invert them. For example a crystal plane with the direction [112] intersects the basis at points $[1,0,0]$, $[0,1,0]$, and $[0,0,1/2]$. This crystal plane then gives the cross section one would see when looking at a crystal in this direction. When denoting a plane directly then one typically uses parenthesis instead of square brackets. Negative directions are denoted by a bar over the respective index. For example $[\bar{k}\bar{k}\bar{k}]$ is exactly the opposite of $[kkk]$. For hexagonal classes such as wurtzite, the basis vectors are shown in Figure 4. It should be mentioned however, that in crystallography

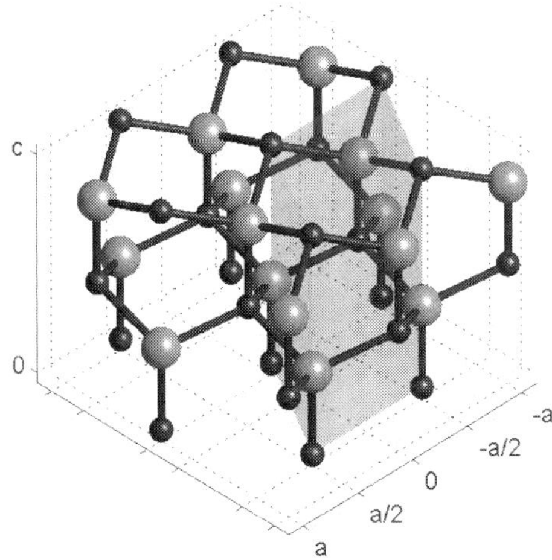

Fig. 2. The wurtzite crystal structure. For clearer impression of the hexagonal symmetry, more than one unit cell is drawn. One unit cell is indicated by the transparent blue box. Also here the atomic bonds are predominantly of ionic nature.

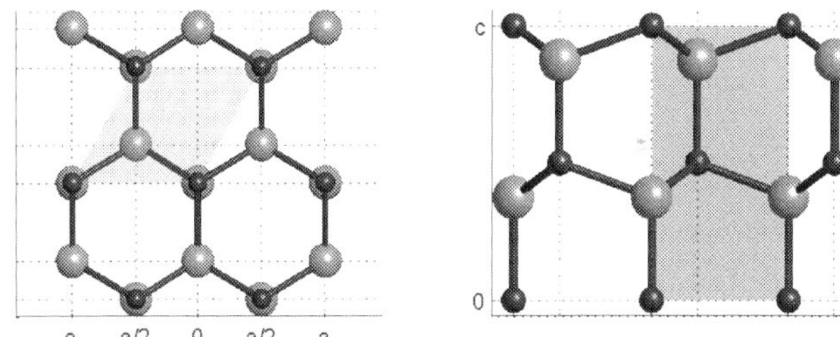

Fig. 3. Top view of the wurtzite structure (left). Here the hexagonal symmetry becomes clear. Side view of the wurtzite structure (right). Note the lack of inversion symmetry which causes both piezoelectric behavior and spontaneous polarization

it is quite common to denote directions in hexagonal crystals by four indices, of which one is redundant. These so called Bravais-Miller indices can be constructed from the standard Miller indices by $[klm] \rightarrow [kl - (h + k)m]$.

3. Fundamental equations

As we are dealing with electromechanical fields, we need fundamental equations for both the mechanical and the electrical part as well as equations for coupling these. In this section we will cover the definitions of linear stress and strain and develop a linear description of the

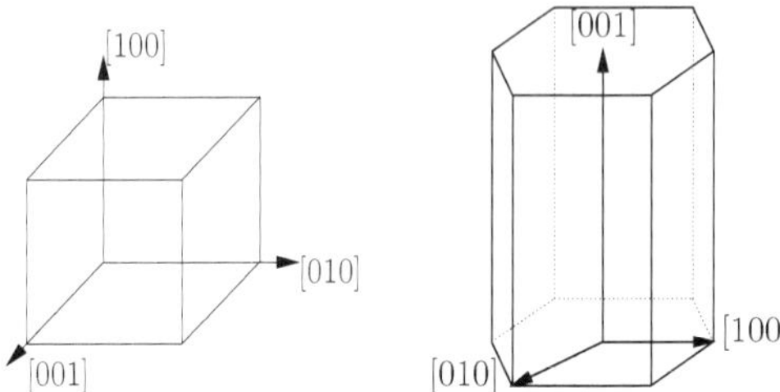

Fig. 4. Basis vectors of the cubic crystal (left) and the hexagonal crystal (right)

piezoelectric effect. Furthermore we will present Maxwell's equations in differential form and derive Navier's law as a linearized form of Newton's Second law. Note that we will stay entirely within the domain of classical physics.

3.1 Stress, strain, and the Piezoelectric effect

The need of a precise definition of deformation of a material should be rather obvious. He we consider material as a continuum consisting of infinitely many, infinitely small elements. Let $\mathbf{r}$ be the position of such an element. Then let $\mathbf{u}$ be the dislocation of the element, i.e. $\mathbf{u} = \mathbf{r} - \mathbf{r}_0$. In general, $\mathbf{u}$ is a function of both space and time, but it does not give a measure of deformation, since it can be non-zero also for undeformed shapes. As an example, taking the gradient of $\mathbf{u}$ will prove insufficient as well, since rotation of an undeformed shape will be non-zero (Landau & Lifshitz, 1986). However, introducing the strain as (Auld, 1990; Landau & Lifshitz, 1986)

$$S_{ij} = \frac{1}{2} \left(\frac{\partial u_i}{\partial r_j} + \frac{\partial u_j}{\partial r_i} + \frac{\partial u_k}{\partial r_i} \frac{\partial u_k}{\partial r_j} \right), \tag{1}$$

allows us to obtain zero if and only if the shape in question is deformed. Assuming that we only work with small $\mathbf{u}$ quantities we can neglect the second order terms to arrive at the definition of linear strain

$$S_{ij} = \frac{1}{2} \left(\frac{\partial u_i}{\partial r_j} + \frac{\partial u_j}{\partial r_i} \right). \tag{2}$$

As we deform a shape of a certain material, obviously we do also create restoring forces. Defining these forces we consider an infinitely small volume element $\delta V = \delta x \delta y \delta z$. Then we can consider traction forces one each of the sides of the element (x, y, and z sides), that is three force vectors:

$$\begin{aligned}
\mathbf{T}_x &= T_{xx}\hat{\mathbf{x}} + T_{xy}\hat{\mathbf{y}} + T_{xz}\hat{\mathbf{z}}, \\
\mathbf{T}_y &= T_{yx}\hat{\mathbf{x}} + T_{yy}\hat{\mathbf{y}} + T_{yz}\hat{\mathbf{z}}, \\
\mathbf{T}_z &= T_{zx}\hat{\mathbf{x}} + T_{zy}\hat{\mathbf{y}} + T_{zz}\hat{\mathbf{z}}.
\end{aligned} \tag{3}$$

The second order tensor consisting of the entries T_{ij} is called the stress tensor and is a measure of pressure.

As long as we stay within the elastic limit, the shape will be restored to its original state once all deformation forces are removed. In other words, the process is reversible and we can treat it thermodynamically. The fundamental thermodynamic relation for deformed bodies reads(Landau & Lifshitz, 1986)

$$dU_{\text{mech}} = \Theta d\sigma + T_{ik}dS_{ik},\qquad(4)$$

where dU_{mech} is the change in internal energy, Θ is the temperature, σ is the entropy, and summation over indices is implicit. However, we have seen that deformations in a piezoelectric material give rise to an electric polarization, stipulating that one needs to take electrical energy into account as well. This contribution reads (Landau et al., 1984)

$$dU_{\text{elec}} = E_i dD_i,\qquad(5)$$

where E_i and D_i denote the ith electric and displacement field component, respectively. Conclusively the total change in internal energy is

$$dU = \Theta d\sigma + T_{ik}dS_{ik} + E_i dD_i.\qquad(6)$$

It will prove convenient however, to look at other thermodynamic quantities as well. Here we chose the electric enthalpy H_{elec}, as this will yield the equation we will use later on. For other interesting quantities and a more detailed derivation of these, see Willatzen (2001). Defining

$$H_{\text{elec}} = U - E_i D_i,\qquad(7)$$

we obtain the differential

$$\begin{aligned}
dH_{\text{elec}} &= dU - d(E_i D_i) \\
&= dU - E_i dD_i - D_i dE_i \\
&= \Theta d\sigma - D_i dE_i + T_{kl}dS_{kl}.
\end{aligned}\qquad(8)$$

Furthermore it is common to assume isentropic operation (Auld, 1990; Willatzen, 2001), even though in principle isothermal operation is possible as well. However, isothermal operation is hard to achieve due to energy dissipation and it has been found experimentally that the isentropic assumption works quaite well (Auld, 1990). In the following we will always assume constant entropy without denoting it specifically.

We will chose to use the thermodynamic state variables **E** and **S**. Then we express the two remaining variables **D** and **T** as

$$D_i = \left(\frac{\partial D_i}{\partial E_j}\right)_S E_j + \left(\frac{\partial D_i}{\partial S_{kl}}\right)_E S_{kl},\qquad(9)$$

$$T_{kl} = \left(\frac{\partial T_{kl}}{\partial E_j}\right)_S E_j + \left(\frac{\partial T_{kl}}{\partial S_{mn}}\right)_E S_{mn},\qquad(10)$$

where the subscript at the partial derivatives denotes the field that is held constant. Furthermore we have used the zero-reference to arrive at these equations.

Notice that we can describe the above derivatives as derivatives of H_{elec} and define a constant, such that (using that order of differentiation is arbitrary)

$$e_{ikl} \equiv \left(\frac{\partial D_i}{\partial S_{kl}}\right)_E = -\left(\frac{\partial^2 H_{\text{elec}}}{\partial S_{kl}\partial E_i}\right) = -\left(\frac{\partial^2 H_{\text{elec}}}{\partial E_i\partial S_{kl}}\right)$$

$$= -\left(\frac{\partial T_{kl}}{\partial E_i}\right)_S = e_{kli}, \tag{11}$$

$$c^E_{klmn} = \left(\frac{\partial T_{kl}}{\partial S_{mn}}\right)_E, \qquad \epsilon^S_{ij} = \left(\frac{\partial D_i}{\partial E_j}\right)_S \tag{12}$$

Note the transposition symmetry $e_{ikl} = e_{kli}$. Similarly it can be shown that $c^E_{klmn} = c^E_{mnkl}$ and $\epsilon^S_{ij} = \epsilon^S_{ji}$. Regarding these constants as material constants we can rewrite the piezoelectric equations as

$$D_i = \epsilon^S_{ij}E_j + e_{ikl}S_{kl} + P^{sp}_i, \tag{13}$$

$$T_{kl} = -e_{kli}E_i + c^E_{klmn}S_{mn}, \tag{14}$$

where the first equation is the direct effect while the latter is called the converse effect. The vector $\mathbf{P}^{sp}$ denotes spontaneous polarization, which is zero for zincblende crystals and has a nonzero z-component for wurtzite. It should be mentioned already here that there are generally two common approaches to these equations when dealing with quantum structures. One is the fully-coupled approach, where the equations above are used as they are, while the other approach is the semi-coupled one where the piezoelectric tensor $\mathbf{e}$ for the converse effect is neglected.

3.1.1 Voigt notation

Due to the symmetry of the tensors involved in the piezoelectric fundamental equations it is possible and useful to contract the index notation. This will transform the nine-element strain and stress tensors to six-element vectors. The subscript contraction for Cartesian coordinates is as follows:

$$xx \to 1, yy \to 2, zz \to 3, \tag{15}$$

$$yz \ \& \ zy \to 4, xz \ \& \ zx \to 5, xy \ \& \ yx \to 6. \tag{16}$$

Note that the piezoelectric stiffness tensor e cannot be fully reduced since it is of third rank. Thus one reduces only the kl-set from equations (13) and (14) such that one can use standard matrix multiplication rules once contracted.

It should also be mentioned that there exist different standards with respect to weights when contraction. One standard is to use the same weight (typically 1) for both stress and strain. Another standard, used by Auld (1990), is to multiply the three shear strains by two while using unit weight for the stress, i.e

$$S_I = 2 \cdot S_{ij} - \delta_{ij}S_{ij}, \qquad\qquad i,j = 1,2,3, I = 1,2,3,4,5,6, \tag{17}$$

$$T_I = T_{ij} \qquad\qquad\qquad i,j = 1,2,3, I = 1,2,3,4,5,6. \tag{18}$$

We will use the latter standard as the resulting contraction operations for the stiffnesses are weightless and thus can be taken directly from several sources containing many of these constants, such as Vurgaftman et al. (2001).

$$c_{IJ} = c_{klmn}, \quad e_{iK} = e_{ikl}. \tag{19}$$

Tensors that already are of second rank are not contracted. Using this contraction, the piezoelectric relations read

$$\mathbf{D} = \varepsilon^S \mathbf{E} + \mathbf{e}\mathbf{S} + \mathbf{P}^{sp}, \tag{20}$$

$$\mathbf{T} = -\mathbf{e}^T \mathbf{E} + \mathbf{c}^E \mathbf{S}, \tag{21}$$

where ε is the 3×3 permittivity matrix, $\mathbf{e}$ is the 3×6 piezoelectric matrix, $\mathbf{c}$ is the 6×6 stiffness, while $\mathbf{D}, \mathbf{E}$ are the 3×1 electric displacement and electric field vectors, respectively, and $\mathbf{T}, \mathbf{S}$ are the 6×1 stress and strain vectors, respectively.

3.2 Material property tensors
For the sake of completeness we give the general form of the material properties of wurtzite and zincblende.

3.2.1 Wurtzite
The wurtzite crystal structure has three independent entries in the piezoelectric tensor, four in the stiffness tensor and two in the purely diagonal permittivity tensor. The matrices read

$$\mathbf{e} = \begin{bmatrix} 0 & 0 & 0 & 0 & e_{x5} & 0 \\ 0 & 0 & 0 & e_{x5} & 0 & 0 \\ e_{z1} & e_{z1} & e_{z3} & 0 & 0 & 0 \end{bmatrix},$$

$$\mathbf{c} = \begin{bmatrix} c_{11} & c_{12} & c_{13} & 0 & 0 & 0 \\ c_{12} & c_{11} & c_{13} & 0 & 0 & 0 \\ c_{13} & c_{13} & c_{33} & 0 & 0 & 0 \\ 0 & 0 & 0 & c_{44} & 0 & 0 \\ 0 & 0 & 0 & 0 & c_{44} & 0 \\ 0 & 0 & 0 & 0 & 0 & \frac{1}{2}(c_{11} - c_{12}) \end{bmatrix},$$

$$\varepsilon = \begin{bmatrix} \epsilon_{xx} & 0 & 0 \\ 0 & \epsilon_{xx} & 0 \\ 0 & 0 & \epsilon_{zz} \end{bmatrix},$$

$$\mathbf{P}^{SP} = \begin{bmatrix} 0 \\ 0 \\ p^{SP} \end{bmatrix} \tag{22}$$

3.2.2 Zincblende
For zincblende there is only one independent entry for the piezoelectric coupling e_{x4} and three independent entries for the stiffness tensor $\mathbf{c}$ while the permittivity in fact can be used as a scalar ϵ_{xx} since all the diagonal entries in the purely diagonal matrix are the same. The

matrices read

$$\mathbf{e} = \begin{bmatrix} 0 & 0 & 0 & e_{x4} & 0 & 0 \\ 0 & 0 & 0 & 0 & e_{x4} & 0 \\ 0 & 0 & 0 & 0 & 0 & e_{x4} \end{bmatrix},$$

$$\mathbf{c} = \begin{bmatrix} c_{11} & c_{12} & c_{12} & 0 & 0 & 0 \\ c_{12} & c_{11} & c_{12} & 0 & 0 & 0 \\ c_{12} & c_{12} & c_{11} & 0 & 0 & 0 \\ 0 & 0 & 0 & c_{44} & 0 & 0 \\ 0 & 0 & 0 & 0 & c_{44} & 0 \\ 0 & 0 & 0 & 0 & 0 & c_{44} \end{bmatrix},$$

$$\varepsilon = \begin{bmatrix} \epsilon_{xx} & 0 & 0 \\ 0 & \epsilon_{xx} & 0 \\ 0 & 0 & \epsilon_{xx}. \end{bmatrix} \tag{23}$$

3.3 Higher order electromechanical effects

This chapter will not cover the handling of nonlinear electromechanical effects. Nevertheless the importance of these effects is being discussed in literature (Voon & Willatzen, 2011) and for this reason we will give a brief introduction to second order piezoelectricity. The other effect, electrostriction, should also be mentioned here. It is a second order effect between strain and the electric field, with the governing equation reading (Newnham et al., 1997)

$$S_{ij} = M_{ijkl}E_k E_l. \tag{24}$$

Note that, opposed to the piezoelectric effect, M_{ijkl} is a tensor of even rank and thus this effect does not need breaking of symmetry as the piezoelectric effect does. However, it is very difficult to estimate the impact of this effect since the electrostrictive tensor is very hard to measure correctly (Voon & Willatzen, 2011).

3.3.1 Second order Piezoelectricity

Instead of assuming a linear relationship between the piezoelectrically generated field and the strain, it is also possible to assume relationships of second order. This yields a piezoelectric tensor that does depend on the strain itself and reads (Bester, Wu, Vanderbilt & Zunger, 2006; Bester, Zunger, Wu & Vanderbilt, 2006)

$$e_{iJ} = e_{iJ}^0 + \sum_{k=1}^{6} B_{iJK}S_K, \tag{25}$$

where e_{iJ}^0 is the piezoelectric tensor at zero strain and B_{iJK} is the fifth rank tensor representing the non-linear piezoelectric effect. Theoretically, the impact of this effect has been shown to be of significance (Bester, Wu, Vanderbilt & Zunger, 2006; Bester, Zunger, Wu & Vanderbilt, 2006), however the agreement with experiments is still disputed (see also discussion in Voon & Willatzen (2011)).

3.4 Navier's and Maxwell's equations

As we deal with coupled electrical and mechanical effects we will need the field equations for both. The electromagnetic field equations consist of the four Maxwell equations

$$\nabla \times \mathbf{E} = -\frac{\partial \mathbf{B}}{\partial t}, \tag{26}$$

$$\nabla \times \mathbf{H} = \frac{\partial \mathbf{D}}{\partial t} + \mathbf{J}_c + \mathbf{J}_s, \tag{27}$$

$$\nabla \cdot \mathbf{B} = 0, \tag{28}$$

$$\nabla \cdot \mathbf{D} = \rho_e, \tag{29}$$

where we have chosen the differential form and $\mathbf{H}$ denotes the magnetic field intensity, $\mathbf{B}$ denotes the magnetic flux density and $\mathbf{J}_c, \mathbf{J}_s$ denote conduction and source currents, respectively, while ρ_e is the free electric charge density. For further details on the origin of these equations the reader is referred to Landau & Lifshitz (1975).

The mechanical field equation of course has its origin in Newton's Second law. Considering a small volume δV with surface δS we have the balance of forces

$$\int_{\delta S} \mathbf{T}\hat{n}\,dS + \int_{\delta V} \mathbf{F}\,dV = \int_{\delta V} \rho_m \frac{\partial^2 \mathbf{u}}{\partial t^2}\,dV, \tag{30}$$

where the first integral is the summation of surface forces, the second is due to the possible presence of a body force $\mathbf{F}\delta V$ and ρ_m is the mass density. In the case where $\delta V \to 0$ the integrands can be regarded as being constant and we obtain Navier's equation

$$\nabla \cdot \mathbf{T} = \rho_m \frac{\partial^2 \mathbf{u}}{\partial t^2} - \mathbf{F}, \tag{31}$$

where $\nabla \cdot \mathbf{T} \equiv \lim_{\delta V \to 0} \frac{\int_{\delta S} \mathbf{T}\hat{n}\,dS}{\delta V}$. In Cartesian coordinates the divergence of stress $\nabla \cdot \mathbf{T}$ reads

$$(\nabla \cdot \mathbf{T})_i = \frac{\partial T_{ij}}{\partial r_j}, \tag{32}$$

where $i, j = x, y, z$. In Voigt notation, the divergence can be expressed by a 3×6 matrix which in Cartesian coordinates reads

$$\nabla \cdot \to \begin{bmatrix} \frac{\partial}{\partial x} & 0 & 0 & 0 & \frac{\partial}{\partial z} & \frac{\partial}{\partial y} \\ 0 & \frac{\partial}{\partial y} & 0 & \frac{\partial}{\partial z} & 0 & \frac{\partial}{\partial x} \\ 0 & 0 & \frac{\partial}{\partial z} & \frac{\partial}{\partial y} & \frac{\partial}{\partial x} & 0 \end{bmatrix} \tag{33}$$

Forms for other coordinate systems can be obtained by considering an elementary volume of this coordinate system and making a linear approximation of the stresses $\mathbf{T}$ on the surfaces on that volume, finally letting the volume approach zero. The divergence for cylindrical and spherical systems are also given in Auld (1990).

With Maxwell's and Navier's equations together with the piezoelectric constitutive relations we have the complete basis for computing electrical and mechanical fields in piezoelectric materials (neglecting second order terms as electrostriction).

3.4.1 Electrostatic approximation

When dealing with problems that are constant in time, the electromagnetic coupling is non-existent and the only equations left to solve are Navier's equation along with the following two of Maxwell's equations

$$\nabla \times \mathbf{E} = 0, \; \nabla \cdot \mathbf{D} = \rho_e. \tag{34}$$

With time varying problems on the other hand, one needs to solve all of Maxwell's and Navier's equations self-consistently. However, in most cases (Auld, 1990) it is possible to use the electrostatic approximation, forcing $\nabla \times \mathbf{E} = 0$ and introducing an insignificant error. Here we will justify this on a simple example and leave further discussions to Auld (1990). Consider an electromechanical plane wave in the z-direction of a bulk piezoelectric material with zero free electric charges of general crystal structure (however, for simplicity considering diagonal terms in ε only as well as assuming non-magnetic properties, as is always the case for either wurtzite or zincblende). In this case we will not be able to assume that also $\mathbf{B}$ and $\mathbf{H}$ depend on z only. After some manipulation using Maxwell's and Navier's equations, one arrives at the dispersion relation

$$\begin{bmatrix} \Gamma_{53}\frac{k^2}{\omega^2} & \Gamma_{54}\frac{k^2}{\omega^2} & \Gamma_{55}\frac{k^2}{\omega^2} - \rho_m & -e_{5x}\frac{k^2}{\omega^2} & -e_{5y}\frac{k^2}{\omega^2} \\ \Gamma_{43}\frac{k^2}{\omega^2} & \Gamma_{44}\frac{k^2}{\omega^2} - \rho_m & \Gamma_{45}\frac{k^2}{\omega^2} & -e_{4x}\frac{k^2}{\omega^2} & -e_{4y}\frac{k^2}{\omega^2} \\ \Gamma_{33}\frac{k^2}{\omega^2} - \rho_m & \Gamma_{34}\frac{k^2}{\omega^2} & \Gamma_{35}\frac{k^2}{\omega^2} & -e_{3x}\frac{k^2}{\omega^2} & -e_{3y}\frac{k^2}{\omega^2} \\ \mu_0 e_{x3} & \mu_0 e_{x4} & \mu_0 e_{x5} & \mu_0 \epsilon_{xx} - \frac{k^2}{\omega^2} & 0 \\ \mu_0 e_{y3} & \mu_0 e_{y4} & \mu_0 e_{y5} & 0 & \mu_0 \epsilon_{yy} - \frac{k^2}{\omega^2} \end{bmatrix} \cdot \begin{bmatrix} S_3 \\ S_4 \\ S_5 \\ E_x \\ E_y \end{bmatrix} = \mathbf{0}, \tag{35}$$

where Γ denotes the piezoelectrically stiffened elastic tensor given by

$$\Gamma = \mathbf{c}^E + \frac{\mathbf{e}^T \mathbf{e}}{\epsilon}^{-1}. \tag{36}$$

Considering standard numerical values of the material properties, i.e. $\Gamma \sim 10^{10}, \rho_m \sim 10^3, e_{iJ} \sim 1, \epsilon \sim 10^{-11}, \mu_0 = 4\pi \cdot 10^{-7}$ it is easily seen that there is only a weak coupling between the submatrices $1 - 3 \times 1 - 3$ and $4 - 5 \times 4x5$. When computing the determinant of this matrix as the sum of all permutation products (negating sign for odd permutations) it becomes clear that the elements with three Γ contributions are dominating in the range of $k^2/\omega^2 = 1/c^2$ with $c \sim 10^3$. The summation terms including three Γ contributions are of order 10^0 while the terms including only two Γ contributions are of order 10^{-6}.

On the other hand, for large ω so we are getting close to electromagnetic wave propagation, we have $c \sim 10^8$. Then the one dominating term becomes $\sim -\rho_m^3 \mu_0 \epsilon_{xx} \mu_0 \epsilon_{yy} \sim 10^{-25}$, with the second largest terms $\sim 10^{-34}$.

Note that in the absence of piezoelectricity we obtain the uncoupled dispersion relations for acoustic and electromagnetic wave propagation, respectively. Thus the piezoelectric coupling terms in the dispersion relation are negligible and one obtains the same dispersion relation as with the electrostatic approximation, where only the $1 - 3 \times 1 - 3$ submatrix will occur in the dispersion and the electromagnetic wave is considered to be purely electromagnetic. Even more, the amplitudes of the strain elements and the electric field elements will also only be perturbed insignificantly.

4. Quantum structures

The key issue for investigating piezoelectric effects in the wurtzite and zincblende crystal structures is their widespread use in optoelectronics and electronics in general. Here we will focus on "clean" quantum structures, i.e. without doping. The major reason for the use of materials such as GaN, AlN and others is their large electronic band gap creating the possibility of large energy transitions as necessary for UV-leds. A basic sketch of a quantum well structure is shown in Figure5

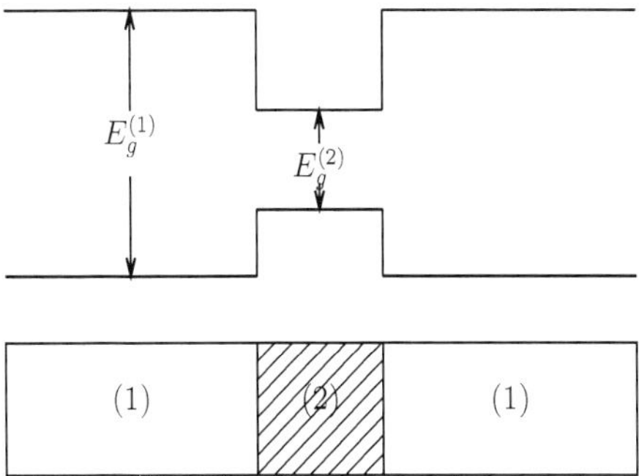

Fig. 5. Basic sketch of a quantum well structure. The indices (1) and (2) denote barrier and well material, respectively. The upper part indicates the conduction and valence band energies for zero electric field.

The three types of quantum structures that differ in the number of confined dimensions are

- Quantum well: one dimension confined
- Quantum wire: two dimensions confined
- Quantum dot: three dimensions confined

One motivation for investigation of these types is that a decrease of dimensionality is reflected in the density of state functions of these structures. The dependency of the density of states (DOS), denoted $N(E)$, on the energy E functions read in a one-band effective model (Singh, 2003)

$$N(E)_{\text{bulk}} = \frac{\sqrt{2}m^{*3/2}\sqrt{E - E_c}}{\pi^2\hbar^3}, \tag{37}$$

$$N(E)_{\text{well}} = \frac{m^*}{\pi\hbar^2}; \ E > E_i (\text{from each subband } i), \tag{38}$$

$$N(E)_{\text{wire}} = \frac{\sqrt{2}m^{*1/2}}{\pi\hbar}(E - E_i)^{-1/2}; \ E > E_i (\text{from each subband } i), \tag{39}$$

$$N(E)_{\text{dot}} = \delta(E - E_i), \tag{40}$$

where E_c is the conduction band energy and m^* is the electron effective mass. Note that the DOS for a quantum dot is discrete, i.e. a quantum dot is treated as a single, isolated particle. A thorough discussion about these three structures can be found in Singh (2003).

The theory presented in this chapter covers electromechanical fields of both well and barrier structures, the latter being used for transistor technology (Koike et al., 2005; Sasa et al., 2006).

5. One-dimensional electromechanical fields in quantum wells

This section contains an example for the application of the above equations on quantum wells. For simplicity we will assume no free charges in the structure as this removes the necessity of solving the Schrödinger equation simultaneously.

The well layer (2) will adapt its lattice constant to the barriers (1) and the strain in the well layer is defined as (Ipatova et al., 1993)

$$\mathbf{S}^{(2)} = \begin{bmatrix} \frac{\partial u_x^{(2)}}{\partial x} - a_{mis} \\[6pt] \frac{\partial u_y^{(2)}}{\partial y} - a_{mis} \\[6pt] \frac{\partial u_z^{(2)}}{\partial z} - c_{mis} \\[6pt] \frac{\partial u_y^{(2)}}{\partial z} + \frac{\partial u_z^{(2)}}{\partial y} \\[6pt] \frac{\partial u_x^{(2)}}{\partial z} + \frac{\partial u_z^{(2)}}{\partial x} \\[6pt] \frac{\partial u_x^{(2)}}{\partial y} + \frac{\partial u_y^{(2)}}{\partial x} \end{bmatrix}, \tag{41}$$

while the strain in layer (1) is defined as usual (see equation (1)). This definition is for wurtzite structures, having two lattice constants a, c. The mismatch a_{mis} is given by $a_{mis} = \left(a^{(2)} - a^{(1)} \right) / a^{(1)}$ and c_{mis} is defined similarly. For use with zincblende, $c_{mis} = a_{mis}$. For the quantum well it is often assumed that all quantities depend exclusively on the z-direction and the x, y-directions are infinite. Note that, since we are working with first order strain, the choice of the denominator for a_{mis} and c_{mis} is arbitrary, as the difference $\left(a^{(2)} - a^{(1)} \right) / a^{(1)} - \left(a^{(2)} - a^{(1)} \right) / a^{(2)}$ is of second order.

5.1 Crystal orientation

As already discussed, the zincblende structure does not exhibit piezoelectric properties upon hydrostatic compression (i.e. no shear). However, as seen in Figure 1 there is reason to believe that a rotation of the crystal structure yields a piezoelectric field upon hydrostatic compression.

The rotation of unit cells is modeled by a rotation of the describing coordinate system transforming coordinates $x, y, z \rightarrow x', y', z'$. The transformation is performed by two subsequent rotations around coordinate axis as shown in Figure 6. The different quantities then transform as

$$\begin{aligned} \mathbf{r}' &= \mathbf{a} \cdot \mathbf{r}, & \mathbf{P}^{SP'} &= \mathbf{a} \cdot \mathbf{P}^{SP}, \\ \mathbf{T}' &= \mathbf{M} \cdot \mathbf{T}, & \mathbf{S}' &= \mathbf{N} \cdot \mathbf{S}, \\ \mathbf{E}' &= \mathbf{a} \cdot \mathbf{E}, & \mathbf{D}' &= \mathbf{a} \cdot \mathbf{D}, \\ \varepsilon' &= \mathbf{a} \cdot \varepsilon \cdot \mathbf{a}^T, & \mathbf{e}' &= \mathbf{a} \cdot \mathbf{e} \cdot \mathbf{M}^T, \\ \mathbf{c}^{E'} &= \mathbf{M} \cdot \mathbf{c}^E \cdot \mathbf{M}^T, \end{aligned}$$

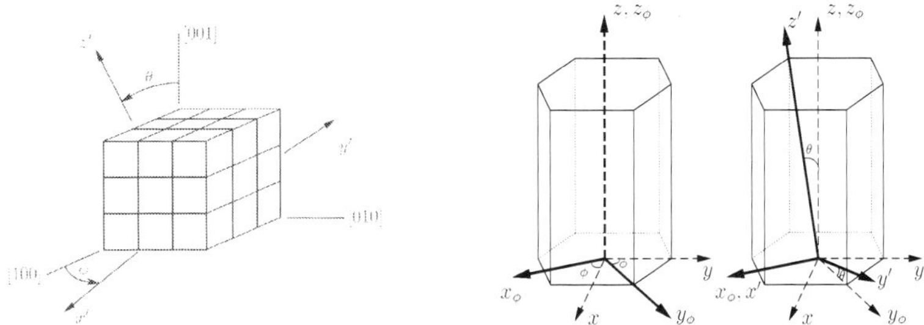

Fig. 6. Subsequent coordinate system rotations - ϕ around z followed by θ around the new x-axis. The cubes to the left indicate the cubic crystal structure while the middle and right figures represent the same operation for hexagonal crystals. Reprinted with permission from Duggen et al. (2008) and Duggen & Willatzen (2010).

where $\mathbf{a}$ is given by (Auld, 1990; Goldstein, 1980)

$$\mathbf{a} = \begin{bmatrix} \cos(\phi) & \sin(\phi) & 0 \\ -\cos(\theta)\sin(\phi) & \cos(\theta)\cos(\phi) & \sin(\theta) \\ \sin(\theta)\sin(\phi) & -\sin(\theta)\cos(\phi) & \cos(\theta) \end{bmatrix}, \tag{42}$$

and the $\mathbf{M}, \mathbf{N}$ matrices are called Bond stress and strain transformation matrices, respectively. They are constructed out of the elements of $\mathbf{a}$ as given in the following (Auld, 1990; Bond, 1943):

$$\mathbf{M} = \begin{bmatrix} a_{11}^2 & a_{12}^2 & a_{13}^2 & 2a_{12}a_{13} & 2a_{13}a_{11} & 2a_{11}a_{12} \\ a_{21}^2 & a_{22}^2 & a_{23}^2 & 2a_{22}a_{23} & 2a_{23}a_{21} & 2a_{21}a_{22} \\ a_{31}^2 & a_{32}^2 & a_{33}^2 & 2a_{32}a_{33} & 2a_{33}a_{31} & 2a_{31}a_{32} \\ a_{21}a_{31} & a_{22}a_{32} & a_{23}a_{33} & a_{22}a_{33}+a_{23}a_{32} & a_{21}a_{33}+a_{23}a_{31} & a_{22}a_{31}+a_{21}a_{32} \\ a_{31}a_{11} & a_{32}a_{12} & a_{33}a_{13} & a_{12}a_{33}+a_{13}a_{32} & a_{13}a_{31}+a_{11}a_{33} & a_{11}a_{32}+a_{12}a_{31} \\ a_{11}a_{21} & a_{12}a_{22} & a_{13}a_{23} & a_{12}a_{23}+a_{13}a_{22} & a_{13}a_{21}+a_{11}a_{23} & a_{11}a_{22}+a_{12}a_{21} \end{bmatrix}, \tag{43}$$

$$\mathbf{N} = \begin{bmatrix} a_{11}^2 & a_{12}^2 & a_{13}^2 & a_{12}a_{13} & a_{13}a_{11} & a_{11}a_{12} \\ a_{21}^2 & a_{22}^2 & a_{23}^2 & a_{22}a_{23} & a_{23}a_{21} & a_{21}a_{22} \\ a_{31}^2 & a_{32}^2 & a_{33}^2 & a_{32}a_{33} & a_{33}a_{31} & a_{31}a_{32} \\ 2a_{21}a_{31} & 2a_{22}a_{32} & 2a_{23}a_{33} & a_{22}a_{33}+a_{23}a_{32} & a_{21}a_{33}+a_{23}a_{31} & a_{22}a_{31}+a_{21}a_{32} \\ 2a_{31}a_{11} & 2a_{32}a_{12} & 2a_{33}a_{13} & a_{12}a_{33}+a_{13}a_{32} & a_{13}a_{31}+a_{11}a_{33} & a_{11}a_{32}+a_{12}a_{31} \\ 2a_{11}a_{21} & 2a_{12}a_{22} & 2a_{13}a_{23} & a_{12}a_{23}+a_{13}a_{22} & a_{13}a_{21}+a_{11}a_{23} & a_{11}a_{22}+a_{12}a_{21} \end{bmatrix}. \tag{44}$$

Note that we have chosen to let the third rotation angle ψ to be zero, as this is a rotation about the z'-axis and does not alter the growth direction. In the following the primes are omitted. It is also noteworthy that calculations for wurtzite show that all the material parameter tensors as well as the misfit strain contributions do not depend on the angle ϕ (Bykhovski et al., 1993; Chen et al., 2007; Landau & Lifshitz, 1986).

5.2 Static case

In the static case the equations to solve in each layer become

$$\nabla \cdot \mathbf{T}^{(i)} = 0, \qquad\qquad \nabla \cdot D^{(i)} = 0, \qquad \nabla \times \mathbf{E}^{(i)} = 0$$

$$\rightarrow \frac{\partial T_3^{(i)}}{\partial z} = \frac{\partial T_4^{(i)}}{\partial z} = \frac{\partial T_5^{(i)}}{\partial z} = 0, \qquad \rightarrow \frac{\partial D_z^{(i)}}{\partial z} = 0, \qquad \rightarrow \frac{\partial E_x^{(i)}}{\partial z} = \frac{\partial E_y^{(i)}}{\partial z} = 0, \qquad (45)$$

where the superscript i denotes the material, as depicted in Figure 5. Usually one would use homogeneous Dirichlet boundary conditions for the electric field $E_x|_{z=z_l,z_r} = E_y|_{z=z_l,z_r} = 0$, corresponding to the case where the two ends are covered by a perfect conductor. As electric coupling conditions force continuity of the tangential components of $\mathbf{E}$ and these components are constant in each layer we obtain $E_x = E_y = 0$ everywhere. Using the definition of strain we find that in each layer

$$\frac{\partial^2 u_x}{\partial z^2} = \frac{\partial^2 u_y}{\partial z^2} = \frac{\partial^2 u_z}{\partial z^2}, \qquad (46)$$

that is, we have linear solutions for the displacement in each layer:

$$u_i = \mathcal{A}_i^{(j)} z' + \mathcal{B}_i^{(j)}. \qquad (47)$$

These coefficients are then found by applying continuity of

$$T_3, T_4, T_5, u_x, u_y, u_z, \text{ and } D_z \qquad (48)$$

at the material interfaces. At the outer boundaries we will assume free ends

$$T_5 = T_4 = T_3 = 0, D_z = D. \qquad (49)$$

The conditions for clamped ends would be $u_x = u_y = u_z = 0$ at the ends. The parameter D is a degree of freedom that in principle corresponds to the application of a voltage across the outer ends (as it changes the electric field and in the static case the electric potential is merely an integration over space). Calculations for a superlattice structure (i.e. a periodic repetition of well and barriers) are exactly the same, with the lattice constants in the well layers adapting to those of the barrier (Poccia et al., 2010).

Calculations for the [111] growth direction of zincblende crystals yields the following analytical expression for the compressional strain in the quantum well (Duggen et al., 2008):

$$S_{zz} = \frac{2\sqrt{3}\frac{e_{x4}^{(2)}}{\epsilon^{(2)}}D + 3\left(c_{11}^{(2)} + 2c_{12}^{(2)}\right)a_{mis}}{4\frac{e_{x4}^{(2)2}}{\epsilon^{(2)}} + c_{11}^{(2)} + 2c_{12}^{(2)} + 4c_{44}^{(2)}} - a_{mis}. \qquad (50)$$

Results for the [111] direction in zincblende quantum wells, with several materials, are given in Table 1. The [111] direction is a rather special case as a compression in the [111] direction yields an electric field in the [111] direction as well and this direction does not couple to the transverse components (i.e. a compression in z-direction does not generate an electric field in x or y directions.) - here zincblende behaves very similar to wurtzite grown along the

[0001] direction. The table also contains a comparison between the fully and the semi-coupled model. The terms S_{semi} and $S_{coupling}$ refer to semi-coupled result and the difference to the fully coupled result, respectively, i.e. $S_{fully-coupled} = S_{semi} + S_{coupling}$.

substrate/QW	S_{semi}	$S_{coupling}$	Deviation	$E'_{z,t}$ $[V/\mu m]$	$E'_{z,e}$ $[V/\mu m]$
GaAs/In$_{0.1}$Ga$_{0.9}$As	0.34%	−0.002%	0.5%	15.56	17 ± 1[a]
GaAs/In$_{0.2}$Ga$_{0.8}$As	0.710%	−0.003%	0.4%	28.63	25[b]
AlN/GaN	1.34%	−0.04%	3.1%	271.6	
GaN/In$_{0.3}$Ga$_{0.7}$N	1.69%	−0.07%	4.4%	355.0	
GaN/InN	7.24%	−0.61%	−9.1%	1441.5	
GaN/AlN	−0.91%	0.04%	−4.7%	−280.3	

[a] Caridi et al. (1990)
[b] J.I.Izpura et al. (1999)

Table 1. Contributions to S'_{zz} in the [111]-grown quantum well layer for different zincblende material compositions with $D = 0$. For GaAs/In$_x$Ga$_{1-x}$As both $E'_{z,t}$ and $E'_{z,e}$, being the theoretical and the experimental electric field in the QW-layer respectively, are listed for comparison

It can be seen that it does not play a role whether one uses the fully-coupled or the semi coupled approach for the nitrides. Note, however, that the electric field generated by the intrinsic strain in the quantum well layer is quite large and will definitely have an influence on the electrical properties.

The same calculations have been carried out for wurtzite quantum wells (and barriers). For the [0001] growth direction, the analytic result for the compressional strain, which is not coupled to the shear strains in this case, reads (Duggen & Willatzen, 2010; Willatzen et al., 2006)

$$S_{xx} = S_{yy} = -a_{mis}, \tag{51}$$

$$S_{zz}^{(1)} = e_{z3}^{(1)} \frac{D - P_z^{(1)}}{e_{z3}^{(1)2} + c_{33}^{(1)} \epsilon_{zz}^{(1)}}, \tag{52}$$

$$S_{zz}^{(2)} = \frac{e_{z3}^{(2)}(D - P_z^{(2)}) + 2a_{mis}(e_{z1}^{(2)} e_{z3}^{(2)} + c_{13}^{(2)} \epsilon_{zz}^{(2)})}{e_{z3}^{(2)2} + c_{33}^{(2)} \epsilon_{zz}^{(2)}}, \tag{53}$$

In principle one can of course find analytic expressions for the general strains as function of the two angles ϕ, θ (for both wurtzite and zincblende). However, these expressions are very cumbersome to comprehend and therefore do not provide additional insight.

Results for the growth direction dependency of a GaN/Ga$_{1-x}$Al$_x$N/GaN well are shown in Figure 7. For this structure the shear strain is negligible and therefore omitted. For other materials, however the shear strain component is significant and there are significant differences between the fully and semi-coupled approach as seen in Figure 8.

Note that for sufficiently large Al-content, the electric field in the GaAlN well becomes zero at two distinct angles. For the MgZnO structures it shows that there even exist up to three distinct zeros (Duggen & Willatzen, 2010). This is of potential importance as it might lead to increased efficiency for the application of white LEDs (Waltereit et al., 2000).

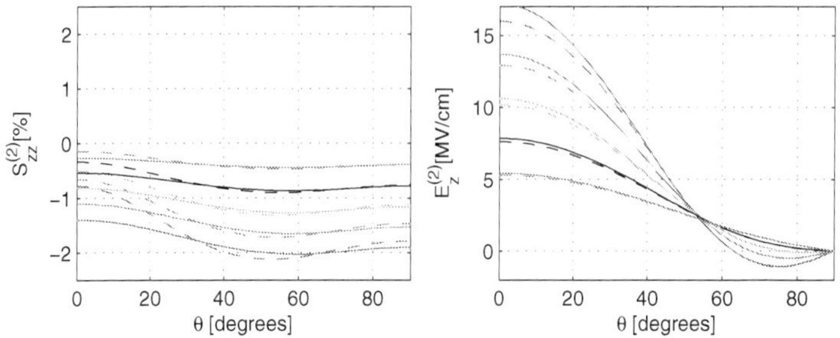

Fig. 7. Compressional strain $S_{zz}^{(2)}$ (left) and electric field $E_z(2)$ (right) for $GaN/Ga_{1-x}Al_xN/GaN$ with several x-values and $D = 0\ C/m^2$. The colors blue, red, green, black, and magenta correspond to $x = 1$, $x = 0.8$, $x = 0.6$, $x = 0.4$, and $x = 0.2$, respectively. Solid (dashed) lines correspond to the semi-coupled (fully-coupled) model. Reprinted with permission from Duggen & Willatzen (2010)

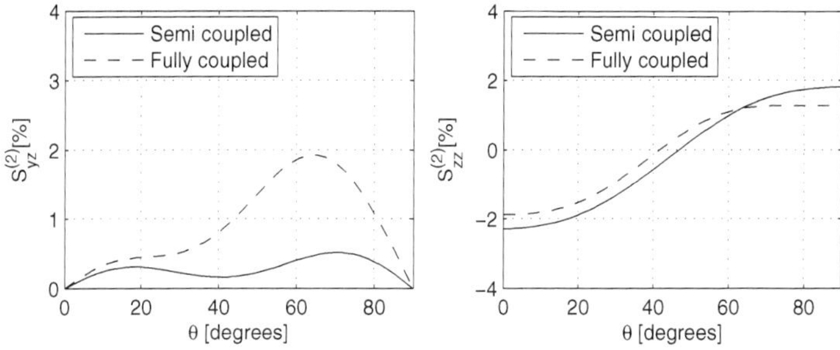

Fig. 8. Shear strain component $S_{yz}^{(2)}$ (left) and compressional strain component S_{zz}^2 (right) in the quantum-well layer of a $Mg_{0.3}Zn_{0.7}O/ZnO/Mg_{0.3}Zn_{0.7}O$ heterostructure for the fully-coupled and semi-coupled models corresponding to $D = 0\ C/m^2$. Reprinted with permission from Duggen & Willatzen (2010)

5.3 Monofrequency case

Both single quantum wells and for superlattice structures might be subject to an applied alternating electric field, which we will model as application of a monofrequent D-field, i.e. we will assume time harmonic solutions $\propto \exp(i\omega t)$, where $\omega = 2\pi f$ and f is the excitation frequency. Here we will limit us to the zincblende case, but the theory is just as well applicable to wurtzite structures, where one needs to take into account the spontaneous polarization P^{SP} as well.

As the coupling conditions are continuity of $\mathbf{T}$, it is convenient to derive the corresponding differential equation for $\mathbf{T}$. As we assume only z-dependency, Navier's equation becomes three equations:

$$\frac{\partial T_I}{\partial z} = \rho_m \frac{\partial^2 u_i}{\partial z},$$

(54)

where I, i are $3, z, 4, y, 5, x$. Furthermore we have that

$$\frac{\partial S_I}{\partial t} = \frac{\partial^2 u_i}{\partial t \partial z},\tag{55}$$

with the same pairs I, i. Differentiating with respect to z and t, respectively, combining and eliminating $\mathbf{u}$ we obtain

$$\frac{\partial^2 T_I}{\partial z^2} = \rho_m \frac{\partial^2 S_I}{\partial t^2}.\tag{56}$$

Then using the piezoelectric fundamental equation along with the electrostatic approximation (forcing $E_x = E_y = 0$ as in the static case) we obtain the set of three coupled wave equations:

$$\Gamma_{33}\frac{\partial^2 T_3}{\partial z^2} + \Gamma_{34}\frac{\partial^2 T_4}{\partial z^2} + \Gamma_{35}\frac{\partial^2 T_5}{\partial z^2} - \rho_m\frac{\partial T_3}{\partial t^2} = \rho_m\frac{e_{3z}^T}{\epsilon^S}\frac{\partial^2 D_z}{\partial t^2},\tag{57}$$

$$\Gamma_{43}\frac{\partial^2 T_3}{\partial z^2} + \Gamma_{44}\frac{\partial^2 T_4}{\partial z^2} + \Gamma_{45}\frac{\partial^2 T_5}{\partial z^2} - \rho_m\frac{\partial T_4}{\partial t^2} = \rho_m\frac{e_{4z}^T}{\epsilon^S}\frac{\partial^2 D_z}{\partial t^2},\tag{58}$$

$$\Gamma_{53}\frac{\partial^2 T_3}{\partial z^2} + \Gamma_{54}\frac{\partial^2 T_4}{\partial z^2} + \Gamma_{55}\frac{\partial^2 T_5}{\partial z^2} - \rho_m\frac{\partial T_5}{\partial t^2} = \rho_m\frac{e_{5z}^T}{\epsilon^S}\frac{\partial^2 D_z}{\partial t^2},\tag{59}$$

where Γ is the piezoelectrically stiffened elastic tensor. Note that the dispersion relation (which is above equations with $D_z = 0$) is the same as in equation (35) with the weak coupling terms removed as is done with the electrostatic approximation.

The general solution to these wave equations consist of forward and backward propagating waves. The solution in each layer for e.g. the x-polarization reads

$$T_5^{(i)} = T_{5A+}^{(i)}\exp(ik_1 z) + T_{5A-}^{(i)}\exp(-ik_1 z) + T_{5B+}^{(i)}\exp(ik_2 z) + T_{5B-}^{(i)}\exp(-ik_2 z)$$

$$+ T_{5C+}^{(i)}\exp(ik_3 z) + T_{5C-}^{(i)}\exp(-ik_3 z) - \frac{e_{5z}^{(i)T}}{\epsilon^{S(i)}}D_z.\tag{60}$$

The other polarizations can then be found by solving the dispersion relation for $T_3(k)/T_5(k)$ and $T_4(k)/T_5(k)$. Thus, when the T_5 amplitudes are known, all amplitudes are known. The coupling conditions between the layers are continuity of stress and continuity of particle velocity (corresponding to continuity of particle displacement in the static case), with the particle velocity $\mathbf{v}$ given by

$$\mathbf{v} = \frac{1}{\rho_m \omega}\frac{\partial \mathbf{T}}{\partial z},\tag{61}$$

where a comment about the dimensionality of $\mathbf{v}$ should be made, since obviously we get elements v_{zx}, v_{zy}, v_{zz}. This is consistent, as the wave has propagation direction z, but three different polarizations x, y, z, i.e. v_5, v_4 describe shear waves while v_3 describes a compressional wave.

The collection of boundary condition equations yields an 18×18 matrix with $\exp(ik_1 z_1)$-like entries. If one would solve for a superlattice consisting of n layers, one would need to solve a $6n \times 6n$ system of equations. As for superlattices this becomes useful when e.g. wanting to compute a macroscopic speed of sound as one can find resonance frequencies and compare to the expression for resonance frequencies of a homeogeneous material. Note that the intrinsic strain will change the bulk speed of sound of the well material, so one cannot simply use a weighted average of the two sound velocities. Furthermore it is expected that operation at resonance strongly influences the properties of the structure (Willatzen et al., 2006).

The first five resonance frequencies for a zincblende AlN/GaN are shown in Figure 9. It is seen that the transversely dominated resonances (only at [111] the, at this direction degenerate, transverse polarizations are uncoupled from the compressional one) are much lower than the compressionally dominated ones, as one would expect. Thus, when computing resonance frequencies it is important not to compute the ideal [111] direction only, but also take into account the significantly lower frequencies as they might occur due to lattice imperfections (Duggen et al., 2008).

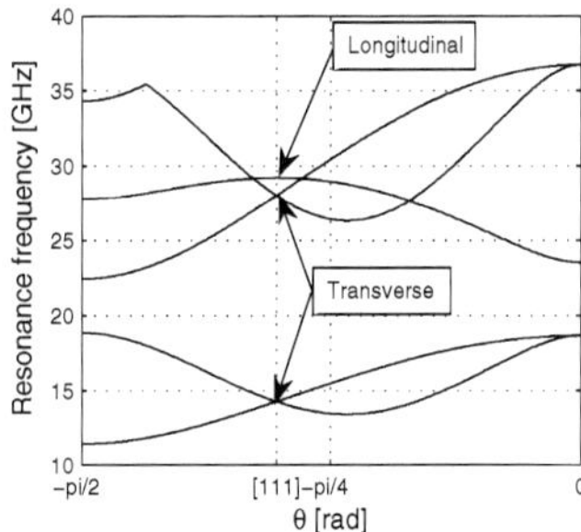

Fig. 9. The first five resonance frequencies for the AlN/GaN structure with $\phi = -\pi/4$. The dimensions of the well-strucure used are 100nm-5nm-100nm. Reprinted with permission from Duggen et al. (2008)

5.4 Cylindrical symmetry of [0001] wurtzite

As we have already noted, the material parameter matrices are invariant under rotation of an angle ϕ around the z-axis. This stipulates investigations of cylindrical structures of wurtzite type. The calculations can, in principle, be done exactly the way described for the quantum well. However, here we consider two degrees of freedom (r, z) which complicates the differential equations and it might not be possible to find analytic solutions anymore. The Voigt notation follows the same standard as for the Cartesian coordinates (including the weight factors) and are

$$rr \to 1, \ \phi\phi \to 2, \ zz \to 3, \ \phi z \to 4, \ rz \to 5, \ r\phi \to 6. \tag{62}$$

The divergence operator becomes

$$\nabla \cdot \to \begin{bmatrix} \frac{\partial}{\partial r} + \frac{1}{r} & -\frac{1}{r} & 0 & 0 & \frac{\partial}{\partial z} & \frac{1}{r}\frac{\partial}{\partial \phi} \\ 0 & \frac{1}{r}\frac{\partial}{\partial \phi} & 0 & \frac{\partial}{\partial z} & 0 & \frac{\partial}{\partial r} + \frac{2}{r} \\ 0 & 0 & \frac{\partial}{\partial z} & \frac{1}{r}\frac{\partial}{\partial \phi} & \frac{\partial}{\partial r} + \frac{1}{r} & 0 \end{bmatrix}, \tag{63}$$

and the material property matrices are transformed in the same manner as for crystal orientation, with

$$\mathbf{a} = \begin{bmatrix} \cos(\phi) & \sin(\phi) & 0 \\ -\sin(\phi) & \cos(\phi) & 0 \\ 0 & 0 & 1 \end{bmatrix}, \tag{64}$$

so since there is cylindrical symmetry, the material parameter matrices remain unchanged. Again using Navier's equation and $\nabla \cdot \mathbf{D} = 0$ one obtains the following linear system of differential equations (with all ϕ-dependencies neglected) (Barettin et al., 2008):

$$L \cdot \begin{bmatrix} u_r \\ u_z \\ V \end{bmatrix} = \begin{bmatrix} -\partial_r \left[(C_{11} + C_{12})a_{mis} + C_{33}c_{mis} \right] \\ -\partial_z \left[2C_{13}a_{mis} + 2C_{13}c_{mis} \right] \\ -\partial_z p^{SP} \end{bmatrix}, \tag{65}$$

$$L = \begin{bmatrix} \partial_r C_{11}\partial_r + \partial_z C_{ee}\partial_z + 1/r\partial r C_{12} + c_{11}\partial_r 1/r \\ \partial_r C_{44}\partial_z + \partial_z C_{13}\partial_r + \partial_z C_{13}/r + c_{44}/r\partial_z \\ \partial_r e_{15}\partial_z + e_{15}/r\partial_z + \partial_z e_{31}\partial_r + \partial_z e_{15}/r \end{bmatrix} \cdot \begin{bmatrix} 1 & 0 & 0 \end{bmatrix}$$

$$+ \begin{bmatrix} \partial_r C_{13}\partial z + \partial_z C_{44}\partial r \\ \partial_r C_{44}\partial_r + \partial_z C_{33}\partial_z + C_{44}/r\partial r \\ \partial_r e_{15}\partial_r + e_{15}/r\partial r + \partial_z e_{33}\partial_z \end{bmatrix} \cdot \begin{bmatrix} 0 & 1 & 0 \end{bmatrix}$$

$$+ \begin{bmatrix} \partial_r e_{31}\partial_z + \partial_z e_{15}/r\partial_r \\ \partial_r e_{33}\partial_z + \partial_z e_{13}\partial_r + e_{15}/r\partial_r \\ -\partial_r \epsilon_{11}\partial_r - \partial_z \epsilon_{33}\partial_z - \epsilon_{11}/r\partial_r \end{bmatrix} \cdot \begin{bmatrix} 0 & 0 & 1 \end{bmatrix}, \tag{66}$$

where ∂_i is short notation for $\partial/\partial i$ and V is the electric potential (thus $E_z = -\partial_z V$). This system can be solved numerically e.g. by using the Finite Element Method. This has been done for a cylindrical quantum dot structure sketched in Figure 10

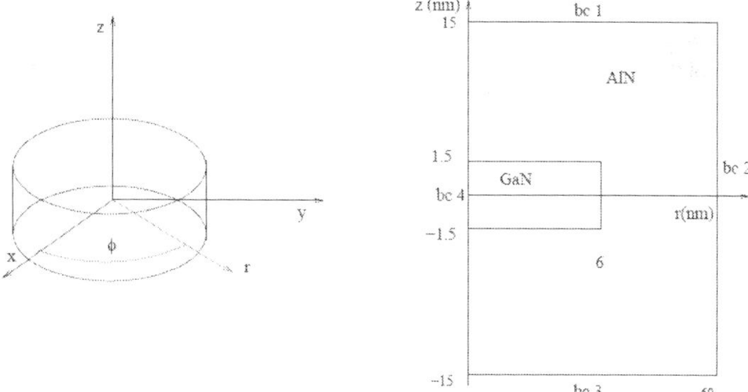

Fig. 10. Geometry of the system under consideration (left) and the two-dimensional equivalent (right). Reprinted with permission from Barettin et al. (2008)

They have found, as can be seen in Figure 11, that the major driving effect for the strain is the lattice mismatch and not the spontaneous polarization.

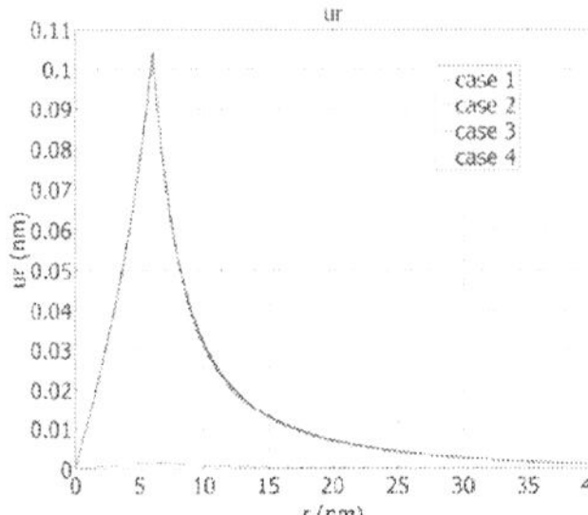
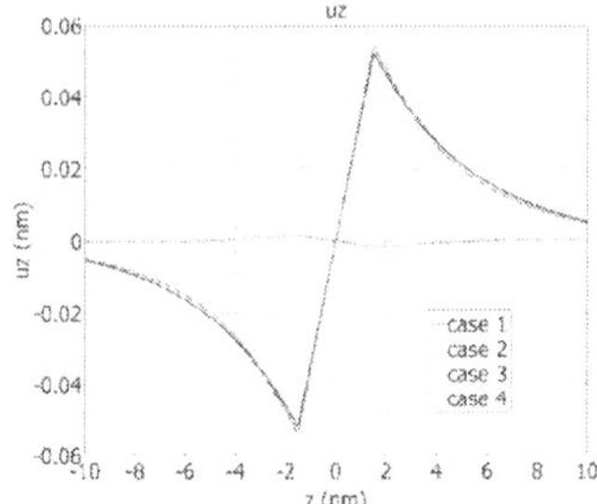

Fig. 11. Displacements u_r at $z = 0$ (left) and u_z at $r = 0$. Four modeling cases are depicted. It suffices to say that only case three does not consider lattice mismatch contributions. Reprinted with permission from Barettin et al. (2008)

Furthermore, using basically the same calculations, Lassen, Barettin, Willatzen & Voon (2008) revealed that calculations in the 3D case can yield a substantially larger discrepancy between semi and fully coupled models, where in the GaN/AlN differences up to 30% were found.

5.5 Other effects

It should be noted that the method described above is by no means secure to be absolutely correct. For example we have disregarded possible free charge densities in order to solve the electromechanical equations self-consistently, without having to solve the Schrödinger equation simultaneously, which would have been necessary otherwise (Voon & Willatzen, 2011). However, it was found by Jogai et al. (2003) that there exists a 2D-electron gas at the interfaces, effectively reducing the generated electric field. Thus the necessity of a fully coupled model is not automatically given, even though calculations as above indicate it.

Also, as already indicated in the piezoelectricity section there might be non-linear effects that are of importance. According to Voon & Willatzen (2011) the effect of non-linear permittivity can be neglected in spite of large electric fields. However, it is not sure whether electrostrictive or second order piezoelectric effects might be of importance. Clearly these questions need further research in order to improve the understanding of electromechanical effects in these structures.

5.6 Alternative: VFF method

As opposed to the above, semi-classical approach there also exist atomistic methods of calculating strains in quantum structures. The are called Valence Force Field (VFF) methods of which Keatings model is the most prominent one (Keating, 1966). Due to limited space we will only present a brief description here, with mainly is taken from Barettin (2009). It should be noted from the start that the piezoelectric effect is not included in this model.

The essence of the model is to impose conditions on the mechanical energy F_s, namely invariance of F_s under rigid rotation and translation as well as symmetries due to the crystal structure. The first condition can be ensured by describing F_s as a function of λ_{klmn}, where

$$\lambda_{klmn} = \left(\vec{u}_{kl} \cdot \vec{u}_{mn} - \vec{U}_{kl} \cdot \vec{U}_{mn} \right) / 2a, \tag{67}$$

where a is the lattice constant, $\vec{U}_{kl} = \vec{X}_k - \vec{X}_l$ with capital $\vec{X}$ denoting nucleus positions in the undeformed crystal and the non-capital $\vec{x}$ denote nucleus positions after deformation. Following assumptions of small deformations and limiting the range of atomic effects to neighboring and second-neighbor terms one arrives at

$$F_s = \frac{1}{2} \sum_{l,l'} \sum_{m,n,m',n'}^{4} B_{mnm'n'}(l-l')\lambda_{mn}(l)\lambda_{m'n'}(l') + O(\lambda^3). \tag{68}$$

where $\lambda_{mn}(l) = \left(\vec{x}_m(l) \cdot \vec{x}_n(l) - \vec{X}_m \cdot \vec{X}_n \right) / 2a$ and l denotes the atom cell index (i.e. the atom which neighbors are considered). Within the harmonic approximation one arrives at

$$F_s = \frac{1}{2} \sum_{l} \left[\frac{\alpha}{4a^2} \sum_{i=1}^{4} \left(x_{0i}^2(l) - 3a^2\right)^2 + \frac{\beta}{2a^2} \sum_{i,j>i,1}^{4} \left(x_{0i}(l) \cdot x_{0j}(l) + a^2\right)^2 \right], \tag{69}$$

where α, β are empirical elastic parameters. The strain is then found by minimizing the elastic energy F_s, fulfilling boundary conditions as e.g. an imposed dislocation of several atoms at an interface between two materials. The VFF method has also been used to determine ground state configurations of lattice mismatched zincblende structures (Liu et al., 2007) as well as non-binary alloys (Chen et al., 2008).

6. Influence of electromechanical fields on optical properties

Since this book covers optoelectronics, we will also have a brief description of the influence of (piezo)electric fields on the optical properties of a quantum well heterostructure. Instead of using the widely used $k \cdot p$ method with eight bands (Singh, 2003) we will limit ourselves to solve the Schrödinger equation for one band, using the effective mass approximation as also has been done by Lassen, Willatzen, Barettin, Melnik & Voon (2008) for investigating a cylindrical quantum dot.

We need to solve the Schrödinger eigenvalue equation, reading

$$H\Psi = E\Psi, \tag{70}$$

where H is the Hamiltonian and is given by Lassen, Willatzen, Barettin, Melnik & Voon (2008)

$$H = \left(k_z \frac{\hbar^2}{m_e^{\parallel}} k_z + k^{\perp} \frac{\hbar^2}{m_e \perp} k^{\perp} \right) + V_{edge} + a_c^{\perp}\epsilon_{zz} + a_c^{\perp}(\epsilon_{xx} + \epsilon_{yy}) - eV, \tag{71}$$

where the m_e denote effective masses, a_c are deformation potentials, e is the fundamental charge, V_{edge} is the band-edge potential. Furthermore, the k-vector is given by $k_j = -i\partial j$ (i being the imaginary unit). Indeed, if one considers a quantum well (i.e. one dimension) there exist analytic solutions to this problem as the Ψ functions can be shown to be linear combinations of Airy functions of first and second kind (Ahn & Chuang, 1986).

The conclusion of the above calculations on a cylindrical quantum dot, performed by Lassen, Willatzen, Barettin, Melnik & Voon (2008) show that the semi-coupled model becomes insufficient when the radius of the quantum dot is comparable or larger than the dot height. In terms of conduction band energy for GaN/AlN the difference between fully and semi-coupled models is up to 36meV which for large radii is comparable to the conduction band energy itself.

	GaN[a]	AlN[a]	ZnO[b]	MgO[c]
$e_{33}[C/m^2]$	0.73	0.97	1.32	1.64
$e_{15}[C/m^2]$	-0.49	-0.57	-0.48	-0.58
$e_{31}[C/m^2]$	-0.49	-0.57	-0.57	-0.58
$c_{11}^E[GPa]$	390	396	210	222
$c_{12}^E[GPa]$	145	137	121	90
$c_{13}^E[GPa]$	106	108	105	58
$c_{33}^E[GPa]$	398	373	211	109
$c_{44}^E[GPa]$	105	116	42	105
$\epsilon_{xx}^S/\epsilon_0$	9.28	8.67	9.16	9.8[d]
$\epsilon_{zz}^S/\epsilon_0$	10.01	8.57	12.64	9.8[d]
$p^{sp}[C/m^2]$	-0.029	-0.081	-0.022[c]	-0.068[d]
$a[10^{-10}\text{m}]$	3.189	3.112	3.20[c]	3.45
$c[10^{-10}\text{m}]$	5.185	4.982	5.15[c]	4.14

[a] Fonoberov & Balandin (2003)

[b] Auld (1990)

[c] Gopal & Spaldin (2006)

[d] Park & Ahn (2006)

Table 2. Material parameters. Data for different materials are taken from references indicated in the first row unless otherwise specified. As Fonoberov & Balandin (2003) we assume $e_{15} = e_{31}$ (except for ZnO) and $\epsilon_{xx} = \epsilon_{zz}$ for MgO due to lack of data. We use linear interpolation to obtain parameters for non-binary compounds.

Material	e_{x4}	$c_{11}^E/10^{10}$	$c_{12}^E/10^{10}$	$c_{44}^E/10^{10}$	ϵ^S/ϵ_0	$a/10^{-10}$	ρ_m
$In_{0.1}Ga_{0.9}As$	0.149[a]	11.82	5.55	5.79	13.13[a]	5.6935	5635[b]
GaAs	0.16[a]	12.21	5.66	6.00	12.91[a]	5.6536	5307[b]
GaN	0.50[c]	29.3	15.9	15.5	9.7[c]	4.50	6150[d]
AlN	0.59[c]	30.4	16.0	19.3	9.7[c]	4.38	3245[d]
InN	0.95[f]	18.7	12.5	8.6	14.86[f]	4.98	6810[e]

[a] Caridi et al. (1990)

[b] Auld (1990)

[c] Fonoberov & Balandin (2003)

[d] Average from Willatzen et al. (2006) and Chin et al. (1994)

[e] Chin et al. (1994)

[f] Davydov (2002)

Table 3. Material parameters for incblende structure materials (in SI units). Parameters from Vurgaftman et al. (2001) if not stated otherwise

7. References

Ahn, D. & Chuang, S. L. (1986). Exact calculations of quasibound states of an isolated quantum well with uniform electric field: Quantum-well stark resonance, *Phys. Rev. B* 34(12): 9034–9037.

Auld, B. A. (1990). *Acoustic Fields and Waves in Solids*, Vol. I, Krieger Publishing Company, Malabar, Florida.

Barettin, D. (2009). *Multiphysics effects in quantum-dot structures*, PhD thesis, University of Southern Denmark.

Barettin, D., Lassen, B. & Willatzen, M. (2008). Electromechanical fields in GaN/AlN Wurtzite quantum dots, *Journal of Physics: Conference Series* 107(1): 012001.
URL: *http://stacks.iop.org/1742-6596/107/i=1/a=012001*

Bester, G., Wu, X., Vanderbilt, D. & Zunger, A. (2006). Importance of second-order piezoelectric effects in zinc-blende semiconductors, *Phys. Rev. Lett.* 96(18): 187602.

Bester, G., Zunger, A., Wu, X. & Vanderbilt, D. (2006). Effects of linear and nonlinear piezoelectricity on the electronic properties of InAs/GaAs quantum dots, *Phys. Rev. B* 74(8): 081305.

Bond, W. L. (1943). The mathematics of the physical properties of crystals, *The Bell System Technical Journal* 22(1): 1–72.

Bykhovski, A., Gelmont, B. & Shur, M. (1993). Strain and charge distribution in GaN-AlN-GaN semiconductor-insulator-semiconductor structure for arbitrary growth orientation, *Applied Physics Letters* 63(16): 2243–2245.
URL: *http://link.aip.org/link/?APL/63/2243/1*

Caridi, E., Chang, T., Goossen, K. & Eastman, L. (1990). Direct demonstration of a misfit strain - generated in a [111] growth axis zinc-blende heterostructure, *Applied Physics Letters* 56(7): 659–661.

Chen, C.-N., Chang, S.-H., Hung, M.-L., Chiang, J.-C., Lo, I., Wang, W.-T., Gau, M.-H., Kao, H.-F. & Lee, M.-E. (2007). Optical anisotropy in [hkil]-oriented Wurtzite semiconductor quantum wells, *Journal of Applied Physics* 101(4): 043104.
URL: *http://link.aip.org/link/?JAP/101/043104/1*

Chen, S., Gong, X. G. & Wei, S.-H. (2008). Ground-state structure of coherent lattice-mismatched zinc-blende $a_{1-x}b_x c$ semiconductor alloys ($x = 0.25$ and 0.75), *Phys. Rev. B* 77(7): 073305.

Chin, V. W. L., Tansley, T. L. & Osotchan, T. (1994). Electron mobilities in gallium, indium, and aluminum nitrides, *Journal of Applied Physics* 75(11): 7365–7372.
URL: *http://link.aip.org/link/?JAP/75/7365/1*

Davydov, S. (2002). Evaluation of physical parameters for the group iii nitrates: BN, AlN, GaN, and InN, *Semiconductors* 36: 41–44. 10.1134/1.1434511.
URL: *http://dx.doi.org/10.1134/1.1434511*

Duggen, L. & Willatzen, M. (2010). Crystal orientation effects on Wurtzite quantum well electromechanical fields, *Phys. Rev. B* 82(20): 205303.

Duggen, L., Willatzen, M. & Lassen, B. (2008). Crystal orientation effects on the piezoelectric field of strained zinc-blende quantum-well structures, *Phys. Rev. B* 78(20): 205323.

Fan, W. J., Xia, J. B., Agus, P. A., Tan, S. T., Yu, S. F. & Sun, X. W. (2006). Band parameters and electronic structures of Wurtzite zno and zno/mgzno quantum wells, *Journal of*

Applied Physics 99(1): 013702.
URL: *http://link.aip.org/link/?JAP/99/013702/1*

Fonoberov, V. A. & Balandin, A. A. (2003). Excitonic properties of strained Wurtzite and zinc-blende GaN/Al$_x$Ga$_{1-x}$N quantum dots, *Journal of Applied Physics* 94(11): 7178–7186.
URL: *http://link.aip.org/link/?JAP/94/7178/1*

Fujita, S., Takagi, T., Tanaka, H. & Fujita, S. (2004). Molecular beam epitaxy of Mg$_x$Zn$_{1-x}$O layers without wurzite-rocksalt phase mixing from x = 0 to 1 as an effect of ZnO buffer layer, *physica status solidi (b)* 241(3): 599–602.
URL: *http://dx.doi.org/10.1002/pssb.200304153*

Goldstein, H. (1980). *Classical Mechanics*, 2nd edn, Addison Wesley, Cambridge, Massachusetts, USA.

Gopal, P. & Spaldin, N. (2006). Polarization, piezoelectric constants, and elastic constants of ZnO, MgO, and CdO, *Journal of Electronic Materials* 35: 538–542. 10.1007/s11664-006-0096-y.
URL: *http://dx.doi.org/10.1007/s11664-006-0096-y*

Ipatova, I. P., Malyshkin, V. G. & Shchukin, V. A. (1993). On spinodal decomposition in elastically anisotropic epitaxial films of III-V semiconductor alloys, *Journal of Applied Physics* 74(12): 7198–7210.
URL: *http://link.aip.org/link/?JAP/74/7198/1*

J.I.Izpura, Sánchez, J., Sánchez-Rojas, J. & Muñoz, E. (1999). Piezoelectric field determination in strained InGaAs quantum wells grown on [111]b GaAs substrates by differential photocurrent, *Microelectronics Journal* 30: 439–444.

Jogai, B., Albrecht, J. D. & Pan, E. (2003). Effect of electromechanical coupling on the strain in AlGaN/GaN heterojunction field effect transistors, *Journal of Applied Physics* 94(6): 3984–3989.
URL: *http://link.aip.org/link/?JAP/94/3984/1*

Keating, P. N. (1966). Effect of invariance requirements on the elastic strain energy of crystals with application to the diamond structure, *Physical Review* 145(2): 637–645.

Kneissl, M., Treat, D. W., Teepe, M., Miyashita, N. & Johnson, N. M. (2003). Ultraviolet AlGaN multiple-quantum-well laser diodes, *Applied Physics Letters* 82(25): 4441–4443.
URL: *http://link.aip.org/link/?APL/82/4441/1*

Koike, K., Nakashima, I., Hashimoto, K., Sasa, S., Inoue, M. & Yano, M. (2005). Characteristics of a zn[sub 0.7]mg[sub 0.3]o/zno heterostructure field-effect transistor grown on sapphire substrate by molecular-beam epitaxy, *Applied Physics Letters* 87(11): 112106.
URL: *http://link.aip.org/link/?APL/87/112106/1*

Landau, L. & Lifshitz, E. (1975). *Course of Theoretical Physics*, Vol. II: The Classical Theory of Fields, Butterworh Heineman, Oxford, UK.

Landau, L. & Lifshitz, E. (1986). *Course of Theoretical Physics*, Vol. VII: Theory of Elasticity, Butterworh Heineman, Oxford, UK.

Landau, L., Lifshitz, E. M. & Pitaevskii, L. (1984). *Course of Theoretical Physics*, Vol. VIII: Electrodynamics of Continuous Media, Butterworh Heineman, Oxford, UK.

Lassen, B., Barettin, D., Willatzen, M. & Voon, L. L. Y. (2008). Piezoelectric models for semiconductor quantum dots, *Microelectronics Journal* 39(11): 1226 – 1228. Papers CLACSA XIII, Colombia 2007.

Lassen, B., Willatzen, M., Barettin, D., Melnik, R. V. N. & Voon, L. C. L. Y. (2008). Electromechanical effects in electron structure for GaN/AlN quantum dots, *Journal of Physics: Conference Series* 107(1): 012008.
URL: *http://stacks.iop.org/1742-6596/107/i=1/a=012008*

Liu, J. Z., Trimarchi, G. & Zunger, A. (2007). Strain-minimizing tetrahedral networks of semiconductor alloys, *Phys. Rev. Lett.* 99(14): 145501.

Nakamura, S., Mukai, T. & Senoh, M. (1994). Candela-class high-brightness InGaN/AlGaN double-heterostructure blue-light-emitting diodes, *Applied Physics Letters* 64(13): 1687–1689.
URL: *http://link.aip.org/link/?APL/64/1687/1*

Newnham, R. E., Sundar, V., Yimnirun, R., Su, J. & Zhang, Q. M. (1997). Electrostriction: Nonlinear electromechanical coupling in solid dielectrics, *The Journal of Physical Chemistry B* 101(48): 10141–10150.
URL: *http://pubs.acs.org/doi/abs/10.1021/jp971522c*

Park, S.-h. & Ahn, D. (2006). Crystal orientation effects on electronic and optical properties of Wurtzite ZnO/MgZnO quantum well lasers, *Optical and Quantum Electronics* 38: 935–952. 10.1007/s11082-006-9007-y.
URL: *http://dx.doi.org/10.1007/s11082-006-9007-y*

Park, S.-H. & Chuang, S.-L. (1998). Piezoelectric effects on electrical and optical properties of Wurtzite GaN/AlGaN quantum well lasers, *Applied Physics Letters* 72(24): 3103–3105.
URL: *http://link.aip.org/link/?APL/72/3103/1*

Poccia, N., Ricci, A. & Bianconi, A. (2010). Misfit strain in superlattices controlling the electron-lattice interaction via microstrain in active layers, *Advances in Condensed Matter Physics* 2010: 261849.
URL: *http://dx.doi.org/doi:10.1155/2010/261849*

Sasa, S., Ozaki, M., Koike, K., Yano, M. & Inoue, M. (2006). High-performance ZnO/ZnMgO field-effect transistors using a hetero-metal-insulator-semiconductor structure, *Applied Physics Letters* 89(5): 053502.
URL: *http://link.aip.org/link/?APL/89/053502/1*

Singh, J. (2003). *Electronic and Optoelectronic Properties of Semiconductor Structures*, Cambridge University Press, Cambridge, UK.

Voon, L. C. L. Y. & Willatzen, M. (2011). Electromechanical phenomena in semiconductor nanostructures, *Journal of Applied Physics* 109(3): 031101.
URL: *http://link.aip.org/link/?JAP/109/031101/1*

Vurgaftman, I., Meyer, J. & Ram-Mohan, L. (2001). Band parameters for IIIŰV compound semiconductors and their alloys, *Applied Physics Review* 89(11): 5815.

Waltereit, P., Brandt, O., Trampert, A., Grahn, H. T., Menniger, J., Ramsteiner, M., Reiche, M. & Ploog, K. H. (2000). Nitride semiconductors free of electrostatic fields for efficient white light-emitting diodes, *Nature* 406: 865–868.
URL: *http://dx.doi.org/10.1038/35022529*

Willatzen, M. (2001). Ultrasound transducer modeling-general theory and applications to ultrasound reciprocal systems, *IEEE Transactions on Ultrasonics, Ferroelectrics, and Frequency Control* 48(1): 100–112.

Willatzen, M., Lassen, B. & Voon, L. C. L. Y. (2006). Dynamic coupling of piezoelectric effects, spontaneous polarization, and strain in lattice-mismatched semiconductor quantum-well heterostructures, *Journal of Applied Physics* 100(2): 024302.

Yoshida, H., Yamashita, Y., Kuwabara, M. & Kan, H. (2008a). A 342-nm ultraviolet AlGaN multiple-quantum-well laser diode, *Nature Photonics* 1(9): 551–554.
URL: *http://dx.doi.org/10.1038/nphoton.2008.135*
Yoshida, H., Yamashita, Y., Kuwabara, M. & Kan, H. (2008b). Demonstration of an ultraviolet 336 nm AlGaN multiple-quantum-well laser diode, *Applied Physics Letters* 93(24): 241106.
URL: *http://link.aip.org/link/?APL/93/241106/1*

22

Optical Transmission Systems Using Polymeric Fibers

U. H. P. Fischer, M. Haupt and M. Joncic
Harz University of Applied Sciences
Germany

1. Introduction

Polymer Optical Fibers (POFs) offer many advantages compared to alternate data communication solutions such as glass fibers, copper cables and wireless communication systems. In comparison with glass fibers, POFs offer easy and cost-efficient processing and are more flexible for plug interconnections. POFs can be passed with smaller radius of curvature and without any mechanical disruption because of the larger diameter in comparison with glass fibers.

The clear advantage of using glass fibers is their low attenuation, which is below 0.5 dB/km in the infrared range (Fischer, 2002; Keiser, 2000). In comparison, POF can only provide acceptable attenuation in the visible spectrum from 450 nm up to 750 nm (Fig. 1). The attenuation has its minimum with about 85 dB/km at approximately 570 nm, which is due to absorption bands of the used Polymethylmetacrylat (PMMA) material (Daum, 2002). For this reason, POF can only be efficiently used for short distance communication up to 100 m. The large core diameter combined with higher Numerical Aperture (NA) results in strong optical mode dispersion, see Fig. 2.

Sources both LEDs and laser diodes in the 650 nm window have been available for some time. It is only recently that LED and Resonant Cavity LEDs (RC-LEDs) sources have become available in the 520 nm and 580 nm windows.

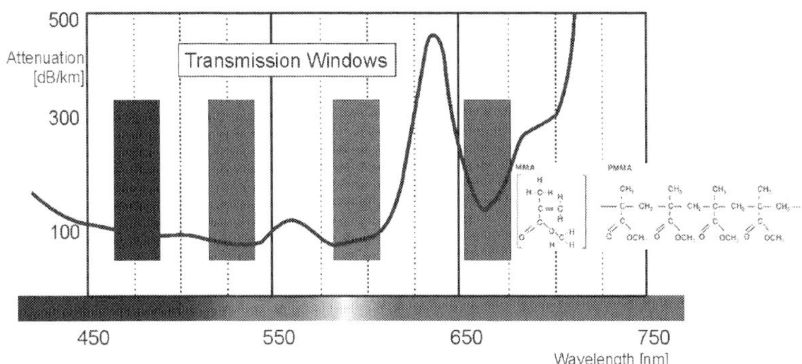

Fig. 1. Attenuation of POF in the visible range, insert: structure of PMMA.

The Numerical Apertur is directly given by the difference of the refractive indices of core and cladding material of the waveguide.

$$NA = (n_1{}^2 - n_2{}^2)^{1/2} \tag{1}$$

$$\Theta = \arcsin (NA) \tag{2}$$

The aperture angle of the waveguide is defined by the arcsin of the NA, which is the amount of input light that can be transferred by the waveguide by total reflection (Senior, 1992). For polymeric fiber systems, the NA calculates to 0.5, which results in the aperture angle of 30°. The difference of the core and cladding refractive indices is in comparison to glass fibers very high : 5%. The numerical aperture NA is correlated to the so-called V-parameter, which gives a correlation to the number of optical modes in the fiber waveguide. The number of the modes allowed in a given fiber is determined by a relationship between the wavelength of the light passing through the fiber, the core diameter of the fiber, and the material of the fiber. This relationship is known as the Normalized Frequency Parameter, or V number. The mathematical description is:

$$V = 2\,\pi\,NA\,a\,/\,\lambda \tag{3}$$

where NA is the Numerical Aperture, a is the fiber radius , and λ is wavelength.

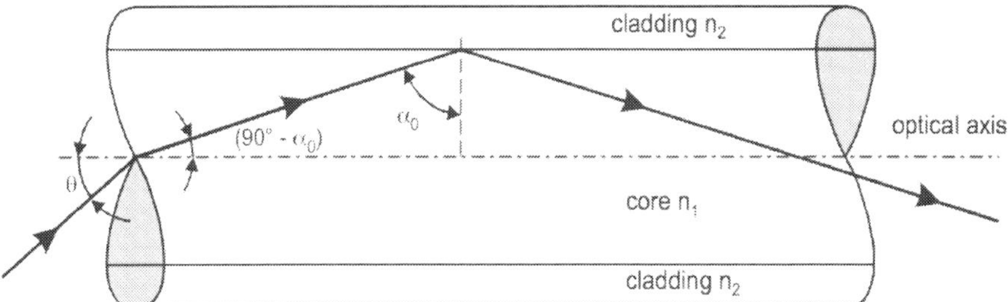

Fig. 2. Optical fiber waveguide.

A single-mode fiber has a V number that is less than 2.405, for most optical wavelengths. It will propagate light in a single guided mode. The multi-mode step index POF has a V number of 2,799, by a given optical wavelength of 550 nm, core radius of 490 μm, and NA of 0.5. This is more than 1000 times larger than for single-mode fiber Therefore the light will propagate in many paths or modes through the fiber. The number of optical modes can be calculated by:

$$N = 0.5\ V^2\ g/(g+2) \tag{4}$$

where g is the index profile exponent, which is infinity for step index fibers. For step index POF the mode number can be calculated to $N \cong V^2/2 = 3.917$ Mio modes. For longer wavelengths the number of modes will reduce to 2.804 Mio modes at 650 nm. The number of modes will reduce the usable bandwidth by mode dispersion, which can be calculated by the difference of the optical path of the mode which is lead through the fiber without reflection t_1 at the core/cladding interface and the path of the mode t_2 which is most reflected due to a high aperture angle of 30°.

$$\Delta t_{mod} = t_1 - t_2 = L_1 \, NA^2 / (2 \, c \, n_2) \tag{5}$$

The skew between the two modes in a POF step index fiber can be calculated to $\Delta t_{mod} \cong 25$ ns for $L_1 = 100$ m and c = velocity of light in vacuum. The bandwidth length product for uniform Gaussian pulses (Ziemann, 2008b)

$$B \, L \cong (0.44 / \Delta t_{mod}) \, L_1 \tag{6}$$

will result in a theoretical bandwidth of 14 MHz for 100 m fiber length. A reduced NA will magnify the bandwidth length product BL up to 100 MHz for a step index POF with a NA of 0.19. To increase the BL product, other types of POF, which are described in detail in chapter 3., are introduced

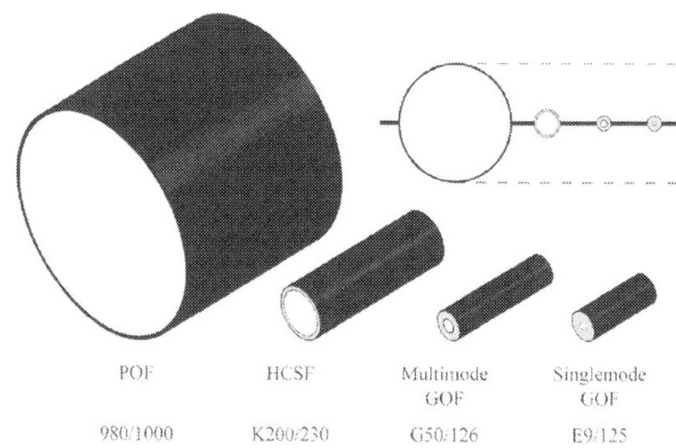

Fig. 3a. Polymeric step index fiber, b. Comparison of the dimension of different optical fiber types.

Like all optical transmission systems, at the beginning of the transmission an electro-optical conversion in a transmitter turns the electrical modulated signals into optical signals (see Fig. 4). This is typically performed by the use of a LED for data speeds up to 150 Mbit/s. For higher data speeds the use of a Laser diode like a VCSEL or edge emitter is necessary.

Modulation format in the existing Fast Ethernet systems is direct modulation by ASK: Non-Return-to-Zero (NRZ). NRZ means that the transmitter switches from maximum level to zero switching with the bit pattern. The advantage is the very easy system set-up. The disadvantage is the large required bandwidth. Usually a minimum bandwidth corresponding to the half of the transmitted bit rate is needed (e.g. 50 MHz for a bit rate of 100 Mbit/s).

For 1 Gbit/s Ethernet direct modulation techniques are not possible for use in POF systems, because of the high mode dispersion of the SI POF. Here, different higher modulation techniques must be implemented:

2. Pulse Amplitude Modulation (PAM)

In pulse-amplitude modulation there are more than two levels possible. Usually 2^n levels are used, with $4 < n < 12$. Due to every symbol transmitting n bits, the required bandwidth and the noise is reduced by $1/n$. A great advantage of PAM is its flexibility and adaptability to the actual signal to noise ratio (Gaudino et al., 2007a, 2007b; Loquai et al., 2010).

2.1 Discrete Multi Tone (DMT)

At DMT the used spectrum is cut into many sub-carriers. Each sub-carrier can now be modulated discrete by quadrature amplitude modulation QAM. Strong signal processing must be implemented with a fast analog-to- digital converter and forward error correction, which makes the overall system expensive. Nowadays, many communication systems like DSL, LTE or WLAN use this method (Ziemann, 2010).

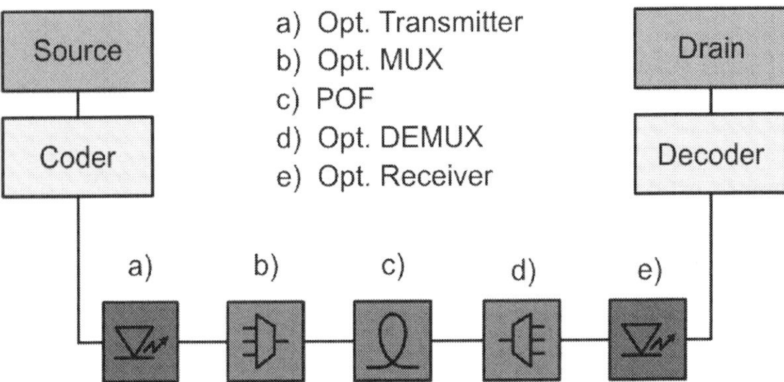

Fig. 4. Basic key elements of an optical transmission line.

At the end of the optical transmission path, an optical/electrical converter must be used. Typically, pin-photo diodes with large active areas are used. In between, the POF medium is situated using multiplexers (MUX) and demultiplexers (DEMUX) for higher effective data rates in the optical pathway. In this paper special optical DEMUX und MUX for wavelength multiplexing are described to extend the data rate of the whole systems for a factor of 4 – 10 in comparison to todays one channel transmission.

The use of copper as communication medium is technically out-dated, but still the standard for short distance communication. In comparison, POF offers lower weight, 1/10 of the volume of CAT cables and very low bending losses down to 20 mm radius. Another reason is the non-existent susceptibility to any kind of electromagnetic interference.

Wireless communication is afflicted with two main disadvantages:

- electromagnetic fields can disturb each other and probably other electronic device,
- wireless communication technologies provide almost no safeguards against unwarranted eavesdropping by third parties, which makes this technology unsuitable for the secure transmission of volatile and sensitive business information.

For these reasons, POF is already applied in various applications sectors. Two of these fields should be described in more detail in the next sections: the automotive sector and the in house communication sector.

2.2 Application areas of POF
2.2.1 Automotive

Since 2000 POF displaces copper in the passenger compartment for multimedia applications, see Fig. 5. The benefits for the automobile manufacturers are clear: POF offers a high operating bandwidth, increased transmission security, low weight, immunity to electromagnetic interference, and ease of handing and installation (Daishing POF Co., Ltd, n.d.). This vehicle bus standard is called Media Oriented Systems Transport (MOST). It is based on synchronous data communication and is used for transmission of multimedia signals over polymer optical fiber (MOST25, MOST50, MOST150) or via electrical conductors (MOST50). The technology was developed, standardized and up to date regularly refined by the MOST Cooperation founded in 1998. MOST was first introduced by BMW in the 7er series in 2001. Since then, MOST technology is used in almost all major car manufacturers in the world, such as VAG Group, Toyota, BMW, Mercedes-Benz, Ford, Hyundai, Jaguar and Land Rover (Wikipedia, 2011). In 2011 there are more than 50 different car types on the market which use the POF in the passenger cabin network structure for multi media data services.

The MOST specification covers all seven layers of the ISO/OSI Reference Model for data communication. On a physical layer polymer optical fiber is used as a media. A light emitting diode (LED) is used for transmission in red wavelength area at 650 nm. PIN photo diode is used as receiver (Grzemba, 2008).

The basic architecture of a MOST network is a logical ring, which consists of up to 64 devices (nods). The logical ring structure is usually implemented on a physical ring, which is however not mandatory. Combined ring, star network or double ring (for critical applications) can also be realised. Plug and play functionality enables easy adding or removing of devices.

In a MOST network one MOST device handles the role of the Timing Master which feeds MOST frames into the ring at a sampling rate of 44.1 kHz (frame is transmitted 44,100 times a second) or 48 kHz. The latest MOST specification recommends sampling rate of 48 kHz. The exact data rate depends on the sampling rate of the system. One after another Timing Slaves on the logical ring receive the signal, synchronize themselves with the preamble, parse the frame, process the desired information, add information to the free slots in the frame and transmits the frame to their successor. Since the MOST system is fully synchronous, with all devices connected to the bus being synchronized, no memory buffering is needed. Each Time Slave contain a fiber optic transceiver - received light signals are converted into electrical domain, processed, converted back into the optical domain and forwarded further.

A MOST frame includes one area for the synchronous transmission of streaming data (audio and video data), one area for the asynchronous transmission of packet data (TCP/IP packets or configuration data for a navigation system), and one area for the transmission of control data. MOST25 frame consists of 512 bits (64 bytes). 60 bytes are used for transmission of data. 6 – 15 quadlets (qualet consists of 4 bytes) of the data can be synchronous data, while the rest of the 60 bytes (0 – 9 quadlets) hold asynchronous data. Two bytes transport the part of the control message which spreads over 16 frames (one block). The first and the last byte of the frame contain the control information for the frame. MOST25 provides a data rate of 22.58 Mbit/s at a sampling rate of 44.1 kHz. This allows up to 15 uncompressed stereo audio channels in CD quality (2x16 bits per channel) / 15 MPEG1 channels for audio-video transmission or up to 60 1-byte connections to be established simultaneously. Maximal data rate is 24.58 Mbit/s at a sampling frequency of 48 kHz.

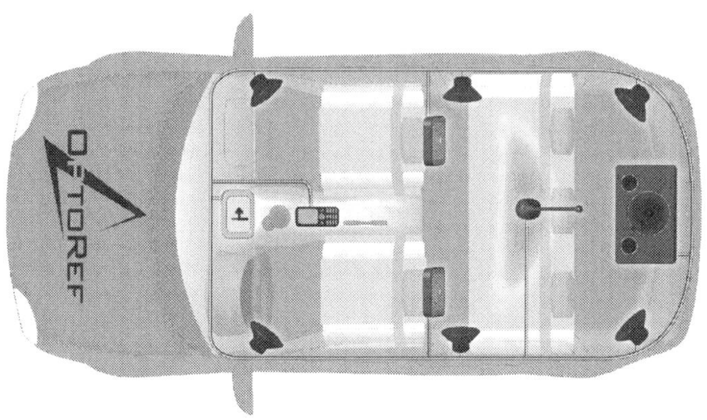

Fig. 5. Multimedia Bus System (MOST-Bus) with POF.

Next MOST generation uses a bit rate of just under 50 Mbit/s for doubling the bandwidth. The name MOST50 derives from this fact. Each frame consists of 1024 bits (128 bytes): 11 bytes for header, which also includes the control channel, and 117 bytes for the payload. The border between synchronous and asynchronous data can be adapted dynamically to the current requirements. The synchronous area can have a width of 0 to 29 quadlets plus one byte (0 to 117 bytes) and the asynchronous area can have a width of 0 to 29 quadlets (116 bytes). Control message consists of 64 bytes.

The latest MOST version (MOST150) was presented in October 2007. MOST150 is designed for high data rate of just under 150 Mbit/s and has a frame of 3027 bits (384 bytes): 12 bytes for header, which also includes the control channel, and 372 bytes for streaming and packet data transfer. It also has access to the dynamic boundary. Both, synchronous and asynchronous areas can have a width in between of 0 and 372 bytes. Besides the three known channels, an Ethernet channel with adjustable bandwidth and isochronous transfer on the synchronous channel for HDTV were introduced. This enables the transmission of synchronous data that require a different frequency than that given by the frame rate of the MOST. MOST150 thus a physical layer for Ethernet in the vehicle (MOST Cooperation, 2010).

Not just multimedia functions can exploit POF. For example, BMW has developed a 10 Mbit/s protocol called ByteFlight, which it uses to support the rapidly growing number of sensors, actuators and electronic control units within cars. Unlike MOST, which employs real-time data transfer, ByteFlight is a deterministic system in which the focus is on making sure that no data is lost (BMW, n.d.). The glass temperature of POF (below 85°C) makes using the fiber in the engine compartment impossible, although this problem might be solved in the foreseeable future. Up to date, a number of different in-car networks for multimedia and security applications has been developed, see Fig. 7.

Fig. 6. MOST applications in the multimedia bus.

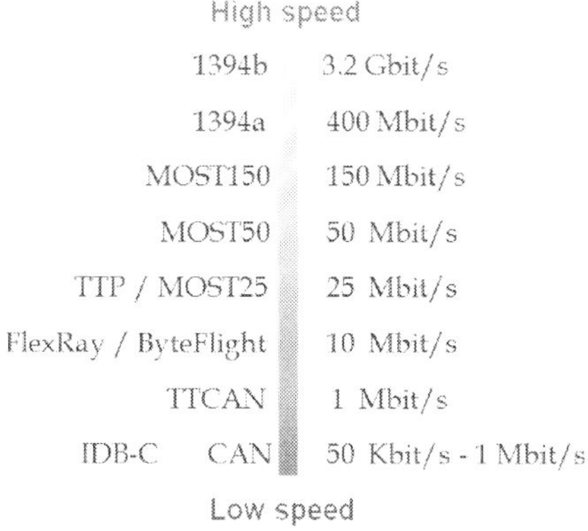

Fig. 7. In-Car network data rates.

2.3 Use of POF in aircraft

To use POF as the transmission media for aircrafts is under the research of different R&D groups due to its specific advantages. The DLR (German Aerospace Center) researches this kind of fiber under the conditions in civil aircrafts. They concluded that "the use of POF multimedia fibers appears to be possible for future aircraft applications" (Cherian et al., 2010). The Boeing Company develops special measurement setups to investigate and analyze POFs for the application under the conditions of daily use in aircrafts. Especially the low weight and the easy and economic handling make this kind of fiber the first choice. But for now the data rates and the temperature range are too low to replace copper for multimedia purposes.

To build aircraft with less weight, all big aircraft manufacturers will use carbon fibers for the aircraft body in all the new aircraft models. Because of its better weight performance, the aviation will loose a lot of its resistance against EMV and outer space radiation. To use optical cables like glass fibers or polymeric fibers is a good approach to bypass the problems of EMV in signal transmission. One coming solution will be the replacement of the electrical copper cables by POF and the application of the bus protocols FlexRay or MOST, which is widely used in the automotive industry (Lubkol, 2008; Strobel, 2010).

In aviation, strong test procedures are introduced for high reliable operation of all system components. High and low temperature operation starting from –60°C up to +130°C must be considered. Also high vibration stability in case of using optical connectors is required. For system relevant usage in the airplane, it is necessary to design the cable in the aircraft for POF use fire- and heat resistant and also waterproof, respectively. Additionally, high temperature POF must be implemented to force stable operation at temperatures in the aircraft up to +130°C, which can occur in the cockpit system unit.

To implement MOST technology in the airplane in the cabin for multimedia usage, the normal standard fiber can be used, because of the not relevant system impact of multimedia provision of the passengers. Up to now, the usage of POF in the airplane is focused in the research area and it will take years to test the reliability for everyday use in the airplane industry.

2.4 In-house

Another sector where POF displaces the traditional communication medium is in-house communication, although the possibilities of application are not confined to the inside of the house itself. In the future, POF will most likely displace copper cables for the so-called last mile between the last distribution box of the telecommunication company and the end-consumer (Koonen et al., 2005, 2009). Today, copper cables are the most significant bottleneck for high-speed Internet.

"Triple Play", the combination of VoIP, IPTV and the classical Internet, is being introduced to the market with force, therefore high-speed connections are essential. It is highly expensive to realize any VDSL system using copper components, thus the future will be FTTH (Fischer, 2007a).

For in-house communications networks data rates between 10 Mbit/s and 100 Mbit/s are typically in use. Copper-cables (Category 5/6) are most widely used in office networks in combination with structured wiring system of DIN EN 50173-1 and DIN EN 50173-2. The 8-core wire in combination with the RJ45 plug can transmit 100/1000 Mbit/s over distances up to 100 meters using Ethernet protocol. Due to the mass-market application of Ethernet (IEEE 802.3), this technique has become very cheap. Most broadband home networks today focus on the combination of Ethernet and RJ45 data cable interface. The disadvantage of this technique depends on the lack of structured cabling in most apartments. The possibilities for re-installation of the thick and inflexible CAT cables are very limited, while most of the wiring has no professional electrical grounding.

In the following the available in-house network technologies are depicted and compared in detail with their specific advantages and disadvantages with POF applications in Table 1:

- Twisted-pair cables belong to the Ethernet standard CAT 5/6 with a star network topology and data rates up to 1 Gbit/s up to 100 m, but due to very thick cables (Ø 7 mm) wide cable channels and complex plug required. They have no electrical isolation, which also leads to a high EMC sensitivity. This disturbing especially in the industrial and automotive environment the transmission.

- Coaxial cables, as they are known from the TV connection, have a diameter of 5 mm and a much higher bandwidth up to 1 GHz for 30 m with large bend radii. However, the electrical isolation from the 230 V power is problematic, which can lead to problems. The EMC problem is related critical as the twisted-pair cable.

- Glass fibers are the media with the highest range and data rate, but expensive compared to alternative techniques, also because of expensive connector assembly and low possible bending radii. Additionally, the small core diameter of 9 microns for single mode fiber is highly vulnerable to pollution. This leads to significant problems in the industrial environment, but without EMC problems.

- Polymer fibers can be easily laid with small bend radii, are very tolerant in terms of buckling and pollution (large core cross-section), without the need of using connectors. It can be shown that POF have a high future potential for increased data rate without having to install additional fibers. Like the glass fiber, POF has a fiber optic to electrical isolation and has a very low EMC sensitivity.

- WLAN is a pure wireless technology with a possible range up to 20 m. Due to absorption by walls, and ceilings the effective range is poor. Furthermore due to interference by third parties, the transmission is not secure. In addition, neighbouring networks will reduce the data rate significantly. This leads especially in the industrial environment to a very large problem, if there are installed WLAN nodes in a very large number. Data rates from 2 up to 100 Mbit/s data rate are possible under optimal conditions, most of the achievable data rates remains well below it.

- Powerline uses the 230 V-house power grid. The range is very limited and depends on the power grid. However, there are only low installation costs, but the high electromagnetic radiation and the uncontrolled distribution over the network are major disadvantages, which makes this network technology for in-house use unattractive.

Technique	Data rate	Range	Security	Costs	Handling	Deployability	Total
Twisted-Pair cable	+	0	0	++	-	0	2+
Coax cable	0	0	0	+	0	0	1+
Glass fiber	++	++	++	--	--	-	1+
POF	0	-	++	+	+	+	4+
WLAN	--	-	--	++	++	++	1+
Powerline	-	-	--	+	+	++	0

Table 1. In-house networks in comparison, division between particularly poor -- and particularly well: ++

In Table 1 an overview is summarized to assess the respective qualities of the alternative networks in view of the most important criteria. It turns out that the most widely used networking technologies such as wireless or twisted pair are leader in the field in terms of costs, but in total the polymer fiber technology shows superior overall properties and combines many advantages of the other transmission media, without their main drawbacks. Keeping these reasons in mind, the further potential of POF seems to be very high.

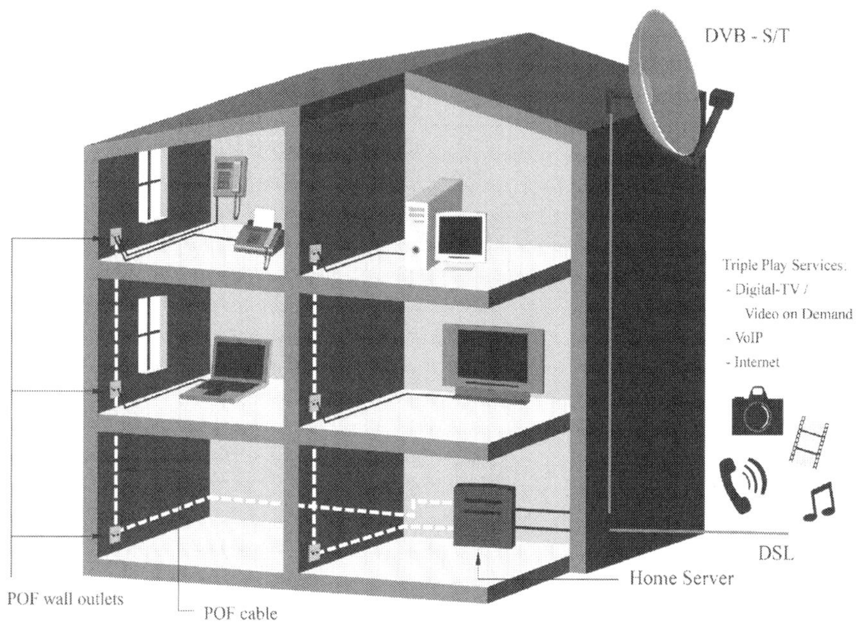

Fig. 8. In-house communication scenario with POF –Ethernet system.

3. State of the art in POF transmission systems

3.1 POF fiber types

Plastic optical fibers for data transmission until recently were limited to step index PMMA fibers that had bandwidths of 38 MHz-100 m (Mitsubishi Eska). More recent results by Mitsubishi with an Eska-Mega fiber shows a three fold increase in bandwidth to 105 MHz-100 m. Increases in bandwidth are also possible with the use of dual step index (DSI), multi-step index (MSI) profiles, multi-core (MC), or combinations of these (Poisel et al., 2003). Here, only the most relevant used POF types are described.

For POF there are in general three fiber types existing, which are on the market available (Table 2).

3.1.1 Step index fibers SI POF

The SI POF (Fig. 9) is already standardized in a IEC 60793-2-40: and IEC 60794-2-40: specification for A4 fiber cables and also in an ETSI recommendation TS 105 175-1. The SI POF fiber is called in the specification as A4.a1 and A4.a2, respectively. Optical properties and also mechanical ones are strictly defined to guaranty a international reliable high level of fabricated fibers. Optical specs are 980/1000 µm diameters, temperature range of -40 °C - +85 °C, and A4a1 is a fiber with attenuation 180 dB/km at 650 nm and bandwidth 4 MHz-km, which is of course the performance of the standard 1 mm, high NA, SI POF. A4a2 refers to the nominally 0.5 mm PF-GI-POF which has an attenuation of less than

100 dB/km in the red, but more significantly, transmits at 850 nm with an attenuation of 40 dB/km. The bandwidth of this fiber is very high, typically of that of GI POF fibers, ~200 MHz-km, but is not in use in cars or in-house networks.

3.1.2 Gradient index fibers GI POF

Here the core-refracting index is distributed in a quadratic behaviour (Fig. 10). This reduces the mode dispersion significantly and relates to a better BL than in step index fibers. BL products of more than 2 GHz-100 m are realized with OM Giga from Optimedia Inc. in Korea. Specifications of different POF types are shown in Table 2:

Specs/Fiber type	SI POF (IEC 60793-2-40)	GI POF (Park 2006)	MC POF (Asahi Chem.)
Diameter (core) mm	1.0 (0.9)	1.0 (0.9)	1.0 (0.9)
Variation of Diameter %	± 5	± 5	± 5
Tensile at break N	> 65	> 65	> 65
Bending Radius mm	> 20	> 35	> 15
Operating Temperature °C	-40 ~ 85	-30 ~ 70	-40 ~ 85
Attenuation dB/km	< 180 at 650 nm	< 200 at 650 nm	< 180 at 650 nm
Bandwidth Mbps	~ 40 @ 100 m	~ 2000 @ 100 m	> 500 @ 100 m

Table 2. Specifications of different POF types.

3.1.3 Multicore fiber MC POF

In this approach, the fiber can be made of many tiny cores, a multi-core POF (Fig. 11). The partition of the core into many individual light guiding areas allows for very small bending radii, helping to ease the installation of the fiber. The numeric aperture and the bandwidth is nearly the same as of the SI-POF. Up to now, there is no international standard available for this new type of POF. In the market available is the fiber from Asahi Chemical, shown in Fig. 11. 19 multi cores of PMMA are introduced in one complete fiber. The NA of this fiber is 0.27 with 1 mm outer diameter and bandwidth length product of 500 Mbit/s-100 m. Operating temperature ranges from –40 °C up to +85 °C. This fiber is an excellent candidate to replace in the next future the SI POF in the mass market, because of its good bandwidth performance and comparable price. On the other hand, standardizing procedures must be realized to make this MC POF acceptable for international network markets.

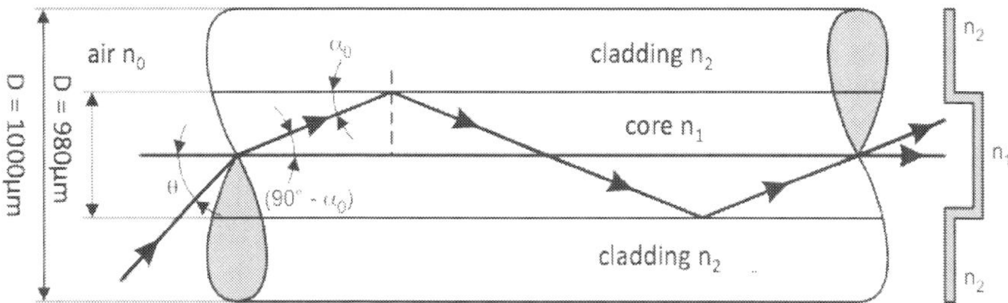

Fig. 9. Step index POF.

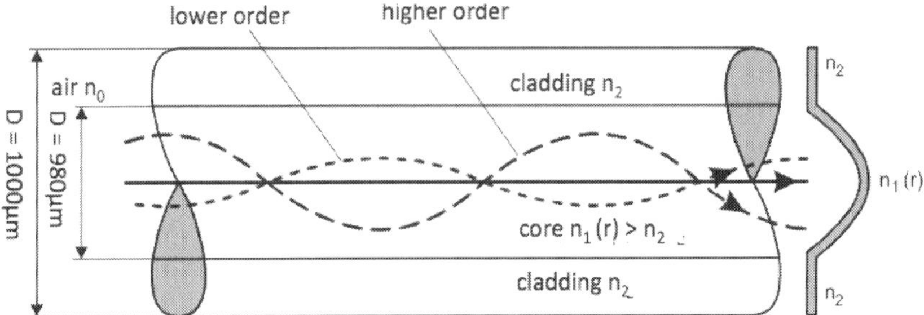

Fig. 10. Gradient index POF.

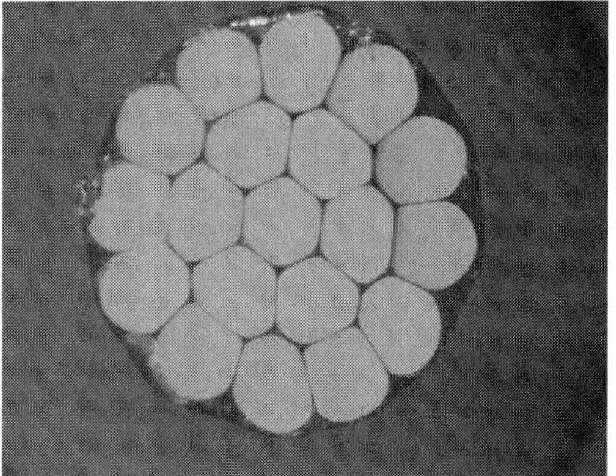

Fig. 11. Multi core POF (Asahi Chemical) 1 mm diameter, 500Mbit/s-100 m.

3.2 Standardization of POF

As shown in this chapter, SI POF is standardized by the International Electrotechnical Commission (IEC) as the A4 category of fibers. In completion, this category contains four types (families A4a-A4d) of SI POF having core diameters ranging from 490 microns to 980 microns for different applications in networks, multimedia sources and sensor systems. This standard also defines other dimensional requirements for these fibers, as well as minimum mechanical and transmission properties (Ziemann 2008). The existing IEC POF standards do not specify any environmental requirements, however.

OFS and Nexans have recently proposed to modify the A4 category fiber standards to include perfluorinated GI POF. According to this proposal, four new fiber families (A4e-A4h) are be added to the A4 category. These families will have core diameters of 500 μm, 200 μm, 120 μm, and 62.5 μm, and are intended to serve a wide variety of applications ranging from consumer electronics to multi-Gb/s data communication.

In Germany, the DKE as the Standardization Division of the VDE Germany has established a POF working group DKE 412.7.1, which is responsible for the international standardisation of Gbit/s POF transmission systems with active and passive elements.

4. WDM over POF

Several sectors will be introduced, where POF offers advantages when compared to the established technologies. Other possible industrial sectors include the aviation or the medical sector. All these applications have one thing in common – they all need high-speed communications systems. The standard communication over POF uses only one single channel. To increase bandwidth for this technology the only possibility is to increase the data rate, which lowers the signal-to-noise ratio and therefore can only be improved in small limitations.

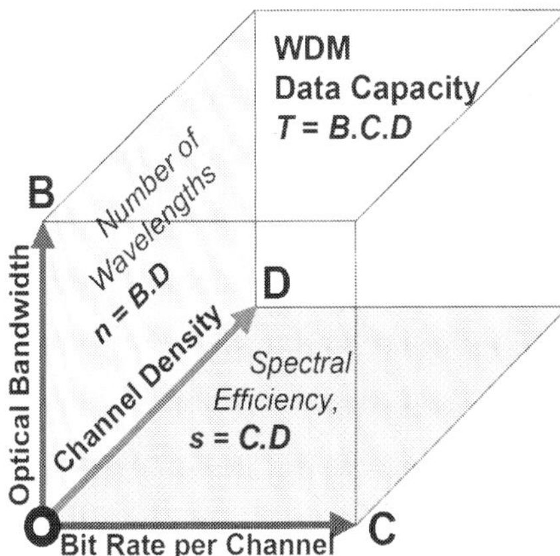

Fig. 12. Advantages of extending the overall transmission capacity using WDM.

Wavelength Division Multiplexing (WDM) is a technique that combines multiple, unique optical signals at different wavelengths (colours) onto a single strand of fiber. At the receiving location, these optical signals are split back out, or demultiplexed, into separate fibers. Essentially, the bandwidth capacity of the fiber is multiplied by the number of wavelengths multiplexed onto the fiber. Fig. 12 conceptually shows the principal benefits of the WDM technique. The WDM system capacity (T) depends on three basic parameters, which are open to modify:

Bit Rate per Channel   (C)
Optical Bandwidth   (B)
Channel Density   (D)

$$T = B \cdot C \cdot D \tag{7}$$

In comparison to single channel transmission systems that only extend the capacity by the help of higher bitrates per channel, WDM will allow to boost up the overall transmission capacity by two additional factors:

- the channel density and
- the optical bandwidth of the system.

Both factors in combination will lead to the total number of wavelength channels which are possible to implement in the whole system. For glass fiber systems the optical bandwidth is characterised by the fibers attenuation curve between 1300 nm and 1650 nm. Here using

POF the bandwidth ist allocated beween 400 nm and 800 nm. Assuming a bandwidth of B = 380 nm and a channel density of D = 1/40 nm, a bit rate per channel of 1 Gbit/s the total capacity will be T = 1 Gbit/s x (380 nm/40 nm) = 9,5 Gbit/s. The application of fixed reference channels for POF systems will be described in chapter 4.6.

For the use of different channel densities, a international system was established for glass fiber systems which defines fixed channel spacing for long distance, metro and short haul networks: this variations of WDM that are commonly used for glass fiber systems: Broad WDM, Coarse WDM, and Dense WDM. Each variation has different capabilities, costs, and operational friendliness.

4.1 Broad WDM

Broad WDM (often just called WDM) utilizes two wavelengths with are parted by more than 200 nm. Broad WDM is very simple to implement. Off-the-shelf optical transmitters without tight control of wavelengths can be used. These applications also utilize low-cost optical multiplexers and demultiplexers with low insertion loss, but are not useful for higher speed systems.

4.2 Coarse WDM

Coarse WDM (CWDM) utilizes multiple wavelengths spaced at 20 nm in the infra red region. The International Telecommunication Union (ITU) in G 694-2 specifies 18 CWDM wavelengths from 1271 nm to 1611 nm for metro networks using optical glass fibers. Transmitters, optical multiplexers, and demultiplexers are at defined wavelengths, but they do not need to be tightly controlled, which translates into lower equipment costs compared to Dense WDM.

4.3 Dense WDM

Dense WDM (DWDM) utilizes many wavelengths spaced narrowly, and they are most commonly located in the C-band, the wavelength range from 1530 nm to 1565 nm. ITU G 694-1 in specifies the center of the DWDM wavelengths. Practical deployments of DWDM today are spaced at 100 GHz frequencies (or approximately 0.8 nm spacing), which allow about 40 wavelengths in the C-band. DWDM requires that the optical transmitters, multiplexers and demultiplexers have very tight control over the wavelength under all operating temperature conditions.

This contribution presents a possibility to open the WDM technique to the POF world. This basic concept can – in theory – also be assigned to POF. However POF shows different attenuation behavior, see Fig. 1. For this reason, only the visible spectrum between 400 nm – 780 nm can be applied when using POF for communication.

For WDM, two key-elements are indispensable: a multiplexer and a demultiplexer. The multiplexer is placed before the single fiber to integrate every wavelength to a single waveguide. The second element, the demultiplexer, is placed behind the fiber to regain every discrete wavelength. Therefore, the polychromatic light must be split in its monochromatic parts to regain the information. These two components are well known for infrared telecom systems, but must be re-developed for POF, because of the different transmission windows.

One technical solution for this problem is available, but it cannot be efficiently utilized in the POF application scenario described here, mostly because this solution is afflicted with high costs and therefore not applicable for any mass production.

4.4 Basic concept of the demultiplexer

As mentioned before, a demultiplexer is essential for WDM (Daum et al., 2008; Chen & Lipscomb, 2000). Several preconditions must be fulfilled to create a functional demultiplexer for POF. First of all, the divergent light beam, which escapes the POF, must be focused. This is done by an on-axis mirror. In the first attempt, a spherical mirror is used. To get perfect results without any spherical aberrations, an ellipsoid mirror should be used.

The second function is the separation of the different transmitted wavelengths (Fischer et al., 2007b). In Fig. 13, this principle is illustrated for three wavelengths (red, green, blue). This is not a limitation for possible future developments, but rather an experimental basis from where to run the various simulations described below. The diffraction is done by a diffraction grating. The diffraction is split into different orders of diffraction. The first order is the important one to regain all information. There a detector line can be installed to detect the signals.

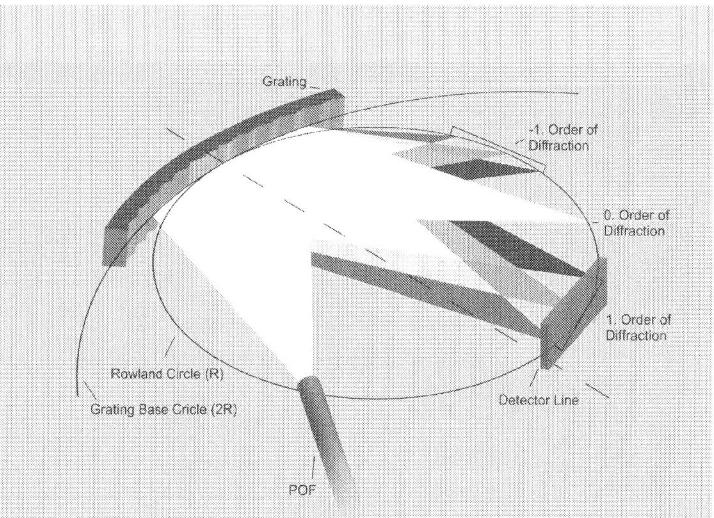

Fig. 13. Principle sketch of a Rowland Spectrometer.

Because the grating is attached to a bended basement only one element can cover both functions, the focusing and the diffracting. Hence the light is not afflicted with any aberrations or attenuations of a focusing lens or other refractive elements, which are necessary for any other setup. The diffraction is split into different orders of diffraction. The first order (z = 1) is the important one for data transmission. The higher the grating constant g of the diffraction grating, the more accurate the shape of the maxima of the different diffraction orders. The diffracted light interferes positively on the detection layer for (Demtröder, 2008):

$$z\lambda = g(\sin\alpha + \sin\beta) \tag{8}$$

with α angle of incidence, β emergent angle and g the grating constant. The following figure illustrates this formula (Fig 14).

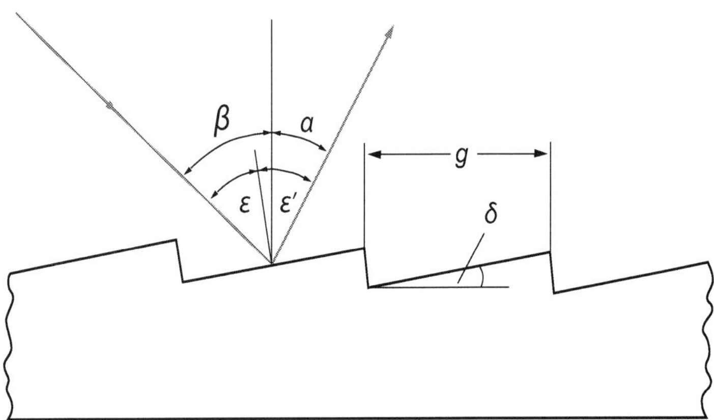

Fig. 14. Grating and behavior of light.

The resolution of the diffraction grating follows the Rayleigh Criterion and depends on the complete number of grating steps N and not on the grating constant (Hecht, 2009):

$$(\lambda/\Delta\lambda) \leq zN \tag{9}$$

This means for the first order of diffraction (z=1) and a number of grating steps N = 3000 (300 lines/mm) that the resolving power is $\Delta\lambda$ = 0.196 nm for λ = 589 nm.
One other characteristic of key elements for POF communication is the three dimensional approach. Key elements of glass fiber communication are usually designed planar. This simplification cannot be adopted for POF communication, because of POF's large numerical aperture and therefore large angle of beam spread.

4.4.1 Results of the simulation
In the following steps, a software program is used to design a demultiplexer based on the general concept outlined above. For the current task, the software OpTaLiX, which is based on the raytracing method, provides all needed functionalities (Blechinger, 2008; Hecht, 2009; Demtröder, 2008). This approach offers different advantages, it is easy to design, analyze and evaluate the simulated results. Also, effective improvements of the configuration can be simulated fast.

4.4.2 Results of the simulation for different line densities
In figure 15, the 2D plot for the reference wavelength (520 nm) of the demultiplexer with an ellipsoid mirror and grating is shown. The multicolored light is emitted by a polymeric fiber. It hits the mirror, where it is focused and diffracted in its monochromatic parts. The light is focused onto a POF- or detector-array.

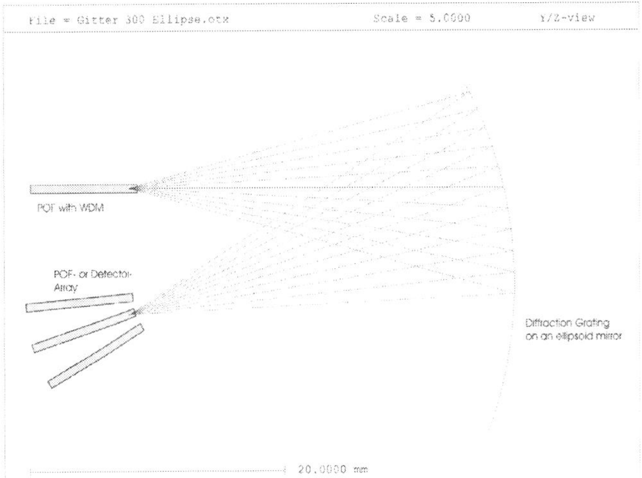

Fig. 15. 2D plot of the demultiplexer.

Without a grating, a perfect point to point mapping (without any aberrations) is possible with an ellipsoid mirror because of the two foci, but there is no separation of the different channels. With a grating stamped on the mirror, the separation of the multicolored light in its monochromatic parts is possible. But this grating distorts the optical path of light dramatically. The first change is the gap of the different colors in the image layer (here the POF- or Detector Array) increases with the line density of the grating. This can be noticed for an ellipsoid mirror (Fig. 16) and for a spherical mirror (Fig. 17) as well. The spherical mirror has the advantage, that the shape can be produced for injection molding easier.

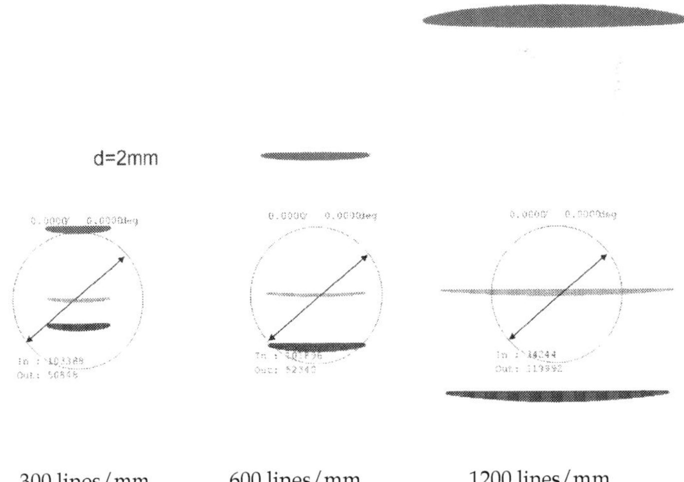

Fig. 16. 2D Plot of the demultiplexer with an ellipsoid mirror and different line densities.

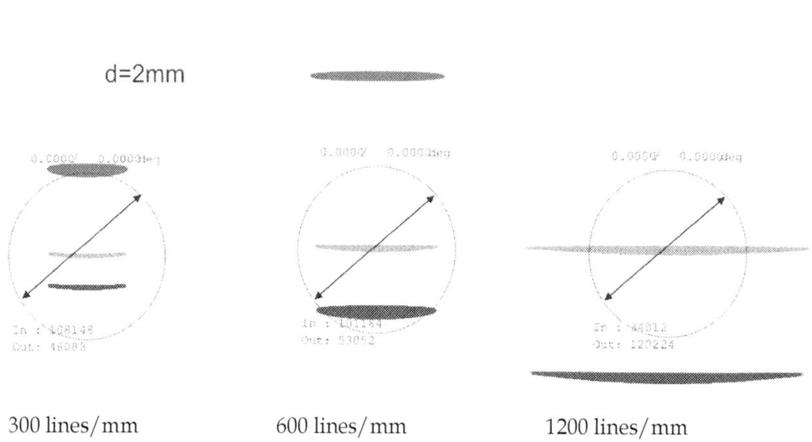

Fig. 17. 2D Plot of the demultiplexer with a spherical mirror and different line densities.

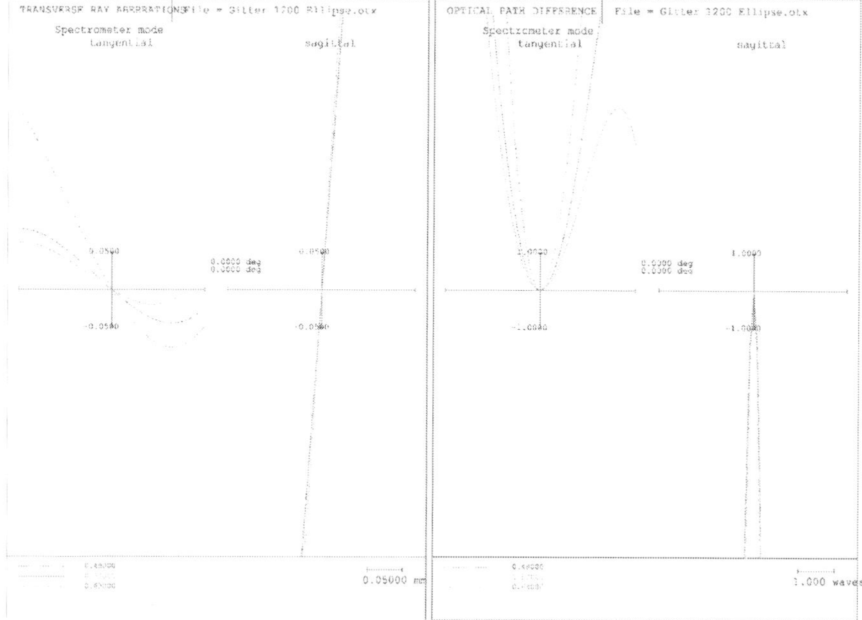

Fig. 18. TRA and OPD for the ellipsoid demultiplexer with 1200 lines/mm.

The second changes are the great aberrations especially for the demultiplexer high line density. To underline this result and to analyze the aberrations in detail, the transverse ray aberration (TRA) and the optical path difference (OPD) in spectrometer mode are shown in figure 18 for the demultiplexer with an ellipsoid mirror and 1200 lines/mm. The chief ray coordinates are irrespective for the TRA and OPD to overlap the different colors.

The TRA shows a slight defocusing for the meridional section, but a very strong defocusing for the sagittal section. The graph of the function in the meridional section exhibits a predominant third order Seidel coefficient. Therefore the slight defocusing in the meridional section compensates the astigmatism. The OPD shows as expected strong deviation from the ideal waveform especially in the sagittal section. This defocusing leads to high losses for the coupling efficiency for the POF- or detector- array in the image layer.

It is obvious that the grating changes the focal length especially of the sagittal section; therefore the shape of the mirror must be improved. It is necessary to change the radius of curvature notable in the sagittal section. Hence the basic shape of the mirror is not longer a sphere or ellipsoid. To meet the demands a higher order shape, which is nearly cylindrical, is used.

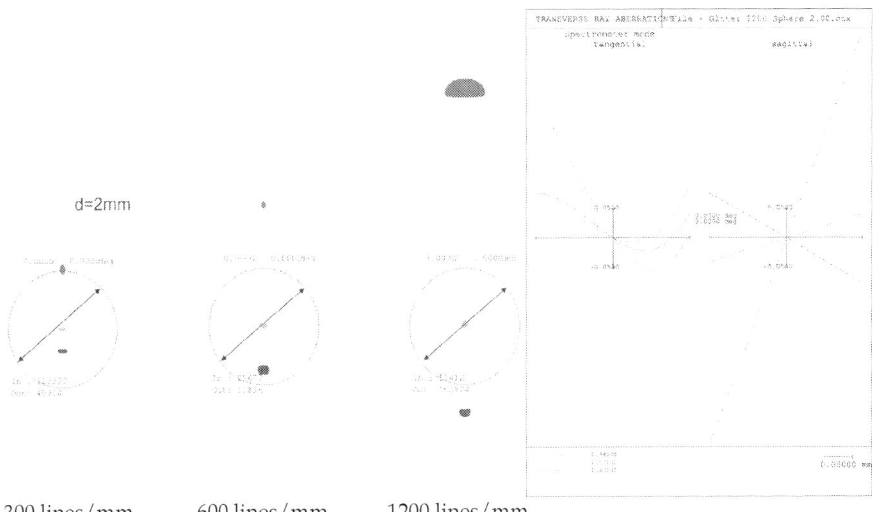

300 lines/mm   600 lines/mm   1200 lines/mm

Fig. 19. Spot Diagram and TRA for the improved DEMUX.

The change of the mirror shape improves the imaging quality substantial. The spot diagram and the TRA for the improved demultiplexer are shown in Fig. 19.

The spot diagram shows three dividable colors. The gap between every color is larger than 2 mm. The TRA shows a marginal shift of the focus of all wavelengths to offset the astigmatism in the meridional section. Because of the spectrometric function of the demultiplexer it is not possible to focus all three colors simultaneously. There is always a combination of over and under correction for the different colors. Hence the radius of the mirror in the sagittal section is optimized to focus the colors completely as much as possible.

This improved demultiplexer can separate three colors with enough space between them to regain the information with a POF- or detector-array. The shapes of the foci feature low coupling losses and the shape of the mirror is easy to produce in injection molding.

4.5 WDM reference comb

Different analyses will be shown in the full chapter including TRA and OPD. The way to the optimized setup will described in detail. Further on a first attempt to standardize the different wavelengths in visible spectrum will be discussed. As described in the previous chapter, WDM has a great chance to expand the overall bandwidth of POF transmission systems. Therefore it is necessary to standardize the WDM channels in frequency or wavelength, based on proven the glass fiber system channel allocation map of ITU recommendation G.694.2-1/2.

4.6 Optical channel allocation map proposal for POF

The usable transmission window in the visible spectrum of POF is located between 400 – 700 nm, which leads to the possible optical bandwidth of 300 nm for POF. Now the ITU proposes a frequency allocation map for WDM in its recommendation G 694.2. Assuming the correlation between wavelength and frequency of electro magnetic waves

$$\lambda\, f = c_{vac} \tag{10}$$

where f = frequency λ = wavelength and c_{vac} = speed of light in vacuum. Calculating the equivalent frequencies to 400 nm and 700 nm will result in 750 THz and 461.5 THz combined with a bandwidth of 320 THz for true WDM transmission. Additionally, a so-called anchor frequency (for glass fiber systems 193.1 THz) will be proposed as 750 THz. The possible transmission windows for WDM channels are dependent on the attenuation of the PMMA based standard SI POF. A possible transmission channel at 490 THz (610 nm) must be omitted because of the attenuation of the OH-Peak at 610 nm at that frequency (see Fig. 20).

The region of low attenuation of less than 90 dB/km is apportioned between 510 THz and 750 THz. In total 9 WDM channels, which are listed in Tab. 3, can be fixed with channel spacing of 40 THz.

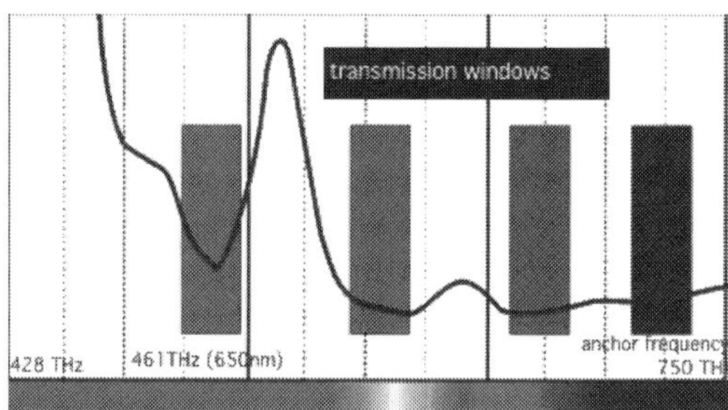

Fig. 20. Proposed optical frequency comb window.

The channel frequencies are calculated by:

$$f_{ch} = 750\text{THz} - n\,(40\text{THz}) \tag{11}$$

where n is the number of the channel.

In total a WDM system data rate of 9 x 2.5 Gbit/s = 22.5 Gbit/s seems to be possible assuming the todays data rates of POF systems using GI POF for transmission medium for a transmission length of 100 m. In this proposal for a new international POF WDM grid, most of the channels are located in the short wavelength region where the attenuation of POF is lower than in the long wavelength region. On the other hand, the "old red" window is already included at channel no 7: f_7 = 470 THz (638 nm).

In Fig. 21 a schematic view of the optical band pass behavior of the WDM filters of the DEMUX/MUX devices are depicted. Supposing a typical X-talk suppression of 30 dB for optical channel separation, a 3 dB filter width of 20 THz for each filter is needed.

Channel	Frequency / THz	Wavelength / nm
0	750	400
1	710	423
2	670	448
3	630	476
4	590	508
5	550	545
6	510	588
7	470	638
8	430	698

Table 3. Proposed optical frequency channels.

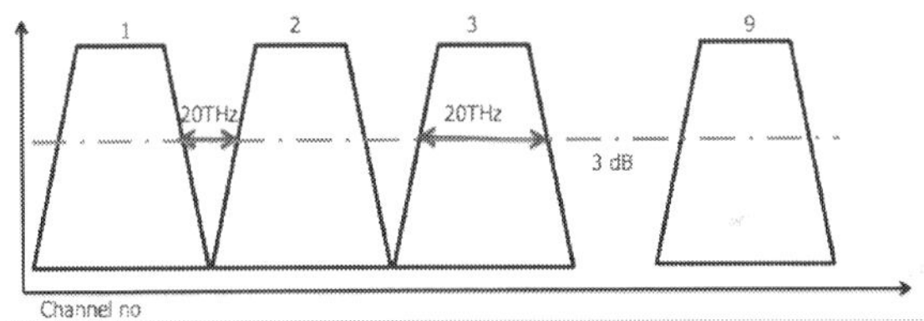

Fig. 21. Proposed optical frequency comb window.

5. Outlook

The simulation results show, that it is possible to build up a mass production convenient demultiplexer for polymeric fiber systems by means of a diffraction grating. A special shape of the mirror is needed to suppress most of the aberrations which results of the grating. The improved demultiplexer can separate all three colors with a gap of 2 mm and crosstalk lower than 30 dB. This demultiplexer has the chance to break through the limitation of standard POF communication also with broad range of usability in optical spectroscopy for sensor systems in automotive and medical applications due to its low cost realization. It can be implemented in combination with all in the market existing POF types like SI POF, MC POF or GI POF with 1 mm outer diameter and a Numerical Aperture of 0.3 – 0.5. The high number of modes in the fibers gives no restriction to optimal function the developed multiplexer for WDM transmission of minimum three different wavelength channels. In the future the device will be extended to multiplex 8 channels.

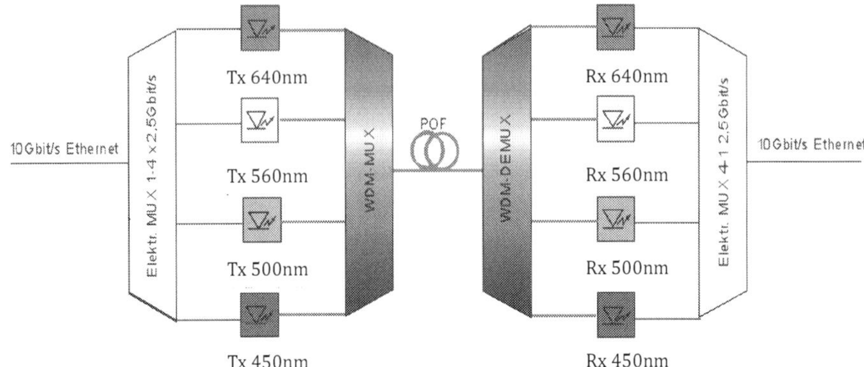

Fig. 22. Possible 10 Gbit/s via POF using WDM technique.

In the next years WDM over POF will expand the total bandwidth of POF transmission systems up to more than 20 Gbit/s. A channel allocation map for 9 WDM channels in the visible range is proposed as an input for the international standardisation organisations IEC and ITU to define a new optical reference standard for POF WDM systems like the ITU recommendation for glass fiber systems G 594.

Soon it would be possible to transmit 10 GbEthernet data via SI POF with the help of the here described WDM over POF technology, shown in Fig. 22. The electrical data stream of 10 GbEthernet will be electrically demultiplexed to four sub data streams of 2.5 Gbit/s. Each of this sub streams will modulate an optical laser diode source with different WDM wavelengths. In a WDM MUX all four colored signals will be combined to be transferred simultaneously via the POF fiber link up to 100 m. At the receiver side an optical DEMUX will spread the optical channels to dedicated photo diodes. The out coming electrical 2.5 Gbit/s data can be electrical multiplexed to the full 10 Gbit/s Ethernet bit stream at the output side.

6. References

Blechinger, H. (2008). *OpTaliX: Optical Design Software*, Available from
       http://www.optenso.com

BMW. (n.d). What is byteflight?, Mar. 2011, Available from
       http://www.byteflight.com/whitepaper/index.html

Chen, R. T. (2000). WDM and Photonic Switching Devices for Network Applications. *Proceedings of SPIE*, Vol. 3949

Cherian, S.; Spangenberg, H. & Caspary, R. (2010). Vistas and challenges for polymer optical fiber in commercial aircraft, *Proccedings of the 19th POF Conference*, Yokohama, Japan, Oct. 19-21, 2010

Colachino, J. (2001). *Mux/DeMux Optical Specifications and Measurements*, Lightchip Inc. white paper, Lightreading Daishing POF Co., Ltd. (n.d). The application of POF in automobile industry, Mar. 2011, Available from
       http://www.dspof.com/en/support-pgdetail-88.html

Daum, W.; Krauser, J.; Zamzow, P. E. & Ziemann 0. (2007). *POF – Polymer Optical Fibers for Data Communication*, Springer, ISBN 3-540-42009-6, Germany

Demtröder, W. (2008). *Experimentalphysik 2: Elektrizität und Optik*, Springer, ISBN 978-3540682103, Germany

DIN EN 50173-1 Information technology - Generic cabling systems - Part 1: *General requirements; German version EN 50173-1:2007/prAB:*, 2009

DIN EN 50173-2 Information technology - Generic cabling systems - Part 2: *Office premises; German version EN 50173-2:2007/prAA:*, 2009

DKE 412.7.1 Standardization Group of POF Gbit/s systems in Germany, April 2011, Available from http://www.vde.com/en/dke/std/projects/POF/Pages/default.aspx

ETSI TS 105 175-1 V1.1.1 Technical Specification *Access, Terminals, Transmission and Multiplexing (ATTM); Plastic Optical Fiber System Specifications for 100 Mbit/s and 1 Gbit/s*, (2010)

Fischer, U. H. P. (2002). *Optoelectronic Packaging*, VDE Verlag GmbH, ISBN 3-8007-2572-X, Germany

Fischer, U. H. P. & Haupt, M. (2007). WDM over POF: the inexpensive way to breakthrough the limitation of bandwidth of standard POF communication, *Proceedings of SPIE*, Vol. 6478

Fischer, U.H.P., M. Haupt & Kragl, H. (2007). Conceptual Design of an Inexpensive POF Demultiplexer, *Proceedings of SPIE*, Vol. 6593

Fraunhofer Institute for Integrated Circuits. (2008). Optical multiplexer for short range communication, Aug. 2008, Available from
http://www.iis.fraunhofer.de/ec/oc/download/demux.pdf

Gaudino, R.; Bosco, G.; Bluschke, A.; Hofmann, O.; Kiss, N.; Matthews, M.; Rietzsch, P.; Randel, S.; Lee, J. &

Breyer, F. (2007). On the ultimate capacity of SI-POF links and the use of OFDM: recent results from the POF-ALL project, *Proceedings of the 16th POF Conference*, Turin, Italy, Sept. 11-13, 2007

Gaudino, R.; Nocivelli, A.; Kragl, H.; Ziemann, O.; Weber, N.; Jaeger, D.; Koonen, T.; Lezzi, C.; Bluschke, A. & Randel, S. (2007). Latest results from the POF-ALL EU Project: Toward Improved Capacity over Large-Core Plastic Optical Fiber, *Proceedings of BBeurope Conference 2007*, Antwerp, Belgium, Dec. 3-6, 2007

Gnauck, A. H.; Chraplyvy, A. R.; Tkach, R. W.; Zyskind, J. L.; Sulhoff, J. W.; Lucero, A.J., et. al. (1996). One terabit/s transmission experiment, *Proceedings of OFC'96*, San Jose, California, USA, Feb. 25 – Mar. 1, 1996

Grzemba, A. (2008). *MOST: The Automotive Multimedia Network*, Franzis Verlag GmbH, ISBN 978-3-7723-5316-1, Germany

Hecht, E. (2009). *Optik*, Oldenbourg, ISBN 978-3486588613, Germany

IEEE 802.3: "Telecommunications and Information Exchange Between Systems - Local and Metropolitan Area Networks - Specific Requirements Part 3: Carrier Sense Multiple Access with Collision Detection (CSMA/CD) Access Method and Physical Layer Specifications Amendment: Physical Layer Specifications and Management Parameters for 10Gb/s Passive Optical Networks".

ITU-T Recommendation G.694.2, *Spectral grids for WDM applications: CWDM wavelength grid* (2011)

ITU-T Recommendation G.694.1, *Spectral grids for WDM applications: DWDM frequency grid* (2011)

KDPOF. (2010). Demonstration of 1Gbps over 50m of low cost SI-POF with KDPOF technology, 02.03.2011, Available from http://www.kdpof.com/Papers.html

Keiser, G. (2000). *Optical fiber communications* (3rd ed.), McGraw-Hill series in electrical engineering, communications and signal processing, ISBN 0-07-232101-6, Singapore

Koonen, A.M.J.; Boom, H.P.A.; Ng'oma, A.; Garcia Larrode, M.; Zeng, J. & Khoe, G.D. (2005). POF Application in Home Systems and Local System, *Proceedings of the 14th POF Conference*, pp. 165-168, Hong Kong, Sep. 19-22, 2005

Koonen, A.M.J.; Pizzinat, A.; Jung, H.-D.; Guignard, P.; Tangdiongga, E. & Boom, H.P.A. (2009). Optimisation of In-Building Optical Networks, *Proceedings of ECOC 2009*, Vienna, Austria, Sept. 20-24, 2009

Last, A. & Mohr, J. (2003). *Fehllicht in LIGA-Mikrospektrometern*, Forschungszentrum Karlsruhe Wiss, Berichte

Loquai, S.; Kruglov, R.; Vinogradov, J.; Ziemann, O.; Swoboda, R. & Atef, M. (2010). Comparison of DMT, PAM and NRZ with optimized receiver bandwidths, *Proccedings of the 19th POF Conference*, Yokohama, Japan, Oct. 19-21, 2010

Lubkoll, J.; Strauss, U.; Pollakis, N. & Strobel, O. (2008). *FlexRay with polymer-clad-silica fiber as transmitting medium in aviation electronics*, ISBN 978-1-4244-3484-8, Marrakech, Dec. 11-13, 2008

Marcou, J. (1997). *Plastic Optical Fibers: Practical Applications*, John Wiley & Sons, ISBN 0471956392, France

MOST Cooperation. (July 2010). Appendix D: Frame Structure and Boundary (informative), In: *MOST Specification Rev. 3.0 E2*, MOST Cooperation, pp. 211-215, MOST Cooperation, Retrieved from
http://www.mostcooperation.com/publications/Specifications_Organizational_Pr ocedures/index.html?dir=291

Nalwa, H. S. (2004). *Polymer Optical Fibers*, American Scientific Publishers, ISBN 1-58883-012-8, USA

Park, Optical network evolution in Korea Park C-W. (Optimedia 2006) available in
http://www.pofac.de/downloads/itgfg/fgt21/FGT21_Oldenburg_Park_GI-POF.pdf, April 2011

Poisel, H.; Ziemann, O. & Klein. K.-F. (2003). Trends in polymer optical fibers. *Proceedings of SPIE*, Vol. 5131, pp. 213-219

Randel, S.; Lee, J.; Spinnler, B.; Breyer, F.; Rohde, H.; Walewski, J.; Koonen, A. M. J. & Kirstädter, A. (2006). 1 Gbit/s transmission with 6.3bit/s/Hz spectral efficiency in a 100m standard 1 mm step-index plastic optical fiber link using adaptive multiple sub-carrier modulation, *Proceedings of ECOC 2006*, Postdeadline Paper, Cannes, France, Sept. 24-28, 2006

Senior, M. J. (1992). *Optical fiber communications: principles and practice* (2nd ed.), Prentice Hall International series in optoelectronics, ISBN 0-13-635426-2, Great Britain

Strobel, O.; Rejeb, R. & Lubkoll, J. (2010). Optical polymer and polymer-clad silica fiber data buses for automotive applications, *Proceedings of 2010 7th IEEE & IET International Symposium on Communication Systems, Networks and Digital Signal Processing (CSNDSP)*, pp. 693-696, ISBN 978-1-4244-8858-2, Newcastle, United Kingdom, July 21-23, 2010

Voges, E. & Petermann, K. (2002). *Optische Kommunikationstechnik*, Springer, ISBN 3-540-67213-3, Germany

Wikipedia. (2011). MOST-Bus, Apr. 2011, Available from
http://de.wikipedia.org/wiki/MOST-Bus

Ziemann, O.; Poisel, H.; Randel, S.; Lee, J. (2008). Polymer optical Fibers for short, shorter and shortest data links, *Proceedings of OFC/NFOEC 2008, San Diego, California, USA, Feb. 24-28, 2008*

Ziemann, O.; Daum, W.; Krauser, J.; Zamzow, P. E. (2008). *POF Handbook: Optical Short Range Transmission Systems*, Springer, ISBN 978-3540766285, Germany

Ziemann, O. (2010). The (hard) Way to a Standard POF Gigabit Interface, *Proceedings of the 19th POF Conference*, Yokohama, Japan, Oct. 19-21, 2010

23

Transfer Over of Nonequilibrium Radiation in Flames and High-Temperature Mediums

Nikolay Moskalenko, Almaz Zaripov, Nikolay Loktev,
Sergei Parzhin and Rustam Zagidullin
Kazan State University of Power
Russia

1. Introduction

Throughout the XX-th century intensive development was received by the high technologies intended for maintenance of stable rates of economic development and global competitive capacity in key industries of manufacture. The contribution of scientific and technical progress in economic growth becomes solving. Now in the developed countries development of high technologies has passed to a stage of the scientific and technical policy directed on introduction of high technologies in sphere of information services, medicine, ecology, power, military-technical manufacture, control of safety of economic activities in any branches of manufacture. Thus the power remains live-providing, a key economic branch in economy of any country and its development should be carried out by advancing rates. On the other hand, the power is a branch in which new scientific and technical achievements take root with high degree of efficiency owing to high level of automation of manufacture and energy transportation.

In the present chapter of the monography basic aspects of a problem of the transfer over of radiation in high-temperature mediums and flames and their decision with reference to problems of remote diagnostics of products of combustion in atmospheric emissions and top internal devices are considered. The special attention is given the account of nonequilibrium processes of radiation which are caused by chemical reactions at burning fuels and photochemical reactions in atmosphere. Radiation of high-temperature mediums is selective in this connection the problem of numerical modeling of spectraradiometer transfer function of atmosphere for non-uniform selective sources of radiation which are flame, combustion products of fuel, torches and traces of aerocarriers, combustion products in top internal chambers is considered. Absence of sharp selection of a disperse phase creates possibility of division of radiation of disperse and gas phases and in the presence of the aprioristic information creates conditions of their remote diagnostics (Moskalenko et al., 2010). The developed measuring complexes (Moskalenko et al., 1980a, 1992b) have allowed to specify substantially the information received earlier under radiating characteristics of products of combustion (Ludwig et al. 1973) and to investigate nonequilibrium processes of radiation in strictly controllable conditions of burning (Kondratyev et al., 2006, Moskalenko et al., 2007a, 2009b, 2010c). The developed two-parametrical method of equivalent mass for functions spectral transmission gas components of atmosphere (Kondratyev, Moskalenko, 1977) has successfully been applied in calculations of radiating heat exchange in high-temperature mediums (Moskalenko, Filimonov, 2001; Moskalenko et al., 2008a, 2009b). The method of numerical modeling of functions spectral transmission on parameters of spectral lines has been used by us for calculations of the transfer over of radiation of torches and traces of aerocarriers in atmosphere and at the decision of return problems of diagnostics of products of combustion by optical methods (Moskalenko & Loktev, 2008, 2009; Moskalenko et al., 2006). Experimental researches of speed radiating cooling a flame are executed by means of calculation of structure of products of combustion (Alemasov et al., 1972) and modeling of radiating heat exchange in chambers of combustion of measuring complexes with control of temperature of a flame by optical methods (Moskalenko & Zaripov, 2008; Moskalenko & Loktev, 2009; Moskalenko et al., 2010).

Measurements of concentration of oxides of nitrogen in flames have shown that their valid concentration much lower in comparison with the data of calculations (Zel'dovich et al., 1947). There was a necessity of finding-out of the reasons causing considerable divergences of theoretical calculations and results of measurements of concentration NO in flames. The reason strong radiating cooling of flames which didn't speak only equilibrium process of their radiation demanded finding-out.

Processes of burning gaseous, liquid and firm fuel have great value in power, and also in technological processes of various industries. At present a principal view of burned fuel in the European territory is gaseous fuel. Partially it is caused by ecological norms and requirements to combustion products. Use of gaseous fuel conducts to reduction of capital expenses at building of thermal stations and boiler installations owing to an exception of expensive filters of clearing of the list of the equipment of station. High heat-creation ability of gas fuel at low operational expenses provides high efficiency of power installations as a whole. A low cost of transportation at use of gas fuel provides its competitiveness in the market. Decrease in losses of heat at its transportation demands creation of small-sized boilers with high efficiency, high thermal stress of top internal space at the raised efficiency that leads to search of optimum design decisions by working out of power installations. Development of rocket technics, creation of space vehicles of tracking their start and support, optimization of systems of detection and supervision demands the data about structural characteristics of torches both spectral and spatial distribution of their radiation which can be received by correct methods of the decision of problems of a transfer over of radiation and radiating heat exchange in the torch. All it has demanded performance of complex researches of processes of radiation at burning and its the transfer over to medium which are discussed more low.

2. Radiating characteristics gas optically active components

Experimental researches radiating optically active components in a range of temperatures $220 \geq T \geq 800K$ have been begun in 1964 for the purpose of reception of the initial data for modeling of radiating heat exchange and spectral and spatial structure of radiation natural backgrounds of the Earth and atmosphere and anthropogenous influences on climate change (Kondratyev & Moskalenko, 1977; Kondratyev et al., 1983; Kondratyev & Moskalenko, 1984). The developed measuring complexes allowed to measure spectra of molecular absorption at pressure from 10^{-3} atm. to 150 atm. That has allowed to parameterized functions of spectral transmission of atmospheric components in a spectral range $0,2 \div 40$ μm at the average spectral permission $\Delta \nu = 2\text{-}10$ cm^{-1}, for atmospheres of the

Earth and other planets. Other direction of researches of radiating characteristics of products of combustion fuels developed in parallel with the first and for the known reasons is poorly reflected in publications. Further we will stop on the analysis of results of researches of radiating characteristics of ingredients a gas phase of products of combustion in a range of temperatures 600÷2500K.

2.1 Measuring devices and results of experimental researches

For the decision of many applied problems connected with the transfer over of radiation of a flame in atmosphere and radiating heat exchange in power installations, data on spectral radiating ability of the various gas components which are products of combustion of flame are required. Independent interest is represented by researches of influence of temperature on formation of infra-red and ultra-violet spectra of absorption or radiation of gas components. Depending on a sort of research problems of spectra of absorption or radiation of gas mediums of measurement it is necessary to carry out or with the average permission Δ =5-20 cm-1, or with the high permission $\Delta\leq$ 0,2 cm-1. In the latter case it is possible to measure parameters of spectral lines and to receive the important information on the molecular constants characterizing vibrational – rotary and electronic spectra of molecules (Moskalenko et el., 1972, 1992). In a range of temperatures 295÷1300 K research of characteristics of molecular absorption it was carried out with use the warmed-up multiple-pass ditches (Moskalenko et el., 1972). Other installation (Moskalenko et el., 1980) allowed to investigate as spectra of absorption and radiation of gases in hydrogen-oxygen, hydrogen-air, the propane-butane-oxygen, the propane-butane-air, methane-oxygen, methane-air, acetylene-oxygen, acetylene-air flames in the field of a spectrum 0,2÷25 µm at temperatures 600÷2500 K, and also to investigate characteristics of absorption of selective radiation of a flame modeled atmosphere of the set chemical composition. Besides, any other component can be entered into a flame, of interest for research.

The Block diagram of experimental installation and design of a high-temperature gas radiator is described (Moskalenko et el., 1972). It includes the lighter, high-temperature absorbing (radiating) to a ditch, system of input of investigated gas and control of their expense, optical system of repeated passage of radiation in a ditch under White's scheme, the block of the gas torches forming two counter streams of a flame in quartz ditch with the heat exchanger for decrease radiating cooling of a flame, coordinating optical prefixes for radiation designing on an entrance crack of spectrometers of reception-registering system with replaceable receivers of radiation PEA – 39A, PEA – 62, BSG – 2, cooled photodetectors with sensitive elements PbS, PbTe, GeCu, GeZn, GeAu, GeAg, germanium bolometer. The spectrum of radiating ability of the high-temperature gas medium is defined by tariroving of a spectrometer on radiation of absolutely black body or normalizing radiation sources. Radiation falling on a reception platform is modulated by the electromechanical modulator with frequency of 11 or 400 Hz (in case of work with PEA and photodetectors). Registration of spectra of radiation was made by spectrometer IRS – 21 or the spectrometers of the high permission collected on the basis of monochromators MDR – 2, DPS – 24, SDL – 1. The last are completed with replaceable diffraction lattices with number of strokes 1200, 600, 300, 150, 75 and the cutting off interferential optical filters providing a working spectral range 0,2 $<\lambda$ <25 µm. The limit of the spectral permission of spectrometers made 0,1÷0,2 cm-1. Spectral radiating ability of the gas medium

$$\varepsilon(v,T) = G(v) \cdot \frac{N^0(v,T_\beta)}{B(v) \cdot N^0(v,T)} , \qquad (1)$$

where T – temperature of the investigated gas medium; G (v), B (v) – recorder indications at registration of radiation from the gas medium (flame) and absolutely black body (ABB); N^0(v, T_v) and N^0(v, T) – spectral brightness ABB at temperatures T_v ABB and T the investigated gas medium.

At work in a mode of absorption of not selective radiation by a flame the radiation modulated by the electromechanical modulator from the lighter is registered. Not modulated radiation of the flame by reception system isn't registered. In the lighter as radiation sources SI lamps – 6 – 100, DVS – 25, globar and ABB with temperature 2500K are used. Radiation from these sources, promodulated by the electromechanical modulator, by means of optical system of the lighter goes in high-temperature absorbing gas to a cell which optical part is collected under White's scheme. The thickness of the absorbing component can change by increase in an optical way at the expense of repeated passage of a beam of radiation between mirrors of system of White. The maximum thickness of the absorbing medium can reach 16 m.

Absorbing (radiating) a cell represents the device executed in the form of established in heat exchanger along an optical axis of the cell two mobile pipes, made of quartz. On a circle of entrance cavities from end faces quartz ditches are located two systems of gas torches (on 6 pieces in everyone) for reception of the hot absorbing (radiating) medium. The internal cavity is filled with two counter streams of a flame. Combustion products leave through a backlash between mobile quartz pipes, the heat exchanger and two unions, located at its opposite ends. Investigated gases can be both combustion products, and other gases entered in a cell and warm flame. For flame creation two various systems of torches are used.

At work about hydrogen-oxygen (hydrogen-air) a flame are used torches of Britske, each of which allows to receive a flame of diffusion type. We will remind that under diffusion flame such flame for which fuel and an oxidizer are originally divided is understood. Fuel and an oxidizer mix up or by only diffusion, or partially by diffusion and partially as a result of turbulent diffusion. For reception the propane-butane-oxygen, the propane-butane-air flame hot-water bottles have been designed and made, each of which allows receiving a flame of Bunsen's type. The flame of Bunsen's type is understood as a flame of preliminary mixed oxidizer and fuel.

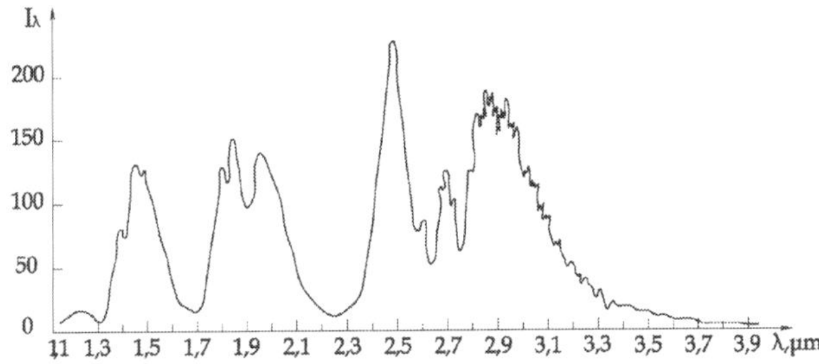

Fig. 1. Radiative spectrum of the hydrogen – oxygen flame at temperature T≈2300K in the range 1,1-4 µm.

Each torch has an adjustable angle of slope of an axis of a torch to an axis of the cell quartz in limits from 20 to 70°. Combustible gases are set fire by a spark. Change of temperature of a flame is reached by change stehiometrical parities of combustible gas and an oxidizer, and also change of combustible gas and oxidizer diluting by buffer gas. Temperature measurement is carried out W – Re and Pt – Po by thermocouples and optical methods.

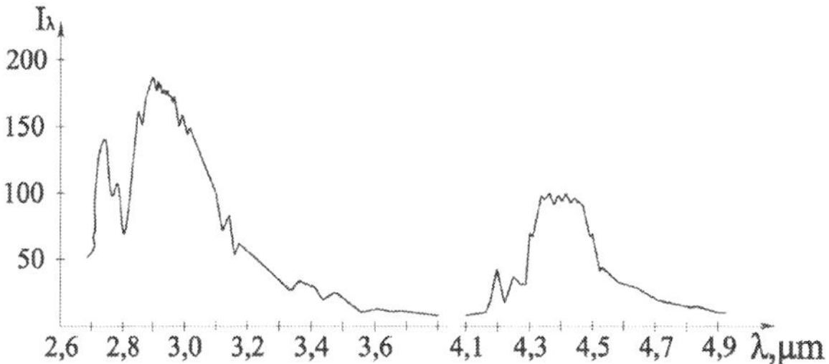

Fig. 2. Radiative spectrum of the hydrogen – oxygen flame in range 2,7 - 5 µm with addition CO_2 in quality of the research gas.

On fig. 1, 2 examples of records of spectra the radiations which have been written down by means of spectrometer IRS–21 are resulted at the average spectral permission at temperature T ≈ 2300K. For oxygen-hydrogen flame radiation bands only water vapor in a vicinity of bands 0,87; 1,1; 1,37; 1,87; 2,7 and 6,3 µm are observed. In ultra-violet spectrum areas are observed electronic spectra of radiation of a hydroxyl OH. With temperature growth considerable expansion of bands and displacement of their centers in red area is observed. At temperatures more 2000K in a flame absence of "windows" of a transparency of a flame, spectral intervals with radiating ability close to zero is observed.
At addition in a flame of gases from a number limit hydrocarbons (methane, ethane, etc.) In radiation spectra bands of carbonic gas (2; 2,7; 4,3; 15 µm) are observed. The similar picture is observed at introduction in a flame and purely carbonic gas. At introduction in flame NO the spectrum of the basic band 5,3 µm NO and a continuous spectrum of radiation NO_2 in a range from 0,3 to 0,8 µm is observed. Data processing of measurements of spectra of radiation of a flame and restoration of a profile of temperature along an axis of an ardent radiator has shown appreciable temperature heterogeneity in zones of an input of a flame in the combustion chamber (Moskalenko & Loktev, 2009) which is necessary for considering at definition of dependence of radiating characteristics of separate components from temperature. This lack has been eliminated in working out of a measuring complex of the high spectral permission (Moskalenko et el., 1992) for research of flames. On working breadboard models of this installation and the experimental sample of this installation the most part of the spectral measurements taken as a principle of parameterization of radiating characteristics of gas components of products of combustion has been executed.
The spectral measuring complex described more low also is intended for registration of spectra of radiation of flames and spectra of absorption of radiation by a flame at the high spectral permission in controllable conditions and has full metrological maintenance. On fig. 3 the block-scheme of this installation is presented. An installation basis make: the block of a high-temperature gas radiator, blocks of optical prefixes 2D-4, intended for increase in an optical way in an ardent radiator and the coordination of fields of vision of the lighter; the

block of a high-temperature radiator of sources of radiation 3 for absolute calibration of a spectrum of radiation of a flame and the Fourier spectrometer of high spectral permission FS – 01. Management of experiment and data processing of measurements by means of software on the basis of measurement-calculation complex IVK – 3. The measuring complex functions in spectral area 0,2–100 µm. Registration of spectra is carried out by means of spectrometers FS – 01, SDL – 1.

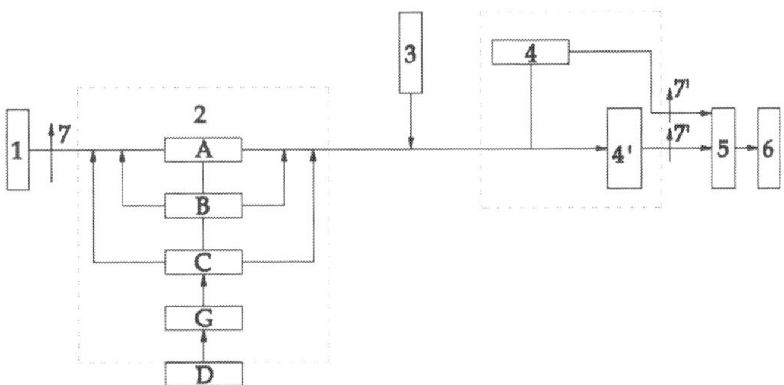

Fig. 3. The experimental installation scheme: 1 – illuminator, 2 – hightemperature gaseous radiator (A – lead – in of research gas system and contrac there expense, B – the mechanism of multiple passing ray thaw a flame, C – the gaseous burner of ascending flow of a flame, G – the gaseous provision system vacuum and control of gaseous expense, D – the system with a water circular pump); 3 – aradiative sources; 4, 4' – optical system for agreement of in trance and exit apertures; 5 – the reception – recording system; 6 – the system of atreatment of measuring data; 7, 7' – electrical mechanical modulators of radiation.

The high-temperature ardent radiator structurally represents the block of a gas radiator closed from above by the water cooled cap with two protective windows, stable in time. Formed at burning of gases flames have a squared shape with a size at the basis 40x20 cm². The torch design allows to investigate hydrogen – oxygen, hydrogen – air and hydrocarbonic flames. Measurements have shown that heterogeneity of a temperature field within a field of vision of optical system makes 3 %. Various variants of optical schemes together with system of repeated passage of radiation constructed under White's scheme, allows to investigate radiation spectra of flames and spectra of absorption of continuous radiation of a flame in a range of lengths of an optical way 0,2÷16 m. The flame temperature is measured by a method of the self-reference of spectral lines in lines of water vapor of bands 1, 38 and 1,87 µm. The average relative error of measurement of temperature of a flame makes ±2 %. Measurement of volume expenses of gases was carried out specially graduated rotameters RS – 5. On a parity of mass fuel consumption and an oxidizer the chemical composition of products of combustion are determined by thermodynamic calculation (Alemasov et al.., 1972). To absolute calibration of spectra of radiation of a flame are applied spectrameasured lamps SIRSh 8,5-200-1 and globar KIM, preliminary graduated on metrology provided standards.

Measurement of spectra of radiation and spectra of absorption of radiation by a flame allow to define spectral factors of nonequilibrium functions of a source of radiation in flames. Such measurements have revealed considerable nonequilibrium source functions in an ultra-violet part of a spectrum of a flame (the factor of nonequilibrium reaches values 20 – 100). At the same time vibrational-rotary spectra of radiation of water vapor in flames remain equilibrium. Nonequilibrium radiations OH in flames is strongly shown in an ultra-violet part of a spectrum and considerably influences radiative transfer over in flames and in vibrational-rotary bands v_1, $2v_1$, $3v_1$, where v_1 – frequency of normal fluctuation OH. The error of measurements of function of a source makes 30 % for an ultra-violet part of a spectrum and 7-10 % in infra-red bands of radiation of a flame. It is found out also nonequilibrium radiations in electronic bands of oxides of nitrogen.

At measurement in a mode of absorption of radiation the flame modulates radiation of the lighter 1. Nonmodulated radiation of a flame doesn't give constant illumination and isn't registered by receiving-registering system. Modulation of radiation of a flame is created by the modulator 7 '. Registration of spectra of radiation of flames in vibrational–rotary bands is carried out by Fourier spectrometer FS – 01 which reception module is finished for the purpose of use of more sensitive cooled receivers of radiation. The major advantage of the Fourier spectrometer in comparison with other spectrometers – digital registration of spectra with application of repeated scanning of spectra and a method of accumulation for increase in the relation a signal/noise. Prominent feature of Fourier spectrometer is discrete representation of the measured spectrum of radiation of a flame with the step equal to the spectral permission. The last has demanded working out of the software for processing of the measured spectra, restoration of true monochromatic spectral factors of absorption and parameters of spectral lines of absorption (radiation), their semiwidth and intensitys. With that end in view measured spectra are exposed to smaller splitting with step $\delta = \Delta/5$, where Δ – the spectral permission of the Fourier spectrometer. Value in splitting points is defined by interpolation.

Reduction of casual noise is reached by smoothing procedure on five or to seven points to splines in the form of a polynom of 5th degree. The spectrum of radiation received in a digital form is exposed to decomposition on individual components of lines.

From the restored contours of spectral lines it is easy to receive intensity and semiwidth of lines. Thus intensity such Lawrence's lines

$$S_m = \int_{-\infty}^{\infty} K_m(v)dv = K_{vm}\pi\alpha_m , \qquad (2)$$

where K_m - absorption factor in the center of a contour of a line, α_m - its semiwidth, K_{vm} - the restored contour of a spectral line. Thus the condition should be met

$$1 - \int_{-\infty}^{\infty} dv \exp\left[-k_m(v)w\right] = \int_{-\infty}^{\infty} A_{Im}(v)dv , \qquad (3)$$

where w - the substance maintenance on an optical way, A_{Im} - the measured function of spectral absorption of such line. Parameters of spectral lines of water vapor can be used for temperature control in a flame (Moskalenko & Loktev, 2008, 2009).

On fig. 4 the example of the measured spectrum of the high spectral permission of radiation of a flame for spectral area $3020 \div 3040$ cm^{-1} is resulted. On fig. 5, 6 spectra of radiating ability of a flame in vibrational–rotary bands of water vapor are illustrated at the average spectral permission Δv.

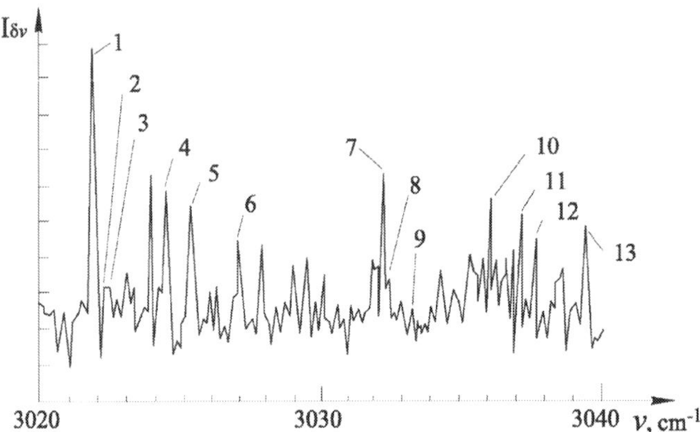

Fig. 4. The record of a high resolution radiative spectrum of the hydrogen – oxygen flame in the range 3020-3040 cm⁻¹. Centers of spectra lines: 1 – 3021,806 (v_1), 2 – 3022,365 ($2v_3$), 3 - 3022,665 ($2v_2$), 4 - 3024,369 (v_1), 5 – 3025,419 ($3v_2 - v_2$), 6 - 3027,0146 (v_1), 7 - 3032,141 (v_3), 8 - 3032,498 ($3v_2 - v_2$), 9 - 3033,538 ($3v_2 - v_2$), 10 - 3036,069 ($3v_2 - v_2$), 11 - 3037,099 ($3v_2 - v_2$), 12 - 3037,580 ($3v_2 - v_2$), 13 - 3039,396 (v_1) cm⁻¹.

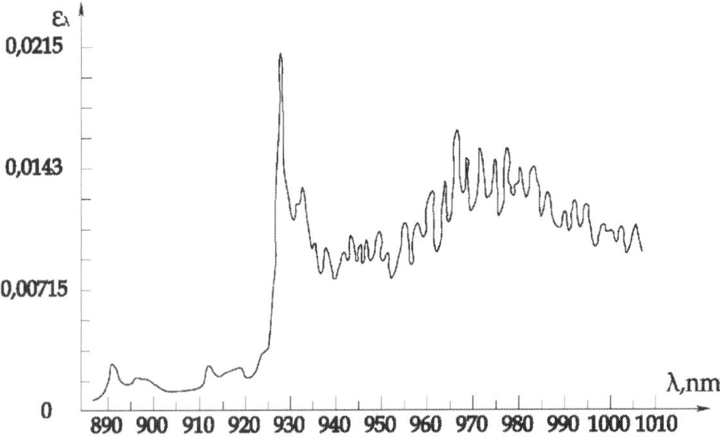

Fig. 5. Spectral emissivity of water vapor at T = 2400K in the band 0,96 μm. ω_{H2O} = 1,59 atm· cm STP, spectral resolution Δv = 10,6 cm⁻¹.

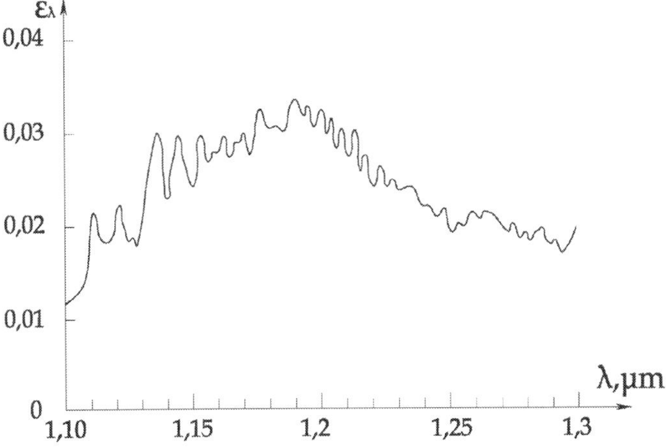

Fig. 6. Spectral emissivity of water vapor in the band 1,14 µm. T = 2400K, ω_{H2O} = 1,59 atm· cm STP, spectral resolution Δv= 15,5 cm^{-1}.

The spectra of radiation of the high spectral permission received in a digital form aren't calibrated on absolute size. Transition from values of relative spectral brightness to absolute radiating ability is carried out on parity

$$A_{\delta v} = \frac{I_{\delta v}\left(1-\bar{\tau}\right)}{\frac{1}{\Delta v}\int I_{\delta v} dv} , \tag{4}$$

where $\bar{\tau}$ – average value of function spectral transmission for the processed site of a spectrum Δv. Data on $\bar{\tau}$ have been received by us earlier for various products of combustion of flames. Further difficult function $A_{\delta v}$ it is decomposed to separate components, using a method of the differentiated moments, according to which

$$A(v) = \sum_{m=1}^{M} A_m \left[\sum_{n=0}^{N} A_{mn}\left(v - v_m^o\right)^n \right]^{-1} , \tag{5}$$

where A_m – a maximum of intensity of such line, A_{mn} - factors of the generalized contour.

$$q_m = \frac{1}{\sum_n A_{mn}\left(v - v_m^o\right)^n} , \tag{6}$$

Characteristics A_m give the full information on separate contours and are defined as decomposition factors abreast Taylor of some function f_m (v), describing such contour:

$$f_m(v) = \frac{1}{m} \sum_{n=0}^{N} A_{mn}\left(v - v_m^o\right)^n . \tag{7}$$

Value A_m is a maximum of amplitude of a contour. The center v_m^0 is defined from a condition of equality to zero of factor A_{m1}. Value of semiwidth of a line turns out from a parity

$$\alpha_m = \sqrt{\frac{\sqrt{A_{m2}^2 - 4A_{m4}} - A_{m2}}{2A_{m4}}}, \qquad (8)$$

Further the profiles received thus are restored on influence of hardware function of a spectrometer. So, we have separate contours of function of absorption $A_m(v)$ from which it is easy to pass to contours of factors of absorption $K_m(v)$:

$$K_m(v) = \sum_{m=1}^{M} K_m\left(v - v_m^0\right), \qquad (9)$$

where M – number of lines in a spectrum, m – line number. On fig. 7 the example of decomposition of function $A_{\delta v}$ on individual contours for oxygen-hydrogen of a flame for a spectrum site 3064÷3072 cm⁻¹, and also comparison (a curve 2) and calculated (a curve 3) on the restored contours of spectral lines of function $A_{\delta v}$ is presented. Integrated intensity of lines were defined from a parity (2). Detailed processing of spectra of radiation of water vapor in flames which has revealed many lines which were not measured earlier has been executed.

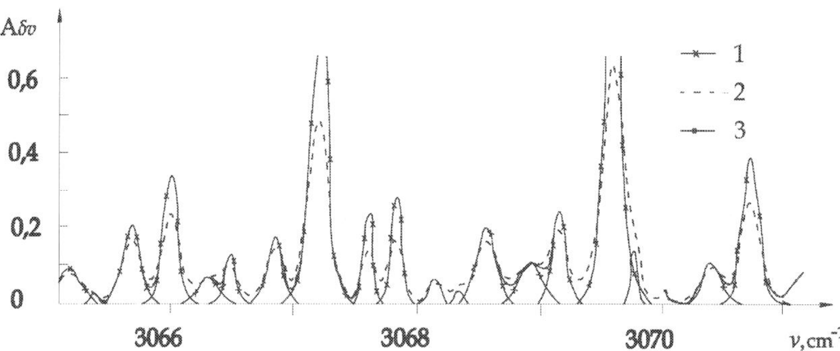

Fig. 7. The expansion of measuring function Aδv on individual contours. 1 – separate components of expansion, 2.3 – function Aδv measuring and calculative by reconstituting parameters of spectral lines accordingly.

In table 1 as an example parameters of spectral lines of water vapor are resulted at temperature T = 2100K for spectral ranges 3271÷3274 and 3127÷3130 cm⁻¹. Recalculation of parameters of lines on other temperatures can be executed under the formula

$$S(T) = S(T_0)\left(\frac{T_0}{T}\right)^{1.5} \frac{Q(T_0)}{Q(T)} \exp\left[-1.439E'\left(\frac{1}{T} - \frac{T}{T_0}\right)\right]. \qquad (10)$$

Statistical sum $Q(T)$ in the ratio (10) is calculated in harmonious approach. That circumstance pays attention that the centers of spectral lines measured at temperatures of spectral lines $T = 2100$ K and temperature $T_0 = 1000$ K don't coincide that is possible, is caused by the displacement of spectral lines caused by pressure, and also temperature displacement of lines. These distinctions in position of the centers of spectral lines surpass often an error of measurements of the centers of lines which in our experiments makes $\pm 0,02$ cm^{-1}. The measured semiwidth of spectral lines of water vapor basically will be coordinated with results of calculations under the theory of the Anderson, executed by us at temperatures $300 \div 3000$ K.

ν , cm^{-1}	S, $atm^{-1}cm^{-1}$	α, cm^{-1}	ν , cm^{-1}	S, $atm^{-1}cm^{-1}$	α, cm^{-1}
3271,731	0,0131	0,075	3127,8714	0,0123	0,129
3271,944	0,00642	0,084	3128,115	0,0015	0,075
3272,101	0,01272	0,080	3128,395	0,0042	0,081
3272,395	0,00408	0,066	3128,600	0,00216	0,076
3272,654	0,00876	0,168	3128,806	0,00277	0,083
3272,811	0,0114	0,080	3129,109	0,00498	0,092
3273,041	0,0236	0,111	3129,273	0,00387	0,091
3273,436	0,033	0,099	3129,589	0,0154	0,105
3273,735	0,0261	0,092	3129,941	0,0130	0,104

Table 1. Parameters of lines of water vapor at T = 2100 K in the hydrogen-oxygen flame for sites of a spectrum 3271 - 3274 and 3127 - 3130 cm^{-1}. STP.

2.2 Device for modeling of the transfer over of selective radiation in structurally non-uniform mediums

The problem of a transfer over of selective radiation of torches and streams of aerocarriers is put in the sixtieth year of XX th century. The transfer over of selective radiation is influenced by following factors: the temperature self-reference of spectral lines of radiation, displacement of spectral lines with pressure, displacement of spectral lines as a result of high speed of aerocarriers (Dopler's effect), the temperature displacement of the spectral lines which have been found out for easy molecules (vapor H_2O, CH_4, NH_3, OH) (Moskalenko et el., 1992). The executed calculations have shown that displacement of spectral lines with pressure in a flame, making thousand shares of cm^{-1}, and doplers displacement of spectral lines in conditions turbulized high-temperature mediums can't render appreciable influence on function spectral transmission. Temperature displacement of spectral lines in a flame make the 100-th shares of cm^{-1} and at high temperatures reach semiwidth of spectral lines and more. It leads to that radiation of a high-temperature kernel of a torch is to a lesser degree weakened by its peripheral layers that strengthens radiating cooling torch kernels. At registration of radiation of a torch of the aerocarrier the effect of an enlightenment of atmosphere is observed more considerably in comparison with the account only the temperature self-reference of spectral lines. If for the temperature self-reference the spectral effect of an enlightenment is observed more intensively for optically thick mediums the effect of an enlightenment of atmosphere at the expense of temperature displacement of spectral lines is shown and for optically thin selective radiators and observed by us earlier at registration of radiation of system «a selective radiator – atmosphere» with the high spectral permission in bands of water vapor.

Earlier the problem of the transfer over of selective radiation was put in interests of the decision of problems of the transfer over of radiation of torches and streams of aerocarriers in atmosphere of the Earth (Moskalenko et el., 1984). It has been found out by numerical modeling that law of the transfer over of radiation in low-temperature and high-temperature mediums considerably differ. Burning and movement of products of combustion in a stream is accompanied by wave processes at which there is high-frequency making (turbulence) and low-frequency (whirls). Thus low-frequency wave processes can make the greatest impact on the transfer over of selective thermal radiation while influence of turbulence on the transfer over of selective radiation can be neglected. At high pressures of the non-uniform medium the thin structure of a spectrum of gas components is greased also with influence of sharp selectivity of spectra of radiation on radiating heat exchange it is possible to neglect.

At low pressure and high temperatures of medium effects of temperature displacement of spectral lines in structurally non-uniform mediums can render the greatest influence on the transfer over of selective radiation, not which account for easy molecules (H_2O, CH_4, NH_3, OH) in settlement schemes can essentially underestimate radiating cooling high-temperature zones of a torch (Moskalenko & Loktev, 2009). On the other hand, sharp selectivity of radiation of the gas medium promotes preservation of heterogeneity a temperature field at movement of products of combustion in a fire chamber owing to decrease in absorption of high-temperature zones of its torch by peripheral low-temperature layers.

Creation non-uniform on temperature of the gas medium in top internal space is promoted also by specificity of radiating heat exchange in top internal space, when speed radiating cooling peripheral zones optically a thick torch above, than in its central part. Even if the burning device forms front of products of combustion homogeneous for temperature in process of movement of gases in a plane, normal to a direction of movement of a stream, there is heterogeneity so heterogeneity of a field of temperature becomes three-dimensional.

Modeling of structurally non-uniform gas mediums is carried out by means of the optics-mechanical device in which the amplitude modeled heterogeneities can varies in a range of temperatures 400÷2500 K. On fig.8 the structure of an optics-mechanical part in section and the top view is shown. Installation contains the lighter with a source of modulated radiation, mirror optical system of repeated passage of a bunch of radiation under White's scheme between which mirrors the block of gas torches mounted on a rack with possibility of change of position and an inclination of a cut of a plane of capillaries of torches concerning a plane of the main sections of mirrors of optical system.

The block of gas torches includes the radiator basis in which branch pipes with capillaries accordingly for combustible and oxidizing gases are built in serially. Cooling of branch pipes with capillaries is carried out by means of radiators of water cooling. Behind a target mirror of optical system are consistently established the mechanical modulator – the breaker of radiation and a spectrometer. The gas torch having possibility of moving on height and a turn in horizontal and vertical planes, together with an optical part make an ardent multiple-pass cell which from above is covered with a metal cap cooled by water.

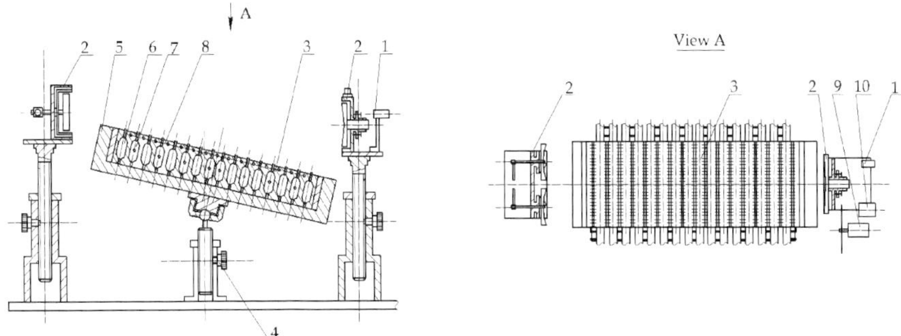

Fig. 8. An optics-mechanical part of a non-uniform radiator: 1 – the lighter; 2 - mirror optical system under White's scheme; 3 - the block of gas torches; 4 – a rack; 5 - the radiator basis; 6 and 7 - capillaries accordingly for combustible and oxidizing gases; 8 – radiators; 9 - the mechanical modulator – the radiation breaker; 10 – a spectrometer.

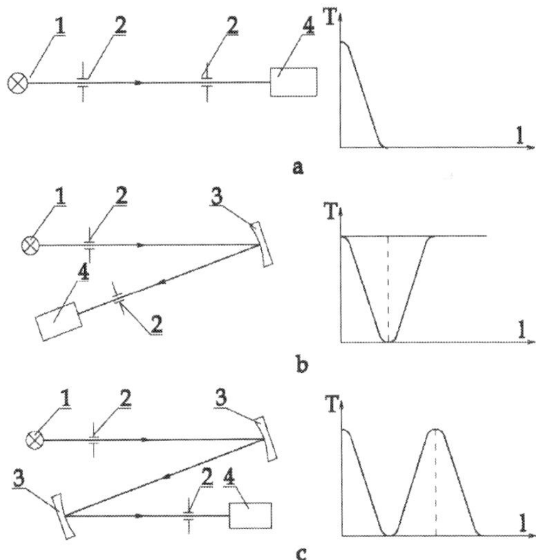

Fig. 9. Formation of profiles of temperature for cases unitary (a), double (b) and triple (c) passages of a beam of radiation through a flame stream. 1 – the lighter; 2 – entrance and target cracks; 3 – spherical mirrors; 4 – the radiation receiver.

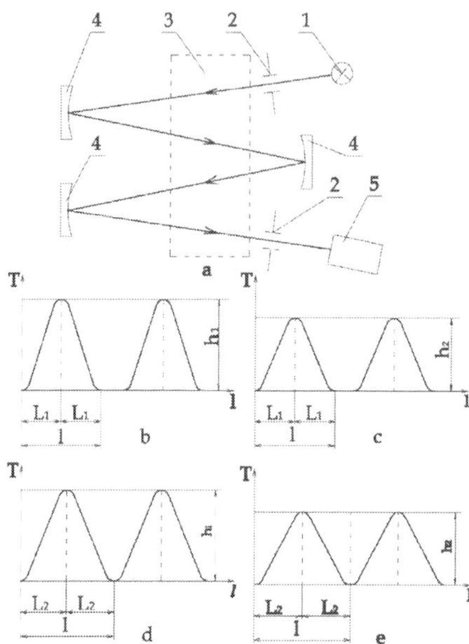

Fig. 10. Formation of profiles of temperature for cases of quadruple passage of radiation through a flame and temperature profiles T corresponding to them on an optical way of radiation l: 1 – a radiation source; 2 – entrance and target cracks; 3 – a flame zone; 4 – mirrors; 5 – a spectrometer.

For preservation of vertical development of a flame at inclined position burning devices on capillaries it is desirable to establish nozzles with a turn corner $(\pi - \alpha)$, where α – a corner of a plane of a cut of capillaries burning devices concerning a horizontal plane. On fig. 9 the kind of temperature profiles for various cases of passage of radiation along an optical way through ardent multiple-pass cell is shown. When the plane of a cut of capillaries of a torch is parallel to a plane of the main sections of mirrors of optical system, the bunch of radiation of the lighter passes through the gas medium homogeneous for temperature. Changing height of position of a gas torch, in this case probably to define distribution of temperature depending on height over a plane of cuts of capillaries. Further this information can be used for definition of a profile of temperature non-uniform on temperature of the gas mediums modeled in installation «a non-uniform gas radiator». Optical schemes are presented in the left part of drawing, and temperature profiles T on an optical way l – in the right part of drawing.

Mirror reflection of these profiles (return temperature profiles) can be received by return turn of a plane of a gas torch concerning a horizontal plane. On fig. 10 the explanatory to formation of profiles of temperature for cases of quadruple passage of radiation through a flame and temperature profiles T corresponding to them on an optical way of radiation l is presented. Radiation from a radiation source through an entrance crack passes a flame zone, is reflected consistently by mirrors after quadruple passage through a flame zone projected on a target crack of receiving-registering system of a spectrometer. Depending on constructive length L zones of a flame along an optical way and height h arrangements gas burning devices over the basis change (to look fig. 10) amplitude of temperature heterogeneity and its half-cycle Δl. On fig. 10 cases are presented, when $h_1 \neq h_2$ and $L_1 \neq L_2$.

For homogeneous system the law of Kirhgof is carried out. In non-uniform medium on structure it is broken, and function spectral transmission in a spectral interval of final width becomes dependent as from thin structure of a spectrum of the radiating volume, and from thin structure of a spectrum of the absorbing medium. Effects of display of sharp selection of spectra of radiating and absorbing mediums on function spectral transmission lead to certain features of radiating heat exchange in a torch and transfer function of distribution of radiation of a torch in medium. So radiation of a kernel of a torch is to a lesser degree weakened by its peripheral layers. In chambers of combustion it leads to increase heat-receptivity by surfaces of heating at the expense of radiating heat exchange, and at distribution of radiation of a torch of the aerocarrier to atmosphere the effect of an enlightenment of atmosphere when atmosphere becomes more transparent for non-uniform high-temperature selective radiators, in comparison with not selective radiators is observed. Consideration of process of the transfer over of selective radiation in atmosphere allows constructing the following scheme of its account through the factors of selectivity defining the relation of function spectral transmission for selective radiation τ_c to function spectral transmission for not selective radiation. If to enter factor of selectivity for a component i:

$$\eta_{i\lambda c} = \frac{\tau_{i\lambda c}}{\tau_{i\lambda n}}, \tag{11}$$

Then full transmission of mediums for selective radiation $\tau_{\lambda c}$ it is represented in a kind:

$$\tau_{\lambda c} = \prod_i \eta_{i\lambda} \cdot \tau_{i\lambda n} \tag{12}$$

As functions $\tau_{i\lambda n}$ are studied, researches $\tau_{\lambda c}$ is reduced to reception of sizes $\eta_{i\lambda}$ as functions of temperature, an optical thickness of radiating and absorbing mediums, and also pressure which can be defined on the basis of experimental researches or the data of numerical modeling of the transfer over of radiation on thin structure of a spectrum of radiating and absorbing mediums (Moskalenko et el., 1984).

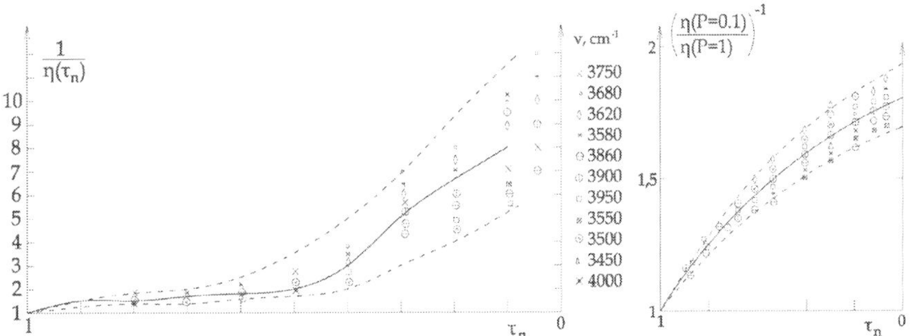

Fig. 11. Dependence of factor of selectivity on function spectral transmission at various frequencies.

The executed experimental researches and results of numerical modeling have shown that sizes $\eta_{i\lambda}$ depend on temperature. At low temperatures of selective radiators, for example streams of turbojets, sizes $\eta_{i\lambda} \leq 1$ and selective radiation is absorbed in atmosphere more intensively than not selective radiation. To calculations of function spectral transmission for not selective radiation it is applied one-parametrical and two-parametrical methods of calculation of the equivalent mass, discussed more low.

Dependence of transfer function on structure of absorbing and radiating mediums is important for considering in problems of remote diagnostics of products of combustion by optical methods and supervision over aerocarriers on their infra-red thermal radiation. The importance of the account of effect of selectivity of radiation on transfer function of atmosphere is illustrated on fig. 11a, on which dependences of spectral factors of selectivity η are presented as function from transmission τ_n for sources of not selective radiation for various sites of a spectrum with the centers v (v – wave number) for optically thin radiator of water vapor. The absorbing medium is atmospheric water vapor. A total pressure P in a selective source and in atmosphere is one atmosphere. The Fig. 11b shows strengthening of display of effect of selectivity with fall of total pressure P to 0,1 atmospheres.

2.3 Functions spectral transmission of vapors H_2O, CO_2 and small components of products of combustion

Let's consider the general empirical technique for calculation of radiating characteristics of a gas phase of products of combustion (Kondratyev & Moskalenko, 1977; Moskalenko et al., 2009), applicable for the decision of problems of radiating heat exchange and the radiation transfer over in torches of aerocarriers, in chambers of combustion of power and power technological units and in the power fire chambers functioning in the conditions of high pressures of a working medium. The developed technique is applicable for function evaluation spectral transmission (the basic radiating characteristic) multicomponent non-uniform on temperature and effective pressure of atmosphere of smoke gases of products of combustion in the chamber of combustion and gas-mains of boilers. A working range of effective pressure $0,01 \leq P_e \leq 100$ atm that provides its use at the decision of problems of radiating heat exchange both in modern boilers, and in perspective workings out of power and power technological units.

Generally at function evaluation spectral transmission $\tau_{\Delta v}$ where v – wave number, Δ – the spectral permission, is necessary to allocate contributions to the absorption caused by wings of remote spectral lines of atmospheric gases $\tau^k_{\Delta v}$, by the induced pressure absorption $\tau^n_{\Delta v}$, selective absorption $\tau^c_{\Delta v}$ by the spectral lines entering into the chosen spectral interval. Then for the set component:

$$\tau_{\Delta v} = \tau^k_{\Delta v} \times \tau^n_{\Delta v} \times \tau^c_{\Delta v}. \tag{13}$$

Function:

$$\tau^k_{\Delta v} \cdot \tau^n_{\Delta v} = \exp\left[-(\beta_{vk}(T) + \beta_{vn}(T))\omega P\right], \tag{14}$$

where $\beta_{vk}(T)$ and $\beta_{vn}(T)$ – factors continual and the absorption induced by pressure, depending on temperature T; ω – the component maintenance; P – partial pressure.

For reception of function spectral transmission $\tau^c_{\Delta v}$ it is offered to use the general parity:

$$\left(\frac{1}{\ln \tau_{\Delta \nu}}\right)^2 = \left(\frac{1}{\ln \tau'^{C}_{\Delta \nu}}\right)^2 + \left(\frac{1}{\ln \tau''^{C}_{\Delta \nu}}\right)^2 + \frac{M}{(\ln \tau'^{C}_{\Delta \nu})(\ln \tau''^{C}_{\Delta \nu})} ,\tag{15}$$

where

$$\tau'^{C}_{\Delta \nu} = \exp[-k_\nu(T)\omega]\tag{16}$$

defines function spectral transmission in the conditions of weak absorption and at elevated pressures ($P \geq 10$ atm) in the conditions of the greased rotary structure of a spectrum of absorption,

$$\tau''^{C}_{\Delta \nu} = \exp[-\beta_{\nu c}(T)\omega^{m}{}_{\nu} P_e^{n}{}_{\nu}]\tag{17}$$

- function spectral transmission at small $P_e < 1$ atm in the conditions of strong absorption. The M parameter characterizes change of growth rate of function transmission at transition from area of weak absorption in area of strong absorption. Parameters k_ν, m_ν, n_ν, $\beta_{\nu c}$, are defined from the experimental data received by means of described above measuring complexes. In conformity with the theory of modeling representation of spectra of absorption $k_\nu = \overline{S}/d$ defines the relation of average intensity to distance between lines, and the size $k_\nu \Delta \nu$ - characterizes intensity of group of the spectral lines located in the chosen spectral interval $\Delta \nu$.

It has been shown that the parity (15) describes any modeling structure of a spectrum, including the law of Buger for a continual spectrum of strongly blocked spectral lines. Really, in this case m=1, n=0, $\beta_{\nu c}$=k_ν, M = − 1. The overshoot of spectral lines is stronger, the it is more parameter m and the less parameter n and the closer parameter $|M|$ to unit. For real spectra parameter $M \in \{0, -1\}$. Continual absorption by wings of lines and the absorption induced by pressure is described by a following set of parameters: m=1, n=1, $k_\nu = \beta_\nu$, M = − 1.

Let's notice that spectra of the absorption induced by pressure submit to other rules of selection in comparison with vibrational-rotary spectra and the bands of absorption forbidden by rules of selection in vibrational-rotary spectra, become resolved in spectra of the absorption induced by pressure. In this connection the account of the absorption induced by pressure can become necessary in radiating heat exchange in power fire chambers. In power fire chambers the account and continual absorption by wings of strong lines and absorption bands is more important.

With temperature growth the density of spectral lines increases and, hence, parameters m_ν, n_ν, M_ν change. In this connection at calculations $\tau^c_{\Delta \nu}$ in the conditions of non-uniform on temperature and pressure of medium average values of these parameters in a certain range of temperatures are used.

For calculation $\tau^c_{\Delta \nu}$ in the conditions of non-uniform on temperature and pressure of medium it is convenient to enter temperature functions:

$$F_{1c}(T) = \frac{K_\nu(T)}{K_\nu(T_0)} , \quad F_{2c}(T) = \frac{\beta_{\nu c}(T)}{\beta_{\nu c}(T_0)} .\tag{18}$$

Then

$$-\ln \tau^{\,C}_{\Delta\nu} = K_{\nu c}(T_0)W_1\,,\quad -\ln \tau^{\,''C}_{\Delta\nu} = \beta_\nu(T)W_2^{\,m_\nu}\,,\tag{19}$$

where

$$W_1 = \int_e \rho(e)F_{1c}[l(T)]dl,\tag{20}$$

$$W_2 = \int_e \rho(e)\left(\frac{P_e(e)}{P_0}\right)^{\frac{n_\nu}{m_\nu}} F_{2c}^{\frac{1}{m_\nu}}[l(T)]dl\,.\tag{21}$$

Here effective pressure:

$$P_e = P_{N_2} + B_{O_2}\cdot P_{O_2} + \sum_{i=1}^{N-1} B_{ik}P_{ik}\,,\tag{22}$$

where P_{N_2} - pressure N_2, P_{O_2} - pressure O_2, B_{ik} – the widening factor (the relation of average semiwidth of lines in the chosen interval of a spectrum for collisions of molecules i-k to average semiwidth of spectral lines in case of impact of molecules of type i with molecules of nitrogen N_2).
Similarly for induced and continual absorption:

$$\beta_{\nu u}(T) = \beta_{\nu u}(T_0)F_u(T)\,,\qquad \beta_{\nu\kappa}(T) = \beta_{\nu\kappa}(T_0)F_\kappa(T)\,.\tag{23}$$

Temperature functions used for calculations $F_u(T), F_\kappa(T), F_{1c}(T), F_{2c}(T)$ can be presented in the tabular form or in the form of simple analytical approximations, for example, in the exponential-sedate form.
It is experimentally shown that for multicomponent atmosphere full function spectral transmission is defined by the law of product of functions on all gas components:

$$\tau_{\Delta\nu} = \prod_i \tau_{i\Delta\nu}\,,\tag{24}$$

where i – component number.
The parity (24) directly follows from static model of spectra and reflects that fact that the thin structure of spectra of each molecule doesn't depend on other molecules. For induced and continual absorption it is a parity it is carried out owing to absence of rotary structure. Numerical modeling of functions spectral transmission on parameters of thin structure of spectra have shown that the parity (24) is carried out with a margin error no more than 1 %.
Parameters K_ν, β_ν, m_ν, n_ν, M are defined from the measured spectra of radiation and radiation absorption by high-temperature gas mediums, modeling with the help heating cells and fiery measuring complexes.
For definition of parameters of functions spectral transmission the data of experimental researches has been added by results of numerical modeling under high-temperature atlases of parameters of the spectral lines prepared with use of the base data, received by means of measuring complexes of the high spectral permission. For an example on fig. 12 spectral

factors of absorption of water vapor K_v, and on fig. 13 – spectral dependences β_v water vapor in bands 1,37, 1,87 and 2,7 μm on experimental data are led. On fig. 14 spectral dependences of factors of absorption CO_2 in band 2,7 μm is given. On fig. 15 spectral factors of absorption K_v in the basic bands CO and NO according to numerical modeling of thin structure of spectra of absorption are illustrated. For vapor H_2O parameters m_v, n_v, M_v poorly depend on length of a wave a range of temperatures 600-2500K and probably to use average values n=0,45, m=0,65, M = −0,2. Strong temperature dependence of spectral factors of absorption in a range of spectrum 10−20 μm pays attention. At growth of temperature from 300 to 2500K increase intensitys the spectral lines entering into the specified interval of a spectrum, in 6600 times is observed.

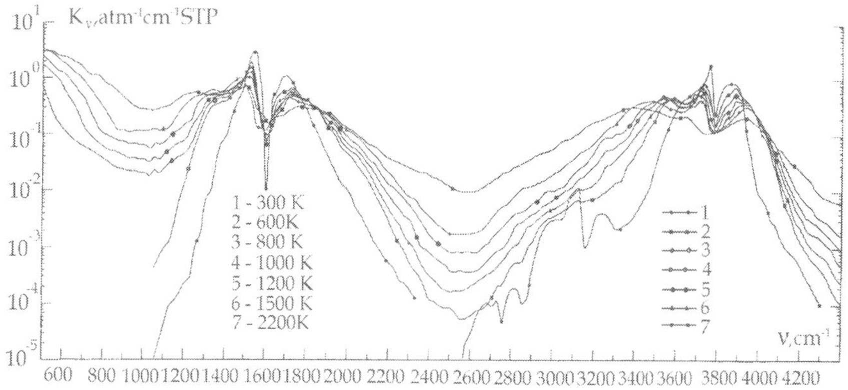

Fig. 12. Spectral factors of absorption K_v water vapor in bands 6,3 and 2,7 μm on experimental data.

Applicability of the received parameterization of functions spectral transmission for the decision of problems of the transfer over of radiation in high-temperature mediums and radiating heat exchange in chambers of combustion with application described above parities for calculations spectral intensitys thermal radiation and nonequilibrium radiation of electronic spectra in non-uniform working mediums under structural characteristics taking into account absorption and scaterring of radiation by a disperse phase has been considered. Main principle of correctness of spent calculations is calculation of equivalent mass on indissoluble trajectories from the radiating volume to a supervision point, including at reflection of radiation from walls and at scattering of radiation by a disperse phase. Streams of thermal radiation on walls of the working chamber are defined by integration spectral intensitys on a spectrum of lengths of waves and a space angle within a hemisphere.

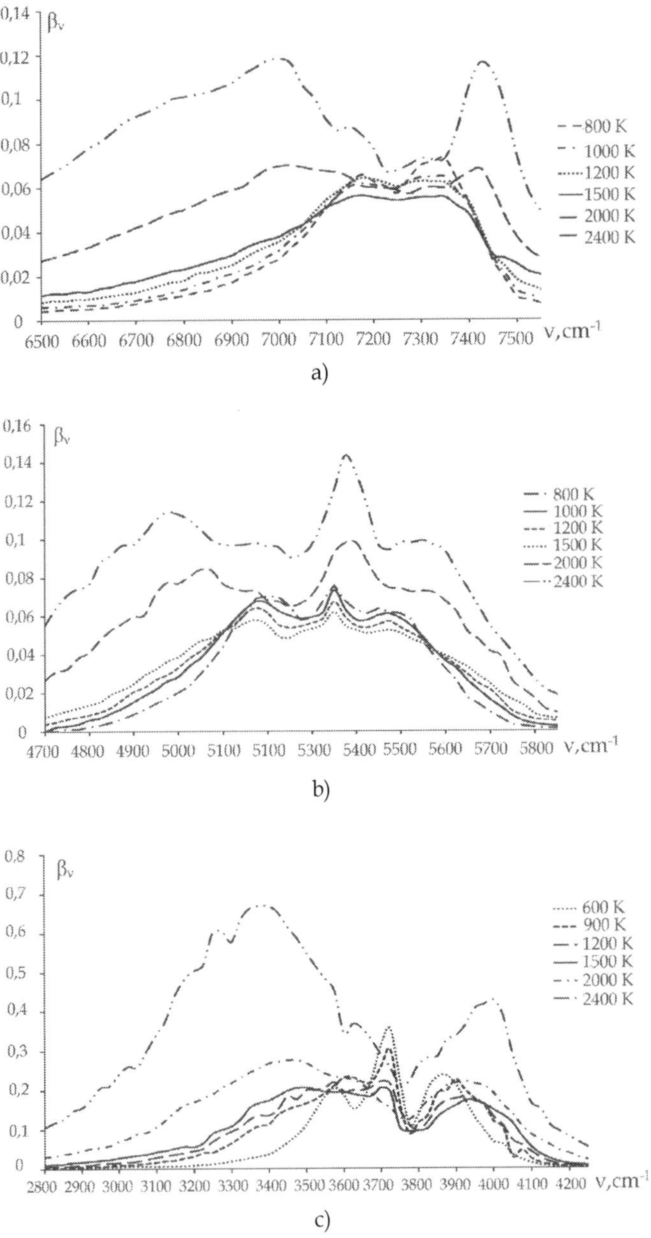

Fig. 13. Spectral dependences of parameter βν in bands 1,37 (a), 1,87 (b) and 2,7 μm (c) water vapor.

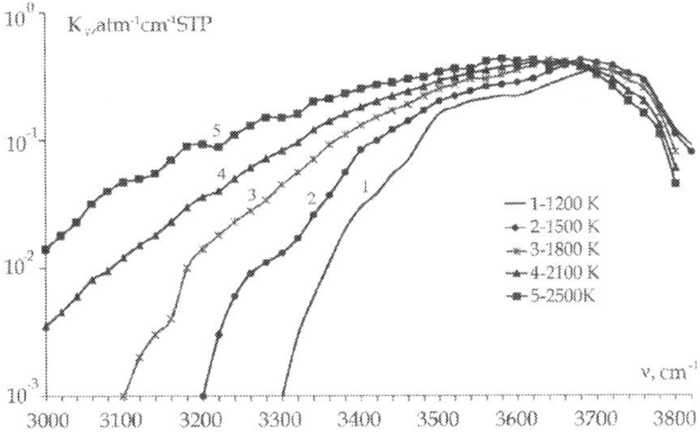

Fig. 14. Spectral dependences of factors of absorption K_v in band 2,7 μm CO_2 on experimental data.

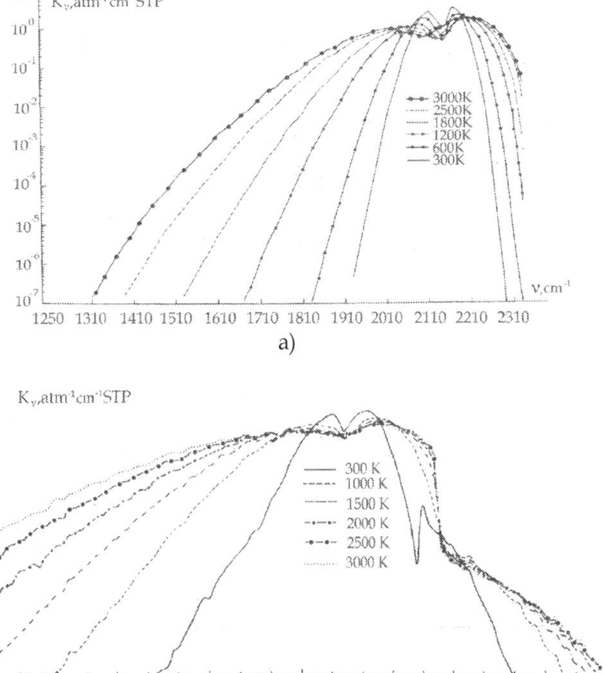

Fig. 15. Spectral dependences of factors of absorption K_v in the basic bands CO (a) and NO (b) by results of numerical modeling of thin structure of a spectrum.

3. Nonequilibrium processes of radiation in flames

Executed spectraradiometry measurements have revealed presence of nonequilibrium radiation in ultra-violet, visible parts of a spectrum as result of display of effect of chemiluminescence in processes of burning of fuel. The contribution of nonequilibrium radiation in radiating cooling a torch has appeared essential. The basic components defining nonequilibrium radiation, are: OH, CH, NO_2, NO, SO_2, CN. Probably also influence of splinters of difficult hydrocarbonic connections which are formed in dissociated process the difficult hydrocarbonic connections which are in wild spirits.

Nonequilibrium radiation is formed by a kernel of a torch and then extends on all volume of the chamber of combustion, being transformed and participating in process of heating of particles of fuel and heatsusceptibility surfaces. Radiating cooling molecules occurs during their relaxation, making $\leq 10^{-4}$ sec, commensurate in due course courses of chemical reaction, and reduces adiabatic temperature of products of combustion in peaks of chemical reactions. Out of zones of chemical reactions radiation is equilibrium.

3.1 Definition nonequilibrium radiating cooling a flame from experimental data

Nonequilibrium radiation is generated by mainly electronic bands of radiation of the raised molecules of products of the combustion, lying in ultra-violet and visible parts of a spectrum and in vibrational-rotary bands. Temperature T of a zone of burning was measured by optical methods with a margin error no more than 2 %. A known chemical composition of gas fuel allows to calculate adiabatic temperature of zones of chemical reactions and to define size $\Delta T = T_a - T$, characterizing radiating cooling zones of active burning. Radiating cooling can be equilibrium and nonequilibrium. Equilibrium radiating cooling ΔT_e it is possible to calculate on absolute spectra of radiation of a flame and on the measured temperature and a chemical composition of products of combustion, speed of the expiration of a stream that allows defining radiating cooling ΔT_n, caused by nonequilibrium radiation. Nonequilibrium radiating cooling $\Delta T_n = \Delta T - \Delta T_e$ is convenient for characterizing in size $\xi = \Delta T_n / T_a$ that which according to our measurements varies in a range of values (0,02-0,13), and increases with growth of temperature T_a.

Features of registration of average temperature of a flame an optical method have demanded working out of a method of definition ΔT_n in conformity with absolute spectra of the radiation registered by the spectral device, allocated optical systems in volumes of the radiating medium.

For definition nonequilibrium radiating cooling of flames results of measurements by optical methods of temperature hydrogen-oxygen, hydrogen-air, the propane-butane-oxygen, the propane-butane-air, acetylene-oxygen flames, formed by burning of gas fuel of a controllable chemical composition in air or oxygen have been used. Specially developed burning devices provided formation homogeneous for temperature flames in optical measuring channels. The temperature of a homogeneous flame was measured by a method of the self-reference of spectral lines of water vapor and on spectral brightness of radiation of a flame in "black" lines of water vapor or CO_2. Really, for a flame homogeneous for temperature and the gas medium which are in thermodynamic balance, spectral brightness of radiation $B_{\Delta\lambda}$ will be defined by a parity:

$$B_{\Delta\lambda} = B_\lambda^{abb}\left(1 - \tau_{\Delta\lambda}\right), \tag{25}$$

where $\tau_{\Delta v}$ – function of spectral transmission a flame for a small range of length of a wave, B_λ^{abb} – spectral brightness of radiation of absolutely black body on length of a wave λ, Δ – semiwidth of hardware function of a spectrometer. If $\tau_{\Delta v} = 0$ (a black line of radiation), $B_{\Delta\lambda} = B_\lambda^{abb}$ and the size $B_{\Delta\lambda}$ unequivocally defines temperature of a homogeneous flame.

The adiabatic temperature of a flame is calculated on a known chemical composition of products of combustion of burned gas and factors of surplus of oxygen and air α. Comparison of calculated values T_a with the measured optical methods in temperatures of flame T has shown that observable temperature T of a flame always more low T_a. The only thing the reason leading to lower value of size T in comparison with in adiabatic temperature T_a, can be effect radiating cooling a flame.

For the homogeneous radiating medium speed radiating cooling will be defined by the formula:

$$\frac{\partial T}{\partial t} = \frac{\oint\limits_S \left(F^{\uparrow}(S) - F^{\downarrow}(S) \right) dS}{\overline{C}p(T)\rho(T)V} , \tag{26}$$

where integration is made on the closed surface S covering all radiating volume V. The size $F^{\downarrow}(S)$ defines the integrated flux of equilibrium radiation entering into radiating volume V in point S; $F^{\uparrow}(S)$ represents an integrated flux of the equilibrium radiation leaving radiating volume V in point S; $\overline{C}p(T)$ - a specific thermal capacity of medium at constant pressure:

$$\overline{C}p(T) = \frac{\sum\limits_i C_{pi}(T)P_i(T)}{\sum\limits_i P_i(T)} , \tag{27}$$

where P_i - partial pressure of i-th component. Summation in the ratio (27) is made on all gas components which are a part of products of combustion; ρ (T) - the density of the gas medium, which dimension is defined by dimension $\overline{C}p(T)$,

$$F^{\uparrow}(S) = \int\limits_0^{\pi/2}\int\limits_0^{2\pi}\int\limits_0^{\infty} J_\lambda^{\uparrow}(S,\theta,\phi)\sin\theta\cos\theta\, d\theta\, d\phi\, d\lambda , \tag{28}$$

$$F^{\downarrow}(S) = \int\limits_0^{\pi/2}\int\limits_0^{2\pi}\int\limits_0^{\infty} J_\lambda^{\downarrow}(S,\theta,\phi)\sin\theta\cos\theta\, d\theta\, d\phi\, d\lambda . \tag{29}$$

In parities (28), (29) antiaircraft corner θ is counted from a normal to the closed surface in a point of supervision S. Boundary conditions at the decision of the equation of the transfer over were set in conformity with the constructional decision of measuring complexes (Moskalenko et al., 1992). Functions spectral transmission were calculated with use of a two-parametrical method of equivalent mass on indissoluble optical ways from the radiating volume to a supervision point (Kondratyev & Moskalenko, 2006).

At data processing of measurements were considered flames horizontal development of two counter streams of the flame surrounded with the quartz heat exchanger, reducing radiating cooling a flame and promoting preservation of uniformity of temperature within an optical way of thermal radiation to the optoelectronic device, and also excluding formation sooty ashes as a result of process of pyrolysis of combustion products at burning hydrocarbonic fuels. The geometrical way of each of counter streams of a flame made 20 cm at speed of a current of a stream ω =8-10 m/s. The geometrical way to devices with vertical development of a flame made 6 cm at speeds of a current ω =10-15 m/s. Researchers have shown that in devices with water cooling of the case of the chamber of combustion (Moskalenko et al., 1992) at burning of hydrocarbonic fuel it is formed sooty ashes. Therefore in these devices as fuel the pure hydrogen excluding possibility of formation sooty ashes in products of combustion was used.

On fig. 16 and in table 2 results of definition of size $\xi=\Delta T_n/T_a$ for hydrogen-oxygen, hydrogen-air, the propane-butane-oxygen, the propane-butane-air, acetylene-oxygen flames are shown. With increase adiabatic temperatures T_a the size ξ increases and in a range of temperatures T_a from 1800 to 3200K varies from 2 % to 13 %.

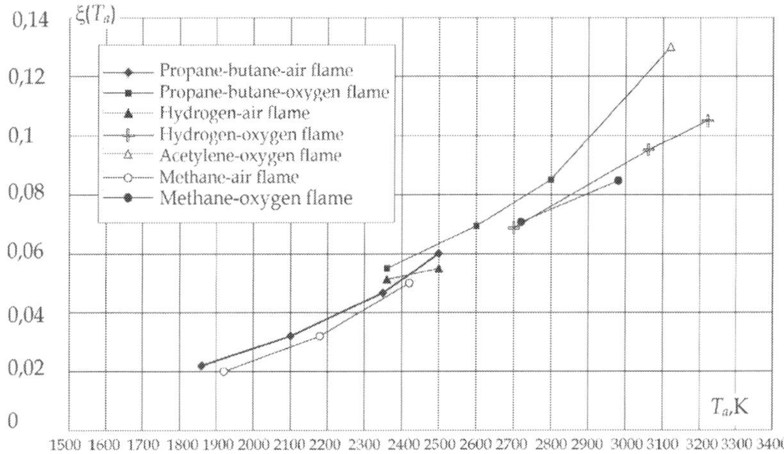

Fig. 16. Dependence of parameter $\xi=\Delta Tn/Ta$ from adiabatic temperatures Ta.

As have shown results of experimental definition of function of source B_λ and factors of nonequilibrium $\eta_\lambda = B_\lambda(T)/B_\lambda^{abb}(T)$, where $B_\lambda^{abb}(T)$ is spectral brightness of absolutely black body (Planck's function), nonequilibrium radiation is formed mainly in electronic spectra of radiation of the molecules located in ultra-violet and visible parts of a spectrum. The effect nonequilibrium radiations in the vibrational-rotary bands lying in infra-red area of a spectrum, is shown considerably only at adiabatic temperature T_a> 2500 K.

The obtained data of experimental researches concerns optically thin torch when influence nonequilibrium is shown in a greater degree. Therefore the data presented on fig. 16 and in tab. 2, can be used for an estimation of the maximum size of energy which is transferred on heatsusceptibility to a surface in case of burning of gaseous fuel. It can be estimated on radiation of gas products of combustion of a kernel of a torch and on change enthalpy of combustion products. In the first case we will use the law of Stefan-Boltzman's defining an integrated hemispherical stream of radiation of absolutely black body,

$$F(T) = \sigma T^4. \tag{30}$$

T_a, K	$\xi = \Delta T_n / T_a$	Type of flame
1860	0,021	propane-butane-air
2100	0,032	propane-butane-air
2350	0,048	propane-butane-air
2500	0,06	propane-butane-air
2360	0,055	propane-butane-oxygen
2600	0,069	propane-butane-oxygen
2800	0,085	propane-butane-oxygen
3120	0,13	acetylene-oxygen
2360	0,051	hydrogen-air
2500	0,056	hydrogen-air
2700	0,068	hydrogen-oxygen
3060	0,095	hydrogen-oxygen
3220	0,105	hydrogen-oxygen
1920	0,020	methane-air
2180	0,033	methane-air
2420	0,050	methane-air
2720	0,071	methane-oxygen
2980	0,094	methane-oxygen

Table 2. Influence of nonequilibrium processes of radiation on radiating cooling a flame. T_a is adiabatic settlement temperature; ΔT_n is radiating cooling the homogeneous flame, caused by nonequilibrium radiation. Factor of surplus of air and oxygen $\alpha > 1$.

Let's enter integrated function transmission:

$$\tau(T) = \int_0^\infty B_\lambda(T)\tau_\lambda(T)d\lambda \bigg/ \int_0^\infty B_\lambda(T)d\lambda. \tag{31}$$

As the range of changes of temperature $(T_a\text{-}T)$ is insignificant, in the specified range of temperatures of a flame $\in \{T_a - T\}$ we will accept τ_n=const. Then a parity:

$$\frac{\Delta F}{F(T_a)} = \frac{F(T_a) - F(T)}{F(T_a)} = 1 - \frac{F(T)}{F(T_a)} = 1 - \frac{F(T_a - \Delta T_n)}{F(T_a)} =$$

$$= 1 - \frac{\sigma(T_a - \Delta T_n)^4}{\sigma T_a^4} = 1 - \left[1 - \left(\frac{\Delta T_n}{T_a}\right)^4\right] \approx \frac{4\Delta T_n}{T_a}. \tag{32}$$

Thus, in a case optically a thin torch on heatsusceptibility surfaces without easing can get from 8 % to 30 % of full radiation of a torch.

Change enthalpy of combustion products will be defined by the formula:

$$\Delta H = \sum_i \left(\int_T^{T_a} C_{pi}(T)\,dT \right) V_i , \tag{33}$$

where $C_{pi}(T)$ - a specific thermal capacity at constant pressure for a component i combustion products, V_i - led to standard temperature and pressure volumes of components of products of combustion (P_{0i}=1 atm, T_{0i}=273K). Summation is carried out on all components i combustion products.

Volumes V_i are calculated on a known chemical composition of gas fuel and value of factors of surplus of air or oxygen. Size estimations $\Delta H/H_a$ yields the results close to sizes $\Delta F/F$ (T_a).

In case of mediums with the big optical density effects of influence nonequilibrium radiations are extinguished owing to suppression processes of chemiluminescence, caused by absorption of nonequilibrium radiation and display radiationless transitions in molecules. Because nonequilibrium radiation is formed in ultra-violet and visible parts of a spectrum, and also in infra-red area of a spectrum, in the transfer over of nonequilibrium radiation it is important to consider radiation absorption thing-dispersion ashy fraction and the hydrocarbonic connections absorbing radiation in spectral area of generation of nonequilibrium radiation. At absorption of nonequilibrium radiation there is a heating of particles of fuel and their ignition, and absorption of nonequilibrium radiation by heatsusceptibility surfaces weakens.

3.2 Definition of spectral factors of nonequilibrium vibrational temperature of components in flames

Quantum-mechanical consideration of a problem of formation of spectra of radiation of flames shows that presence nonequilibrium in radiation in electronic spectra (Moskalenko et al., 2009) it is accompanied by the nonequilibrium mechanism of radiation and in vibrational-rotary spectra as cores, overtone and composition bands, as have confirmed results of our measurements. The mathematical model of the transfer over of radiation is developed for the analysis of spectra of radiation flames in nonequilibrium radiating biphasic mediums in approach of unitary scattering of radiation (Moskalenko et al., 2009; Moskalenko & Loktev, 2009). The Chemical composition of products of combustion was calculated (Alemasov et al., 1973). Measurements were carried out on the spectral installations described in section 2.1 in spectral area 0,2÷25 µm in a range of temperatures 1500-2900 K. Hydrogen, methane, the propane-butane and acetylene in oxygen and in air were burned. The spectral resolution made Δ=3÷5 cm^{-1}. Besides, in the field of a spectrum 2,5÷3,5 µm were used results of the measurements executed with the high spectral resolution Δ=0,05 cm^{-1} by means of the Fourier spectrometer that has allowed to reveal influence nonequilibrium a flame in the basic band OH 2,8 µm. Results of the executed researches have shown that radiation of water vapor in vibrational-rotary bands is equilibrium. Nonequilibrium character of radiation CO_2 starts to be shown at thermodynamic temperatures of flame T_t>2100 K. Nonequilibrium character of radiation in electronic spectra of oxides of nitrogen in a spectral range 0,3÷0,7 µm was detected. Diffused radiation NO_2 has poorly expressed maximum of radiation in a spectrum range 0,55÷0,58 µm. At burning CH_4, propane-butane, C_2H_2 there is a formation of particles of soot to optical density on length of a wave $\lambda=0,55\mu m$ accordingly equal 0,098; 0,2; 0,4 m^{-1} in a flame. Spectral dependences of factors of absorption and easing sooty ashes are used by us for restoration of its microstructure with the subsequent calculation of matrixes light scattering on a radiation spectrum (Moskalenko et al., 2010). We will more low stop on the analysis of spectra of nonequilibrium radiation of hydroxyl OH.

For research of radiating characteristics of a hydroxyl it is expedient to use hydrogen-oxygen a flame in order to avoid overlapping of bands of radiation OH other radiating components. Also it is expedient to choose modes of burning with factors of surplus of an oxidizer $\alpha_{ox}<1$. Spectra of radiation H_2-O_2-flame are measured on installation SUVGI (Moskalenko et al., 1980; Moskalenko et al., 1992) and laboratory breadboard models of this installation at temperatures 1500-2500K. Hydrogen-oxygen flames, used in the present researches, have is bluish-white coloring with less contrast external zone of orange color. The flame height strongly depends on a mode of burning and on the average makes 30 cm. At heights more than 25 mm over hydrogen and oxygen capillaries ardent streams see height more merge, forming a continuous radiating torch. The maximum concentration of the basic radiating components of the hydrogen-oxygen flame: water vapor and a hydroxyl makes accordingly 58,46 % (at $\alpha_{ox}=1$) and 12,21 % (at $\alpha_{ox}=1,3\div1,5$). Thermodynamic parameters of products of combustion of the hydrogen-oxygen flame, calculated at pressure $P=1atm$, are presented in table 3. In spectra of radiation of a flame the most intensive bands are observed in a range $261\div360$ nm. In flames with α_{ox} <0,5 in combustion products molecular oxygen is absent and in spectra of radiation of a band of Schuman-Runge of hot molecular oxygen weren't observed.

Unlike a propanee-butane-oxygen flame, a hydrogen-oxygen flame is transparent, since its emissivity in the visible is $\varepsilon_\lambda \leq 0.003$. Weak emission bands belonging to water vapor are observed in the neighborhoods of $\lambda = 577.5, 579.7, 693, 718, 766$, and 813 nm. The Na doublet was also detected from the slight concentration of sodium in the hydrogen. The emission peak of nitrogen dioxide (NO_2) is observed in the range of 550–580 nm. Members of the Delandres scheme of the ground state of the OH bands make a small contribution to the visible emission from the hydrogen-oxygen flames.

Table 3 lists the main thermodynamic parameters of the combustion products of the hydrogen-oxygen flame, including the thermodynamic temperature T of the flame and the chemical composition of the combustion products in the form of the volume concentrations of the components, as functions of the oxygen excess coefficient α_{ox}.

The OH radical is formed in all combustion processes for hydrogen-containing fuels and plays a significant role in the chemical chain reaction mechanism (Kondratyev, 1958; Kondratyev & Nikitin, 1981). It is a component of the combustion products in the exhaust plumes of ballistic rockets. Intense UV emission of the OH group is an important signifier of the launching of rockets and the firing of artillery systems. Large amounts of the OH groups are produced during nuclear explosions in the atmosphere.

The emission and absorption spectra of OH in hydrogen-oxygen flames has been studied for several decades. Small sized premixed flames have been used for this. The results indicate a disequilibrium over the internal degrees of freedom in the reaction zone of the flame. The emission from hydrogen flames at low pressures is essentially pure chemiluminescence. The most important criterion for chemiluminescent radiation is that the absolute intensity of the OH group radiation must exceed its equilibrium value because of a nonequilibrium vibrational and rotational energy distribution which leads to anomalously high vibrational and rotational temperatures. Increasing the pressure and introducing inert gases into the combustion zone cause quenching of the chemiluminescence on the part of hydroxyl groups. At low pressures the emission intensity of the OH groups in hydrogen low-temperature flames is many orders of magnitude greater than the equilibrium emission intensity. Here we measure the UV emission and absorption of OH in a hydrogen-oxygen flame. This flame can be used to model the temperatures in the exhaust plumes of rocket engines.

α_{ox}	T, K	K_1	H_2	H_2O	H	OH	O	O_2
0,1	1089	0,79	0,9	0,1	-	-	-	-
0,15	1442	1,19	0,85	0,15	-	-	-	-
0,2	1763	1,59	0,7998	0,2	0,002	-	-	-
0,25	2047	1,98	0,7483	0,2497	0,0019	-	-	-
0,3	2272	2,38	0,6917	0,2977	0,0099	0,0006	-	-
0,35	2475	2,78	0,6354	0,3444	0,0181	0,0019	-	-
0,4	2620	3,18	0,5756	0,3871	0,0379	0,0051	0,0001	-
0,45	2733	3,57	0,5166	0,4256	0,0468	0,0102	0,0005	0,0001
0,5	2822	3,97	0,4604	0,4595	0,0606	0,0174	0,0014	0,0005
0,6	2945	4,76	0,3617	0,5129	0,0809	0,0368	0,0049	0,0027
0,7	3017	5,56	0,2842	0,5483	0,0896	0,0584	0,011	0,0083
0,8	3055	6,35	0,2256	0,5694	0,0895	0,0784	0,0186	0,0183
0,9	3071	7,14	0,1818	0,5804	0,0843	0,0946	0,0261	0,0326
1,0	3074	7,94	0,1486	0,5846	0,0768	0,1065	0,0328	0,0504
1,1	3068	8,73	0,1231	0,5843	0,0688	0,1146	0,0382	0,0708
1,2	3057	9,52	0,1031	0,5809	0,0609	0,1196	0,0422	0,0931
1,3	3043	10,32	0,0871	0,5755	0,0536	0,1221	0,045	0,1164
1,5	3008	11,91	0,0636	0,5609	0,0413	0,1221	0,0476	0,1643
2	2907	15,87	0,0317	0,5171	0,0212	0,107	0,0435	0,2792
3	2699	23,81	0,0093	0,4357	0,0055	0,0664	0,0245	0,4582
5	2291	39,69	0,0007	0,3215	0,0002	0,0172	0,0038	0,6563
10	1546	79,37	-	0,1817	-	0,0002	-	0,8181

Table 3. Thermodynamic parameters of the combustion products of hydrogen-oxygen flames at a pressure of P=1atm.

The UV spectrum of the OH is formed by the electronic transition $A^2\Sigma - X^2\Pi$ consisting of the four most intense sequences $\Delta v = 0, \pm 1, \pm 2$ lying in the region 261–360 nm. The 0–0 band is the most intense. The nonequilibrium OH emission of the flame is characterized by the spectral nonequilibrium coefficient of the radiation defined η_λ as the ratio of the intensity of the nonequilibrium emission to the equilibrium intensity at wavelength λ. The spectral brightness of the equilibrium OH emission $B_{\lambda e}^{OH}$, for a uniform medium is given by:

$$B_{\lambda e}^{OH} = B_\lambda^0 \left(1 - \tau_\lambda^{OH} \right), \tag{34}$$

where $\tau_\lambda{}^{OH}$ is the spectral transmission function of the OH groups in the flame; $B_{\lambda e}^{OH}$ is the absolute spectral brightness of the OH emission expressed in energy units, and is the spectral brightness of an absolute black body at the thermodynamic temperature of the flame.

In accordance with the definition, the spectral nonequilibrium coefficient is written in the form:

$$\eta_\lambda = B_{\lambda n}^{OH} / B_{\lambda e}^{OH} = B_{\lambda n}^{OH} \Big/ B_{\lambda e}^{0}\left(1-\tau_\lambda^{OH}\right), \tag{35}$$

Where $B_{\lambda n}^{OH}$ is the spectral brightness of the nonequilibrium emission from the OH groups.

An examination of the theory of nonequilibrium radiation transfer (Moskalenko et al., 2006) shows that, under the given conditions for formation of the flare, η_λ depends on the optical thickness of the emitting medium and reaches a maximum for small optical thicknesses of OH, when the spectral transmission function $\tau_\lambda{}^{OH}\rightarrow 1$. If η_λ is obtained at $\tau_\lambda{}^{OH}$, then $\eta_{\lambda\max}^{OH} = \eta_\lambda^{OH}\Big/\tau_\lambda^{OH}$.

The quantity $\eta_{\lambda\max}$ shows up in the computational integro-differential equations for modelling radiative transfer in nonequilibrium radiating media (Moskalenko & Zaripov, 2008). Studies of the integral effects of nonequilibrium radiative processes on radiative transfer have shown that the effect of the nonequilibrium in the radiative processes increases with the adiabatic temperature T_a of the flame; this is also confirmed by the present spectral measurements of the nonequilibrium OH emission. The values of $\eta_{\lambda\max}$ can be used for determining the spectra (wavelength dependence) of the vibrational temperatures (Moskalenko et al., 2010).

In the faint nonequilibrium emission bands, the spectral transmission $\tau_\lambda\rightarrow 1$ and small values of $(1-\tau_\lambda)$ cannot be observed experimentally. In that case, the vibrational temperatures obtained for the fainter bands can be used to find the spectral absorption coefficients and integral intensities of the faint emission bands whose absolute emission spectra have been recorded by measurement devices.

In order to determine the degree of nonequilibrium of the OH group emission, their transmission spectra were measured under the same conditions as those for measuring the emission spectra. The structure of the transmission spectra correlates with the structure of the emission spectra. The resulting spectral variations in η_λ and $\eta_{\lambda\max}$ (fig. 17) are rapidly varying. The degree of variation in the 281–296 nm range is slightly higher than in the 306–328 nm range. The maximum value of $\eta_{\lambda\max}$ for T=2380 K is 100. In the 306-328 nm band, $\eta_{\lambda\max}$ reaches 50 (fig. 17).

It should be noted that the spectral dependences of η_λ for different temperatures are essentially different. From our standpoint, the reason for this is that it is impossible to make the conditions for obtaining the emission and transmission spectra completely identical, since they have were not recorded simultaneously. The existence of a spectral variation in $\eta_{\lambda\max}$ indicates that the vibrational temperature may depend on the vibrational and rotational quantum numbers. In addition, the shape of the spectrum lines may have an effect on the value of η_λ. Turbulent inhomogeneity during burning of the flame may also have some effect on η_λ. For practical calculations it is appropriate to introduce the average degree of nonequilibricity $\bar{\eta}_{\lambda\max}$ which for, say, the sequence is $\bar{\eta}_{\lambda\max}(\Delta\upsilon=0)=18$, and for the sequence $\Delta\upsilon=+1$, is $\bar{\eta}_{\lambda\max}(\Delta\upsilon=+1)=28$ at T=2380 K. According to the theory of formation of nonequilibrium emission process $\bar{\eta}_{\lambda\max}(\Delta\upsilon=-1)=28$ for the sequence $\Delta\upsilon=-1$. This value of $\bar{\eta}_{\lambda\max}$ and the absolute spectral intensities of the nonequilibrium emission from the hydroxyl group have been used to determine the spectral absorption coefficients and the integrated absorption intensity of the OH for the $\Delta\upsilon=-1$ sequence (Fig.18).

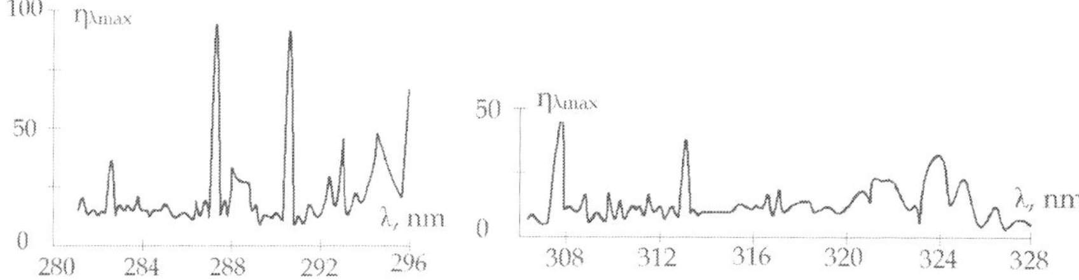

Fig. 17. Spectral dependence of the maximum degree of nonequilibrium in the emission from OH groups at a temperature of 2380 K.

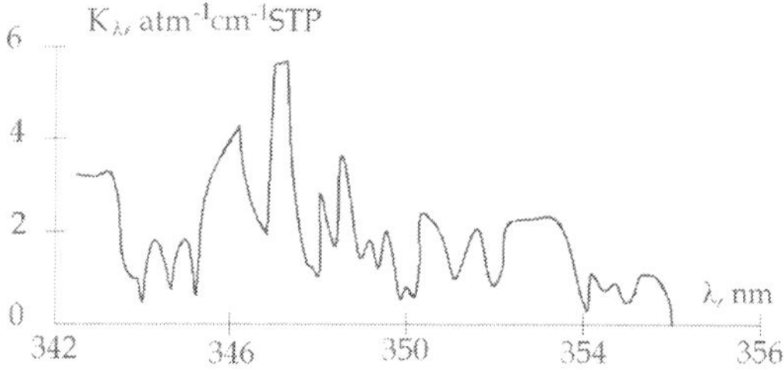

Fig. 18. Spectral dependence of the absorption coefficient for the emission from OH groups in the 342–356 nm range at 2380 K obtained from spectra of the nonequilibrium flame emission.

Our measurements of the absolute intensity spectrum can be used to determine the OH vibrational temperature by the method of (Broida and Shuler, 1952) using the formula:

$$0{,}624\cdot\frac{E_1-E_2}{T}=4\cdot\lg\frac{\lambda_2}{\lambda_1}+\lg\frac{P_1}{P_2}+\lg\frac{J_2}{J_1}\,,\tag{37}$$

where E_1 and E_2 are the energies of the upper vibrational levels of the chosen emission sequences; λ_1 and λ_2 are the
locations of the centers of the bands or of their edges; P_1 and P_2 are the vibrational transition probabilities; J_2 and J_1
are the emission intensities of the individual vibrational bands of the chosen sequences.
The values of J_2 and J_1 are determined experimentally after energy calibration of the emission spectra. The position of the edges of the bands can be determined fairly accurately after spectral calibration of the measured emission spectra. The vibrational transition

probabilities P_1 and P_2 differ considerably according to the data of various authors. Thus, it has been proposed (Moskalenko and Loktev, 2009) that the vibrational temperatures be measured from $\eta_{\lambda\max}$, and the average vibrational temperature, from $\bar{\eta}_{\lambda\max}$. It should be noted that the integrated intensities of the electronic absorption spectra for the $\Delta\upsilon=0,\pm1$ sequences can be used in an experimental determination of the transition probabilities P for these sequences. When Eq.(37) is used legitimately, the vibrational temperatures found by the two different methods (Broida & Shuler, 1952; Moskalenko & Loktev, 2009) should coincide.

The results of this determination of the vibrational temperatures indicate a nonequilibrium distribution of the energy released during the combustion reaction over the vibrational levels of the electronically excited state $A^2\,\Sigma^+$. The integrated intensities of the OH group bands for the sequences $\Delta v = 0, +1, -1$ are equal, respectively, to 30592, 14472, and 1980 atm^{-1}·cm^{-2} at STP. According to the present measurements, the total intensity of all the absorption bands of the OH groups in the 281–333 nm range is 47044 atm^{-1}·cm^{-2} at STP. The intensities are in the proportion $S(\Delta v = 0) : S(\Delta v = +1) : S(\Delta v = -1) = 1:0.49:0.065$.

Figure 19 shows the measured spectral dependence of the absorption coefficients of the hydroxyl groups and Fig. 20 the spectral dependence of the vibrational temperature $T_{\lambda k}$ for the $\Delta v = 0, +1$ sequences of the hydroxyl bands according to the formula (Moskalenko & Loktev, 2009):

$$T_{\lambda k} = \frac{C_2}{\lambda}\left\{\ln\left[\frac{\exp(C_2/\lambda T)-1}{\eta_{\lambda\max}}+1\right]\right\}^{-1}, \tag{38}$$

where λ is the wavelength, T is the flame temperature, and C_2 is the second constant in the spectral brightness of a black body (Planck function).

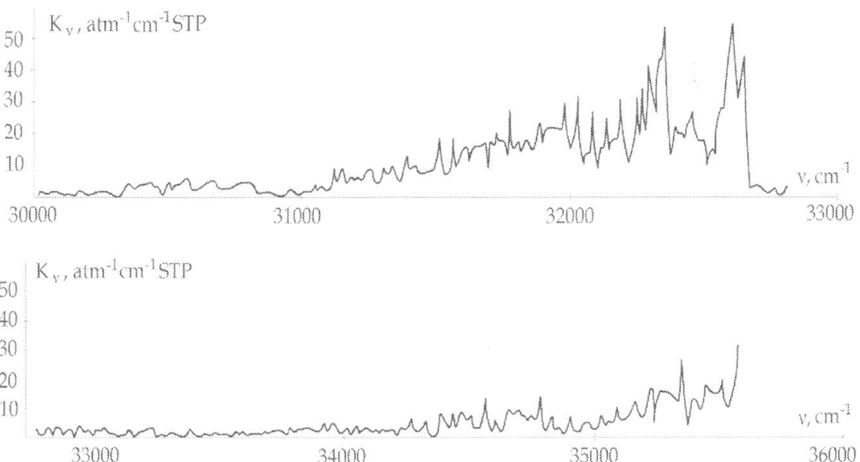

Fig. 19. Spectral dependence of the absorption coefficient $K\nu$ for OH groups at 2450 K in the 281–333 nm range (30000–36000 cm^{-1}).

The data obtained here can be used to study the mechanisms responsible for the nonequilibrium character of the emission from hydrogen-oxygen flames, calculating the spectral brightness of the UV emission from rocket exhaust plumes, solving radiative heat transfer problems in power generation furnaces, and in atmospheric nuclear explosions. The present study of the effect of nonequilibrium processes on the formation of the emission spectra of hydrogen-oxygen flames applies to a total pressure of $P = 1{\cdot}10^5$ Pa (1 atm) in the flames. The radiative nonequilibrium coefficients are higher at lower flame pressures, as confirmed by experiments on the effect of nonequilibrium radiative processes on radiative cooling of flames (Moskalenko and Zaripov, 2008). These data show that nonequilibrium radiative processes must also have an effect on the vibrational-rotational emission bands of OH, and the degree of this influence will depend on the relaxation time of hydroxyl from an excited electronic state into an equilibrium state and will show up in the form of additional bands that are shifted toward the IR relative to the main transition. Because of overlap of the main vibrational-rotational band of OH with the intense emission spectrum of water vapor, it will be necessary to measure the emission spectra of flames with high spectral resolution in order to establish the effect of nonequilibrium processes on the formation of the IR emission spectrum of OH. However, the present analysis of the emissivity spectra of hydrogen-oxygen flames has revealed the existence of continuum absorption in the 4.1–4.4 μm range owing to vibrational-rotational transitions of OH groups in excited electronic states (See Fig. 21).

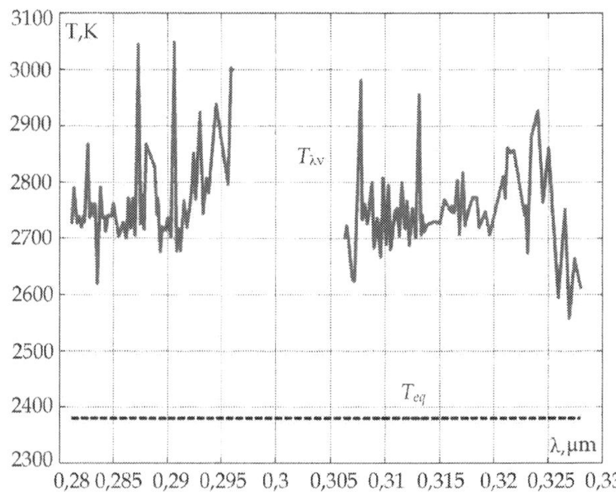

Fig. 20. Spectral dependence of the equilibrium thermodynamic temperature T_{eq} and the vibrational temperature $T_{\lambda v}$ for the $\Delta v = 0$, +1 sequences of the electronic bands of the hydroxyl group.

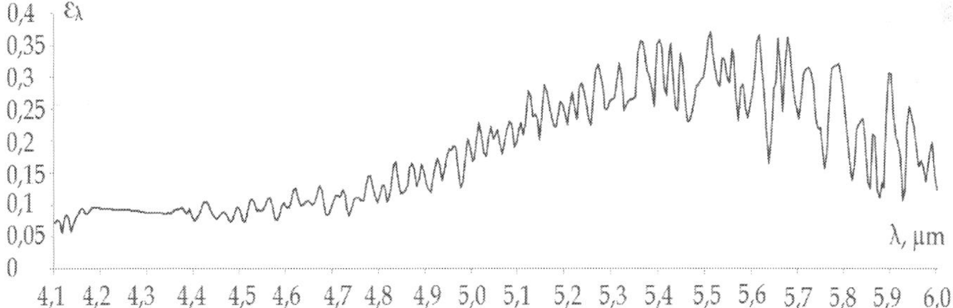

Fig. 21. Spectral emissivity of a hydrogen-oxygen flame with continuum emission of the hydroxyl group in an excited electronic state in the 4.1–4.4 μm range.

The analysis of spectra of radiation of the high spectral permission the hydrogen-oxygen flame in a spectral range of 3000-3500 cm^{-1} received at the spectral permission Δ = 0,1 cm^{-1} (Moskalenko et al., 1992), has revealed presence of nonequilibriumnes in the basic band v_1 hydroxyls OH. In a considered range of a spectrum radiation is formed by vibrational-rotary bands of vapor H_2O and hydroxyl OH. The spectrum of radiation of a flame was energetically not normalized. Besides, because of presence in experimental installation not blown and anonvacuum a site of an optical way the true spectrum of radiation of a flame is transformed by lines cold (T=291 K) water vapor and carbonic gas. For definition of vibrational temperature of hydroxyl OH the true spectrum of radiation corrected from influence of hardware distortion and atmospheric absorption of radiation by vapor H_2O and CO_2 at first is restored. Are most informative for definition of vibrational temperature of a Q-branch of band OH which vibrational temperature surpasses equilibrium temperature of a flame in a spectrum of wave number (vibrational temperature of vapor H_2O in a flame) on size ΔT_v=350÷850 K at average size ΔT_v=650K. This value ΔT_v is recommended to be used at definition of the contribution of radiation OH in radiating heat exchange in high-temperature mediums, including chambers of combustion of boiler installations. We will notice that nonequilibrium spectra of radiation OH in the basic vibrational-rotary band 0 – 1 it is caused by formation as a result of chemical reactions of molecules OH in the basic electronic state with superfluous population of the raised vibrational levels in comparison with equilibrium and transitions of molecules OH with the nonequilibrium populated raised electronic state to selection rules on vibrational and rotary quantum numbers. Definition of vibrational temperature from spectra of radiation is more preferable, as nonequilibrium radiation amplifies not only at the expense of increase in intensity of lines with growth of vibrational temperature ΔT_k, but also growth of function of source B_λ (T_v) with increase T_v.

In spectra of radiation of a flame influence nonequilibrium radiations OH in vibrational-rotary bands of the first and second overtones of the basic and raised electronic states is revealed. Their influence on radiating heat exchange high-temperature of flames can be more essential in connection with growth integrated intensitys bands $2v_1$, $3v_1$ OH with increase in temperature and weaker overshoot of spectra OH lines of water vapor. Bands $2v_1$ and $3v_1$ OH the raised electronic state settle down in a vicinity of lengths of waves 2,1 and 1,4 μm, and for the basic electronic state they are in a vicinity of lengths of waves 1,43 and 1 μm.

4. Results of definition of a microstructure of sooty ashes and its optical characteristics

Sooty ashes it is generated at burning of any hydrocarbonic fuel, including wood and an industrial production waste. The nobility suffices for the decision of the scalar equations of the transfer over of radiation spectral factors of absorption, scattering and indicatryss of scattering ashes and their spatial distribution on the radiating volume. The last depends on a chemical composition and a microstructure sooty ashes which are defined by a kind of fuel and a mode of its burning. The least concentration sooty ashes are observed at fuel burning in oxygen with surplus factor $\alpha \geq 1$ and at most high temperatures of a flame (Moskalenko et al., 2010). Also more thin microstructure ashes in this case takes place. At burning firm power fuel in fire chambers of boiler installations and power technological units the microstructure ashes in top internal volume changes in process of combustion of particles and is multimodal in connection with simultaneous generation sooty ashes in the course of fuel combustion. Thus a roughly-dispersion the fraction ashes is defined by characteristics of grinding devices. For gaseous fuel optical characteristics strongly depend on fuel structure. At hydrogen burning sooty ashes this kind of fuel isn't formed also is non-polluting at its burning in oxygen atmosphere. At burning of gas hydrocarbonic fuel a microstructure sooty ashes and its mass concentration in a flame depends on structure of gas fuel and can change largely (Moskalenko et al., 2010).

4.1 Results of definition of a microstructure of sooty ashes at burning of gas fuel

Variety used fuels and technologies of their burning studying of influence of a chemical composition of fuel, a microstructure of its scattering, a way of giving of fuel and an oxidizer (air), a fire chamber design, burning cause of devices on radiating heat exchange. Last is defined by a chemical composition, a microstructure and a phase condition of hot particles of fuel, a chemical composition of a gas phase of products of combustion and radiating characteristics ashes which is generated in the course of burning of fuel or as a result of ionic nucleation from a gas phase of products of combustion. Process of burning of fuel is accompanied by simultaneous radiating heat exchange and mass exchange combustion products on volume of a fire chamber and a heat transfer to evaporating surfaces. In the present work possibilities of definition of microstructural characteristics of ashes from data on easing of radiation by products of combustion and from radiation optical characteristics of flames which are included into the decision of the scalar equation of the transfer over of radiation are considered and are necessary for modeling of radiating heat exchange in high-temperature diphasic mediums, for example, in chamber fire chambers of power and power technological units. Thereupon optical characteristics industrial ashes are analysed, including sooty particles of the most probable not spherical forms. In case of spherical and cylindrical particles calculations were carried out under the theory of Mi. In case of not spherical particles optical characteristics were calculated in approach of geometrical optics and effective sections of absorption of scattering and scattering indicatryss are received by their averaging on space within a space angle 4π in the assumption that spatial orientation depending on a radiation direction is equiprobable. This assumption is proved enough in case of the decision of problems of radiating heat exchange in combustion chambers where strong turbulence of products of combustion and presence of whirlwinds is observed. For the analysis results of measurements of spectra of radiation of radiation and radiation easing flames, received on measuring complexes (Moskalenko et al., 1980, 1992) and their breadboard models.

Microstructure restoration ashes from the given spectral measurements of radiation of radiation of a flame and easing of radiation by a flame is a return task of remote sounding and demands for the decision of data on spectral dependences of optical characteristics ashes from distribution of particles in the sizes. Ashe of combustion products is polydisperse and on the modeling chosen by us it is represented superposition of separate fractions, microstructure N (r) which is defined by parity (Kondratyev & Moskalenko, 1983):

$$N(r) = \sum_{i=1}^{N} N_i(r) = \sum_{i=1}^{N} A_i \cdot r_a^{a_i} \cdot \exp\left[-b \cdot r^{c_i}\right], \tag{39}$$

where A_i, a_i, b_i, c_i – parameters of i-th fraction, N – number of fractions. As well as in (Kondratyev & Moskalenko, 1983) decisions on the restored spectral factors of easing and radiation absorption ashes we will represent from them normalized for length of a wave λ =0.55 µm, and the mass maintenance ashes on an optical way we will express in an optical thickness $|-\ln \tau|$, where τ – function spectral transmission on length of a wave λ =0.55 µm. The essence of a problem of restoration of a microstructure ashes consists in definition of weight functions $N_i(r)$ which in the best way will describe spectral dependence of the measured factors of absorption and radiation easing ashes (see fig. 22).
It is necessary to notice that flames form strongly absorbing gas medium so lengths of waves for the decision of a return task should be chosen so that radiation of gases in them was minimum. Inclusion in processing of lengths of waves with $\lambda > 4.2$ µm only increases errors of restoration of a microstructure of ashes.

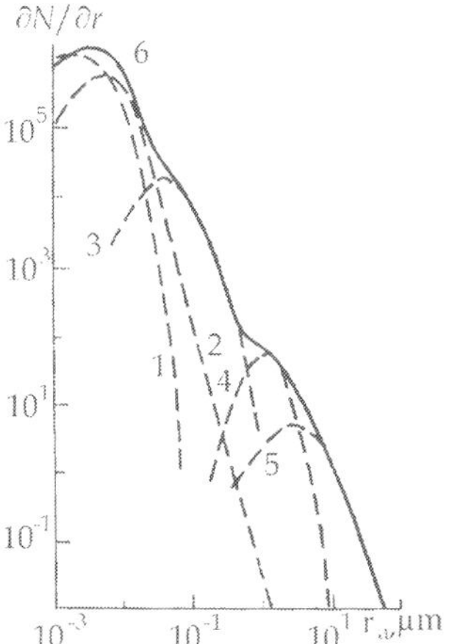

Fig. 22. An example of construction of model of microstructural characteristics (a curve 6) sooty ashes in the form of superposition of definitely weighed fashions curves 1, 2, 3, 4, 5.

Experience of restoration of a microstructure sooty ashes in flames has shown that in flames is present thin-dispersion sooty primary ashes which is generated in flames from a gas phase of products of combustion (Moskalenko et al., 2009) as a result of ionic nucleation, the microstructure and which optical density strongly depend on a chemical composition of gas fuel. The most probable modal radius of particles of this fraction ashes $r_m=0.003$ μm is received at methane burning in air and in oxygen. A population mean of optical density $(\partial\bar{\tau}/\partial L)=0.099$ m⁻¹. At propane-butane burning in air the increase in optical density, so $(\partial\bar{\tau}/\partial L)=0.2$ m⁻¹ is observed. The highest values $(\partial\bar{\tau}/\partial L)=0.4$ m⁻¹ is observed in flames at burning of acetylene C_2H_2 when the microstructure sooty ashes is observed most roughly-dispersion. At burning C_2H_2 higher value unburning fuel is observed also.

Microstructure restoration sooty ashes on spectral dependence of factors of radiation and easing of radiation by a flame demands inclusion in consideration of lengths of waves in an ultra-violet part of a spectrum, visible and near infra-red ranges. For definition of a microstructure of huge particles of mineral fraction ashes the additional information on scattering indicatryss of radiation is required. Mineral ash in flames is present at a liquid phase, the complex which indicator of refraction remains to unknown persons that complicates reception of matrixes of scattering of radiation for the purpose of their further use at definition of a microstructure mineral ash in flames. It is important to remember that mineral ash in flames and gas-mains boiler installations it is enriched by carbon (soot) and its radiating properties strongly depend on a subtlety of crushing of firm fuel. Than is thinner highling crushing, the disperse phase of products of combustion has especially high volume factors of absorption (radiation). In process of burning out of particles of firm fuel their size decreases and the specific density of particles that leads to reduction of their radiating ability increases. However, as a result breaking the thaw in the course of them degassing, is generated more a thin dispersion fashion mineral ash with modal radius $r_m{\approx}0.3\text{-}1$ μm, strengthening radiating ability of products of combustion in visible and near infra-red spectrum ranges. Optical characteristics of some equivalent ensemble of spherical particles are defined by a complex indicator of refraction $n=m\text{-}ix$, where m and x – its valid and imaginary parts which for a multicomponent chemical composition of particles are described by a parity:

$$m=\frac{\sum\limits_i m_i\cdot\rho_i}{\sum\limits_i\rho_i}\ ;\ x=\frac{\sum\limits_i x_i\cdot\rho_i}{\sum\limits_i\rho_i}\ ,\tag{40}$$

where ρ_i – mass concentration of connection i in a particle. Summation extends on all connections which are a part of a particle, including air for friable particles, and also a moisture at the influence account condensation processes on optical characteristics of a disperse phase in atmosphere. In table 4 parameters of models of a microstructure of the particles used as the aprioristic information at restoration of a microstructure of a disperse phase of products of combustion are resulted.

For microstructure definition sooty ashes in flames it is necessary to restore at first its spectral optical density. Function spectral transmission gas components of products of combustion for the chosen optical channels of sounding (optimum for sounding of lengths of waves) and their contribution to spectral intensity of radiation of a flame is for this purpose calculated. Further function of spectral easing of radiation ash and function spectral transmission ashes $\tau^a_{\lambda a}$, caused only is defined by radiation absorption. On the measured spectral functions of easing of radiation τ^a_λ and spectral radiating ability ashes $(1\text{-}\tau^a_{\lambda a})$ it is defined then spectral factors of easing σ^a_λ and absorption (radiation) $\sigma^a_{\lambda a}$ ashes. The restored spectra σ^a_λ and $\sigma^a_{\lambda a}$ serve for microstructure definition ash by optimum adjustment under spectral dependences σ_λ^a and $\sigma_{\lambda a}^a$. In table 5 the measured spectra σ_λ^a and $\sigma_{\lambda a}^a$ for methane-oxygen, the propane-butane - oxygen and acetylene-air flames are resulted. The restored distributions of a microstructure sooty layer in considered flames are presented in table 6.

№	a	b	c	r_m, µm
1	1	50	0.5	$1.6 \cdot 10^{-3}$
2	0.2	6	0.5	$4.4 \cdot 10^{-3}$
3	1	9	0.5	$4.9 \cdot 10^{-2}$
4	1	7.5	0.5	$9.0 \cdot 10^{-2}$
5	2	12	0.5	$8.9 \cdot 10^{-2}$
6	1.5	6	1	$2.5 \cdot 10^{-1}$
7	0.1	0.5	2	0.3
8	1	1	2	0.3
9	3	6	0.5	1.0
10	3	4	0.5	2.25

Table 4. Parameters of models of a microstructure of particles modified scale - distributions.

п/п	λ, µm	0.287	0.304	0.338	0.55	0.82	1.08	2.62	4.15	9.1
A	σ_λ^a, m^{-1}	0.161	0.154	0.144	0.099	0.072	0.058	0.029	0.019	0.008
	$\sigma_{\lambda a}^a$, m^{-1}	0.105	0.099	0.092	0.065	0.049	0.041	0.022	0.015	0.007
B	σ_λ^a, m^{-1}	0.234	0.230	0.225	0.204	0.174	0.164	0.103	0.082	0.044
	$\sigma_{\lambda a}^a$, m^{-1}	0.126	0.126	0.120	0.115	0.098	0.091	0.065	0.052	0.031
C	σ_λ^a, m^{-1}	0.385	0.398	0.405	0.400	0.380	0.372	0.271	0.225	0.156
	$\sigma_{\lambda a}^a$, m^{-1}	0.190	0.197	0.208	0.210	0.200	0.196	0.158	0.135	0.096

Table 5. The restored spectral optical density sooty ashes σ_λ^a and $\sigma_{\lambda a}^a$ for methane-oxygen (A), the propane-butane - oxygen (B) and acetylene-air (C) flames.

Results of the present measurements of a microstructure sooty ashes are used in calculations of optical characteristics, including spectral factors of easing, scattering, absorption and scattering indicatryss (Moskalenko & Loktev, 2009), in radiating heat exchange and problems of remote sounding.

Sooty ashes in combustion products define unburning fuel and reduce thermal efficiency.

On a microstructure sooty ash probably to define volume of particles ashe:

$$V = A \cdot \int_r \frac{4}{3} \cdot \pi \cdot r^3 \cdot \frac{\partial N}{\partial r} dr , \qquad (41)$$

where A – normalizing multiplier which is defined on optical density on length of a wave λ =0.55 μm. Knowing density sooty ashes ρ, it is possible to define chemical unburning of fuel $\rho_c \cdot V$.

r, μm	$\partial N / \partial r$			r, μm	$\partial N / \partial r$	
	A	B	C		B	C
0.001	$7 \cdot 10^5$	$1 \cdot 10^5$		0.15	$5 \cdot 10^1$	$2 \cdot 10^3$
0.002	$1.1 \cdot 10^6$	$3 \cdot 10^5$		0.2	$1 \cdot 10^1$	$7 \cdot 10^2$
0.003	$1 \cdot 10^6$	$5 \cdot 10^5$		0.3	$2 \cdot 10^0$	$1.6 \cdot 10^2$
0.004	$9 \cdot 10^5$	$6 \cdot 10^5$	$1 \cdot 10^3$	0.4	$8 \cdot 10^{-1}$	$4.0 \cdot 10^1$
0.006	$8 \cdot 10^5$	$6.5 \cdot 10^5$	$2 \cdot 10^3$	0.6	$3 \cdot 10^{-1}$	$1.5 \cdot 10^1$
0.008	$6 \cdot 10^5$	$6 \cdot 10^5$	$3 \cdot 10^3$	0.8	$1 \cdot 10^{-1}$	6
0.01	$4 \cdot 10^6$	$5 \cdot 10^5$	$5 \cdot 10^3$	1	$3 \cdot 10^{-2}$	2
0.015	$8 \cdot 10^5$	$3 \cdot 10^5$	$1 \cdot 10^4$	1.5		0.6
0.02	$1 \cdot 10^4$	$1 \cdot 10^5$	$1.3 \cdot 10^4$	2		0.2
0.03	$1 \cdot 10^3$	$1.6 \cdot 10^4$	$2 \cdot 10^4$	3		0.06
0.04	$1 \cdot 10^2$	$3 \cdot 10^3$	$1.3 \cdot 10^4$	4		0.02
0.06	$1 \cdot 10^1$	$1 \cdot 10^3$	$8 \cdot 10^3$	6		0.01
0.08	$1 \cdot 10^0$	$6 \cdot 10^2$	$5 \cdot 10^3$	8		0.003
0.1	0.1	$1.5 \cdot 10^2$	$3 \cdot 10^3$	10		0.001
				15		0.0001

Table 6. The restored distributions of a microstructure sooty ashes in methane-oxygen (A), the propane-butane - oxygen (B) and acetylene-air (C) flames.

Concentration of sooty ashes in combustion products strongly depends on structure and a mode of burning of gas fuel and changes in a range of values $4 \times 10^{-7} \div 2.6 \times 10^{-4} / m^3$. The minimum concentration of soot is observed in flame methane-oxygen at factor of surplus of oxygen α =1,03. The maximum concentration is measured for flame acetylene-air.

The received data on microstructure ashes for various flames are used for calculations of matrixes light-scatterig and spectral dependences of factors of easing, absorption and scattering, scattering indicatryss on a spectrum of lengths of waves with their subsequent use in calculations of radiating heat exchange in chambers of combustion of boilers and power technological units. Normalizing optical characteristics sooty ashes it is desirable to carry out with a binding on optical density on length of a wave λ =0.55 μm that allows to use the measured values received in the present work, at calculations of radiating heat exchange in projected objects. Thus it is possible to believe that burning of various components of fuel occurs independently. The weight of each microstructure in optical density sooty ashes is defined by a share of each component in weight of burned fuel.

The microstructure of sooty ashes at movement of products of combustion changes in connection with its burning out thin-dispersion fraction and capture of small particles of soot by large particles in the course of coagulation. These effects can be considered by change of weight of optical density of various fractions sooty ashes on fire chamber volume (Moskalenko et al., 2009).

Trustworthy information on concentration and a microstructure of ash can be received also by the analysis of the selected tests of products of combustion at lower temperatures with use warmed up multiple-pass cell as the working chamber (Moskalenko et al., 2010). Spectral dependence of easing of radiation bears the information on a microstructure ashes and, hence, about mass concentration ashes.

The knowledge of distribution of mass concentration and fields of temperatures of a disperse phase is of great importance by optimization and heat exchange intensification in top internal devices at coal and black oil burning.

5. Spectral and spatial distribution of thermal radiation in multichamber fire chambers

In the present section possibility of creation of boiler installations with the multichamber fire chambers is considered, allowing essentially to intensify radiating heat exchange in fire chambers and to raise vapor-productivity boiler installations. Now there are two possibilities of increase of efficiency of radiating heat exchange in fire chambers: increase in number of torches in a fire chamber with creation of strong radiating contrasts separate flames on the general thermal background of products of combustion in a fire chamber; creation of multichamber fire chambers where the increase in efficiency radiating cooling is reached at the expense of preservation of dimensions of the top internal chamber by a multichamber design.

The second variant is more effective and supposes wide variety of constructional decisions and allows to design vapor boilers with natural and compulsory circulation, and also to apply the mixed forms of circulation. Multichamber fire chambers can be used for burning firm hydrocarbonic fuels. In this case application in boiler installations of burning of fuel in the cyclonic prefire chambers is perspective, allowing to clear products of combustion of a cindery component. The top internal chamber in this case is actually radiating heat exchanger which for increase of efficiency radiating cooling combustion products is carried out by the multichamber.

The design of a multichamber fire chamber with ascending movement of products of combustion in a fire chamber and vertical development of a flame is most effective (Moskalenko & Zaripov, 2008; Moskalenko et al., 2008) with matrix burning device, the general gas collector for oxidizer giving (air or oxygen) and a collector for giving of gas fuel to multirow to torches. Burning devices are expedient for the transfer out with a radiator for its cooling by water on an independent circulating contour. At use the hearth burning devices providing maximum time of stay of products of combustion in a fire chamber. For the given height of a fire chamber and its sizes increase vapor-productivity the boiler can be reached by increase in speed radiating cooling $\partial T/\partial t$ which is reached by decrease in cross-section section of separate chambers in a multichamber fire chamber.

On fig. 23 dependence of the relation of speed radiating cooling $\partial T(L)/\partial t$ by the speed $(\partial T(L)/\partial t)_{max}$ from distance L between walls of a multichamber fire chamber is presented.

From the presented drawing it is visible that speed $\partial T(L)/\partial t$ reaches the maximum value at small distances L between fire chamber walls when function spectral transmission aspires to unit. With increase in distance L between walls of a fire chamber speed radiating cooling $\partial T(L)/\partial t$ decreases and, hence, for achievement of temperature T_{uh}=const on an exit from a fire chamber time $\Delta t(L)$ decreases with increase vapor-productivity a boiler. The less size L in a multichamber fire chamber, the at smaller height H from hearth fire chambers is reached value T_{uh}. The full value radiating cooling $\Delta T = T_a - T_{uh}$ depends on an arrangement of the torches forming the ascending stream of a flame.

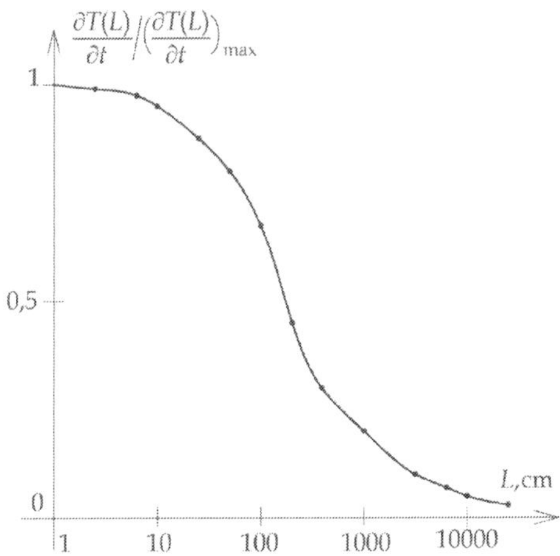

Fig. 23. Relative dependence of average speed radiating cooling $\dfrac{\partial T(L)}{\partial t}\Big/\Big(\dfrac{\partial T(L)}{\partial t}\Big)_{max}$ from

distance L between walls heatsusceptibility surfaces. Fuel – natural gas, the hearth burning device of matrix type with multirow torches, factor of surplus of air α =1,03.

For formation short flames it is expedient to use multirow burning devices the height of an ardent zone in which depends on diameter of tubes. For definition of height of an ardent zone in the presented calculations the theory of similarity and really observable heights of a flame was used. At diameter of tubes D=8 mm through which fuel (diameter of an individual nozzle) height of an ardent zone of products of combustion arrives makes 70 cm. With increase in diameter D the height of an ardent zone increases and, hence, its speed nonequilibrium radiating cooling decreases.

At modernization of existing steam-boilers with their transfer into an operating mode of multichamber fire chambers it is possible to increase in 2-3 times them vapor-productivity. Application of multichamber fire chambers in case of use of gas fuel allows to reduce steam-boiler dimensions (at the same vapor-productivity).

Designing of multichamber fire chambers demands consideration of technological aspects of their manufacturing about economic efficiency of manufacture. From our point of view manufacturing of a boiler of the cylindrical form of tower type with multichamber fire chambers and rectangular multichamber fire chambers should be rather effective. High thermal loadings of the bottom part heatsusceptibility surfaces of a steam-boiler demand optimization of loading of circulating contours of the heat-carrier and designs heatsusceptibility the surfaces providing their steady hydrodynamic characteristics.

It is necessary to notice that in a homogeneous medium of the infinite sizes of radiating heat exchange doesn't exist.

5.1 Mathematical model of the transfer over of nonequilibrium radiation in flames and high-temperature diphasic mediums

Errors of calculations at the decision of problems of radiating heat exchange are defined by function spectral transmission in structurally non-uniform multicomponent mediums.

Generally at function evaluation spectral transmission $\tau_{\Delta v}$ it is necessary to allocate contributions to the absorption, caused by wings of the remote spectral lines of the various atmospheric gases $\tau_{\Delta v}^k$, the absorption induced by pressure $\tau_{\Delta v}^n$, selective absorption of the spectral lines entering into the chosen spectral interval (owing to distinctions in these cases of function spectral transmission $\tau_{\Delta v}$ from the maintenance absorbing (radiating) gas, effective pressure P, and temperature T). Then for the set component function spectral transmission is defined as product of three considered above functions. Similar division allows providing universality of the description $\tau_{\Delta v}$ for any almost realized atmospheres of top internal devices of the present and the future workings out.

For multicomponent atmosphere $\tau_{i\Delta v}$ it will be defined as product $\prod\limits_i \tau_{i\Delta v}$, where i - component number. Legitimacy of this law is checked up experimentally and follows from independence of thin structure of spectra of the various absorbing (radiating) components which are a part of torches and oven atmosphere.

Let's believe that structural characteristics of the top internal chamber are known. For the account of nonequilibrium processes of radiation in a torch we will express function of a source for nonequilibrium radiation of a component i a torch in a kind

$$B_{i\lambda}(T) = B_\lambda^{abb}(T) \cdot \eta_{i\lambda}(T), \tag{42}$$

Where $\eta_{i\lambda}(T)$ - factor of nonequilibrium radiations for a component i.

Let's consider at first the elementary case of the absorbing medium: radiation scattering is absent or radiation scattering is neglected. We will assume that the temperatures of walls T_g is known, distribution of temperature T on volume of the top internal chamber and a field of concentration of gas and disperse components are set. Let O - a supervision point in the top internal chamber, K - a point of intersection of a vector of supervision l with a surface of the top internal chamber. A vector of scanning of volume of space from point K we will designate L. We will assume also that a wall surface is Lambert's. Then spectral intensity of thermal radiation in a direction l will be defined by a parity:

$$J_{\Delta\lambda}(l) = \int\limits_0^{L_k} \sum_i B_\lambda\big[T(l)\big] \eta_{i\lambda}\big[T(l)\big] \frac{d\tau_{i\Delta\lambda}}{dl} \prod_{k\neq i} \tau_{k\Delta\lambda}(l)\,dl + \left(1-\delta_\lambda^k\right) B_\lambda\left(T^k\right) \tau_{\Delta\lambda}\left(l_0^k\right) +$$

$$+ \frac{\delta_\lambda^k}{\pi} \int\limits_0^{2\pi} \int\limits_0^{L_k} B_\lambda\big[T(L)\big] \sum_i \eta_{i\lambda}\big[T(L)\big] \frac{d\tau_{i\Delta\lambda}}{dl} \left[L(\theta,\varphi)+l_0^k\right] \cdot \prod_{k\neq i} \tau_{k\Delta\lambda}\left[L(\theta,\varphi)+l_0^k\right] dL d\Omega + \tag{43}$$

$$+ \frac{\delta_\lambda^k}{\pi} \int\limits_0^{2\pi} \left[1-\delta_\lambda^g\right] B_\lambda\big[T(L_g)\big] \tau_{\Delta\lambda}\left(l_0^k + l_k\right) d\Omega,$$

where $T(l)$ - temperature of medium along an optical way l; $B_\lambda\big[T(l)\big]$ - spectral brightness of radiation of absolutely black body at temperature T in a point l; δ_λ - spectral factor of reflection of a wall; l_0^k - an optical way between points O and K; T^k - temperature in point K; $\tau_{\Delta\lambda}(l)$ - function spectral transmission for an optical way l in a spectral interval in width $\Delta\lambda$; λ - length of a wave; $T(L_g)$ - temperature in a point of intersection of vector L with a wall surface; $d\Omega$ - a space angle element; θ, φ - antiaircraft and azimuthally corners, accordingly; $\tau_{\Delta\lambda}(L+l)$ means function spectral transmission along an optical way $(L+l)$; the index «g» means wall border.

In the ratio (43) product undertakes on all components $k \neq i$, including ashes,

$$\tau_{\Delta\lambda} = \prod_i \tau_{\Delta\lambda i},\qquad(44)$$

where $\tau_{\Delta\lambda i}$ - function of spectral transmission for i-th component as gas, so disperse phases of top internal atmosphere. For gas components function $\tau_{\Delta\lambda i}$ are calculated on a two-parametrical method of equivalent mass, considered in section 2.2.

For the account of absent-minded radiation, we will choose the beginning of coordinates at the bottom of a fire chamber. An axis of coordinates z we will choose in conformity with symmetry of an ascending stream of products of combustion. We will enter polar system of coordinates. We will designate a supervision point z_n with antiaircraft θ_0 and azimuthal φ_0 supervision corners; θ, φ – flowing antiaircraft and azimuthally corners of integration on space. Then any point in fire chamber space will be characterized by height z concerning a bottom of a fire chamber and corners θ, φ, and a surface limiting space of a fire chamber – coordinates z_g, θ, φ. The radiation going to the top hemisphere from a point of supervision z_n, we will name ascending with intensity $J\uparrow$. The radiation going to the bottom hemisphere with intensity $J\downarrow$ we will name descending. The corner of scattering of radiation $\Psi(\theta_0, \varphi_0, \theta, \varphi)$ depends as on a supervision direction θ_0, φ_0, and current corners of integration θ, φ of absent-minded radiation. We will assume further that the fire chamber surface has temperature $T(z_g, \theta, \varphi)$ and spectral factor of reflection $\delta_\lambda(z_g, \theta, \varphi)$, λ - length of a wave of radiation.

Let's enter further scattering indicatryss $f(z, \Psi)$ in such a manner that

$$\int d\phi \int f(z,\Psi)\sin\Psi d\Psi = 1 .\qquad(45)$$

Let $\tau_{\Delta\lambda a}(l)$ - function spectral transmission at the expense of absorption of radiation of a gas phase of top internal atmosphere and its disperse phase, $\tau_{\lambda s}^a(l)$ - function spectral transmission (easing) only at the expense of scattering of radiation of a disperse phase of top internal atmosphere, $\tau_{\lambda a}^a(l)$ - function spectral transmission at the expense of absorption of radiation by aerosols for which following parities are fair:

$$\tau_{\lambda a}^a(l) = \exp\left[-\sigma_{\lambda a}^a \int_l \sigma_0^a(l)dl\right],\qquad(46)$$

$$\tau_{\lambda s}^a(l) = \exp\left[-\sigma_{\lambda s}^a \int_l \sigma_0^a(l)dl\right],\qquad(47)$$

$$\tau_{\Delta\lambda a}^a(l) = \tau_{\lambda a}^a(l)\prod_i \tau_{\Delta\lambda i}^a(l),\qquad(48)$$

where l - an optical way which runs radiation beam, $\sigma_{\lambda a}^a, \sigma_{\lambda s}^a$ - spectral normalizing volume factors of absorption and aerosol scattering, $\sigma_{\lambda 0}^a(l)$ - volume factor of easing of radiation by an aerosol on length of a wave λ =0,55 μm, $\tau_{\Delta\lambda i}(l)$ - function spectral transmission for i-th component of a gas phase of atmosphere for spectral intervals in width Δ with the center λ which are calculated on a two-parametrical method of equivalent mass. Let's choose a supervision direction l which will cross borders of a surface of a fire chamber for the top hemisphere $z_g^+(\theta_0, \phi_0)$ and for the bottom hemisphere $z_g^-(\theta_0, \phi_0)$ concerning a horizontal plane $z = z_n$.

Then for intensity of ascending radiation $J_{\Delta\lambda}^\uparrow$ in approach of unitary scattering in a direction θ_0, ϕ_0, in a point z_n

$$J_{\Delta\lambda}^\uparrow = J_{1\Delta\lambda}^\uparrow + J_{2\Delta\lambda}^\uparrow + J_{3\Delta\lambda}^\uparrow + J_{4\Delta\lambda}^\uparrow + J_{5\Delta\lambda}^\uparrow , \tag{49}$$

and for intensity of descending radiation

$$J_{\Delta\lambda}^\downarrow = J_{1\Delta\lambda}^\downarrow + J_{2\Delta\lambda}^\downarrow + J_{3\Delta\lambda}^\downarrow + J_{4\Delta\lambda}^\downarrow + J_{5\Delta\lambda}^\downarrow , \tag{50}$$

Where $J_{1\Delta\lambda}^\downarrow$ - own descending radiation of the medium of the top internal chamber in a supervision point; $J_{2\Delta\lambda}^\downarrow$ - radiation of a wall of the top internal chamber in the supervision direction, weakened by top internal atmosphere; $J_{3\Delta\lambda}^\downarrow$ - disseminated in a direction of supervision the radiation which is starting with volume of top internal atmosphere (from point volume); $J_{4\Delta\lambda}^\downarrow$ - absent-minded radiation of all walls of the fire chamber, reflected from a point l^g on a wall in a supervision direction; $J_{5\Delta\lambda}^\downarrow$ - own radiation of all walls of the chamber, weakened by oven atmosphere and reflected from a point on a wall in a supervision direction. The physical sense of components intensitys in the ratio (49) for ascending radiation is similar.

For nonequilibrium radiation source function is various for various radiating components and can change within top internal volume and on a spectrum of lengths of waves of electron-vibrational transitions of molecules. If sizes $\eta_{i\lambda}$ for components i are known, in the intensity equations it is necessary to enter summation of radiations on components i under the badge of integrals, having replaced size $B_\lambda(T)$ in size $B_\lambda(T)\cdot\eta_{i\lambda}$. Then for intensity of ascending radiation:

$$J_{1\Delta\lambda}^\uparrow = \int\limits_{z_g^-}^{z_n} \sum_i \frac{B_\lambda\left[T(z),\theta_0,\phi_0\right]\cdot\eta_{i\lambda}\left[T(z),\theta_0,\phi_0\right]\tau_{\lambda s}^a(z,\theta_0,\phi_0)\dfrac{\partial\tau_{i\Delta\lambda a}(z,z_n,\theta_0,\phi_0)}{\partial z}}{\prod\limits_{k\neq i}\tau_{i\Delta\lambda a}^a(z,z_n,\theta_0,\phi_0)}\, dz , \tag{51}$$

$$J_{2\Delta\lambda}^\uparrow = B_\lambda\left[T_g(z_g^-,\theta_0,\phi_0)\right]\tau_{\lambda s}(z_g^-,\theta_0,\phi_0)\tau_{\Delta\lambda a}(z_g^-,\theta_0,\phi_0) , \tag{52}$$

$$J_{3\Delta\lambda}^\uparrow = \int\limits_0^{2\pi}\partial\phi\int\limits_{-\pi/2}^{\pi/2}\sin\theta\,\partial\theta\int\limits_{z_g^-}^{z_n} f_\lambda\left[z,\psi(\theta,\phi,\theta_0,\phi_0)\right]\frac{\partial}{\partial z}\tau_{\lambda s}\left[z,z_g,(\theta,\phi),\theta,\phi;z,z_n,\theta_0,\phi_0\right]\cdot$$

$$\cdot \int_{z_g^+}^{z} \left\{ \sum_i B_\lambda[T(z,\theta,\phi)] \eta_{i\lambda}[T(z,\theta,\phi)] \frac{\partial}{\partial z} \tau_{i\Delta\lambda\alpha}\left[z,z',\theta,\phi;z_n,z,\theta_0,\phi_0\right] \prod_{k\neq i} \tau_{k\Delta\lambda\alpha}\left(z,z',\theta,\phi;z_n,z,\theta_0,\phi_0\right) \right\} dz' dz, \tag{53}$$

$$J_{4\Delta\lambda}^{\uparrow} = \frac{\delta_{\lambda g}\left(z_g^-,\theta_0,\phi_0\right)}{\pi} \frac{2\pi}{\int_0^{} d\phi} \int_{-\pi/2}^{\pi/2} \sin\theta d\theta \int_{z_g^-}^{z_n} f_\lambda\left[z,\Psi\left(\theta,\phi,\theta_0,\phi_0\right)\right] \frac{\partial}{\partial z} \tau_{\lambda s}\left[z,z_g,(\theta,\phi),\theta,\phi;z,z_n,\theta_0,\phi_0\right] \cdot \tag{54}$$

$$\cdot \tau_{\Delta\lambda a}\left[z,z_g,(\theta,\phi),\theta,\phi;z,z_n,\theta_0,\phi_0\right] \cdot \left[1-\delta_{\lambda g}\left(z,\theta_0,\phi_0\right)\right] B_\lambda^g[T(\theta,\phi)] dz,$$

$$J_{5\Delta\lambda}^{\uparrow} = \frac{\delta_\lambda\left(z_g^-,\theta_0,\phi_0\right)}{\pi} \int_0^{2\pi} d\phi \int_{-\pi/2}^{\pi/2} \left\{1-\delta_\lambda\left(z_g,(\theta,\phi)\right)\right\} \cdot B_\lambda^g[T(\theta,\phi)] \tau_{\lambda s}\left[z_g^-,z_g(\theta,\phi),\theta,\phi;z_g^-,z_n,\theta_0,\phi_0\right] \cdot \tag{55}$$

$$\cdot \tau_{\Delta\lambda a}\left[z_g^-,z_g(\theta,\phi),\theta,\phi;z_g^-,z_n,\theta_0,\phi_0\right] d\theta,$$

where summation is carried out on all components i, and product – on all components $k \neq i$; z_g^- means that fire chamber borders are located below supervision height z_n; $\delta_{\lambda g}^+$ means reflection factor on border of the fire chamber located at height $z_g > z_n$.

For intensity of descending radiation $J_{1\Delta\lambda}^{\downarrow}$ it is easy to write parities, similar (50-54),

$$J_{1\Delta\lambda}^{\downarrow} = \int_{z_g^+}^{z_n} \sum_i B_\lambda[T(z),\theta_0,\phi_0] \cdot \eta_{i\lambda}[T(z),\theta_0,\phi_0] \tau_{\lambda s}^a(z,\theta_0,\phi_0) \frac{\partial \tau_{i\Delta\lambda a}\left(z,z_n,\theta_0,\phi_0\right)}{\partial z} \prod_{k\neq i} \tau_{i\Delta\lambda a}^a\left(z,z_n,\theta_0,\phi_0\right) dz, \tag{56}$$

$$J_{2\Delta\lambda}^{\downarrow} = B_\lambda\left[T_g\left(z_g^+,\theta_0,\phi_0\right)\right]\left[1-\delta_\lambda\left(z_g^+,\theta_0,\phi_0\right)\right] \tau_{\lambda s}\left(z_g^+,\theta_0,\phi_0\right) \tau_{\Delta\lambda a}\left(z_g^+,\theta_0,\phi_0\right), \tag{57}$$

$$J_{3\Delta\lambda}^{\downarrow} = \int_0^{2\pi} d\phi \int_{-\pi/2}^{\pi/2} \sin\theta d\theta \int_{z_g^+}^{z_n} f_\lambda\left[z,\Psi\left(\theta,\phi,\theta_0,\phi_0\right)\right] \frac{\partial}{\partial z} \tau_{\lambda s}\left[z,z_g,(\theta,\phi),\theta,\phi;z,z_n,\theta_0,\phi_0\right] \cdot \tag{58}$$

$$\cdot \int_{z_g^+}^{z} \left\{ \sum_i B_\lambda[T(z,\theta,\phi)] \eta_{i\lambda}[T(z,\theta,\phi)] \frac{\partial}{\partial z} \tau_{i\Delta\lambda\alpha}\left[z,z',\theta,\phi;z_n,z,\theta_0,\phi_0\right] \prod_{k\neq i} \tau_{k\Delta\lambda\alpha}\left(z,z',\theta,\phi;z_n,z,\theta_0,\phi_0\right) \right\} dz' dz, \tag{59}$$

$$J_{4\Delta\lambda}^{\downarrow} = \frac{\delta_{\lambda g}\left(z_g^+,\theta_0,\phi_0\right)}{\pi} \int_0^{2\pi} d\phi \int_{-\pi/2}^{\pi/2} \sin\theta d\theta \int_{z_g^+}^{z_n} f_\lambda\left[z,\Psi\left(\theta,\phi,\theta_0,\phi_0\right)\right] \frac{\partial}{\partial z} \tau_{\lambda s}\left[z,z_g,(\theta,\phi),\theta,\phi;z,z_n,\theta_0,\phi_0\right] \cdot \tag{60}$$

$$\cdot \tau_{\Delta\lambda a}\left[z,z_g,(\theta,\phi),\theta,\phi;z,z_n,\theta_0,\phi_0\right] \cdot \left[1-\delta_{\lambda g}\left(z,\theta_0,\phi_0\right)\right] B_\lambda^g[T(\theta,\phi)] dz,$$

$$J_{5\Delta\lambda}^{\downarrow} = \frac{\delta_\lambda\left(z_g^+,\theta_0,\phi_0\right)}{\pi} \int_0^{2\pi} d\phi \int_{-\pi/2}^{\pi/2} \left\{1-\delta_\lambda\left(z_g,(\theta,\phi)\right)\right\} B_\lambda^g[T(\theta,\phi)] \cdot \tau_{\lambda s}\left[z_g^+,z_g(\theta,\phi),\theta,\phi;z_g^+,z_n,\theta_0,\phi_0\right] \cdot \tag{61}$$

$$\cdot \tau_{\Delta\lambda a}\left[z_g^+,z_g(\theta,\phi),\theta,\phi;z_g^+,z_n,\theta_0,\phi_0\right] d\theta.$$

Processes of nonequilibrium radiation at burning hydrocarbonic fuels practically aren't developed also their influence on radiating cooling a torch of top internal space practically isn't studied. From the most general reasons of formation of electron-vibrational spectra it is possible to draw a conclusion that the greatest influence on process of radiating heat exchange in fire chambers nonequilibrium renders radiations at burning of gaseous hydrocarbonic fuel and black oil which incorporate S-contents and N-contents the components forming nonequilibrium radiation of a high-temperature kernel of a torch.

At the decision of problems of radiating heat exchange in boilers operate integrated intensity thermal radiation which are defined by integration spectral intensitys thermal radiation on a spectrum of lengths of waves λ:

$$J^{\uparrow}\left(z_n,\theta_n,\phi_n\right) = \int_0^{\infty} J_{\lambda}^{\uparrow}\left(z_n,\theta_n,\phi_n\right) d\lambda\,, \tag{62}$$

$$J^{\downarrow}\left(z_n,\theta_n,\phi_n\right) = \int_0^{\infty} J_{\lambda}^{\downarrow}\left(z_n,\theta_n,\phi_n\right) d\lambda\,. \tag{63}$$

Knowing sizes $J^{\uparrow\downarrow}\left(z_n,\theta_n,\phi_n\right)$, it is possible to define streams of thermal radiation on any direction including on heatsusceptibility surfaces, having executed spatial integration $J^{\uparrow\downarrow}$ within a space angle 2π. In particular, for streams of descending and ascending radiation

$$F^{\uparrow\downarrow}\left(z\right) = \int_0^{2\pi} J^{\uparrow\downarrow}\left(z,\theta,\phi\right) d\Omega\,, \tag{64}$$

where $d\Omega$ – a space angle element. Radiating change of temperature will be defined from a parity

$$\frac{\partial T\left(z,\theta,\phi\right)}{\partial t} = \frac{1}{\rho\left(z,\theta,\phi\right)C_p\left(z,\theta,\phi\right)} \cdot \frac{\partial F\left(z,\theta,\phi\right)}{\partial z}\,, \tag{65}$$

where $\rho\left(z,\theta,\phi\right), C_p\left(z,\theta,\phi\right)$ - accordingly density and a thermal capacity in a local point with coordinates z, θ, φ,

$$F\left(z,\theta,\varphi\right) = F^{\uparrow}\left(z,\theta,\varphi\right) - F^{\downarrow}\left(z,\theta,\varphi\right).$$

If heat exchange process is stationary, $dT\left(z,\theta,\phi\right)/dz = const$ for any local volume with coordinates z, θ, φ. If heat exchange process is not stationary there are time changes of temperature in the local volumes which time trend can be calculated by application of iterative procedure of calculations on each time step i so

$$T_{i+1}\left(z,\theta,\phi\right) = T_i\left(z,\theta,\phi\right) + \Delta t \frac{dT_i\left(z,\theta,\phi\right)}{dt}\,. \tag{66}$$

However thus it is necessary to take into consideration and influence of other mechanisms of heat exchange: diffusion, turbulent diffusion, convective heat exchange.

Most intensively radiating cooling it is shown in a torch kernel, in this connection its temperature always below theoretical on 15-20 %. The last means that during combustion of fuel the torch considerably cools down as a result of radiating cooling. Degree radiating cooling a torch is maximum, if the stream expires in free atmosphere. In the closed volume of a fire chamber radiating cooling increases with growth of temperature of a torch, degree of its blackness at the expense of absorption of radiation by gas and disperse phases of products of combustion and decreases at rise in temperature heatsusceptibility surfaces and their factors of reflection. In cold zones of a fire chamber can take place and radiating heating if in them active components contain optically. If there are temperature inversions in temperature distributions in zones of temperature inversions radiating heating or easing radiating cooling also can be observed.

Full radiating cooling combustion products in a fire chamber depends on time of their stay in top internal volume and, hence, from speed of movement of products of combustion $V(z)$ in a fire chamber which can change on fire chamber height. Full radiating cooling combustion products ΔT it is defined by the formula:

$$\Delta T = \int_{0}^{H} \frac{1}{V(z)} \cdot \frac{\partial T}{\partial z} dz , \tag{67}$$

where H is fire chamber height.

Let's analyze the physical processes proceeding in the top internal chamber under the influence of nonequilibrium short-wave radiation which is generated in ultra-violet and visible parts of a spectrum as a result of a relaxation of the raised molecules formed at burning of fuel. If the difficult molecule is formed in wild spirits and dissociates on unstable short-living splinters also its splinters will be in wild spirits and to generate nonequilibrium short-wave radiation. Owing to small time of life of these connection spectral lines of nonequilibrium radiation will be much wider, than for equilibrium radiation, and can create the diffuse spectra of radiation which are not dependent from widening of pressure. Functions spectral transmission for the nonequilibrium medium submits to the law of Buger:

$$\tau_{\Delta \upsilon}(L) = \int_{L} \exp\left[-k_{\upsilon}(L)\right] dL , \tag{68}$$

where $k_{\upsilon}(L)$ - absorption factor, υ - the wave number, and integration is carried out along optical way L to a torch kernel. Vibrational and rotary structures of a spectrum of nonequilibrium radiation it will be washed away and poorly expressed. There is a basis to believe, as nuclear spectra of elements also can be nonequilibrium that proves to be true on an example of nuclear spectra of the sodium which lines of radiation have appeared nonequilibrium and at high temperatures can't be used for definition of temperature of a flame. Hence, probably to expect presence of photochemical reactions under the influence of the short-wave radiation, products of combustion essentially influencing a chemical composition in the top internal chamber.

Feature of nonequilibrium processes of radiation is considerable cooling zones of chemical reactions in time $\approx 10^{-4}$ sec, commensurable in due course courses of chemical reactions. In this connection the equilibrium temperature of a flame considerably decreases that leads to much lower concentration of a monoxide of nitrogen NO. Really, it agree (Zel'dovich et al., 1947)

$$[NO]_{max} = 4.6\sqrt{C_{N_2}C_{O_2}}\exp\left(\frac{-21500}{RT_{max}}\right), \tag{69}$$

where at fuel burning in air

$$C_{O_2} = 21(\alpha - 1)V_0 / (1 + \alpha V_0)$$
$$C_{N_2} = 79\alpha V_0 / (1 + \alpha V_0)$$

(70)

Here T_{max} - the maximum absolute temperature in peaks of volumes of chemical reactions, R - a gas constant, V_0 - theoretically necessary quantity of air for fuel burning, α - factor of surplus of air. At high temperatures real concentration NO in combustion products on an order and lower, than intended under formulas (69, 70) that from our point of view is caused nonequilibrium radiating cooling peaks of chemical reactions. And real concentration NO can depend on depth of turbulence burning and a spectrum of whirls.

By consideration of radiating heat exchange in the top internal chamber with torch burning of firm fuel in the twirled streams it is necessary to take into consideration the phenomenon of separation of particles when the largest particles are taken out in peripheral zones of a fire chamber where, settling, can grasp sooty ashes, formed as a result of pyrolysis in cold zones of a fire chamber, and then to flow down in a cold funnel.

Considering dependence of absorption of nonequilibrium radiation by combustion products, we will pay attention to strengthening of absorption with increase in capacity of the top internal chamber. Hence, with increase in capacity of a fire chamber nonequilibrium radiation in a greater degree passes in thermal energy of particles of fuel and thermal energy of products of combustion. Nonequilibrium radiating cooling decreases also and concentration NO_x increases with increase in capacity of a fire chamber that is really observed by results of statically provided supervision.

Let's pay attention to results of measurements of a chemical composition of products of combustion of wood (Moskalenko et al., 2010) when raised concentration NO_2 have been found out. If at burning of black oil and gases the relation of concentration $C(NO_2)/C(NO) \approx$ 0.1, at burning of wood the relation of concentration $C(NO_2)/C(NO) \approx 1/3$. It means that the increase in concentration NO_2 causes increase in intensity of the nonequilibrium radiation reducing temperature of a flame, and, hence, leads to reduction of concentration NO. Considering optical properties of a disperse phase depending on a microstructure of liquid or firm fuel at chamber burning, it is necessary to notice that concentration NO will increase in smoke gases with increase in a subtlety of scattering of liquid fuel and crushing of firm fuel. From the point of view of ecological influence of atmospheric emissions on flora and fauna expediently chamber burning of fuel of rough crushing and scattering. Besides, from the point of view of minimization of anthropogenous influences on medium it is expedient to burn fuel at lower pressure as nonequilibrium radiating cooling amplifies with pressure decline in a fire chamber (process of suppression of a chemical luminescence with pressure decline it is weakened).

Presence chemical unburning leads to formation of heavy hydrocarbons in combustion products (especially benzologies) that causes suppression of nonequilibrium radiation in a fire chamber. The last can be formed in interfaces of the top internal chamber and weaken heatsusceptibility of screens owing to strong absorption of ultra-violet radiation.

Presence of connections of sulfur in fuel leads to occurrence of nonequilibrium radiation SO_2 in the field of a spectrum $\lambda < 0.4$ μm which reduces flame temperature, and, hence, and concentration NO_x.

5.2 Modelling of radiating heat exchange in multichamber fire chambers

Let's consider results of modeling of radiating heat exchange of multichamber fire chambers taking into account nonequilibrium processes of radiation (Moskalenko et al., 2009), section 5.1 executed on algorithms for diphasic structurally non-uniform medium of top internal space of the chamber of combustion.

On fig. 24 for an example results of calculations of vertical profiles of speed radiating cooling $\partial T(z)/\partial t, \partial T(z)/\partial z$ and stationary distribution of temperature $T(z)$ from fire chamber height z over cuts of capillaries matrix burning devices are illustrated. Fuel is natural gas of a gas pipeline of Shebalovka-Brjansk-Moskva, the size of horizontal section of a cell of a multichamber fire chamber 1,25x1,6 m².

Speed of giving of products of combustion on an initial site of a fire chamber makes values υ_0=25 m/s and υ_0=20 m/s at pressure in a fire chamber $1 \cdot 10^5$ Pa. Height of an ardent zone Δz = 0,7 m. In calculations are considered equilibrium and nonequilibrium processes of radiation on the algorithms considered above. It is supposed that process of burning of various components of gas fuel occurs independently at optimum value of factor of surplus of air α =1,03.

The microstructure sooty ashes is measured at burning (to look section 4) methane, propane-butane and acetylene (Moskalenko et al., 2010). Optical characteristics sooty ashes are calculated for the measured microstructures of a disperse phase of products of combustion. Volume factors of easing, absorption and scattering normalized on the measured values of optical density ash (Moskalenko et al., 2009).

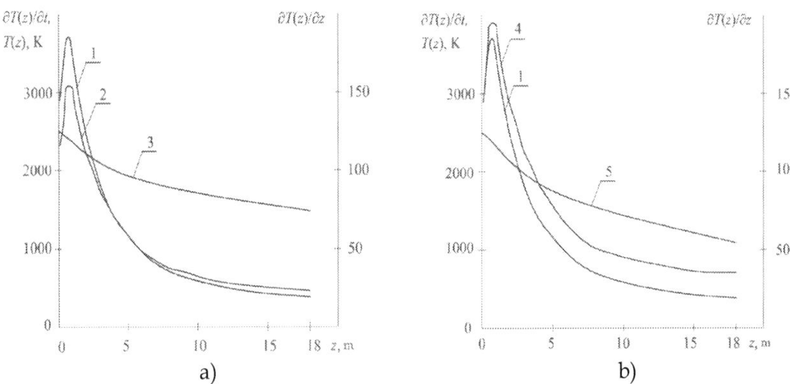

Fig. 24. Results of calculation of radiating heat exchange in a multichamber fire chamber with the size of horizontal section of a cell 1,25x1,6 m² for initial average speed of a current of products of combustion of 25 m/s (a) and 20 m/s (b). $\partial T(z)/\partial t, \partial T(z)/\partial z$ - speeds radiating cooling, T (z) – a temperature profile of average on section of temperature depending on height z over cuts of capillaries multirow torches. 1– $\partial T(z)/\partial t$; 2 – $\partial T(z)/\partial z$ for initial average speed of a current of products of combustion of 25 m/s; 3 – T (z) for initial average speed of a current of products of combustion of 25 m/s; 4 – $\partial T(z)/\partial z$ for initial average speed of a current of products of combustion of 20 m/s; 5 – $T(z)$ for initial average speed of a current of products of combustion of 20 m/s.

On fig. 25 examples of spectral and spatial distributions of thermal radiation on heatsusceptibility surfaces of a cell of a multichamber fire chamber by results of the closed modeling of process of radiating heat exchange with calculation of speed radiating cooling products of combustion and their temperature depending on height over cuts of capillaries multirow a torch forming ascending streams of a flame are resulted. Horizontal section of a cell of a multichamber fire chamber – a square with the party of 1,4 m. Fuel – natural gas of a gas pipeline of Shebalovka-Brjansk-Moskva, factor of surplus of air α = 1,03. Average initial speed of products of combustion makes 25 m/s. Pressure in a fire chamber – 10^5 Pa. The executed calculations of heatsusceptibility surfaces show that the greatest thermal loading the bottom part of lateral screens and heatsusceptibility is exposed to a surface hearth of fire chambers. So, on the central axis of the lateral screen at heights 1, 7, 17 meters from a cut of capillaries of a torch falling streams of heat make accordingly 260,313; 99,709; 48,387 kW/m². For the center hearth of fire chambers the falling stream of heat answers value of 249,626 kW/m², and the ascending stream of heat at height h = 18 m on an axis of a cell of a fire chamber makes 41,115 kW/m². A full stream

$$F = \int_s F(S)dS = \sum_i V_i \int_0^h C_{ip}\,[t(z)]dz , \qquad (71)$$

where C_{ip}, V_i – accordingly a thermal capacity at the constant pressure, answering to temperature t in a point z and volume for a component i combustion products. This condition at the closed modeling of heat exchange is carried out with a margin error 1 %. In approach of "gray" radiation when calculations are carried out under the law of Buger, overestimate heatsusceptibility on 15 % is observed. The account of effective pressure reduces an error of calculation full heatsusceptibility by 5-6 %. At use of a two-parametrical method of equivalent mass in calculations of function spectral transmission at modeling of a disperse phase of products of combustion in the present calculations it is supposed that burning of each component of fuel occurs independently that allows to use optical density sooty ashes by results of measurements on ardent measuring complexes. For methane, propane-butane, acetylene the optical density on length of a wave 0,55 µm is accepted according to equal 0,1; 0,2; 0,4 m⁻¹ in an ardent zone. Above an ardent zone it is observed exponential recession of numerical density thin-dispersion ashes with height in connection with its burning out. More rougly-dispersion fractions 2,3 sooty ashes don't burn out, and their distribution doesn't depend on height. The contribution of each fraction ashes is normalized according to volume concentration CH_4, propane-butane, C_2H_2.

On fig. 26 distribution of an integrated stream of the radiation calculated taking into account absorption (radiation) by basic optically by active components of products of combustion on lateral walls of a cell of a multichamber fire chamber depending on height of a fire chamber in case of weak approximation is illustrated. On fig. 27 distribution of an integrated stream of radiation to lateral walls of a cell of a multichamber fire chamber depending on height the fire chambers calculated with use of function spectral transmission on a two-parametrical method of equivalent mass is presented. On fig. 28 distribution of an integrated stream of the radiation calculated taking into account absorption (radiation) by basic optically active components of products of combustion, but without effective pressure is resulted. For the given design of a multichamber fire chamber the contribution of nonequilibrium radiation to radiating heat exchange makes 7,5 % from a full stream. Absence of the account of

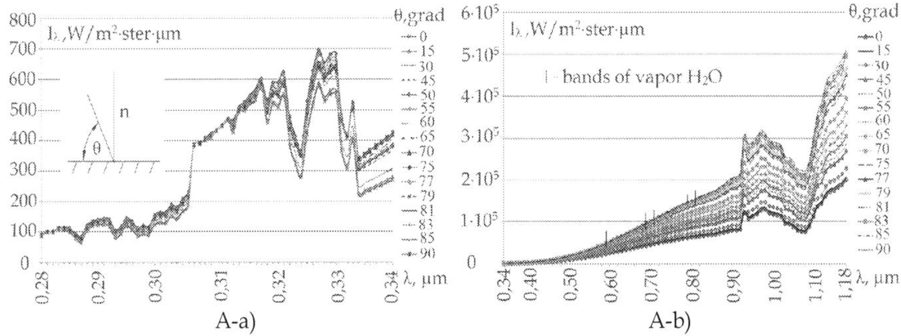

A-a)

A-b)

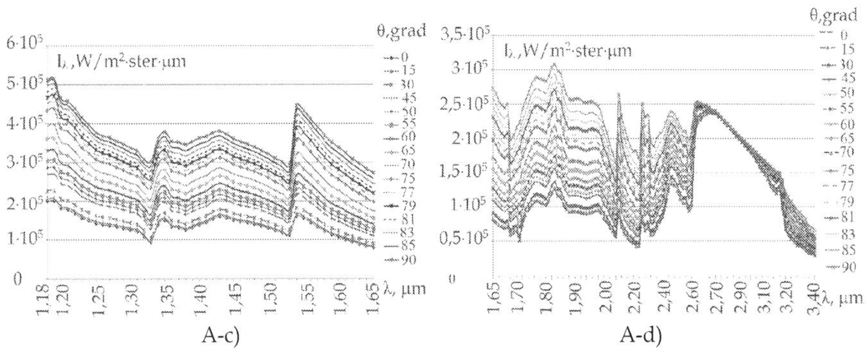

A-c)

A-d)

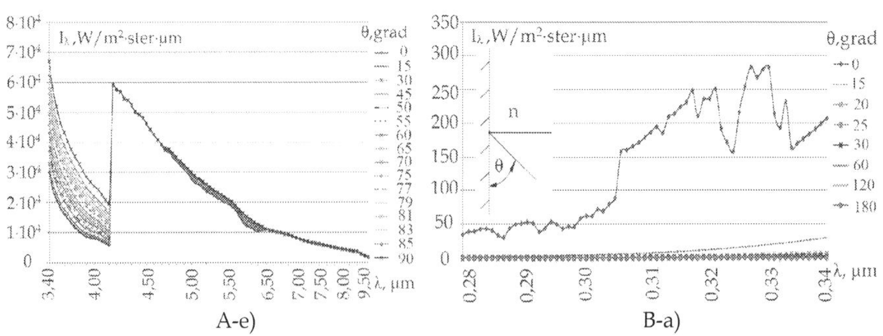

A-e)

B-a)

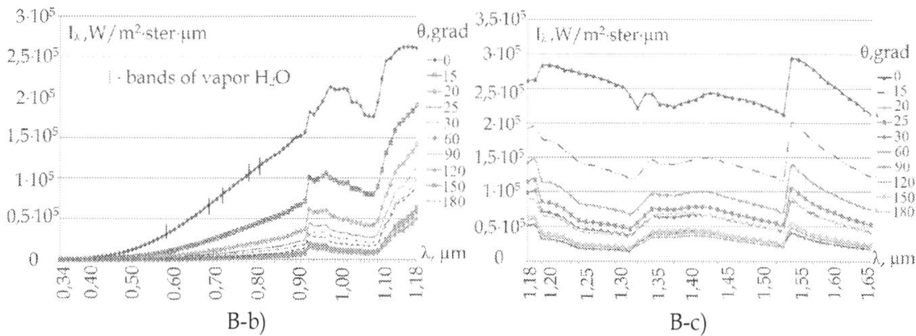

B-b)

B-c)

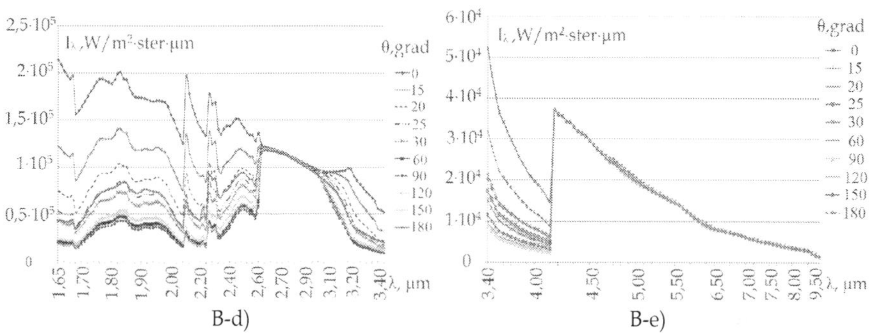

B-d)

B-e)

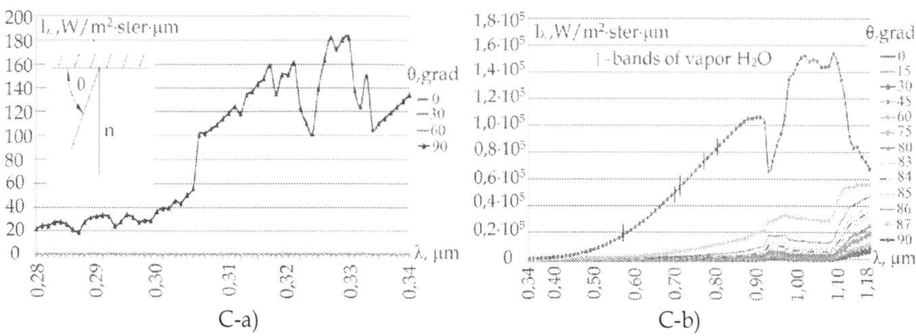

C-a)

C-b)

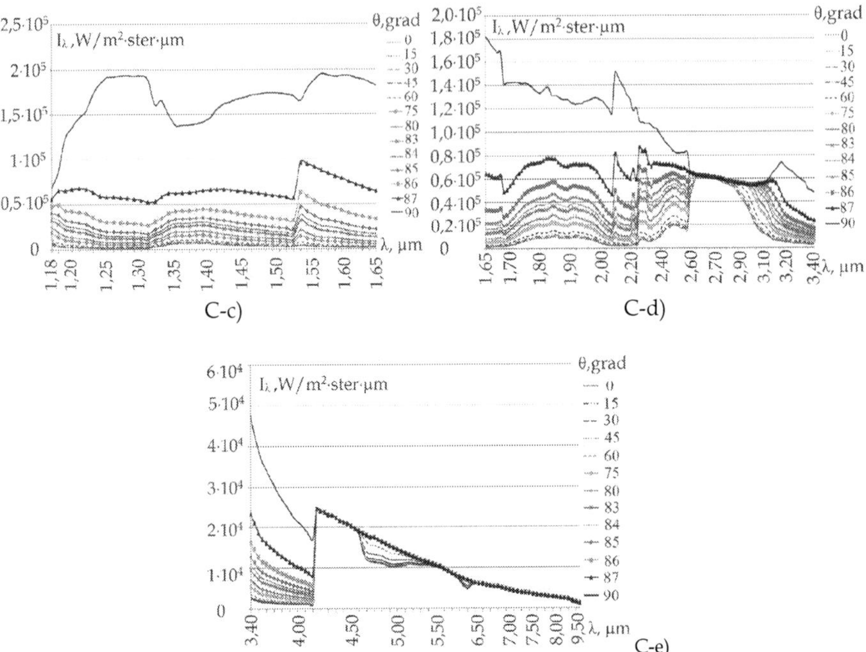

Fig. 25. Spectral and spatial distribution of thermal radiation in spectrum ranges: a) 0,28÷0,34 μm; b) 0,34÷1,18 μm; c)1,18÷1,65 μm; d) 1,65÷3,4 μm; e)3,4÷9,5 μm. A − descending radiation on a hearth heatsuscebility surface, B − falling radiation on lateral screens of a cell of a multichamber fire chamber at level 7 meter from a cut of capillaries multirow torches, C − ascending radiation at level of 18 meter from a cut of capillaries multirow torches of a cell of a multichamber fire chamber

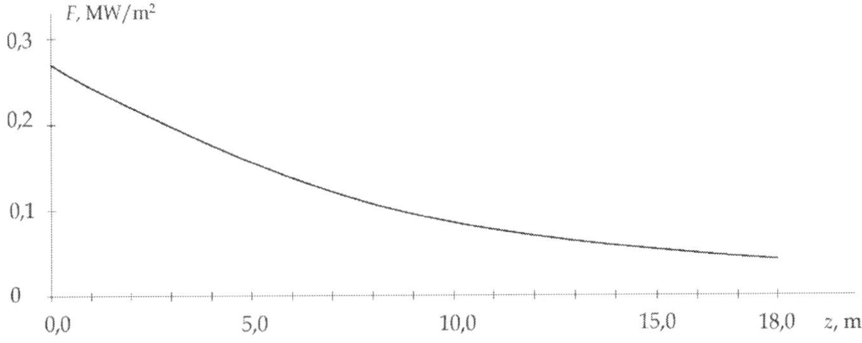

Fig. 26. Distribution of an integrated stream of the radiation depending on height of a fire chamber in case of weak approximation.

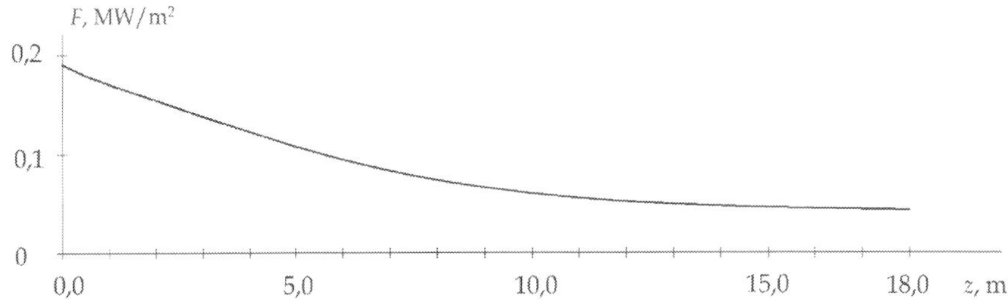

Fig. 27. Distribution of an integrated stream of the radiation depending on height the fire chambers calculated with use of function spectral transmission on a two-parametrical method of equivalent mass.

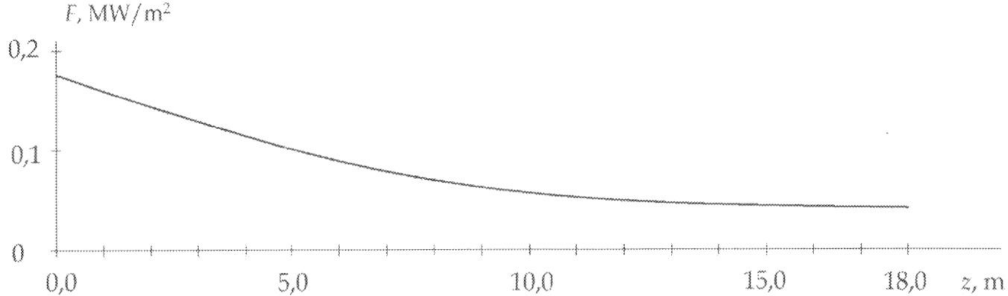

Fig. 28. Distribution of an integrated stream of the radiation calculated without effective pressure.

effective pressure in functions spectral transmission gas components underestimates radiating heat exchange on 5-6 %. The disperse phase of products of combustion influences radiating heat exchange at the expense of radiation ashes. Radiation scattering ashes poorly influences radiating heat exchange in strongly absorbing top internal atmosphere. Reflection of radiation from walls of the top internal chamber leads to reduction of speed radiating cooling in top internal volume.

Generally at heat exchange calculations it is necessary to consider the transfer over of heat at the expense of recirculation of products of combustion in a fire chamber and massexchange owing to diffusion which influence temperature distribution on volume of the chamber of combustion. With the advent of supercomputers there is possible an application of numerical methods of the decision of problems of the transfer over of radiation (Marchuk & Lebedev, 1981; Surgikov, 2004; Moskalenko et al., 1984) which restrain insufficient reliability of data on parameters of spectral lines of gas components of products of combustion.

6. Conclusion

In the conclusion we will stop on the basic results received in the course of the present work.
1. The developed measuring optic-electronic complexes for research of optical characteristics of high-temperature mediums and flames have allowed to spend registration of spectra of absorption and spectra of radiation various flames with the average and high spectral

permission at various lengths of an optical way from 0,2 to 16 m. Uniformity of temperature flames provided possibility of measurement of their temperature by optical methods with a margin error ±2 %. The method of definition nonequilibrium radiating cooling a flame from experimental data on its temperature is developed. Data on a role of nonequilibrium processes on radiating cooling optically the thin torch are received, allowing estimating influence of nonequilibrium processes of radiation in ultra-violet, visible and infra-red parts of a spectrum on radiating heat exchange of torches of aerocarriers and in top internal chambers.

2. The analysis of results of long-term measurements of radiating characteristics of gas and disperse phases of products of combustion is made and radiating characteristics various optically active components of products of combustion, including the cores (vapor H_2O and CO_2) and small components are received. Data on a microstructure sooty ashes and to its optical characteristics are received at burning of various gas hydrocarbonic components in oxygen and in air. Strong dependence of a microstructure sooty ashes from molecular structure of gas fuel and a burning mode is observed. Mass concentration sooty ashes is minimum at burning of methane CH_4 and is maximum at burning of acetylene C_2H_2. The microstructure sooty ashes at black oil burning is close to its microstructure at acetylene burning. Parameterization of gas components of products of combustion is executed on a two-parametrical method of equivalent mass.

3. The method of modeling of the transfer over of thermal radiation in nonequilibrium to radiating multicomponent non-uniform atmosphere under structural characteristics of top internal space is developed. The design of multichamber fire chambers with ascending movement of products of combustion in a fire chamber and vertical development of a flame of the hearth multirow torches forming uniform for all chambers of a multichamber fire chamber burning device of matrix type with the general gas collector for giving of gas fuel and a collector for giving of an oxidizer (air or oxygen) is offered. The burning device is expedient for the transfer out with a radiator for cooling by its water on an independent circulating contour. The design of a multichamber fire chamber at use of gas fuel allows to raise efficiency on 2-3 % and to increase it vapor-productivity in 2-3 times at preservation of parameters of vapor and boiler dimensions.

4. The closed modeling of radiating heat exchange in the chamber of combustion of a multichamber fire chamber with horizontal section of a cell of a boiler 1,25x1,6 m and 1,4x1,4 m is executed at factor of surplus of air α =1,03 and average initial speed of a current of products of combustion of 25 m/s and 20 m/s. Data in the speeds radiating cooling $\partial T(z)/\partial t, \partial T(z)/\partial z$ and to temperature profile T (z) depending on height z over cuts of capillaries matrix burning devices are received. Calculations heat susceptibility on heatsusceptibility to surfaces of the top internal chamber is executed. Full stream F of thermal radiation on a fire chamber surface will be coordinated with change enthalpy on an exit from the top internal chamber with a margin error 0,3%. Nonequilibrium radiating cooling makes 7,5 %. The account of effective pressure in a fire chamber leads to growth of a full stream of radiation F on 5÷6 %.

5. Consideration of optical properties of gas and disperse phases of products of combustion hydrocarbonic fuels and the offered algorithms of numerical modeling allows to draw following conclusions:

- nonequilibrium radiation reduces concentration of harmful component NO_x;
- nonequilibrium radiation leads to heating of particles of fuel and accelerates their ignition by that more intensively, than more small particles and then more their section of absorption of radiation;

- in case of burning of gas fuel nonequilibrium radiation practically isn't transformed by products of combustion and without easing is absorbed by walls of the chamber of combustion (screens);
- in case of presence of a disperse phase nonequilibrium radiation is absorbed by aerosols (soot and fuel particles) and its role in processes of radiating heat exchange is weakened, as energy of nonequilibrium radiation as a result of absorption passes in thermal energy of particles;
- nonequilibrium radiation (especially rigid ultra-violet radiation) can to initiate photochemical reactions in processes of combustion and to influence radiating heat exchange through changes of radiating properties of products of combustion.

6. The basic component defining nonequilibrium radiation in flames is hydroxyl OH. Factors of absorption OH in ultra-violet and infra-red areas of a spectrum are defined. Quantum-mechanical consideration of formation of spectra of nonequilibrium radiation shows that nonequilibrium radiation is shown both in electronic, and in vibrational-rotary spectra of molecules OH which is in raised and basic electronic conditions: bands v_1, $2v_1$, $3v_1$, where v_1 – frequency of normal vibration. Nonequilibrium radiation OH is revealed in a vicinity of lengths of waves 1; 1,43; 2,1; 2,7; 4,1 μm in flame hydrogen-oxygen. The method of definition of vibrational temperature in radiation spectra flames is developed. Presence of spectral structure of vibrational temperature testifies to its dependence on vibrational and rotary quantum numbers.

7. For a homogeneous mediums the law of Kirchhoff is carried out. In non-uniform medium on structure it is broken also function spectral transmission becomes depending as from thin structure of a spectrum of the radiating volume, and thin structure of a spectrum of the absorbing medium, and differs from function spectral transmission for sources of not selective radiation which are measured in laboratory experimental researches. The transfer over of selective radiation is influenced by following factors: the temperature self-reference of spectral lines of radiation, displacement of spectral lines with pressure, the temperature displacement of the spectral lines which have been found out for easy molecules (vapor H_2O, CH_4, NH_3, OH). Till now influence of last two factors wasn't investigated. Quantummechanics calculations of displacement of spectral lines with pressure make thousand shares of cm^{-1} and in conditions turbulized atmosphere can't render essential influence on function spectral transmision. Temperature displacement of spectral lines in a flame make the 100-th shares of cm^{-1} and at high temperatures reach semiwidth of spectral lines and more. It leads to that radiation of a high-temperature kernel of a torch is weakened to a lesser degree by its peripheral layers that increases heatsusceptibility surfaces of heating at the expense of radiating heat exchange. At registration of radiation of a torch of the aerocarrier in atmosphere the effect of an enlightenment of atmosphere in comparison with the account only the temperature self-reference of spectral lines of radiation of a torch is observed more considerably.

8. The analysis of radiating heat exchange between gas and disperse phases of products of combustion gaseous fuels shows that the temperature sooty particles should be below thermodynamic temperature of gases that weakens influence of a disperse phase on radiating heat exchange in the top internal chamber. On the other hand, absorbing properties of sooty ashes define its role in radiating heat exchange, forming a field of thermal radiation in space of the top internal chamber. Scattering of radiation by particles of a disperse phase of products of combustion shows weak influence on distribution of streams of radiation on heatsusceptibility surfaces of the top internal chamber. Mass concentration of

sooty ashes and its microstructure considerably depend on structure of gas fuel and a burning mode. At performance of calculations of radiating heat exchange the disperse phase of products of combustion is supposed multicomponent and is defined by various mechanisms of its formation. Each fraction of an aerosol has the optical characteristics, normalized on easing factor at length of a wave λ =0,55 μm. Spectral factors of easing, absorption, scattering and indicatryss of scattering are calculated for polydisperse ensemble of spherical particles of the set chemical compound. The electronic database includes three fractions of sooty ashes (primary thin-dispersion sooty ash, fraction of average dispersion, and coagulation fraction of soot of smoke gases), flying fraction ashes and roughly-dispersion fraction of products of combustion of firm fuel. As structural characteristics optical density on length of a wave λ =0,55 μm for various fractions of a disperse phase of products of combustion acts. The real spectral optical characteristics entering into settlement formulas are calculated on an electronic database in the assumption that burning of each component of fuel occurs independently that allows using optical density and a microstructure of sooty ashes by results of measurements on ardent measuring complexes.

7. References

Alemasov, V.E. & Dregalin, A.P. et al. (1972). *Thermodynamic and Physical Properties of Combustion Products*, VINITI, Moscow, Russia

Broida, H.P. & Shuler, K.E. (1952). Kinetics of OH Radical from Flame Emission Spectra. IV. A Study of Hydrogen-Oxygen Flame. *Journ. Chem. Phys.*, Vol.20, No.1, pp. 168-174

Ludwig, C.B. & Malkmus, W. et al. (1973). *Handbook of Infrared Radiation from Combustion Gases*, NASA, Washington, USA

Kondratyev, V.N. (1958). *Kinetics of Gaseous Chemical Reaction*, AN SSSR, Moscow, Russia

Kondratyev, K.Ya. & Moskalenko, N.I. (1977). *Thermal Emission of Planets*, Gidrometeoizdat, Leningrad, Russia

Kondratyev, B.N. & Nikitin, E.E. (1981). *Chemical Process in Gases*, Nauka, Moscow, Russia

Kondratyev, K.Ya.; Moskalenko, N.I. & Pozdnyakov B.N. (1983). *Atmospheric Aerosols*, Gidrometeoizdat, Leningrad, Russia

Kondratyev, K.Ya. & Moskalenko, N.I. (1984). *Hoti-house Effect of Atmosphere and Climate (Vol.12)*, VINITI, Moscow, Russia

Kondratyev, K.Ya.; Moskalenko, N.I. & Nezmetdinov R.I. (2006). Role Nonequilibrium Process of Radiative Growing Cold of Combustion Products on Content of Nitrous Oxides in Atmospheric Emission, *Dokl. AN*, Vol.14, pp. 815-817

Marchuk, G.I. & Lebedev, W.I. (1981). *Numerical Methods in Theory Neutrons of Transfer*, Atomizdat, Moscow, Russia

Moskalenko, N.I. & Mirumyanth, S.O. et al. (1976). Installation for Complex Research of Characteristics Molecular Absorption of Radiation by Atmospheric Gases, *Journ. Appl. Spectrosc.*, Vol.19, No.4, pp. 752-756

Moskalenko, N.I. & Cementhov, C.A. et al. (1980). Spectral Installation for Research of Molecular Absorption and Radiation by Gases in Hightemperature mediums, *Journ. Appl. Spectrosc.*, Vol.54, No.2, pp. 377-382

Moskalenko, N.I.; Ilyin, Yu.A. & Kayumova G.V. (1992). Measuring Complex of High Spectral Permission for Research of Flame, *Journ. Appl. Spectrosc.*, Vol.56, No.1, pp. 122-127

Moskalenko, N.I. & Filimonov, A.A. (2001). Modeling of Heat Emission Transfer in Hightemperature mediums, *Problems of Energetic,* No.11-12, pp. 27-41

Moskalenko, N.I. & Chesnokov, S.P. (2002). Thin Parameterization of Gaseous Components Radiative Characteristics of Hydrocarbonful Fueles, *Problems of Energetic,* No.1-2, pp. 10-19

Moskalenko, N.I.; Loktev, N.F. & Zaripov, A.V. (2006). Diagnostics of Flames and Combustion Products by Optical Methods, *Proc. IV-th Russian National Conference on Heat Transfer,* pp. 277-280, Moscow, Russia, October 23-27, 2006

Moskalenko, N.I.; Zaripov, A.V.; Loktev N.F. & Nezmetdinov R.I. (2007). Research of Role of Nonequilibrium Process in Radiative Growing Cold, *Problems of Gas Dynamics and Heatmassexchange,* Vol.2, pp. 47-50, Sankt-Petersburg, Russia, May 21-27, 2007

Moskalenko, N.I. & Zaripov, A.V. (2008). Research of Role of Nonequilibrium Process in Radiative Crowing Cold of Combustion Products of Firing Chamber, In: *Current Problems in Modern Science,* S.S. Chernov, (Ed.), No.3, 45-73, SIBPRINT, ISBN 978-5-94301-044-6, Novosibirsk, Russia

Moskalenko, N.I. & Loktev, N.F. (2008). Thing Parameters of Radiative Characteristics of Combustion Products and its Application in Tasks of Remote Diagnostics, *Materials from International Conference "Energy – 2008: Innovation, Solutions, Prospects",* pp. 2224-2230, Kazan, Russia, September 15-19, 2008

Moskalenko, N.I.; Zagidullin, R.A. & Kuzin, A.F. (2008). Manny Firing Chambers and Heatexchangers as Means Increase of Effectiveness in Heat Engineering, *Materials from International Conference "Energy – 2008: Innovation, Solutions, Prospects",* pp. 230-234, Kazan, Russia, September 15-19, 2008

Moskalenko, N.I. & Loktev, N.F. (2009). Numerical Modeling in Tasks of Remote Diagnostics of Combustion Firing Fuels and Technological Mediums, In: *Technics and Technology in XXI-th Century: Modern Conditions and Prospects of Development: Monograph,* S.S. Chernov, (Ed.), Vol.4, 13-47, SIBPRINT, ISBN 978-5-94301-068, Novosibirsk, Russia

Moskalenko, N.I.; Zaripov, A.V. & Zagidullin, R.A. (2009). Emission Spectrums and Radiative Heatexchange Mediums, Flames and Firing Chamber, In: *Technics and Technology in XXI-th Century: Modern Conditions and Prospects of Development: Monograph,* S.S. Chernov, (Ed.), Vol.4, 48-87, SIBPRINT, ISBN 978-5-94301-068, Novosibirsk, Russia

Moskalenko, N.I. & Loktev, N.F. (2009). Methods of Modeling Selective Radiation Transfer in Structure – nonhomogeneous Mediums, *Thermal Process in Technique,* Vol.1, No.10, pp. 432-435

Moskalenko, N.I.; Loktev, N.F.; Safiullina, Ya.S. & Sadykova, M.S. (2010). Ingredients Identification and Determination of Ingredient Composition of Atmospheric Emission and Combustion Products by Means of Fine Structure Spectrometry Method, *International Journal of Alternative Energetic and Ecology,* Vol.8, No.2, pp. 43-54

Moskalenko, N.I.; Rodionov, L.V. & Yakupova F.S. (1984). Modeling of Transfer over of Touch Radiation of Differet Carriers, *Problems of Special Engineering,* Ser.1, No2, pp. 54-58

Moskalenko, N.I.; Zaripov, A.V. & Ilyin, Yu.A. (2010). Investigation of Nonequilibrium Hydroxyl Emission Spectra, *Russ. Phys. Journ.,* Vol.53, No.2, pp. 107-113

Moskalenko, N.I.; Zaripov, A.V.; Loktev, N.F. & Ilyin, Yu.A. (2010). Emission Characteristics of Hydrogen-Oxygen Flames, *Journ. Appl. Spectrosc.*, Vol.77, No3, pp. 378-385

Surgikov, S.T. (2004). *Thermal Radiation of Gases and Plasma*, MGTU, Moscow, Russia

Young, S.J. (1977). Evolution of Nonithothermal Band Models for H_2O, *Journ. Quant. Spectrosc. Radiat. Transfer.*, Vol.18, No.1, pp. 29-45

Zachor, A.S. (1968). General Approximation for Gaseous Absorption, *Journ. Quant. Spectrosc. Radiat. Transfer.*, Vol.8, No.2, pp. 771-784

Zel'dovich, Ya.B.; Sadovnikov, P.Ya. & Frank-Kamentsky, D.A. (1947). Oxidizing of Nitrogen to Firing, *AN SSSR*, Moscow, Russia

24

Photopolarization Effect and Photoelectric Phenomena in Layered GaAs Semiconductors *

Yuo-Hsien Shiau
Graduate Institute of Applied Physics, National Chengchi University
Taiwan, Republic of China

1. Introduction

The studies of transport properties in semiconductors have made great progresses for the past decades. This is mainly due to the advanced technologies for development of new materials and the application of nonlinear dynamics to the fundamental well-known materials. In particular, the discipline of nonlinear dynamics grows fast, which is due to the cooperation of theoretical background and experimental findings. Among systems considered, semiconductors represent interesting and highly productive examples of the experimental investigation of nonlinear dynamics. One of the typical findings observed in nonlinear semiconductors is the dynamics of propagating electrical solitary waves which could be periodic or chaotic. Many of these phenomena have been studied in bulk semiconductors as well as superlattices, and can be successfully explained by means of theoretical as well as numerical approaches (Amann & Schöll, 2005; Bonilla & Grahn, 2005; Cantalapiedra et al., 2001; Gaa & Schöll, 1996; Wack, 2002). Of particular interest is that GaAs semiconductors have been shown to generate microwave radiation. The generation was attributed to propagating space-charge waves (or high-field domains). The domain shape parameters such as the maximum fields and the domain size are controllable with changing the concentration of ionized donors that are doped in the semiconductor substrate.

It is known that nonlinear electro-optic characteristics can be observed in an n^+-n^--n-n^+ GaAs sandwich structure under optical excitation, where potential applications including optical control of microwave output, ultrafast electric switches, memory cells and other areas. The key factor in such a system is that propagating space-charge waves (SCWs) were formed at the cathode and destroyed at (or before) the anode being due to a balance of the diffusion of carriers and the nonlinearity in the velocity-field characteristic, where it can be realized that propagating SCWs are equivalent to the case of the laser beam propagation in Kerr-type nonlinear optical media. Besides, the notch profile (i.e., the n^- layer) will be strongly influenced by the optical illumination, which will result in the tuning traveling-distance of SCWs. Owing to that, optical control of microwave output can be expected, and this phenomenon is related to the photopolarization effect. In addition, the interesting phenomena including optically induced hysteresis and long-lived transient behaviors can be observed in a layered semiconductor. In the meanwhile, the development of multiple sandwich structures has been known to be helpful for the high-power microwave generation; however, electro-optic characteristics are less known in this system. Concerning on multiple sandwich

*This work was partially supported by the National Science Council of the Republic of China (Taiwan) under Contract Nos. NSC 98-2112-M-004-001-MY3

structures without laser illumination, it would be expected that coherent/identical SCWs initiated from different doping notches will show up, but it has never been reported that persistent photoelectric phenomena can be observed in multiple GaAs sandwich structures due to the photopolarization effect. In this Chapter, using 10-ns-duration pulse of a Nd:YAG laser to generate electron-hole pairs is considered. Interestingly, both exponential and non-exponential photoelectric relaxations can be numerically observed via the well-accepted drift-diffusion model; therefore, the switching time for non-exponential relaxations would be much higher than that of exponential relaxations. Thus, photo-induced persistent charge transport can be discovered in the present multiple GaAs sandwich structures, which is believed to be important properties of opto-electronic and transport processes in layered semiconductors.

The remainder of this Chapter is organized as follows. Optically induced hysteresis, the photopolarization effect, and non-exponential photoelectric relaxations will be introduced in Sec. 2. Sec. 3 will provide a numerical evidence of long-lived transient behaviors under the consideration of optical stochasticity. Concluding remarks will be given at the end of this Chapter.

2. Photoelectric phenomena

The physical basis for hysteretic switching between low- and high- conducting states in nonlinear semiconductors is usually related to the exhibition of S-shaped negative differential conductivity (SNDC) on the current-density-field characteristic. Up to now, several mechanisms have been proposed to induce SNDC. For example, two-impurity-level model with impact ionization (II) for n-GaAs at 4.2 K (Schöll, 1987), the interband breakdown for n-GaAs at room temperature (Gel'mont & Shur, 1970), and generic N-shaped NDC characteristics connected with a large load resistance (LR) (Döttling & Schöll, 1992; Shiau & Cheng, 1996), i.e., inverted SNDC, etc. Theoretical analyses and predictions, under the assumption of spatial homogeneity, are usually based on the local or global bifurcation schemes around the operating points. However, in whatever theoretical models II or LR is a key factor to induce SNDC. In this section we numerically demonstrate, even without consideration of II and LR, the hysteretic switching in an n^+-n^--n-n^+ GaAs sandwich structure (Oshio & Yahata, 1995) under local optical excitation. It is interesting to find that quenched and transit modes can coexist at the same laser intensity. And the transition between these two dynamical states is hysteretic, i.e., optically induced hysteresis. These results also indicate using a layered semiconductor as an inverter of optical input to microwave output and this electro-optic phenomenon shall be potentially useful for applications. Moreover, in realistic situations the electric field in SCWs could become strong enough to generate electron-hole pairs due to the II effect. Therefore, considering the influence of the II effect on this hysteresis is also performed. The numerical results show that this hysteretic switching still can be maintained but the transition regions between these two dynamical states will be perturbed. Thus, the II effect is not a primary factor in our system. This is why we called optically induced hysteresis, a novel nonlinear electro-optic characteristic, between two different conducting states in a semiconductor device.

Before going to introduce our computational model, it shall be noted that the function of this doping notch is to establish local space-charge field for dipole-domain nucleation. In order to study the bipolar transport, we further consider local optical excitation which is close to the doping notch. We expect the domain dynamics originally determined by the doping notch and external dc bias will be influenced by the optical intensity. The motivation of consideration of local optical excitation in the active region is to redistribute the space-charge field around the doping notch via optical generation of hole carriers. More precisely speaking, local optical excitation is to make doping notch as a collector of electrons. Then the internal field in doping notch will become stronger, which can speed up electrons and influences the dipole-domain nucleation. Therefore, if the optical intensity is large enough, even at lower dc bias, the quenched domain could be controlled and possibly becomes the transit domain. Thus the so-called optical control of electrical propagations in a GaAs-based Gunn device is expected.

As the sandwich structure we consider a GaAs-based semiconductor device with the doping profile $N_D(x)$, denoted as n$^+$(8.0 μm)-n$^-$(4.0 μm)-n(40.0 μm)-n$^+$(8.0 μm), shown in Fig. 1. The active region is sandwiched between the highly-doped n$^+$ cathode and anode regions. A 4 μm doping notch, at the beginning of the active region, is to initiate the dipole domain near the cathode. In addition, the laser illumination with 3 μm width is considered and denoted as $I(x)$. The detailed model equations and the GaAs parameters used for numerical simulation are listed in Tables I and II, respectively (Shiau & Peng, 2005). In the following the results obtained from direct numerical simulation with the consideration of the drift-diffusion model and fixed boundary conditions will be demonstrated in detail.

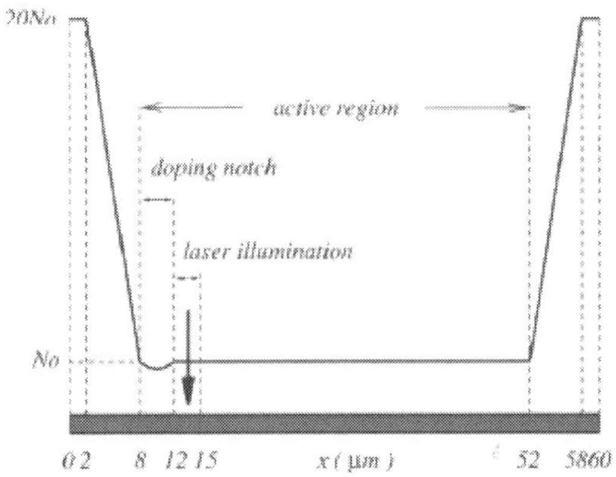

Fig. 1. Schematic illustration of device doping profile $N_D(x)$ and local laser illumination $I(x)$. The sample length is 60 μm.

Without laser illumination and II effect, the simulated model shall display traditional quenched or transit domains which are dependent of the external dc bias (Sze, 1969). When external bias is 12 V, the dynamical characteristics of the quenched mode are clearly exhibited via spatiotemporal behaviors of the electric field and time-dependant total current density in Fig. 2. A high-field domain (Fig. 2(a)) grows from the doping notch then annihilate before reaching the anode. The waveform of oscillating current density J is shown in Fig. 2(b) and the associated oscillating frequency f_0 is around 20 GHz. Of course, in this case the hole density shall be zero in the whole sample. If external bias is increased to 20 V, the transit mode, also clearly shown in Fig. 2, cyclical propagation from the cathode to the anode is obtained. And the oscillating current frequency is around 2 GHz. As the case of the quenched domain, p is still zero in the whole computational domain. Therefore, the simulated model is well to describe the traditional electrical-induced domain formation and propagation. Now we consider local laser illumination $I(x)$ applied to the semiconductor device which is operated

Continuity equations for the carrier densities:

$$\frac{\partial n}{\partial t} + \frac{\partial J_n}{\partial x} = gI(x) + G_{ii} - \gamma np,$$

$$\frac{\partial p}{\partial t} + \frac{\partial J_p}{\partial x} = gI(x) + G_{ii} - \gamma np, \quad I(x) : \text{local optical excitation.}$$

Poisson equation:

$$\frac{\partial^2 \phi}{\partial x^2} = \frac{e}{\epsilon}\left[p + N_D(x) - n \right], \quad N_D(x) : \text{doping profile.}$$

Particle current densities for electrons (J_n) and holes (J_p):

$$J_n = nv_\text{n}(E) - D_n \frac{\partial n}{\partial x},$$

$$J_p = -pv_\text{p}(E) - D_p \frac{\partial p}{\partial x}, \quad D_n(D_p) : \text{diffusion coefficient of electrons (holes).}$$

Drift velocities for electrons (v_n) and holes (v_p):

$$v_\text{n}(E) = \frac{\mu_1 + \mu_2 R \exp\left(-\frac{\Delta E}{kT_e}\right)}{1 + R \exp\left(-\frac{\Delta E}{kT_e}\right)} E, \quad T_e(T_L) : \text{electron (lattice) temperature.}$$

$$v_\text{p}(E) = \begin{cases} \mu_p E & (E < E_p), \\ \mu_p E_p & (E \geq E_p). \end{cases}$$

Energy balance equation:

$$E^2 = \frac{3k}{2e\tau_e}(T_e - T_L) \frac{1 + R \exp\left(-\frac{\Delta E}{kT_e}\right)}{\mu_1 + \mu_2 R \exp\left(-\frac{\Delta E}{kT_e}\right)}, \quad \tau_e : \text{the energy relaxation time.}$$

Impact ionization:

$$G_{ii} = g_0 \exp\left[-\left(\frac{E_{th}}{E}\right)^2 \right] \left(|J_n| + |J_p| \right).$$

Boundary conditions:

$$n(0,t) = n(L,t) = 20N_0, p(0,t) = p(L,t) = 0,$$

$$\phi(0,t) = 0, \phi(L,t) = \text{external dc bias.}$$

Dynamical variables:

n, p : electron and hole densities.

ϕ, E : electric potential and electron field.

Parameters:

$\mu_1(\mu_2), \mu_p$: electron mobility at the lower (upper) valley and hole mobility.

g_0, E_{th} : impact ionization rate and threshhold field for impact ionization.

g, γ : the rate for generation and recombination of electron-hole pairs.

E_p : threshhold field for the saturated hole velocity.

ΔE : the energy difference between the two valleys.

R : the ratio of density of states for the upper valley to the lower valley.

Table 1. Model equations.

at 12 V. When the applied laser intensity is 192 kW/cm^2, a transit domain with a smaller domain width (compared to the upper portion of Fig. 2(a)) is shown in Fig. 3. Not surprising, the stationary and nonuniform hole distribution is obtained around the notch region (Fig. 3). The oscillating current frequency in this case is also around 2 GHz. As we already mentioned that the optical generation of hole carriers is to control the space-charge field of the doping notch. When the suitable laser intensity is applied, the transition from the quenched mode to the transit mode at lower dc bias still can be observed. Therefore, the optical control of electrical propagations is numerically demonstrated. Besides, these two different electrical propagations can coexist at the same laser intensity. Fig. 4 illustrates the formation of quenched and transit modes when the laser intensity is 96 kW/cm^2. Thus this kind of bistable characteristic is due to the optical excitation, which is little known in n$^+$-n$^-$-n-n$^+$

parameter	value	parameter	value
D_n	200 cm^2/s	R	94
D_p	20 cm^2/s	k	8.6186×10^{-5} eV/K
ϵ	1.17×10^{-12} F/cm	ΔE	0.31 eV
e	1.61×10^{-19} C	T_L	300 K
g	4.4×10^{18} cm^{-1}sec^{-1}W^{-1}	τ_e	10^{-12} s
γ	10^{-10} sec^{-1}cm^3	N_0	5×10^{15} cm^{-3}
μ_1	8500 cm^2/V·s	g_0	2×10^5 cm^{-1}
μ_2	50 cm^2/V·s	E_{th}	550 kV/cm
μ_p	400 cm^2/V·s	E_p	40 kV/cm

Table 2. The fundamental constants and GaAs parameters for numerical simulation.

semiconductor devices. We carefully check the transition in between quenched and transit modes, and find that this transition is hysteretic. The transition regions corresponding to the quenched mode to the transit mode and vice versa are observed at 132 kW/cm^2 and 82 kW/cm^2, respectively. The detailed electro-optic characteristic of this two different electrical propagations is illustrated in the upper portion of Fig. 5 via oscillating current frequency f versus laser intensity I plot. The circular and triangular symbols denote, respectively, quenched and transit domains. It is clear to see that the oscillating current frequency of the quenched domain gradually decreases when I increases. Moreover, the oscillating current frequency of the transit domain is independent of the laser intensity, which means that the traveling time of the transit domain is only related with the bulk property and not influenced by the local optical excitation. In addition, the inset in the upper portion of Fig. 5 is the corresponding hysteretic J_{min}-I curve when I is slowly increased (top) and decreased (bottom), where J_{min} is the extreme minimum current density. Moreover, the calculated electric field of quenched and transit domains is much small than the E_{th} ($= 550$ kV/cm) (Hall & Leck, 1968), i.e., at least less than one order of magnitude. Therefore, the II process to generate electron-hole pairs in high-field domains can be neglected. Nevertheless, in the following we still investigate the influence of the II effect on the hysteretic f-I and J_{min}-I plots. The lower portion of Fig. 5 clearly shows that the influence of the II effect is only to perturb the transition regions which become at 122 kW/cm^2 and 77 kW/cm^2. Compared to the upper portion the threshold laser intensities for domain transition are all decreased. The reasonable explanation is that in addition to the local optical excitation the internal field in doping notch will become stronger due to the generation of hole carriers via the nonlinear amplification of the II effect. Therefore, the laser intensity needed for domain transition will decrease. The inset in the lower portion of Fig. 5 is also the corresponding hysteretic J_{min}-I curve.

Here, we would like to give a more detailed explanation for Fig. 5. It is well known that the notch profile affects strongly on the shedding of the high-field domain. In this study the doping notch illustrated in Fig. 1 can be changed due to the local optical excitation. This result we called the photopolarization effect. At low illumination, a stationary hole profile is established nearby the n$^-$ layer. Therefore, the original dipole field owing to the n$^-$ layer will be enhanced by the present hole distribution. In other words, the length and associated profile of the n$^-$ layer are *effectively* as well as *perturbatively* changed. This phenomenon finally will result in the tuning traveling-distance of the quenched domain. At high illumination, the hole distribution can extend to the cathode emitter (i.e., the n$^+$ layer) and leads to a significant enhancement of the dipole field. Consequently, the *effective* length and associated profile of the n$^-$ layer have a dramatic change, which result in the transit dynamics even at a lower dc

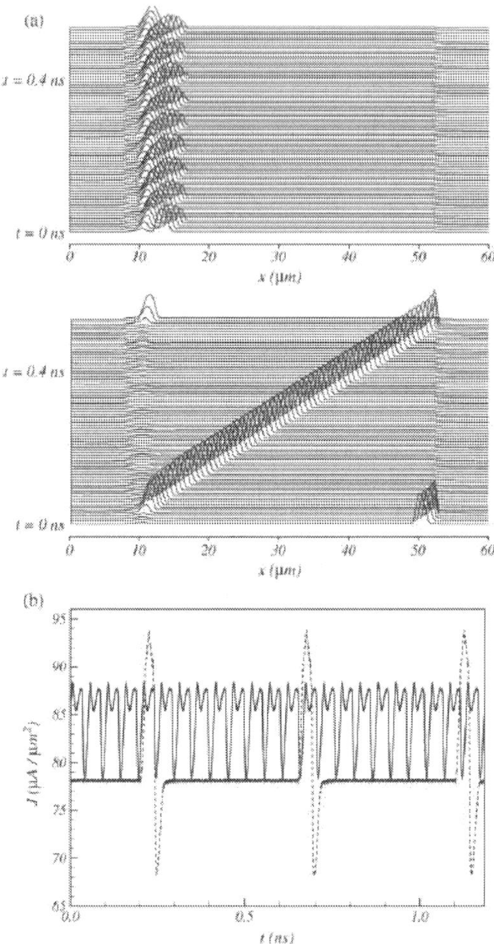

Fig. 2. Without consideration of local laser illumination and the II effect, the dynamical characteristics of the quenched and transit modes for dc bias, respectively, being equal to 12 V and 20 V. (a) the quenched domain in upper portion and the transit domain in lower portion. (b) the time evolution plot of the total current density for the transit mode (dashed line) and the quenched mode (solid line). The maximum values of electric fields in (a) are 84.0 kV/cm for the transit domain and 35.5 kV/cm for the quenched domain.

bias. In addition to the mention above, it is also interesting to find that there is a hysteretic transition of the hole distribution when laser illumination is varied from low to high and vice versa. Therefore, the nonlinear electro-optic characteristic in Fig. 5 can be explained as a result of the photopolarization effect. The detailed quantitative analysis of the *effective* notch profile

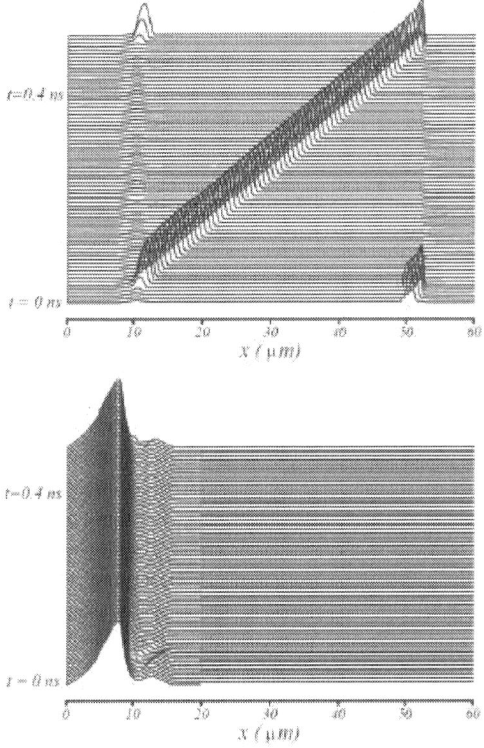

Fig. 3. Illustrations of dynamical characteristics for the transit mode when this semiconductor device is operated at 12 V and illuminated with 192 kW/cm^2. The spatiotemporal behaviors of electric fields and hole densities are shown in, respectively, upper portion and lower portion. The maximum values of electric fields and hole densities are, respectively, equal to 49.7 kV/cm and 2.77aÑ1015 cm^{-3}.

influenced by the lager illumination was reported via the cross-correlation matrix method (Shiau, 2006).

Concerning on multiple sandwich structures, our system is a series combination of two identical n$^+$(4.0 μm)-n$^-$(2.0 μm)-n(20.0 μm)-n$^+$(4.0 μm) layered semiconductors, which is biased at 12 V. Without the consideration of optical illumination, coherent/identical SCWs initiated from different doping notches will show up. The drift velocity, traveling distance, and cyclic period for each of SCWs would be approxmatively equal to 0.8×10^7 cm/s, 2 μm, and 0.025 ns, respectively. In addition, the potential drop across the first (or second) layered semiconductor is 6 V. However, when the locally illuminated n region adjacent to the doping notch (i.e., 1.5 μm illumination) is considered, it would be interesting to find non-identical SCWs in these two layered semiconductors. Fig. 6 depicts that the potential drop ϕ across the second layered semiconductor is a function of time, where 10-ns-duration pulse of a Nd:YAG laser is switched on at $t = 0$, and the dashed (solid) line is resulted from laser

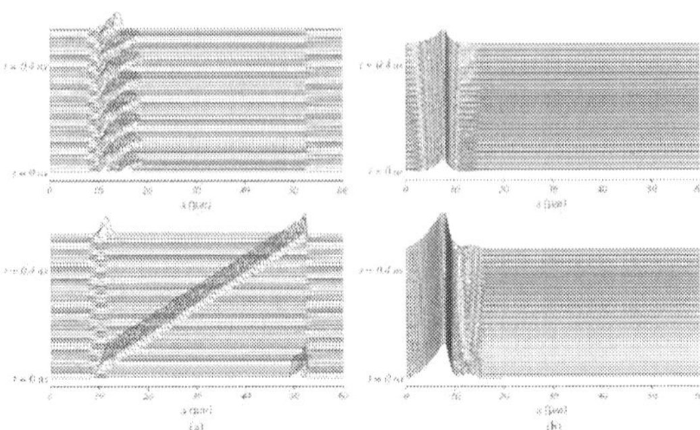

Fig. 4. Coexistence of different electrical propagations when the laser intensity is decreased to 96 kW/cm^2 and the dc bias is still kept at 12 V. (a) the quenched domain in upper portion and the transit domain in lower portion. (b) upper portion: hole distribution for the quenched mode and lower portion: hole distribution for the transit mode. The maximum values of electric fields in (a) are 49.7 kV/cm for the transit domain and 36.8 kV/cm for the quenched domain. As for the maximum values of hole densities in (b) are 2.77aÑ10^15 cm^{-3} for the transit mode and 0.98aÑ10^15 cm^{-3} for the quenched mode.

illumination on the second (first) layered semiconductor. During 10 ns illumination, ϕ exhibits complicated dynamical response including exponential decrease (or exponential increase) as well as oscillating patterns, where the oscillating period is roughly equal to 0.045 ns (see the inset shown in Fig. 6). When the illumination is switched off at $t = 10$ ns, exponential and non-exponential photoelectric relaxations will show up.

The exponential decrease (increase) during 10 ns illumination reflects the discharge (charge) process of the second layered semiconductor, and the oscillating pattern is associated with the formation of new SCWs. Moreover, non-identical SCWs in these two layered semiconductors can be expected. After 10 ns illumination, the exponential (non-exponential) photoelectric relaxation represents a recharge (discharge) process. In particular, the discharge process displays a non-monotonic behavior, where the potential drop ϕ quickly crosses the equilibrium value 6 V and slowly goes back to 6 V. Thus, photo-induced persistent charge transport can be discovered in the present multiple GaAs sandwich structures

Finally, we would like to discuss the validity of the present approach. The simulated model is a combination of the well-known drift diffusion equations, a single electron temperature model, and fixed boundary conditions. To our knowledge, it can successfully describe the domain dynamics for longer devices (Shaw et al., 1992). The parameters used in this simulation for GaAs are taken from Refs.(Yeh, 1993; Sze, 1969), while the coefficients of impact ionization are from Ref.(Hall & Leck, 1968). Another important issue we would like to discuss is the laser properties. the interesting phenomena in this study occur at laser intensity with the order of magnitude from 10^4 W/cm^2 to 10^5 W/cm^2. As far as we know, such conditions are experimentally feasible since two wave mixing experiments in GaAs have been performed with light at more than 10 MW/cm^2 intensity (Disdier & Roosen, 1992; Duba et al., 1989). And the laser source we considered is 10-ns-duration pulse of a Nd:YAG laser with wavelength 1064 nm, which ensures our simulation results can be observed in this pulse laser. As for the technique of 3 μm (or 1.5 μm) laser illumination also can be easily performed via GRIN (Chen, 1996), i.e., graded refractive index lenses. Therefore, we believe that our numerical results could be observed experimentally.

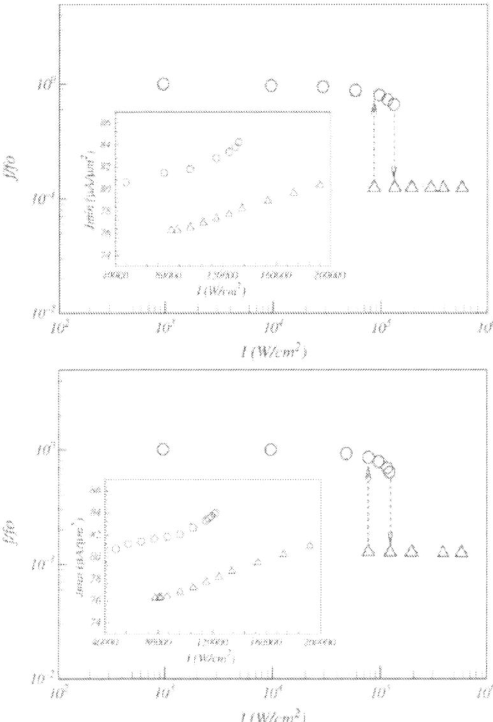

Fig. 5. The upper portion: the nonlinear electro-optic characteristic when exclusion of the II effect. The corresponding J_{min}-I hysteretic curve is shown in the inset. The lower portion: the same illustrations as in the upper portion but inclusion of the II effect.

3. Long-lived transient behaviors with optical stochasticity

In this section, we report the electrical response with long-lived transient characteristic, i.e., lock-on effect, in the above-mentioned GaAs-based Gunn device under stochastic stimuli. Our results can be considered as a numerical evidence for the new development of nonlinear theory which describes the so-called noise delayed decay (NDD) of unstable dynamical states (Agudov & Malakhov, 1999; Horsthemke & Lefever, 1984). Furthermore, this nonequilibrium decay process in our system can be used to control the response time of an ultrafast microwave switch (i.e., a nanosecond switch). To the best of our knowledge, the nonequilibrium decay rate in realistic semiconductors influenced by the optical stochasticity has never been reported. Moreover, the concept of NDD can be generally applied to other semiconductor systems with unstable dynamical states. For example, semiconductors with persistent photoconductivity can be used as many important applications. However the locking time of the dark current (i.e., an unstable dynamical state) is fundamentally related with material preparation (Joshi et al., 1999). According to the NDD effect, the locking time can possibly be prolonged and

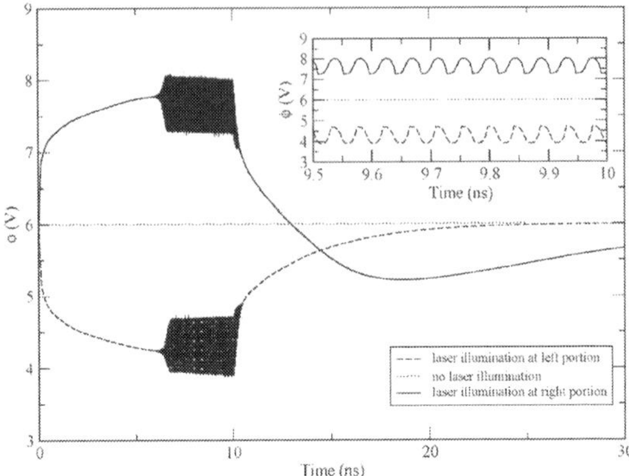

Fig. 6. Exponential and non-exponential photoelectric relaxations in multiple sandwich structures.

controlled. Therefore, the NDD effect developed by the nonlinear theory may play a crucial role for the semiconductor industry.

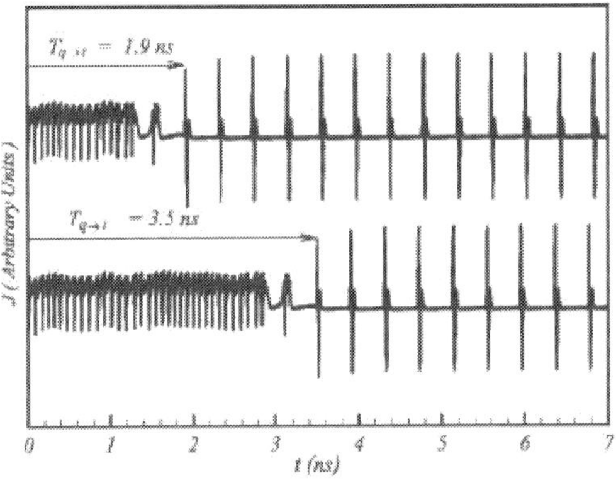

Fig. 7. Time evolution plot of the transient J values for laser intensity beginning at 115 kW/cm^2 and then suddenly increased to 125 kW/cm^2. The upper portion: without the additional randomness, the switching time $T_{q \to t}$ is 1.9 ns. The lower portion: when A is 10^{-2}, $T_{q \to t}$ is increased to 3.5 ns.

The upper portion of Fig. 7 illustrates the transient current oscillation from the optically-induced quenched domain to transit domain with the laser intensity initially at 115 kW/cm^2 then increased suddenly to 125 kW/cm^2. The time location of the first extreme maximum current density for optically-induced transit domain is defined as the switching time $T_{q \to t}$ from the optically-induced quenched to transit domain. The switching time in this case is around 1.9 ns. This is why a GaAs-based Gunn device can be regarded as an ultrafast microwave switch. If we further consider the additional weak noise in the laser intensity, the local uniform laser illumination $I(x)$ shall be replaced by $I(x)\,[1 + A\eta(t)]$ with an additional requirement $1 + A\eta(t) \geq 0$. Here A is the strength of the noise, and $\eta(t)$ must satisfy the white noise conditions: $\langle \eta(t) \rangle = 0$ and $\langle \eta(t)\eta(t') \rangle = \delta(t - t')$. The symbol $\langle \rangle$ denotes the ensemble average. The lower portion of Fig. 7 demonstrates that the switching time is increased to 3.5 ns with $A = 10^{-2}$. It shall be noted that the additional randomness is just given at laser intensity being equal to 125 kW/cm^2 for the artificial control of electrical response. The corresponding spatiotemporal dynamics of Fig. 7 are shown in Fig. 8 via contour plots. The white area represents the higher electric-field values and intermediate shading from gray to black shows different levels of lower electric-field values. It is seen clearly that the optically-induced quenched domain temporally locks in the space, which is strongly dependent of the additional randomness. The detailed relation between $T_{q \to t}$ and A is plotted with squared symbols in Fig. 9. It is found that not only the prolonged switching time but also the nonlinear $T_{q \to t}(A)$ spectrum. When A exceeds 10^{-2}, the switching time is fixed at 3.5 ns, while the NDD effect disappears as $A \leq 10^{-3}$. Now we discuss the switching time $T_{t \to q}$ from the optically-induced transit to quenched domain with laser intensity initially at 87 kW/cm^2 then decreased suddenly to 76 kW/cm^2. Similarly, the additional randomness is given at which the laser intensity equals 76 kW/cm^2. The $T_{t \to q}(A)$ spectrum, as shown by circle symbols in Fig. 9, displays no NDD effect in this transition branch. The switching time is fixed at 1.3 ns when A is increased from 0 to 10^{-1}.

The findings shown in Fig. 9 can be simply realized via classical attractor dynamics. If the initial conditions were fixed, the phase trajectory from the initial state to the final state and the associated relaxation time, i.e., switching time, can be determined. It is also noted that the dynamical characteristic of the model system, i.e., the phase trajectory and the relaxation time, are not influenced by a small strength of randomness (e.g., A is smaller than 10^{-3}). However, $T_{q \to t}(A)$ and $T_{t \to q}(A)$ exhibit different behaviors with $10^{-3} \leq A \leq 10^{-1}$. Such a result indicates that the stability of the relaxation path in this system is quite complex. The phase trajectory from the transit mode to the quenched mode still maintains its stability and is not perturbed by the strength of white noise. Nevertheless, it's a different story for the quenched mode to the transit mode. The unstable quenched mode tends to be temporally locked in phase space due to the additional randomness, i.e., noise-enhanced stability for unstable dynamical states. Naively, stronger noise can drive the dynamical system promptly into the convergent regime of the final stable state. Therefore, noise-enhanced stability should be gradually decreased as A is increased. Surprisingly, in the interval of $10^{-2} \leq A \leq 10^{-1}$ our calculations indicate that no sign of increase of $T_{q \to t}$ with respect to A. We believe that the NDD effect shown in $T_{q \to t}(A)$ spectrum is still an unsolved problem in stochastic processes. However, these different $T_{q \to t}(A)$ and $T_{t \to q}(A)$ spectrums possibly can be explained via local as well as global bifurcation scenarios around the transition points. According to the bifurcation analysis, catastrophic scenarios can be numerically observed both in quenched state to transit state and vice versa. A common situation for catastrophe to arise is where the system undergoes a crisis at which an attractor collides with the basin boundary separating it and another coexisting attractor (Thompson & Stewart, 1993). If the attractor is chaotic, after the crisis, the chaotic attractor is destroyed and converted into a nonattracting chaotic saddle. A dynamical trajectory then wanders in the vicinity of the chaotic saddle for a period of time before it approaches to the other attractor. This phenomenon is known as chaotic transient (Ditto et al., 1989; Grebogi et al., 1987) which can be realized via the transition from chaotic quenched state to periodic transit state in the upper portion of Fig. 8. In order to illustrate the formation of chaotic quenched domains near the transition region with laser intensity being equal to 122 kW/cm^2, the bifurcation diagram of current density with local minimum

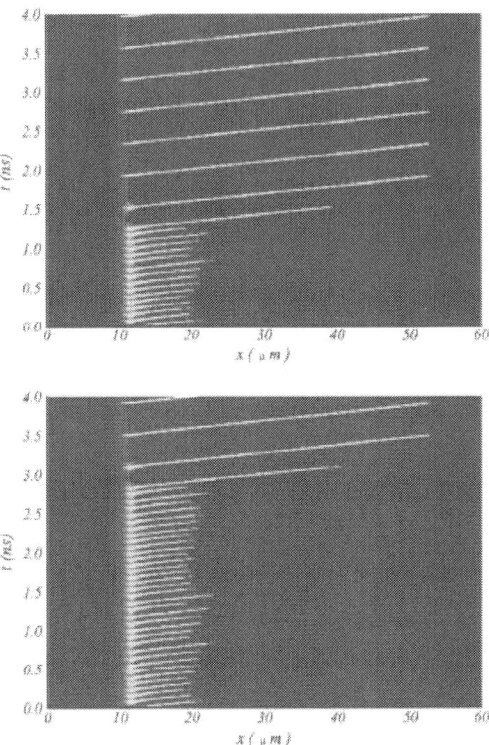

Fig. 8. The contour plots of the spatiotemporal behaviors in Fig. 7 for A being equal to 0 (the upper portion) and 10^{-2} (the lower portion).

values J_{bifur} versus laser intensity is illustrated in Fig. 10 which exhibits an incomplete period-doubling scenario, where J_{bifur} is detected when the associated values is smaller than 85 $\mu A/\mu m^2$. If the laser illumination with randomness is considered, the wandering time in the vicinity of the chaotic saddle will be prolonged and can be realized in the lower portion of Fig. 8. Therefore, the NDD effect in the nonlinear $T_{q \to t}(A)$ spectrum is strongly related with the dynamical structure nearby the chaotic saddle. If the attractor is periodic, after the crisis, the periodic attractor becomes unstable and quickly approaches to another coexisting attractor. The wandering time in this case should be shorter than that of the case in the vicinity of the chaotic saddle. Moreover, the dynamical structure in this case should not be perturbed by the additional weak noise. Therefore, the underlying physics for the $T_{t \to q}(A)$ spectrum with fixed and shorter values also can be interpreted as catastrophe from periodic transit state to periodic quenched state.

In summary, we introduce light-triggered spatiotemporal dynamics in layered semiconductors. Interesting phenomena shown in this Chapter include optically induced hysteresis, the photopolarization effect, exponential and non-exponential photoelectric relaxations, and long-lived transient behaviors under the consideration of optical stochasticity.

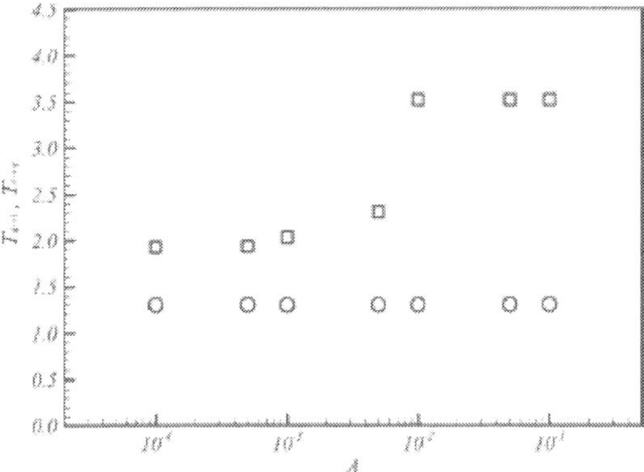

Fig. 9. Illustration of nonlinear $T_{q \to t}(A)$ and flat $T_{t \to q}(A)$ spectrums. Please note that the values of $T_{q \to t}(A)$ and $T_{t \to q}(A)$ are obtained from the ensemble average. Each switching time is averaged over 20 different realizations.

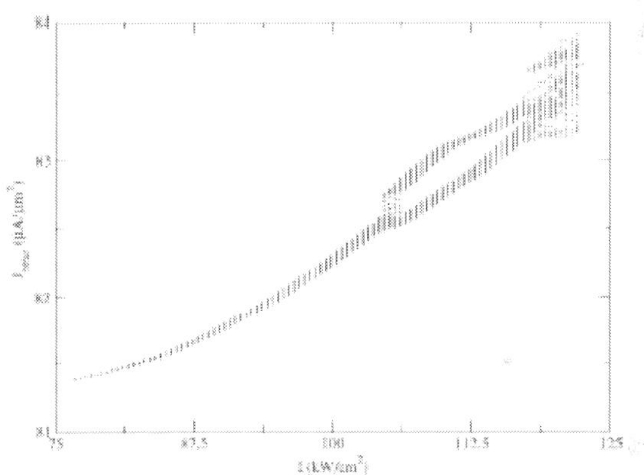

Fig. 10. The bifurcation diagram of J_{bifur} versus I represents the transition from periodic to chaotic quenched domains when laser intensity is increased.

These findings would be valueable to the development of new optoelectronics. In addition, with the assistance of nonlinear theory the operation of opto-electronic devices would be far beyond capabilities of the traditional operation.

4. References

Agudov, N.V. & Malakhov, A.N. (1999). Decay of unstable equilibrium and nonequilibrium states with inverse probability current taken into account, *Phys. Rev. E* Vol. 60: 6333-6342.

Amann, A & Schöll, E. (2005). Bifurcations in a system of interacting fronts, *J. Stat. Phys.* Vol. 119: 1069-1138.

Bonilla, L.L. & Grahn, H.T. (2005). Nonlinear dynamics of semiconductor superlattices, *Rep. prog. Phys.* Vol. 68: 577-683.

Cantalapiedra, I.R., Bergmann, M.J., Bonilla, L.L. & Teitsworth S.W. (2001). Chaotic motion of space charge wavefronts in semiconductors under time-independent voltage bias, *Phys. Rev. E* Vol. 63 (No. 056216): 1-7.

Chen, C.L. (1996). *Elements of Optoelectronics & Fiber Optics*, Irwin, Chicago.

Disdier, L. & Roosen, G. (1992). Nanosecond four-wave mixing in semi-insulating GaAs, *Opt. Commun.* Vol. 88: 559-568.

Ditto, W.L., Rauseo, S., Cawley, R., Grebogi, C., Hsu, G.H., Kostelich, E., Ott, E., Savage, H.T., Segnan, R., Spano, M.L. & Yorke, J.A. (1989). Experimental observation of crisis-induced intermittency and its critical exponent, *Phys. Rev. Lett.* Vol. 63: 923-926.

Döttling, R & Schöll, E. (1992). Oscillatory bistability of real-space transfer in semiconductor heterostructures, *Phys. Rev. B* Vol. 45: 1935-1938.

Dubard, J., Smirl, A.L., Cui, A.G., Valley, G.C. & Boggess, T.F. (1989). Beam amplification by transient energy transfer in GaAs and Si, *Phys. Status Solidi (b)* Vol. 150: 913-919.

Gaa, M. & Schöll, E. (1996). Traveling carrier-density waves in n-type GaAs at low-temperature impurity breakdown, *Phys. Rev. B* Vol. 54: 16733-16741.

Gel'mont, B.L. & Shur, M.S. (1970). S-type current-voltage characteristic in highly doped Gunn diodes, *Sov. Phys. Solid State* Vol. 12: 1304-1308.

Grebogi, C., Ott, E., Romeiras, F. & Yorke, J.A. (1987). Critical exponents for crisis-induced intermittency, *Phys. Rev. A* Vol. 36: 5365-5380.

Hall, R. & Leck, J.H. (1968). Avalanche breakdown of gallium arsenide p-n junctions, *Int. J. Electron.* Vol. 25: 529-537.

Horsthemke, V. & Lefever, R. (1984). *Noise Induced Transitions*, Springer, Berlin.

Joshi, R.P., Kayasit, P., Islam, N., Schamiloglu, E., Fleddermann, C.B. & Schoenberg, J. (1999). Simulation studies of persistent photoconductivity and filamentary conduction in opposed contact semi-insulating GaAs high power switches, *J. Appl. Phys.* Vol. 86: 3833-3843.

Oshio, K.I. & Yahata, H. (1995). Non-periodic current oscillations in the Gunn-effect device with the impact-ionization effect, *J. Phys. Soc. Jpn.* Vol. 64: 1823-1836.

Schöll, E. (1987). *Nonequilibrium Phase Transitions in Semiconductors*, Springer, Berlin.

Shaw, M.P., Mitin, V.V., Schöll, E. & Grubin, H.L. (1992). *The Physics of Instabilities in Solid State Electron Devices*, Plenum, New York.

Shiau, Y.H. & Cheng, Y.C. (1996). Static and dynamic hysteresis in deep-impurity doped n-GaAs under d.c. bias voltage, *Solid-State Electron.* Vol. 39: 205-210.

Shiau, Y.H. & Peng, Y.F. (2005). Long-lived transient behavior in an n^+-n-n^+ semiconductor device with optical stochasticity, *Phys. Rev. E* Vol. 71 (No. 066216): 1-6.

Shiau, Y.H. (2006). Detection of hidden structures in a photoexcited semiconductor via principal-component analysis, *Solid-State Electron.* Vol. 50: 191-198.

Thompson, J.M.T. & Stewart, H.B. (1993). *Nonlinear Dynamics and Chaos*, Wiley, New York.

Wacker, A. (2002). Semiconductor superlattices: A model system for nonlinear transport, *Phys. Rep.* Vol. 357: 1-111.

Yeh, P. (1993). *Introduction to Photorefractive Nonlinear Optics*, John Wiley & Sons, New York.

Sze, S.M. (1969). *Physics of Semiconductor Devices*, Wiley, New York.

25

Optoelectronics in Suppression Noise of Light

Jiangrui Gao, Kui Liu, Shuzhen Cui and Junxiang Zhang
State Key Laboratory of Quantum optics and Quantum optics Devices
Institute of Opto-Electronics, Shanxi University, Taiyuan
China

1. Introduction

High quality light, whenever coherent light and nonclassical light, is more and more widely used in many filed. The coherent state behave minimum classical noise and better coherence, is mostly important light in quantum information, precise measurement and quantum metrology etc. In quantum information and quantum optics, a coherent light is used to entanglement generation (Ou, Z. et al. 1992), quantum teleportation (Furusawa, A. et al. 1998), quantum computation (Menicucci, N. et al. 2006) and quantum secret sharing (Lance, A. et al. 2004). The quiet laser beam, shot noise limited laser beam is important in above application.

In the past 20 years there has been much interest in the generation and application of various nonclassical light sources, such as squeezed-state entanglement, quantum-correlated twin beams, and sub-poissonian light, for which improved signal-to-noise ratios and measurement precision beyond the quantum-noise limit are anticipated. In particular, sub-poissonian light is a promising light source in quantum information and optical communication because of its communication capacity (Tapster, P. et al. 1988; Teich, M. & Saleh, B. 1985; Machida, S. et al. 1987,1989; Richardson, W. et al. 1991; Li, R. et al. 1995).

In 1990 Mertz et al. (Mertz, J. et al. 1990) obtained sub-Poissonian light by using the feed-forward technique, and in the year of 1997, P. K. Lam et al. (Lam, P. et al. 1997) experimental demonstrate noiseless signal amplification of using positive optoelectronic feed forward. The technique of optoelectronics is applied into noise suppression of light and noiseless signal amplification, and it is extended to other displacement transformation, such as quantum teleportation (Furusawa, A. et al. 1998), quantum cloning (Andersen, U. 2005; Braunstein, S. et al. 2001); furthermore, the purification of nonclassical light recently (Hage, B. et al. 2008; Dong, R. et al. 2008).

In this paper, we introduce the application of opto-electronics in noise suppression. It includes two parts, section II is noise suppression of fiber laser to SNL by optoelectronic feed forward and section III is preparation of sub-poissonian state by opto-electronic feedforward.

2. Suppression intensity noise of fiber laser

As we all known, a general laser beam is near the coherent light, the fiber laser is the new type laser, it behave some advantageous: high output power, small size and good frequency stability, easy to coupled with the fiber, but the light is not complete coherent light, especially at the low sideband frequency, the light have higher noise than shot noise limit (SNL). It hinders fiber laser application in the quantum information, so suppression the noise of fiber laser is a significant issue. Intensity noise of fiber laser include beating noise, environmental perturbations, thermal noise, and fluctuations in pump power (Cheng, Y. et al. 1995).Various methods for partially reducing output noise have been proposed. High frequency beating noise from fiber ring laser can be reduced by intracavity spectral filtering (Sanders, S. et al.1992). Intensity noise of fiber laser at the relaxation oscillation frequency was suppressed by integrating a negative feedback circuit (Spiegelberg, C. et al. 2004; Ball, G. et al. 2008).

Optoelectronic feed forward, which is easy and practical, is a method to suppression noise of light easily. In the section we introduce an optoelectronic feed-forward loop used to suppress the intensity noise of the fiber laser. Selecting the best delay time and feed-forward gain, the intensity noise can be suppressed near to SNL at sideband frequency. The maximum noise reduction 22dB was obtained at 6.0MHz. Furthermore, altering the delay time and filter band of optoelectronic system, the light can near to shot noise at differ frequency point. After the noise suppression, the fiber laser can be widely applied into the filed of quantum information.

The experiment setup of laser noise suppression is shown in fig.1. It is based on the noise correlation between reflection light and transmission light.

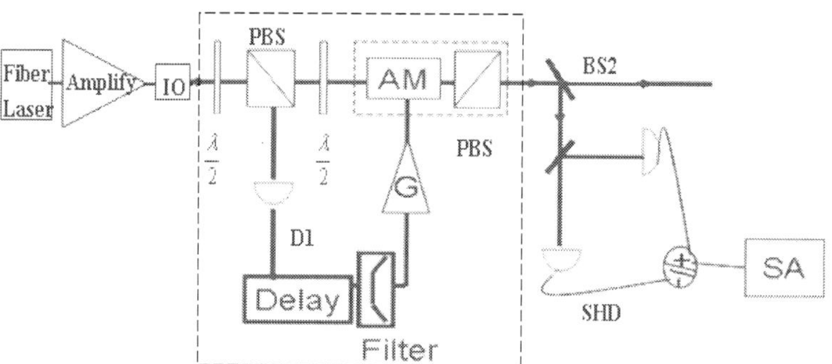

Fig. 1. The experiment setup of suppression of intensity noise of fiber laser with OptoElectronic feedforward. PBS: polarization beam splitter; BS: beam splitter; AM: amplitude modulator; D1; detector; SHD: homodyne detectors; SA: spectrum analyzer.

As shown in fig.1, the laser output from fiber laser is divided into two parts by beam splitter which is composed by polarization beam splitter (PBS) and half wave plate. The detector D1 detects the reflected light and the measurement of D1 is taken as the signal of feed forward. After passed through delay, filter and amplifier, the signal is acted on amplitude modulator (AM). As the result, the noise of transmission is suppressed, and the noise is measured by homodyne detectors.

2.1 Theory of optoelectronic feed forward in noise suppression

The output field of fiber laser can be expressed:

$$\hat{A}_{in}(t) = A_{in} + \delta\hat{A}_{in}(t) \tag{1}$$

Here, $\hat{A}_{in}(t)$ is the annihilate operator of input field, A_{in} is the average value of input field, $\delta\hat{A}_{in}(t)$ is fluctuation of input field.

After passed through AM, the light field can read as (Lam, P. et al. 1997):

$$\hat{A}_{out}(t) = \sqrt{\eta_3}[\sqrt{1-R}\,A_{in} + \sqrt{1-R}\,\delta\hat{A}_{in}(t) + \sqrt{R}\,\delta\hat{v}_1(t) + \delta\hat{r}(t)] + \sqrt{1-\eta_3}\,\delta\hat{v}_3(t) \tag{2}$$

Where $\hat{v}_1$ is vacuum noise from the first BS; $\hat{v}_3$ is vacuum noise corresponding to imperfect of homdyne detector system, η_3 is the efficiency of homdyne detectors. $\delta\hat{r}$ is fluctuation from feedback loop; η_2 is efficiency of detector D1.

$$\delta\hat{r}(t) = \int_{-\infty}^{\infty} \kappa(\tau)\sqrt{R\eta_2} \times A_{in}[\sqrt{R\eta_2}\,\delta\hat{X}_{Ain}(t-\tau)$$
$$-\sqrt{(1-R)\eta_2}\,\delta\hat{X}_{v1}(t-\tau) + \sqrt{R}\,\delta\hat{X}_{v2}(t-\tau)]d\tau \tag{3}$$

Here $\kappa(\tau)$ is response function of feedback loop; $\delta\hat{X}_{in} = \delta\hat{A}_{in} + \delta\hat{A}_{in}^+$, $\delta\hat{X}_{vi} = \delta\hat{A}_{vi} + \delta\hat{A}_{vi}^+$ are amplitude quadrature and phase quadrature, respectively.

The amplitude quadrature of output field can be deduced from ref. (1) and (2)

$$\delta\hat{X}_{out}(t) = \delta\hat{A}_{out}(t) + \delta\hat{A}_{out}^+(t)$$
$$= \sqrt{\eta_3}[\sqrt{1-R}\,\delta\hat{X}_{in}(t) + \sqrt{R}\,\delta\hat{X}_{\hat{v}_1}(t) + \delta\hat{X}_r(t)] + \sqrt{1-\eta_3}\,\delta\hat{X}_{\hat{v}_3}(t) \tag{4}$$

Here

$$\delta\hat{X}_r(t) = \delta\hat{r}(t) + \delta\hat{r}^+(t)$$
$$= \int_{-\infty}^{\infty} G(\tau)[\sqrt{R\eta_2}\,\delta\hat{X}_{Ain}(t-\tau) - \sqrt{(1-R)\eta_2}\,\delta\hat{X}_{v1}(t-\tau) + \sqrt{R}\,\delta\hat{X}_{v2}(t-\tau)]d\tau$$

$$G(\tau) = \kappa(\tau)\sqrt{R\eta_2} \times A_{in}$$

The equation (4) can be easily solved by taking the Fourier transform into the frequency domain:

$$\delta\hat{X}_{out}(\omega) = \sqrt{\eta_3}\{\sqrt{1-R}\,\delta\hat{X}_{in}(\omega) + \sqrt{R}\,\delta\hat{X}_{\hat{v}_1}(\omega) + G(\omega)[\sqrt{R\eta_2}\,\delta\hat{X}_{in}(\omega)$$
$$-\sqrt{(1-R)\eta_2}\,\delta\hat{X}_{v1}(\omega) + \sqrt{R}\,\delta\hat{X}_{v2}(\omega)]\} + \sqrt{1-\eta_3}\,\delta\hat{X}_{\hat{v}_3}(\omega) \tag{5}$$

Here G(ω) is the gain of feedback, $G(\omega) = \kappa(\omega)\sqrt{R\eta_2} \times A_{in}$.

We can obtain the noise spectrum of output filed:

$$V_{out}(\omega) = \left\langle \left|\delta X_{Aout}\right|^2\right\rangle$$
$$= \eta_3\left|\sqrt{(1-R)} + G(\omega)\sqrt{R\eta_2}\right|^2 V_{in}(\omega) + \eta_3\left|\sqrt{R} - G(\omega)\sqrt{(1-R)\eta_2}\right|^2 V_1 \tag{6}$$
$$+ \eta_3\left|G(\omega)\sqrt{(1-\eta_2)}\right|^2 V_2 + (1-\eta_3)V_3$$

Here V_1, V_2, V_3 are vacuum noise spectrum and $V_1 = V_2 = V_3 = 1$; V_{in} is the intensity noise spectrum of input field.

To infer from eq.(6), when $G(\omega) = G(\omega)_{opt}$, the intensity noise suppression is best, and then the minimum output noise $V_{out}(\omega)$ is:

$$V_{out}^{opt}(\omega) = \frac{[R\eta_2 + (1-R)\eta_3](V_{in}(\omega)-1)+1}{(V_{in}-1)R\eta_2+1} \quad , \tag{7}$$

$$G(\omega)_{opt} = \frac{-(V_{in}-1)\sqrt{(1-R)R\eta_2}}{V_{in}R\eta_2 + \eta_2(1-R)+(1-\eta_2)} \quad , \tag{8}$$

When η_2=0.95, η_3=0.20, the optimum output noise V_{out}^{opt} as the function of input noise V_{in} and BS reflection R is shown in fig.2. With the same R, higher the input noise is, higher the output noise is, and with the same input noise, higher the reflection R is, lower the output noise V_{out} is. Just only when $R=1$, the output noise V_{out}^{opt} reach at shot noise limit (SNL). In general, the reflection R is always smaller than 1 in order to get effective intensity of output field, the method can suppress main noises, but not all noises.

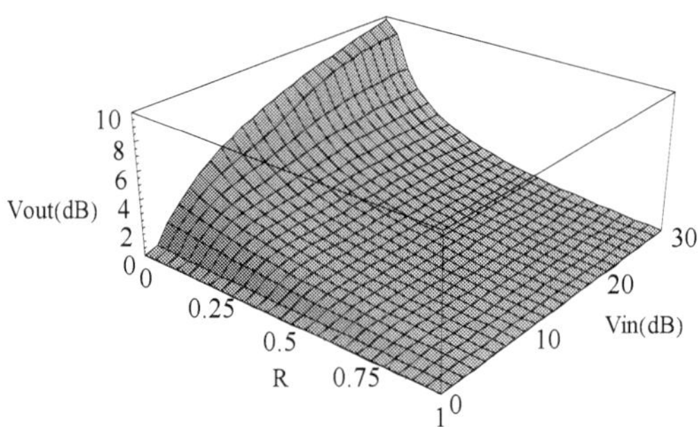

Fig. 2. The optimum output noise V_{out}^{opt} as the function of input noise V_{in} and BS reflection R, η_2=0.95, η_3=0.20.

2.2 Noise suppression

The relationship of residual output noise and reflection R is shown in fig. 3. at 6 MHz, the dot is experiment data; the dash line is theory result as eq. (7). The detector D1 can be in saturation when the R is high, we make $\eta_2 \times R$ = constant to correct the theory of dash line, and the solid line is the corrected theoretical result. In the experiment, $\eta_2 = 0.95$, C=0.24 and $V_{in} / SNL = 24dB$.

Because of the more information we can be got from laser beam when the reflection is higher, that means the signal and noise ratio is better, the noise suppression will be better as shown in fig.3. When reflection R is large enough, the increase of noise suppression efficiency become flat. In order to obtain both the output power and noise suppression efficiency simultaneously, we select the reflection $R = 0.3$.

With best selecting delay time, bandwidth filter and gain of feedback loop, the noise of suppression vs analysis frequency is shown in fig. 4 when the reflection $R = 0.3$. The maximum noise reduction 22dB is obtained at 6MHz, and the noise is near to SNL. The output noise only can be most suppressed at a frequency point with limited bandwidth due to limitation of bandwidth of optoelectronic feedback loop. In order to suppress noise in large range of frequency, we can use a high finesses cavity as model cleaner to suppress wideband noise effectively. We have obtained the shot noise limit laser from 3MHz by mode cleaner [Liu, K. et al. 2009].

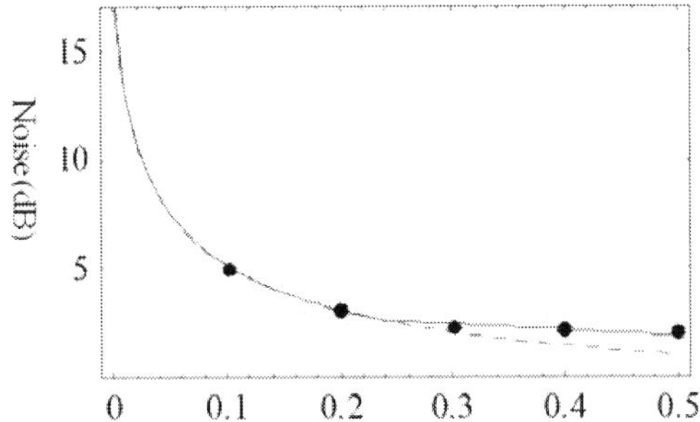

Fig. 3. Residual Noise Vs Reflectivity. ● The results of experiment; --- The results of theory , $\eta_2 = 0.95$; — The results of theory , $\eta_2 \times R = 0.24$.

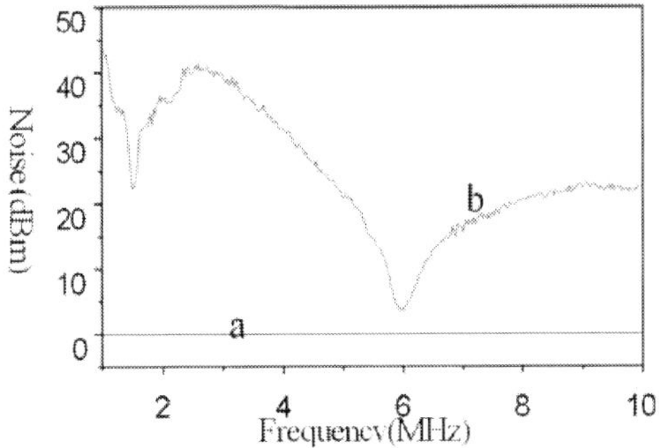

Fig. 4. The best noise suppression of Fiber Laser with Opt-Electronic feedbak at 6.2 MHz . a) the electronics noise. b) the SNL.

3. Generation of sub-poissonian state

Sub-Poissonian light is an important light source in studies of quantum information and optical communication. Sub-Poissonian light is also the light with the largest communication capacity in known light fields (Yamamoto, Y. & Haus, H. 1986). There are three techniques for generation of sub-Poissonian light: direct conversion, a feedback technique, and a feed-forward technique. In 1990 Mertz et al. (Mertz, J. et al. 1990) obtained sub-Poissonian light by using the feed-forward technique and a nondegenerate optical parametric oscillator (NOPO) pumped by the 528 nm line of an argon-ion laser; the observed noise reduction was as much as 24% below the shot-noise limit (SNL).Subsequently, Kim and Kumar (Kim, C. & Kumar, P. 1992) analyzed theoretically the generation of tunable sub-Poissonian light with an intensity feed-forward scheme for application in precision absorption spectroscopy. Then, pulses sub-Poissonian light was obtained experimentally also by traveling wave optical parametric deamplification (Li, R. et al. 1995). In 2006 Zou, H. obtained noise reduction of 45% wavelength tunable sub-Poissonian light by using the feed-forward technique (Zou, H. et al. 2006).

Zhang *et al.* (Zhang, Y. et al. 2002) presented statistics on twin beams emerging from a nondegenerate OPO. They recorded the twin beams' photocurrent difference fluctuation and demonstrated sub-Poissonian distributions of the twin beams' intensity difference. Laurat et al. (Laurat, J. et al. 2003) reported their experimental demonstration of the conditional preparation of a continuously variable nonclassical state of light from twin beams by a data-acquisition system; this is a kind of postselected sub-Poissonian light.

In this section we demonstrate the achievement of a frequency-tunable high-intensity sub-Poissonian state with optoelectronic feed-forward and quantum-correlated twin beams. A maximum noise reduction of 2.6 dB (45%) below the SNL was measured, and a wavelength tunable range of 7.4 nm was demonstrated. Sub-Poissonian distribution with twin beams was also demonstrated by direct measurement of the sub-Poissonian state.

3.1 The principle of generation sub-Poissonian light by optoelectronics

Generation of quantum-correlated twin beams is a prerequisite to our scheme. It is well known that a NOPO running above threshold is one of the best choices for intensity-correlated twin beams. Several groups of scientists have demonstrated experimentally quantum correlation of the intensity of twin beams. The intensity feed-forward scheme is shown schematically in Fig. 5.

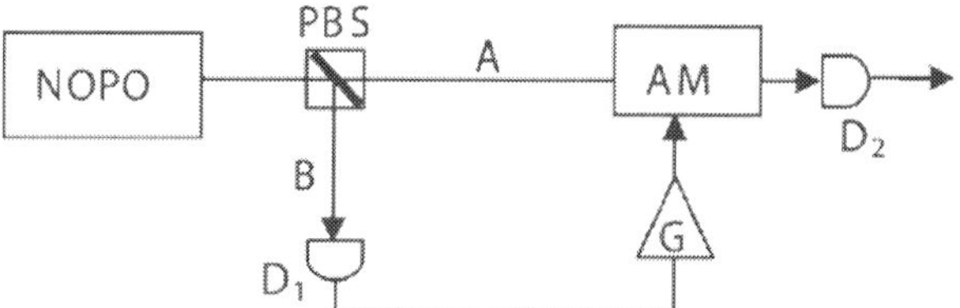

Fig. 5. Schematic of the feed-forward scheme. The quantum-correlated signal and idler beams from the NOPO are separated into beams A and B. Beam B is directly detected by detector D1 to correct beam A by modulation. PBS: polarization beam splitter; G:amplifier; AM: amplitude modulator; NOPO: nondegenerate optical parametric oscillator.

Beams A and B from the NOPO are quantum-correlated twin beams (signal and idler). Beam B is detected directly to correct beam A. After optimum correction, intensity noise S_A^{opt} of beam A at analysis frequency becomes (Mertz, J. et al. 1990)

$$S_A^{opt} = 2S_{A-B}(\Omega)[1 - \frac{S_{A-B}(\Omega)}{2S(\Omega)}]\tag{9}$$

where $S(\Omega)$ is the intensity noise spectrum of a single beam (beam A or beam B) and $S_{A-B}(\Omega)$ is the noise spectrum of the intensity difference between the twin beams, which characterizes the quantum correlation between the twin beams. Both $S(\Omega)$ and $S_{A-B}(\Omega)$ are normalized to their respective SNL. The intensity noise of a single beam is decided mainly by the OPO's state of operation. While the OPO cavity operates close to oscillator threshold, each beam generated by the OPO in general has a large amount of excess noise. Under the circumstances, the intensity noise spectrum of corrected beam A can be written approximately as

$$S_A^{opt} = 2S_{A-B}(\Omega) \tag{10}$$

It is obvious that only when the intensity difference squeezing is more than 3 dB ($S_{A-B} < 0.5$) can the noise power of the corrected signal beam be below the shot-noise level, that is, sub-Poissonian light.

3.2 Experiment and results

A schematic of the experimental setup is shown in Fig. 6. A homemade intracavity frequency-doubled and frequency-stabilized cw ring neodymium:yttrium aluminum perovskite laser serves as the light source. The output second-harmonic wave at 540 nm is used to pump the semimonolithic NOPO. The oscillation threshold is less than 120 mW, and an output power of 40 mW is obtained at a pump power of 170 mW.

The output infrared beams are reflected by dichroic mirror M and then separated by a polarizing beam splitter (PBS) into detection arms A and B. The detectors (PA$_1$, PA$_2$, PB$_1$, and PB$_2$) are Epitax 500 p-i-n photodiodes, with quantum efficiencies of about 0.94. The displacement operation is performed by use of an electro-optical modulator and a highly reflecting (99/1) mirror with the help of a local beam for reducing loss.

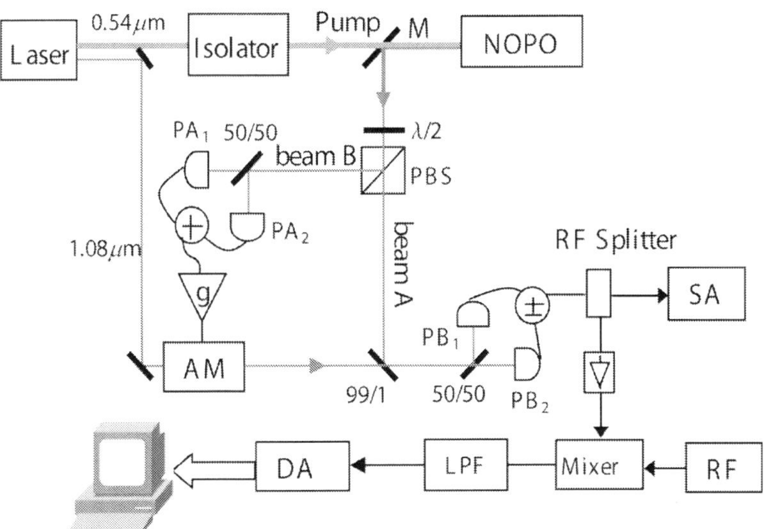

Fig. 6. (Color online) Experimental setup: PA$_i$, PB$_i$ (i=1,2): photodetectors; AM: amplitude modulator; RF Splitter: radio-frequency power splitter; SA: spectrum analyzer; RF: radio-frequency signal; LPF: low-pass filter; DA: data acquisition card; PBS: polarization beam splitter; M: mirror.

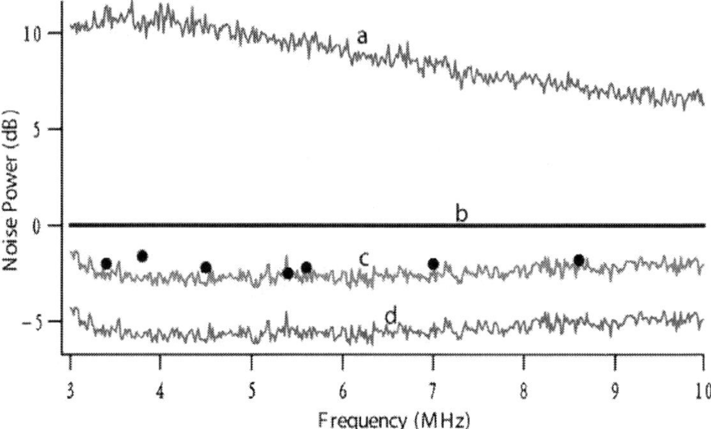

Fig. 7. (Color online) Noise spectra from 3 to 10 MHz. All noise power is normalized to the corresponding SNL. Curve *a*, noise without feed-forward correction; curve *b*, the SNL. Curve *d*, noise spectrum of the intensity difference between the signal and the idler fields. Curve *c*, obtainable noise of beam A [according to Eq. (9)] with data curve *a* and curve *d*. Filled circles, experimentally measured results.

Experimental results are shown in Fig. 7. All the noise power spectra are normalized to their respective SNL. Curve a is the noise power spectrum of beam A without feed-forward correction $S(\Omega)$; line b is the normalized SNL. It is clear that the single beam has a large mount of excess noise. The excess noise is typically 6–10 dB above the SNL in the measurement 3–10 MHz. Curve d is the noise power spectrum of the intensity difference between the twin beams $S_{A-B}(\Omega)$, the quantum correlation between beams A and B. The curve exhibits a more than 5 dB intensity difference squeezing in the entire range, and the maximum value reaches 5.7 dB near 4.5 MHz. Curve c is the obtainable minimum noise power of the sub-Poissonian beam (corrected beam A) according to Eq. (9) with data curve a $S(\Omega)$ and curve d $S_{A-B}(\Omega)$. Because there is much excess noise in each beam from the OPO, curve c is approximately 3 dB higher than curve d [Eq. (10)]. The filled circles in Fig. 7 are the directly measured noise power of prepared sub-Poissonian light at different analysis frequencies. It is clear that the maximum noise reduction is 2.6 dB (45%) below shot-noise level at 5.5 MHz frequency, and the experimental result accords well with the expected values of curve c.

We tuned the sub-Poissonian wavelength roughly by changing the KTP crystal temperature in the OPO. A wavelength range from 1078.9 to 1083.8 nm is covered for a signal beam, and the noise of a sub-Poissonian field is reduced to more than 2 dB below the SNL throughout all the wavelength-tunable range, as shown in Fig. 8. If arm A is measured to control arm B, a range from 1076.4 to 1081.2 nm for the sub-Poissonian field will be demonstrated. One can also obtain continuous frequency tuning by tuning the laser frequency

Additionally, the distribution of sub-Poissonian light is acquired as shown in Fig. 9. In the experimental setup (Fig. 6), a part of the signal after rf splitting is mixed with a sinusoidal local oscillator at 5.5 MHz. After a 100 kHz low-pass filter, the signal is collected at a sampling rate of 500 kHz by a 12 bit acquisition card. The statistical distribution of the photocurrent fluctuation is shown in Fig. 9, with 200,000 points for each curve. The points in the figure are the experimental results, and the solid curves are Gaussian fits of the probability distribution. Curve a represents the probability distribution of the prepared sub-Poissonian field, curve b corresponds to a coherent state (the SNL), and curve c corresponds to single-beam field without correction. It is shown that the sub-Poissonian distribution of light fluctuation is narrower than a standard Gaussian distribution of the coherent state. The uncorrected single-beam fluctuation distribution is a super-Poissonian and is much broader than the standard Gaussian distribution. The photocurrent fluctuation of the sub-Poissonian field can also be compared with the standard Gaussian distribution. A noise reduction of 1.2 dB below the SNL is calculated from average half-widths (Fig. 9) and does not accord well with what we observed with the spectrum analyzer because of the narrow bandwidth of the prepared sub-Poissonian field and a nonideal low-pass filter. The calculated photocurrent fluctuation of a single beam is 9 dB above the SNL, which accords well with what we observed with the spectrum analyzer.

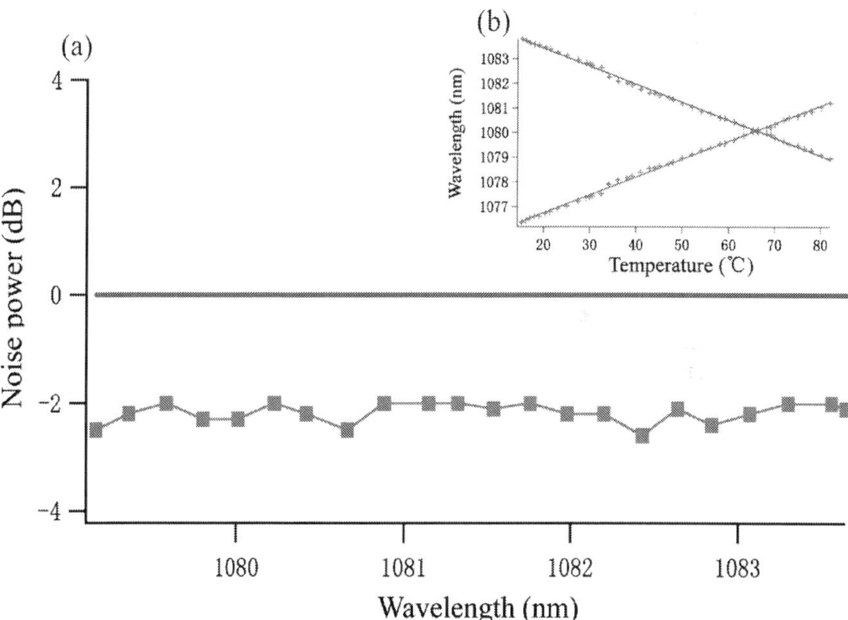

Fig. 8. (Color online) (a) Normalized sub-Poissonian light noise from 1079 to 1083.7 nm. (b) Wavelength of twin beams versus temperature of the crystal in the OPO.

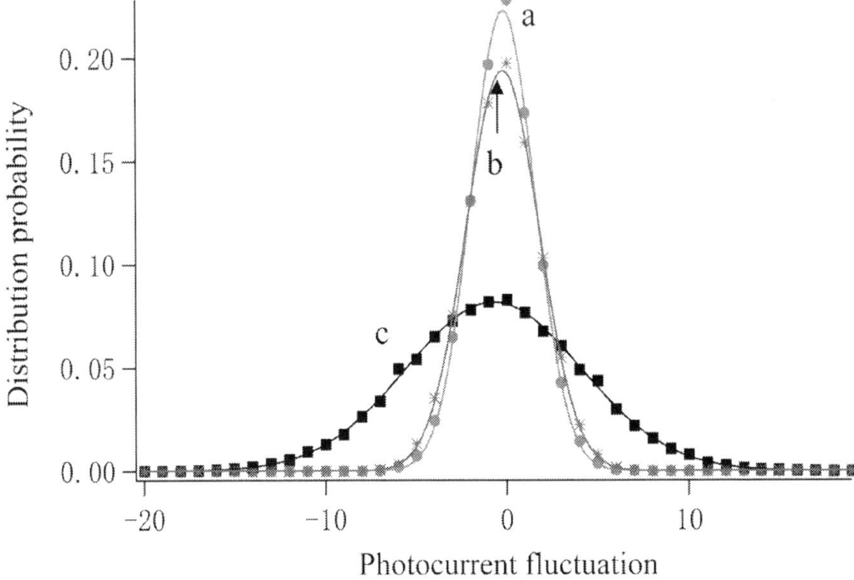

Fig. 9. (Color online) Intensity fluctuation distribution at 5.5 MHz. Curve *a*, prepared sub-Poissonian field; curve *b*, coherent light; curve *c*, single beam from the NOPO (beam A).

4. Conclusion

We introduce the application of opto-electronics feed-forward in noise suppression, including both classical noise (fiber laser noise suppression) and quantum noise (preparing sub-Poissonian) suppression. The technique of opto-electronics has been widely applied and will be more and more significant in the field of quantum optics and quantum information.

5. References

Andersen, U.; Josse,V. & Leuchs, G. (2005). Unconditional Quantum Cloning of Coherent States with Linear Optics. Phys.Rev. Lett. Vol. 94, No. 24, pp.240503, ISSN:1079-7114.

Ball, G.; Hull-Allen,G. & Holton, C. (2008). Low noise single frequency linear fiber laser. Electronics Letters. Vol. 29, No. 18, pp. 1623-1625, ISSN: 0013-5194.

Braunstein, S. Nicolas, J.; Iblisdir, S.; Loock, P. & Massar, S. (2001). Optimal Cloning of Coherent States with a Linear Amplifier and Beam Splitters. Phys.Rev. Lett. Vol. 86, No. 21, pp.4938-4941, ISSN:1079-7114.

Cheng, Y.; Kringlebotn, J.; Loh, W.; Laming, R. & Payne, D. (1995). Stable single-frequency travelling-wave fiber loop laser with integral saturable-absorber-based tracking narrow-band filter. Opt. Lett. Vol. 20, No. 8, pp. 875-877, ISSN:0146-9592.

Dong, R.; Lassen, M.; Heersink, J.; Marquardt, C.; Filip, R.; Leuchs, G. & Andersen, G. (2008). Experimental entanglement distillation of mesoscopic quantum states. Nature Phys. Vol. 4, No. 2 November, pp.919-923, ISSN:1745-2473.

Furusawa, A.; Sørensen, J.; Braunstein,S.; Fuchs, C.; Kimble, H. & Polzik, E. (1998). Unconditional Quantum teleportation. Science. Vol.282, No.5389, pp.706-709, ISSN:1095-9203.

Hage, B.; Samblowski, A.; DiGuglielmo, J.; Franzen, A.; Fiurášek, J. & Schnabel, R. (2008). Preparation of distilled and purified continuous-variable entangled states. Nature Phys. Vol. 4, No. 2 November, pp.915-918, ISSN:1745-2473.

Kim, C. & Kumar, P. 1992. Tunable sub-Poissonian light generation from a parametric amplifier using an intensity feedforward scheme. Phys.Rev. A. Vol. 45, No. 7, pp.5237-5242, ISSN:1050-2947.

Lam, P.; Ralph, T.; Huntington, E. & Bachor, H. (1997). Noiseless Signal Amplification using Positive Electro-Optic Feedforward. Phys.Rev. Lett. Vol. 79, No. 8, pp.1471-1474, ISSN:1079-7114.

Laurat, J.; Coudreau, T.; Treps, N.; Maitre, A. & Fabre, C. (2003). Conditional Peraration of a Quantum State in the Continuous Variable Regime: Generation of a sub-Poissonian State from Twin Beams. Phys.Rev. Lett. Vol. 91, No. 21, pp.213601, ISSN:1079-7114.

Li, R.; Choi, S. & Humar, P. (1995). Generation of sub-Poissonian pulses of light. Phys.Rev. A. Vol. 51, No. 22, pp.R3429-R3432, ISSN:1050-2947.

Liu, K., Cui S., Zhang, H. Zhang, J. and Gao J. R. et al. 2011. Chin. Phys. Lett. Vol. 28, No.7, pp.074211, ISSN:1741-3540.

Machida, S.; Yammmoto, Y. & Itaya, Y. (1987). Observation of amplitude squeezing in a constant- current- driven semiconductor laser. Phys.Rev. Lett. Vol. 58, No. 10, pp.1000-1003, ISSN:1079-7114.

Machida, S. & Yamamoto, Y. (1989). Observation of amplitude squeezing from semiconductor lasers by balanced direct detectors with a delay line. Opt. Lett. Vol. 14, No. 19, pp. 1045-1047, ISSN:0146-9592.

Menicucci, N.; Loock, P.; Gu, M.; Weedbrook, C.; Ralph, T. & Nielsen, M. (2006). Universal Quantum Computation with Contiunous-Varible Cluster States. Phys.Rev. Lett. Vol. 97, No.11, pp.110501, ISSN:1079-7114.

Mertz, J.; Heidmann, A.; Fabre, C.; Giaocobino, E. & Reynand, S. (1990). Observation of high-intensity sub-Poissonian light using an optical parametric oscillator. Phys.Rev. Lett. Vol. 64, No. 24, pp.2897-2900, ISSN:1079-7114

Ou, Z.; Pereira, S. F.; Kimble, H. J. & Peng, K. C. (1992). Realization of the Einstein-Podolsky-Rosen paradox for continuous variables.Phys.Rev. Lett. Vol.68,No. 25,pp.3663-3666, ISSN:1079-7114.

Richardson, W.; Machida, S. & Yamamoto, Y. (1991). Squeezing photon-number noise and sub-Poissonian electrical partition noise in a semiconductor laser. Phys.Rev. Lett.Vol. 66, No. 22, pp.2867-2870, ISSN:1079-7114.

Sanders, S.; Park, N.; Dawson, J. W. & Vahala, K. J. (1992). Reduction of the intensity noise from an erbium-doped fiber laser to the standard quantum limit by intracavity spectral filtering. Appl. Phys. Lett. Vol. 61, pp. 1889-1891, ISSN: 0003-6951.

Spiegelberg, C.; Geng, J. H. & Hu, Y. D. (2004). Low-Noise-Narrow-linewidth Fiber Laser at 1550nm. Journal of Lightwave Technology. Vol. 22, No. 1, pp.57, ISSN: 0733-8724

Tapster, P. et al. 1988. Use of parametric down-conversion to generate sub-poissonian light. Phys.Rev. A. Vol. 37, No. 8, pp.2963-2967, ISSN:1050-2947.

Teich, M. & Saleh, B. 1985. Observation of sub-Poisson Franck-Hertz light at 253.7nm. J.Opt.Soc.Am.B. Vol. 2, No. 2, pp.275- 282, ISSN:1520-8540.

Yamamoto, Y. & Haus, H. 1986. Peaparation, measurement and information capacity of optical quantum states. Rev. Mod. Phys. Vol. 58, No. 4, pp. 1001-1020, ISSN:0034-6861.

Zhang, Y.; Kasai, K. & Watanable, M. (2002). Investigation of the photon-number statistics of twin beams by direct detection. Opt. Lett. Vol. 27, No. 14, pp. 1244-1246, ISSN:0146-9592.

Zou, H.; Zhai, S.; Guo, J.; Yang, R. & Gao, J. R. (2006). Preparation and measurement of tunable high-power sub-Poissonian light using twin beams.Opt. Lett. Vol. 31, No. 11, pp. 1735-1737, ISSN:0146-9592.

26

Anomalous Transient Photocurrent

Laigui Hu[1] and Kunio Awaga[2]
[1]Department of Applied Physics, Zhejiang University of Technology,
[2]Department of Chemistry and Research Center for Materials Science, Nagoya University,
[1]China
[2]Japan

1. Introduction

The operating principle in conventional optoelectronic devices is based on steady-state photocurrent. In these devices, photogenerated carriers have to travel long distances across the devices. Various dissipation mechanisms such as traps, scattering and recombination dissipate these carriers during transport, and decrease device response speed as well as optoelectronic conversion efficiency, especially in organic devices (Forrest & Thompson, 2007; Pandey et al., 2008; Saragi et al., 2007; Spanggaard & Krebs, 2004; Xue, 2010). Such organic devices have received considerable attention due to their potential for of large-area fabrication, combined with flexibility, low cost (Blanchet et al., 2003), and so on. Efforts to substitute inorganic materials by organic ones in optoelectronics have encountered a serious obstacle, i.e., poor carrier mobility that prevents photogenerated carriers from travelling a long distance across the devices.

Typically, exciton diffusion length in organic materials is approximately 10-20 nm (Gunes et al., 2007). Internal quantum efficiency decreases with the increase in film thickness (Slooff et al., 2007) since recombination will occur prior to exciton dissociation if photogenerated excitons are unable to reach the region near the electrodes. Therefore, though a thicker film can result in an enhanced light harvesting, collecting carriers using electrodes becomes difficult. In addition, the poor mobility of organic materials always triggers the formation of space charges in thin film devices, and the space charges additionally limit the photocurrent (Mihailetchi et al., 2005).

In this chapter, we introduce an anomalous transient photocurrent into optoelectronics based on Maxwell's theory on total current, which consists of conduction and displacement current. In contrast to organic optoelectronic devices based on conduction photocurrent, which sufferrs from poor carrier mobility, the anomalous photocurrent can contribute to optoelectronic conversion and "pass" through an insulator. Though such anomalous photocurrent, or photoinduced displacement current, has received previous attention (Andriesh et al., 1983; Chakraborty & Mallik, 2009; Iwamoto, 1996; Kumar et al., 1987; Sugimura et al., 1989; Tahira & Kao, 1985), its mechanism and characteristics are still largely unresolved. We systematically explained this phenomenon based on our theoretic analyses and experiments on an organic radical 4'4-bis(1,2,3,5-dithiadiazolyl) (BDTDA) (Bryan et al., 1996) thin film device. A double-layer model was introduced, and a new type of device with structure of metal/blocking layer/semiconductor layer/metal was developed to reproduce the anomalous photocurrent (Hu et al., 2010b). The photocurrent transient is observed to involve polarisation in the materials, and stored charges within the phtocells can be released by the time-dependent conduction photocurrent. The formulae derived for this phenomena are promising for the characterisation of carrier transport in organic thin films.

In this chapter, we firstly demonstrate the anomalous photocurrent and steady-state photocurrent in the BDTDA photocells with a structure of ITO/BDTDA (300 nm)/Al (Hu et al., 2010a; Iwasaki et al., 2009). The anomalous photocurrent in the BDTDA films is observed to involve a large polarisation current induced by the formation of space charges near the electrodes. Subsequently, a series of formulae based on the total current equation for a double-layer system have been developed to fit experimental data. The theoretical ideas behind this formula are discussed as well.

Based on the analyses, the metal/blocking layer/semiconductor layer/metal photocell is demonstrated using different organic materials, including insulators and semiconductors, to reproduce the anomalous photocurrent. We introduce the enhancement of anomalous photocurrent by employing a transparent dielectric polymer with a larger dielectric constant (as a blocking layer) since larger polarisation current can be produced. Fast speed can be achieved since the performance is mainly limited by the fast dielectric relaxation (Kao, 2004). These are promising for high-speed operation in optoelectronics. Afterward, the properties of anomalous photocurrent, including light intensity dependencies, are demonstrated.

Finally, we briefly introduce a new method for mobility measurements based on the double-layer model. Unlike the time of flight technique and field effect transistor measurements, this method can be used for an ultra-thin organic semiconductor to check carrier transport along the directions perpendicular to electrodes in photocells. Furthermore, we demonstrate that the technique can be utilised to check the dominant carrier types in a semiconductor. The final section includes the summary and proposals.

2. Anomalous photocurrent in BDTDA photocells

Anomalous transient photocurrent has been independently revealed in organic materials and amorphous inorganic materials. In extant literatures, mechanisms such as trapping/detrapping or electron injection from electrodes were adopted to interpret this behaviour in different materials. A common understanding from previous reports is that the transient photocurrent comes from organic or amorphous materials with poor carrier mobility or large thickness. However, the effects of the dielectric properties on related materials were seldom studied in detail. Moreover, we observed the anomalous transient photocurrent in a radical BDTDA thin film device with a significant imbalance of carrier transports. As a model material, behaviour in the BDTDA devices will be introduced in this section, as well as the physical properties of the pink BDTDA thin films.

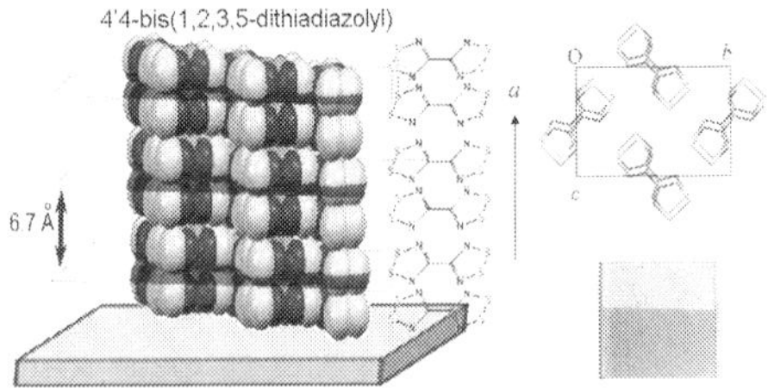

Fig. 1. Molecular π-stacking along the monoclinic *a* axis of BDTDA, a photograph of a thin film on ITO, and the molecular packing in the *bc* plane for this material

2.1 Characteristics of BDTDA thin films
2.1.1 Film structures

BDTDA is a disjoint diradical. Molecular orbitals for the two unpaired electrons are localised to separate five-membered rings, and exchange interactions between the two radical centres are very small. Its crystal structure consists of a face-to-face BDTDA dimer, indicating that intermolecular interaction is stronger than intradimer interaction. These dimers show π-stacking along the monoclinic a axis. Packing of dimeric stacks produces a herringbone-like motif with electrostatic $S^{\delta+}...N^{\delta-}$ contacts, in which all the molecular planes of BDTDA are parallel to the bc plane. It is notable that BDTDA films consist of alternating 1-dimensional π-stacking with molecular planes parallel to the substrates, as shown in Fig. 1 (Iwasaki et al., 2009; Kanai et al., 2009). Therefore, π-stacking can bridge the distance between bottom and top electrodes, which aids photoconduction between the electrodes.

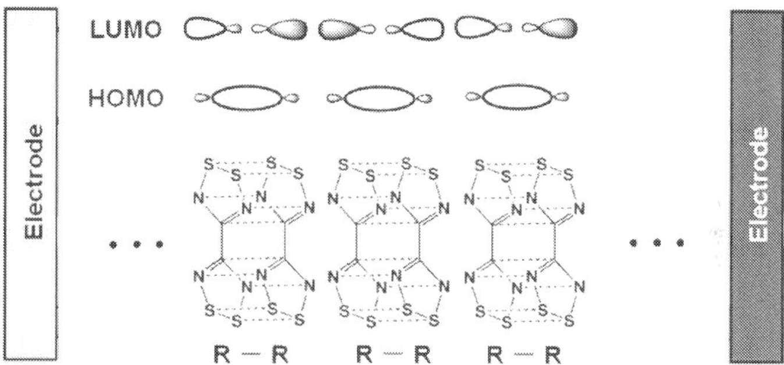

Fig. 2. Bonding and antibonding supramolecular orbitals of π–radical dimer.

2.1.2 Imbalance of carrier transport in BDTDA films

Considering that two π-radical BDTDA molecules exhibit face-to-face overlap, a bonding supramolecular orbital and an antibonding supramolecular orbital are developed (Fig. 2). The population of the bonding supramolecular orbital is concentrated at the centre of the dimer, while that of the antibonding supramolecular orbital spreads outside along the R−R axis. Since these radical dimers create stacking chains with π-π interactions, the antibonding supramolecular orbitals are expected to form a wide band through a large interdimer overlap; population of the lowest unoccupied molecular orbital (LUMO) spreads towards the outside of the dimer. By contrast, the highest occupied molecular orbital (HOMO) forms a narrow band. Therefore, a significant imbalance of carrier transport can be expected, specifically high photoconductivity by the electron migration in the wide LUMO band and poor hole mobility in the narrow HOMO band. In addition, the valence bond image (Iwasaki et al., 2009) suggests that the photoexcited state includes a character of charge transfer, namely, R:R $\rightarrow$ R⁺R⁻, where R is a radical. In other words, electrons will be directly promoted from one molecule to another by photons, which can be regarded as a precursor stage of charge separation. These characteristics are promising for developing photoactivities.

2.1.3 Space charge limited current in BDTDA films

To characterise the diradical film, photocells with a structure of ITO/BDTDA (300 nm)/Al were prepared (Fig. 3) and current-voltage (J-V) characteristics were recorded. BDTDA was prepared as described in a previous report (Bryan et al., 1996), and was thermally evaporated onto ITO glasses. As a top electrode, Al was also thermally evaporated onto the thin films. The effective area of this photocell was approximately 0.02 cm². The sample was then fixed into a cryostat with a pressure below 1 Pa. During measurement, the Al electrode was grounded, and bias polarity was defined as plus when a positive bias voltage was applied to ITO.

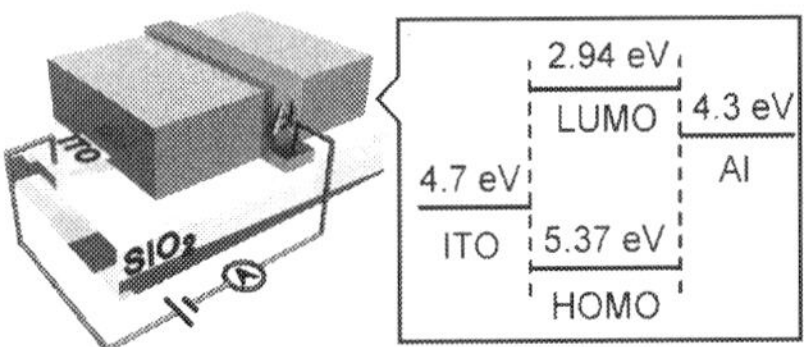

Fig. 3. Schematic views and an energy diagram of BDTDA photocells.

J-V characteristics were investigated using a picoammeter/voltage source under dark conditions and the bias voltage was scanned from -3 V to 3V. As shown in Fig. 4(a), the J-V curve exhibits a rectification behaviour, and rectification rate is approximately 10^2 at ±2 V. This behaviour is reasonable, as the work functions of the two electrodes are different, and non-injecting (see energy diagram of electrodes and BDTDA in the inset of Fig. 3). The applied bias V was corrected (van Duren et al., 2003) to compensate for the built-in voltage ($V_{bi} \approx 0.4$ V) that arises from work function difference between the two electrodes. Voltage drop across the series resistance of BDTDA devices was ignored, as its value was negligibly small.

Figure 4(b) exhibits the log (J)-log (V) plots for the data in Fig. 4(a). This curve consists of two regions with a crossover point at ~0.8 V, below which the J-V curve demonstrates Shockley behaviour that is ascribed to the injection limited current. At higher voltages (V > 0.8 V), the J-V curve shows a linear dependence, and its slope can be estimated as ~4.9. This value indicates that space charge limited current dominates the curve, though the dependence does not satisfy Child's law ($J \propto V^2$) (Coropceanu et al., 2007; Karl, 2003). This is a bulk limited current ascribed to a trap-controlled space charge limited current or a space charge limited current with a field dependence of carrier mobility (Blom et al., 1997; Sharma, 1995). Therefore, space charges are easily generated in this thin film devices, mainly due to significant imbalance of carrier transport and relatively large thickness (300 nm).

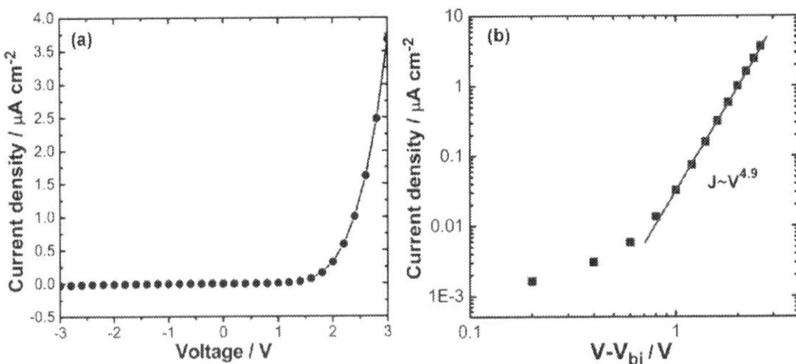

Fig. 4. J-V characteristics of a BDTDA photocell under a dark condition; (a) linear plot of J versus V; (b) log (J)-log(V) plot for the data in (a).

2.2 Photoresponses of BDTDA films

To measure the photocurrent of the photocells, a monochromated light, and green laser (532 nm) that can produce a stronger illumination, were employed as light source to irradiate the samples. To match the absorption band of BDTDA thin films, light with a wavelength of 560 nm was chosen for weak illumination to the transparent ITO electrode. We adopted lock-in techniques or an AC method (Ito et al., 2008) for normalised photocurrent-action spectra.

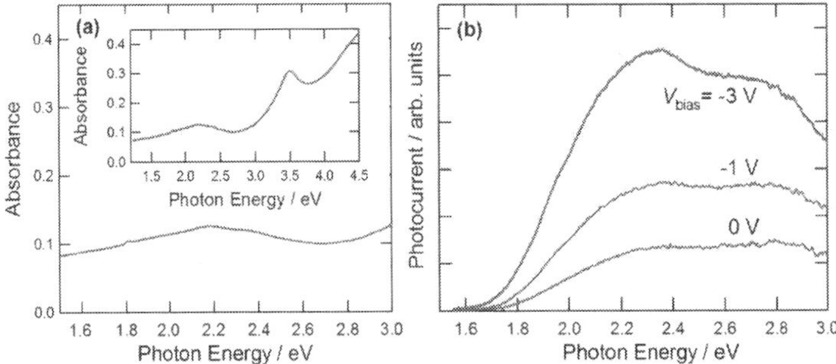

Fig. 5. (a) Absorption spectrum of BDTDA thin film; the inset shows the whole data within the range of 1.4-4.5 eV; (b) photocurrent-action spectra

2.2.1 Steady-state photocurrent of BDTDA films

To determine the optical properties of BDTDA thin films for photocurrent measurements, absorption spectrum of the BDTDA thin film (100 nm) on a quartz substrate within the range of 1.5-3.0 eV was recorded, as shown in Fig. 5(a). The inset shows data in the whole range of 1.2-4.5 eV. It is notable that there is a broad band around 2.1 eV that covers the whole visible range. The molecular orbital calculations indicate that this broad band is a complex of various electronic transitions, including intramolecular-, intradimer-, and interdimer transitions, allowed in the dimeric structure of this disjoint diradical. Subsequently, we examined the photoresponse of ITO/BDTDA (300 nm)/Al sandwich-type photocells.

Figure 5(b) shows the plots of photocurrent versus the photon energy (photocurrent-action spectra) measured by a lock-in technique with bias voltages V_{bias}= -3, -1 and 0 V. Photocurrent is obtained in the whole range of visible light (1.8-3.0 eV), while it shows a quick decrease below 2.2 eV. This decrease is possibly caused by the fact that absorptions below this energy are due to intramolecular excitations. The wide-range response, shown in Fig. 5(b), is advantageous for practical application as photodetectors.

Figure 6 is the photocurrent induced by green laser light illuminating from the ITO side with a small reverse bias voltage V_{bias} of -3V. Upon illumination, conductivity is enhanced with an on/off gain of 1.8×10^2 under an excitation light intensity of 1.59 mW/cm². The corresponding photoresponsivity (R_{res}) was calculated to be approximately 3.5 mA/W based on the relation $R_{res} = (I_{ph})/IA$, where A is the effective device area; I_{ph} and I are the photocurrent and the incident light intensity, respectively. The on/off ratio increases with the light intensity, and its maximum value observed in our experiments is approximately 10^3. Meanwhile, the photoresponsivity demonstrates an inverse behaviour, and changes from 10^{-1} to 10^{-4} A/W, which is comparable to that of the most advanced organic polymer photodetectors for visible region (Hamilton & Kanicki, 2004; Narayan & Singh, 1999; O'Brien et al., 2006; Xu et al., 2004).

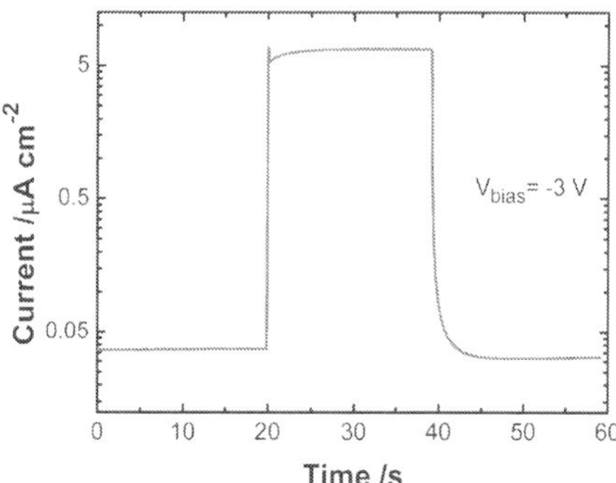

Fig. 6. On/off switching properties of the BDTDA photocell.

It is notable that the ITO/BDTDA/Al cells produce a photocurrent even at V_{bias}= 0 V, due to the potential difference of the electrodes, specifically ITO (4.8 eV) and Al (4.3 eV). This photovoltaic behaviour is consistent with the energy scheme in Fig. 3 taken by UPS/IPES measurements (Iwasaki et al., 2009). It is possible that the charge separation character in the photoexcited state, namely R+R-, contributes to this photovoltaic behaviour.

2.2.2 Anomalous transient photocurrent of BDTDA films

Figure 7(a) shows the photoresponses of an ITO/BDTDA/Al photocell with a bias voltage of 0 V. Upon illumination, a large anomalous transient photocurrent followed by a steady-state photocurrent was observed. Upon removal of illumination, a negative anomalous transient photocurrent was detected. Both the anomalous transient photocurrent and steady-state photocurrent increase with increases in light intensity. Figure 7(b) demonstrates the short circuit photoresponses under a reverse bias voltage of -2 V. Note that the anomalous transient photocurrent can be dramatically suppressed by applying a bias voltage. In particular, the negative current is nearly eliminated, while the steady-state current is increased. It is notable that anomalous transient photocurrent values under the zero bias can be comparable to those of the steady-state photocurrent under a bias voltage V. Positive anomalous transient photocurrent with weak excitation light intensity ($\leq$ 0.57 μW/cm^2) decreases exponentially with time, and decay time of the positive anomalous transient photocurrent shows light-intensity dependence. As shown in Fig. 7(a), a stronger illumination causes faster decay. Meanwhile, for the light intensity of > 0.57 μW/cm^2, positive anomalous transient photocurrent cannot fit well with a single exponential simulation. This indicates that anomalous transient photocurrent is a superposed signal with different mechanisms.

Quantum efficiencies for steady-state photocurrent and anomalous transient photocurrent were calculated by neglecting reflection losses at the device surfaces. Figure 8(a) shows the intenal quantum efficiency (Pettersson et al., 2001) versus photon energy plots for the peak values of the positive (red curve) and negative anomalous transient photocurrent (blue curve) and for steady-state photocurrent under monochromatic illumination with weak intensity from a halogen lamp. Intenal quantum efficiency values for the positive and negative anomalous transient photocurrent show increases with an increase in photon energy, and their values are considerably higher than that of the steady-state photocurrent (black curve). It is notable that the transient intenal quantum efficiency for the positive anomalous transient photocurrent reaches an extremely high value of 65% at the photon energy of 2.8 eV, and its root mean square value is estimated to be ~30%; intenal quantum efficiency values of steady-state photocurrent are ~6%, corresponding to an external quantum efficiency of ~2%.

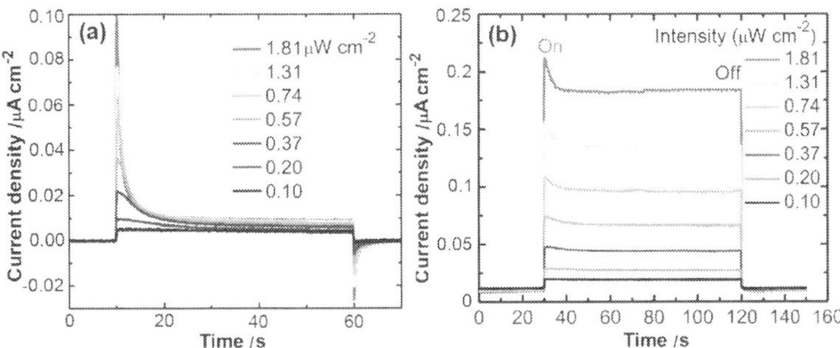

Fig. 7. Photoresponses of a BDTDA photocell with an illumination of 560 nm; (a) photoresponses under different light intensities with a zero bias voltage; (b) photoresponses under different light intensities with a bias voltage of -2 V.

To explore the recombination processes and mechanisms for anomalous transient photocurrent, we examined the light intensity dependence of the positive anomalous transient photocurrent and steady-state photocurrent. The results are shown in Fig. 8(b), where both axes are in a logarithmic scale. Both anomalous transient photocurrent and steady-state photocurrent obey a power law: $J \propto I^{\alpha}$, with α = 0.93 for the former or α = 0.28 for the latter. The former value suggests that monomolecular or geminate recombination (Binet et al., 1996) plays a role in the process. The latter α value suggests that the steady state suffers from higher order recombination processes, such as Auger (Wagner & Mandelis, 1996) and quadrimolecular recombinations (Marumoto et al., 2004). Considering that the α value is close to 0.25, quadrimolecular recombinations are more likely; adjacent photogenerated R+R- pairs may interact with each other and recombine simultaneously.

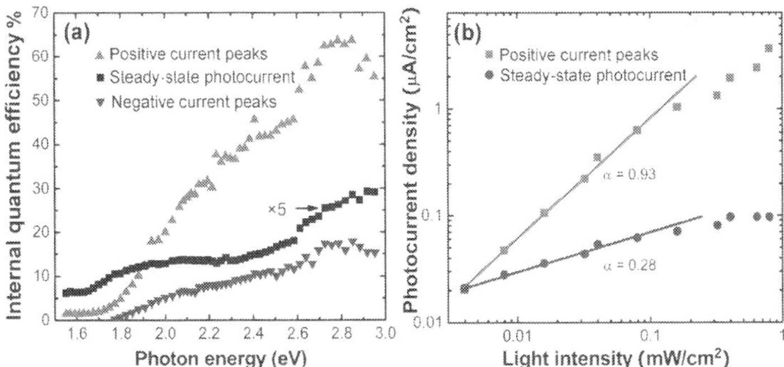

Fig. 8. (a) Intenal quantum efficiency values of the anomalous transient photocurrent and steady-state photocurrent for a BDTDA photocell; (b) light intensity dependence of the positive anomalous transient photocurrent (red points) and the steady-state photocurrent (blue points) induced by the green laser.

3. Mechanisms of anomalous photocurrent in BDTDA

Due to imbalance of carrier transports and the energy scheme of photocells, the junction at the Al/BDTDA interface plays the dominant role for the transient photoresponse (Hu et al., 2010a) if the BDTDA film is fully depleted. On the contrary, ITO/BDTDA with a larger barrier plays a main role if the film is not depleted. The junction acts as an active region (dark pink region in Fig. 9), which makes a different contribution to the anomalous transient photocurrent compared with the bulk region as blocking region (shallow pink region). The thick film can be treated as a double-layer system with widths of d_a and d_b (Hu et al., 2010b). Due to the large thickness and an imbalance of carrier transport, space charges are accumulated in the active layer. The built-in electric field will be changed, which may lead to the generation of polarisation current in the film.

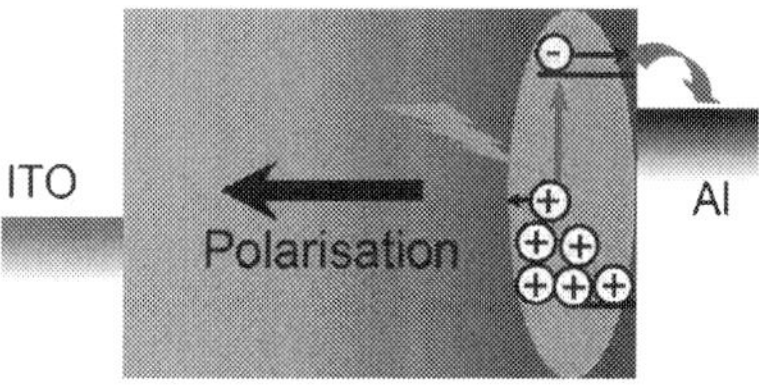

Fig. 9. A schematic view of BDTDA photocells.

3.1 Total current in a double-layer model

Theoretic analyses were performed to explore the mechanisms. To simplify the related theoretic analyses, electric fields in both regions are regarded as uniform and thus the BDTDA films can be separated into a double-layer film. Moreover, the thickness of both layers is assumed to be constant.

3.1.1 Theoretic analyses for a double-layer model

Based on the total current equation (Guru & Hiziroğlu, 2004), the current density j through the double layers is written as follows:

$$j = \left(\sigma_{b0} + \sigma_b\right)E_b\left(t\right) + \varepsilon_0\varepsilon_b\frac{dE_b\left(t\right)}{dt} = \left(\sigma_{a0} + \sigma_a\right)E_a\left(t\right) + \varepsilon_0\varepsilon_a\frac{dE_a\left(t\right)}{dt}, \tag{1}$$

where σ_{b0}, σ_b, ε_b, and $E_b(t)$ pertain to dark conductivity, photoconductivity, relative dielectric constant, and the time t dependence of the uniform electric field, respectively, in the bulk blocking region. Meanwhile, σ_{a0}, σ_a, ε_a, and $E_a(t)$ express the corresponding quantities in the active junction region. The constant ε_0 denotes the dielectric constant of free space. The first terms in both sides of Eq. (1) are conduction current, while the second terms are displacement current. All parameters for conductivity are assumed to be time independent and dark conductivities were ignored. Taking the bias voltage ($V = E_b d_b + E_a d_a$) and the boundary condition $\varepsilon_b E_b(0)=\varepsilon_a E_a(0)$ into account, we can resolve Eq. (1), and the time dependence of E_b, E_a and j can be written as:

$$E_b\left(t\right) = \frac{\sigma_a V}{d_b\sigma_a + d_a\sigma_b} + \left(\frac{\varepsilon_a}{d_b\varepsilon_a + d_a\varepsilon_b} - \frac{\sigma_a}{d_b\sigma_a + d_a\sigma_b}\right)Ve^{-t/\tau}, \tag{2a}$$

$$E_a(t) = \frac{\sigma_b V}{d_b \sigma_a + d_a \sigma_b} + \left(\frac{\varepsilon_b}{d_b \varepsilon_a + d_a \varepsilon_b} - \frac{\sigma_b}{d_b \sigma_a + d_a \sigma_b} \right) V e^{-t/\tau} , \tag{2b}$$

$$j = \frac{\sigma_b \sigma_a}{d_b \sigma_a + d_a \sigma_b} V + \frac{\left(\varepsilon_a \sigma_b - \varepsilon_b \sigma_a \right)^2 d_b d_a V}{\left(d_b \varepsilon_a + d_a \varepsilon_b \right)^2 \left(d_b \sigma_a + d_a \sigma_b \right)} e^{-t/\tau} , \tag{3}$$

where

$$\tau = \frac{\varepsilon_0 \left(d_b \varepsilon_a + d_a \varepsilon_b \right)}{d_b \sigma_a + d_a \sigma_b} . \tag{4}$$

As shown in Eq. (4), the physics of decay time τ relates to the extraction speed for the free carriers by electrodes and dielectric property/polarisation in the films. Subsequently, we can estimate the total collected charges at time t in the active side electrode, which is given by the following:

$$Q(t) = \int_0^t Aj(t)dt - \int_0^t i(t)dt , \tag{5}$$

where $i(t)$ is the discharging current in the external circuit. We take into account that the voltage drop across load resistor R is equal to that across the photocell, particularly the following:

$$i(t)R = \frac{\int_0^t Aj(t)dt - \int_0^t i(t)dt}{C} , \tag{6}$$

where C is the capacitance of the photocell. Since the initial current $i(0) = 0$, Eq. (6) can be resolved and the external discharging current $i(t)$ (i.e., anomalous transient photocurrent) is expressed as follows:

$$i(t) = \frac{S\xi_0}{(\tau - RC)} \left(e^{-\frac{t}{\tau}} - e^{-\frac{t}{RC}} \right) + S\varsigma (1 - e^{-\frac{t}{RC}}) , \tag{7}$$

where

$$\xi_0 = \frac{\varepsilon_0 \left(\varepsilon_a \sigma_b - \varepsilon_b \sigma_a \right)^2 d_b d_a V}{\left(d_b \varepsilon_a + d_a \varepsilon_b \right) \left(d_b \sigma_a + d_a \sigma_b \right)^2} , \quad \varsigma = \frac{\sigma_b \sigma_a}{d_b \sigma_a + d_a \sigma_b} V$$

3.1.2 Simplified analyses for BDTDA photocells

In general, photogenerated excitons in organic materials can be dissociated only at donor-acceptor interfaces, or by a strong local electric field (Nicholson & Castro, 2010). If the film thickness is considerably larger tthan exciton diffusion length and carrier drift length, the excitons and carriers far from the electrodes cannot contribute to the photocurrent. In other words, the photoconductivity σ_a, which is proportional to carrier mobility μ and density n in the junction (active region), is considerably larger than that in the bulk region, as well as the dark conductivity. Therefore, other conductivities (σ_{b0}, σ_b, and σ_{a0}) can be ignored. Eqs. (2) and (7) can therefore be expressed as follows:

$$E_b(t) = \frac{V}{d_b} + \left(\frac{\varepsilon_a}{d_b\varepsilon_a + d_a\varepsilon_b} - \frac{1}{d_b} \right) V e^{-t/\tau} , \tag{8a}$$

$$E_a(t) = \frac{\varepsilon_b}{d_b\varepsilon_a + d_a\varepsilon_b} V e^{-t/\tau} , \tag{8b}$$

$$i(t) = \frac{S\xi}{(\tau - RC)} \left(e^{-\frac{t}{\tau}} - e^{-\frac{t}{RC}} \right) \tag{9}$$

with

$$\xi = \frac{\varepsilon_0\varepsilon_b^2 d_a V}{(d_b\varepsilon_a + d_a\varepsilon_b) d_b} \tag{10}$$

and

$$\tau = \frac{\varepsilon_0(d_b\varepsilon_a + d_a\varepsilon_b)}{d_b\sigma_a} . \tag{11}$$

Obviously, photogenerated carriers in the junction region that can be collected by electrodes will be exhausted if photoconductivity of the blocking region is extremely small. The space charge will be accumulated in the film and thus the electric field can be changed, as shown in Eq. (9). This naturally leads to a polarisation current. Two mechanisms, including τ and RC time constant, are responsible for the decay of the anomalous transient photocurrent. The derivative calculation was performed for Eq. (9), and a rise time τ_R can be obtained. In particular, after a time

$$t = \tau_R = \frac{RC\kappa}{RCI - \kappa} \ln \frac{RCI}{\kappa} , \tag{12}$$

with $\kappa = \tau I = \varepsilon_0(d_b\varepsilon_a + d_a\varepsilon_b)/d_b e a \mu$, the largest current density J_m can be achieved, which is expressed as follows:

$$J_m = \frac{\xi}{(\tau - RC)} \left(\frac{RC}{\tau} \right)^{-\frac{RC}{RC-\tau}} \left(1 - \frac{RC}{\tau} \right) . \tag{13}$$

Since $\sigma_a = e a I \mu$, where a is quantum efficiency and aI means carrier density with a light intensity of I, and e means the elementary charge, the decay time τ can be written as follows:

$$\tau = \frac{\varepsilon_0(d_b\varepsilon_a + d_a\varepsilon_b)}{e d_b a I \mu} , \tag{14}$$

which suggests a relationship of $\tau \propto I^{-1}$. This relation fits the experimental data well. We consider the situation of a weak illumination, which will lead to a large τ. If $\tau \gg RC$, the discharging current density in Eq. (9) will be

$$J \approx J_m e^{-\frac{t}{\tau}},\tag{15}$$

with a maxima $J(t)$ value J_m

$$J_m = \frac{d_a Ve\alpha I\mu}{\left(d_b\,\varepsilon_a/\varepsilon_b + d_a\right)^2} \propto I .\tag{16}$$

Equation (15) suggests that the anomalous transient photocurrent exhibits exponential decay under weak irradiation and/or with a very small RC time constant in the circuit, which fits well with the experimental behaviour in Fig. 7(a). It is notable that $J_m \propto \varepsilon_b^{\,2}$ in Eq. (16), indicating the effects from the dielectric constant of the bulk region. On the contrary, stronger illumination triggers a smaller τ, which is related to the dielectric constants of the materials and photoconductivity in the junction region. If $\tau \ll RC$, the time constant in the circuits will dominate the decay, and shows resistance dependence as well as an exponential relationship.

3.2 Discussions

Both Eqs. (7) and (9) indicate that the anomalous transient photocurrent is a superposed signal with two mechanisms, namely, electron extraction from the junction region, and discharging process in the external circuit with a time constant of RC. It is clear that the thickness of our BDTDA films (300 nm) is excessively large, exceeding the exciton diffusion length and carrier drift length. Upon illumination, photogenerated electrons near the cathode are extracted as conduction current, while electrons on the other side cannot move across the thick film to compensate. This induces the transient conduction current.

The capacitance and dielectric constant in the equations involve polarization mainly in the bulk region triggered by photogenerated space charges in the films. The dielectric property of BDTDA strongly influences the anomalous transient photocurrent. Based on theoretic analyses, it is natural that the anomalous behaviour is universal for the thin films with large polarity, poor mobility and relatively large thickness. Though the carriers in organic materials cannot withstand a long trip due to various means of dissipation, including traps and recombination, displacement or a polarisation current can generate a large anomalous transient photocurrent without experiencing a long trip. Fast generation of this photocurrent is possible because the photoinduced polarisation current allows localised charges to oscillate around their equilibrium states. This is promising for high-speed organic photodetectors.

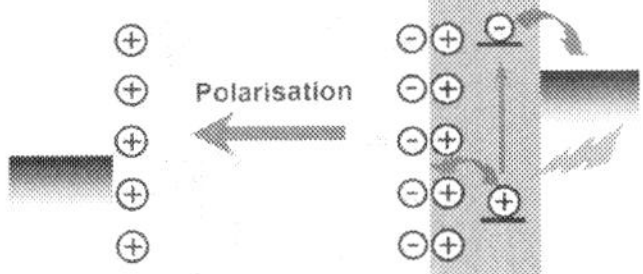

Fig. 10. A schematic display of an anode/ blocking layer /active layer/cathode photosensor.

4. Metal/insulator/semiconductor/metal type photocells

Based on the double-layer model, we developed a device to confirm the theoretic analyses in Section 3. A transparent thick organic insulator layer as a blocking layer was adopted to substitute the bulk region in BDTDA photocells, and an organic semiconductor thin layer as an active layer was chosen to substitute the junction region. Figure 10 demonstrates the photocell with a structure of metal/organic insulator/organic semiconductor/metal, which may be utilised for light detection as well. The thickness of the semiconductor layer is targeted around 20 nm, which is equivalent to the carrier drift length. The organic double layers between the metals induce an imbalance of carrier transports; in particular, only one type of carrier can be collected by the electrodes. These will facilitate accumulation of the other type of carriers as space charges at the interface of the blocking layer and active layer. In this structure, the dielectric property of the insulator layer will strongly influence the signals.

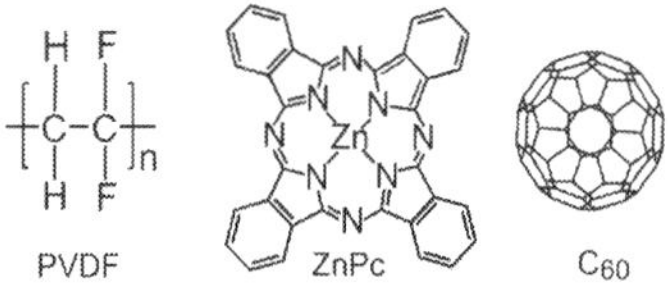

Fig. 11. Chemical structures of PVDF and ZnPc:C_{60} donor-acceptor systems.

4.1 Photoresponses of ITO/PVDF/ZnPc:C_{60}/Al

To check the photoresponse of this kind of photocell, an equivalent metal/blocking layer/semiconductor layer/metal photocell was fabricated with ITO and Al electrodes. A well-known transparent polymer, polyvinylidene fluoride (PVDF, 8 wt% in dimethyl formamide), was adopted for the blocking layer and spin-coated onto a hot ITO glass slide (100 °C). Thickness was estimated to be ~1 μm by cross-sectional SEM images. At the top of the blocking layer, a 30-nm active layer with a high charge-separation efficiency was prepared with zinc phthalocyanine (ZnPc) and fullerene (C_{60}) (molar ratio: 1:1, see Fig. 11 for their molecular structures) by co-deposition. Subsequently, the Al cathode was thermally evaporated onto the blend film. Photocurrent measurements were conducted under an illumination from a green laser (532 nm) controlled by a multifunction synthesiser. Photoresponses across a load resistor of 10^5 Ω were recorded on an oscilloscope.

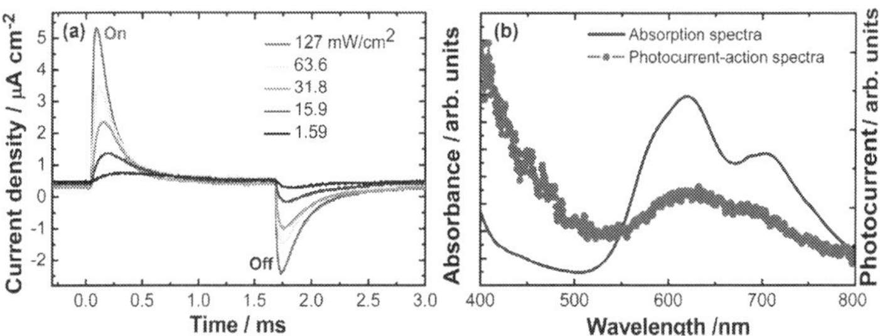

Fig. 12. (a) Photoresponses of an ITO/PVDF/ZnPc:C_{60}/Al photocell under an illumination of different intensities; (b) a comparison between the absorption spectra of the blend films (blue curve) and photocurrent-action spectra of the photocell (red curve).

4.1.1 Photoresponses

Figure 12 shows the photoresponses with various light intensities. Upon laser illumination, a large anomalous transient photocurrent similar to that in the BDTDA photocells is observed, and a negative anomalous transient photocurrent appears just after the illumination. Both the positive and negative anomalous transient photocurrent increase with increases in light intensity, and a faster decay can be obtained under a stronger illumination, which fits the expectation of Eq. (12). Absorption spectra of the blend films and photocurrent-action spectra (Fig. 12(b)) were collected for comparison. The peaks in these spectra are in agreement, indicating that the active layer does play a primary role in the production of this anomalous transient photocurrent. It is notable that no signals were obtained in the ITO/PVDF/Al structure, suggesting that only the active layer was the sensitive component. In addition, the relationship between anomalous transient photocurrent and weak light intensity was observed to exhibit linearity.

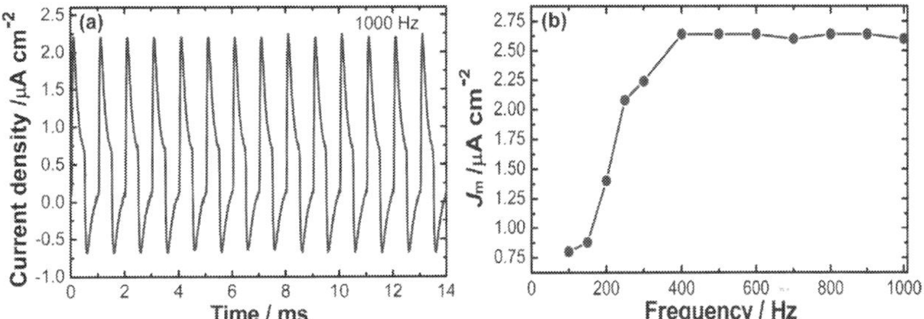

Fig. 13. (a) Photoresponses of an ITO/PVDF (1 μm)/ZnPc:C_{60} (30 nm)/Al photocell with a light modulation of 1 kHz (31.8 mW/cm²). (b) Frequency dependence of the photoresponses.

We examined the reproducibility of the anomalous transient photocurrent as well. Continuous current oscillation induced by frequency modulation is stably observed without degeneration (Fig. 13(a)). Evidently, the effective current will be increased as modulation frequency increases, as more current peaks can be generated in a fixed time period. It is notable that the values of the anomalous transient photocurrent peaks increase with increases in modulation frequency, and saturation is subsequently achieved after a certain modulation frequency, as shown in Fig. 13(b).

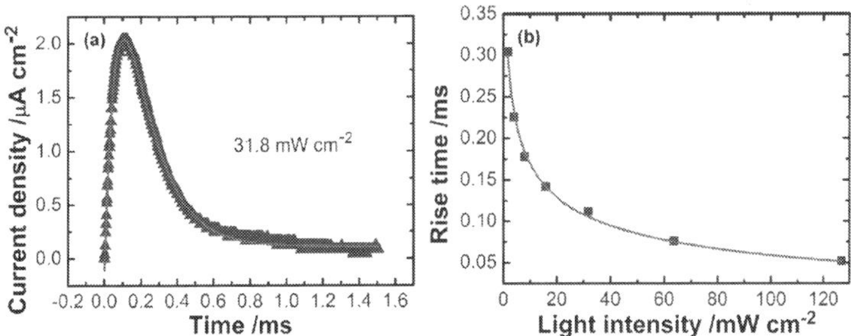

Fig. 14. (a) Simulations for the positive anomalous transient photocurrent based on (a) Eq. (7) and (b) Eq. (12) at 100 Hz.

4.1.2 Theoretic analyses for the transient photocurrent

We performed theoretic simulations for the anomalous transient photocurrent from the metal/blocking layer/semiconductor layer/metal photocells based on Eqs. (7) and (12), as shown in Fig. 14. The blue triangles in Fig. 14(a) show the time dependence of the current density of positive anomalous transient photocurrent obtained under an illumination of 31.8 mW/cm^2. The solid red curve in this figure shows the theoretical simulations from Eq. (7). The RC time constant was extracted from the simulation to be 6.8×10^{-5} s, which is considerably close to the experimental value (4.2×10^{-5} s, experimentally determined for the present circuit by an LCR meter at 100 Hz). τ was estimated to be $\sim1.6\times10^{-4}$ s, during which $1-(1/e)$ of the photogenerated carriers that can be extracted will be collected by electrodes. The blue squares in Fig. 14(b) depict the dependence of the rise time τ_R on light intensity. This behaviour is reproduced by Eq. (12) (solid curve) as well. The RC time constant and τ are estimated to be 8.3×10^{-5} s and 1.4×10^{-4} s under an illumination of 31.8 mW/cm^2, respectively. Both simulated values from Eqs. (7) and (12) are in approximate agreement with each other, suggesting that the established double-layer model is reasonable for the explanation of anomalous transient photocurrent.

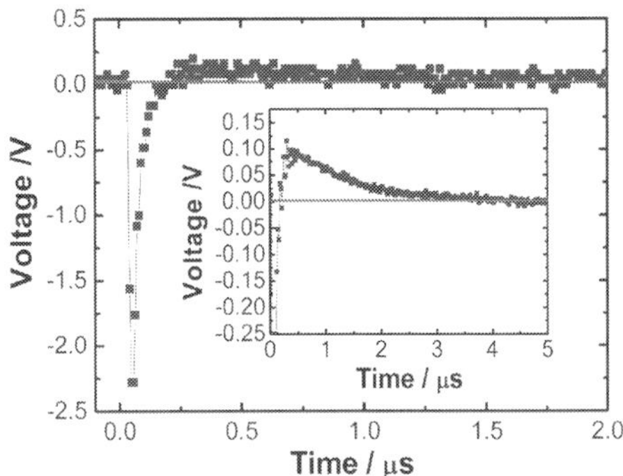

Fig. 15. Impulse response of the ITO/PVDF/ZnPc:C$_{60}$/Al photocell under a zero bias voltage; the inset is a magnified version of the recovery process.

4.1.3 Impulse response

To evaluate the lifetime of anomalous transient photocurrent, an impulse response was examined with a nanosecond laser beam (600 nm) from an optical parametric oscillator pumped by a Nd:YAG laser (10 Hz; pulse width: ~6 ns; power: ~1.08 µJ/pulse). A digital oscilloscope and a dc 300-MHz amplifier were used to collect voltage response with an input resistance of 50 Ω. A photocell with a structure of ITO/polystyrene (1 µm)/ZnPc:C$_{60}$ (20 nm)/Al was prepared for comparison with the ITO/PVDF (1 µm)/ZnPc:C$_{60}$ (20 nm)/Al photocells. The fabrication method for the polystyrene blocking layer was the same as that for PVDF.

Figure 15 shows the impulse response of the photocell with a PVDF blocking layer, which consists of rise, decay, and recovery processes. This behaviour is similar to that of the pyroelectric detectors with slower rise, decay, and recovery times (Odon, 2005; Polla et al., 1991), though their mechanisms are quite different. The RC constant in this circuit was estimated to be ~5 ns. Rise and decay time of the PVDF photocell can be observed as ~15 and 100 ns, respectively. Both the rise and decay times show an RC constant dependence; they increase along with increases in the RC constant (not shown). However, the recovery time exhibits a long time scale of ~2.5 µs (see inset of Fig. 15) and is independent of the RC constant.

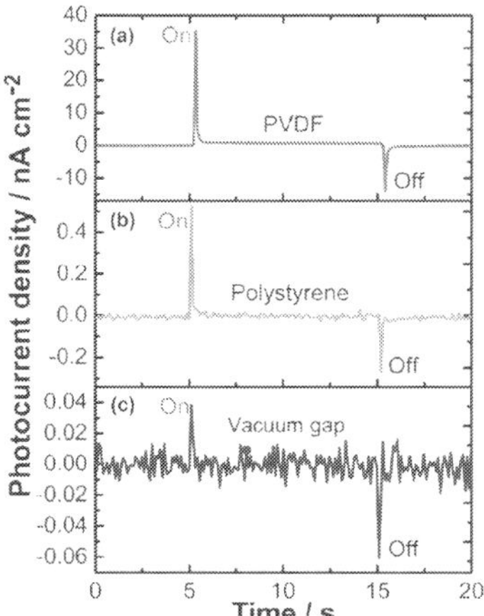

Fig. 16. Dielectric constant dependence of the anomalous transient photocurrent under an illumination (532 nm) of 160 mW/cm^2; (a), (b) and (c) show the short-circuit anomalous transient photocurrent in the metal/blocking layer/semiconductor layer/metal photosensor with PVDF, polystyrene, and vacuum gap as blocking layers, respectively.

Faster response can be achieved by decreasing the dielectric constant ε of blocking layer. For example, substitution of the PVDF layer ($\varepsilon \approx 7$-13) (Kerbow & Sperati, 1999) by polystyrene ($\varepsilon \approx 2.6$) (Cullen & Yu, 1971) brings about a considerably faster rise (~5 ns) and decay (~8 ns) times at 0 V, but recover time remains to be ~1 µs. Slow recovery time could be ascribed to an energy barrier between the donor-acceptor and/or semiconductor-metal interfaces. Considering that the polarisation current is proportional to the variation rate of E_b triggered by the photogenerated space charges, a faster generation of space charges by a sharper light pulse can bring about a larger anomalous transient photocurrent, even when only a small number of space charges are generated. Therefore, device speed is mainly determined by the rise and decay time, even though the system does not completely recover.

4.1.4 Dielectric influences

We examined the relation between the dielectric constant ε_b of the blocking layer and the quantum efficiency of anomalous transient photocurrent. Photocells with three different blocking layers (1 µm), namely, with vacuum gap ($\varepsilon = 1$), polystyrene, and PVDF were prepared. Thickness of all the active layers is approximately 20 nm. Figure 16 demonstrates the short-circuit photoresponses of the three photocells against a strong illumination (160 mW/cm^2). The values of the anomalous transient photocurrent dramatically increase with ε_b as predicted in Eq. (16). As such, we can control the transient conversion efficiency by changing the ε value of blocking layer. It is notable that the positive anomalous transient photocurrent of the PVDF photocell is ~8×10^2 times larger than that of the vacuum-gap photocell, though a rough estimation based on Eq. (16) suggests a difference of two orders of magnitude. The internal quantum efficiency of anomalous transient photocurrent in this PVDF cell under a weak illumination (0.2 µW/cm^2; 560 nm) from a halogen lamp is calculated to be approximately 34% (root mean square, rms). The photoresponsivity at 560 nm (0.2 µW/cm^2) reaches 10 mA/W (rms) even without applying a bias voltage, which is comparable to those of conventional organic photodetectors operated by a bias voltage (Iwasaki et al., 2009; Narayan & Singh, 1999; O'Brien et al., 2006). It is believed that more charges stored in the PVDF photocell with a larger ε were released upon illumination, when compared with the polystyrene and vaccum gap photocells.

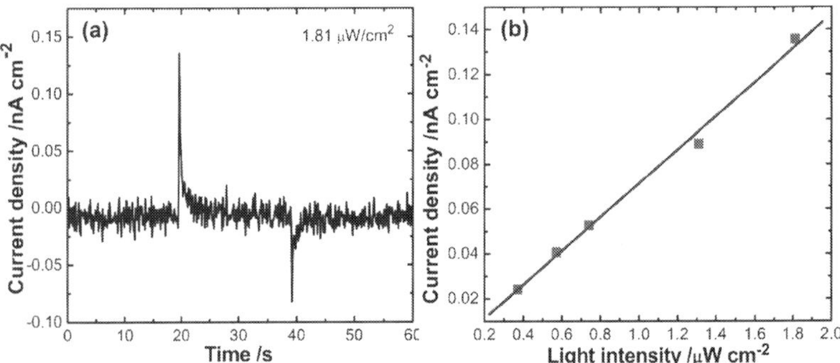

Fig. 17. (a) Photoresponse of ITO/ ZnPc/polystyrene (1 µm)/ Al photocells under a zero bias voltage; (b) light intensity dependence of peak anomalous transient photocurrent value.

4.2 Photoconductivity dependence

Based on Eq. (14), decay time τ is inversely proportional to photoconductivity $\sigma_a = en\mu$, where $n = aI$ is the photogenerated free carrier density. Therefore, larger carrier mobility and density will induce faster decay and a larger signal if the RC time constant in a circuit is very small. To check this relationship, ZnPc (30 nm) as the active layer was utilised in the metal/blocking layer/semiconductor layer/metal structure by thermal evaporation with a speed of 1 Å/s. Polystyrene layer (1 µm) by spin coating was adopted as the blocking layer. Two types of photocells with different structures for this material were produced, namely, ITO/polystyrene/ZnPc/Al and ITO/ZnPc/polystyrene/Al. As expected, the former does not exhibit signals since ZnPc is an excellent donor material. The latter shows a signal (see Fig. 17) and only holes are collected by the ITO electrode, which can be judged by the current direction. However, compared to those from the blend film (or bulk-heterojunction) devices, the signal from ZnPc photocells is considerably weaker due to a lower charge separation efficiency, which leads to a smaller carrier density. We likewise examined the light intensity dependence of anomalous transient photocurrent. As predicted in Section 3, intensity dependence of anomalous transient photocurrent does exhibit linearity (Fig. 17(b)) under weak illumination with a monomolecular or geminate recombination.

4.3 Discussions

Photoresponses from the metal/blocking layer/semiconductor layer/metal structure even with a vacuum gap is promising, indicating potential for pulse light detection. As we know, metal/semiconductor/metal type organic thin film device usually exhibits a large dark current due to pin holes, which leads to a small photocurrent. The employment of a blocking layer hampers the formation of pin holes and results in an extremely small dark current. It is possible now to utilise ultrathin films only with the highest internal quantum efficiency for light detection. Compared with ideal metal/semiconductor/metal photocells with the same thickness in which conduction photocurrent J_{ph} can be written as follows:

$$J_{ph} = \frac{Ve\alpha I\mu}{d_b + d_a},$$ (17)

the metal/blocking layer/semiconductor layer/metal structure may exhibit a larger anomalous transient photocurrent current if we select an appropriate blocking layer and active layer with a small value of d_b/d_a, and small dielectric constant ratio ($\varepsilon_a/\varepsilon_b$). These can be judged from Eq. (16) [1]. Therefore, it is possible to release more stored charges in the capacitor type photocells upon illumination in addition to the photogenerated free carriers. Equation (14) likewise indicates a promising method for characterisation of carrier transport. As we can see from this equation, the anomalous transient photocurrent signal is related to carrier mobility and density. Therefore, mobility can be estimated if we extract the decay time τ from theoretic simulations, and obtain the photogenerated carrier density by other methods, such as light-induced electron paramagnetic resonance technique. In addition, we can utilise the metal/blocking layer/semiconductor layer/metal structure to determine the carrier type in an organic semiconductor, as described in Section 4.2.

5. Conclusion

In summary, we analysed the anomalous transient photocurrent in the BDTDA photocells based on a double-layer model. Resuts indicate that the dielectric property of organic materials will strongly influence the anomalous behaviour. For instance, a large dielectric constant will induce a larger anomalous photocurrent. This was confirmed in equivalent devices, such as ITO/PVDF or polystyrene/ZnPc:C_{60}/Al, in which the PVDF and polystyrene layer act as the bulk region, and the blend film acts as the junction region in the BDTDA photocells. Both theoretic and experimental results fit well with each other, suggesting that polarisation and fast extraction of photogenerated carriers in the blend film play significant roles in this behaviour. The theoretic analyses likewise indicate that the anomalous transient photocurrent may achieve a larger value if proper conditions are satisfied, compared to the conventional metal/semiconductor/metal photocells with the same total thickness. It is notable that the metal/blocking layer/semiconductor layer/metal structure is immune from pin-hole effects which usually exist in ultrathin conventional devices. Stored charges in the metal/blocking layer/semiconductor layer/metal capacitor photocells can be released upon illumination, which is quite different from the conventional principles for light detection or harvesting. These indicate potential optoelectronic conversion for pulse light detection in various fields, including communications, remote control, and image sensors. The obtained theory may also play a role in the characterisation of carrier transport along the directions perpendicular to the electrodes in the metal/blocking layer/semiconductor layer/metal photocells.

---

[1] For comparison, Eq. (16) can be changed to the following:

$$J_m = Ve\alpha I\mu / \left(d_a + 2d_b\,\varepsilon_a/\varepsilon_b + d_b^{\,2}\,\varepsilon_a^{\,2}/d_a^{\,2}\varepsilon_b^{\,2} \right).$$

6. Acknowledgments

The authors are indebted to Prof. Hiroshi Ito for his technical supports and constructive suggestions. Also acknowledged are the research group members who contributed to this work through useful discussions and provision of materials. This research was supported by a Grant-in-Aid for Scientific Research from the Ministry of Education, Culture, Sports, Science, and Technology (MEXT) and by CREST, JST. Dr. Hu also thanks the National Natural Science Foundation of China (No. 11004172 and 10804098) and the Zhejiang Provincial Natural Science Foundation of China (No. Y607472).

7. References

Andriesh, A. M. *et al.* (1983), Anomalous Transient Photocurrent in Disordered Semiconductors - Theory and Experiment. *Solid State Communications*, Vol. 48, No. 12, pp. 1041-1043, ISSN 0038-1098.

Binet, F. *et al.* (1996), Mechanisms of recombination in GaN photodetectors. *Applied Physics Letters*, Vol. 69, No. 9, pp. 1202-1204, ISSN 0003-6951.

Blanchet, G. B. *et al.* (2003), Large area, high resolution, dry printing of conducting polymers for organic electronics. *Applied Physics Letters*, Vol. 82, No. 3, pp. 463-465, ISSN 0003-6951.

Blom, P. W. M. *et al.* (1997), Electric-field and temperature dependence of the hole mobility in poly(p-phenylene vinylene). *Physical Review B*, Vol. 55, No. 2, pp. R656-R659, ISSN 0163-1829.

Bryan, C. D. *et al.* (1996), Preparation and characterization of the disjoint diradical 4,4'-bis(1,2,3,5-dithiadiazolyl) $[S_2N_2C\text{-}CN_2S_2]$ and its iodine charge transfer salt $[S_2N_2C\text{-}CN_2S_2]$. *Journal of the American Chemical Society*, Vol. 118, No. 2, pp. 330-338, ISSN 0002-7863.

Chakraborty, A. K. & Mallik, B. (2009), Photoinduced anomalous current changes in some organometallic materials. *Current Applied Physics*, Vol. 9, No. 5, pp. 1079-1087, ISSN 1567-1739.

Coropceanu, V. *et al.* (2007), Charge transport in organic semiconductors. *Chemical Reviews*, Vol. 107, No. 4, pp. 926-952, ISSN 0009-2665.

Cullen, A. L. & Yu, P. K. (1971), Accurate Measurement of Permittivity by Means of an Open Resonator. *Proceedings of the Royal Society of London Series a-Mathematical and Physical Sciences*, Vol. 325, No. 1563, pp. 493-509, ISSN 0950-1207

Forrest, S. R. & Thompson, M. E. (2007), Introduction: Organic electronics and optoelectronics. *Chemical Reviews*, Vol. 107, No. 4, pp. 923-925, ISSN 0009-2665.

Gunes, S. *et al.* (2007), Conjugated polymer-based organic solar cells. *Chemical Reviews*, Vol. 107, No. 4, pp. 1324-1338, ISSN 0009-2665.

Guru, B. S. & Hiziroğlu, H. R. (2004), *Electromagnetic Field Theory Fundamentals*, Cambridge Univ. Press, Cambridge. ISBN 0521830168.

Hamilton, M. C. & Kanicki, J. (2004), Organic polymer thin-film transistor photosensors. *Ieee Journal of Selected Topics in Quantum Electronics*, Vol. 10, No. 4, pp. 840-848, ISSN 1077-260X.

Hu, L. G. *et al.* (2010), Highly efficient alternating photocurrent from interactive organic-radical dimers: A novel light-harvesting mechanism for optoelectronic conversion. *Chemical Physics Letters*, Vol. 484, No. 4-6, pp. 177-180, ISSN 0009-2614.

Hu, L. G. *et al.* (2010), Optoelectronic conversion by polarization current, triggered by space charges at organic-based interfaces. *Applied Physics Letters*, Vol. 96, No. 24, pp. 243303, ISSN 0003-6951.

Ito, H. *et al.* (2008), Photocurrent of regioregular poly(3-alkylthiophene)/fullerene composites in surface-type photocells. *Thin Solid Films*, Vol. 516, No. 9, pp. 2743-2746, ISSN 0040-6090.

Iwamoto, M. (1996), Transient current across insulating films with long-range movements of charge carriers. *Journal of Applied Physics*, Vol. 79, No. 10, pp. 7936-7943, ISSN 0021-8979.

Iwasaki, A. *et al.* (2009), Interactive Radical Dimers in Photoconductive Organic Thin Films. *Angewandte Chemie-International Edition*, Vol. 48, No. 22, pp. 4022-4024, ISSN 1433-7851.

Kao, K.-C. (2004), *Dielectric Phenomena in Solids* Elsevier Academic Press, Amsterdam. ISBN 978-0-12-396561-5.

Karl, N. (2003), Charge carrier transport in organic semiconductors. *Synthetic Metals*, Vol. 133, No. pp. 649-657, ISSN 0379-6779.

Kumar, A. *et al.* (1987), Anomalous Decay of Photocurrent in Amorphous Thin-Films of $Ge_{22}Se_{78}$. *Physical Review B*, Vol. 35, No. 11, pp. 5635-5638, ISSN 0163-1829.

Marumoto, K. *et al.* (2004), Quadrimolecular recombination kinetics of photogenerated charge carriers in regioregular poly (3-alkylthiophene) / fullerene composites. *Applied Physics Letters*, Vol. 84, No. 8, pp. 1317-1319, ISSN 0003-6951.

Mihailetchi, V. D. *et al.* (2005), Space-charge limited photocurrent. *Physical Review Letters*, Vol. 94, No. 12, pp. 126602, ISSN 0031-9007.

Narayan, K. S. & Singh, T. B. (1999), Nanocrystalline titanium dioxide-dispersed semiconducting polymer photodetectors. *Applied Physics Letters*, Vol. 74, No. 23, pp. 3456-3458, ISSN 0003-6951.

Nicholson, P. G. & Castro, F. A. (2010), Organic photovoltaics: principles and techniques for nanometre scale characterization. *Nanotechnology*, Vol. 21, No. 49, pp. 492001, ISSN 0957-4484.

O'Brien, G. A. *et al.* (2006), A single polymer nanowire photodetector. *Advanced Materials*, Vol. 18, No. 18, pp. 2379-2383, ISSN 0935-9648.

Odon, A. (2005), Voltage Response of Pyroelectric PVDF Detector to Pulse Source of Optical Radiation. *Measurement Science Review*, Vol. 5, No. pp. 55-58, ISSN 1335-8871

Pandey, A. K. *et al.* (2008), Size effect on organic optoelectronics devices: Example of photovoltaic cell efficiency. *Physics Letters A*, Vol. 372, No. 8, pp. 1333-1336, ISSN 0375-9601.

Pettersson, L. A. A. *et al.* (2001), Quantum efficiency of exciton-to-charge generation in organic photovoltaic devices. *Journal of Applied Physics*, Vol. 89, No. 10, pp. 5564-5569, ISSN 0021-8979.

Polla, D. L. *et al.* (1991), Surface-Micromachined Pbtio3 Pyroelectric Detectors. *Applied Physics Letters*, Vol. 59, No. 27, pp. 3539-3541, ISSN 0003-6951.

Saragi, T. P. I. *et al.* (2007), Spiro compounds for organic optoelectronics. *Chemical Reviews*, Vol. 107, No. 4, pp. 1011-1065, ISSN 0009-2665.

Sharma, G. D. (1995), Electrical and photoelectrical properties of Schottky barrier devices using the chloro aluminium phthalocyanines. *Synthetic Metals*, Vol. 74, No. 3, pp. 227-234, ISSN 0379-6779.

Slooff, L. H. *et al.* (2007), Determining the internal quantum efficiency of highly efficient polymer solar cells through optical modeling. *Applied Physics Letters*, Vol. 90, No. 14, pp. 143506, ISSN 0003-6951.

Spanggaard, H. & Krebs, F. C. (2004), A brief history of the development of organic and polymeric photovoltaics. *Solar Energy Materials and Solar Cells*, Vol. 83, No. 2-3, pp. 125-146, ISSN 0927-0248.

Sugimura, A. *et al.* (1989), Anomalous Photoinduced Current Transients in Nematic Liquid-Crystals. *Physical Review Letters*, Vol. 63, No. 5, pp. 555-557, ISSN 0031-9007.

Tahira, K.&Kao, K. C. (1985), Anomalous Photocurrent Transient in Polyethylene. *Journal of Physics D-Applied Physics*, Vol. 18, No. 11, pp. 2247-2259, ISSN 0022-3727.

van Duren, J. K. J. *et al.* (2003), Injection-limited electron current in a methano-fullerene. *Journal of Applied Physics*, Vol. 94, No. 7, pp. 4477-4479, ISSN 0021-8979.

Wagner, R. E. & Mandelis, A. (1996), Nonlinear photothermal modulated optical reflectance and photocurrent phenomena in crystalline semiconductors .1. Theoretical. *Semiconductor Science and Technology*, Vol. 11, No. 3, pp. 289-299, ISSN 0268-1242.

Xu, Y. F. *et al.* (2004), Photoresponsivity of polymer thin-film transistors based on polyphenyleneethynylene derivative with improved hole injection. *Applied Physics Letters*, Vol. 85, No. 18, pp. 4219-4221, ISSN 0003-6951.

Xue, J. G. (2010), Perspectives on Organic Photovoltaics. *Polymer Reviews*, Vol. 50, No. 4, pp. 411-419, ISSN 1558-3724.

Part 6

Nanophotonics

27

Nanophotonics for 21st Century

S. K. Ghoshal[1], M. R. Sahar[1], M. S. Rohani[1] and Sunita Sharma[2]
[1]Advanced Optical Material Research Group, Department of Physics,
Faculty of Science, Universiti Teknologi Malaysia,
Skudai, Johor,
[2]Materials Science Group, Inter-University Accelerator Centre,
Aruna Asif Ali Marg, New Delhi,
India

1. Introduction

Recent advances in the synthesis and understanding of material properties at nanoscale in addition to the development of the nanofabrication techniques has enabled researchers to control and manipulate photons at nanometer scales. These have given the birth of an emerging hybrid technology with multi-facets applied interests, popularly known as *nanophotonics*. Nanophotonics can be defined as the science and engineering of light-matter interactions. These interactions, which take, place, on the one hand, within the light wavelength and sub-wavelength scales and, on the other hand, are determined by the physical, chemical and structural nature of artificially or natural nanostructured matter. It is envisaged that nanophotonics has the potential to provide ultra-small optoelectronic components, high speed and greater bandwidth. Nanophotonics has significant potential applications in the field of science and technology. Some of them are sensors, lasers, optoelectronic chips, optical communications, optical microscopy (by overcoming the usual diffraction limit), bio-imaging, targeted therapy, barcodes, harvesting solar energy, etc., to cite a few (Kalele et al., 2007; Prasad, 2004; Shen et al., 2000). Nanophotonics are classified into three main branches as illustrated in the block diagram (figure 1) depicting various process and techniques in nanophotonics including plasmonics.

As mentioned, the field nanophotonics deals with a number of interestingly important topics in photonics and materials structures at nm length scales and their applications in general. Waves in the form of electromagnetic and quantum mechanical, and materials as semiconductors and metals are the focus. Different approaches to confine these waves and devices employing such confinement are the key issue in nanophotonics. Localization of light and applications to metallic mirrors, photonic crystals, optical waveguides, micro-resonators, plasmonics are gaining tremendous applied interest. Localization of quantum mechanical waves in low dimensional structures such as quantum wells, wires and dots has been demonstrated. Devices incorporating localization of both electromagnetic and quantum mechanical waves, such as resonant cavity quantum well lasers and micro-cavity-based single photon sources are on the way of commercialization. Some system-level applications of the introduced concepts, such as optical communications, biochemical sensing and quantum cryptography are targeted for the near future.

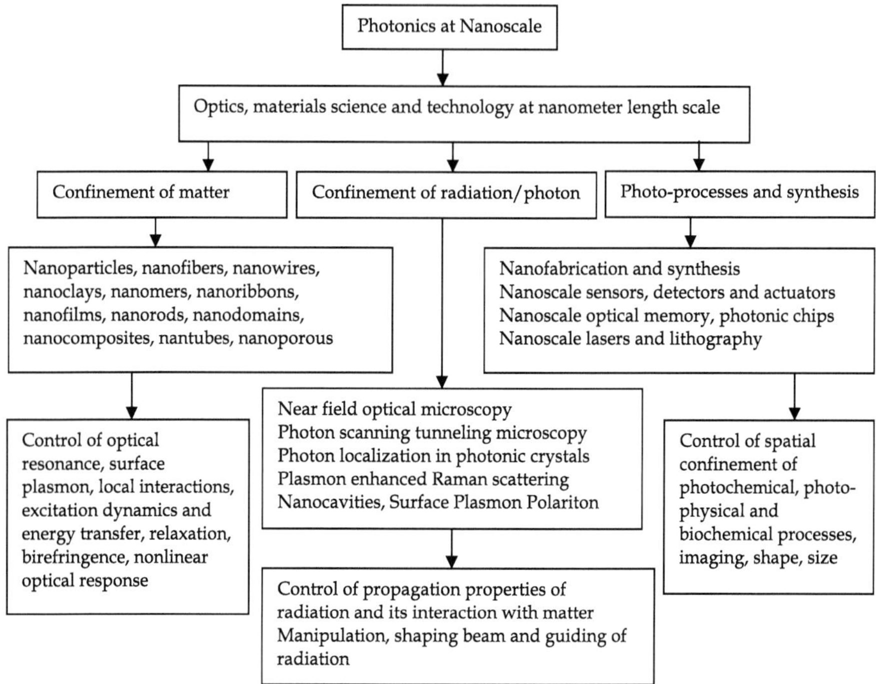

Fig. 1. A block diagram representing various major processes and techniques available in nanophotonics.

The main objectives of nanophotonics research is to control the optical energy and its conversion on the nanometer scale by combining the properties of metal, organic, semiconductor, organo-metallic, polymers and dielectric materials to create new, combined states of light and matter often called meta-materials. Specific examples of scientist's targets are:

- *Controlled quantum coupling at the nanoscale*: The ability to prepare coupled nanostructures presents tremendous opportunity to induce and control the interactions of photons, plasmons, polarons, polaritons and excitons, thereby producing new elementary excitations that have no bulk counterpart. Basic scientific research on these excitations is performed, with application to many disciplines such as solar energy conversion, nanoscale photonic devices, new photochemical processes, communications, sensing, imaging and quantum logic.

- *Understanding ultrafast processes at ultra-small length scales*: The outcome of ultrafast processes can be very different in nanoscale versus bulk materials, with potentially great impact on the physical properties and photochemical products of the nanoscale system. Research has been continuously performed to understand, manipulate, optimize and control these differences.

- *New routes to functional nanophotonic materials*: The development of novel optical materials via three main approaches are:

- Advanced colloidal synthesis,
- Lithographically assisted synthesis,
- Advanced near-field optical lithography techniques for generating hybrid nanoscale structures over large areas,
- Ion-beam milling and sputtering,
- Chemical vapor deposition, and
- Laser ablation
- *Efficient energy transport in plasmonic nanostructures*: Some research is ongoing for significant improvement of:
  - The range of plasmon propagation minimizing loss,
  - The manipulation of localized surface plasmon and surface plasmon polariton,
  - The spectral bandwidth that can be supported by plasmonic structures, and
  - The minimum lateral dimension for efficient plasmon propagation for guiding.

The art of research is the basis for entirely new, efficient solar concentrators, lasers or all optical nanoscale circuitry.

To achieve these objectives a four-pronged experimental and theoretical approach is often utilized:

- *Materials generation via physical and chemical synthesis and lithography,*
- *Optical instrumentation development for advanced characterization, fabrication,*
- *Materials modeling and rigorous numerical simulations, and*
- *Optimization and functionalization of devices via computer aided design*

Some of the leading equipments that are used to achieve this target are Near-field scanning optical microscope, Con-focal Raman microscope, Nanolithography, Ultra-fast transient absorption spectroscopy, Ultra-fast microscope, Chemical vapor deposition, Chemical synthesis, Atomic force and Scanning tunneling microscopy, Size-selected cluster facility and cluster-based nanomaterials for nanophotonics and nano(photo) catalysis (Sharma et al., 2005). Nanosphere lithography provides a very simple yet powerful way to fabricate nanoparticle arrays with precise control over size, shape and inter-particle spacing that is employed to fabricate sensors. They are made from mono-dispersed negatively charged silica or polystyrene nanospheres deposited on a negatively charged glass substrate that helps the nanospheres to diffuse freely until they reach their minimum energy configuration. The capillary forces draw these nanospheres together and finally form a close packed hexagonal pattern on the substrate. These self-assembled nanospheres thus form a mask through which a metal can be deposited by physical evaporation. The nanosphere mask is removed by sonicating the entire sample to obtain a triangular pattern of metal. The optical properties of these sensors can be tuned in the visible range by changing the size of nanospheres.

There is a great need to understand how electromagnetic waves behave in the presence of periodic or nearly periodic arrays of molecules and nanosized metal clusters, i.e. how to achieve optical frequencies in X-ray wavelength scales? Now there are the tools, to map topology, size and shape with optical properties. In the past, these tools were not available, and the tools still needed are the subjects of active research. It is the locally modified electromagnetic field, which allows one to make use of molecular photonics, taking nanophotonics to the molecular scale of a few nanometers. This, in turn, opens up possibilities not just for further miniaturization but also for radically enhanced light-matter interactions. Presently, some of the very genuine engineering concerns relates to when producing ultra-compact, low power and high-sensitivity optical devices, towards the level of single photon detection and emission, and onwards to computation by molecules.

Classes of materials, that inhibit the flow of light, within optical band frequencies, are called photonic band gap crystals or photonic band gaps for short. An array of closed packed nanospheres is the simplest realization of photonic band gap materials, made of polystyrene or silicon dioxide called opals. The range of the inhibited light depends mainly on both the size of the spheres and the number of layers. Another more advanced photonic band gap structure called an inverse opal structure can realize a photonic band gap around 1.5 µm that is made by using the nanospheres as a template. The spaces around the spheres are filled with silicon, and then the spheres themselves are removed. Photonic crystals or photonic band gap crystals are periodic dielectric or metallo-dielectric nanostructures that are designed to affect the propagation of electromagnetic waves in the same way as the periodic potential in a semiconductor crystal affects the electron motion by introducing allowed and forbidden electronic energy bands. Researchers in the field of photonic crystals are just beginning to understand the behavior of light in periodic media. Currently, there is an incipient understanding of what happens in periodic structures and, at the other end, optical processes in single molecules. Particular interest lies in materials with negative refractive indices and meta-materials. However, the gap in between is not bridged yet and requires in-depth investigations. The key issue is what happens when one has nearly periodic, or quasi-periodic, molecular or cluster arrays or clusters? The description of their optical properties is in rapid progress as is the behavior of light in complex media. For example, existing knowledge base is not enough to design novel devices, the associated platforms and system configurations.

Nanophotonics is the field of nanotechnology concerned with discovering and developing nanomaterials that can control the flow of light and in some cases localize or confine it within a volume. Intuitively, we view light as rays, which either propagate in a single direction, being absorbed or reflected to some extent by any object on which it impinges. However, the propagation of light through a material is itself a quantum effect, involving the excitation and relaxation of electrons in the material. Therefore, creating a material with structural and compositional features on a length scale comparable to the wavelength of light i.e. 300 - 750 nm for the visible light enables us to guide light in any direction of our desire.

There are two major challenges to reach the molecular scale and develop mass production techniques. In molecular photonics the aim is to achieve sufficient control over light-matter interactions through nanoscale physical, chemical and structural modification interactions using a few or single molecules. Today, optical spectroscopy of single molecules provides a tool to sense the local environment, which complements and extends ensemble-averaged measurements performed in conventional bulk structures. The molecular-scale variables range from signal strength, orientation, emission spectrum and energy transfer to neighboring molecules. This has undeniable potential in areas requiring ultra-sensitive detection without suffering from random electrical noise. Therefore, molecular photonics holds the promise to key advances in quantum sources of light, efficient photon harvesting systems and optical tuning and switching, among others.

Nanophotonics comprises the fabrication of nanomaterials and their interaction with light, using near-field optics, and thereby controlling the spatial confinement of photo-physical and photochemical process. One way to induce interactions between light and matter on a nanoscale is to confine light to nanodimensions that are much shorter than the wavelength of light (Li et al., 2004; Rosa et al., 2005; Huang et al., 2008). The second approach is to confine matter to nanodimensions, thereby limiting interactions between light and matter to nanoscopic dimensions that are synthesized using nanotechnology lithographic techniques as indicated in figure 1. The last way is nanoscale confinement of photo-processes by inducing photochemistry or a light-induced phase change. This approach provides methods for nanofabrication of photonic structures and functional units. The three main applied areas of nanotechnology and their technical concepts are shown in figure 2, whereas the widespread utility in every sphere of basic and applied sciences as well as engineering is illustrated in figure 3. Similar to the periodic electron-crystal lattice, one can fabricate photonic-crystal lattice. The refractive index varies with a much larger period of around 200 nm. The similarity between electronic and photonic crystals is represented in figure 4. The following characteristics are essential for the materials for making nanophotonics devices:

- Nanoscale quantum dots can be formed in high density.
- Quantum dots are in isolated condition, when not in photo-excited state.
- Nanorods or nanowires large quantum yield and high aspect ratio.
- Highly excited level energy can be controlled by material mixture ratio and dimension.
- Light element excitation is stable at room temperature, and high-excited level does not degenerate.

The brilliant, yet simple result is that one can treat photons in a similar manner as one does with electrons, and begin to envision 'optical' circuit components as existed electronic circuit components.

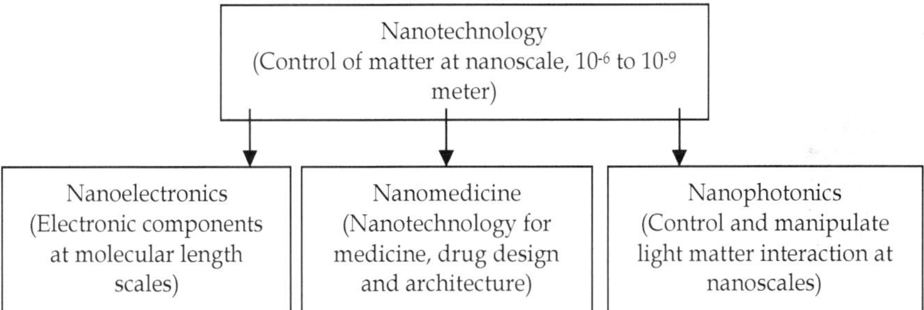

Fig. 2. Block diagram depicting three main branches of nanotechnology.

It is realized that an increasing number of novel materials, techniques, concepts and processes continue to be discovered in the context of nanotechnology and nanophotonics. There is an urgent need to develop the required understanding and to evaluate this new knowledge from the perspective of both novel science and potential applications. Understanding the fundamentals of nanophotonics is important because to fabricate a test structure it is necessary to home in on a window of parameters. The permutations on parameterization, optimization and functionalities are infinite, because there are millions of molecules, hundreds of different atoms and structures one can attach to them, and arranged them to realize a functional device. A window of parameters always helps to target the nanostructures and molecular laboratory samples into research objects, which can give information on device-relevant properties. The obvious boundary conditions are power consumption and cost that needs to be minimized for commercialization. The best option that is recently adopted is to work with molecules embedded in matrices; they are not only efficient but also relatively easier to handle and to manufacture.

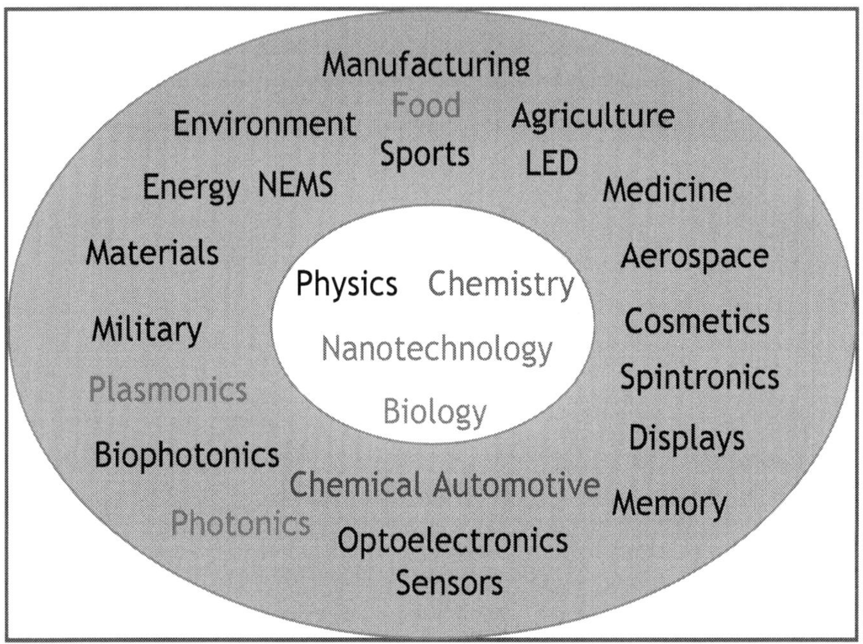

Fig. 3. Schematics representing various branches of nanotechnology and related applied fields in science and engineering.

Nanoscale photo-processes are comprised of nanoscale lithography, fabrication of nanoscale structures, and nanoscale optical memories. Foundations for nanophotonics require basic equations describing propagation of photons in dielectrics that has some similarities to the propagation of electrons in crystals (see figure 4). There are similarities between photons and electrons. Wavelength of light (photon) is given by

$$\lambda = \frac{h}{p} = \frac{c}{v} \tag{1}$$

Wavelength of electron is given by

$$\lambda = \frac{h}{p} = \frac{h}{mv} \tag{2}$$

The Maxwell's electromagnetic equations for the light (photon) describes the allowed frequencies of light and in a similar way Schrodinger's equation for the electrons describes allowed energies of electrons. Free space solution of the wave equation for both the photon and the electron are plane wave (see figure 5). Interaction potential in a medium for propagation of light affected by the refractive index of the dielectric medium and the propagation of electrons are affected by the Coulomb potential. In this case, both electrons and photons have propagation through classically forbidden zones. Photon tunnels through classically forbidden zones, the electric, and the magnetic fields decay exponentially that results k-vector is imaginary. Electron wave function decays exponentially in forbidden zones and has finite tunneling probability as shown in the figure 6.

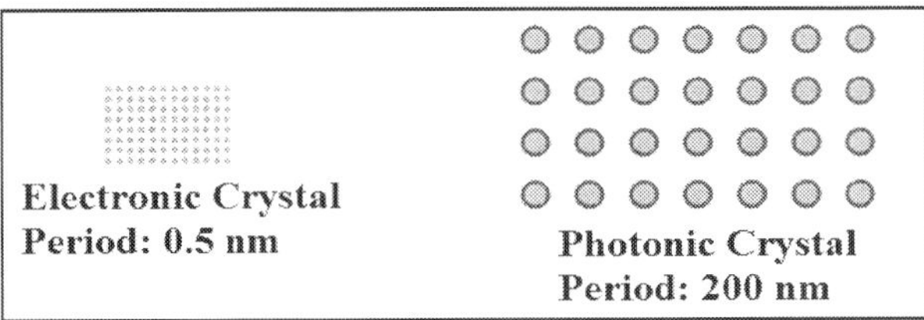

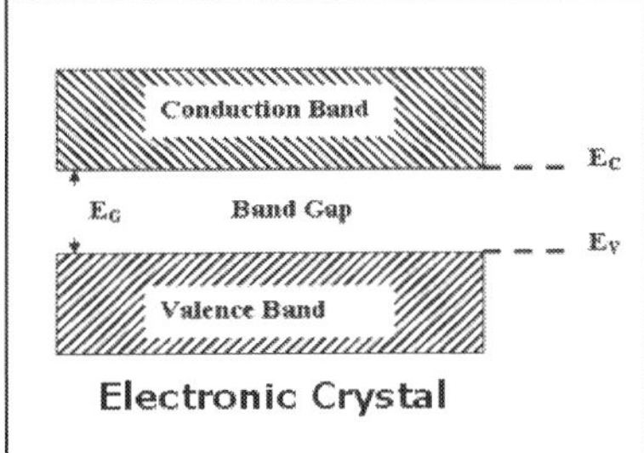

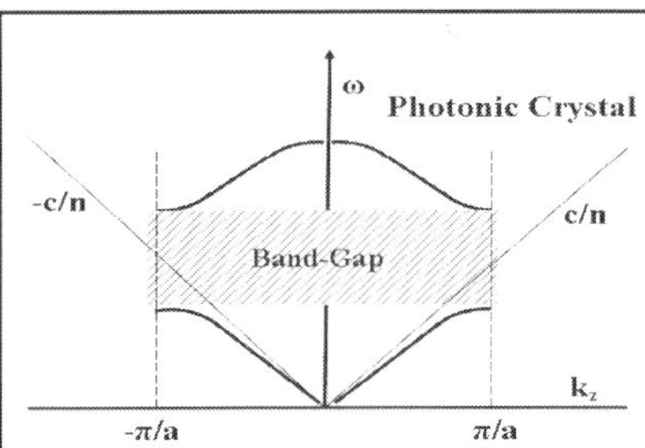

Fig. 4. The typical band structures of electronic and photonic crystals.

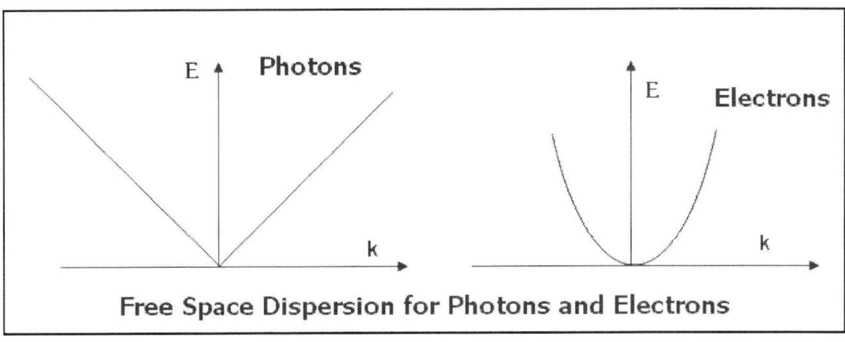

Fig. 5. Free space dispersion relations for photons (left) and electrons (right).

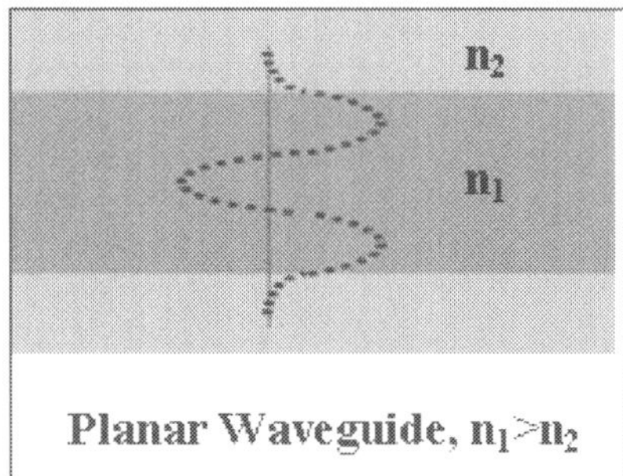

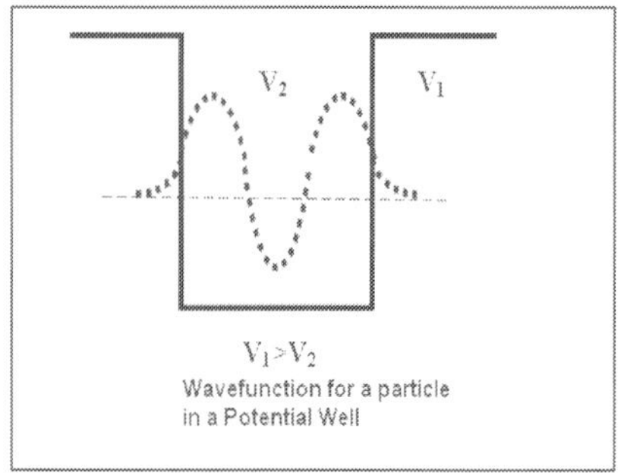

Fig. 6. Tunneling of photon (left) and electron (right) in the classically forbidden region.

Quantum confinement (see figure 7) causes manifestations of new optical effects due to the change in the nature of density of states from continuous to discrete structures as the dimension is reduced from three to zero. It is the density of states that guides optical and electrons correlation effects in quantum confined nanostructures. Size dependence of optical properties due to confinement produces a blue shift of the band-gap in semiconductors, and increases the oscillator strengths and in turn increases the optical transition probability. Location of discrete energy levels depends on the size and nature of the confinement. The oscillator strength increases with increasing confinement from bulk to quantum well to quantum wire to quantum dot. Confinement produces sub-bands within the conduction and valence bands, enabling intra-band optical transitions, which are not allowed in the bulk structures. These infrared transitions have applications in making new quantum cascade lasers and detectors. The oscillator strengths increase as the width of the quantum well decreases. The most important optical feature of these structures is that absorption/emission spectra shift to shorter wavelengths as the size become smaller. In case of metallic nanostructure, somewhat spectacular things happen due to which they show strong luminescence and plasmon absorption bands.

Artificial structures or meta-materials with the optical equivalent of the energy gap in semiconductors promise a wealth of new devices that could satisfy the demand for ever-faster computers and optical communications (Ozbay et al., 2004). Presently, the availability of personal computers that operate at 1 GHz (10^9 Hz), are really very impressive, but what is the likelihood of a 100 GHz desktop computer appearing on the market at affordable price? Indeed, with the current understanding of semiconductor technology, even producing a 10 GHz personal computer would seem to be difficult. However, by transmitting signals with light rather than electrons (a domain of optical computing), it might be possible to build a computer that operates at hundreds of terahertz (10^{12} Hz) which is of the order of the processing speed of the fastest earth simulator computer.

Researchers now believe that such an awesome processing engine could be built from optical components made from so-called photonic crystals and quasi-crystals. These materials possess structures with periodically modulated refractive indices that can be designed to control and manipulate the propagation of light. They are attracting more and more attention not only in the field of photonics but also in the fields of chemistry, physics, and microelectronics. This is because the photonic crystals enable us to control the flow of photons by means of photonic band gaps, which is a photonic analogue of energy band gaps in semiconductors. In recent years, much of the developments on the photonic band gap materials are made using insulators and semiconductors. However, metals or metallo-organic structures can also be used for the photonic band gap in the form of photonic surfaces using the surface plasmon resonance. The photonic band gaps can be introduced by texturing metal surfaces with three-dimensional patterns using spheres or any other shape with a spatial period roughly half of the free-space wavelength of light. Excitation of surface plasmons leads to the formation of standing waves and opening of the stopgap useful for channeling, guiding and controlling photons.

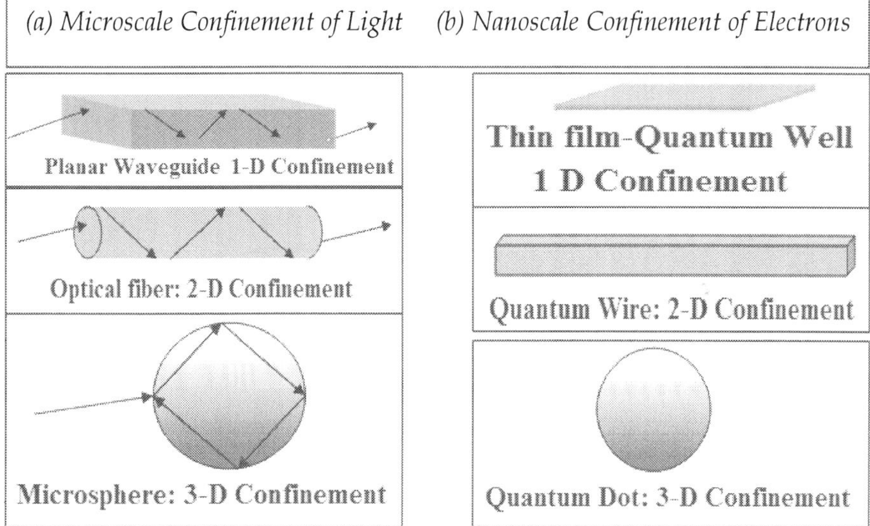

Fig. 7. Nanoscale confinements of photon (left) and electron (right) in different spatial dimensions.

The structures of traditional optical waveguides cannot be smaller than $\lambda/2$. Furthermore, wave guiding along bent guides causes dramatic loss. The problem associated with loss is overcome in photonic band gap structures, a potential candidate for nanophotonics device as already mentioned. However, the dimensions of the structures are still limited by the wavelength of light. This restriction can be overcome by wave guiding of the plasmonic excitation in closely placed metal nanoparticles. The photonic band gap crystals (see figure 8) and photonic band gap fibers involve periodic variation of dielectric constant over wavelength-scale. In common, electronic crystals like sodium chloride, the periodicity is of the order of one nm. Photonic crystals are produced by periodically varying refractive index in one, two or three dimensions and the period is comparable to the wavelength of light. Thus, the field of photonic crystals can be thought as microphotonics. However, in order to fabricate photonic crystals with micron-scale period, the fabrication technique must have nanoscale resolution. Therefore, it is appropriate to include photonic crystals in the domain of nanophotonics.

Basic electromagnetic equations describing propagation of photons in dielectrics have some similarities to propagation of electrons in crystals. The inspiration and very idea of photon localization in photonic band gap materials is drawn from nature. Undoubtedly, nature has demonstrated its advance capability in synthesizing nanomaterials to a level of sophistication and functionality far beyond our own. This inspiration comes from the glittering appearance of butterfly's wing and peacock's feathers, which has a highly ordered periodic structure at the nanoscale. The key to confining and guiding light within a material lies in the periodicity of the structures. An even more recent discovery is the capability of the weevil beetle to produce opals, a precious silica base gem, which have been synthetically produced to make three-dimensional photonic crystals. This discovery opens many doors for accessing the advanced molecular machinery of nature for fabricating photonic materials. Figure 9 illustrates an inverse opal photonic band gap structure that has been realized.

Fig. 8. A photonic band gap structure used for channeling light scaled in microns.

It is well known fact that electronics is the aspect ratio and is limited in speed, whereas, photonics is the diffraction that is limited in size. Plasmonics can go beyond the sub-diffraction limit and is not limited my any means because the surface plasmon wavelengths can reach to nanoscale at optical frequencies. Plasmonics will enable an improved synergy between electronic and photonic devices by naturally interfacing with similar size of electronic components and similar operating speed photonic network. Surface plasmon resonance has already been used by biochemists to detect the presence of a molecule on a surface. In recent years, plasmons are being exploited in many applications by manipulating and guiding light at resonant frequencies. There is a need to understand the origin and the fundamentals of physics associated with surface plasmon resonance before they are employed to many revolutionary applications. Rapid progresses in materials synthesis and nanofabrication techniques over the last decade have led to variety of applications. Some of the examples of plasmon-assisted nanophotonics are, high-resolution plasmon printing, laser shaping, solar concentrators, nanoscale waveguides, sensing, bio-detection at the single-molecule level and enhanced transmission through sub-wavelength apertures, to cite a few.

Since the inception of surface plasmon optics there has been a gradual transition from fundamental research to more applied oriented research. Presently, the surge in surface plasmon-based studies is happening at a time when crucial technological areas such as advanced optical lithography, optical data storage, and high-density electronics manufacturing are approaching fundamental physical limits. Several current technological challenges may be overcome by utilizing the unique properties of surface plasmons. A wide range of plasmon-based optical elements and techniques have now been developed, including a variety of passive waveguides, active switches, biosensors, lithography masks, and more. These developments have led to the notion of *plasmonics*, the science and technology of metal-based optics and nanophotonics.

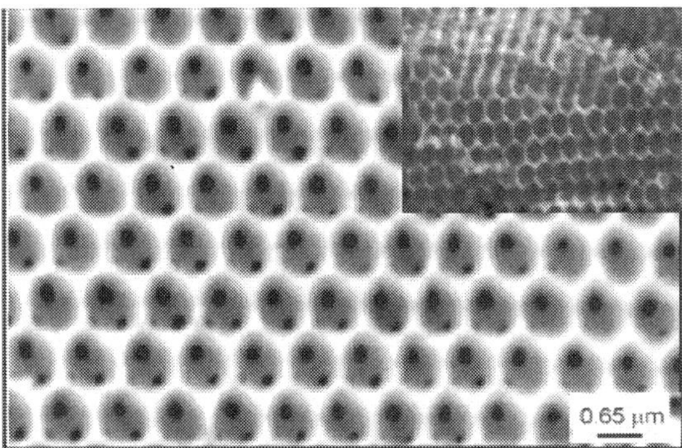

Fig. 9. An inverse opal photonic crystal structure (inset shows dielectric and air holes).

In photonics, metals are not usually thought of as being very useful, except perhaps as mirrors. In most cases, metals are strong absorbers of light, a consequence of their large free-electron density. However, in the miniaturization of photonic circuits, it is now being realized that the metallic structures can provide unique ways of manipulating light at length scales smaller than the wavelength. Surface plasmons, the light-induced excitations of electrons on metal surfaces, may provide integration of electronics and optics on the nanoscale. The optical manipulation of micron/submicron sized particles and bio-molecules through plasmonics have been investigated by Miao et al., and the schematic diagram of the experimental set-up is shown in figure 10. In recent years, a new horizon in nanophotonics has opened up called 'plasmonics' that consists of interaction of light wave with metal nanoparticles (Gopinath et al., 2009). Electromagnetic point of view metals are plasmas, comprising fixed, positive ion cores and mobile conduction electrons. At the plasma frequency, the real part of the dielectric constant changes sign. Therefore, the miniaturization of photonic circuits, allows metallic structures to manipulate light at length scales smaller than the wavelength. The typical thinking about metals is either conductors in electronics or reflectors in optics; however, metals as mirrors are not always desirable. Their residual absorption and low damage threshold makes them unsuitable for lasers and gyroscopes. Plasmonics has given photonics the ability to go to the nanoscale. Plasmonics utilizes the collective oscillations of conduction electrons in metallic nanostructures called plasmons, excited by the incident electromagnetic waves. In the following section, we will discuss in detail the basics of plasmonics, which includes the origin of surface plasmon resonance (see figure 11(a)), their propagation characteristics (figure 11(b)) and their exploitation in innumerable applications. Some of the interesting phenomena in this field are:

- Generation of the surface plasma waves and the dependence of plasmon resonance on metal particle size, and geometry.
- Dependence of the plasmon resonance condition on the dielectric adjacent to the metal film.

- Enhancement of the electromagnetic field close to metal nanoparticles.
- Effects of the metal nanoshells on plasmon resonance.
- Propagation of high-frequency electromagnetic waves along sub-wavelength-wide metal wave-guides.
- Effect of metal surface on radiative decay of molecules.

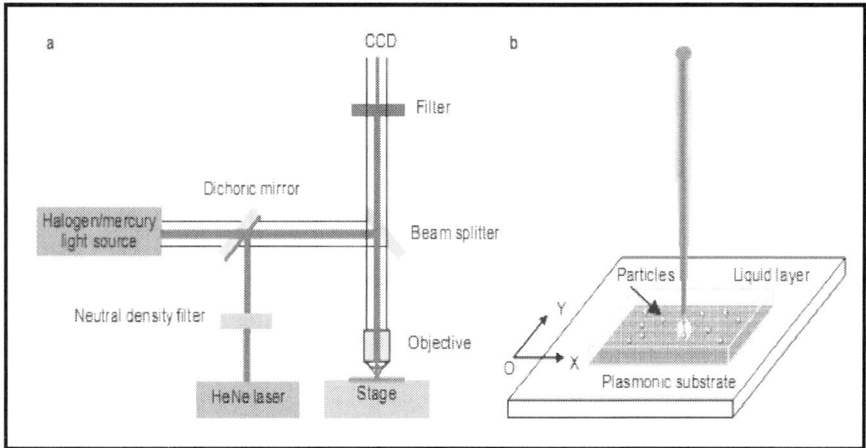

Fig. 10. a) Experimental set up for the trapping and concentration experiments on the plasmonic substrate. b) Detailed configuration of the sample chamber.

The published scientific literatures containing the words 'surface plasmon' over last decade clearly indicates the exponential growth of the field of plasmonics. This rapid growth is stimulated by the development and commercialization of advanced sample preparation methods, powerful electromagnetics simulation codes, nanofabrication techniques and physical analysis techniques providing scientists and engineers with the necessary tools for designing, fabricating, and analyzing the optical properties of metallic nanostructures. A major boost to the field was given by the development of a commercial surface plasmon resonance based sensor in the end of last century. Presently, an estimated fifty percent of all publications on surface plasmons involve the use of plasmons for bio-detection and bio-photonics applications.

The chapter is organized as follows. In Section 2, we review the salient features of photonic crystals and photonic band gaps. A birds' eye view of the recent development in plasmonics and its future prospects is presented in Section 3. The recurring theme of Section 4 is the main challenges in nanophotonics research and the upcoming market strategies. The physics behind the quantum confinement and the behavior of quantum-confined materials are discussed in Section 5 and 6. Section 7 describes the possibility of widespread exciting applications and the need of further research. Conclusions and further outlook are summarized in Section 8.

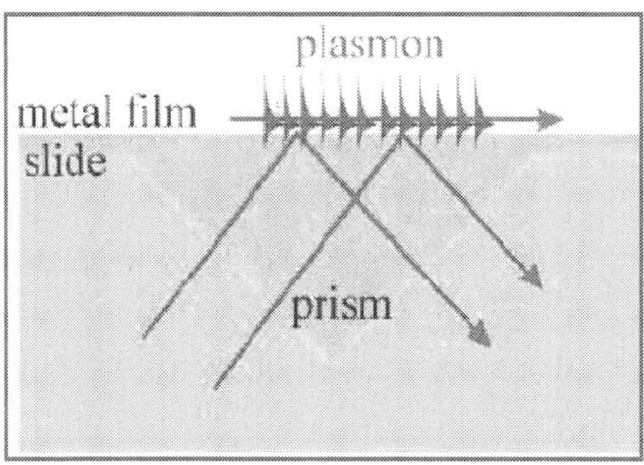

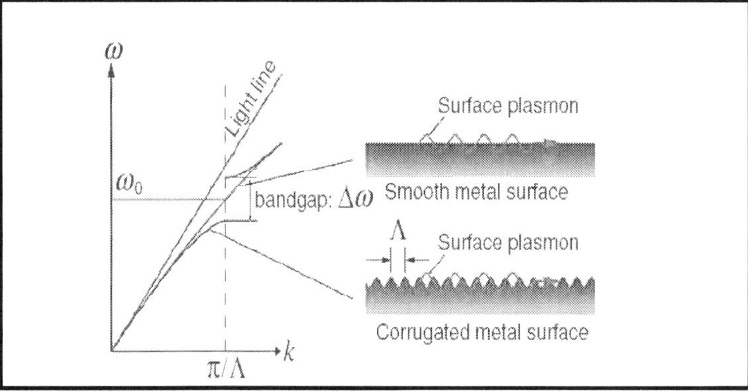

Fig. 11 (a) The origin of surface plasmon resonance at metal dielectric interface (Left panel). (b) The dispersion relation of the surface plasmon travelling on a flat metal surface and a corrugated metal surface (Right panel).

2. Photonic crystals: an amazing material for ultrafast technology

This section focuses on the importance of the photonic crystals called photonic band gap materials. In an age when technology influences society on daily basis, there is an increased demand for new ways to process and exchange information quickly and efficiently. One solution is to implement optical circuits, which use light instead of electricity to relay and process information. However, with the current methods of building optical circuits, it is difficult to manipulate light without losing efficiency. In fact, nanotechnology is helping the photonics to grow at rapid space. Nanotechnology will not just be about electrons moving through circuits but will go beyond our imagination. Information has been transmitted for many years using electromagnetic waves. Most of the communication technology developments exploited radio- and micro-waves and in more recent time information, technology is using much shorter wavelength of light in optical fibers. The natural progression here is towards what are called photonic devices, in which light is channeled through microscopic tubes or lattices of tubes, which are now a topic of ever-increasing active research and debate.

The solution to the problem of rapid information processing and safe transfer at cheaper rates is to use photonic crystals as waveguides in optical circuits. They are small-fabricated crystals, which can efficiently manipulate the direction in which light travels. Photonic crystals are usually viewed as an optical analog of semiconductors that modify the properties of light similar to a microscopic atomic lattice that creates a semiconductor band-gap for electrons. It is therefore believed that by replacing relatively slow electrons with photons as the carriers of information, the speed and bandwidth of advanced communication systems will be dramatically increased. The present century's exciting application possibilities of these materials in science and technology, its present status and future scope are briefly explored in this section.

The various issues addressed here related to the field of photonic crystals are:

- What are photonic crystals?
- What is meant by photonic band gap materials?
- What are the methods used to analyze theory of photonic crystals?
- How do they look like?
- What physics make them so potential for devices?
- Is there any use of these crystals so far?
- What are their fabrication processes?
- What potential applications do they have?
- What are challenges and difficulties?
- How promising are they for future technology?

Microelectronics has allowed by miniaturization of components, such as transistors, resistors, and capacitors on one single chip. On the other hand, photonics has not achieved the same level of miniaturization yet while there are demands for cheaper, faster, and more compact laser-based communication systems in modern time. The integration of photonic components such as lasers, detectors, couplers, and wave-guides is still at a very primitive stage, compared with that of microelectronics. This is due to difficulties with implementing integrated optical components smaller than certain sizes. For instance, small bends or curves of wave-guides lead to the leakages of optical signals, so the bends have to be bigger than certain critical length. Photonic crystals can then be used as they offer a potential to provide solutions to the challenge of photonic component integration by means of its nature to confine photons within their structures. Moreover, intriguing phenomena in photonic crystals can be exploited, which are not achievable by conventional isotropic media. Photonic crystals can be a breakthrough in photonic technology in the near future (Miao et al., 2008).

The absence of allowed propagating electromagnetic modes inside the structures in a range of wavelengths called a photonic band gap, gives rise to distinct optical phenomena such as inhibition of spontaneous emission, high-reflecting omni-directional mirrors and low-loss-wave guiding amongst others. Since the basic physical phenomenon is based on diffraction, the periodicity of the photonic crystal structure has to be in the same length-scale as half the wavelength of the electromagnetic waves ~300 nm for photonic crystals operating in the visible part of the spectrum (Rayleigh, 1888). The periodicity, whose length scale is proportional to the wavelength of light in the band gap, is the electromagnetic analogue of a crystalline atomic lattice. This periodicity acts on the electron wave function to produce the familiar band gaps as in the case of semiconductors in condensed matter physics. The study of photonic crystals is likewise governed by the Bloch-Floquet theorem, and intentionally introduced defects in the crystal. The later is similar to electronic dopants give rise to localized electromagnetic states in linear waveguides and point-like cavities. The crystal can thus form a kind of perfect optical insulator that can confine light without loss around sharp bends, in lower-index media, and within wavelength-scale cavities, among other novel possibilities for control of electromagnetic phenomena (Joannaopoulos et al., 2008). The periodicity of the photonic crystals can be in one, two, and three dimensions that allow interesting properties such as bending light at 90° around corners as shown in figure 12.

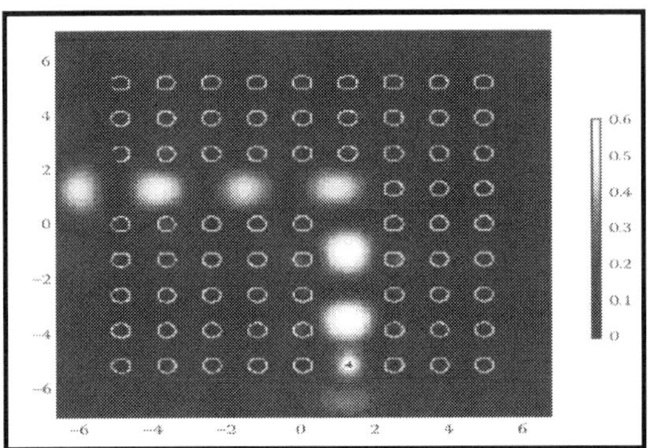

Fig. 12. Bending of light at 90° around corners in the photonic crystals.

One-dimensional periodic system continued to be studied extensively, and appeared in applications from reflective coating to distributed feedback diode lasers. In the former case, the reflection band corresponds to the photonic band gap and for the later, a crystallographic concept is inserted in the photonic band gap to define the laser wavelength. Yablonovitch and co-workers (Yablonovitch, 1987) produced the first photonic crystal by mechanically drilling holes a millimeter in diameter into a block of material with a refractive index of 3.6. The material, which became known as Yablonovite, prevented microwaves from propagating in any direction and exhibited a 3-dimensional photonic band gap. Other structures that have band gaps at microwave and radio frequencies are currently being used to make antennae that direct radiation away from the heads of mobile-phone users (Sajeev, 1987; Lodahl, 2004; Kim, 2008; Sonnichsen, 2005). Later on, photonic crystals of semiconducting colloidal particles were fabricated for realizing photonic band gaps in the visible region of the electromagnetic spectrum. They are also fabricated by the spontaneous self-organization of mono-disperse colloidal spheres such as silica or polystyrene to form a three-dimensional crystal having long-range periodicity. As mentioned, the photonic crystals are materials with periodically varying relative permittivity and are optical equivalents of semiconductors. However, the true potential of these materials lies in manipulating light of wavelength comparable with their lattice parameter. The voids between the particles form regions of low relative permittivity, while the spheres form regions of high relative permittivity, i.e. periodically varying refractive indices (see figure 13). The refractive index variation contrasts for photons in a similar manner to the periodic potential that an electron experiences while traveling in a semiconductor. For sufficiently large contrast, the creation of a complete photonic band gap may occur that results a frequency range where light cannot propagate inside the photonic crystal. This is the underlying principle by which a colloidal photonic crystal blocks certain wavelengths in the photonic band gap, while allowing other wavelengths to pass. The photonic band gap can be tuned by changing the size, shape and symmetry of the particles and the geometry of voids. Using core-shell particles similar photonic crystals are prepared with a large contrast in the refractive indices of the core and shell materials, where the photonic band gap are tuned from the visible to the infrared ranges by changing the refractive indices contrast. It has taken over a decade to fabricate photonic crystals that work in the near infrared (780 - 3000 nm) and visible (300 - 750 nm) regions of the spectrum. The main challenge has been to find suitable materials and processing techniques to fabricate structures that are about a thousandth the size of microwave crystals (Kalele et al., 2007; Sajeev, 1987).

One of the most important features in photonic crystals is the photonic band gap, which is analogous to band gaps or energy gaps for electrons traveling in semiconductors. In case of semiconductors, a band gap arises from the wave-like nature of electrons. Electrons as waves within a semiconductor experience periodic potential from each atom and are reflected by the atoms. Under certain conditions, electrons with certain wave vectors and energy constitute standing waves. The range of energy, named 'band gap', in which electrons are not allowed to exist. This phenomenon differentiates semiconductors from metals and insulators. In the similar manner, standing waves of electromagnetic waves can be formed within a periodic structure whose minimum features are about the order of the wavelength. In this case, the medium expels photons with certain wavelengths and wave vectors. Such a structure acts as an insulator of light, and this phenomenon is referred to as photonic band gap ((Yablonovitch, 1987; Sajeev, 1987; Lodahl, 2004). The origin of photonic band gap in photonic crystals can be explained with the help of Maxwell's equations.

It is well known that in a silicon crystal, the atoms are arranged in a diamond-lattice structure in which the electrons moving through this lattice experience a periodic potential while interacting with the silicon nuclei via the Coulomb force, that results in the formation of allowed and forbidden energy states. No electrons can be found in the forbidden energy gap or simply the band gap for pure and perfect silicon crystals. However, for real materials with defects the electrons can have energy within the band gap due to the broken periodicity caused by a missing silicon atom or by an impurity atom occupying a silicon site, or if the material contains interstitial impurities. Now, consider a situation in which the photons are moving through a block of transparent dielectric material that contains a number of tiny air holes arranged in a regular lattice pattern. The photons will pass through regions of high refractive index of the dielectric intersperse with regions of low refractive indexed air holes. In case of a photon, this contrast in refractive index looks just like the periodic potential that an electron experiences traveling through a silicon crystal. Indeed, if there is large contrast in refractive index between the two regions then most of the light will be confined within either the dielectric material or the air holes. This confinement results in the formation of allowed energy regions separated by a forbidden region, photonic band gap. As the wavelength of the photons is inversely proportional to their energy, the patterned dielectric material will block light with wavelengths in the photonic band gap, while allowing other wavelengths to pass freely (Mia et al., 2008). It is also possible to create energy levels in the photonic band gap by changing the size of a few of the air holes in the material. This is the photonic equivalent to breaking the perfect periodicity of the crystal lattice. The diameter of the air holes is a critical parameter, in addition to the contrast in refractive index throughout the material. Photonic band gap structures can also be made from a lattice of high-refractive-index material embedded within a medium with a lower refractive index (core-shell for example). A naturally occurring example of such a material is opal. However, the contrast in the refractive index in opal is rather small, and hence the appearance of a small band gap (Kalele et al., 2007).

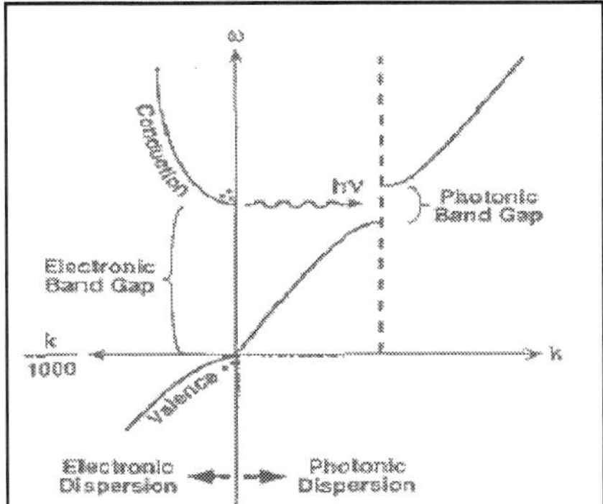

Fig. 13. Schematic representing the electronic and photonic band gaps in the Brillouin zone.

Let us consider the simplest one-dimensional (1D) structure in order to describe the phenomenon of formation of photonic band gap in the photonic crystals that has alternating layers of two dielectrics. The incident wave in entering a periodic array of dielectric sheets is partially reflected at the boundaries of the dielectric layers. If the partially reflected waves are in phase and superimposed, they form a total reflected wave, and the incident wave is unable to enter the medium, as depicted in figure 14. The range of wavelengths in which incident waves are reflected is called a 'stop band'. A structure that exhibits stop bands to every direction for given wavelengths, the stop bands are considered a 'photonic band gap'. On the other hand, when the wavelength of an incident wave does not lie within the band gap, destructive interferences occur and partially reflected waves cancel one other. Consequently, the reflection from the periodic structure does not happen and the light passes through the structure as illustrated in figure 15.

For two-dimensional (2D) structure, the condition in which such reflections occur at the interfaces of two dielectrics and a photonic band gap arises from the superposition of partial reflected waves are somewhat complex. To realize an effective photonic band gap, back-scattered waves should be in phase, forming one reflected wave in which the Bragg's condition has to satisfy, the same condition has to be satisfied for incident waves from every direction to attain a photonic band gap. An intuitive idea regarding the nature photonic crystal structure obtained from Bragg's law indicates that the distance from one lattice point to neighboring ones should be same so that scattered waves are superimposed and in phase at any point of the structure. Moreover, the structure should possess symmetry to as many directions as possible so that scattered waves from one lattice point experiences the same orientation of neighboring lattice points. The same concept can be extended to three-dimensional (3D) periodic structure, where the incident waves turned into partially reflected waves and the transmitted waves at boundaries between the two media. If the partially reflected waves are in phase, the scattered waves add up to a net reflected wave, resulting in a stop band. The condition for Bragg's law must be satisfied at each lattice point that can be either a dielectric material or an air hole surrounded by a dielectric. If the stop bands exist for every direction and those *'stop bands'* overlap within certain wavelength

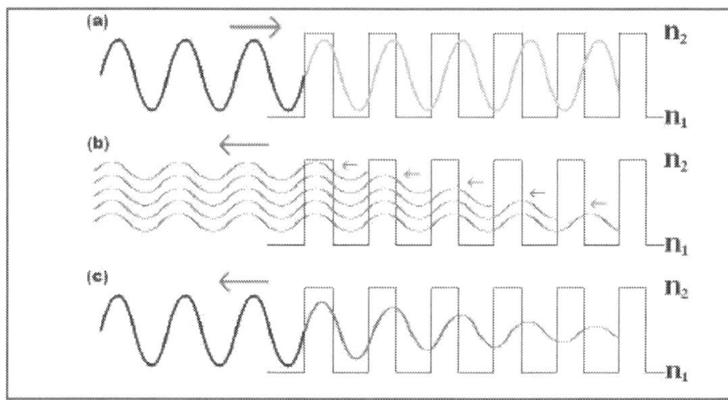

Fig. 14. The constructive interference for the photonic band gap in one dimension. (a) An incident wave within the photonic band gap enters the periodic structure with two different refractive indices n_1 and n_2. (b) The incident wave is partially reflected by the boundary of the structure. (c) The incident wave is totally reflected when each reflected wave is in phase, and is unable to penetrate the structure.

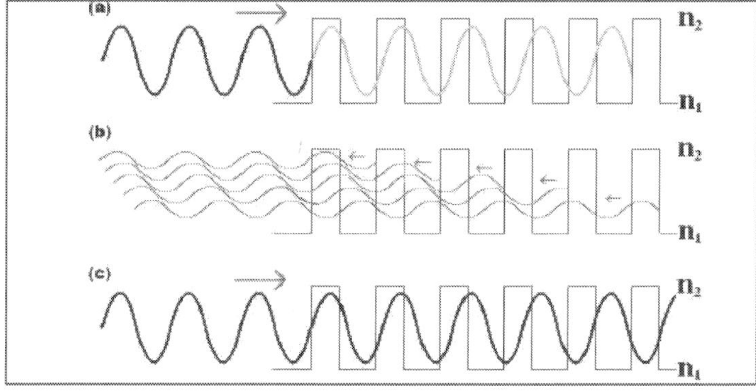

Fig. 15. The destructive interference. (a) An incident wave outside the photonic band gap enters the periodic structure. (b) The incident wave is partially reflected by the boundary of the structure, but each reflected wave is out of phase and interfere destructively. (c) The incident wave penetrates the structure without being reflected.

region then a complete photonic band gap arises in three-dimension. Photonic band gaps results from the net interferences of scattered incident light waves from lattice points of a periodic structure. It is important to note that high refractive index contrasts of the periodic structures play pivotal role for the photonic band gaps to occur or to become more pronounced for a given structure (Joannaopoulos et al., 2008).

There are two reasons for the importance of high refractive index contrasts. First, each photonic crystal structure has a threshold value of refractive index contrast to exhibit a photonic band gap as depicted in figure 16. This phenomenon is attributed to the fact that interfaces of two dielectrics with higher contrast of refractive indices tend to scatter waves from any direction, so stop bands to any direction, a photonic band gap, are more likely to take place. Second, the higher the refractive index contrast is, the fewer layers are necessary to have sufficient photonic band gap effects. As explained in figure 14, each layer or lattice of photonic crystal partially reflects the propagating wave. Consequently, if each layer reflects more waves due to a higher refractive index contrast, sufficient net reflections can be achieved by fewer layers of lattices than a structure with the same configuration but with a lower refractive index contrast. This condition helps us to choose materials such as semiconductors for photonic crystals (Mia et al., 2008; Rayleigh, 1888; Yablonovitch, 1987).

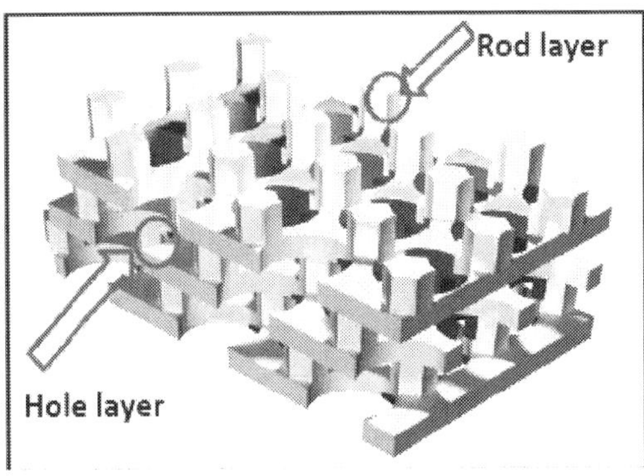

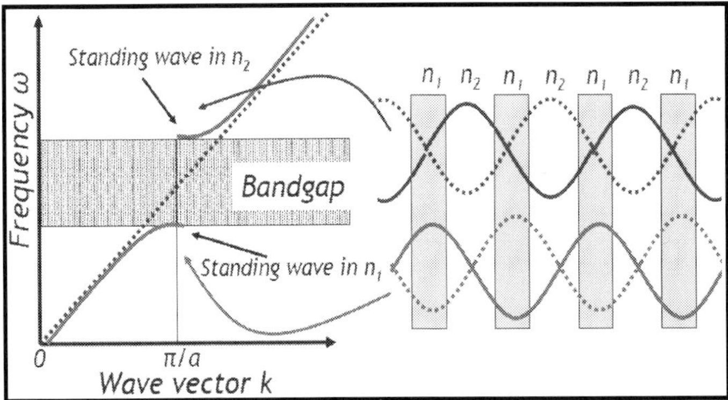

Fig. 16. (Left) A 3D photonic crystal consisting of an alternating stack of triangular lattices of dielectric rods in air and holes in dielectrics (courtesy Yablonovitch). (Right) Projected band diagrams and the band gap for a finite-thickness slab of air holes in dielectric with the irreducible Brillouin zone.

By combining Maxwell's equations with the theorems of solid-state physics a surprising and simple result emerges, that explain the phenomena of light bouncing among infinity of periodic scatterers. Like electrical insulators, which keep the currents in the wires where they belong, one can also build an optical insulator, a photonic crystal to confine and channel photons. The emergence of photonic crystals is due to the cooperative effects of *periodic* scatterers that occur when the period is of the order of the wavelength of the light.

They are called 'crystals' because of their periodicity and 'photonic' because they act on light i.e. photon. Once such a medium is obtained, impervious to light, one can manipulate photons in many interesting ways. By carving a tunnel through the material, an optical 'wire' can be achieved from which no light can deviate. Even more interesting things can happen by making a cavity in the center of the crystal, an optical 'cage' can be created in which a beam of light could be caught and held, because the very fact that it cannot escape would render it invisible. These kinds of abilities to trap and guide light have many potential applications in optical communications and computing (Joannaopoulos et al., 2008). A typical photonic crystal slab structure with tunnels and cavities that are made to confine and control light is presented in figure 17.

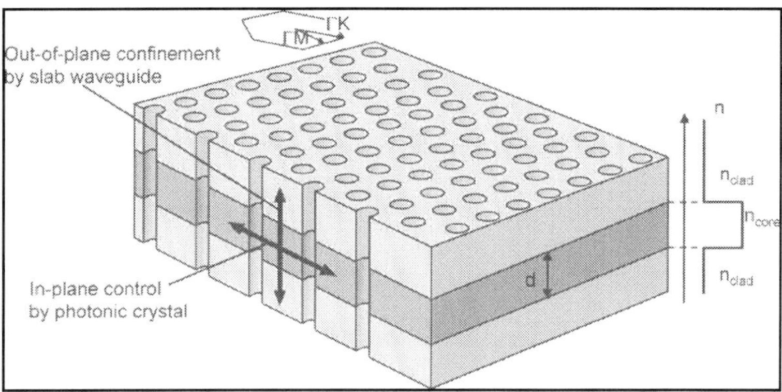

Fig. 17. A 2D photonic crystal slab. In-plane, light is controlled by the photonic crystal, while in the vertical direction it is confined by the layer with the higher refractive index.

To achieve a large band gap, the dielectric structure should consist of thin, continuous veins/membranes along which the electric field lines can run. This way, the lowest band(s) can be strongly confined, while the upper bands are forced to a much higher frequency because the thin veins cannot support multiple modes (except for two orthogonal polarizations). The veins must also run in all directions, so that this confinement can occur for all wave vectors and polarizations, necessitating a complex topology in the crystal. Furthermore, in two or three dimensions one can only suggest rules of thumb for the existence of a band gap in a periodic structure. The design of 3D photonic crystals is a trial and error process (Sanjeev, 1987). The typical band structure for photonic crystals for transverse electric and transverse magnetic mode is shown in figure 18. Interestingly, the 2D systems exhibit most of the important characteristics of photonic crystals, from nontrivial Brillouin zones to topological sensitivity to a minimum index contrast, and can also be used to demonstrate most proposed photonic-crystal devices (Yablonovitch, 1987).

The numerical computations are the crucial part of most theoretical analyses for photonic band gap materials due to their complexity in high index-contrast directional dimensionality of the systems. Computations are typically fall into the following three categories:

1. The time-evolution of the fields with arbitrary initial conditions in a discretized system are modeled and simulated by the time-domain 'numerical experiments' using finite difference method.

2.	The scattering matrices are computed in some basis to extract transmission/reflection through the structure (mainly eigenvalues) and the definite-frequency transfer matrices can be achieved.
3.	The frequency-domain methods can directly extract the Bloch fields and frequencies by diagonalizing the eigenoperator.

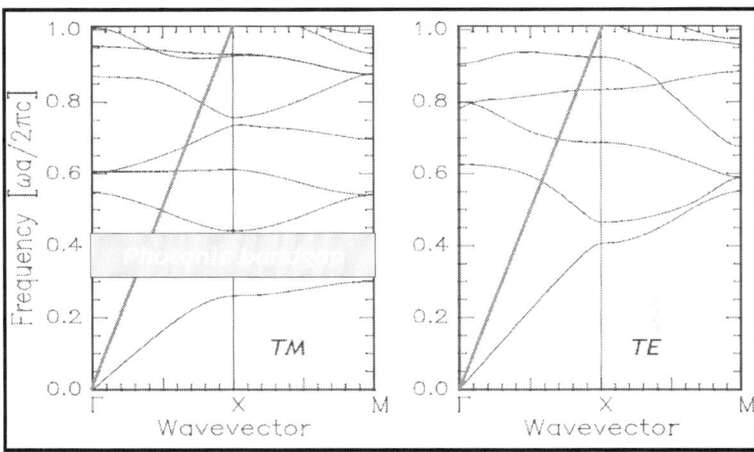

Fig. 18. Band diagrams and photonic band gaps for the two polarizations TE/TM (electric field parallel/perpendicular to plane of periodicity).

The directly measurable quantities such as transmission can be obtained intuitively from the first two categories. The third category is more abstract, yielding the band diagrams that provide a guide to interpretation of measurements as well as a starting point for device design and semi-analytical methods. For many systems, several band diagrams are computed by the frequency-domain method.

Photonic-crystal slabs have two new critical parameters that influence the existence of a gap. Firstly, it must have mirror symmetry in order that the gaps in the even modes and odd modes can be considered separately. Such mirror symmetry is broken in the presence of an asymmetric substrate. In actual practice, the symmetry breaking can be weak if the index contrast is sufficiently high so that the modes are strongly confined in the slab. Secondly, the height of the slab must not be too small that weakly confines the modes or not too large so that higher-order modes will fill the gap. The required optimum height must be around half a wavelength relative to an averaged index that depends on the polarization (Joannaopoulos et al., 2008). The photonic-crystal slabs are one way of realizing 2D photonic-crystal effects in 3D. A 3D periodic crystal is formed by an alternating hole-slab/rod-slab sequence by stacking of bi-layers that has a 21 % plus complete gap for $\varepsilon = 12$, forbidding light propagation for all wave vectors and all polarizations (Sanjeev, 1987). This kind of crystal slabs confines light perfectly in 3D, because its layers resemble 2D rod/hole crystals, it turns out that the confined modes created by defects in these layers strongly resemble the TM/TE states created by corresponding defects in 2D. Therefore, it can be used for direct transfer of designs from two to three dimensions while retaining omni-directional confinement (Joannaopoulos et al., 2008).

Over the years, it is realized that the fabrication of photonic crystals can be either easy or extremely difficult depending upon the desired wavelength of the band gap and the level of dimensionality. Lower frequency structures that require larger dimensions are easier to fabricate because the wavelength of the band gap scales directly with the lattice constant of the photonic crystals. At microwave frequencies, where the wavelength is of the order of 1 cm, the photonic crystals are decidedly macroscopic and simple machining techniques or rapid prototyping methods can be employed in building the crystals. Moreover, at the optical wavelengths, photonic band gaps require crystal lattice constants less than 1 μm and are difficult to fabricate. Building photonic band gaps in the optical regime requires methods that push current state-of-the-art micro and nanofabrication techniques. Since 1D photonic band gaps require periodic variation of the dielectric constant in only one direction, they are relatively easy to build at all length scales compare to 3D one (Sanjeev, 1987; Lodahl et al., 2004; Kim et al., 2008; Sonnichsen et al., 2005; Joannaopoulos et al., 2008). The 1D photonic band gap mirrors commonly known as distributed Bragg reflectors that have been used in building optical and near-infrared photonic devices for many years. Two common examples of devices that have been realized using 1D photonic band gaps are distributed feedback lasers and vertical-cavity surface-emitting lasers. The 2D photonic band gaps require somewhat more fabrication, but relatively ordinary fabrication techniques can be employed to achieve such structures. There are several examples of 2D photonic band gaps operating at mid- and near-IR wavelengths. Clearly, the most challenging photonic band gap structures are fully 3D structures with band gaps in the IR or optical regions of the spectrum. As mentioned above, the fabrication of 3D photonic band gaps is complicated by the need for large dielectric contrasts between the materials that make up the photonic band gap crystal, and the relatively low filling fractions that are required. The large dielectric contrast demands dissimilar materials, and often the low-dielectric material is air with the other material being a semiconductor or a high-dielectric ceramic. The low dielectric filling fraction ensures that the photonic band gap crystal has mostly air, while the high dielectric material must be formed into a thin network or skeleton. Combining these difficulties with the need for micron or sub-micron dimensions to reach into the optical region, the fabrication becomes extremely difficult indeed (Sanjeev, 1987; Lodahl et al., 2004).

The deep x-ray lithography and other techniques are useful to fabricate the photonic band gaps structures in which the resist layers of polymethyl methacrylate are irradiated to form a 'three-cylinder' structure. The holes in the polymethyl methacrylate structure are usually filled with ceramic material due to their low value of dielectric constant not favorable for the formation of a photonic band gap. A few layers of this structure can be fabricated with a measured band gap centered at 2.5 THz. The layer-by-layer structure can be fabricated by laser rapid prototyping using laser-induced direct-write deposition from the gas phase. The photonic band gap structure consisted of oxide rods and the measured photonic band gap is centered at 2 THz. The measured transmittance shows band gaps centered at 30 and 200 THz, respectively. In this way, it is possible to overcome very difficult technological challenges, in planarization, orientation and 3D growth at micrometer length scales. Finally, the colloidal suspensions have the ability to form spontaneously the bulk 3D crystals with submicron lattice parameters. In addition, 3D dielectric lattices have been developed from a solution of artificially grown mono-disperse spherical SiO_2 particles. However, both these procedures give structures with a quite small dielectric contrast ratio (< 2), which is not enough to give a full band gap. A lot of effort is being devoted to find new methods in increasing the dielectric contrast ratio. Several groups are trying to produce ordered macro-porous materials from titania, silica, and zirconia by using the emulsion droplets as templates around which material is deposited through a sol-gel process (Xing-huang et al., 2008). Subsequent drying and heat treatment yields solid materials with spherical pores left behind the emulsion droplets. Another very promising technique in fabricating photonic crystals at optical wavelengths is 3D-holographic lithography (Miao et al., 2008).

Materials with photonic band gaps could speed up the internet by improving the transmission of long-distance optical signals. One drawback with conventional optical fibers is that different wavelengths of light can travel through the material at different speeds. Over long distances, time delays can occur between signals that are encoded at different wavelengths. This kind of dispersion is worse if the core is very large, as the light can follow different paths or 'modes' through the fiber. A pulse of light traveling through such a fiber broadens out, thereby limiting the amount of data that can be sent. These problems could be solved by an extremely unusual 'holey fiber' as show in figure 19. The fiber has a regular lattice of air-cores running along its length and transmits a wide range of wavelengths without suffering from dispersion. It is made by packing a series of hollow glass capillary tubes around a solid glass core that runs through the centre. This structure is then heated and stretched to create a long fiber that is only a few microns in diameter. The fiber has the unusual property that it transmits a single mode of light, even if the diameter of the core is very large. This fiber can be produced even in a better way by removing the central solid glass core to form a long air cavity. In this case, the light is actually guided along the low-refractive-index air core by a photonic-band-gap confinement effect. Since the light is not actually guided by the glass material, very high-power laser signals could potentially be transmitted along the fiber without damaging it.

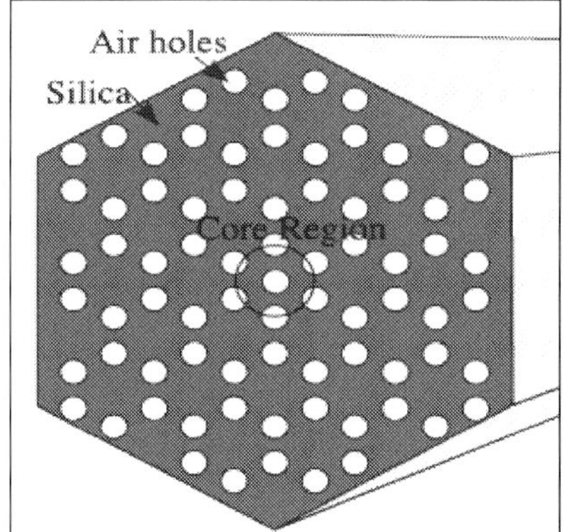

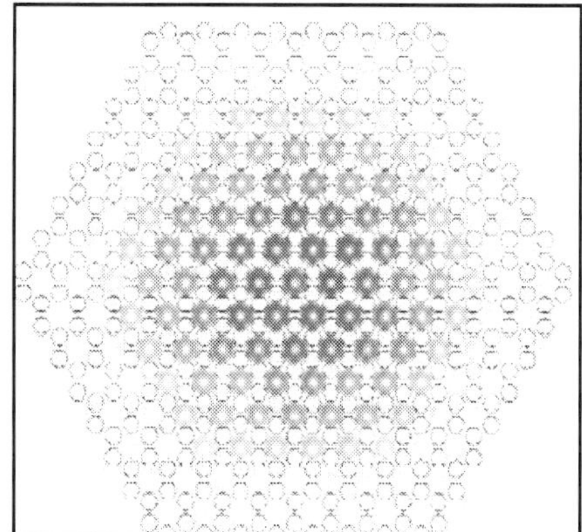

Fig. 19. Air-core photonic crystal fibers. Arrangements of voids and dielectric media (left) and light propagations through holes (right).

Defects in photonic band gap structure allow designing small, but highly efficient micro lasers. A point defect in the crystal gives rise to a resonant state with a defined resonant frequency in the band gap. Light is trapped in this cavity as the photonic band gap prevents it from escaping into the crystal. The photonic crystals built from the photo emissive materials, such as III-V semiconductors and glasses doped with rare-earth atoms, can also be used to make narrow-line width lasers that could potentially be integrated with other components in an optical-communications system. These lasers are made by introducing a small number of holes that are slightly smaller or larger than the other holes in the photonic-crystal lattice. These 'micro cavities' generate a narrow defect mode within the photonic band gap. While the material emits light in a wide spectral range, only the wavelength that matches the wavelength of the defect mode is amplified because it can propagate freely through the material. The laser cavity is formed either by the crystal surface or by external mirrors that surround the glass. The intensity of the propagating light increases as it undergoes successive reflections and travels back and forth through the photonic crystal. Meanwhile, light at other wavelengths are trapped within the photonic crystal and cannot build up. This means that the laser light is emitted in a narrow wavelength range that is directly related to the diameter of the micro cavity divided by the diameter of the regular holes. Moreover, the line width can be reduced further by using unusual geometries of the photonic-crystal lattice (Sanjeev, 1987). Such micro cavities are also much more efficient at trapping light than the cavities formed in semiconductor diode lasers since there are fewer directions in which the photons can escape from the cavity. The rate of photoemission in an active medium can be greatly increased by maintaining a high optical flux density. As micro cavities act as light traps, they provide a good method of enhancing the rate of photoemission in light emitting diodes, and are crucial for the operation of lasers. Moreover, the increased rate of photoemission means that micro cavity light emitting diodes and photonic-crystal lasers can be switched on and off at far greater speeds compared with conventional devices, which could lead to higher data-transmission rates and greater energy efficiency.

Preliminary experiments have been performed at microwave frequencies on defect structures within photonic crystals made from 'passive' materials that do not emit light. Photonic-crystal micro cavities that are fabricated from passive materials, such as silicon dioxide and silicon nitride, could also be used to create filters that only transmit a very narrow range of wavelengths. Such filters could be used to select a wavelength channel in a *'dense wavelength division multiplexing'* communications system (Lodahl, 2004). Indeed, arrays of these devices could be integrated onto a chip to form the basis of a channel de-multiplexer that separates and sorts out light pulses of different wavelengths. Figure 20 shows a photonic-crystal device that works as a simple filter. This is made by growing a thick layer of silicon dioxide on the surface of a silicon substrate, followed by a layer of silicon nitride. The positions of the holes were defined by patterning the top surface of the waveguide with electron-beam lithography. The underlying silicon dioxide was then etched away to create a freestanding porous silicon-nitride membrane that blocks light over the wavelength range 725 - 825 nm. Similar devices can also be fabricated with band gaps at shorter visible wavelength. Miniature wave-guides that could be used to transmit light signals between different devices are a key component for integrated optical circuits. However, the development of such small-scale optical interconnects has so far been inhibited by the problem of guiding light efficiently round very tight bends.

Conventional optical fibers and waveguides work by the process of total internal reflection. The contrast in refractive index between the glass core of the fiber and the surrounding cladding material determines the maximum radius through which light can be bent without any losses. For conventional glass waveguides, this bend radius is about a few millimeters. However, inter-connects between the components on a dense integrated optical circuit require bend radii of 10 μm or less. It is possible to form a narrow-channel waveguide within a photonic crystal by removing a row of holes from an otherwise regular pattern. Light will be confined within the line of defects for wavelengths that lie within the band gap of the surrounding photonic crystal. Since a porous material has no available modes at this wavelength, an optical quantum well forms in the waveguide region and traps the light. Under these conditions, we can introduce a pattern of sharp bends that will either cause the light to be reflected backwards or directed round the bend.

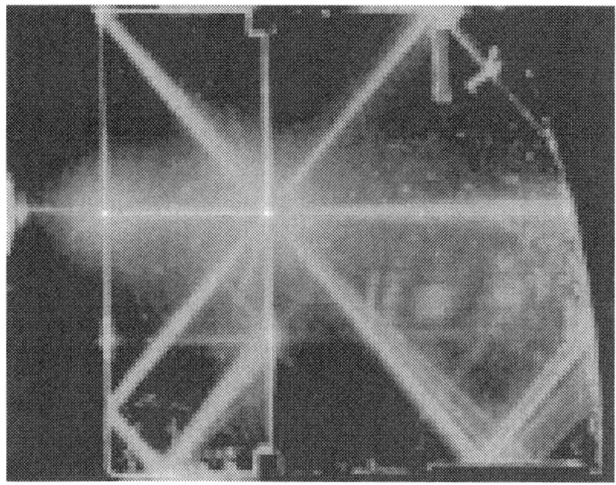

Fig. 20. A photonic crystal devices that work as a simple filter.

We conclude with the note that the original innovative research into photonic crystals/photonic band gap materials is necessary to achieve immediate commercial applications, but without intense research, it would not have been possible to set into these new classes of structures or a whole host of other tangential pursuits. The most important and useful thing that comes out of the research is *new ideas and paths of investigation*. Research breeds more research, which will eventually lead to something that genuinely be commercialized. Though the field of nanophotonics and nanotechnology is growing up exponentially and newer applications are coming at rapid space, however, more focused research is needed to get position in the market by defeating the existing technology.

3. Plasmonics: a new avenue of nanoscale optics

The term 'plasmonics' refers to the science and technology dealing with the manipulation of the electromagnetic signals by coherent coupling of photons to free electron oscillations at the interface between a conductor and a dielectric. Plasmons are electrons density waves and is created when light hits the surface of a metal at the precise frequency. Because these density waves are generated at optical frequencies, very small and rapid waves, they can theoretically encode a lot of information; more than what is possible for conventional electronics (Kim et al., 2008). Surface plasmons are optically induced oscillations of the free electrons at the surface of a metal. Plamonics is thought to embody the strongest points of both optical and electronic data transfer. Optical data transfer, as in fiber optics, allows high bandwidth, but requires bulky 'wires', or tubes with reflective interiors. Electronic data transfer operates at frequencies inferior to fiber optics, but only requires tiny wires. Plasmonics, often-called 'light on a wire', would allow the transmission of data at optical frequencies along the surface of a tiny metal wire, despite the fact that the data travels in the form of electron density distributions rather than photons (Sonnichsen, 2005). We would like to address the following relevant issues in plasmonics:

- What is plasmonics and plasmon resonance?
- How to get materials for plamonics applications?
- Why research is necessary in plasmonics?
- What is the present status for commercialization?
- Why are they so interesting?
- What are challenges and difficulties in plasmonics?
- How promising are they for future technology?

Since the middle of nineteenth century, after the first demonstration of stable dispersion of gold nanoparticles by Michael Faraday the scientific insight and queries on the interaction of light with matter has intrigued scientists. Without invoking the word nano in ancient time, artists have been exploiting sparkling red, yellow and green colors exhibited by metal nanoparticles especially of gold and silver as colorants in glasses for the decoration of windows and doors of many cathedrals, palaces, mosques and temples. Faraday concluded that metal nanoparticles having size much smaller than the wavelength of light exhibit intense colors that has no bulk counterpart. Gustav Mie in 1908 successfully explained the origin of such colors of dispersion using Maxwell's theory of classical electromagnetic radiation in which the phenomena was attributed to strong absorption and scattering of light by dispersion of metal nanoparticles (Kalele et al., 2007). However, during last two decades a series of noble-metal particles fabricated using advanced nanotechnology route showed a strong absorption band in the visible region of electromagnetic spectrum, arising from a resonance between collective oscillations of conduction electrons with incident electromagnetic radiation. Consequently, scientists are interested to guide, manipulate and control such strong absorption band associated with plasmon and hence the genesis of plasmonics. The formation of electric dipoles originates from the interaction of incident electromagnetic field that induces strong polarization of conduction electrons and weaker polarization to the immobile heavier ions. The net charge difference between the electrons and the ions acts as a restoring force that can be visualized as simple harmonic oscillator in the Lorentz model. The plasmon resonance is the resonance between the frequency of oscillation of the electrons and the frequency of the incident photon and is characterized by a strong absorption band. For nanoscale matter, the surface by volume ratio is high and most of the optical and electronic structure properties are dominated by the surface rather than the bulk. In this case, since a net charge difference is felt at the surface of a nanoparticle, the resonance is also known as *surface plasmon resonance*. The pictorial representation of surface plasmon resonance on the metal dielectric interface and on an array of two gold nanoparticle is shown in figure 21 (left panel) and (right panel) respectively.

The generation of *surface plasmon* is like *'an ocean of light'*. Dropping a piece of stone into a quiet lake one creates the ripples that spread out across its surface. The same thing happens when a photon hits the surface of a metal, where the 'ripples' consist of collective oscillations of electrons and have wavelengths of the order of nanometers. During such oscillations these 'surface plasmons' can pick up more light and carry it along the metal surface for comparatively large distances. Using plasmons light can not only be focused into the tiniest of spots but can also be directed along complex circuits or manipulated it many different ways. It is possible to achieve all of this at the nanoscale that is several orders of magnitude smaller than the wavelength of light (Pendry, 2000). This nanoscale is far below the resolution limits of conventional optics. Due to this reason, plasmonics has occupied a place in naophotonics in its own right. Several potential applications such as lasers, sensors, memory, communications, solar cells, biochemical sensing, optical computing and even cancer treatments are widely explored. Some of the exciting features of this field will be explored in this Section.

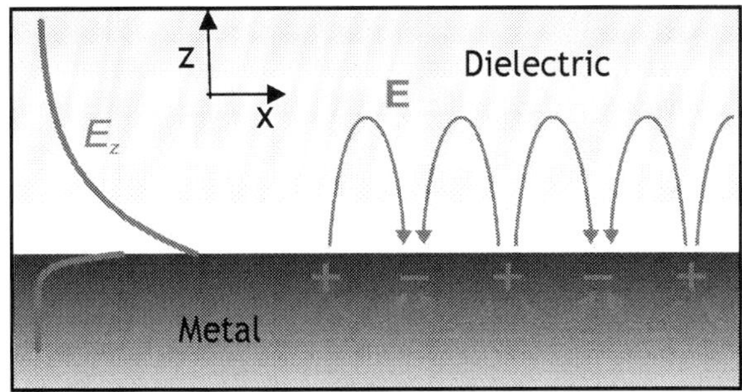

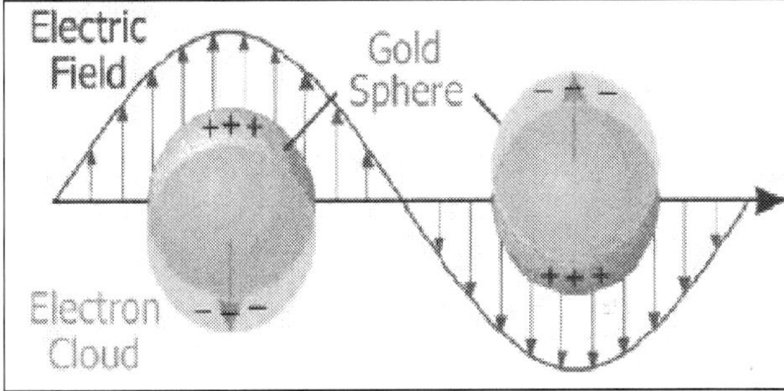

Fig. 21. The surface plasmon resonance: EM wave at metal-dielectric interface (left), and in gold nanoparticles (right).

The optical extinction properties of small metal particles have been studied for many years. Noble metal nanoparticles embedded in a dielectric exhibit a strong absorption peak due to a collective motion of free electrons, that is, a surface plasmon resonance. For isolated spherical particles, the resonance peak occurs generally in the visible part of the spectrum. The particular frequency depends on the particle size, and the dielectric constants of the metal and of the surrounding medium. For particle ensembles, however, electromagnetic coupling between neighboring particles shifts the plasmon absorption bands. Numerical calculations have demonstrated that nanoparticle size, nearest neighbor spacing, the overall ensemble size and shape have a critical effect on extinction spectra. The extinction coefficient that is the sum of the absorption and the scattering cross-sections is a useful parameter for surface plasmon resonance to occur in metals. The field plasmonics, the optical properties of metal structures at the nanoscale has made rapid development due to the ability of engineering metal surfaces and particles at the nanoscale. Advanced techniques like, electron beam lithography, chemical vapor deposition, and deep-UV lithography, focused ion beam milling and self-assembly has provided routes to engineer complex arrays of metal nanostructures. These chains of metal nanoparticles are exploited to excite, control, guide, direct and manipulate plasmons. The plasmons are attractive because they can effectively confine the optical excitation in a nanoscale volume and thereby mediate strong optical interactions. In addition, the wavelength at which these phenomena are observed can be tuned by varying the metal nanostructure shape, size and dielectric environment. This in turn, provides a broad domain with flexibility from which it is possible to choose the desired optical properties for an application (Kim, 2008; Sonnichsen, 2005; Prodan, 2003).

The coupling of light with electronic surface excitations, specifically, surface plasmon polaritons offers the opportunity to bridge the orders of magnitude difference in sizes between optical and electronic carriers. To develop schemes for coupling and transporting surface plasmons around a chip, the determination of their propagation lengths is particularly important. Researchers have already excited surface plasmons using a focused beam of electrons and then detected the luminescence emitted as the plasmons decayed. Based on these cathode-luminescence intensity decay profiles, they could determine propagation lengths as a function of wavelength. Gold and silver thin films on silicon and quartz substrates respectively were patterned with gratings to direct the emission, allowing the measurement of propagation lengths as short as several hundred nanometers. However, the resolution of the technique is limited by the excitation volume, which in principle, would increase as the film thickness decrease (Sonnichsen, 2005). Using surface plasmon we can obtain ultra-small, wavelength-sensitive directional sensors or detectors. The resonant coupling between the nanoparticles can concentrates light into well-defined hot spots and acts as antennas by suitably engineering the metal nanostructures (Waele et al., 2007). Coupling metal nanoparticle arrays to optical emitter's directional emitters may be achieved. In order to provide the control over the color, directionality and polarization of light-emitting diodes the enhanced optical density of states near the surface of metal nanoparticles can be used. The enhancement of optical density of surface states is highly efficient for the large-scale applications of solid-state lighting, bio imaging, sensing and solar concentrators. Recent calculations and experiments confirms that light scattering from metal nanoparticle arrays can effectively fold the path of sunlight into the layer and thereby strongly enhance its effective absorption (Pillai et al., 2007).

It is known from Maxwell's equations that an interface between a dielectric (e.g. silica glass) and a metal (e.g. silver or gold) can support a surface plasmon. A surface plasmon is a coherent electron oscillation that propagates along the interface together with an electromagnetic wave. These unique interface waves result from the special dispersion characteristics (dependence of dielectric constant on frequency) of metals. What distinguishes surface plasmons from 'regular' photons is that they have a much smaller wavelength at the same frequency. For example, a He-Ne laser, whose free-space emission wavelength is 633 nm, can excite a surface plasmon at a silicon/silver interface with a wavelength of only 70 nm. When the laser frequency is tuned very close to the surface plasmon resonance, surface plasmon wavelengths in the nanometer range can be achieved. The short-wavelength surface plasmons enable the fabrication of nanoscale optical integrated circuits, in which light can be guided, split, filtered, and even amplified using plasmonic integrated circuits that are smaller than the optical wavelength (Kim, 2008; Loo et al., 2005). The reduction in wavelength comes at a price and as a result, surface plasmons are often having loss. One way to achieve long propagation lengths is to use very thin metal films. In this case, surface plasmons on both surfaces of the metal film interact, and both a symmetric and an asymmetric field distribution can exist. One of these modes has low loss and, for metal films as thin as 10 nm, the centimeter propagation lengths can be achieved for surface plasmons in the infrared. At a given frequency, the surface plasmon wavelength is strongly dependent on the metal thickness. Thus, the plasmonic integrated circuit engineer has an extensive toolbox, including choice of metal (dispersion), metal thickness, and excitation frequency (Loo et al., 2005).

When a light source such as a luminescent quantum dot or dye molecule is placed close to a metal, it can excite a surface plasmon through a near-field interaction. With a light-emitting diode embedded in a plasmonic structure, surface plasmons can be electrically excited. Such surface plasmons may serve as an alternative to overcome the information bottlenecks presented by electrical interconnects in integrated circuits. Coupling to surface plasmons can also enhance the extraction efficiency of light from light emitting diodes. Metallic nanoparticles have distinctly different optical characteristics than surface plasmons at planar interfaces. Nanoparticles show strong optical resonances, again because of their large free-electron density. As a result, a plane wave impinging on a 20 nm diameter silver particle is strongly 'focused' into the particle, leading to a large electric field density in a 10 nm region around the particle. Ordered arrays of nanoparticles can possess even further enhanced field intensities because of plasmon coupling between adjacent particles. By varying nanoparticle shape or geometry, it is possible to tune the frequency of surface plasmon resonance over a broad spectral range. For example, gold ellipsoids or silica colloids covered with a gold shell show resonances that coincide with the important telecommunications wavelength band. The ability to achieve locally intense fields has many possible applications, including increasing the efficiency of light emitting diodes, (bio) sensing, and nanolithography. The light-carrying phenomenon when light falls on a thin film of metal containing millions of nanometer-sized holes shows some surprising results. Interestingly, the film was found to be more transparent than expected, and thus generate many applied research possibilities. The holes were much smaller than the wavelength of visible light, which should have made it almost impossible for the light to get through at all. When the incoming photons struck the metal film, they excited surface plasmons, which picked up the photons' electromagnetic energy and carried it through the holes, re-radiating it on the other side and giving the film its transparency (Ebbesen, 1998).

Arrays of metal nanoparticles can also be used as miniature optical waveguides. In linear chain arrays of nanoparticles, a plasmon wave propagates by the successive interaction of particles along the chain. The propagation length is small (~100 nm), but may be increased by optimizing particle size and anisotropy. The effect of quantum confinement make these nanoparticle array waveguides attractive as they provide confinement of light within ~50 nm along the direction of propagation, a 100-fold concentration compared to dielectric waveguides. A very peculiar effect occurs in metal films with regular arrays of holes, in which, local field enhancements are predicted to occur along the holes. This effect leads to much larger optical transmission through the holes than expected, based on consideration of their geometric areas. The precise role of surface plasmons in these effects is still the subject of lively scientific debate, but applications of the enhanced transmission characteristics in nanoscale optical storage appear promising (Prodan et al., 2003).

Clearly, there is a plenty of plasmonic concepts still waiting for exploration. The clinical studies are ever increasing and encouraged with promising results (Loo et al., 2005). The applications spanning from (bio) sensing, optical storage, solid-state lighting, interconnects and waveguides. Indeed, it appears that metals can shine a bright light toward the future of nanoscale photonics. Most of the early work in plasmonics focused on the study of resonances and electromagnetic field enhancements in individual metal nanoparticles and particle assemblies (Prasad, 2004; Rayleigh, 1888; Pendry, 2000). It is possible to form nanoscale hot spots through plasmon coupling within arrays of metal nanoparticles. In these hot spots, the intensity of light from an incident beam can be concentrated by more than four orders of magnitude that lead to a large improvement in sensing techniques that use optical radiation, such as Raman spectroscopy, with potential applications in medical diagnostics (Polman, 2008). In phenomena that are nonlinear in light intensity the effect of light concentration via plasmons are robust. This has recently been demonstrated by the on-chip generation of extreme-ultraviolet light by pulsed laser high harmonic generation (Kim et al., 2008). This opens up a new avenue in lithography or imaging at the nanoscale with soft x-rays. The methodology of fluorescence energy transfer that is routinely used in biology is limited in length scale (Sonnichsen et al., 2005). Using the highly sensitive plasmonic interaction between metal nanoparticles this can be overcome. Due to the very high sensitiveness to nanoparticles separation, precise measurements of the plasmon resonance wavelength of metal particle assemblies functionalized with bio-molecules can be used as a molecular-scale ruler that operates over a much larger length scale. Practical applications of this concept in systems biology, such as imaging of the motion of molecular motors, bio labeling and bio sensing are being exploited (Polman, 2008). The standard commercial pregnancy tests and the detection of bio-molecules are based on the measurement principle of plasmonic resonance shifts. The possibility of using of particles composed of a dielectric core and a metallic shell in future cancer treatments is underway. The injected shell-core nanoparticles are selectively bound to malicious cells and then laser irradiation at a precisely engineered plasmon resonance wavelength is focused to heat the particles and thereby destroy the cells (Atwater et al., 2009).

One of the main challenges of present plasmonic research is to shrink visible wavelength regime into the soft x-ray wavelength regime. The long distance propagation of surface plasmons along metal waveguides using plasmonic structures based on metal nanoparticles is a new paradigm of research. Using the tools of nanotechnology one can precisely controls material structures and geometry that allows the wave-guiding properties to be controlled in ways that cannot be achieved with regular dielectric waveguides. Particularly, extremely short wavelengths can be achieved at optical frequencies using plasmonic waveguides. A recent experiment demonstrate that light with a free-space wavelength of 651 nm can be squeezed to only 58 nm in a metal-insulator-metal plasmonic waveguide (Miyazaki et al., 2006). The propagation speed of plasmons can be further reduced well below the speed of light by suitably engineering the structures of plasmonic waveguide. Integrating nanoholes in metal films that acts as efficient color filters a more efficient plasmonic waveguide structures have been fabricated. In some complex geometry by tailoring the plasmon waveguides, a negative refractive index for the guided plasmon has been observed. This is very interesting because the two-dimensional negative refraction in these plasmonic waveguides may be useful for plasmonic lens and high resolution imaging (Lezec et al., 2007). The research on planar plasmon propagation is targeted to the design of plasmonic integrated circuits. Using these plasmonic integrated circuits optical information can be generated, manipulated, switched, amplified, guided and detected within dimensions much smaller than the free space wavelength of light. The dream is the integration of optics with nanoscale semiconductor integrated circuit technology. So far, it seems plasmo-eletronic integration is impossible because of the different length scales of optics and electronics. It is hoped that in these devices of nanoscale dimensions a relatively small propagation lengths could be tolerated despite of plasmons decay during their propagation. The plasmo-electronic technology may open a wealth of prospects in designing plasmon laser or amplifier of nanodimension.

As mentioned before, optical 'meta-materials' with artificially engineered permittivity and permeability will fulfill the ever-growing market demand of advanced materials for optoelectronics and nanophotonics circuitry. The fabrication of metallic nanoresonators in 2D and 3D arrangements employing meta-materials is a step forward in this direction. A stack of metallic 'fishnet' structures shows negative index of refraction at near infrared light wavelength (Valentine, 2008). It is possible to achieve sub-wavelength optical imaging due to the peculiar nature of light refraction in the materials of negative refractive indices (Pendry, 2000). Surprisingly, precisely engineered geometries with negative refractive indices may even act as invisibility cloaks for visible wavelengths. It is needless to mention that, the field plasmonics has grown from an embryo with fundamental insights to a vast field with important applications and commercialization. To shape up the plasmonic research, several novel basic and applied research topics are undertaken, including the femto- and atto-second dynamics and coherent control of plasmons, 4D imaging, plasmo-electronic integration, lasing spacers cloaking using novel geometries (Engheta, 2008) and quantum mechanical effects at the sub-nanoscale level. These studies are very exploratory with innovations and enriched with novel scientific thoughts as well. Many new exciting applications of plasmonics are waiting to capture the market. These efforts, in turn, have benefited greatly from the flowering of nanotechnology in general over the past decade, which brought with it a proliferation of techniques for fabricating structures at the nanoscale, exactly what plasmonics needed to progress from laboratory curiosity to practical applications (Brongersma et al., 2007).

The plasmonics made a breakthrough in the field of the solar cell design using semiconductors to enhance the efficiency. In this route, gold nanoparticles on the surface of semiconductors are fabricated that act as reflectors and focus light into the semiconductor and thereby increase the absorption efficiency by concentrating more light (see figure 22). The other route in which, tiny gold nanoantennas could redirect sunlight vertically allows it to propagate along the semiconductor rather than passing straight through the surface. In both the approach, the cell could get by with a much thinner semiconductor layer and acts as a superior concentrator of light. Using plasmonic techniques, not only the cost of the solar cells is decreasing but the efficiency at extracting the available energy from sunlight also drastically improving. An optimistic model calculation and theoretical estimate shows that the use of plasmonics in photovoltaics could increase the absorption two to five times and commercialization of such solar cells look promising. The amorphous silicon based solar cells available in the market have efficiencies of around 10–15 % and the predicted enhancements could translate into efficiencies of about 20 %. Currently available crystalline silicon solar cells have efficiencies around 21 % and the new figure could approach the theoretical maximum of about 30 %. The large scale and low-cost commercial applications are facing the challenges of developing workable device designs, architecture and fabrication techniques for mass production (Atwater et al., 2009).

The beauty of plasmonics is that it can bring the optics closer in size to the transistor, which can offer optical pathways on virtually the same scale as the silicon structures found in advanced microchips. The design of chip with the integration of metals is possible to distribute light over an integrated circuit by surface plasmons. Structures of gold and silver nanowires, nanorods, nanodots (Verhagen, 2009) or grooves are etched into metal surfaces

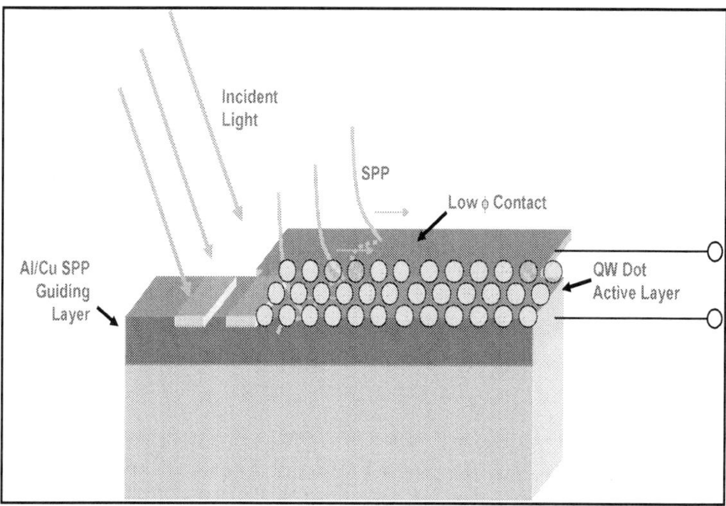

Fig. 22. Light manipulation in plasmonic quantum dot solar cell; the surface plasmons are generated to help direct light using nanoantennas in devices such as solar cells.

(Bozhevolnyi, 2006). They are expected to provide pathways that guide light across a chip irrespective of the designer's directionality. The only trade-off here is with the smallness of the structure, in which forcing the plasmons to travel through too narrow channels can cause leak out from the sides and thereby gets lost. However, guiding surface plasmons over distances of more than 100 µm is possible, which is roughly a thousand times bigger than the features on a current generation microchip. This research has opened new possibilities for plasmonic nanocircuits to carry information using light waves along complex paths and through many processing steps. Recent progress in the laser miniaturization has shown the promise to fabricate plasmonic waveguides. Moreover, plasmonics offers the possibility of integration of plasmo-electro chip at the nanoscale, at lengths much shorter than the wavelength of laser light. Rather than amplifying light in a conventional laser cavity, a plasmonic *'spaser'* would amplify it with the help of plasmons and the first experimental evidence for such plasmon-based lasing has already been reported (Noginov, 2009; Oulton 2009).

The full integration of these plasmon lasers into standard micro-circuitry, however, needs a suitable way to trigger the spasers using standard electrical currents. *'SPASER'* the Surface Plasmon Amplification by Stimulated Emission of Radiation is a new device that has been introduced very recently. In a spaser, a surface plasmon plays the same role as a photon in a laser. A plasmon enters the resonator as a nanoparticle embedded in a gain material containing chromophores such as semiconductor nanocrystals or dye molecules. The gain medium must be capable of producing population inversion, which allows it to lase or *'spase'* in this case. Spacers are ultrafast nanoplasmonic chips with high degree of integration. In addition to creating light and guiding it across, spacers communicate and control each other through their near fields or are connected with nanoplasmonic wires and perform ultrafast microprocessor functions (Noginov, 2009). The plasmonics based optical computing requires a series of bits in a digital data stream that can be obtained by turning the flow of plasmons on and off at high speeds. A plasmonic modulator using silicon technology has been realized and the working principle of this device is based on the use of an electric field to control the propagation of surface plasmons through the device (Dionne, 2009). They are not only much smaller in size compared to conventional optical counterparts but their operation frequency can easily reach tens of terahertz that is much above the gigahertz limit of modern computers.

One of the niche areas in plasmonics is surface-enhanced Raman spectroscopy. One can enhance the signal by several orders of magnitude larger and is strong enough to detect a single molecule (Fleischmann, 1974; Nie, 1997). The surface-enhanced Raman spectroscopy is very useful in the biochemical and materials sciences for providing information on the chemical composition of molecules at very small concentrations and detail microstructures. Surface-enhanced Raman spectroscopy is a plasmonic effect in which silver/gold nanoparticles act as nanontennas to gather the incoming laser light and, through their surface plasmons, concentrate it. In this case, a dual amplification results gigantic signal enhancement by concentrating the light first and then scattered by nearby molecules and amplified again by the silver/gold nanoparticles on the way back out (Atwater et al., 2009). Presently, the surface-enhanced Raman spectroscopy faces some problem for commercialization. This is due to formidable difficulties in achieving highly accurate control over the surface nanostructures and their mass production. Other sensing techniques such as localized surface plasmon resonance may be a suitable alternative in which the surface is covered with nanostructures in the shape of rods or triangles plays important role. The plasmonic properties depend strongly on the properties of the surrounding medium and the changes to the refractive index lead to experimentally measurable changes to the wavelength of surface plasmon resonance (Anker et al., 2008). Surprisingly, the huge sensitivity of localized surface plasmon resonance based devices can reach to the limit of single-molecule detection, and can even focus to destroy cancer cells! For cancer treatments, gold nanoparticles can be injected and guided to the tumor by antibodies bound to the particles' surface. By illuminating the area near nanoparticles with a low dose, using infrared laser light gets absorbed to create plasmons in the gold and burn the infected cells and leaves healthy tissue undamaged. The cancer cells are finally killed by heating up the nanoparticles with accumulated energy through localized surface plasmons (Hirsch et al., 2003). This kind of cancer therapy has successfully been tested on mice for complete elimination of the tumors and waiting for the human clinical trials with patients having head and neck cancers.

Exponential rise in nanophotonics research provided amazing data processing and signal transport capabilities that have the potential to enhance computer performance remarkably. However, to realize this objective much powerful integration techniques for newly emerging nanophotonic devices with conventional nanoelectronics components are urgently required. Undoubtedly, a natural choice for an ideal platform for the marriage of these distinct technologies would be the silicon. Consequently, the lack of an intrinsic source of surface plasmon polaritons compatible with silicon-based complementary metal-oxide semiconductor fabrication techniques slowed down the growth of the integration of plasmonic components with silicon. Presently, complementary metal-oxide semiconductor has reached to true nanoscale devices composed of complex and intertwined dielectric, semiconductor and metallic structures. The impressive developments and availability in computer aided circuit design, lithography, Monte Carlo method, electronic and photonic-device simulations, an increasingly wide variety of integrated optoelectronic functionalities are making the silicon-based technology more robust (Hryciw et al., 2010). Plasmonics is playing major role in the design of future silicon-based optoelectronic and plasmo-electronic chips based on the manipulation of surface plasmon polaritons. The plasmonics research began with passive routing of light in waveguides with diameters much smaller than the wavelength of the light. The surface plasmon polariton waveguides was not perceived as a superior alternative to high-index dielectric waveguides as the propagation length in such high-confinement is limited to a few tens of μm. It is important to keep in mind that the size of dielectric waveguides is limited by the fundamental laws of diffraction, which is much larger than the electronic devices on a chip. However, the sub-wavelength dimensions of plasmonic devices are uniquely capable of reconciling the size mismatch and bridge the gap between dielectric micro-photonics and nanoelectronics. The passive waveguides and light-concentrating structures are the two exciting outcome of the plasmonic studies.

Using surface plasmons, by channeling and concentrating light on sub-wavelength structures miniaturized photonic circuits with waveguides having nanometer length scales have been fabricated. This photonic circuit first converts the incident light to a surface plasmon wave that propagates and eventually converts back to light. These waveguides are realized by depositing gold stripes on a dielectric surface. It is possible to channel the electromagnetic energy using a linear chain of gold and silver nanoparticles over a distance of ~200 nm without any significant loss. In this geometry, each nanoparticle with dimension much smaller than the wavelength of incident light acts as an electric dipole and thereby produces surface plasmon. The inter-particle spacing in the array plays an important role in deciding the interactions. The near-field electric-dipole interactions dominates when the inter-particle separation become much smaller than the wavelength of incident light. This is highly desirable for the wave guiding application of arrays of gold or silver nanoparticles. Active plasmonic devices are designed to switch and detect light in ultra-compact geometries that may exceed the stringent requirements of complementary metal-oxide semiconductor technology (Walters et al., 2010). A crucial ingredient called complementary metal-oxide semiconductor-compatible plasmonic sources can now be added through surface plasmon polaritons. These surface plasmon polariton emitters will play a crucial role in chip-scale optical information links useful for novel integrated bio-sensing applications. A silicon-based source for active Plasmon waveguide using Si nanocrystals as the active medium whose operation principle is similar to other device are created. This device is fabricated using atomic layer deposition and low-pressure chemical vapor deposition processes occurred at around room temperature to be compatible with complementary metal-oxide semiconductor processing (Pavesi, 2003).

The silicon microelectronics world is currently defined by length scales that are many times smaller than the dimensions of typical micro-optical components, the process scaling driven by Moore's law. The size mismatch poses severe challenge to integrate photonics with complementary metal-oxide semiconductor electronics technology. One promising solution is to fabricate optical systems at metal/dielectric interfaces, where surface plasmon polaritons offer totally new and unique opportunities to confine and control light at length scales below 100 nm. Many passive components developed using plasmonics suggests the potential of surface plasmon polaritons for applications in sensing and communication. Active plasmonic devices based on III–V materials and organic materials and an electrical source of surface plasmon polaritons using organic semiconductors have been reported. It is established that a silicon-based electrical source for surface plasmon polaritons can be fabricated using low temperature micro technology processes that are compatible with back-end complementary metal-oxide semiconductor technology (Hryciw et al., 2010). The highly confined modes of metal–dielectric–metal called metal–insulator–metal plasmonic waveguides dramatically alter the light-emission properties of optical emitters located between the metals (Jun et al., 2008). Moreover, there exist an efficient electromagnetic decay pathway for the surface plasmon polariton emission thereby the radiative decay rate of excited emitters can be increased order of magnitude, a direct consequence of the Purcell effect (Hryciw et al., 2010). The small size of the surface plasmon polariton mode that is directly translates to a strong coupling to the surface plasmon polariton emitter is primarily responsible for the large modification of the decay rate in these plasmonic structures. Other high-confinement metal oxide semiconductor and silicon slot waveguides shows similar beneficial effects and lay the foundation of an entire new set of silicon-based sources (Hryciw et al., 2009; Galli et al., 2006; Jun et al., 2009). The enhancements in high-confinement waveguides are very broadband in nature that allows effective use of emitters across the entire visible and near-infrared spectrum to achieve power-efficient incoherent light sources. Even for poor emitters the reduced radiative lifetime is beneficial in increasing the efficiency that allows faster source modulation. The other important benefit of metal-dielectric-metal waveguides is that they only support a single propagating mode and provide low-loss dielectric waveguides (Veronis et al., 2007). The wave guiding based on high-confinement sources is altogether a new class of chip scale devices. They combine efficient charge injection and facile photon extraction by an electrically pumped, plasmon-enhanced light source that inspires new way of designing truly nanoscale photonic devices and circuits for future miniaturization (Brongersma et al., 2007).

Surface plasmon polaritons are quasi-two-dimensional electromagnetic excitations. They propagate along a dielectric-metal interface in which the field components decay exponentially into both neighboring media. The field of a plane surface plasmon polariton comprises a magnetic field component, which is parallel to the interface plane and perpendicular to the propagation direction. It has two electric field components, of which the main one is perpendicular to the interface. The numerical simulations shows that nanometer sized metal rods can support extremely confined surface plasmon polariton modes that propagates over hundreds of nanometers. Similar observations have been made for the electromagnetic excitations supported by chains of metal nano-spheres. Metal stripes of finite width are employed for lateral confinement of the surface plasmon polariton along the stripes. In conventional integrated optics based on dielectric waveguides, the problem of miniaturization is approached. This is achieved by making use of the photonic band gap effect that is essentially a manifestation of Bragg reflection of waves propagating in any direction because of periodic modulation of the refractive index (Maier, 2007; Brongersma et al., 2007).

The efficient wave guiding along straight and sharply bent line defects in 2D photonic band gap structures has been demonstrated for light wavelengths inside the photonic band gap. It became clear that these photonic band gap structures, when properly designed and realized, might be advantageously used for miniature photonic circuits allowing for an unprecedented level of integration. Furthermore, one can conjecture that other (quasi) 2D waves, e.g., surface plasmon polaritons, might be employed for the same purpose. The surface plasmon polariton photonic band gap effects for all directions in the surface plane of a silver/gold film having a 2D periodic surface profile has also been demonstrated. It should be emphasized that the interaction of surface plasmon polariton with a periodic surface corrugation, similarly to the interaction of a waveguide mode with a periodic array of holes, produces inevitably scattered waves propagating *away* from the surface. This unwanted process results in the additional propagation loss and has to be taken into account when considering the surface plasmon polariton photonic band gaps structures. The surface plasmon polariton guiding along line defects in surface plasmon polariton photonic band gap structures with ~45 nm high and 200 nm wide gold bumps arranged in a 400 nm period triangular lattice on the surface of a 45 nm thick gold film is also reported. The efficient surface plasmon polariton reflection by such an area and surface plasmon polariton guiding along channels free from scatterers was observed, as well as significant deterioration of these effects at ~800 nm, indicating the occurrence of the surface plasmon polariton photonic band gaps effect in these structures (Veronis et al., 2007; Kalele et al., 2007).

There are many uses of gold and silver nanoparticles and nanorods from cancer-cell diagnostics, cancer-cell imaging and photo-thermal therapy. In the plasmonics applications of bio imaging or drug delivery, mostly the nanoparticles of gold and silver are used as it offers highly favorable and biocompatible optical and chemical properties. Moreover, metal nanoshells having the same volume as metal nanoparticles show much stronger and sharper surface-plasmon-resonance bands due to its enhanced surface area. Therefore, the nanoshells are preferred for the detection of macromolecules, DNA, proteins and microorganisms. The integration of biology and the materials science at nanoscale has the potential to revolutionize many fields of science and technology. The relevance of nanometer scale stems from the natural dimensions of bio-molecules, such as, DNA, proteins, viruses and sub-cellular structures as they fall in the length scale of 1 to 1000 nm. Gold nanoparticles are mostly exploited for bio imaging and therapeutic applications due to their strong properties of light scattering. In addition, the scattered intensity depends on the size and shapes of nanoparticles and their aggregation states (Hryciw et al., 2009; Kalele et al., 2007).

The non-photo-bleaching character of gold is suitable for detecting very low concentration and can be used as contrast agents in various biomedical imaging techniques. It is demonstrated that the antibody-conjugated gold nanoparticles bind specifically to the surface of malignant cells with much more affinity than healthy cells; also, malignant cells required half the energy to be destroyed photo-thermally than healthy cells. Gold nanocages have been employed in optical coherence tomography by using scattered light for noninvasive imaging to detect cancer at an early and treatable stage. The materials structures fabricated using nanosphere lithographic technique can be used for chemosensing and biosensing by realizing through shifts in the surface Plasmon resonance peak. The mechanism of the shift can be attributed to the changes in the local relative permittivity as well as charge transfer interactions between the adsorbed analyte and the metal. The wavelength shifts are more reliable rather than the intensity changes in biophotonics. Some spectacular observation on the shift of surface-plasmon absorption bands has been made in recent years using nanosphere lithography-deposited silver and gold nanotriangles and nanorods having size ~50 nm (Kalele et al., 2007; Walters et al., 2010).

Despite of many roadblocks remain to the commercialization of plasmonic technologies ranging from the plasmo-electronic integration on a single microchip to device issues there is renewed research interests. The future challenge would be to minimize the losses in the metal nanostructures, and he smart design of the plasmonic structures has been attempted to reduce the losses to acceptable levels. Although, the plasmonics research has already made remarkable progress but the understanding of physics very close to the metal surface still far from being fully understood. Current commercially available optical devices are too large and show rather high losses in the optical signal strength. Plasmonic-based devices will perhaps overcome this problem because a light beam could in principle, relay information through the chip on more channels and at a higher speed than conventional integrated circuitry can handle. Nevertheless, the plasmonics has given to photonics the ability to go to the nanoscale and properly take its place among the nanosciences.

4. Challenges of nanophotonics

Nanophotonics has the potential to improve optoelectronic products in a wide array of new applications. There are no yardsticks or a clear roadmap yet regarding the stability, efficiency, tolerances, longevity and large-scale production. However, the promises of technologies and applications are diverse and multidisciplinary, in early stages of development, and the opportunities are spread throughout the value chain. There are different challenges for nanophotonics (Ghoshal et al., 2007; Chu et al., 2005). Some of them are

- Market strategy

Nanophotonics has the potential to improve optoelectronic products in a wide array of new applications, including multi target markets each worth few billions of dollars. Making market strategy is a greater challenge than the technology. These findings are presented in most of the studies in recent time.

- Single –molecule addressing (pre-requisite for architectures work)
- Optical nanoscopy of molecules
- Designing plasma-optic chip
- Assessment of nanowires, nanoparticles and nanoarrays in nanophotonics
- Assessment of metal nanoparticles, nanoarrays and nanorods
- Hierarchy of interactions with other quasi particles
- Energetically sound
- Amplification and gain
- Integration, costs, standards, etc.

Although, these are the main challenges but there are expected benefits from nanophotonics. Some of them are:

- Bridge the gap between current photonic systems and future approaches bringing in example:
  - access to further integration
  - lower noise
  - mass production techniques and accurate fabrication
  - plasma-electro integration
  - cheaper and efficient devices
- Moving towards molecular photonics as the probable limit of integration which
  - will dissipate less energy
  - will occupy less volume
  - will require lower input signal
  - will probably rely on self assembly
  - will be more lasting
  - will be more flexible
  - will be more sensitive

- Molecules might compute, sense, act and serve building block of more complex structures
  - time scale
  - length scale
  - input-output schemes, algorithms.

Majority of the present fabrication techniques are based on expensive technologies and processes belonging to the very high technology pursuits at an expensive category, such as e-beam lithography, projection lithography, ion-beam milling and extreme ultraviolet lithography. There is great interest and need for approaches as well as strategies for large-scale production, which are more cost-effective and environment friendly. The development of the emerging nanopatterning methods that is recently established may overcome certain barriers related to mass production. The parallel-processes including self-assembly for polymeric materials with matrices can be tested with few milligrams to optimize a novel optical materials. Moreover, getting devices within the next one or two decades would require to produce kilograms of nanophotonic and molecular photonic materials and manufacture them cost efficiently. Thus, preparation techniques are needed that can be scaled up. The criteria on tolerances, efficiency of energy transfer and longevity are other significant issues that manufacturer have to meet when it comes to material purity because this will be an important cost factor, as will lateral size control. Nevertheless, much of these techniques have to be well established with the relevant industrial standards and free of hazards. Until now, it is far from being understood that how efficiently the energy can be converted from a molecular transition into an electronic transition. How efficient a photonic device will be to convert photons to electrons and finally back to photons? The stability of such processes is major concern to the researchers.

5. Nanoscale optical confinement

Size quantization in materials is the manifestation of quantum-confinement in low dimensional quantum structures. Materials at nanometer length scale having large surface to volume ratio possess different electronic and optical density of states and shows many emerging properties those absent in bulk structures. The change in the nature of optical band gaps and the opening up the gap are the manifestations of optical confinement due to which plasmon can be localized and photons can be confines in nanostructures. They have quantum optical properties that are absent in the bulk material due to the confinement of electron-hole pairs (called excitons) in a region of a few nanometres. For example, bulk metals was never thought of as useful candidates for photonics applications, and due to their high reflection and absorption coefficients they have been generally overlooked as elements to guide, focus and switch light at visible and infrared wavelengths. However, at the nanoscale the intriguing guiding and refractive properties of metal structures can be realized since the metal components become semi-transparent due to their small size (Pavesi, 2007; Ghoshal et al., 2007).

Light can be localized and manipulated in appropriately designed metallic and metallo-dielectric nanoparticle array structures. Interesting phenomena occur near the plasmon frequency where optical extinction is resonantly enhanced due to the effect of quantum confinement. Recent interest exploits the collective oscillations of the conduction electrons of this plasma in arrays of metal nanoparticle can also be used as miniature optical waveguides in linear chain arrays of nanoparticles. A plasmon wave propagates by the successive interaction of particles along the chain. The propagation length is small (~100 nm) but may increase by optimizing particle size and anisotropy. A nanoparticle array waveguides is attractive because they provide confinement of light within ~50 nm along the direction of propagation and a 100-fold concentration compared to dielectric waveguide (Koller et al., 2008; Stockman et al., 2007). The confinement of photons in a nanoscale optical cavity offers other possibilities. The spontaneous emission rate depends on cavity properties, increasing with quality factor and decreasing with mode volume. Photonic-crystal cavities can be made with model volumes smaller than the cube of the wavelength (measured as the wavelength λ in air divided by refractive index n of the medium) and can be fabricated with high quality factors. That greatly enhances the spontaneous-emission rate and the fraction of those photons coupled into a single-cavity mode. This allows a single nanocavity to operate as a laser with a very low threshold (Kobayashi et al., 2002).

The surface plasmon resonance associated with array of nanoparticles of noble metals within the size range ~ 10 to 100 nm can be localized to each nanoparticle known as *localized surface plasmon resonance.* On the other hand, the surface plasmon resonance associated with a metallic thin film of thickness in the 10 - 100 nm range can travel across the metal dielectric interface called *propagating surface plasmon resonance.* Although light is not able to propagate through a bulk metal, plasmons are able to propagate at a metal–dielectric interface over distances as large as several centimeters. Furthermore, the plasmon wavelength can be tuned by proper control of the metal film's thickness, size, shape and geometry of nanoparticles. The fabrication of plasmonic waveguide with wavelength shorter than that in free space is in fact the practical realization of this concept. The nanostructures made of gold and silver have been studied much more extensively due to their unique optical and electronic structure properties. In addition to the size and the shape of nanoparticles, surface plasmon resonance also greatly influenced by diverse phenomena such as Rayleigh scattering from nanoparticles, aggregation of nanoparticles, charge transfer interactions, changes in local refractive index, presence of defects, etc. These kinds of interactions are highly useful and are often explored for the detection of bio-molecules using metal nanostructures. The detection process involves monitoring the changes associated with the surface plasmon of the metal nanostructures on addition of the bio-molecules. Surface plasmon resonance can also be tuned over a very wide spectral range using novel nanostructure such as nanoarrays, nanorods and nanoshells (Prasad, 2004; Shen et al., 2000; Kalele et al., 2007).

The optical properties of silicon nanocrystallites of known sizes, present in super-critically dried porous silicon films of porosities as high as 92 %, have been measured by a variety of techniques. The band gap and luminescence energies have been measured as a function of size for the first time. The band gap increases by more than 1 eV due to quantum confinement (Bettoti et al., 2003; Pavesi, 2003). The peak luminescence energy, which also shifts to the blue, is increasingly Stokes shifted with respect to the band gap, as the size decreases. The measured band gap is in agreement with realistic theories and the Stokes-shift between band gap and luminescence energies coincides with the exciton binding energy predicted by these theories. These results demonstrate unambiguously and quantitatively the role of quantum confinement in the optical properties of this indirect gap semiconductor (Behren et al., 1998). Optical confinement effects in nanostructured materials enable new innovative device concepts that can radically enhance the operation of traditional semiconductor devices. A larger fraction of the solar spectrum can be harnessed while maximizing the solar cell operating voltage by using quantum wells, quantum wires and quantum dots embedded in a higher band-gap barrier material.

Nanostructured devices thus provide a means to decouple the usual dependence of short circuit current on open circuit voltage that limits conventional solar cell design. Ultra-high conversion efficiencies are predicted for solar cells that collect both low and high-energy photons from the solar spectrum while maintaining high voltage operation. Unprecedented opportunities are arising for re-engineering existing products. For example, cluster of atoms (nanodots, macromolecules), nanocrystalline structured materials (grain size less than 100 nm), fibres less than 100 nm in diameter (nanorods and nanotubes), films less than 100 nm in thickness provide a good base to develop further new nanocomponents and materials in nanophotonics. The buckyball (C_{60}) has opened up an excellent field of chemistry and material science with many exciting nanophotonic applications because of its ability to accept electrons. Carbon nanotubes (CNTs) have shown a promising potential in the safe, effective and risk free storage of hydrogen gas in fuel cells, increasing the prospects of wide uses of fuel cells and replacement of internal combustion engine. In addition, one can exploit the potential of CNTs in oil and gas industry. Nanotechnology offers a myriad of applications for production of new gas sensors, optical sensors, chemical sensors, and other energy conversion devices to bio implants. Nanoporous oxide films of TiO_2 to enhance photo voltaic cell technology are used. Nanoparticles are perfect to absorb solar energy and that is useful in very thin layers on conventional metals to absorb incident solar energy. New solar cells based on nanoparticles of semi conductors, nanofilms and nanotubes by embedding in a charge transfer medium is an active field of research. Films formed by sintering of nanometric particles of TiO_2 (diameter 10-20 nm) combine high surface area, transparency, excellent stability and good electrical conductivity and are ideal for photovoltaic applications. Nonporous oxide films are highly promising material for photovoltaic applications (Zhiyong et al., 2009).

Nanoscale slab-slot-waveguides that provides high optical confinement have found abundant applications in silicon photonics. After developing an analytical mode solver for general asymmetric slot waveguides, the confinement performance of symmetric as well as asymmetric geometries was systematically analyzed and compared. For symmetric structures, 2D confinement optimization by varying both low-index slot and high-index slab width revealed a detailed saturation trend of the confinement factor with the increase of the studied width. Furthermore, simple design rules on how to choose the slot and slab width for achieving optimal confinement was obtained. For asymmetric structures, we demonstrated that the confinement performance was always lower than the 2D optimized confinement of the symmetric structures providing the two high-index slab layers and the two cladding layers have same refractive indices, respectively. In addition, the sensitivity of the confinement to the degree of asymmetry was studied, and we found that the fabrication tolerance on the material and structural parameters might be reasonably large for symmetric structures designed at optimal confinement (Ma et al., 2009). A change in shape of nanoparticles can produces much larger shifts in the surface-plasmon resonance band due to spatial confinement effect. The surface-plasmon resonance band often splits in to two bands due to the elongation of nanoparticle along some particular axis of aisotropy. The shorter-wavelength band termed as the transverse plasmon resonance band corresponds to the oscillations of electrons along any minor axis, while the longer-wavelength band termed as the transverse plasmon resonance band corresponds to the oscillations of the electrons along the major axis. In some semiconductor nanoparticles different kinds of excitations such as bi-exciton, dark and bright exciton, polaron, bi-polaron and polariton etc. emerges due to the effect of quantum confinement in contrast to their bulk counterpart.

6. Quantum-confined materials

There is much current interest in the nonlinear optical properties of quantum confined semiconductor nanocrystals, particularly for their potential use in nonlinear photonic devices. Glass is an attractive host matrix for the nanocrystals in such applications (Dvorak et al., 1995). Quantum-confined materials refer to structures, which are constrained to nanoscale lengths in one, two or all three dimensions that possess density of states entirely different from bulk structure as shown in figure 23. The length along which there is quantum confinement must be smaller than de Broglie wavelength of electrons for thermal energies in the medium. The thermal energy is given by $E = mv^2/2 = k_BT$ and de Broglie wavelength is $\lambda = h/mv$. For effective quantum-confinement, one or more dimensions must be less than 10 nm. Quantum-confined structures show strong effect on their optical properties. Artificially created structures with quantum-confinement on one, two or three dimensions are called, quantum wells, quantum wires and quantum dots respectively (Botez et al., 2005). Low-dimensional semiconductors and other materials can modify the density of states by restricting their size to the order of an electron wave function and thereby show many fascinating effects. The quantized energy levels in quantum structures became the driving force for next-generation solar cells and other nanophotonic chip design due to the mitigation of the thermal scattering as well as the control of the light absorption range.

The quantum efficiency of the photoluminescence and electroluminescence from a bulk noble metal is very low ($\sim 10^{-10}$) compared to the corresponding nanosystems, which is several orders of magnitude higher ($\sim 10^{-4}$). However, in the case of silver and gold nanorods or nanoparticles, the luminescence efficiency is enhanced by six to seven orders of magnitude, and in addition, a red (blue) shift for the luminescence maximum is observed with increasing (decreasing) aspect ratio of the nanorods or diameter of nanoparticles. A linear increase in the photoluminescence efficiency with square of aspect ratio (diameter) is remarkable. The enhancement of the luminescence efficiency is also observed for rough surfaces of gold and silver. The rough surface can be modeled as an ensemble of randomly oriented hemispheroids of nanometer size, and the surface plasmon resonance arises from them amplifies the local fields, thereby resulting in the enhancement of photoluminescence often called lightening rod effect. All these are primarily attributed to quantum confinement effects.

Nanoparticulate metals and semiconductors that have atomic arrangements at the interface of molecular clusters and 'infinite' solid-state arrays of atoms have distinctive properties determined by the extent of confinement of highly delocalized valence electrons. At this interface, the total number of atoms and the geometrical disposition of each atom can be used significantly to modify the electronic and photonic response of the medium. In addition to the novel inherent physical properties of the quantum-confined moieties, their 'packaging' into nanocomposite bulk materials can be used to define the confinement surface states and environment, inter-cluster interactions, the quantum-confinement geometry, and the effective charge-carrier density of the bulk. Current approaches for generating nanostructures of conducting materials, especially the use of three-dimensional crystalline super-lattices as hosts for quantum-confined semiconductor atom arrays (such as quantum wires and dots) with controlled inter-quantum-structure tunneling (Sucky et al., 1990) are getting momentum. Quantum-confined nanocrystallites of GaAs are fabricated in porous glass and the bound electronic nonlinear refractive index, the two-photon absorption coefficient, and the refraction from carriers generated by two-photon absorption are simultaneously determined using the Z-scan method and compared to those of bulk GaAs. The quantum-confined Stark effect in single cadmium selenide (CdSe) nanocrystallite quantum dots has also been widely studied due to their potential application in displays and renewable energy. The electric field dependence of the single-dot spectrum is characterized by a highly polarizable excited state (~10^5 cubic angstroms, compared to typical molecular values of order 10 to 100 cubic angstroms), in the presence of randomly oriented local electric fields that change over time. These local fields result in spontaneous spectral diffusion and contribute to ensemble inhomogeneous broadening. Stark shifts of the lowest excited state more than two orders of magnitude larger than the line width were observed, suggesting the potential use of these dots in electro-optic modulation devices (**Empedocle et al., 1997**).

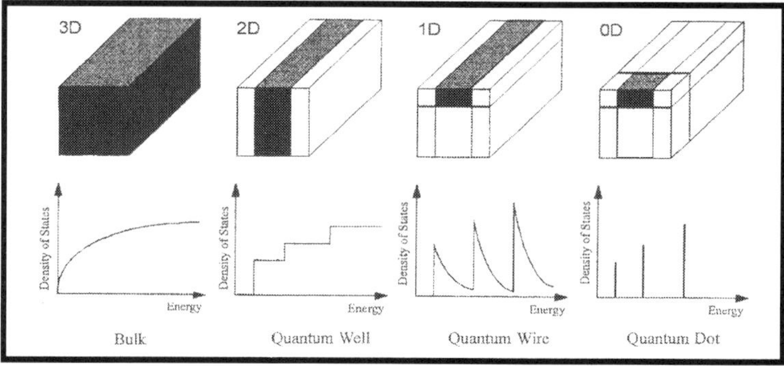

Fig. 23. The change in the density of states with confinement dimensions. Bulk structure suffers no confinement but 0D structure has confinement in all three spatial dimensions and hence density of states is discrete delta functions.

The carrier contribution to elastic constants in quantum confined heavily doped non-linear optical compounds has been studied. The basis of this newly formulated electron dispersion law take into account the anisotropies of the effective electron masses and spin orbit splitting constants together with the proper inclusion of the crystal field splitting in the Hamiltonian within the framework of k.p formalism. It has been found, taking different heavily doped quantum confined materials that, the carrier contribution to the elastic constants increases with increase in electron statistics and decrease in film thickness in ladder like manners for all types of quantum confinements with different numerical values, which are entirely dependent on the energy band constants. This contribution is greatest in quantum dots and least in quantum wells together with the fact the heavy doping enhances the said contributions for all types of quantum-confined materials (Baruah et al., 2007).

To understand the size quantization on the luminescence (see figure 24) from the metal nanoparticles a simplified model for the band structure is developed. This model includes the s-p conduction band and two sets of occupied d bands. The luminescence from metal nanoparticles can be attributed to the transition of electrons from a completely filled d band (highest occupied molecular orbital) to the unoccupied levels in the sp band (lowest unoccupied molecular orbital). The excitation involves a transition from states in the upper d band to the unoccupied levels in s-p band at and above the Fermi level. The emission arises from the direct recombination of conduction electrons with holes in the d bands of the quantum confined structures. The emission bands either appear due to radiative inter-band recombination of electrons and holes in the s-p and d bands, or originate from radiative intra-band transitions within the s-p band across the band gap. This model has been employed for describing luminescence from silver, gold and other nanoparticles (Link et al., 2002).

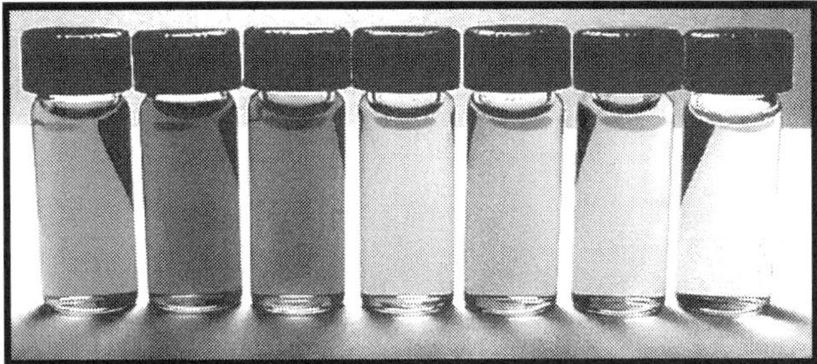

Fig. 24. The effect of size quantization: light emission from gold colloids with varying size of nanoparticles , smallest particles emit violet and the largest emit red color.

7. Applications

Nanophotonics is a unique field because it combines scientific challenges with large variety of near- term applications (Koller et al., 2008; Stockman et al., 2007). Fundamental research on nanophotonics leads to applications in communications technology, lasers, solid-state lighting, data storage, lithography, biosensors, optical computers, imaging, solar cells, light-activated medical therapies, displays, and smart materials to cite a few (**Empedocles et al., 1997;** Baruah et al., 2007; Rahmani et al., 1997; Nezhad et al., 2007; Shen et al., 2000; Levy et al., 2004). Some of the important areas that will find large market in next ten to fifteen years from now are identified and represented in figure 25.

Two-dimensional photonic crystal lasers have been fabricated on III-V semiconductor slabs. Tuning of the spontaneous emission in micro and nanocavities has been achieved by accurate control of the slab thickness. Different structures, some of them of new application to photonic crystal lasers, have been fabricated like the Suzuki-phase or the coupled-cavity ring-like resonators. Laser emission has been obtained by pulsed optical pumping. Optical characterization of the lasing modes has been performed showing one or more laser peaks centred around 1.55 µm. Far field characterization of the emission pattern has been realized showing different patterns depending on the geometrical shape of the structures. These kinds of devices may be used as efficient nanolaser sources for optical communications or optical sensors (Postigo et al., 2007).

Nanoscale confinement of matter is achieved on nanoscale crystals and they are used for: (i) optical up-conversion of radiation in rare-earth nanocrystals, (ii) size-dependent emission properties of nanoscale semiconductor crystals (quantum dots), and (iii) size-dependent absorption and confinement is exploited in plasmonic solar cells and biosensing. Metal nanoparticles and nanotips used in surface enhanced Raman spectroscopy. Photonic band gap crystals and photonic band gap fibers (photonic fibers) involve periodic variation of dielectric constant over wavelength-scale. They find applications in fabrication of Micro-Opto-Electro Mechanical Systems, Micro-Optics, i.e. Micro-lasers, Directional Couplers between waveguides, Bio-photonic Chips etc. Optical wires (wave-guides), filters, and transistors are now possible, and all of them could harness the speed of light for optical communications, sensing, data processing and storage.

Silicon photonics offers high-density integration of individual optical components on a single chip. Strong light confinement enables dramatic scaling of the device area and allows unprecedented control over optical signals. Silicon nanophotonic devices have immense capacity for low-loss, high-bandwidth data processing. Fabrication of silicon photonics system in the complementary metal–oxide–semiconductor-compatible silicon-on-insulator platform also results in further integration of optical and electrical circuitry. Following the Moore's scaling laws in electronics, dense chip-scale integration of optical components can bring the price and power per a bit of transferred data low enough to enable optical communications in high performance computing systems (Bettoti et al., 2003; Pavesi, 2003).

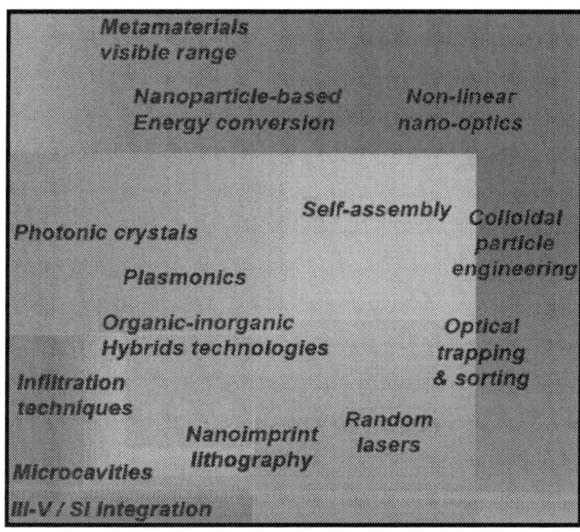

Fig. 25. Schematic diagram for the required technology and materials development in next fifteen years from now in every five years periods as shaded by different colors.

Photonic crystals and photonic band gap materials have a myriad of uses. These uses range from computing to telecommunication and from lasers to circuits that can also be found in almost anything digital. Light-emitting diodes play a key role in optical-communication systems. These devices are made from so-called photo emissive materials that emit photons once they have been excited electrically or optically. These photons are typically emitted in many different directions and have a range of wavelengths, which is not ideal for communications applications. A type of light emitting diode can be created that only emits light in the forward direction by placing a reflector behind the photo emissive layer. However, the efficiency of such a device is limited by the efficiency of the reflector.

Some of important applications of the photonic crystals include, perfect dielectric mirror in which the reflectivity of photonic crystals derives from their geometry and periodicity, not a complicated atomic-scale property (unlike metallic components mirror). The only demand on such materials is that for the frequency range of interest, these should be essentially lossless. Such materials are widely available all the way from the ultraviolet regime to the microwave. The nonlinear effects of the materials exploits the non-linear properties for construction of photonic crystal lattices open new possibilities for molding the flow of light. In this case, the dielectric constant is additionally depending on intensity of incident electromagnetic radiation and any non-linear optics phenomena can occur. In wave-guides and junctions, the existence of guided modes with different parities is an important factor that deeply influences the transmission through the Y junction. The cavity now produces a skewed field that can be understood as a superposition of modes with odd and even parities.

Photonic crystals could be used to design a mirror that reflects a selected wavelength of light from any angle with high efficiency. Moreover, these could be integrated within the photo emissive layer to create a light emitting diode that emits light at a specific wavelength and direction. Ideally, one needs to build a truly 3-D lattice structure to gain complete control of the light in all three dimensions. Fortunately, 2-D periodic lattices exhibit some of the useful properties of a truly 3-D photonic crystal, and are far simpler to make. These structures can block certain wavelengths of light at any angle in the plane of the device, and can even prevent light entering from certain angles in the third dimension (i.e. perpendicular to the surface). Thus, 2-D photonic crystals are a good compromise for many applications and are easily incorporated within planar waveguides.

Improving the existing concepts of light sources is a hot topic in nanophotonics, particularly, flat-panel displays through photo-detectors. However, the ambitions of nanophotonics and molecular photonics go well beyond this simple and straightforward demand. In the field of displays one needs to produce images using liquid crystals or with fluorescent molecules in a polymer matrix. In the near future, the target would be to obtain the internet screen being displayed in the living rooms on a larger screen, with electronic wallpaper acting as the display that is, realizing a component of ambient intelligence. Therefore, nanophotonics can bridge the gap between existing optoelectronic devices and future optics circuitry by incorporating optically active nanostructures and molecules.

Predictions are always difficult to make, however, the future for photonic-crystal circuits and devices looks certain. Within few years, a number of basic applications will start making an appearance in the market place. Among these will be highly efficient photonic-crystal lasers and extremely bright light emitting diodes. It will be possible to implement optical manipulation, signal processing and electronic circuitry all in one chip. Photonic crystals may also be used for high-speed computers of the next generation. Progress has already been made towards replacing the slow copper connections in computers with ultra-fast optical interconnects. Then photons, rather than electrons, will pass signals from board to board, from chip to chip and even from one part of a chip to another. Imagine a computer that operates at the speed of light!

Metal nanostructures have received considerable attention for their ability to guide and manipulate electromagnetic energy in the form of surface plasmon polaritons. The unique properties of surface plasmon polaritons may enable an entirely new generation of chip-scale technologies, known as plasmonics. Such plasmonic devices could add functionality to the already well-established electronic and photonic device technologies. The patterned metal structures are used as tiny wwaveguides to transport electromagnetic energy between nanoscale components at optical frequencies. Interestingly, the surface plasmons and surface plasmon polaritons, which are collective oscillations of electrons in small particles (nanosystems) or on metal dielectric interfaces induced by an incident electromagnetic radiation, is a topic of tremendous interest with ever-growing surprises and promises. In recent years, the potential of plasmonics has been realized in making highly sensitive and very small photonic devices (nanolasers as an example). A wide range of applications emerging from controlling and manipulating the light at the nanoscale such as high-resolution optical imaging below the diffraction limit, enhanced optical transmission through sub-wavelength apertures, bio-detection at femto-mole level, nanoscale wave guiding, THz communications, optical computing and surface-enhanced Raman scattering have been established. Many futuristic applications regarding plasmonic materials rendering invisibility are also inspiring researchers. Surface plasmon polaritons can be tightly bound to the metal surface, penetrating ~100 nm into the dielectric and ~10 nm into the metal. This feature implies the possibility of using surface plasmon polaritons for miniature photonic circuits and optical interconnects and has attracted a great deal of attention to surface plasmon polaritons.

Plasmonic nanoparticles have tremendous technological importance and found applications in therapy and imaging. Beyond biosensing, plasmonic nanoparticles have further applications in biology and medicine. For example, they simultaneously give contrast in electron and optical microscopy such that a labeled cell can be imaged in optical and subsequently in electron microscopy. A modified version uses fluorophore-gold nanoparticle conjugates for simultaneous contrast in electron and fluorescence microscopy. Gold nanoparticles are also promising in hypothermal cancer therapy. In these experiments, nanoparticles that rae attached to cancer cells are heated by the absorption of light until the cancer cell overheats and dies (Shalaev et al., 2007). Plasmonics have wide applications in the energy sector as of improving lighting technologies by improving the efficiency of light emitting diodes up to 14 times. In medicine, for selectively destroying cancer cells by radiation absorption it is very useful. It is even possible in theory to make an object invisible by using plasmonics with metaparticles who have extraordinary optical properties for electronic waves. In computing technology for processing information with speed of light, it is recommended. If it can be combined with graphene, a newly emerged transparent conducting material, innovations beyond our wildest dreams may be achieved. Applications of surface plasmons in solid-state lighting and lasing are just appearing, but it may be that traffic lights are composed of surface plasmon light emitting diodes in a few years time!

Recent advancements in laser science using nanophotonics have tackled physical limitations to achieve higher power, faster and smaller light sources. The quest for ultra-compact laser that can directly generate coherent optical fields at the nanoscale, far beyond the diffraction limit of light, remains a key fundamental challenge. Microscopic lasers based on photonic crystals, carbon nanomaterials, micro-disks, semiconductor quantum structures, metal clad cavities and nanowires can now reach the diffraction limit, which restricts both the optical mode size and physical device dimension to be larger than half a wavelength (Bonaccorso et al., 2010). While surface plasmons are capable of tightly localizing light, ohmic loss at optical frequencies has inhibited the realization of truly nanoscale lasers. Progress in theory and modeling has paved the way to reduce significantly the plasmonic loss while maintaining ultra-small modes by using a hybrid plasmonic waveguide. Experimental demonstration of nanoscale plasmonic lasers producing optical modes 100 times smaller than the diffraction limit, utilizing a high gain CdS semiconductor nanowire atop a silver surface separated by a thick insulating gap has been reported. Direct measurements of emission lifetime reveal a broadband enhancement of the nanowire's spontaneous emission rate by up to 6 times due to the strong mode confinement and the signature of apparently *threshold-less lasing*. Since plasmonic modes have no cut-off the downscaling of the lateral dimensions of both device and optical mode are possible. As these optical coherent sources approaches molecular and electronics length scales, *plasmonic lasers* offer the possibility to explore extreme interactions between light and matter, opening new avenues in active photonic circuits, bio sensing and quantum information technology. Let us turn our attention to the wonder material graphene and see its future promises for plasmonic-based devices for photonics and optoelectronics (Zouhdi et al., 2009; Jablan et al., 2009).

Amongst many, the novel material *'graphene'* has become something of a celebrity material due to its unusual conductive, thermal, electronic, plasmonic and optical properties, which could make it useful in a range of sensors, lasers and semiconductor devices. In addition to flexibility, robustness and environmental stability it possesses high mobility and optical transparency. Most of the recent studies mainly focused on fundamental physics and electronic devices. Undoubtedly, its true potential lies in nanophotonics and optoelectronics, where the combination of its unique optical and electronic properties can be fully exploited, even in the absence of a band gap, and the linear dispersion of the Dirac electrons enables ultra-wideband tunability. Graphene being a potential candidate of advanced plasmonic materials is a rapidly rising star on the horizon of materials science and condensed-matter physics (Bonaccorso et al., 2010; Murray et al., 2007). Graphene's ultra-wide broadband capability will be useful for future nonlinear optical devices including high speed, transparent and flexible photosensitive systems, which could be further functionalized to enable chemical sensing. With an ever-growing interest in the widespread applications of graphene, ultrafast and tunable lasers have become a reality. The combination of graphene photonics with plasmonics could lead to a wide range of advanced devices. The charge carriers in graphene are negative electrons and positive holes, which in turn are affected by plasmons. The *'plasmaron'* is a composite particle; a charge carrier coupled with a plasmon has recently been observed. The rise of graphene in photonics and optoelectronics is shown by several recent results, ranging from solar cells and light-emitting devices to touch screens, photo-detectors and *ultrafast lasers*. Electroluminescence has also been reported recently in pristine graphene. This observation could lead to new light emitting devices based entirely on graphene. In this communication, we will briefly discuss the state-of-the-art in this emerging field (Maier, 2007; Brongersma et al., 2007).

Optical properties of plasmons in graphene are in many relevant aspects similar to optical properties of surface plasmons propagating on dielectric-metal interface, which have been drawing a lot of interest lately because of their importance for nanophotonics. It is reported that plasmons in doped graphene simultaneously enable low-losses and significant wave localization for frequencies below that of the optical phonon branch. Large plasmon losses occur in the inter-band regime via excitation of electron-hole pairs that can be pushed towards higher frequencies for higher doping values. For sufficiently large dopings, there is a bandwidth of frequencies where a plasmon decay channel via emission of an optical phonon together with an electron-hole pair is non-negligible. There are certain frequencies in which plasmons have the low losses and that makes graphene potentially interesting for nanophotonic applications (Zouhdi et al., 2009).

Nanohole arrays have emerged from an interesting optical phenomenon to the development of applications in photo-physical studies, photovoltaics, in imaging chips for digital cameras, and as a sensing template for chemical and biological analyses. Numerous methodologies have been designed to manufacture nanohole arrays, including the use of focus ion-beam milling, soft-imprint lithography, colloidal lithography and modified nanosphere lithography. Hole-arrays in plasmonics are increasingly finding their way into applications as selective filters for color sensors and channeling light into optical devices. Hole-arrays can be made into highly selective filters for sensors that depend on detecting specific colors, or for efficiently extracting monochromatic light from light emitting diodes and lasers. For commercialization NEC laboratory is focusing on prototypes of plasmon-enhanced devices for displays and telecommunications. To reduce pixel noise and improve camera sensitivity hole arrays are placed on top of individual pixels that helps to capture incoming light more efficiently. Another plasmonic technique for channeling light into a device is to sprinkle its surface with nanoscale particles made of a metal such as gold. These nanoparticles function like an array of tiny antennas: incoming light is taken up by plasmons and then redirected into the device's interior.

With nanosphere lithography or colloidal lithography, the experimental conditions control the density of the nanosphere mask and, thus, the aspect of the nanohole arrays. Low surface coverage of the nanosphere mask produces disordered nanoholes. Ordered nanohole arrays are obtained with a densely packed nanosphere mask in combination with electrochemical deposition of the metal, glancing angle deposition or etching of the nanospheres prior to metal deposition. The optical properties of nanoholes have interesting applications in analytical chemistry. In particular, applications of these novel plasmonic materials are demonstrated as substrates for a localized surface plasmon resonance, surface plasmon resonance, surface enhanced Raman spectroscopy and in electrochemistry with nano-patterned electrodes (Polman, 2009). Perhaps, the most promising application of nanohole-arrays called nanoantennas is in the improvement of solar cell efficiency because the currently used solar cells made from silicon has reached to saturation and is expensive. However, to harvest the wide range of wavelengths from the solar spectrum, particularly in the red and infrared part of the spectrum, the semiconductor layer has to be relatively thick and that makes it bulky up to 350 μm thick. Other solar-cell materials except few recently developed nanosystems and organic materials have the same problem and the efficiency is low. Nanoantennas based on surface-enhanced Raman spectroscopy solar cell can overcome this difficulty. Some applications of surface-enhanced Raman spectroscopy have reached the market. For example, in specifically prepared colloids of gold nanoparticles, a clustering of these nanoparticles is triggered by the presence of pregnancy hormones. This leads to a color change induced by plasmonic effects that has been widely commercialized in pregnancy tests.

The other important application of surface plasmons in nanophotonics is the enhanced optical transmission through sub-wavelength- apertures. For light transmission through an aperture having lateral dimensions smaller than $\lambda/2$, the efficiency is extremely low as a result light cannot propagate freely because transmission is possible only via the highly inefficient photon tunneling mechanism. Surprisingly, a substantial enhancement of transmitted light at resonant wavelengths could be achieved for nanoscale apertures (diameter ~200 nm and period ~500 nm) in the form of arrays of holes or slits fabricated on a metal film. This enhancement of the transmission intensity which is much larger than that predicted by classical diffraction theory is primarily due to the coupling of the incident light (photons) with surface plasmons. This finding on enhanced transmission through a large number of nanoholes suggests that even the photons impinging between the nanoholes can be transmitted. Furthermore, the evanescent waves that originate from the diffraction of the incident radiation by the nanoholes diffract while tunneling through the nanoholes and produce interference with the incident waves. Eventually, the surface plasmons enhance the field associated with the evanescent waves, thereby resulting in the enhancement of the transmitted light intensity.

Nanophotonics have potential to make significant impact on military and warfare systems, in both symmetric and asymmetric warfare. The technology parameters identified are real and can be translated into military applications. It is very likely that nanotechnology insertion into disruptive technologies that could pose threats is difficult to anticipate by the very nature of such threats and the sometimes-embryonic state of research in nanophotonics relative to other technologies. Some applications may take substantial investments and time before they are realized. Nanophotonics definitely requires a fabric of supporting and enabling technologies that must be smart, fast and environment friendly. This technology set-up at least in part does not exist today, which further complicates our further outlook. Provoked by the computer games and software development strong emphasis has been placed on information technology as a possible 'game changer,' with nanophotonics. Certainly, it could potentially enable a much more ubiquitous and pervasive data processing and sensing capability than is currently available.

8. Conclusions & future outlook

This new frontier called nanophotonics offers numerous opportunities for both fundamental research and application. The information technology based on the nanotechnology is varied from the targets that can be realized in the near future, to those with long-term goals such as quantum info-communication and quantum computing technology. In the quantum dot area, quantum dot laser, quantum dot optical amplifier, quantum dot nonlinearity devices, and quantum dot light detectors are about to emerge into the real world. The expectations are high for optical circuit of the optical router and like which will be the basic component of the router in the future photonic network and quantum dot amplifier for application to optical 3R relay generator will play a key role. The expectation is that the quantum dot be used for this single photon generator. The expectation is also high for the realization of nanophotonics based on the new principle. The nanophotonic element is an ultimate element that uses strongly connected condition of the light and electron that include the polariton laser. Investment in nanophotonics should lead to realization of low energy consumption society and promotion of heals of the people. There will come an age when nanophotonic elements will play the key roles including the fusion of quantum dot and photonic crystal. The quantum dot will be the basic component of future quantum computing elements in the future and expectations are very large for emergence of new 'non-continuous' technology based on the quantum dot technology.

Since photonic crystals offer prospects for numerous applications: low-threshold lasing, optical-power limiting, chemical and bio sensing, and optical switching. It will play a major role in further development of this exciting field. Photonic components such as fiber optic cables can carry lot of data but are bulky compared to electronic circuits. Electronic components such as wires and transistors carry less data but can be incredibly small a single technology that has the capacity of photonics and the smallness of electronics would be the best bridge of all. *So 'Light on a Wire,' Is Circuitry Wave of Future.* Nanophotonics, according to world leaders, has the potential to transform telecommunications technologies in a way even more radical than the emergence of electron-based integrated circuits. People are talking about components such as single photon sources, single photon detectors in the quest for low-noise signals. Components like waveguides, which carry light from one part of the optical circuit to another, and switches are essential. There are good prospects of switching functionality, because optical properties of some molecules are highly non-linear, which is an essential condition for fast switching devices. Therefore, there are very promising device-relevant properties here, suitable for future optical circuits and systems.

The main limitation to plasmonics today is that plasmons tend to dissipate after only a few millimeters, making them too short-lived to serve as a basis for computer chips, which are a few centimeters across it. For sending data even longer distances, the technology would need even more improvement. The key is using a material with a low refractive index, ideally negative, such that the incoming electromagnetic energy is reflected parallel to the surface of the material and transmitted along its length as far as possible. There exists no natural material with a negative refractive index, so nanostructured materials must be used to fabricate effective plasmonic devices. For this reason, plasmonics is frequently associated with nanotechnology. Before all-plasmonic chips are developed, plasmonics will probably be integrated with conventional silicon devices. Plasmonic wires will act as high-bandwidth freeways across the busiest areas of the chip. Plasmonics has also been used in biosensors. When a particular protein or DNA molecule rests on the surface of a plasmon-carrying metallic material, it leaves its characteristic signature in the angle at which it reflects the energy. Therefore, it is necessary to develop smart integration techniques to fabricate plasmo-electronic chips.

Future commercial opportunities of nanophotonics lay ahead in each of the following fields: solar energy, biophotonics, solid-state lighting and displays, computing, telecommunications, sensors and security, etc, to cite a few. This field is expected to revolutionize many aspects of modern life. The commercial potential is frankly overwhelming and mind boggling. Nanophotonics promises solar cells that provide cheap, clean power wherever it is needed. Solid-state lighting systems those are much more efficient and versatile than current light bulbs. Flexible electronic displays that allow cell phones will be the size of credit cards. The computer screens as thin and flexible as a sheet of heavy paper is possible. Computers that connect to the internet at the speed of light are targeted. The new medical devices that give doctors the power to detect and treat diseases in novel ways are up coming. This is just the beginning! The movement from nanophotonics to molecular photonics provides a link to developments in molecular electronics. Ultimately, there is a need to bring together the advantages of molecular photonics with those of molecular electronics, based on the common strategies of preparation, handling and integration. It may well be that molecular electronics is the electronics of the future and molecular photonics, the photonics of the future.

In conclusion, this chapter on 'nanophotonics for 21st century' focuses on the paramount importance and tremendous technological potential of this field for nanoscale device miniaturization at low cost and with maximum efficiency. Nanophotonics is newly developed very vast and expanding paradigm of nanoscience, and is a unique part of physics/chemistry/materials science because it combines a wealth of scientific challenges with a large variety of near-term applications. Nanophotonics is the use of materials nanotechnology in photonics or the use of photonic materials in nanotechnology. The exciting application possibilities of nanophotonics related upcoming areas in science and technology, its present status and future scope are explored in this presentation. Nanophotonics could well revolutionize the fields of telecommunications, computing, imaging and sensing in particular. This presentation argues, why is the research into nanophotonics important? Nanophotonics is all about the manipulation and emission of light in both the far field and near-field using materials at nanoscale. It is the study of the behavior of light on the nanometer scale and their interactions in different material media. The ability to fabricate devices in the nanoscale that has been developed in recent years called nanotechnology is the main thrust. Nanotechnology is defined as a novel and multidisciplinary use of materials or processes at the nanometer scale (below 100 nm). However, we attempt to give a panoramic overview of nanophotnics, the science of light-matter interaction at nanometer scale (< 1 μm to ≥ 1 nm), which deals with optical processes at the much smaller length scale than the wavelength of optical radiation. The nanoscale matter-radiation interaction includes nanoscale confinement of radiation, nanoscale confinement of matter, and nanoscale photo-physical or photochemical transformation, offer numerous opportunities for both fundamental research and technological ramifications. Quantum-confined materials are structures that can constrain to nanoscale lengths in one, two or all three dimensions are employed for implementations of devices in nanophotonics. Confinement of light results in field variations similar to the confinement of electron in a potential well. For light, the analogue of a potential well is a region of higher refractive index bounded by a region of lower refractive index.

It is important to note that the overview on nanophotonics provided here cannot do justice to all the different research directions in the field of photonic crystals and plasmonics. Without attempting to be exhaustive, we have tried to present a short but somewhat complete perspective including the fundamentals and some of the current burning research topics. Together, these represent the true state-of-the art in the field of nanophotonics. Several applications are highlighted. In addition, the importance in understanding the electromagnetic wave propagation in photonic crystals/photonic band gap materials and meta-materials are briefly discussed. The chapter concludes with the future of nanophotonics that seems bright and close to reality. The ever-ending research activities aimed in this field are expected to revolutionize many aspects of modern life and will remain fertile ground for basic and applied research.

9. Acknowledgements

S. K. Ghoshal is indebted to Prof. K. P. Jain (IIT, Delhi) and Prof. Sir Roger Elliott (Oxford University) for introducing this field and educating him with many critical concepts. S. K. G gratefully acknowledges the financial support from Universiti Teknologi Malaysia through the research grant (VOTE 4D005/RMC and Q.J130000.7126.00J39/GUP).

10. References

Alu, A. & Engheta, N. (2008). Multifrequency Optical Invisibility Cloak with Layered Plasmonic Shells. *Phys. Rev. Lett.*, 100, 113901(1-4), ISSN 0031-9007.

Anker, J. N.; Hall, W. P.; Lyandres, O.; Shah, N. C.; Zhao, J. & Van Duyne, R. P. (2008). Biosensing with plasmonic nanosensors. *Nature Mater.*, 7(6), 442–453, ISSN 1476-1122.

Atwater, H. & Polman, A. (2009). Surfing the Waves, a Report by Joerg Heber. *Nature*, 461, 720-722, ISSN 0028-0836.

Baruah, D.; Choudhury, S.; Singh, K. M. & Ghatak, K. P. (2007). Influence of quantum confinement on the carrier contribution to the elastic constants in quantum confined heavily doped non-linear optical and optoelectronic materials: simplified theory and the suggestion for experimental determination, *Journal of Physics: Conference Series*, 61, 80-84, ISSN 0953-8984.

Behren, J. V.; Buuren, T. V.; Zacharias, M.; Chimowitz, E. H. & Fauchet, P. M. (1998). Quantum confinement in nanoscale silicon: The correlation of size with bandgap and luminescence. *Solid State Communication*, 105(5), 317-322, ISSN 0038-1098.

Bettoti, P.; Cazzanelli, M.; Dal Negro, L.; Danese, B.; Gaburro, Z.; Oton, C. J.; Vijaya Prakash, G. & Pavesi, L. (2003). Silicon nanostructures for photonics. *J. Phys.: Cond. Mat.*, 14, 8253-8265, ISSN 1361-648X.

Bonaccorso, F.; Sun, Z.; Hasan, T. & Ferrari, A. C. (2010). Graphene Photonics and Optoelectronics. *Nature Photonics*, 4, 611-614, ISSN 1749-4885.

Botez, D. (2005). Intersubband Quantum-Box Semiconductor Lasers: High-Power, Efficient CW Sources for the 3–10µm Wavelength Range. *Laser Physics*, 15(7), 966-974, ISSN 1054-660X.

Bozhevolnyi, S. I.; Volkov, V. S.; Devaux, E.; Laluet, J. Y. & Ebbesen, T. W. (2006). Channel plasmon subwavelength waveguide components including interferometers and ring resonators. *Nature*, 440, 508–511, ISSN 0028-0836.

Brongersma, M. L. & Kik, P. G. (2007). *Surface Plasmon Nanophotonics*. Springer, ISBN 978-1-4020-4349-9, UK.

Chu, T.; Ishida S. & Arakawa, Y. (2005). Compact 1 × N thermo-optic switches based on silicon photonic wire waveguides. *Opt. Express*, 13(25), 10109-10114, ISSN 1094-4087.

Dionne, J. A.; Diest, K.; Sweatlock, L. A.; Atwater, H. A. & Plasmostor, A. (2009). Metal−Oxide−Si Field Effect Plasmonic Modulator. *Nano Lett.*, 9(2), 897–902, ISSN 1530-6984.

Dvorak, M.D.; Justus, B.L.; Gaskill, D.K & Hendershot, D.G. (1995). Nonlinear absorption and refraction of quantum confined InP nanocrystals grown in porous glass. *Appl. Phys. Lett.*, 66 (7), 804-806, ISSN 0003-6951.

Ebbesen, T. W.; Lezec, H. J.; Ghaemi, H. F.; Thio, T. & Wolf, P. A. (1998). Extraordinary optical transmission through sub-wavelength hole arrays. *Nature*, 391, 667–669, ISSN 0028-0836.

Empedocles, S. A. & Bawendi, M. G. (1997). Quantum-Confined Stark Effect in Single CdSe Nanocrystallite Quantum Dots. *Science*, 278 (5346), 2114-2117, ISSN 0036-8075.

Fleischmann, M.; Hendra, P. J. & Mc Quillan, A. (1974). Raman spectra of pyridine adsorbed at a silver electrode. *J. Chem. Phys. Lett.*, 26, 163–166, ISSN 0009-2614.

Galli, M.; Politi, A.; Belotti, M.; Gerace, D.; Liscidini, M.; Patrini, M.; Andreani, L. C.; Miritello, M.; Irrera, A.; Priolo, F. & Chen, Y. (2006). Strong enhancement of Er3+ emission at room temperature in silicon-on-insulator photonic crystal waveguides. *Appl. Phys. Lett.*, 89, 241114-241116, ISSN 0003-6951.

Ghoshal, S. K.; Mohan, D.; Kassa T. T. & Sharma, S. (2007). Nanosilicon for Photonic Application. *Int. J. Mod. Phys. B*, 21(22), 3783-3796, ISSN 0217-9792.

Gopinath, A.; Boriskina, S. V.; Svetlana, V.; Reinhard, B. M. & Bjoern, M. L. (2009). Deterministic aperiodic arrays of metal nanoparticles for surface-enhanced Raman scattering (SERS). *Opt. Express*, 17(5), 3741-3753, ISSN 1094-4087.

Hirsch, L. R.; Stafford, R. J.; Baukson, J. A.; Sershen, S. R. & Rivera, B. (2003). Nanoshell-mediated near-infrared thermal therapy of tumors under magnetic resonance guidance. *Proc. Natl. Acad. Sci. USA*, 100, 13549–13554, ISSN 0028-0836.

Hryciw, A.; Chul, Y. J. & Brongersma, M. L. (2010). Electrifying plasmonics on silicon. *Nature Mater.*, 9, 3-4, ISSN 1476-1122.

Hryciw, A.; Jun, Y. C & Brongersma M. L. (2009). Plasmon-enhanced emission from optically-doped MOS light sources. *Opt. Express*, 17, 185–192, ISSN 1094-4087.

Jablan, M.; Buljan, H. & Soljačić M. (2009). Plasmonics in graphene at infrared frequencies. *Phys. Rev. B.*, 80, 245435- 245439, ISSN 1094-1622.

Joannaopoulos, J. D.; Johnson, S. D.; Winn, J. N. & Meade R. D. (2008). *Photonic Crystals Molding the Flow of Light*. Princeton University Press, ISBN 978-0-691-12456-8, Princeton, New Jersy.

Jun, Y. C.; Briggs, R. M.; Atwater, H. A. & Brongersma. M. L. (2009). Broadband enhancement of light emission in silicon slot waveguides. *Opt. Express*, 17, 7479–7490, ISSN 1094-4087.

Jun, Y. C.; Kekatpure, R. D.; White, J. S. & Brongersma M. L. (2008). Nonresonant enhancement of spontaneous emission in metal-dielectric-metal plasmon waveguide structures. *Phys. Rev. B*, 78, 153111(1-4), ISSN 1550-235.

Kalele, S. A.; Tiwari, N. R.; Gosavi, S. W. & Kulkarni, S. K. (2007). Plasmon-assisted photonics at the nanoscale. *J. Nanophotonics*, 1, 012501-012520, ISSN 1934-2608.

Kim, S.; Jin, J.; Kim, P; Young-Jin Kim, Y.; Kim & Seung-Woo (2008). High-harmonic generation by resonant plasmon field enhancement. *Nature*, 453, 757-760, ISSN 0028-0836.

Kobayashi, M.; Kawazoe, T.; Sangu, S. & Yatsui, T. (2002). Nanophotonics: design, fabrication, and operation of nanometric devices using optical near fields. *J. Selected Topics in Quantum Electronics*, 8, 839-862, ISSN 1077-260X.

Koller, D. M.; Hohenau, A.; Ditlbacher, H.; Galler, N.; Reil, F.; Aussenegg, F. R.; Leitner, A. N.; List, E. J. W. & Krenn, J. R. (2008). Organic plasmon-emitting diode and references therein. *Nature* Photonics, 2, 684–687, ISSN 1749-4885.

Levy, U. C.; Nezhad, H.; Nakagawa, W.; Tetz; Chen, K.; Pang, L. & Fainman, Y. (2004). *Quantum Sensing And Nanophotonic Devices Book Series: Proceedings of The Society of Photo-Optical Instrumentation Engineers*, Ed.(s): M. Razeghi, G. J. Brown 5359 (SPIE) pp. 126-144.

Lezec, H. J. ; Dionne, J. A. & Atwater, H. A. (2007). Negative Refraction at Visible Frequencies. *Science*, 316, 430-432 *ISSN* 0036-8075.

Li, X.; Liu, H.; Wang, J.; Zhang, X. & Cui, H. (2004). Preparation and properties of YAG nano-sized powder from different precipitating agent. *Opt. Mater.*, 25, 407-412, ISSN 0925-3467.

Link, S.; Beeby, A; Fitzgerald, S.; El-Sayed, M. A.; Schaaff, T. G. & Whetten, R. L. (2002). Visible to infrared luminescence from a 28-atom gold cluster. *J. Phys. Chem.*, 106, 3410-3415, ISSN 1089-5647.

Lodahl, P.; van Driel, A. F.; Nikolaev, I. S.; Irman, A.; Overgaag, K.; Vanmaekelbergh, D. & Vos, W. L. (2004). Controlling the dynamics of spontaneous emission from quantum dots by photonic crystals. *Nature,* 430, 654-657, ISSN 0028-0836.

Loo, C.; Lowery, A.; Halas, N.; West, J. & Drezek, R. (2005). Immunotargeted Nanoshells for Integrated Cancer Imaging and Therapy. *Nano Lett.,* 5(4), 709-711, ISSN 1530-6984.

Ma, C.; Zhangand, Q. & Van Keuren, E. (2009). Analysis of symmetric and asymmetric nanoscale slab slot waveguides. *Optic Communication,* 282(2), 324-328, ISSN 0030-4018.

Miao, X.; Wilson, B. K.; Pun, S. H. & Lin, L.Y. (2008). Optical manipulation of micron/submicron sized particles and biomolecules through plasmonics. *Opt. Express,* 16(18), 13517-13525, ISSN 1094-4087.

Maier, S. A. (2007). *Plasmonics: Fundamentals and Applications.* Springer, ISBN 978-0-387-33150-8, Germany.

Miyazaki, H. T. & Kurokawa, Y. (2006). Squeezing Visible Light Waves into a 3-nm-Thick and 55-nm-Long Plasmon Cavity. *Phys. Rev. Lett.,* 96, 097401(1-4), ISSN 0031-9007.

Murray, W. A. & Barnes, W. L. (2007). Plasmonic Materials. *Adv. Mater.,* 19, 3771-3776, ISSN 1521-4095.

Nezhad, M.; Abashin, M.; Ikeda, K.; Pang, L.;. Kim, H. C.; Levy, U.; Tetz, K.; Rokitski, R. & Fainman, Y. (2007). *Optoelectronic Integrated Circuits IX Book Series: Proceedings of The Society of Photo-Optical Instrumentation Engineers,* Ed.(s): L. A. Eldada, E. H. Lee 6476 (SPIE), pp. A4760-A4760.

Nie, S. & Emory, S. R. (1997). Probing Single Molecules and Single Nanoparticles by Surface-Enhanced Raman Scattering. *Science,* 275, 1102–1106, ISSN 0036-8075.

Noginov, M. A.; Zhu, G. & Belgrave, A. M. U. (2009). Demonstration of a spaser-based nanolaser. *Nature,* 460, 1110–1112, ISSN 0028-0836.

Oulton, R. F.; Sorger, V. J.; Zentgraf & Thomas, X. (2009). Plasmon lasers at deep subwavelength scale. *Nature,* 461, 629–632, ISSN 0028-0836.

Pavesi, L. (2003). Will Silicon be the photonics material of the third millennium. *J. Phys.: Cond. Mat.,* 15, R1169-1196, ISSN 1361-648X.

Pendry, J. B. (2000). Negative Refraction Makes a Perfect Lens. *Phys. Rev. Lett.,* 85, 3966-3969, ISSN 0031-9007.

Pillai, S.; Catchpole, K. R.; Trupke, T. & Green, M. A. (2007). Surface plasmon enhanced silicon solar cells. *J. Appl. Phys.,* 101, 093105(1-8), ISSN 1089-7550.

Polman, A. (2008). Plasmonics Applied. *Science,* 322, 868-869, ISSN 0036-8075.

Postigo, P. A.; Alija A. R.; Martinez, L. J.; Dotor, M. L.; Golmayo, D.; Sanchez-Dehesa, J.; Seassal, C.; Viktorovitch, P.; Galli; Politi, M. A.; Patrini, M. & Andreani, L. C. (2007). *Photonics and nanostructures,* 5(2-3), 104, ISSN 1569-4410.

Prasad, P. N. (2004). Polymer science and technology for new generation photonics and biophotonics. *Current Opinion in Solid State and Materials Science,* 8, 11–19, ISSN 1359-0286.

Prodan, E; Radloff, C; Halas, N. J. & Nordlander, P (2003). A Hybridization Model for the Plasmon Response of Complex Nanostructures. *Science,* 302, 419-422, ISSN 0036-8075.

Rahmani, A.; Chaumet, P. C.; Fornel, F.de & Girard, C. (1997). Fluorescence lifetime calculations in a confined geometry field propagator of a dressed junction. *Phys. Rev. A,* 56, 3245-3254, ISSN 1050-2947.

Rayleigh, J. W. S. (1888). On the remarkable phenomenon of crystalline reflexion described by Prof. Stokes. *Phil. Mag.*, 26, 256–265, ISSN 1478-6443.

Rosa, E. De La; Diaz-Torres, L. A.; Salas, P.; Arrendondo, A.; Montoya, J. A. C. & Rodriguez, A. (2005). Low temperature synthesis and structural characterization of nanocrystalline YAG prepared by a modified sol–gel method. *Opt. Mater.*, 27, 1793-1799, ISSN 0925-3467.

Sajeev, J. (1987). Strong localization of photons in certain disordered dielectric superlattices. *Phys. Rev. Lett.*, 58, 2486-2489, ISSN 0031-9007.

Shalaev, V. M. & Kawata, S. (2007). Nanophotonics with surface plasmons *(Advances in Nano-optics and Nano-photonics)*, 1-340, Elsevier Science, ISBN 0444528385, UK.

Sharma, A.; Dokhanian, M;. Kassu, A. & Parekh Atul N. (2005). Photoinduced grating formation in azo-dye-labeled phospholipid thin films by 244-nm light. *Opt. Lett.*, 30(5), 501-503, ISSN 0146-9592.

Shen, Y.; Friend, C. S.; Jiang, Y.; Jakubczyk, D.; Swiatkiewicz, J. & Prasad, P. N. (2000). Nanophotonics: Interactions, Materials and Applications. *J. Phys. Chem. B*, 104, 7577-7587, ISSN 1089-5647.

Sonnichsen, C.; Reinhard, B. M.; Liphardt, J. & Alivisatos, A. P. (2005). A molecular ruler based on plasmon coupling of single gold and silver nanoparticles. *Nat. Biotechnol.*, 23(6), 741-745, ISSN 1087-0156.

Stockman, M. I.; Kling, M. F.; Kleineberg, U. & Krausz, F. (2007). Attosecond Nanoplasmonic Field Microscope. *Nature* Photonics, 1, 539-542, ISSN 1749-4885.

Stucky Galen D. & Mac Dougall James, E. (1990). Quantum Confinement and Host/Guest Chemistry: Probing a New Dimension. *Science*, 247(4943), 669-678, ISSN 0036-8075.

Valentine, J.; Li, J.; Zentgraf, T.; Bartal G. & Zhang, X. (2008). Three-dimensional optical metamaterial with a negative refractive index. *Nature*, 455, 376-380, ISSN 0028-0836.

Verhagen, E.; Spasenović, M.; Polman, A. & Kuipers, L. (2009). Localization in Correlated Bilayer Structures: From Photonic Crystals to Metamaterials and Semiconductor Superlattices. *Phys. Rev. Lett.*, 102, 203904(1-4), ISSN 0031-9007.

Veronis, G. & Fan, S. (2007). Theoretical investigation of compact couplers between dielectric slab waveguides and two-dimensional metal-dielectric-metal plasmonic waveguides. *Opt. Express*, 15, 1211–1221, ISSN 1094-4087.

Waele, R.; de Koenderink, A. F. & Polman, A. (2007). Tunable Nanoscale Localization of Energy on Plasmon Particle Arrays. *Nano Lett.*, 7(7), 2004-2008, ISSN 1530-6984.

Walters, R. J.; van Loon, R. V. A.; Brunets, I.; Schmitz J. & Polman, A. A. (2010). Silicon-based electrical source of surface plasmon polaritons. *Nature Mater.*, 9(1), 21–25, ISSN 1476-1122.

Xing-huang, Y; Song-sheng, Z.; Rui-min, Y.; Jing, C;. Zhi-wei, X.; Chun-jia, L. & Xue-tao L. (2008). Preparation of YAG:Ce3+ phosphor by sol-gel low temperature combustion method and its luminescent properties. *Trans. Nonferrous Met. Soc. China*, 18, 648-653, ISSN 1003-6326.

Yablonovitch, E. (1987). Inhibited Spontaneous Emission in Solid-State Physics and Electronics. *Phys. Rev. Lett.*, 58, 2059-2062, ISSN 0031-9007.

Zhiyong, F.; Ruebusch, D. J.; Rathore, A. A.; Kapadia, R.; Ergen, O.; Leu, P. W. & Javey, A. (2009). Challenges and Prospects of Nanopillar-Based Solar Cells. *Nano. Res.*, 2, 829-843, ISSN 1998-0124.
Zouhdi, S.; Sihvola, A. & Vinogradov, A. P. (2009). *Metamaterials and Plasmonics: Fundamentals, Modeling, Applications.* Springer, ISBN 978-1-4020-9405-7, Germany.